The Author

Jerram L. Brown is Professor of Biology and Professor of Brain Research at the University of Rochester. He received his A.B. and M.S. degrees from Cornell University, and his Ph.D. from the University of California at Berkeley. He also spent two years as a postdoctoral fellow in neurobiology at the *Physiologisches Institut* of the University of Zurich.

Professor Brown's present research interests involve the behavioral ecology of social systems, with emphasis on altruism and aggression.

THE EVOLUTION OF BEHAVIOR

BROWN, Jerram L. The evolution of behavior. Norton, 1975. 761p il tab bibl 74-32453. 15.95. ISBN 0-393-09295-X. C.I.P.

An outstanding book that covers the field of animal behavior from an ecological, physiological, and evolutionary standpoint — the latter being the unifying factor throughout. Brown, a neurobiologist at the University of Rochester whose research interest lies in the behavioral ecology of social systems, has organized the book into six sections. Opening with a very lucid explanation of the phylogeny of behavior, he follows with a treatment of the genetic basis of behavior. By far the longest two sections are those dealing with social organization and animal communication. Physiological and developmental aspects of behavioral evolution are treated in the last two sections. Each chapter is written in crisp scientific style with very generous illustrations. The graphs, tables, line drawings, and black-and-white photographs are all of high quality and supplement the text nicely. A 57-page bibliography indicates the degree to which the subject was researched by the author.

Jerram L. Brown

UNIVERSITY OF ROCHESTER

The EVOLUTION OF BEHAVIOR

W · W · NORTON & COMPANY · INC ·

NEW YORK

THIS BOOK was composed in Linofilm Melior. Composition, printing, and binding were done by the Kingsport Press, Inc.

Library of Congress Cataloging in Publication Data
Brown, Jerram L
The evolution of behavior.
Bibliography: p.
Includes index.
1. Animals, Habits and behavior of. 2. Evolution.
I. Title. [DNLM: 1. Behavior. 2. Behavior, Animal.
BF671 B878e]
QL751.B726 591.5 74-32453

ISBN 0 393 09295 X

2 3 4 5 6 7 8 9 0

Dedicated to my wife
Esther R. Brown

Contents

Preface xv

I THE PHYLOGENY OF BEHAVIOR 1

Diversity and Constancy. The Analysis of Comparisons.

1. Patterns of Diversity in Behavior among Animal Species 4

Correlation of Behavior with Phylogeny. Discovery of Natural Groupings through Behavioral Analysis. Discovery of New Species through Behavioral Analysis. Behavioral Consistency within Species. The Adaptive Value of Behavior. Adaptive Radiation in Behavior. Domestication and Behavior. Evolutionary Convergence and Parallelism in Behavior. Progressive Evolutionary Change in Behavior. Fossil Records of Behavior. Clines in Behavior. CONCLUSIONS

II THE GENETIC BASES OF BEHAVIOR 21

2. Behavior, Genetics, and Evolution 22

Demonstration of the Genetic Basis of Behavioral Differences. Single-Gene Differences. Heritability of Polygenic Behavioral Traits. Heritability: Comparison of Inbred Lines. Heritability: Selection for Behavioral Traits. Genetic Polymorphism. Interspecific Hybrids. Developmental Genetics of Behavior. Innateness. CONCLUSIONS

III SOCIAL ORGANIZATION: THE SOCIAL CONTEXT OF BEHAVIOR AND NATURAL SELECTION 39

The Units of Natural Selection. Fitness. Kin Selection. Population Selection.

3. Social Units and Their Patterns of Dispersion 47

Patterns of Dispersion. The Units of Dispersion. Temporal Changes in Dispersion. Dispersal and Spacing Behavior. Home Range. CONCLUSIONS

4. The Comparative Study of Spacing Behavior: An Introduction to Territoriality and Other Spacing Phenomena 58

Phylogenetic Incidence of Territoriality. Definition. Difficult and Borderline Cases. Diversity of Territories. The Central Role of Territory Defense. The Ontogeny of Territorial Behavior. Artificial Territories. CONCLUSIONS

5. Animals in Groups 72

TYPES OF GROUPS / FORCES OF ATTRACTION WITHIN GROUPS / THE SPACE AROUND AN INDIVIDUAL Nearest-neighbor Distance. Individual Distance. SOCIAL DOMINANCE IN GROUPS History. Methods of Observation and Dominance Criteria. Dominance Matrices in Natural Populations of Birds. Theoretical Considerations. Peck-Right and Peck-Dominance. Natural Selection for Status-Indicating Behavior. Survival Value of Dominance. Social Roles. CONCLUSIONS

6. Behavior and Populations 97

EXCLUSION EFFECTS Effects on Breeding Density. Effects of Territoriality on Reproduction. Effects on Survival. The Regulation of Density by Aggressive Behavior. SUBORDINATION EFFECTS Responses to Population Increases: Emigration. Survival of Emigrants and Transients. Flexibility in Social Organization in Response to High Densities. Effects of High Densities on Recruitment. Physiology as a Function of Population Density. Neuroendocrine Physiology as a Function of Defeat. Evolutionary Fate of Emigrants. CONCLUSIONS

7. The Evolution of Diversity in Spacing Systems 124

THE EVOLUTION OF TERRITORIALITY Early Hypotheses. The Mating-Territory Hypothesis. The Food-Territory Hypothesis. A General Theory. Population Selection and Territorial Behavior. THE EVOLUTION OF COLONIALITY Clumped versus Regular Dispersion in Blackbirds. An Energy-Budget Approach. A Survival Approach. Social Stimulation. THE EVOLUTION OF SURVIVAL GROUPS Some Benefits of Survival Groups. THE EVOLUTION OF OVERLAPPING HOME RANGES IN RODENTS / CONCLUSIONS

8. Mating Systems and Sexual Selection 151

TYPES OF MATING SYSTEMS / SEXUAL SELECTION / TYPES OF SEXUAL SELECTION / MALE DOMINANCE AND COMPETITIVE MATING The Fur Seal. Red and Roe Deer. Other Mammals. General Characteristics. MALE ADORNMENTS AND FEMALE CHOICE Isolated Mating Territories. Communal Mating Grounds. Sage Grouse and Other Lek Species. All-purpose Territories. General Characteristics. ECOLOGY OF MATING SYSTEMS Ecological Models of Sexual Selection. Ecological and Behavioral Factors Influencing Female Choice. Habitat and Diet. GENETICS AND SEXUAL SELECTION Fisher's Concept of Sexual Preference. Genetic Models of Sexual Selection. Frequency-dependent Sexual Selection. Genetics of Male Courtship. Genetics of Female Selectivity. CONCLUSIONS

9. Aid-giving Behavior: Diversity, Ecology, and Evolution 186

Diversity and Phylogenetic Incidence. THE EVOLUTION OF REPRODUCTIVE RATES AND PARENTAL CARE Clutch Size. Evolution and Environment. Uncrowded Environments. r_{max}-Selection. Colonizing Species. Crowded Environments. K-Selection. Parental Care and K-Selection. THE EVOLUTION OF ALTRUISTIC BEHAVIOR IN VERTEBRATES What Is Altruism? Where Does Operational Altruism Occur? Sketches of Some Communal Breeders. Is Kin Selection Involved? Are Altruists Basically Selfish? Reciprocal Altruism. Other Theories. Symbiosis or Mutualism. Types of Altruism. CONCLUSIONS

10. The Evolution of Sociality in Bees 214

The Spectrum of Bee Societies. The Honeybee. Haplodiploidy and Altruism. CONCLUSIONS

11. Social Parasitism 224

Polyphyletic Origins. SOCIAL PARASITISM IN INSECTS Ants. Inquilinism via Slavery. Inquilinism via Temporary Parasitism. Inquilinism via Food Parasitism. Evolution of Inquilinism. Nest Parasites of Bumblebees. Social Parasitism of Solitary Wasps. SOCIAL PARASITISM IN BIRDS Brood Parasitism and Egg Mimicry in Cuckoos. Nestling Mimicry in the Viduines. Some Intermediate Stages in the Evolution of Brood Parasitism. The Advantage of Being Parasitized. Other Brood Parasites. The Evolution of Brood Parasitism. CONCLUSIONS

12. Diversity in Primate Social Organizations 242

MATING SYSTEMS Solitary Species. Monogamous Pairs. Single-Male Harems. Age-Graded-Male Troops. Multi-Male Troops. Troop Size and Sexual Selection in Baboons. Sexual Dimorphism and Sexual Selection. THE ROLE OF AGONISTIC BEHAVIOR IN PRIMATE SOCIAL ORGANIZATION Dominance Phenomena within Groups. Agonistic Behavior between Groups. COOPERATION, AID-GIVING, AND KINSHIP Parental Behavior. Kinship and Dominance. Cooperation. Kin Selection. ENVIRONMENTAL INFLUENCES ON SOCIAL ORGANIZATION Group Size and Carrying Capacity. Density and Territoriality. Food. CONCLUSIONS

IV THE ADAPTEDNESS OF BEHAVIOR IN ANIMAL COMMUNICATION 267

The Problem: The Adaptive Nature of Species Differences. Concepts of Adaptation. Communication Exemplifies Environmental Adaptation. What Is Animal Communication? Brain Size and Communication. Motivation and Communication. Sensory Modes.

13. Visual Communication: Contexts and Measurement 271

CONTEXTS OF VISUAL COMMUNICATION The Demonstration of Visual Communication. SOME TERMS AND CONCEPTS / THE MEASUREMENT OF COMMUNICATION / CONCLUSIONS

14. The Evolutionary Origins of Displays 282

THE CONTEXTS OF DISPLAYS The Spatial Context of Threat Displays. The Temporal Context of Threat Displays. Quality of Opponent. THE POSTURAL COMPONENTS AND ORIENTATION OF THREAT DISPLAYS / DECISION-MAKING The Adaptive Value of Displays: Tipping the Balance. ORIGINS OF DISPLAYS Thermoregulation. Respiration. Intention Movements of Locomotion. Protective Movements. Redirected Attacks. Out-of-Context Behavior. CONCLUSIONS

15. Selection Pressures on Displays 301

DISCRETE AND GRADED DISPLAYS / INTRASPECIFIC DISTINCTIVENESS OF DISPLAYS / INTERSPECIFIC DISTINCTIVENESS IN VISUAL SIGNALS / TYPES OF EVOLUTIONARY CHANGE IN DISPLAYS

Changes of the Whole Display. Exaggeration of Coloration. Exaggeration of Structure. Exaggeration of Movement. Transfer of Function. THE PHYSICAL ENVIRONMENT AND THE DIRECTION OF SELECTION Nest Sites. Display Perches. Aquatic Displays. Ground-feeding Species. Visibility. SOCIAL ENVIRONMENT AND DISPLAYS Mating Systems. Agonistic Systems. Aid-giving Systems. INTERSPECIFIC DISPLAYS / CONCLUSIONS

16. The Evolution of Auditory Communication 329

PHYLOGENETIC OCCURRENCE / EVOLUTIONARY ORIGINS OF SOUND-PRODUCING MECHANISMS From Respiratory Structures. From Beating a Substrate. From Rubbing of Appendages. BASIC CHARACTERISTICS OF SOUND SIGNALS IN DIFFERENT ANIMAL GROUPS Crickets. Frogs and Toads. Birds. Mammals. BEHAVIORAL EFFECTS OF SOUND PRODUCTION: CONTEXTS AS CLUES Crickets. Fishes. Amphibians. Reptiles. Birds. Mammals. EXPERIMENTAL AND ANALYTICAL STUDIES OF AUDITORY COMMUNICATION Insects. Amphibians. Bird Songs. Recognition within the Family: Birds. PROPERTIES OF SOUNDS IN RELATION TO USES Alarm Calls. Advertisement Songs. Species Diversity. CONCLUSIONS

17. Chemical Communication 370

INTERSPECIFIC CHEMICAL COMMUNICATION / SEX ATTRACTANTS Insects. Primates. ALARM SUBSTANCES / TRAIL MARKING / REPERTOIRE DIVERSITY Social Insects. Deer. Mice. Other Mammals. OPTIMIZATION OF PHEROMONAL PROPERTIES Divergence of Mating Signals in Allopatric Populations: Visual Signals. CONCLUSIONS

18. The Origin of Species and of Species-specific Behavior: Behavioral Isolating Mechanisms 400

SPECIES-SPECIFIC BEHAVIOR Extinction. Unique Selection Pressures in Unique Environments. Species Differences Per Se. THE BIOLOGICAL SPECIES / ISOLATING MECHANISMS Types. Coaction of Isolating Mechanisms. ALLOPATRIC SPECIATION / SYMPATRIC SPECIATION / MATING SIGNALS AS ISOLATING MECHANISMS Visual Signals. Auditory Signals. Chemical Signals. EVOLUTIONARY DIVERGENCE OF BEHAVIOR IN ALLOPATRY Divergence of Mating Signals in Allopatric Populations: Auditory Signals. Homogamic Mating in Allopatric Populations. Evolutionary Divergence of Behavior in Sympatry. Character Displacement. Mating Calls and Songs.

Experimental Selection for and against Pre-Mating Isolation.
CONCLUSIONS

19. The Dances of Bees 449

Methods. Some Chemical Cues. The Dances. Population Differences in Dancing. Other Bees. Other Insects.
CONCLUSIONS

V THE PHYSIOLOGICAL BASIS OF SPECIES CONSTANCY AND SPECIES DIVERSITY IN BEHAVIOR 465

20. Reflexes in Simple and Complex Animals 468

NERVE NETS / SPECIES DIVERSITY OF BEHAVIOR BASED ON NERVE NETS / REFLEXES UNDER CENTRAL CONTROL / THE DEFINITION OF A REFLEX / REFLEXES AS INTEGRATIVE MECHANISMS / CONCLUSIONS

21. Stereotypy in the Mechanisms of Coordination of Motor Patterns 490

SOME CHARACTERISTICS OF SPECIES-TYPICAL, STEREOTYPED MOTOR PATTERNS / GIANT NEURONS AND FIXED BEHAVIOR PATTERNS The Importance of Giant Neurons. Giant Neurons in Cephalopods. Giant Axons in Crayfish. CENTRAL PATTERN GENERATORS / CONCLUSIONS

22. Sensory Systems: Feature Detection 516

RECOGNITION OF BIOLOGICALLY SIGNIFICANT AUDITORY STIMULI Moths. Frogs. Other Animals. RECOGNITION OF BIOLOGICALLY SIGNIFICANT VISUAL STIMULI / TYPES OF RECOGNITION / CONCLUSIONS

23. Sensory Systems: Orientation, Echolocation 543

TYPES OF ORIENTATION IN INVERTEBRATES / ECHOLOCATION BY ANIMALS History. Functions and Characteristics of Echolocation Systems. Who Echolocates? ECHOLOCATION BY BATS Species Diversity in Foraging and Flying. Prey Capture. Capture Sequence. EL Sounds. Echo Processing. Adaptions. ECHOLOCATION BY PORPOISES / CONCLUSIONS

24. Long-distance Migrations 557

THE PHENOMENA OF MIGRATION Distance. Accuracy. Length of Nonstop Flights. Natural Barriers and Pathways. Time of Day. Altitude. Weather. INTERACTION OF LEARNING AND INHERITANCE IN MIGRATION / NAVIGATIONAL CUES Sun. Stars. Nonvisual Cues: Magnetic Fields. CONCLUSIONS

25. The Programming of Behavior: Biological Rhythms, Attention, and Motivation 588

BIOLOGICAL RHYTHMS Short-term Rhythms. Circadian Rhythms in Some Invertebrates. Circadian Rhythms in Vertebrates. ATTENTION / MOTIVATION / CONCLUSIONS

VI THE DEVELOPMENT OF SPECIES-TYPICAL BEHAVIOR 607

26. The Beginning of Behavior in Embryos 612

The First Behaviors. Spontaneity. The Functions of Motor Neurons. Evolutionary Adaptations. The Program of Development. Individuation versus Reflex Integration. Inheritance and Environment in the Growth of Nerve Cells. CONCLUSIONS

27. The Naive Young Animal: Imprinting and Species Recognition 627

EXPERIMENTAL STUDIES OF SEXUAL IMPRINTING / ANALYSIS OF THE IMPRINTING PROCESS Critical Period. DEVELOPMENT OF SPECIES RECOGNITION / SEXUAL IMPRINTING IN EVOLUTION Disruption of Isolating Mechanisms via Imprinting. Effects of Sexual Imprinting on Gene Frequencies. CONCLUSIONS

28. The Development of Songs and Calls 653

IN WHAT SPECIES IS VOCAL LEARNING IMPORTANT? / GENETIC FACTORS IN SONG DEVELOPMENT / TRANSITIONS IN THE DEVELOPMENT OF VOCALIZATION / SPECIES DIVERSITY IN SONG LEARNING Little or No Evidence of Learning. Learning from Neighboring Conspecifics. Learning from the Father. Learning from Other Species in Normal Song Development. Learning from the Mate. Dialects. THEORIES OF SONG-LEARNING / THE EVOLUTION OF VOCAL LEARNING / CONCLUSIONS

Literature Cited 677

Appendix: Testing Kinship Theories Ecologically 735

Subject Index 737

Author Index 741

Species Index 752

Preface

THIS IS A BOOK about behavior written with the central, unifying theme of *biological evolution*. The factual material that forms the basis of this treatment consists of the vast body of descriptive and experimental studies on behavior of all kinds of animals. In the real sense of the word this is a *comparative* study of behavior. Here, *comparative* is not used, as it has sometimes been, to mean *pertaining to animals other than man*. This book is principally concerned with comparisons of behavior between different kinds of animals, including man—but comparisons are only the starting point.

The comparative study of behavior inevitably raises certain questions. Some species are found to differ from others with respect to certain behavioral traits, while other species may closely resemble each other. Why do these differences and similarities exist? This question is part of a central problem in biology. Similarities and differences among different kinds of organisms are the data with which theories of biological evolution are concerned. The "answers" to questions raised by the comparative study of behavior must be sought within the framework of evolutionary biology. Three principal approaches to behavior—ecological, physiological, and developmental—can all be interpreted profitably within the framework of evolutionary biology and brought to bear on the common problem of the patterns of behavioral similarities and differences among different kinds of animals. To accomplish this task is the goal of this book.

It is hoped that this book will be of use to individuals in all areas of behavior study who wish to acquaint themselves with an evolutionary approach to behavior. It is also intended to be useful as a college textbook in comparative animal behavior—in fact it follows very closely my lectures on that subject at the University of Rochester. As a text it makes a fundamental departure from the presently available books. First of all it focuses explicitly on a single but broad and basic problem area, rather than attempting to survey the whole field of animal behavior. Secondly, it makes no attempt to come up with a "synthesis of ethology and psychology" or to concentrate on the "common ground" of these fields. Material is selected on the basis of suitability to the task. As a result it is drawn from various disciplines, but far more from the various areas of biology than from psychology.

I feel that adherence to the theme of biological evolution in this book is to the advantage of both biologically and psychologically trained behaviorists, for this procedure allows a clearer distinction between two approaches to behavior. Psychology, and its subdivisions such as physiological psychology, developmental psychology, and psychophysics, are basically concerned with the *individual* and processes that go on within the individual. On the other hand, the central concepts in this book are concerned with *populations*. It is in some senses a treatise on certain aspects of the population biology of behavior. As such it draws on comparative studies by zoologists, evolutionists, geneticists, ecologists, ethologists, and psychologists – persons whose fields of interest overlap greatly.

The time seems ripe now for formal recognition of the distinctions between these two branches of behavior study. The methods, approaches, problems, and concepts of behavior study at the population level are fundamentally different from those at the organismic and lower levels. Recognition of the reasons for this basic difference will be healthy for both divisions. On the one hand, evolutionary and population biologists can approach behavior with a clearer delineation and fuller appreciation of their own field; and on the other, physiologically and developmentally oriented behaviorists can gain by viewing their old problems with the perspective of the population biologist and perhaps by discovering new ones. *Vive la différence!*

This book attempts to introduce the reader to the evolutionary branch of behavior study, but it is hoped that the beginning student of behavior will acquaint himself or herself with both branches. It is anticipated that most students will have had some introductory material in behavior, usually in psychology, before taking a comparative behavior course, and that they will also take the basic courses in the mechanisms – the physiology and development – of behavior. A general knowledge and appreciation of evolutionary biology is helpful in using this book; I have assumed the reader to have as much knowledge in that area as can be obtained in a good course in introductory biology.

The content of recent books on animal behavior has varied greatly for reasons that usually are not stated. Most have given considerable attention to the mechanisms of behavior per se and have given quite broad coverage to behavioral topics of a variety of sorts. In the present work a clear departure is made from this across-the-board but eclectic coverage; topics that concern physiological mechanisms are largely omitted. Consequently, studies on sensory processes, nervous systems, drive, motivation, regulatory behavior, and related topics are avoided. The book covers physiological topics not for their intrinsic interest, but for the insight they give to the evolution of behavior. Similarly the abundant literature on the ontogeny of behavior is not given full coverage. Instead, ontogeny is treated only from the standpoint of the comparative behaviorist and evolutionist.

A word of explanation may be necessary here because some behaviorists think many of the omitted subjects are important—or traditional—parts of a comparative course on behavior. In the first place, using animals other than man does not per se make the topic "comparative." Rats are used in psychological experiments not for the purpose of comparing results with those for other species, but mainly as a convenient mammal with which to investigate phenomena thought to be common to mammals generally and therefore to man. In the second place, many comparative studies have not been carried far enough to justify the erection of evolutionary hypotheses. The observed differences between species may be of dubious validity, and evolutionary interpretations may depend on weak data. Or even if valid these studies may not elucidate the evolution of behavior. The comparative study of learning, for example, although a subject of great interest to behaviorists, has been particularly subject to methodological difficulties, and even the presumed superiority of primates in this general area has been difficult to demonstrate objectively (Warren 1965; Rumbaugh and McCormack 1967).

The topics included have been chosen because they help to explain the evolution of behavior. Social behavior and social organization have received special attention because they greatly influence the frequencies of genes in populations. Furthermore, social phenomena reveal the importance of various kinds of natural selection. The area of animal communication has received emphasis because it is a rich source of comparative material and because it displays so well the concept of evolutionary adaptedness. Examples in the physiology of behavior illustrate critical concepts of how species-typical behavior is mediated physiologically. Examples in the development of behavior clarify problem areas that have been foci of comparative studies.

Acknowledgments

I WOULD LIKE to thank the following persons who kindly read and commented on the manuscript: Ernst W. Caspari, for a review of Chapter 2; John F. Eisenberg, for his comments on Chapter 12; Gordon H. Orians, for a critique of Chapter 8; S. T. Emlen, for suggestions on Chapter 24; and especially Jack P. Hailman, who read the entire manuscript. All errors and imperfections are my own and in no way attributable to these readers. My wife, Esther, has been invaluable in many ways, particularly in proofreading, typing, indexing, and obtaining illustrations and reproduction permissions. I thank warmly those persons who freely contributed material for illustrations. I thank also Christopher P. Lang and Katherine L. Hyde of W. W. Norton for their patience and skill in shepherding this manuscript through the various stages from infancy to publication.

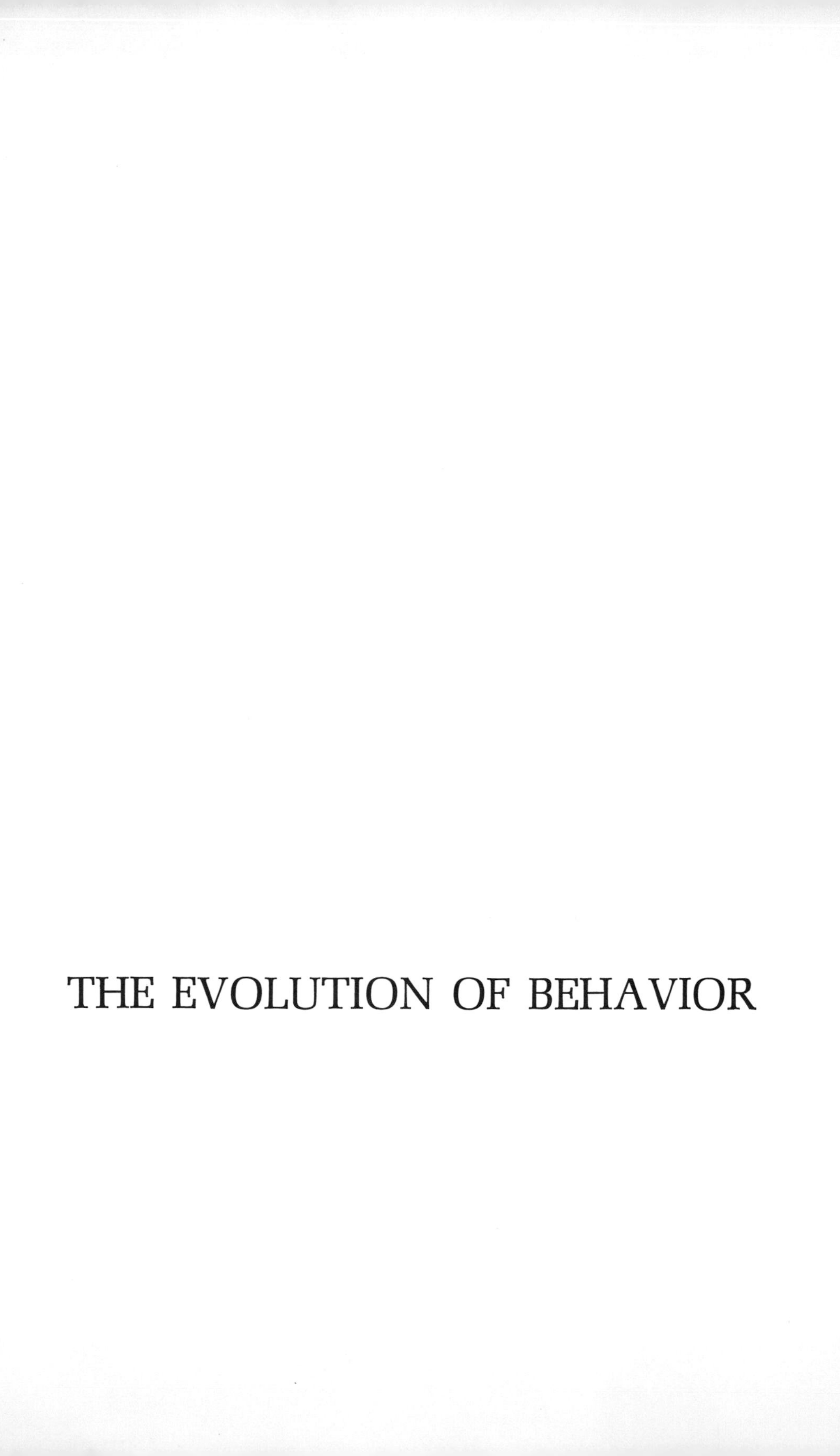

THE EVOLUTION OF BEHAVIOR

THE PHYLOGENY OF BEHAVIOR

Diversity and Constancy / The diversity of life is truly astounding. With roughly a million species of animals already discovered, not to mention plants and unicellular organisms, and more being described every year, the diversity of kinds of living matter must be considered one of the fundamental characteristics of the living world—the explanation of which constitutes one of the central problems of biology.

Not only are many species of animals to be found, but these tend to persist virtually the same for generation after generation. Indeed, the fossil record provides dramatic proof that the basic features by which many present-day groups of animals are differentiated have existed for millions of years. The constancy of the various forms of life and their resistance to change from generation to generation must also be regarded with awe as characteristic of all forms of life on this planet and a central concern of biology.

These two properties of life, diversity and constancy, have inspired some of man's greatest thoughts and scientific accomplishments. The study of diversity has become today's discipline of *evolutionary biology*. One of the great and universal generalizations of biology is Darwin's theory of evolution through natural selection. Today's science of *genetics* might be conceived as the study of constancy. Another great generalization of biology is the theory of particulate inheritance that was launched by Gregor Mendel and has matured into today's study of the molecular biology of inheritance.

These two branches of biology, evolution and genetics, apply to all attributes of life—to structure, to physiology, and to behavior. This book represents an approach to two universal characteristics of life, diversity and constancy, as they are manifested in behavior. It brings to the study of behavior the approaches of evolutionary biology and of genetics.

When we consider for a moment that nearly all of the species of animals differ from one another in some aspect of their behavior, we come face to face with the biological phenomena that the comparative behaviorist attempts to understand. The questions confronting him are many. What differences in behavior between species are to be found? Do they fall into recognizable patterns, and are these patterns explainable according to natural laws that can be discovered and tested? What are the physiological and anatomical properties that are responsible for the persistence of these differences within the lifetimes of individuals and from generation to generation? How can such behavioral diversity among species be maintained in view of the great plasticity in behavior due to the processes of learning and memory? At what stages in ontogeny do differences between species arise, and what is the role of environmental differences in bringing them about? What are the various means by which genetic factors bring about differences in behavior between individuals, populations, and species? These are the principal questions dealt with in this book.

The Analysis of Comparisons / The starting point in a study of diversity of behavior patterns in a group of species often consists of a set of comprehensive descriptions of the characteristic behavior patterns of these species. Each of these descriptions may be called an *ethogram*. Ethograms for all species are important, but some attract special attention. For example, Schaller's (1963) pioneering study of the mountain gorilla marked a turning point in the study of primate behavior by showing that the great apes were not deadly gargantuan monsters but could be studied quite safely under natural conditions, at close range, using the traditional methods of vertebrate zoologists. Ethograms of rare and endangered species, such as the studies on the California condor, ivory-billed woodpecker, and whooping crane, also attract interest. Scientifically, ethograms of bizarre or inaccessible species in a group already well studied are particularly important; an example is the study of the Galapagos swallow-tailed gull (Nelson 1968*b*; Hailman 1965; Harris 1970; Snow and Snow 1968).

When the behavior patterns of many species are known, differences and similarities can be observed. From the patterns of similarities and differences, hypotheses about their evolution can be created and observations designed to test them can be made. The basic sequence, greatly simplified, is then (1) description, (2) comparison, (3) evolutionary hypothesis, (4) testing, and (5) evaluation.

This procedure will be referred to as the *analysis of comparisons*. The hypotheses to be tested involve such concepts as biological function, adaptedness, selection pressures, environmental influences, and phylogenetic relationship.

Perhaps it is useful to distinguish between the analysis of comparisons, the broad topic of this book, and the *comparative method* as commonly used in evolutionary biology. Since the time of Whitman (1899), zoologists have used what has come to be called the comparative method in the study of behavior. This method has been used with success by Lorenz and his students (Lorenz 1950; Hinde and Tinbergen 1958; Baerends 1958), and it has come to be associated traditionally with the science of ethology. In the hands of ethologists it has been used primarily in the consideration of possible homologies; other questions raised by comparisons, such as identification of selection pressures, have received much less attention from these workers. In this book these other questions will receive considerably more attention than was given to them by the classical ethologists. To stretch the meaning of the comparative method to include all aspects of the analysis of comparisons as employed in this book would be an injustice to a time-honored concept.

1 Patterns of Diversity in Behavior among Animal Species

Instinct and organs are to be studied from the common viewpoint of phyletic descent.

Charles Otis Whitman, 1899

Correlation of Behavior with Phylogeny / When zoologists began looking at the phylogenetic distribution of behavior patterns, they made some basic observations that pointed the way toward future work on the evolution of behavior. Probably the most important of these early observations can be phrased as follows: The distribution of behavioral similarities and differences in a group of species tends to be correlated with the phylogenetic relationships within the group.

The basic truth of this generalization has been demonstrated in numerous comparative behavioral studies by many different authors working with many different animal groups (Mayr 1958). Among the first to recognize it were Charles Otis Whitman (1899, 1919), who worked on the behavior of Columbidae (pigeons and doves) at Woods Hole, Massachusetts, and Oscar Heinroth (1911), who studied many species at the Berlin zoo, but especially the family Anatidae (ducks, geese, and swans). Although these workers came independently to similar conclusions at the turn of the century, it was not until many years later that the European and American groups became well aware of each others' findings.

Whenever fairly large natural groups of species were studied by a person familiar with both their behavior and their phylogenetic relationships, as determined morphologically, the same general finding was reached. Certain kinds of behavior could be used to characterize natural groups of species. Some appreciation of this kind of relationship between behavior and phylogeny (inferred from morphology) may be obtained from Figure 1–1, which is based on a comparative study of birds of the order Pelecaniformes by Van Tets (1965). The first division, between the Pelecani on the one hand and the Phaëthontidae and

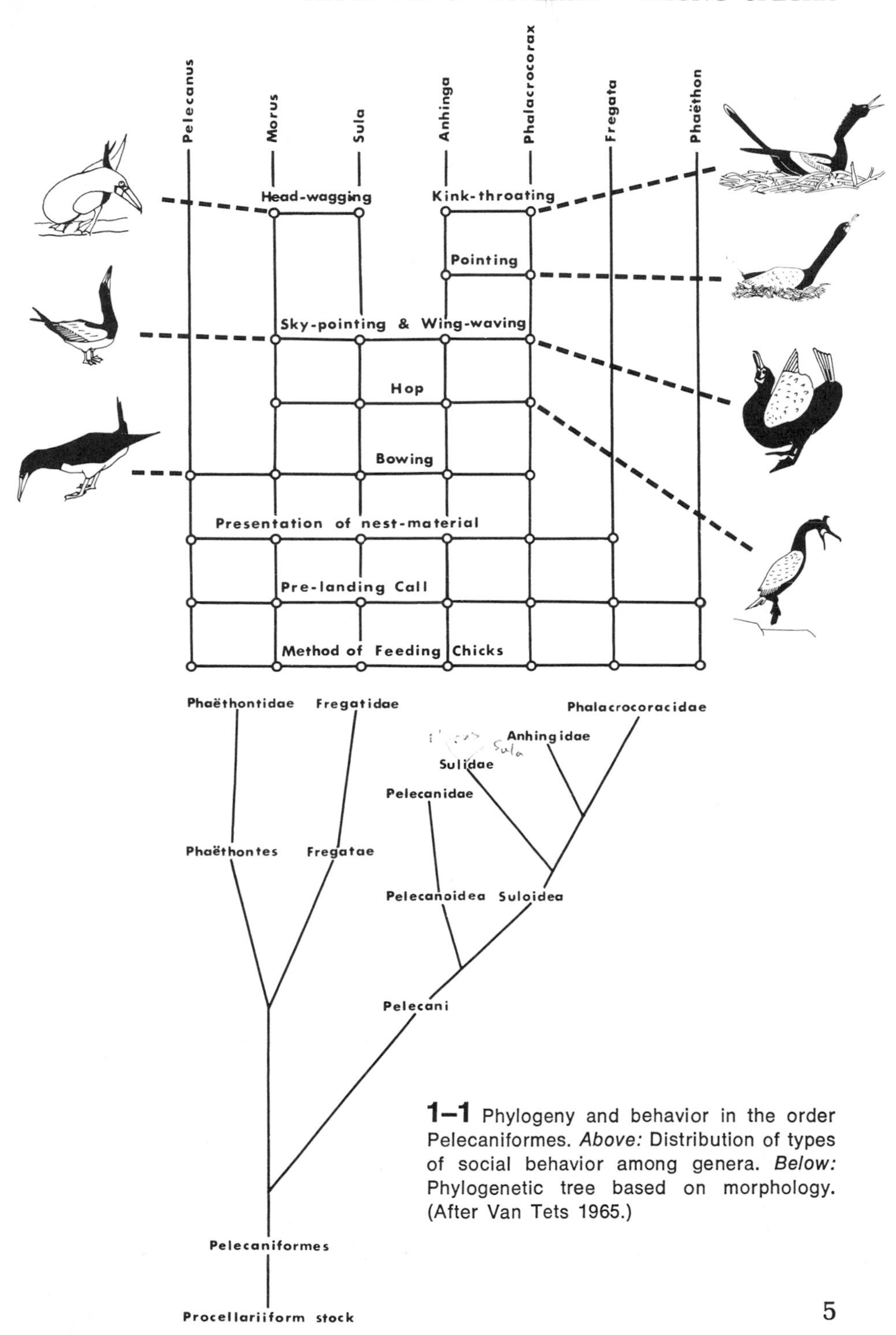

1–1 Phylogeny and behavior in the order Pelecaniformes. *Above:* Distribution of types of social behavior among genera. *Below:* Phylogenetic tree based on morphology. (After Van Tets 1965.)

Fregatidae on the other, is characterized behaviorally by the presence of bowing in the Pelecani but not in the latter groups. The Pelecanidae are distinguished by the absence of several behaviors found in the other groups. The Sulidae (*Morus* and *Sula*) are characterized by head-wagging, and the Anhingidae and Phalacrocoracidae by kink-throating and pointing. It is interesting that the distribution of behavioral traits conforms better to a classification based on morphology than to one based on biochemical traits as well. Is this because the morphology influences the behavior? For example, the absence of kink-throating in the Sulidae and its presence in the Phalacrocoracidae might be due to the morphological specializations of the neck for plunge-diving in the Sulidae and for surface-diving in the Phalacrocoracidae. Similar explanations can be imagined for several of the other behavioral differences.

Among other animal species, the family Anatidae was among the first to be thoroughly studied from a behavioral point of view, and it probably remains the most thoroughly studied large family of birds. Through the studies of Heinroth (1911), Lorenz (1941), Delacour and Mayr (1945), Johnsgard (1961, 1965), and others, the behavior of roughly 90 percent of the 150 species of Anatidae in the world had been described by 1966 (Johnsgard 1967). Through the work of Tinbergen and his students, over half of the roughly 35 species of the tribe Larini (gulls) have been studied ethologically (Tinbergen 1959; Moynihan 1959). Studies of insect behavior have been especially useful in this respect; the works of Evans (1953, 1962, 1966*a*, *b*, *c*; Evans and Eberhard 1970) on solitary wasps have been outstanding. In a review of 101 species and subspecies of the fruit fly *Drosophila*, Spieth (1952) reached similar conclusions about the relationship between behavior and phylogeny. The usefulness of behavioral traits for characterizing natural groups of animals has been reviewed by Mayr (1958), Wickler (1961), Amadon (1959), and others. Many papers, too numerous to list, treat the usefulness of certain types of behavior in taxonomy and systematics or of behavior in general in the taxonomy of a particular phylogenetic group.

Discovery of Natural Groupings through Behavioral Analysis / After the usefulness of behavioral traits in phylogenetic classifications had become apparent, it was not long before many behaviorists began to discover new relationships and confirm previously uncertain ones.

The Ptilonorhynchidae (bowerbirds and catbirds), a family of birds in New Guinea and Australia, provides one such example. Stresemann (1953) and Marshall (1954) found that two principal groups could be recognized. Catbirds are monogamous and do not build bowers, and the male parent helps to feed the young. Bowerbirds are polygamous and build bowers, and the male does not feed the young. Furthermore, the bowerbirds can be divided into two types on the basis of the courtship behavior of males. "Maypole-builders" clear and decorate a circular display area around a central sapling; "avenue builders" clear and decorate a narrow rectangular or oval space on the ground (see Figure 15–11).

Mayr (1958) has reviewed many more such cases in which the classification of individuals into families, subfamilies, or genera has been modified through the use of behavioral characters.

The use of behavioral characteristics in the working out of phylogenetic relationships depends on the concept of common ancestry. This has been formalized as the concept of *homology*. Terming behavioral patterns in different species homologous implies that certain similarities of these patterns are explainable on the hypothesis of common ancestry. Such assertions can, of course, never be "proven" or "established," as some ethologists have maintained, since events in the past can in theory not be known with absolute certainty. At best one can show that a particular hypothesis of common ancestry is highly convincing or highly probable. In theory the concept of homology is relatively simple; yet in practice and in its fine points it often becomes controversial. The student who wishes to look into the methodology for investigating phylogenetic groupings should consult the recent literature of evolutionary morphology and of numerical taxonomy, since it is in these fields that the methods find their most frequent use and most searching analyses. Later it may be useful to read less sophisticated critiques such as that of Atz (1970) and earlier authors (listed by Atz) on their application to behavior.

Discovery of New Species through Behavioral Analysis / In some cases visible morphological differences between species may be nearly or completely undetectable. Such groups of species are known as *sibling species* (Mayr 1963). The discovery of sibling species has been greatly aided by behavioral studies. One such new species was discovered by Adriaanse (1947) while watching wasps then known as *Ammophila campestris*. On close inspection it was observed that there were two types that could be distinguished according to the nesting behavior described in Table 1–1. The second form was subsequently discovered to have consistent morphological differences from the first and was named *A. adriaansei*.

New species of fireflies have been discovered by their patterns of

TABLE 1–1 Behavioral differences between two sibling species of wasps (*Ammophila*). (After Mayr 1958.)

Behavioral trait	*A. campestris*	*A. adriaansei*
Material used to fill nest hole	carried on foot from nearby "quarry"	flown in from distant source
Food	sawflies	caterpillars
Provisioning sequence	egg, then prey	prey, then egg
Breeding season	earlier, to August	later, to mid-September

flashing (Barber 1951), and new species of frogs, crickets, and birds by their calls and songs. In most cases correlated morphological differences are found on close examination, but in the frog *Hyla versicolor* no such differences proved reliable. In this species complex, two types of mating calls were discovered, and it was shown that females consistently chose the same one when the calls were played to them through two loudspeakers (Johnson 1966). Both males and females were identifiable only behaviorally. It was also found that crosses between the two forms showed a high degree of incompatability (Johnson 1959, 1963). For further material about sibling species discovered by their behavior, see the discussion of behavioral isolating mechanisms (Chapter 18).

Behavioral Consistency within Species / If behavioral traits are to be used in classification, it is desirable that all members of the species having a given trait show it consistently. Absolute differences are preferable to average differences. An example of a behavioral trait that is essentially invariable in some species but variable in others is head-scratching in birds. It was pointed out by Heinroth (1930) that birds scratch their heads in one of two ways, with one wing partially extended and the leg passing over it to reach the head (indirect), or with both wings folded (direct), as in Figure 1–2. That species differ consistently in this respect you can easily verify while sitting on a park bench. Pigeons (*Columba livia*) will be seen to scratch their heads only in the direct fashion, while house sparrows (*Passer domesticus*) use only the indirect method. Some species have been reported to use either method (e.g., Nice and Schantz 1959; Ficken and Ficken 1958, 1968; Dunham 1963); nevertheless, the consistency of this behavior in most species has made it useful in classification (Simmons 1957, 1961; Brown 1959; Brereton and Immelmann 1962).

Another simple behavioral trait that usually is completely con-

1–2 Head-scratching movements. *Left:* The direct method, with wing folded and leg under it (Nashville warbler, *Vermivora ruficapilla*). *Right:* The indirect method, with wing partially extended and leg passing over it (ruby-crowned kinglet, *Regulus calendula*).

sistent within a species but differs between families is the method of drinking. This too you can verify from a park bench.* Most species of birds drink in one of two ways. Pigeons drink like mammals, in one respect: they suck up water into their throats, keeping the bill somewhat deep in the water while swallowing. But house sparrows seem unable to pump water upward into their throats and swallow while the head is down. They first raise the head and neck so the water can flow by gravity into the throat. As a result, they alternate repeatedly between taking water in their bills and throwing their heads back. The sucking method of drinking is one of the behavioral traits that characterize most members of the family Columbidae (pigeons and doves). In this family it seems to be related to the use of crop milk in feeding the young. It is also found in a few species of waxbills (Poulsen 1953) and colies (Cade and Greenwald 1966), where it may be adaptive in desert conditions. The frequently cited use of this character to ally the Pteroclidae (sand grouse) with the Columbidae is an error, since it has been observed that sand grouse do not drink the way pigeons do (Schönholzer 1959: 388; Cade, Willoughby, and Maclean 1966).

The Adaptive Value of Behavior / In the early applications of behavioral study to the problem of phylogeny, interest centered on the usefulness of behavior patterns as taxonomic characters, and little attention was devoted to the environments and selection pressures that

* Methods of drinking among zoo animals have been reviewed by Schönholzer (1959).

more efficient means of obtaining the same types of insect prey. It is notable that stick-poking is not known among mainland species; but a somewhat similar type of behavior has been described in one population of brown-headed nuthatches in the southern United States (Morse 1968). Stick-probing has also been found in the insular New Caledonian crow, *Corvus moneduloides* (Orenstein 1972).

The Hawaiian honeycreepers (Drepaniidae) show adaptive radiation in an even more extreme form, and their structural modifications and colorful plumages are much more bizarre than the drab Galapagos finches. Although adaptive radiation in structure has been studied in Hawaiian honeycreepers (Baldwin 1953; Bock 1970), opportunities for study of the behavior of these species are diminishing. Several species have become extinct or are now surviving perilously due to encroaching civilization. Of course, many families and orders of animals provide good examples of adaptive radiation, but its occurrence on archipelagos is especially impressive and instructive.

Domestication and Behavior / Adaptive radiation can also take place within a single species under the conditions of domestication in which mating between the breeds is prevented. In their reviews of the effects of domestication on the dog, Scott and Fuller (1965; see also Scott 1967, 1968) pointed out that most breeds of dogs were developed for special purposes. Dogs excelling in a given task were selected for breeding. Therefore, dogs that could stop and point silently at game birds without making them fly away were selected as pointers; beagles were selected for success in rabbit hunting; terriers, for their ability to go down fox and badger burrows; sheep dogs, for shepherding; and greyhounds, for rapid chases over open country. There was even developed in England at one time a "thievish dog" that, because it was essentially barkless, was used by game poachers who hunted mainly at night and wished to avoid detection.

Many unusual behaviors have been selected for in other species. The roller canary is a variety famed for its unique song (Marler 1959). Siamese fighting fish and Asian gamecocks have been bred for success in fights. Various breeds of pigeons have been selected for homing, aerial acrobatics (tumblers), display (pouters), and vocalizations (Levi 1965). Differences in gait of horses have been bred selectively.

Probably the most common behavioral change resulting from domesticiation is a loss of aggressiveness toward man. This is evident in laboratory rats, cattle, horses, pigs, goats, sheep, water buffalo, zebra finches, chickens, turkeys, and possibly in almost every domesticated species of animal except those species bred for fighting, such as bulls, fighting fish, cocks, and Chinese crickets.

Domesticated species provide many good examples of the correlation between behavior and phylogeny at a level below that of the species. Behavioral differences among breeds are conspicuous in most domesticated species. In these cases it is certain that the similarities within a breed are due mainly to common ancestry, or in other words to genetic factors. The role of genetic factors has been extensively demonstrated for four breeds of dogs by breeding experiments (Scott and Fuller 1965). The behavior of domesticated animals and the process of domestication have been reviewed by Hale (1962).

Evolutionary Convergence and Parallelism in Behavior / When a trait has evolved similarly in two or more clearly unrelated groups, the case is commonly interpreted as convergent evolution (Figure 1–5). Many such traits are known. Sociality has developed independently in several groups of Hymenoptera (wasps and ants), and in Isoptera (termites). Both bats and most birds fly. Certain squirrels, flying lemurs, lizards, and frogs can glide (Figure 1–6). Several unrelated types of birds such as swallows, flycatchers, and wrens use old woodpecker holes for nest sites and have evolved specialized behavior adapted to them. Polygamy and monogamy have evolved independently in many animal groups. Cleaning symbioses have evolved separately in unrelated species of fishes. Several small passerines in different families have similar "seee"–alarm calls. Such cases are thought to have come about through similar selection pressures acting on different types of animals.

When a trait has evolved similarly, presumably under similar

1–5 Evolutionary convergence and parallelism. In convergence, g and h are similar but evolved from different phyletic lines. In parallelism, n, o, p, and q are similar but evolved from a common ancestor, z, from which they all now differ considerably. (From V. H. Heywood, *Plant Taxonomy*—Studies in Biology—Edward Arnold (Publishers) Ltd.)

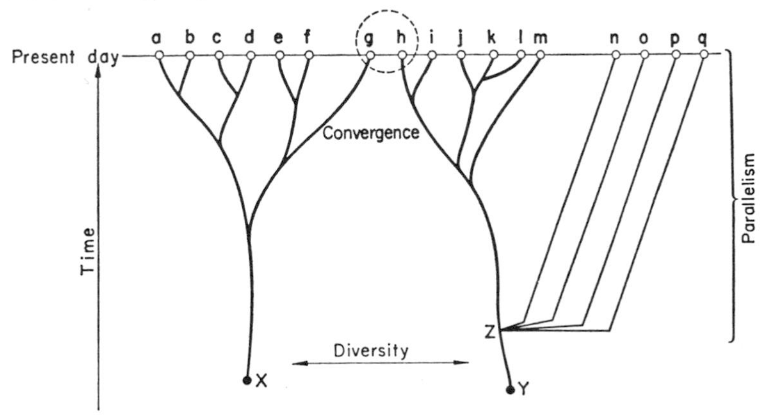

1–6 Evolutionary convergence in gliding behavior. These animals inhabit forests of Malaya. The colugo has the most extensive skin flaps of any gliding animal and can travel 210 feet with a drop of only 40 feet (11 degrees). The arboreal flying frog is more like a parachutist, the large webs acting mostly to slow its rate of fall. *Draco* and the gecko are lizards.

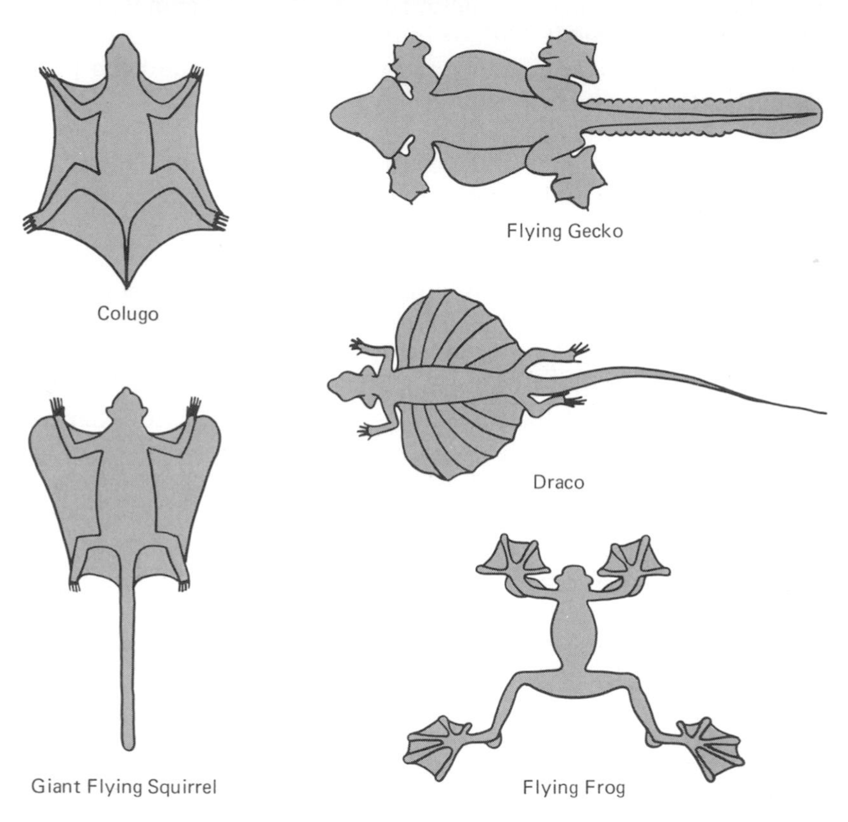

selection pressures, in two or more groups having a common ancestor that lacked the trait, parallel evolution is said to have occurred (Figure 1–5). Little attention has been given to parallelism in the evolution of behavior, and no convincing examples have been described. (However, longitudinal contraction, used for escape by several genera of worms, might be described as parallelism, since it is mediated by different neuronal circuits in different genera—see Figure 21–3.) Nevertheless, the condition is known for some nonbehavioral traits; and it deserves some attention by ethologists.

Progressive Evolutionary Change in Behavior / When comparative studies show that a set of species can be arranged to form a progression in one or more characters from one extreme to another, it can be concluded that since intermediate forms are possible and still exist, evolution could have proceeded through a similar series (but usually not the identical one).

The question of which extreme evolved from which often cannot be answered unless there exists either a fossil record or a logical argument showing that one of the two possible directions is unlikely. For example, the balloon flies described by Kessel (1955) form a series with intermediate species in a range between one in which the male presents a silken ball to the female during courtship (Figure 1–7) and the more numerous species that lack this trait.

There seems at first glance to be no apparent advantage of the silk-ball courtship over the normal type; both accomplish the same end. But comparative study of the intermediate types reveals a logical reason for the evolutionary change and for concluding that the silken-ball type evolved from the normal type and not vice versa. The female balloon fly is voracious and may eat the male during courtship. This danger is minimized in one species by having the male present an insect prey to the female during the courtship preliminaries. But even this has its dangers if the female finishes the prey too soon. Another species prolongs the female's consumption of the gift by secreting a web of silken fibers around it. This trend is carried further by a species in which the prey is relatively small and in which the web itself becomes more important. From here it is only a small step to the terminal stage, in which the female gets an empty web. In summary, using P for prey and W for web, we have the following progression of phenotypes: $P^-W^- \rightarrow P^+W^- \rightarrow P^+W^+ \rightarrow P^-W^+$.

1–7 Presentation of silk ball by male to female in the courtship of *Empis*. (From Meisenheimer 1921.)

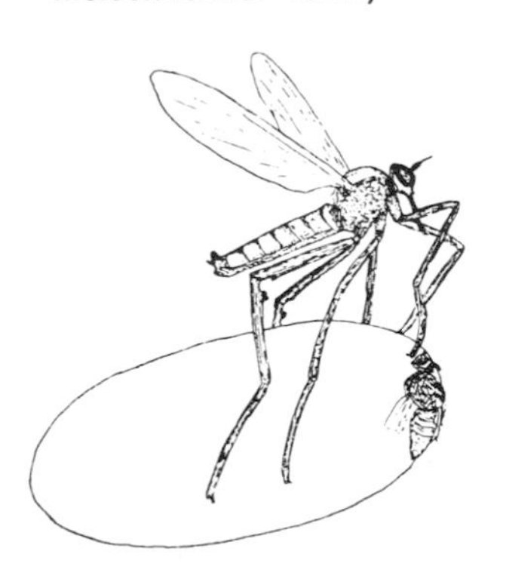

The prey-carrying behavior of wasps, which likewise exhibits a series, has been studied comparatively by Evans (1962). Background material on wasps and excellent illustrations of prey carriage are provided by Evans and Eberhard (1970). Figure 1–8 shows the distribution of the various types of prey carriage among the major groups of wasps.

The ancestors of the familiar Aculeate wasps were probably parasitoids, represented today by ichneumons, chalcids, and gall wasps. In

1–8 Distribution of types of prey carriage on phylogenetic tree of wasps. **A,** modified abdominal tergite; **B,** on sting; **C,** hind legs; **D,** flies with prey in legs; **E,** flies with prey in mandibles; **F,** walks forward, prey in mandibles; **G,** walks backward, prey in mandibles. (After Evans 1962.)

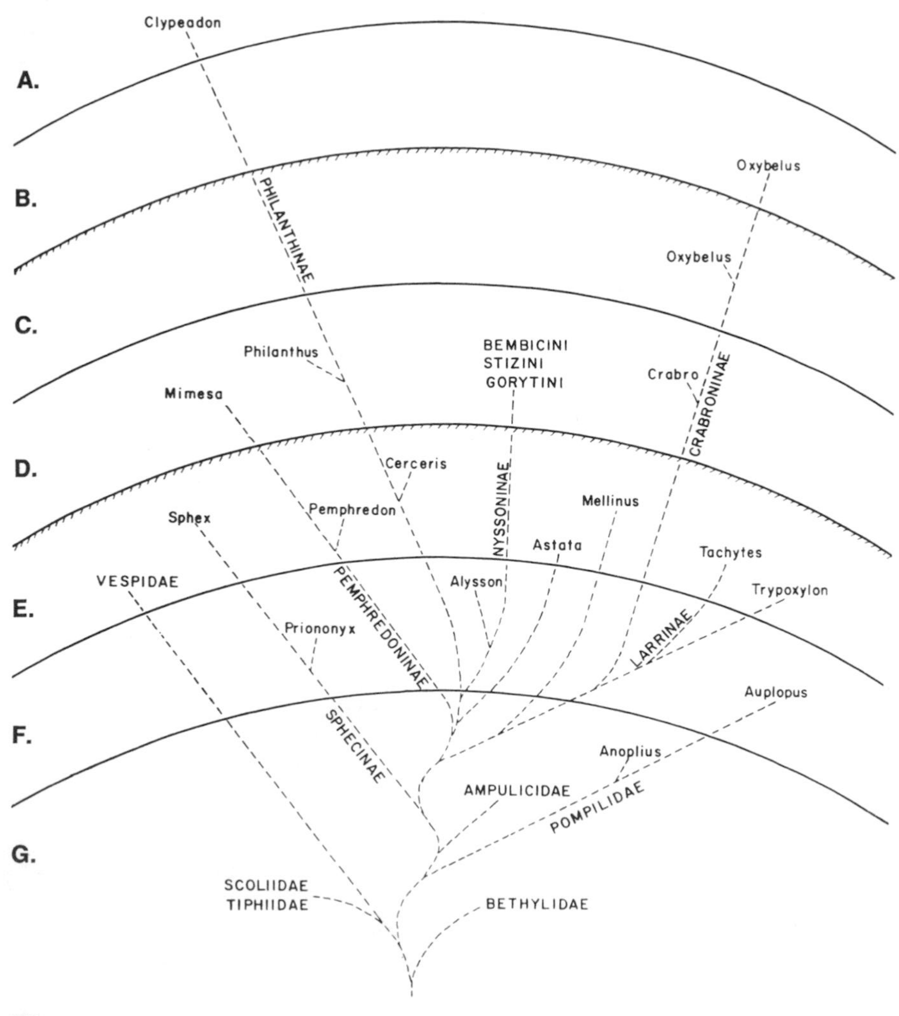

these groups, females do not have a nest or a sting and do not carry prey; they lay several eggs on living prey such as caterpillars and grubs. The eggs hatch and the parasitic larvae develop on the host until it finally dies. A member of the Tiphiidae, *Methoca*, differs from this pattern: it stings its prey and lays only one egg per host. The victim is left in its burrow in the ground (Figure 1–9) or dragged backwards to a suitable hole. The spider wasp *Auplopus* (Figure 1–10) drags her prey *frontwards* by the mandibles to a definite nest, allowing the mother to see where she is going. Species utilizing smaller prey may fly back to their nests carrying the prey in their mandibles (several genera are listed in Figure 1–8). The disadvantage of carrying the prey in the mandibles is that the female must leave the prey exposed on the ground while she opens or digs her burrow. During this relatively brief time, various cleptoparasitic flies or wasps may lay their eggs on the prey item. The eggs of the cleptoparasite hatch before those of the wasp, and the wasp egg or larva subsequently dies. By carrying the prey in the middle or hind legs, the wasp can open the burrow with her forelegs while she protects her prey from cleptoparasites. In this series the prey is carried progressively more posteriorly, starting at the front of the wasp and finishing at the rear. In the last stage of the series, the wasps carry their prey either on their sting, which is sometimes barbed in these species (e.g., *Oxybelus quadrinotatum*), or on a terminal abdominal tergite modified to function as an ant clamp, as in *Clypeadon laticinctus* (shown in Figure 1–10). We know enough about wasp phylogeny to say that in this series the progression was from mandibular to abdominal carriage, rather than vice versa.

Many other cases of series or progressions are known. Species that use *external constructions* are often particularly useful for studies of this sort because the constructions persist long after the behavior that built them, and they can be collected and directly compared together. Evolutionary progressions in external constructions have been suggested by the webs of spiders (Kaston 1964; Kullman 1972), the houses of caddisfly larvae (Ross 1964), and the nests of bees (Michener

1–9 A primitive method of provisioning the nest in wasps. A spider wasp, *Sericopompilus apicalis,* drags a spider backward to her nest. (From Evans 1953.)

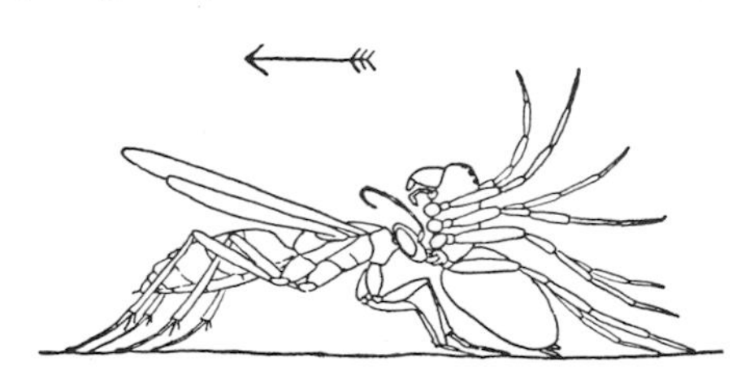

1–10 A specialized method of prey carriage in wasps. *Left:* A female *Clypeadon* digs into her nest entrance while holding an ant on the tip of her modified abdomen. *Right:* Terminal abdominal segment (tergite) of a typical digger wasp, *Auplopus* (*above*), and of *Clypeadon* (*below*), showing modification in shape for carrying ants. (From Evans 1962. Photo by André Steiner.)

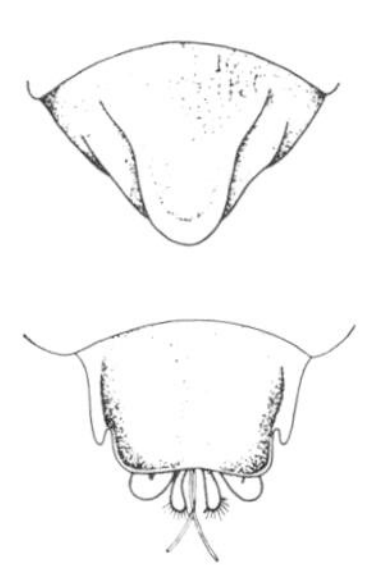

1964), termites (Schmidt 1964), and birds (Collias 1964; Collias and Collias 1963, 1964; Crook 1963). The construction of nest burrows by digger wasps has been investigated by Evans (1966*a*, *b*, *c*). Wickler (1962) has done a beautiful comparative study of mouth breeding in fishes. The methods of carrying nest materials by lovebirds form another fascinating progression (Dilger 1960, 1962).

Fossil Records of Behavior / Fossils are rarely of use to behaviorists. One can sometimes make certain deductions about how extinct species probably behaved on the basis of fossil skeletons, teeth, and certain specialized parts, such as horns (Colbert 1958). In a few cases evolutionary progressions have been recognized in the fossilized tracks of worms in successive sedimentary strata (Seilacher 1964, 1967; Raup and Seilacher 1969).

Clines in Behavior / In evolutionary studies the term *cline* usually refers to a pattern of geographic variation (within a species) in which a character changes gradually from one part of the range of the species to another. Clines in body size and appearance are often associated with geographic differences in climate (e.g., Allen's Rule, Gloger's Rule; see Mayr 1963). Clines in behavior have received little attention. Clinal variation in a structure used as a part of a display has been analyzed in jays (see Figure 15–13; Brown 1963*a*). Clinal patterns of geographic variation among populations probably also occur in the chemical cue preferences for feeding responses in garter snakes (Burghardt 1970) and in nesting behavior of mice of the genus *Peromyscus* (King et al. 1964; Layne 1969).

Since behavioral differences within a species tend to be rather subtle and since the role of environmental factors in their determination is usually uncertain, they have not been extensively investigated for most types of behavior. Yet, they are useful in evolutionary studies for showing the kinds of phenotypic (and presumably genetic) differences within one species upon which natural selection may act to modify behavior. Examples are known of clines and other types of geographical variation in the calls and songs of insects, frogs, and birds, and in the displays of lizards. These will be discussed later in relation to auditory communication and the evolution of isolating mechanisms.

CONCLUSIONS

This brief survey of the patterns of variation in behavior among and within species has revealed that many behavioral traits vary in patterns identical to the classical ones found by evolutionary biologists using morphological traits. Evolutionary convergence, adaptive radiation, clines, progressive series, and domestication may be demonstrated for behavioral traits as well as for morphological ones. The correspondence between natural groupings of species suggested by behavioral traits and those suggested by morphological and other nonbehavioral traits is impressive. All of these considerations reaffirm the general conclusion that the behavior of every species of animal, including man, has been profoundly influenced by its evolutionary history even down to seemingly insignificant details of behavior.

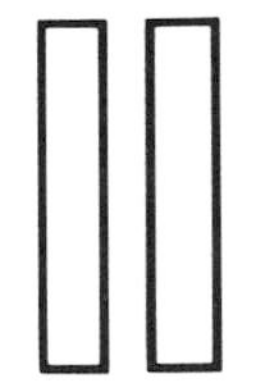

THE GENETIC BASES OF BEHAVIOR

THE FACT THAT patterns of behavioral similarities and differences among species are correlated with phylogeny strongly suggests that evolution has been the primary determinant of these patterns. If so, it follows that behavioral differences and similarities among species must be largely under genetic control and that there must be intraspecific genetic variability in behavior on which selection may act. Experimental verification of these conclusions is abundant, but only a tiny fraction of it can be mentioned here. In the following chapter we shall (1) examine some studies that demonstrate the genetic bases of behavior, (2) introduce some quantitative concepts for the analysis of the relative contributions of genetic and environmental determinants of variation, and (3) consider briefly the role of genes in the development of behavior. This chapter is not intended as a review of the field of behavior genetics; only a few topics useful to the theme of the evolution of behavior are included here. Other aspects of behavior genetics are covered in the chapters on mating systems, motor patterns, and speciation. The reader interested in a more comprehensive introduction or review of behavior genetics is referred to Fuller and Thompson (1960), Parsons (1967), Hirsch (1967), Manosevitz et al. (1969), Thiessen (1971), and McClearn and DeFries (1973).

2 Behavior, Genetics, and Evolution

THROUGHOUT this book concepts are employed that are predicated on genetic determination of behavior in populations. However, only in this chapter will we deal with actual experimental verification of such assumptions per se. There are basically only three methods of demonstrating genetic determination conclusively: (1) crossing experiments, (2) selection experiments, and (3) comparison of identical and fraternal twins. We will discuss here the first two, which are useful in working with animals.

An elegant way to verify that behavior can be determined genetically is to find behavioral differences among individuals that by previous crossing experiments are known to differ genetically at only a single gene locus. In normal animals, however, detectable effects of genetic variation at a single locus are rare. In cases when effects of single genes cannot be isolated, statistical methods are necessary to estimate the roles of inheritance and environment. We will examine examples of both the single-gene and the statistical approaches. Our brief consideration of the quantitative methods for analysis of such variation should also aid in understanding the processes of natural selection. It is convenient to introduce the effects of different types of selection here in the same context.

Genetic variation in natural populations is most conspicuous in genetic polymorphisms and in species differences. These are briefly discussed. Finally, innateness and the roles of genetic and environmental factors in development are briefly discussed.

Demonstration of the Genetic Basis of Behavioral Differences / One of the two basic experimental methods used to reveal the genetic basis of differences in behavior is the crossing experiment, in which individuals that differ in a particular kind of behavior are crossed and their progeny examined. In all such crosses it is of the utmost importance that the environment be constant or well controlled.

The types of behavioral differences that can be studied through crosses are partially determined by the nature of the parental stocks. In order of genetic complexity, the crosses may involve individuals differing by (1) a single gene, (2) a block of genes, such as a chromosomal inversion or a whole chromosome; individuals of (3) different inbred strains that are almost purely homozygous, (4) different breeds (for example, of dogs or of chickens), (5) different natural populations within one species, or (6) different species, genera, and higher categories. As a general rule, except for abnormal mutations, the complexity of the differences in a given behavioral trait that may be found between individuals tends to be correlated with the complexity of the genetic differences between them. For example, individuals of *Drosophila pseudoobscura* differing only in their chromosomal inversions do not differ in behavior as much as individuals from different species and subgenera of *Drosophila*.

Single-Gene Differences / One of the more convincing methods of establishing genetic effects on behavior is through the study of single-gene mutations. The known examples of these almost invariably represent gross abnormalities, since single-gene differences with small effects within the normal range of phenotypic variation are difficult to discover except by special techniques (e.g., Benzer 1967; Konopka and Benzer 1971). Nevertheless, in species such as *Drosophila melanogaster*, in which many single-gene mutants are readily available, good opportunities exist for studying the effects of single-gene differences on behavior. One of the first studies of this type was done on the *yellow* mutant of *Drosophila melanogaster*, which was known to be less successful than the wild type in competitive mating. Bastock (1956) confirmed that the deficiency existed and showed that it was attributable in part to the smaller amount of time that *yellow* males spent in the wing-vibration phase of courtship (see Figure 8–10).

A more spectacular example of a behavioral difference with an apparently simple genetic explanation was found in the hygienic behavior of honeybees by Rothenbuhler (1964, 1967). The Van Scoy strain of honeybees is susceptible to a bacterial infection known as American foulbrood, while the Brown strain is resistant. Resistance is due in large part to the cleaning behavior of worker bees, which remove dead larvae in the Brown strain (hygienic), thus preventing spread of the infection, but not in the Van Scoy strain (nonhygienic). Crosses between the two strains showed that the F_1 (first filial generation) workers were nonhygienic and that the genes conferring resistance were therefore recessive. In backcrosses to the homozygous recessive Brown line, roughly a quarter (6/29) of the colonies were hygienic, suggesting that the trait was controlled by two loci. Examination of the nonhygienic colonies revealed three types: (1) one-third of the colonies performed

2–1 Genetics of hygienic behavior in honeybees. *Above:* Experiments and their results. Number of observations refers to colonies rather than individual workers (see Chapter 10). The nonhygienic colonies could be divided into those that remove the dead larvae if the cells are first uncapped and those that do not. *Below:* Genetic hypothesis offered in explanation of responses to brood killed by American foulbrood in 63 colonies of bees. (From Rothenbuhler 1967.)

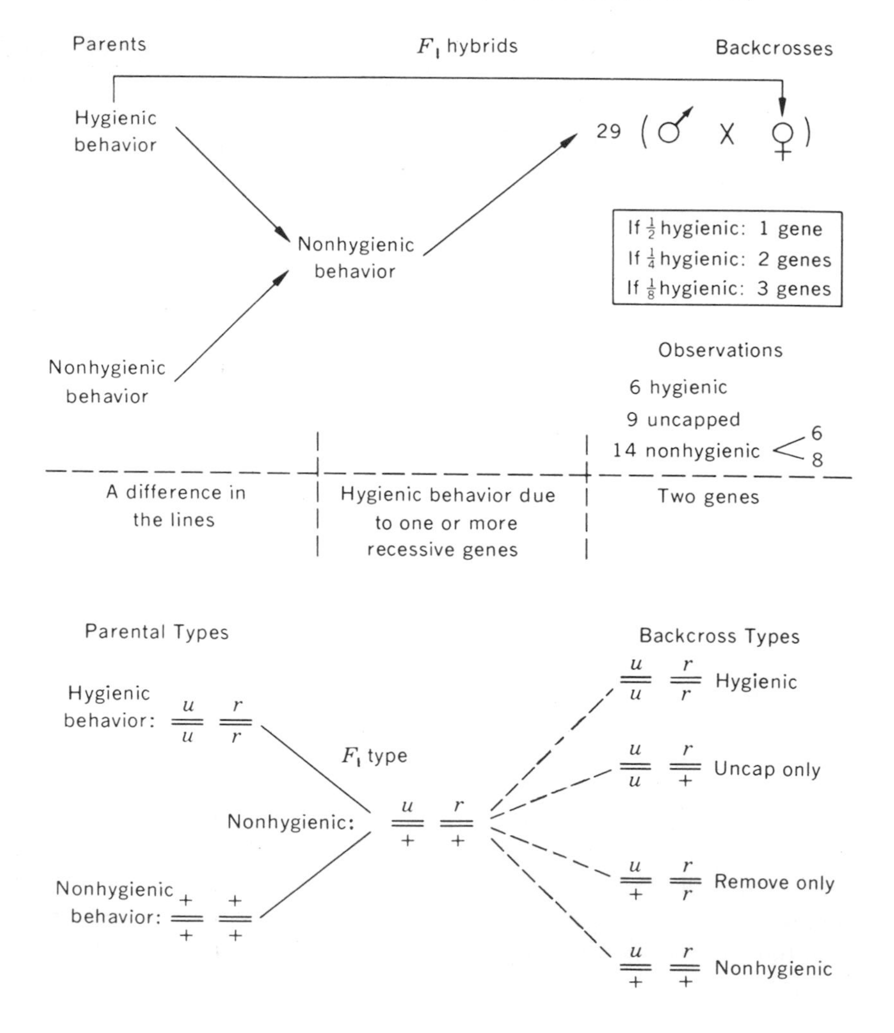

the uncapping of the cells of dead larvae but failed to remove them; (2) one-third would not remove the caps of the cells themselves, but when caps were removed for them they would remove the dead larvae; and (3) one-third would neither uncap cells nor remove dead larvae from opened cells. This result was consistent with the hypothesis that one of two loci affected uncapping behavior, the other affected removal behavior, and that both together controlled the difference in hygienic behavior between the two strains (Figure 2–1). There was no evidence of one genotype learning from another.

Heritability of Polygenic Behavioral Traits / Although many morphological phenotypes have been found that are attributable to genetic variation at a single locus and that can be studied by the classical Mendelian methods based on segregation of distinct types, few such genetically simple behavioral phenotypes are known. Genes responsible for such traits are known as *major genes*. More commonly, a behavioral trait varies along a continuum in which the influence of a single gene is not detectable. This situation arises when genes at many loci have small, additive effects, and when the influence of the environment on variability in the trait is large in relation to the influence of single loci. Genetic determination of variation in such traits is termed *polygenic,* and the genes are known as *polygenes.*

Since polygenic systems are so widespread in behavioral traits, we shall introduce briefly some of the concepts used for their analysis. For details the reader is referred to Parsons (1967), Hirsch (1967), and Falconer (1960). The principal statistical concepts to be employed are the mean, $\bar{x}$,

$$\bar{x} = \frac{\Sigma x_i}{n},$$

the variance, V,

$$V = \frac{\Sigma(x_i - \bar{x})^2}{n - 1},$$

and the standard deviation, SD,

$$SD = \sqrt{V}.$$

Here x_i is an individual measurement and n is the number of measurements. The principal statistical method used is known as *analysis of variance;* it is described in most textbooks on statistical analysis. For the present purposes it is not necessary to know how to perform an analysis of variance; it is only necessary to understand that it is an objective method for partitioning variance into its components. Basically, the sample variance in the phenotype, V_P, of a behavioral trait is divided into a part due to genetic effects, V_G; a part due to environmental effects, V_E; and a part due to the interaction of genotype with environment, if present, $COV_{G,E}$, that is, the covariance of genotypic values and environmental deviations. This may be written

$$V_P = V_G + V_E + 2\ COV_{G,E}.$$

One of the tasks of behavioral genetics is to evaluate variation within and among populations by means of these concepts.

Heritability: Comparison of Inbred Lines / One of the approaches to the problem of estimating the components of V_P makes use of inbred strains. These strains have been sib-mated (brother × sister) for many generations, a procedure that is expected to cause a fall in the proportion of heterozygotes of about 19 percent per generation. In theory, such strains should ultimately be completely homozygous. Consequently, within one inbred strain we assume $V_G = 0$.

An example of this approach has been described by Parsons (1967) for *Drosophila melanogaster*. In five strains sib-mated for at least 140 generations, mean durations of copulation ranged from 16.7 to 21.4 min. Analysis of variance for these data revealed the following, where M_2 is the within-strains variance and M_1 is the between-strains variance: $M_2 = 14.00$; neglecting $COV_{G,E}$, M_2 can be taken as V_E because there was no within-strains genetic variability. In this instance $V_E = M_2 = 14.00$. The expected variance between strains has both environmental and genetic components. $M_1 = V_E + rV_G = 172.75$, where r is the number of replicates for each strain. Rearranging and substituting,

$$V_G = \frac{M_1 - V_E}{r} = \frac{M_1 - M_2}{r} = \frac{172.75 - 14.00}{52} = 3.05.$$

The *heritability*, h^2, can then be calculated as

$$h^2 = \frac{V_G}{V_G + V_E} = \frac{3.05}{3.05 + 14.00} = 0.18.$$

This is also known as the *degree of genetic determination*. It varies between 0, in the case of no genetic variability, and 1, in the case of no environmental variability.

It is important to avoid thinking of heritability in developmental terms or as a property of an individual. It is not an estimate of the importance of genetic factors in the ontogenetic development of behavior in an individual. It is a characteristic of the genetic variability in a sample *population*, for the kind of behavior observed, and for the precise method of observation used.

The population aspects of heritability (no matter how estimated) are worth considering. The greater the differences among the means of the inbred strains, the greater will be M_1, V_G, and h^2. Consequently, estimates of these factors depend on the investigator's choice of strains. If he wishes to maximize V_G and h^2, he would choose strains that are as different as possible in respect to the behavioral trait being analyzed. Since the amount of variability among laboratory stocks of the standard experimental animals such as *Drosophila*, mice, and rats is related to the amount of effort expended by geneticists to preserve variations that have been discovered, it is evident that single values for h^2 derived from such samples have little meaning. The method of comparison of inbred strains is most meaningful when the strains have been drawn randomly from a natural population, but even here it has severe limitations due to the necessarily small sample that can be drawn and due to the necessary absence of lethals in the selected line.

Heritability: Selection for Behavioral Traits / Artificial selection can be used to alter population phenotypes in several ways, depending on which fractions of the population are chosen for succeeding generations (Mather 1953; see Figure 2–2). In *stabilizing selection*, also known as normalizing selection, breeders are chosen from the center of the

2–2 The three basic modes of selection. Note the differences in means and distributions characteristic of each. The ordinate indicates the frequency of individuals in the population; the abscissa shows a metric scale for any quantitative character being considered.

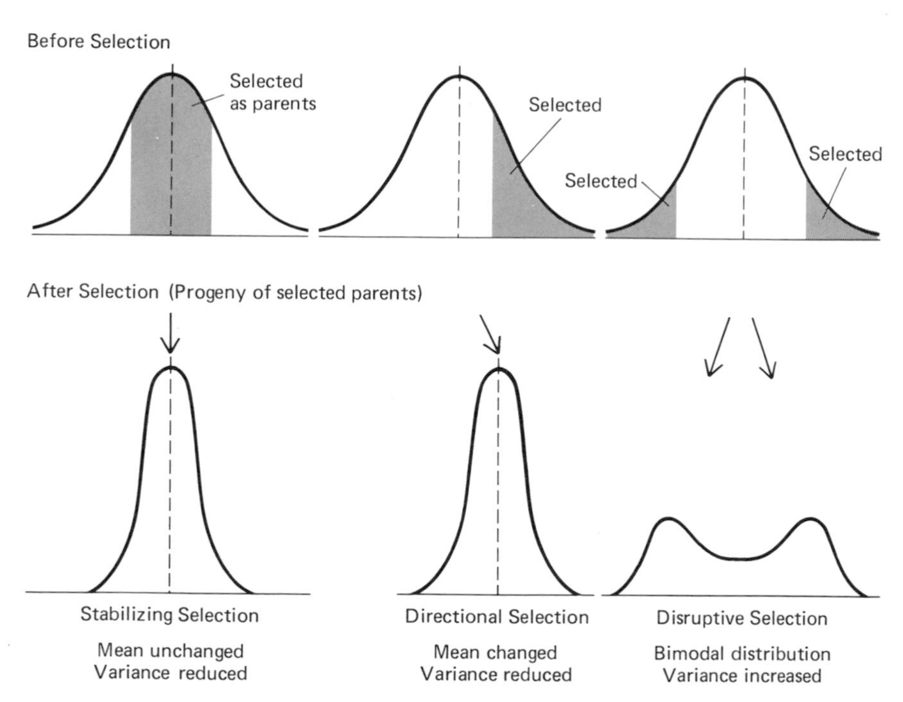

distribution and extremes are rejected. The effect is primarily to reduce genetic variability in the selected line and secondarily to preserve the same mean value.

In *directional selection* breeders are chosen from one end of the distribution. The primary effect is to shift the mean of the population in the direction of the extreme chosen; secondarily, genetic variation in the selected line is reduced.

In *disruptive selection* breeders are chosen from both extremes of the distribution and the central part is rejected. The primary effect is to produce a bimodal distribution. Disruptive selection is thought to be involved in the origin and maintenance of genetic polymorphisms (Mather 1955; Thoday 1959, 1965). Dobzhansky (1968) has termed this diversifying selection in reference to its important role in natural populations inhabiting diverse environments.

It is worth considering briefly the relationship between heritability and selection (Figure 2–3). The *selection differential,* S, reflects the mean phenotypic value of the fraction chosen as parents and is expressed as a deviation from the mean phenotypic value for the whole population. The magnitude of the *response* to selection, R, or the difference between the means of the parental and filial generations, depends upon S and h^2:

$$R = h^2 S.$$

2–3 Estimation of realized heritability in a selection experiment. In this case $h^2 = 0.5$. What would the curve for the filial generation look like if $h^2 = 0$? What would it look like if $h^2 = 1.0$?

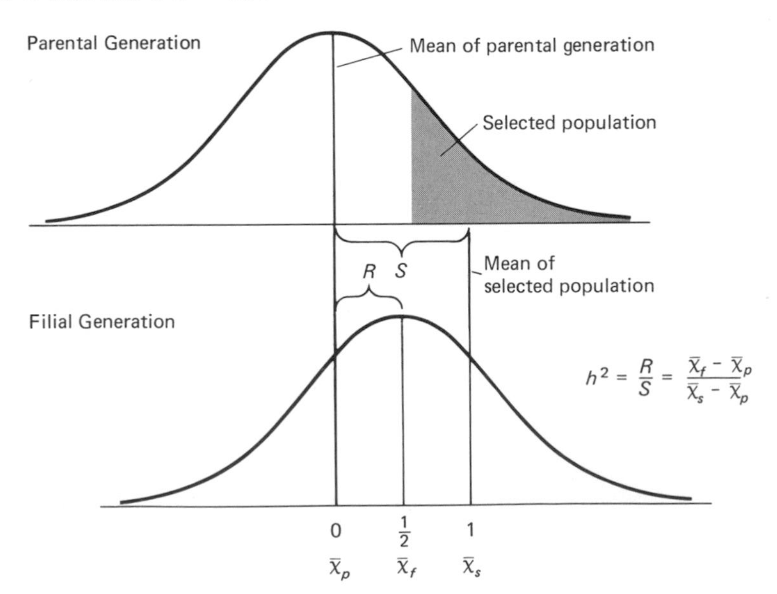

Thus, when h^2 is 0, the mean of the filial population should not differ significantly from the mean of the parental population, regardless of the selection differential. This should happen when directional selection is attempted on an inbred, homozygous strain – since all variation is attributable to the environment in such cases. When h^2 is 1, the mean of the filial generation should approximate the mean of the selected sample. In most cases, of course, with intermediate values of h^2, the response to selection would lie between these extremes.

It can be seen from the above equation that selection experiments can be used to determine heritability, since R and S can be determined empirically. The value so obtained is termed *realized heritability* (Roberts 1967).

$$\text{realized heritability} = h^2 = \frac{R}{S}$$

The results of selection experiments, whether with behavioral or structural traits, tend to vary in detail with the trait chosen and the animals involved. With minor exceptions, however, the results of selection for behavior parallel results for structure. This is not surprising since all behavior must have a structural basis (in the broad sense). With directional selection the mean can usually be shifted and the variance may or may not be reduced.

Mating behavior is of considerable importance as a direct component of fitness and as a factor in the origin of species. Many selection experiments have been performed on various aspects of mating behavior in *Drosophila* (reviewed by Petit and Ehrman 1969; Spiess 1970), and some of these will be discussed in the chapters on sexual selection and on the evolution of behavioral isolating mechanisms. Experiments by Manning (1961) on mating speed (latency to copulation) in *Drosophila melanogaster* serve to illustrate the results of artificial selection on behavior and provide background for our return to the genetics of mating behavior in later chapters. Using 50 pairs of virgin flies, four to six days old, placed together in a half-pint milk bottle, Manning selected the first and last 10 pairs formed to initiate his fast and slow lines. Thereafter, the first 10 pairs in the fast lines and the last 10 in the slow lines were selected in each generation.

Results for two fast and two slow lines of repeated selection and for controls are shown in Figure 2–4. Considerable divergence between fast and slow lines was attained in the first 7 generations, but was not greatly increased by further selection to 25 generations. A heritability of approximately 0.30 was estimated for mating speed for this population. Fluctuations in all the lines from generation to generation tended to parallel each other, suggesting the influence of unidentified environmental factors.

Crossing fast and slow lines in both directions gave intermediate F_1 mating speeds, while mating two fast lines against each other and the two slow lines against each other yielded fast and slow maters, respectively, thus indicating that both sexes were affected. The nature of the behavioral change was also investigated. In general activity, fast maters were remarkably sluggish and slow maters reacted strongly to any stimulus by flying and running. This conclusion from casual observation was verified quantitatively by counting the number of grid squares entered by a fly in the first minute after introduction to the grid chamber. The effect of activity on mating speed was to increase the length of disturbance following introduction of the flies into the test bottle, thereby delaying the onset of courtship. Selected males placed with unselected females responded by pairing appropriately rapidly or slowly. The frequency of licking (terminal phase in courtship just preceding copulation) shown to unselected females was higher in the males of the fast than slow lines. Females of slow lines showed lowered receptivity to control males, and those of fast lines showed increased receptivity. The results of selection could be explained through effects on both general and sexual activity. Fast maters had reduced general activity and increased sexual activity. Slow maters had increased general activity and reduced sexual activity. These experiments illustrate a com-

2–4 Selection for fast and slow mating speed in *Drosophila melanogaster. Mating time* is the time from the release of flies into the container until copulation begins. There were two fast lines, *FB, FA,* and two slow lines, *SB, SA.* Mean mating time is plotted on a logarithmic scale. Selection was relaxed in generations 11, 14, and 21. (From Manning 1961.)

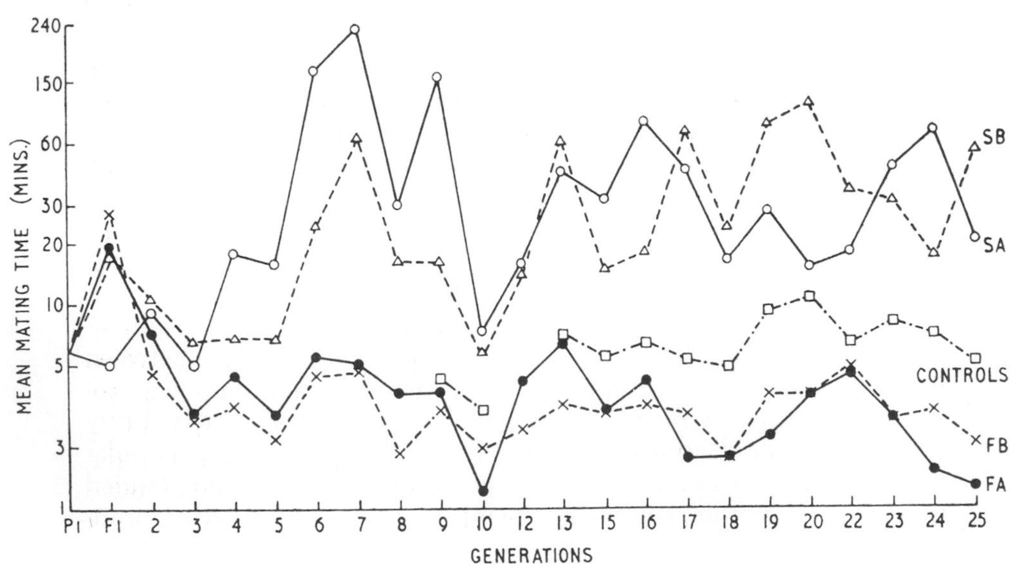

mon result of selection for behavioral traits: A genetic change in behavior is often achieved through relatively nonspecific effects instead of, or in addition to, specific effects on the trait selected. Thus mating speed can be affected by general activity and probably also by other nonspecific factors, such as phototaxis.

Other behavioral traits that have been studied in selection experiments include phototaxis and mating choice in *Drosophila,* and aggressiveness, emotionality (defecation in an open space), and audiogenic seizures in mice (see references in Hirsch 1967; Parsons 1967; Manosevitz et al. 1969). A more dramatic response to selection than was observed for mating speed was obtained for geotaxis in *D. melanogaster.* Compare the response for geotaxis (Figure 2–5) with that for mating speed (Figure 2–4). The observation that selection continued to

2–5 A striking demonstration of the effectiveness of continued artificial selection for a behavioral trait. In over 60 generations of selection for positive or negative geotaxis (tendency to go downward or upward, respectively, in a vertical maze) in *Drosophila melanogaster,* genetic variation for the behavioral trait was still not exhausted, since responses to selection were continuing. (From Erlenmeyer-Kimling et al. Copyright 1962 by the American Psychological Association. Reprinted by permission.)

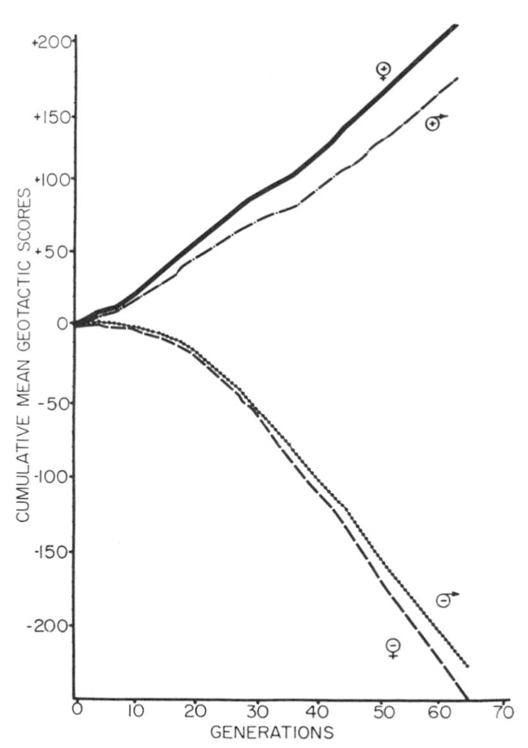

be effective even after 60 generations reflects the large number of genes that affect geotaxis.

Genetic Polymorphism / When qualitatively distinct genotypes are maintained in a population at a higher frequency than could be expected from mutation alone, the condition is termed *genetic polymorphism.* The examples best known to naturalists include the color phases or morphs of lady beetles, Batesian mimic butterflies, arctic foxes, screech owls, various herons, blue and snow geese, and red-backed salamanders. At least 3 percent of the roughly 700 species of birds and 2 percent of the 650 species of mammals in North America north of Mexico are known to show such visible polymorphism. To geneticists the best known examples are the chromosomal inversion types in *Drosophila.* There has been continuing interest in finding behavioral differences between these types, but few examples have been discovered. The inversion karyotypes (chromosomal types) will be considered later in the context of sexual selection.

One clear example of genetic polymorphism in behavior is the difference in choice of substrate color between normal (light-colored) and melanic (dark-colored) nocturnal moths. This polymorphism in *Biston betularia* is controlled mainly by alleles at a single locus. It has been shown experimentally under "natural" conditions that avian predators take proportionately fewer of the melanic forms in the heavily polluted regions where dark roosting substrates predominate (Kettlewell 1956; Ford, 1964). Thus, selection tends to favor the color form that best matches the background color of the substrate it settles on for the day.

Sargent (1966) collected several North American species at night, including two color forms of one species (*Catocala ultronia*), and placed them in an experimental box with sides of four shades of gray. He found that the moths tended to settle on the shades of gray which best matched the reflectances of their forewings. Similar results had been obtained by Kettlewell (1955) in England for *Biston betularia.* A visually guided direct response to substrate reflectance seems to be involved (Sargent 1968). A similar situation has been found in limpets (Giesel 1970).

Interspecific Hybrids / The richest source of genetic and behavioral variability is undoubtedly to be found among different species; and since genetic and behavioral variability are prerequisites for most behavior-genetics research, interspecific hybrids deserve serious consideration. Unfortunately, such hybrids are difficult to obtain, since many interspecies crosses do not yield viable progeny. Even with such closely related sibling species as the mating-call types within the *Hyla* *versicolor* complex (see Chapters 1, 18), genetic incompatibility prevents genetic analysis. Backcrosses and F_2s, of course, are generally even rarer among species. Until ways are developed to overcome these difficulties, the use of interspecific hybrids in studying behavior differences will be limited mainly to those few cases in which pairings are interfertile.

There is, despite these limitations, one kind of genetic information that can be obtained in virtually no other way than by using hybrids: the genetic nature of conspicuous interspecific differences in complex behaviors, those commonly called reflexive or instinctive. Many such cases have been studied in animals as diverse as crickets (Hörmann-Heck 1957), poeciliid fishes (Clark et al. 1954), and grouse (Evans 1966). Some types of behavior in hybrids are intermediate with respect to the parents, but many tend to resemble one or the other parent species to varying degrees. Some behaviors may disappear altogether. Or in some cases hybrids may show behavior of both species. It has even been found that a hybrid duck has shown a display lacking in either parent species, but present in other species of the family (Lorenz 1958; Wall 1963).

Commonly a behavioral pattern is so complex and involves so many components that intermediacy is difficult to estimate. The calls of the dove species studied by Lade and Thorpe (1964) differ from each other in duration of certain notes, rhythm, timing, pitch, and number of notes. In the hybrids most of these elements are so interdependent that degree of resemblance to one or the other parent cannot be accurately measured (Figure 2–6).

The breeding of interspecific hybrids beyond the F_1 is rarely possible, but it appears to be practical in some ducks (Sharpe and Johnsgard 1966), doves (Davies 1970), the swordtail-platy hybrids (Clark et al. 1954), some cricket hybrids (Hörmann-Heck 1957), and in *Drosophila* spp. (Ewing 1969). In the study of crickets, results indicated single-gene inheritance in certain species-typical behaviors. These and other studies, such as Rothenbuhler's (1967) work on bees, show that a variety of genetic phenomena can be found to underlie species-typical behavior. The ontogenetic development of behavior in hybrids appears to be a promising field. Dilger's (1962) studies of nest-building in lovebirds have revealed complex interactions between genetic factors and learning. Few other studies of this phenomenon in relation to species-typical behavior have been made.

Developmental Genetics of Behavior / The pathways between genes and a given type of behavior are invariably complex and involve all structural levels of organization in animals, from molecules through cells, organs, and systems to the whole animal, and sometimes to interacting groups of animals. Clearly genes may affect behavior in many indirect ways.

Gross abnormalities in metabolism, the inborn errors of metabolism, such as phenylketonuria, have been a field of special interest in man (Hsia 1967). In *Drosophila melanogaster* gross somatic abnormalities due to single-gene mutations are known to have a detrimental effect on courtship behavior. These include such mutants as *Bar* and *white,* which have reduced visual capacity because eye function and structure are affected (Connolly et al. 1969); *forked* and *hairy,* whose bristle sense organs are affected; and *vestigial* and *dumpy,* which have malformed wings (Ewing and Manning 1967; Manning 1968). Other mutants in *Drosophila,* such as *yellow,* show less conspicuous effects. The recent biochemical work on *Drosophila* in population genetics

2–6 Sound spectrograms of the coos of some dove species and their hybrids. Dove songs develop normally in the species-specific manner even when the eggs are cross-fostered between species; learning the coo pattern from a parent or foster parent is unimportant. (After Lade and Thorpe 1964.)

(e.g., Hubby and Lewontin 1966; Lewontin and Hubby 1966) suggests that a biochemical approach to the action of genes on behavior in *Drosophila* is not necessarily as difficult as it might appear.

The nervous system may be a crucial link in the gene-behavior pathways. Some progress in this area has been made through study of neurological mutants in mice having correlated behavioral alterations (Sidman et al., 1965), and by selection experiments for neurochemical traits correlated with behavioral plasticity (Bennett et al. 1964).

Another approach is direct comparison of the brains of different species or breeds of animals. Comparative neuroanatomy has many examples of gross and subtle species differences in brain structures that are correlated with behavioral differences. The highly developed auditory pathways in porpoises (Kruger 1966; Jansen and Jansen 1969) and echolocating bats (Mann 1963; Findley 1969) provide clear examples. Even between breeds of rats differences in brain structure have been found; Smith (1928) found that the olfactory bulbs of wild-type Norway rats had significantly more granule cells and fewer mitral cells than those of albino rats, but the behavioral significance of this difference is unknown.

Future advances in this area would seem to be dependent on developmental neurochemistry and neurochemical genetics.

Innateness / The term *innate* has two principal meanings (Lehrman 1970). As used by many comparative behaviorists it means or implies "genetically determined." In this sense it is clear that many *differences* between individuals in the same or different species are innate. Innateness is best demonstrated by means of genetic experiments; however, the "isolation experiment" may also be strongly suggestive of the importance of genetic factors (see Part VI).

Although the essence of genetic experimentation consists of the analysis of *differences*—of variations in traits—geneticists for convenience sometimes refer to a *trait* as being genetically controlled, even though strictly speaking this is incorrect. For example, since there are in *Drosophila melanogaster* many mutant genes that affect eye color, eye color may be said to be genetically controlled. This is an oversimplification, however, since it is really the differences in eye color that are referred to. The development of eye color requires certain anatomical and biochemical substrates where appropriate enzymatic chemical reactions can occur; thus normal development requires environmental factors. In fact, the development of any trait requires both genetic and environmental factors. Consequently, when a biologist refers to a trait as being innate he means that certain *differences* are genetically determined; he does not mean that the "trait itself" is wholly genetically determined or that environmental influences on the development of the trait are lacking. Such a position would clearly be absurd.

Innate has also been applied to behavior to indicate that the behavior develops normally in an isolation or deprivation experiment (see Part VI for further discussion of this point). Such experiments are not genetic experiments and should not be interpreted in terms of genetics; they are developmental experiments and should be interpreted strictly in terms of development. The dual connotations of *innate* are legitimate, but they have led to much misunderstanding and needless debate (reviewed in Lehrman 1970); it is to the discredit of both ethology and psychology that such foolishness persists.

CONCLUSIONS

To investigate genetic determination of behavior the methods of selection experiments and crossing experiments are often used. It is convenient to investigate the behavioral effects of a single gene if mutant strains are available. An early study of this sort was done on the *yellow* mutant in *Drosophila melanogaster* by Bastock (1956). A good example of behavioral differences attributable to variation at few loci is the hygienic behavior of bees. Such differences are due to the effects of *major genes.* More commonly, behavioral differences are due to the cumulative effects of many *polygenes,* each with a small effect. For these behaviors the concepts of phenotypic variance and its partitioning by statistical methods into genetic and environmental effects are useful.

The fraction of the phenotypic variance attributable to genetic variation is known as *heritability.* This fraction may be estimated in various ways. The concept of heritability is useful in understanding the effects of natural and artificial selection upon populations.

The mean and variance of a behavioral trait are typically altered in predictable ways by the three main selection regimes. In *stabilizing selection,* the parents chosen for the next generation tend to have phenotypic values close to the mean of the population; consequently, their progeny also tend to resemble the mean. In *directional selection,* the parents chosen for the next generation have values that differ in one direction from the mean; consequently, the mean of the progeny will be likely to differ from the mean of the population in the previous generation. In *disruptive selection,* parents are chosen from both extremes but not from near the mean, so that the population tends to become bimodal.

The effectiveness of selection in any of these regimes depends upon the existence of genetic variation correlated with the phenotypic variation. If the correlation is high, heritability will be high, and selection will be most effective. If the correlation is low, heritability will be low and selection will be least effective. In the case of directional selection this relationship may be used to calculate realized heritability. Artificial selection is likely to be effective on any behavioral trait. Detailed selection experiments have been done with *Drosophila* on geotaxis and on mating speed, the results being more spectacular for the former.

Genetic morphs, such as the dark and light forms of certain moths, may also show correlated behavioral differences. The genetics of interspecific differences in behavior are sometimes difficult to investigate because of barriers to the exchange of genes between species. However, such studies are valuable for the understanding of the nature of species differences.

The developmental pathways by which genes affect behavior are complicated and diverse, but all have a chemical basis and many affect the nervous system or sense organs. Where behavioral differences are attributable to genetic differences, the term *innate* has been used. This term has also been used for behaviors that develop without the aid of learning from other individuals. Since the two meanings are operationally different, some confusion has resulted.

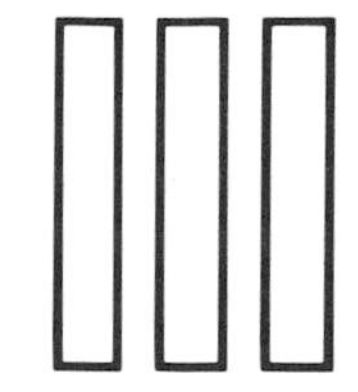

SOCIAL ORGANIZATION: THE SOCIAL CONTEXT OF BEHAVIOR AND NATURAL SELECTION

ALTHOUGH THE diversity of social organizations among animals is fascinating in itself, we shall focus in this section on the relevance of social organization to the processes of natural selection. Since we are dealing with the whole animal kingdom we shall use the terms *social behavior* and *social organization* in their broadest senses.

Social behavior for our purposes will be taken to include all behavior directly related to potential or actual encounters between individuals, regardless of whether or not the behavior is gregarious or cooperative ("social" in the vernacular sense). This is customary ethological usage. It allows us to consider all behavioral relationships among individuals of a species in one sphere of discourse without prejudging the beneficial or harmful effects of such behavior to the society.

When discussing social behavior in relation to natural selection it is convenient to recognize three major determinants of social organization: (1) agonistic behavior and spacing systems, (2) sexual behavior and mating systems, and (3) aid-related behavior and systems for care of offspring.

Agonistic behavior primarily includes acts of attack, escape, threat, defense, and appeasement (Scott and Fredericson 1951). It is convenient to group these behaviors together because (1) they are functionally related to intraspecific, competitive situations and, consequently, to the dispersion pattern of the species, (2) they are intricately related motivationally and physiologically, and (3) they tend to occur together in time and space.

Spacing behavior includes all types of behavior that tend to increase the space between individuals; *grouping behavior* includes all types of behavior that tend to decrease it. Together they determine the species-typical *dispersion patterns.*

Sexual behavior forms a self-evident category; it refers to courtship and copulation and to the competition for mates (here also involving agonistic behavior).

Aid-related behavior includes behavior involving giving or asking for aid actively or passively. As used here it includes parental care of young, solicitation of aid by the young or by adults, and other kinds of behavior that are cooperative (mutually beneficial) or unilaterally beneficial to the receiver.

Natural selection acts somewhat differently on these three determinants of social organization; there are different sets of theories for each. It will be convenient to stress sexual selection in relation to sexual behavior, kin-selection when discussing aid-related behavior, and the competitive aspects of individual selection in relation to agonistic behavior.

The term *social organization,* or *social system,* will be used to refer to the totality of social relationships among all members of a species or other sample population. Consequently, it includes all relationships of a sexual, agonistic, or aid-related nature within the group. To specify completely the typical social organization of a sample population and the variability within it, the investigator must know all the types of social relationships that exist among all the members of the sample—a difficult task in a large group of individuals.

The term *social organization* does not imply absence of randomness or the presence of "organization." It is merely an expression of convenience. Complete randomness in all social behavior would be nearly impossible to show in practice anyway. Consequently, every species has a social organization of some sort. Species differ in the details of structure of their social organizations, as will be seen in the discussions that follow. The term *social system* is sometimes used in reference to a particular type of social organization of a species.

The Units of Natural Selection / Social organization affects natural selection in a variety of ways. To realize this fully it is first necessary to consider the units upon which natural selection acts. These may be variously classified, but basically they include at least the following: (1) individuals, (2) families or kinship groups, (3) spatially defined populations including local populations known as *demes*, and (4) species. Variation in behavior can be found at any of these levels.

It has been generally recognized since Darwin that natural selection acts differentially on the variation among individuals. Much more recently the works of Wright (1945), Kalela (1954), Wynne-Edwards (1962, 1963), MacArthur and Wilson (1967), and others have raised the possibility of selection among demes, or local populations. Similarly, selection may act differentially on families, such as the colonies of bees or ants, (Hamilton 1964), or the families of swifts or starlings (Lack 1954*a*). Competition and other ecological relationships between species have long been known, and the fossil record proves that some species survive (are "selected") while others become extinct. Evolution at the level of ecological communities has also been considered, but this will not concern us here. Social organization is directly related to each of these levels of selection. The details of these relationships will be brought out in the following chapters. For further discussion of the units of selection see Lewontin (1970).

Fitness / The two processes by which natural selection is mediated at the levels of the individual and kinship groups are *differential natality* and *differential mortality*. These together determine genotype *relative fitness*, which may be thought of as a fraction, *W*, that can be multiplied by the frequency of a genotype in the present breeding generation to give the frequency of that genotype in the next breeding generation (assuming nonassortative mating, a large population, and the absence of differential migration or mutation and other disturbing factors). Relative fitness is a mathematical and biological concept that indicates the ability of a genotype, relative to other genotypes in the population, to perpetuate itself.

The concept of *relative fitness* is basic to nearly all evolutionary thinking. Because of its importance to the evolution of behavior, and especially the material in Part III, it is worthwhile to delineate it more precisely before proceding.

Before considering what happens to a genotype that has a competitive advantage in natural selection (superior fitness), it is helpful to consider the case in which no such advantage is present. The ability in man to taste phenylthiocarbamide (PTC) serves as a convenient example. It is inherited as a simple dominant; consequently, tasters may be of two genotypes: *TT*, homozygous, and *Tt*, heterozygous. Nontasters, *tt*, are homozygous recessive.

Assume (1) a population of 100 individuals, consisting of 20 of the genotype *TT*, 40 of *Tt*, and 40 of *tt*; (2) that p is the frequency of *T* and q the frequency of *t* in both sexes; (3) that no mutation at this locus occurs; (4) that the members of this population mate at random with respect to PTC-tasting ability; and (5) that neither genotype and neither gametic type has a selective advantage in survival or reproduction. The gene frequency of *T* is then

$$p = \frac{\text{number of } T \text{ genes in the population}}{\text{number of } T \text{ and } t \text{ genes}}$$
$$= \frac{2(20) + 40}{2(20) + 2(40) + 2(40)} = \frac{80}{200} = 0.4$$

Since $p + q = 1$, the frequency of *t* must then be

$$q = 1 - p = 0.6$$

What will be the frequencies of *T* and *t* in the next generation if the above assumptions are observed? The table below shows a method of

Genotypic frequencies in a Hardy-Weinberg equilibrium.

		♂ Parents		
		TT 0.2	*Tt* 0.4	*tt* 0.4
	TT = 0.2	*TT* = .04	*TT* = .04 *Tt* = .04	*Tt* = .08
♀ Parents	*Tt* = 0.4	*TT* = .04 *Tt* = .04	*TT* = .04 *Tt* = .08 *tt* = .04	*Tt* = .08 *tt* = .08
	tt = 0.4	*Tt* = .08	*Tt* = .08 *tt* = .08	*tt* = .16

Equilibrium totals *TT* = 0.16
Tt = 0.48
tt = 0.36
Note that still $p = 0.4$,
$q = 0.6$

finding them. The male and female parents are shown according to the frequency of each genotype. All possible matings are shown together with the frequencies of the genotypes among the offspring. Particularly important is the observation that the gene frequencies remain the same even though the genotypic frequencies change. If these progeny were taken as the parents for the next generation, both genic and genotypic frequencies would remain the same. In subsequent generations this would hold true indefinitely, provided the original assumptions were met.

Such a condition is called a state of *genetic equilibrium.* This particular kind is known as a Hardy-Weinberg equilibrium, after the men who first pointed out the relevance of the binomial theorem to population genetics. Using this theorem the operations in the table above can be summarized differently, as follows:

Calculation of genotypic frequencies from gene frequencies in a Hardy-Weinberg equilibrium.

♀ \ ♂		p 0.4	q 0.6	$p^2 + 2pq + q^2 = 1$
p	0.4	$p^2 = 0.16$	$pq = 0.24$	$0.16 + 0.48 + 0.36 = 1$ TT Tt tt
q	0.6	$pq = 0.24$	$q^2 = 0.36$	

This is a useful equation, because if a Hardy-Weinberg equilibrium exists and p or q is known, all genotypic frequencies for the equilibrium condition can be calculated. Similarly, if any one of the genotypic frequencies is known, the corresponding gene frequencies can be calculated (since p^2 = frequency of *TT*, q^2 = frequency of *tt*, $2pq$ = frequency of *Tt*). For our purposes, however, the most important conclusion to be drawn is that *genic and genotypic frequencies tend to remain the same from generation to generation*—in the absence of complicating influences, such as selection, nonrandom mating, differential mutation, differential additions or subtractions to the population, or chance effects. Consequently, genetic variability in a population is likely to remain at the same level unless modified by these influences.

Of the factors that tend to alter an equilibrium, selection is of primary importance. The ability to taste PTC seems neither beneficial nor harmful in modern life; but for the sake of a hypothetical example, assume that in primitive man PTC was a constituent of various potential food plants that caused severe illness if eaten in moderate quantity. Individuals who could detect PTC in the food could learn to avoid the illness simply by taste. Since ability to taste PTC may confer longer survival, under such conditions p may increase and q decrease. The general conditions of change in p and q during selection will now be considered.

Let the genotypes have the *fitness values* (W) given below, where S and s are the *selection coefficients* for TT and tt, respectively. The coefficient of selection is a measure of the intensity of selection against the genotype concerned.

$$W_{TT} = 1 - S$$
$$W_{Tt} = 1$$
$$W_{tt} = 1 - s$$

It is convenient here to take $W_{Tt} = 1$, but this is completely arbitrary and even misleading in certain theoretical contexts (Wallace 1968*a*, *b*).

At the end of the next generation, in this example, the number of individuals (new population size) will be less because selection has eliminated some; consequently, the denominators in the frequency expressions must be less than one. The new population size is obtained by multiplying the genotypic frequencies by their respective fitness values and summing the products:

$$(1 - S)p^2 + 2pq + (1 - s)q^2 = 1 - Sp^2 - sq^2$$

The genotypic frequencies in the next generation are as follows:

GENOTYPE	FREQUENCY
TT	$\frac{(1 - S)p^2}{1 - Sp^2 - sq^2}$
Tt	$\frac{2pq}{1 - Sp^2 - sq^2}$
tt	$\frac{(1 - s)q^2}{1 - Sp^2 - sq^2}$

The new frequency of T can be obtained by adding the frequency of TT individuals to half the frequency of Tt individuals (substituting $1 - p$ for q). The new frequency of T will then be

$$p_{n+1} = \frac{p - Sp^2}{1 - Sp^2 - sq^2}$$

From the above one can derive an expression for the change in the frequency (Δp) of a gene such as that for PTC-tasting (T),

$$\Delta p = p_{n+1} - p_n = \frac{pq(sq - Sp)}{1 - Sp^2 - sq^2}$$

Different results are obtained depending on the values of S and s. Four cases may be recognized.

1) If $S = s = 0$ there is no selection; equilibrium remains unchanged.

2) If S and s are both positive, the heterozygotes are most fit, and a stable equilibrium results at

$$\hat{p} = \frac{s}{S+s}, \hat{q} = \frac{S}{S+s}$$

The carets over p and q designate the equilibrium frequencies.

3) If S and s are both negative, either *TT* or *tt* eventually becomes eliminated, although an unstable equilibrium may result temporarily if S and s are equal.

4) If S and s are of different sign or if one is zero, either *TT* or *tt* eventually becomes eliminated.

In terms of our example, if the fitness of nontasters were less then the fitness of the tasters, nontasters would be selected against and *t* would eventually disappear from the population.

Further details may be obtained from Haldane (1959), Li (1955), or Wilson and Bossert (1971).

From these manipulations it can be seen that fitness is a relative concept, not an absolute one. The fitness of a genotype depends on the other genotypes with which it is present and on the environmental conditions. Consequently, values for W for a particular genotype may vary greatly depending on the environment and the genotypes present.

Fitness has also been called relative fitness, survival value, and adaptive value. It is basically an expression of the ability of a genotype to replace itself in successive generations. As such, the concept is extremely useful in evolutionary thinking and it will be employed in the chapters to follow. Although we shall not calculate fitness values, the mathematical expressions have been given to make clear exactly what is meant by fitness and to emphasize that the concept is no longer as vague as it was in Darwin's time, but is actually quite precise and tractable to mathematical or experimental analysis.

Kin Selection / An improvement of the classical concept of fitness, which had been based upon the inherent properties of individuals, was made by considering how fitness values are modified by social interactions with other individuals (Hamilton 1964). Social interactions help to characterize groups of individuals: some groups may be composed of selfish members; others, of relatively nonaggressive or even altruistic individuals. If such groups, which may be as simple as a mother and her dependent offspring, also differ genetically, selection may act to modify the social behavior in the direction that improves the long-term reproductive performance of the group. A group is more likely to be genetically homogeneous if it is composed of close relatives (kin) than distant relatives. The process of selection acting among groups that are defined on the basis of kinship is known as *kin selection* (Smith 1964; Brown 1966). The role of kin selection in the evolution of behavior will be explored in Chapters 9 and 10.

Population Selection / The concept of differential long-term reproductive success among natural units has been applied above to individuals and to groups defined on the basis of kinship. It can also be applied to groups defined geographically or micro-geographically. Populations living in islands of suitable habitat in a sea of unsuitable surroundings can in theory also survive and reproduce. To the extent that genetic factors are responsible for differential survival and reproduction of spatially defined populations, selection may be effective. This type of selection has been termed *population selection* or *deme selection* (also interpopulation and interdemic selection: Smith 1964; Brown 1966; Lewontin 1970). The term *deme* has been used in connection with species that inhabit patchy environments in which the overall population of a geographic area is divided up into smaller, more natural units by habitat barriers; these natural, local populations have become known as demes (following Mayr 1963:137).

Whether or not population selection is of any real importance in evolution has been hotly debated and remains to be clarified. Its importance seems to have been vastly exaggerated by its principal proponents. There do not appear to be any behavioral phenomena that cannot be explained more convincingly on the basis of individual or kin selection than on the basis of population selection. Nevertheless, the effects of social behavior on population dynamics are important, regardless of their evolutionary explanation, and they will be considered in Chapter 6.

To summarize: in the most important kind of selection, the unit of selection is the *individual*. It is the individual that carries the genes, survives, reproduces, and dies. This will be termed individual selection when it is necessary to distinguish it from population or kin selection. In the latter types the units of selection are the *population* (spatially defined) and the *lineage group or family* (defined on the basis of genetic relationship). These are the units that survive, reproduce, and die. It is of the utmost importance that the differences between these different kinds of selection be clearly understood. The student should be able to recognize which of these three processes is involved in any particular evolutionary hypothesis, especially when it is not explicitly mentioned.

3 Social Units and Their Patterns of Dispersion

UNDERLYING all social behavior of a species is the distribution in space of its individuals in their preferred habitats. This distribution, known as the *dispersion pattern*, is determined mainly by the behavior of individuals toward each other. It is important because (1) it places restrictions on the opportunities for social behavior of various types; (2) consequently, it also restricts the directions in which natural selection is likely to change the communicative behavior of a species; and (3) it plays an integral role in the population dynamics of the species.

The distribution of individuals in two-dimensional space can be roughly categorized in terms of the units of dispersion and the dispersion patterns of these units. Describing the patterns of aquatic and aerial species in space requires the use of three dimensions and is more difficult. Three-dimensional patterns will not be considered here, but an introduction to the subject can be found in the literature on schooling in fishes (Breder 1959; Breder and Rosen 1966; Cullen et al. 1965; Shaw 1970).

Patterns of Dispersion / Every animal species has a characteristic pattern of dispersion in its natural habitat. Some live in dense colonies while others are solitary and widely spaced. A first step in the analysis of the social organization of a population is to describe its characteristic dispersion pattern. We may simplify the problem of describing the dispersion of individuals in a population by analyzing the spatial distribution of all individuals at one point in time. This can be done through aerial photos, quadrat sampling, and other methods. With appropriate statistical analyses (Andrewartha and Birch 1954; Greig-Smith 1964; Iwao 1968; Pielou 1969) it is possible to demonstrate that a population is distributed randomly in the available space or more commonly, that it departs from randomness, in the direction of clump-

3–1 The three basic patterns of dispersion. Each dot represents a social unit, which may be an individual, pair, family, or other type of group.

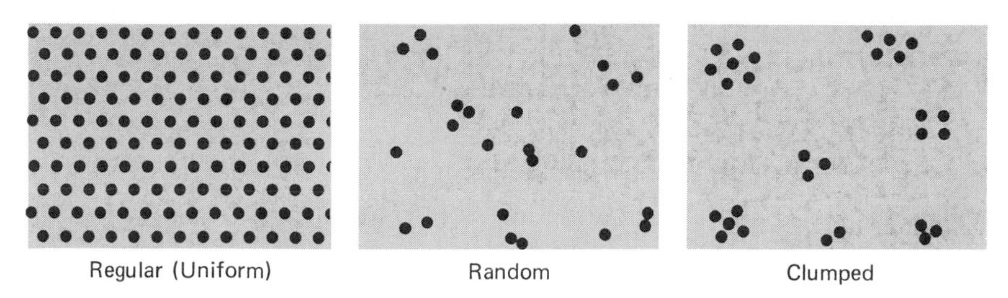

ing or in the opposite direction, regularity. Figure 3–1 shows these three conditions.

In nature *random* dispersion is rare; it has been found in some species of sessile, bivalve clams (Jackson 1968), in groups of kangaroos, *Megaleia rufa* and *Macropus cangaru* (Caughley 1964), and in the mobile, deposit-feeding mollusc, *Nucula proxima* (Levinton 1972). *Clumped* or aggregated patterns are common in many invertebrates and vertebrates (e.g., colonies, herds, flocks, troops, and schools). *Regular* or uniform spacing is conspicuous in many vertebrates and in some invertebrates. Regular dispersion is frequently caused by territorial behavior, but can arise from other causes as well.

The Units of Dispersion / The principal unit of dispersion is the individual. In many species, however, individuals may be found in groups whose dispersion pattern differs from that of the individuals. Groups may have a regular, random, or clumped dispersion pattern of their own. For example, groups of baboons tend to have a regular dispersion, while family groups of Canada geese tend to flock together outside the breeding season. In many species such groups (pairs, families, or clans), rather than individuals, are the *primary social units* (Brown and Orians 1970): the smallest social units in which the members can survive and reproduce normally. In such species the isolated individual often has a much reduced fitness compared with its group-living peers. In species in which individuals live alone the individual is the primary social unit, and the fitness of isolated individuals would be normal for the species.

In an analysis of dispersion it is important to determine through behavioral observation the nature of the primary social units. There may be secondary social units in addition to the individual. In gelada baboons (*Theropithecus gelada;* Crook 1966) and hamadryas baboons (*Papio hamadryas;* Kummer 1968*a*, *b;* Kummer and Kurt 1963), males excluded from harem groups, which are the primary social units, may

associate in all-male bands. Birds unable to obtain territories and consequently unable to pair may associate in nonbreeding flocks while the primary social units, pairs, breed—for example, the red grouse, *Lagopus scoticus* (Jenkins et al. 1963) and the Australian magpie, *Gymnorhina tibicen* (Carrick 1963).

Temporal Changes in Dispersion / Although some species maintain their typical dispersion pattern all year, others undergo seasonal shifts. Migratory birds are conspicuous in this respect. For example, the various species of junco (e.g., *Junco hyemalis*) have a regular dispersion during the breeding season, when the primary social unit is the breeding pair, and a clumped dispersion for the remainder of the year, when they occur mainly in flocks. The importance of the flock as a social unit is shown by the observation that isolated individual juncos in winter suffer a higher mortality than flocked ones (Fretwell 1969).

Dispersion shifts may also occur on a daily basis. Baboon troops, which tend to be regularly spaced during the day, may roost together on the same cliff at night. Colonially roosting bats and birds, and some schooling fishes, also change their spacing pattern daily.

Dispersal and Spacing Behavior / The characteristic dispersion pattern of a species in the breeding season is achieved in many animals by a process that often can be divided into two phases, dispersal and spacing.

Dispersal refers to the movements of animals from a source, such as a roost, an area of high population density, or from a birthplace (Johnston 1961; Berndt and Sternberg 1968). Dispersing animals typically lack territories or stable home ranges; they cover longer distances than would settled individuals. Sometimes dispersal movements are dramatic. In Eurasia and North America certain birds, such as the Bohemian waxwing (*Bombycilla garrula*), the snowy owl (*Nyctea scandiaca*), the red crossbill (*Loxia curvirostra*), and certain titmice, are known to have erratic population irruptions in which large numbers suddenly appear in a region where they are normally absent or rare. The lemmings (*Lemmus lemmus*) of Norway may have massive dispersal movements in which the topography of fjords and valleys naturally funnels many lemmings into smaller areas (Curry-Lindahl 1962, 1963; Archer 1970). When local population densities are unusually high and food and shelter are scarce, dispersal to other areas may result in some improvement in survival rates, both for those remaining behind and for those dispersing.

Dispersal in vertebrates and marine invertebrates typically occurs in the larval or immature stages of the life cycle, during or shortly after the breeding season, when the number of individuals is near its peak for the year. In insects it is typically the winged adult that disperses. During the process many animals die, and dispersal behavior can be significant in the population dynamics of a species (Lidicker 1962).

By the time the breeding season arrives, dispersal usually has ended. At this time sexually mature individuals begin looking for a place to settle and breed. This final settling process determines the breeding dispersion pattern and is known as *spacing behavior.* (Spacing behavior may also occur in other contexts and other seasons.) It is the process by which the available habitat is divided up among the potential claimants. Individuals that have found a place to settle may, by their behavior or their mere presence, cause other individuals to look elsewhere. The movements of individuals in search of a settling area are known as spacing movements. They are typically short and local in comparison with dispersal and migratory movements, which tend to be much longer. Unlike dispersal movements, they are largely guided by behavioral interactions with neighbors.

Home Range / Although the discussion above assumes that individuals, or other units of dispersion, remain at fixed points, this is necessarily an oversimplification for mobile animals. Further analysis of dispersion patterns in mobile animals requires techniques for plotting and analyzing location data that derive from repeated samplings of the same individuals at various times. If the points where an individual has been observed over a specified time period are plotted on a map, they can be used to provide an estimate of the home range of the individual. The home range is simply the area in which an animal normally lives, exclusive of migrations, emigrations, dispersal movements, or unusual erratic wanderings. Home range is defined and estimated operationally and without reference to the presence or absence of particular types of behavior (such as territoriality) and without reference to the home ranges of other individuals. Use of the term home range implies neither the absence nor presence of territoriality and neither the absence nor presence of other individuals. Although this is self-evident from the definition, the term has been misused in these ways.

Several methods are available for deriving home range areas from a set of points plotted on a map (reviewed by Mohr and Stumpf 1966; Jennrich and Turner 1969; Metzgar 1972). A graphic method developed by Odum and Kuenzler (1955) is instructive because it illustrates the problems of sampling and comparison. In this method (Figure 3–2) the location of an animal is sampled at specified time intervals (e.g., 5 or 10 min. for a bird or a squirrel); the points are numbered successively and plotted on a map. The area enclosed by the outermost points of the sample (the largest possible polygon) is then plotted as a function of the number of observations. The result is known as an *observation-area curve*. Figure 3–2 shows that the area increases as sample size increases but tends to reach a limit. Consequently, to be strictly comparable,

3–2 Determination of an observation-area curve for a wood pewee. Observed positions (*above*) were plotted at five-minute intervals; the maximum areas enclosed after each increment of ten observations are enclosed by solid lines. The dashed line encloses the calculated home range size—10.8 acres at the 1-percent level (see text), as shown in the observation area curve, *below*. (From Odum and Kuenzler 1955. Reprinted by permission of the American Ornithologists' Union.)

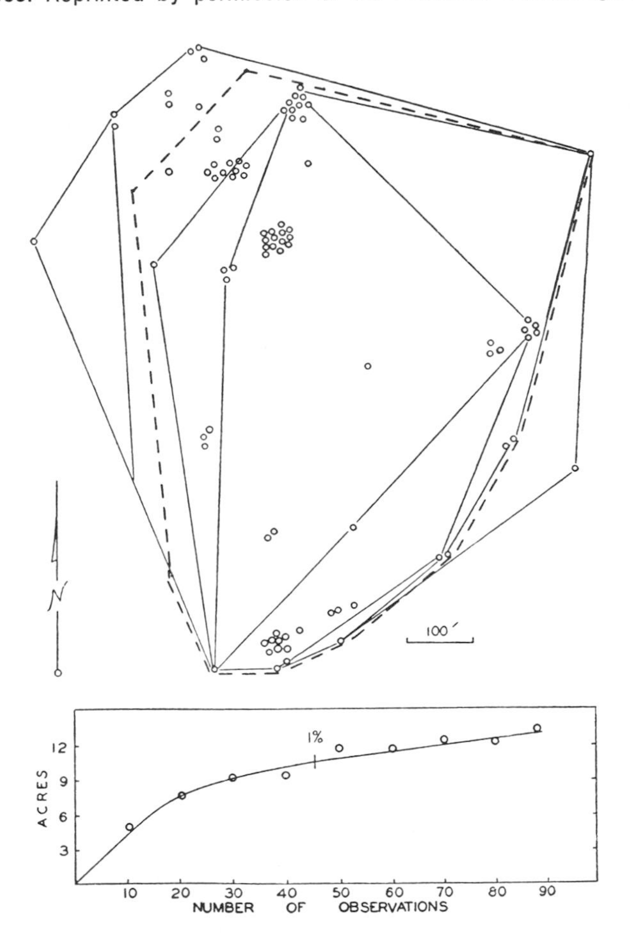

home range estimates should be made from samples of large and comparable size. For comparison one can choose that point on the curve where a specified increase in sample size (e.g., 1 percent) results in a specified increase in home range area (e.g., 1 percent).

Home range has been shown to vary with sex (e.g., L. E. Brown 1966; Fitch and Shirer 1970), age (Dhondt and Huble 1968; Stefanski 1967; Ralph and Pearson 1971; Fitch and Shirer 1970), time of day (Weeden 1965), and stage of the breeding cycle (Odum and Kuenzler 1955; Stenger and Falls 1959; Weeden 1965; Stefanski 1967). Weeden used observation-area curves to show that the home range of a male tree sparrow (*Spizella arborea*) varied with the stage of the nesting cycle (Figure 3–3).

Other methods for comparing home ranges make use of the mean radius of the plotted points from their geometric mean (Tinkle 1965) or of the determinant of the covariance matrix of the capture points (Jennrich and Turner 1969). The greater precision and comparability of the newer methods of home range determination should make possible a clearer understanding in the future of the factors that influence home range.

Animals do not frequent all parts of the home range equally.

3–3 Observation-area curves for one male tree sparrow at different stages of his breeding cycle. (From Weeden 1965.)

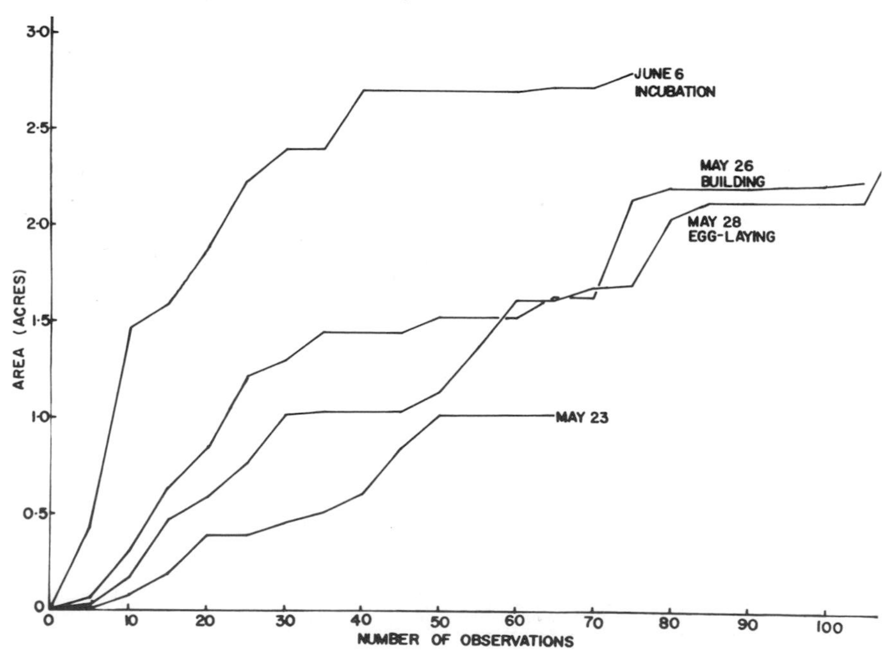

Typically, the area around a nest, burrow, or favored feeding or resting area will receive heavier use. The differential use of parts of the tree sparrow's home range has been shown by Weeden (Figure 3–4). The most heavily used areas tend to be toward the center of the home range, and so are sometimes called *core areas*.

The methods for obtaining location data for individual animals are as diverse as the species studied (Giles 1969; Bub 1967–1969).

3–4 Home ranges of tree sparrows (*Spizella arborea*) showing "core areas," variations in frequency of use of different parts of the home range, and maximum-polygon estimates of home range. In this species males fight but females are not territorial. Each square of the grid is 66 x 66 ft. (1/10 acre). The observer watched a single pair continuously from 3:00–7:00 A.M., plotting all locations of the male and female. The data from four to eight of these four-hour periods were combined for each pair to give the distributions shown here. Observations were made in May and June after the courting period. Intensive use = 16 or more observations in a grid square; moderate = 6–15; slight = 1–5. (From Weeden 1965.)

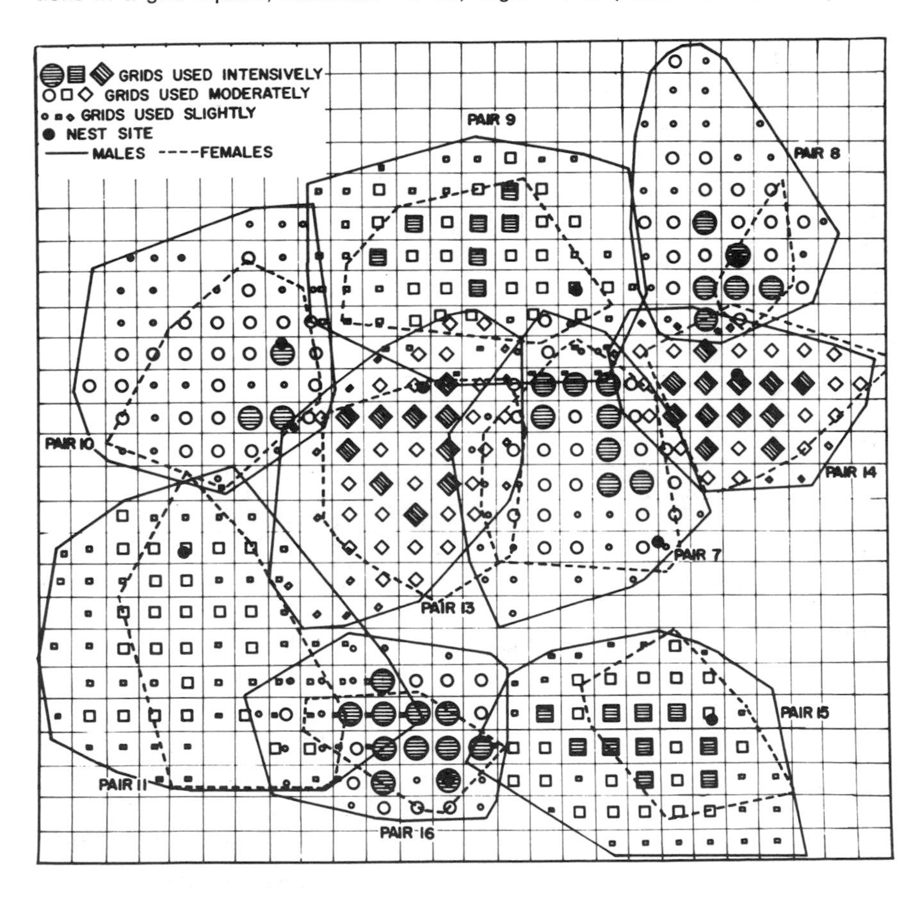

The first requirement is a means of recognizing individuals. This usually requires capturing individuals, marking, and releasing them. Subsequently the individual is observed visually, by radio, with a Geiger counter, by recapture, or by some other ingenious method.

3–5 Home ranges of three red foxes determined by radiotelemetry during May 6–June 3, 1964 in Minnesota. Lines between points connect consecutive locations separated by no more than a one-hour difference. (From A. B. Sargeant, 1972. "Red fox spatial characteristics in relation to waterfowl predation." *J. Wildl. Manage.* 36(2):225–236.)

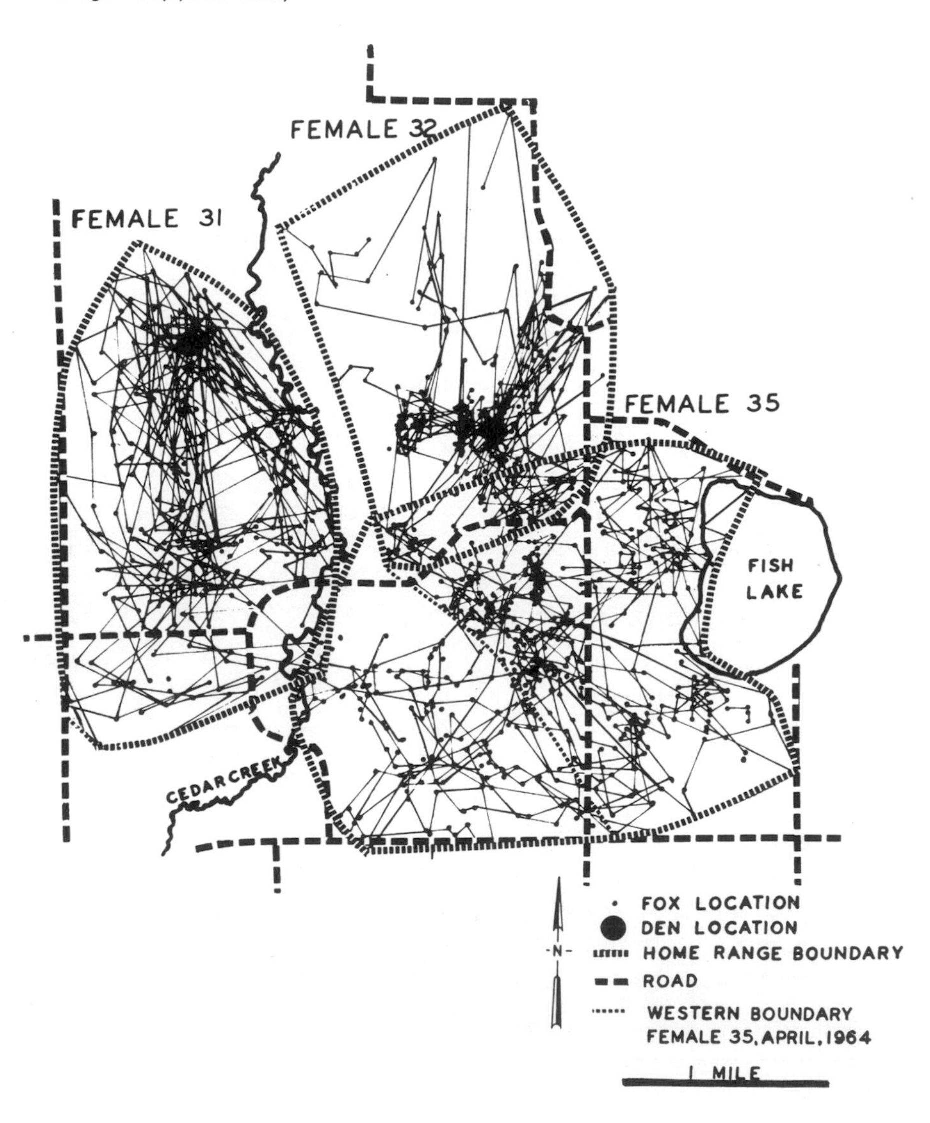

Southern and Lowe (1968) were able to plot the home ranges of tawny owls from the numbered metal tags used to mark mice in the same area; knowing the home range of the mice, they could plot the areas where the owls had been hunting by the numbers on the metal tags, which were regurgitated in pellets by the owls along with the bones and other remains of the prey.

Visual and radio-tracking methods (Mackay 1970) for plotting locations of marked animals provide more information than those requiring recapture. The home ranges of some red foxes (*Vulpes fulva*) that had been tagged with radio transmitters and monitored remotely are shown in Figure 3–5. Such data, together with the use of computers, allow more detailed analyses than have been made in the past. For example, it has become feasible to construct three-dimensional use-intensity maps and formulate an index of the overlap of use-intensity (Adams and Davis 1967), to compute the probability that neighbors will meet by chance (Jorgensen 1968), and to monitor large numbers of animals continuously and simultaneously (Cochrane et al. 1965). These technical advances provide the potential for greater precision and complexity in future studies of local movements and social behavior in natural populations.

The variation in home range area among various species is tremendous, as the small sample of estimates from the literature reveals in Table 3–1. Much of this variation can be accounted for by the size of the animal alone. As body size increases, daily energy requirements also increase, and, other things being equal, a larger home range is necessary to fill them.

The relationship of home range area, A, in acres, to body weight,

TABLE 3–1 Some estimates of area of home range for individuals of various species. (From Schoener 1968; Jewell 1966; and L. E. Brown 1966.)

Species	Area of home range (in acres, except where noted)
MAMMALS	
Meadow vole, *Microtus pennsylvanicus*	0.02–0.58
Deer mouse, *Peromyscus maniculatus*	0.74–1.67
Whitetail deer, *Odocoileus virginianus*	160
Mountain gorilla, *Gorilla gorilla*	8–15*
BIRDS	
Song sparrow, *Melospiza melodia*	0.4
Scrub jay, *Aphelocoma coerulescens*	5.3
Great horned owl, *Bubo virginianus*	525
Red-tailed hawk, *Buteo jamaicensis*	1,050

* Sq. mi.

W, in kilograms, can be described by equations of the form $A = aW^b$, as in Table 3-2. Values of *b* for mammals, birds, and lizards, respectively, have been determined by McNab (1963), Armstrong (1965), Schoener (1968), and Turner et al. (1969), and are shown in Table 3-2. The same type of equation can be used to describe the relationship between basal rate of metabolism, *M*, and body weight, as shown in Table 3-2. The equations for home range and metabolic weight for mammals have similar values of *b*, but in birds and lizards they do not. If there is a direct relationship between home range and energy requirements, as the similarity in mammals suggests, it does not extend to the other two groups.

TABLE 3-2 Regressions of metabolic rate $(M)^a$ and home range size $(A)^b$ on body weight $(W)^c$ in three groups of vertebrates. (From F. B. Turner et al., *Ecology*. Reprinted by permission of the publisher. Copyright 1969, Ecological Society of America.)

Group	Relationship	Function
Mammals	basal metabolism and body weight	$M = 70W^{0.75}$
	home range and body weight	$A = 6.76W^{0.63}$
Birds	basal metabolism and body weight	$M = kW^{0.69}$
	home range and body weight	$A = kW^{1.16}$
Lizards	standard (30°C) metabolism and body weight	$M = 0.82W^{0.62}$
	home range and body weight	$A = 171.4W^{0.95}$

[a] kcal/day for mammals and birds; cm^3 O_2/hr for lizards
[b] acres for mammals and birds; m^2 for lizards
[c] kg for mammals and birds; g for lizards

This relationship may also differ among species with different diets, as suggested for birds in Figure 3-6. It has been claimed for both mammals and birds that home range increases as a function of body weight more rapidly for predators than for grazing herbivores. This is probably because the food of predators tends to be more sparsely distributed than the food of grazers and requires more energy to obtain.

The methodological difficulties in comparing home range size, particularly where so many different techniques of gathering and analyzing data have been used, are considerable; and the differences postulated between birds and mammals and between herbivores and carnivores should perhaps be regarded as tantalizingly tentative. Nevertheless, some safe generalizations relating home range area to such factors as diet, social organization, habitat, and physical aspects of the environment can probably be made in the future.

3–6 Relationship between size of home range (in acres) and body weight (in grams) for birds in various feeding categories. Omnivores (10–90 percent animal food) are shaded; herbivores, half-shaded; predators, clear. N = nuthatch species. (From T. W. Schoener, *Ecology.* Reprinted by permission of the publisher. Copyright 1968, Ecological Society of America.)

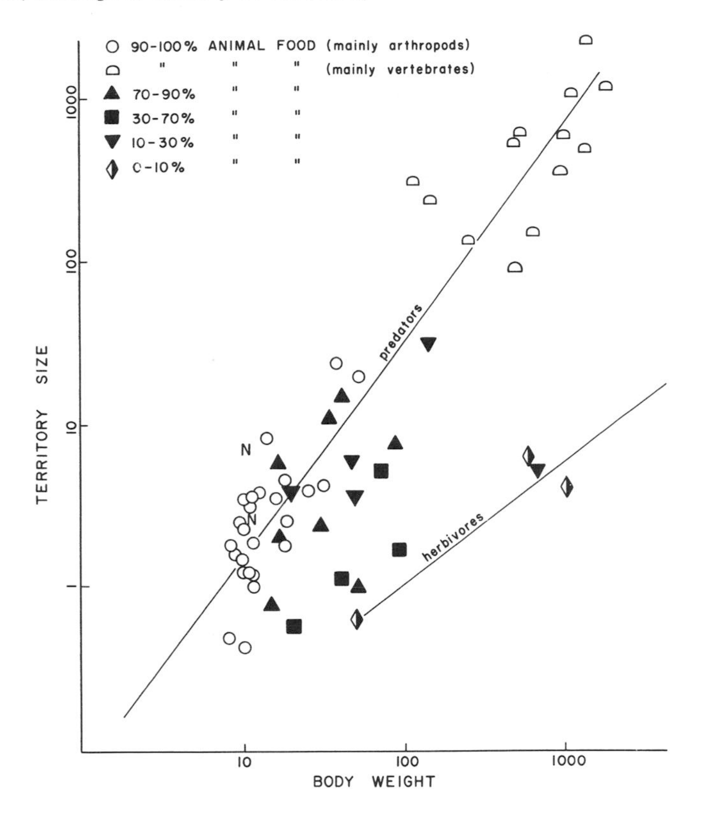

CONCLUSIONS

In this chapter some basic concepts for discussing animal social organization have been described. Dispersal, spacing movements, dispersion patterns, and home range are important aspects of the life history of an individual. In the following chapters we shall examine the behaviors that bring about different dispersion patterns and the problems an individual faces in establishing a home range. We can then examine the adaptive value of species differences in social behavior and their reflection in different types of social organization.

4 The Comparative Study of Spacing Behavior: An Introduction to Territoriality and Other Spacing Phenomena

MANY TYPES OF behavior affect the distances between individuals and consequently influence dispersion. Virtually any form of social behavior either increases or decreases the distance between the participating individuals. Some of these behaviors will be examined in the next several chapters. Agonistic behavior is of prime importance in spacing animals. Its effects on dispersion can be divided for convenience into the spacing out of individuals and groups as separate social units, which will be discussed in this chapter, and the spacing and dominance of individuals or groups *within* a group, which will be discussed in Chapter 5.

Several spacing patterns that result from various combinations of agonistic behavior and site attachment are listed in Table 4–1. In some of these a fixed area is involved, while in others a moving resource is defended. In some, the home ranges of rival individuals are mutually exclusive, while in others they overlap broadly or completely. The size of the social unit also varies.

The best known case is that of territoriality. A *territory* may be conceived as a fixed area from which rivals are driven (excluded) by the active efforts, or "*defense*," of the owner. The owner is commonly an adult male, but territories may also be defended by females, pairs, families, or larger groups. Usually defense is limited to rivals of the same species (intraspecific), but in a minority of cases members of other species, usually similar in appearance, are also excluded or threatened (interspecific territoriality: Orians and Willson 1964; Murray 1971).

Phylogenetic Incidence of Territoriality / Territoriality is widespread in the animal kingdom but is characteristic of the more complex and mobile groups. It is nearly universal in some form in birds and frequent in mammals and fishes. Among reptiles it occurs in lizards but not usually in the less mobile snakes and turtles. Territorial behavior among amphibians is mainly confined to the frogs, in which defense of a mating area occurs in some species. Among the Arthropoda, territorial behavior has been found in some butterflies, wasps, and crabs. Studies of these groups are referred to in the review of Brown and Orians (1970).

Definition / The conventional definition of a territory as a *"defended area"* dates from Noble (1939). It has the virtues of simplicity, flexibility, and long use. There are several potential sources of confusion, however. The following discussion applies to all groups of animals; the concepts are the same regardless of whether they are applied to mammals, birds, or butterflies.

The area defended is delineated by plotting points of defense on a map (see Figure 4–1). Inherent in the concept of territoriality and in the operational procedure by which a territory is delineated is the *fixity* of the area. The area is defined with reference to a map, not by referring to the location of an individual wherever he may be. The concept of individual space or individual distance (see Chapter 5) is thereby excluded. Similarly, defense of a moving object, such as of a female mammal by a male (Ewer 1968), or of a raft of eider ducks by a gull who steals food from them (Ingolfasson 1969), should probably also be excluded.

Some persons have attempted to define territory as an area of dominance (e.g., Emlen 1957; Willis 1967; Murray 1969; Wolf 1970). This is misleading; although a territory owner is always dominant in his territory, this never has been the criterion of the definition, nor can anything be gained by changing the established definition. The idea of defense has traditionally implied, first, the use of attack and/or threat toward intruders, and second, that the intruder is repulsed from the area. In other words, the territory is more than an area of dominance; it is also nearly an *exclusive area* in which exclusion is the result of the aggressive behavior of the owner. Areas of dominance from which submissive individuals are *not* excluded can be called *dominions* to distinguish them from territories (see Chapter 5). Territory should not be defined simply as an exclusive area, as Pitelka (1959) and Schoener (1968) have suggested, because (1) there are cases of exclusive areas that are not simply due to the aggressive behavior of the owner, and

TABLE 4–1 Some spacing patterns in animal societies.

Spacing pattern	Individual	Social unit: Pair	Family	Large group
ALL OR MOST OF HOME RANGE EXCLUSIVE				
Defended = all-purpose territory	European hamster (*C. cricetus*) European robin in fall	European robin in spring	lar gibbon (*Hylobates lar*)	African lion (*Panthera leo*) Mexican jay (*Aphelocoma ultramarina*)
Undefended = undefended exclusive area	nesting female grouse	?	?	most baboons (*Papio*) Juncos (*Junco*) in winter
PART OF HOME RANGE DEFENDED AND EXCLUSIVE				
Mating territory	lek birds and mammals	?	?	?
Nesting territory	mice voles	herring gull (*Larus argentatus*)	?	?
Feeding territory	pectoral sandpiper (*Erolia melanotos*) and hummingbirds in fall	?	?	?
LITTLE OR NONE OF HOME RANGE EXCLUSIVE				
Defense of moving resource				
Drifting food source	glaucous gull (*Larus hyperboreus*)	?	?	?
Females	male hamadryas baboon (*Papio hamadryas*)	?	?	?
Maintenance of free space around a social unit (individual distance, individual space)	chaffinch (*Fringilla coelebs*) in winter	partridge (*Perdix perdix*)	partridge	?
Regular dispersion due to dominance of owner (dominion)	Steller's jay (*Cyanocitta stelleri*)	?	?	?

(2) a behavioral concept such as territoriality should be defined at least partly on the basis of behavior, not by dispersion criteria alone.

Finally, some differences between the concepts of home range and territory should be emphasized. The two concepts are not identical, nor are they alternatives in any way. Any individual of any species can be said to have a home range of some sort, if it has settled in one area; yet few of these home ranges are territories. A home range may or may not be defended in part or in whole; and it may or may not overlap with the home ranges of other individuals. It is useful to compare species in terms of the degree to which the home ranges of individuals overlap. A continuum can be recognized, extending from those having no overlap to those that overlap greatly (Table 4–1). A most unfortunate tendency resulting from such comparisons (Burt 1943) was the custom of referring to the species with little overlap as "territorial species" and those with much overlap as "home-range-type species." Since territory and home range are defined in different terms and by different operations, they cannot logically be placed on the same continuum.

In summary, several concepts bear some resemblance to that of a territory—namely, individual space, area of dominance (dominion), exclusive area, and home range. Each of these shares certain properties with the others. In some species most of these concepts pertain to the same area; in others, they do not. Each of these concepts is valuable for certain purposes. All are included in the category of spacing behavior, which is a general term for all types of behavior that bring about dispersion patterns. To make any of these terms synonymous with any of the others would lead to confusion of the basic concepts. Perhaps the inherent properties of Noble's definition of a territory as a defended area would be more clear if the definition were understood to mean the following: A territory is a fixed area from which intruders are excluded by some combination of advertisement (e.g., scent, song), threat, and attack.

Difficult and Borderline Cases / One complication that arises in applying the concept of territory to mammals is the *interdigitation of pathways* that may occur. As Ewer (1968) emphasized, the home ranges of many species of mammals consist of many narrow pathways between the parts most used; the result is more a system of lines than a circular area of homogeneous usage. The lines often radiate from a burrow, den, or core area. With this type of home range it is possible for much interdigitation of neighboring home ranges without any actual overlap. The operational technique of connecting the outermost points of observation with a line, yielding the largest-possible-polygon home range estimate, would be misleading in this instance, since it would be likely to suggest overlapping of home ranges when in fact none occurred. This pitfall can be avoided by observing the animals directly.

Often the exact location of the territorial border is obvious, but in some cases it is not. The border of the territory may be a zone (as in Figure 4–1) rather than a line. Another complication shown in Figure 4–1 is caused by the "trespassing" of neighbors. Although neighbors are driven out of the territory, or excluded, their occurrence within the territory as trespassers means that objectively the area is not totally exclusive. The important point for establishing the location of the border is that trespassers are driven out, not that they never enter.

A frequent borderline case is the concurrence of aggressive behavior, uniform dispersion, and a moderate to large overlap in home range. This combination of factors is found in chipmunks (Dunford 1970), domestic cats (Leyhausen and Wolff 1959), *Octopus* (Yarnall 1969), Steller's jays (Brown 1963*b*), and ant-tanagers (Willis 1967), among others. The occurrence of some spacing out due to agonistic behavior and core areas of dominance has led some authors to describe such situations as territorial. However, the large and normal overlap in home range requires that such areas not be termed territories. Unfortunately, there is no single accepted term adequate to designate such cases and to distinguish them from territories; however, they might be termed *areas of dominance* or *dominions*. Since in these species subordinates may tend to avoid contact with dominants, different individuals use the same areas but at different times (as with domestic cats). Dominance is discussed further in Chapter 5.

A rather unusual intermediate situation was found in cheetahs (Eaton 1970). Males in families or adult groups exclude neighboring families from each other's home ranges, but the home ranges tend to shift a little from day to day so that the area defended is constantly changing. This is not the same as a *family space* (see Chapter 5), since the areas are scent-marked like the territories of many other mammals. Perhaps such areas could legitimately be termed *moving territories*.

Another difficult case is that of *undefended exclusive areas*. In baboons the home ranges of neighboring troops may or may not overlap. It is common for them to inhabit nearly exclusive areas but without defending them overtly. Red foxes (*Vulpes fulva*) also seem to have undefended exclusive areas, as shown in Figure 3–5 (Sargeant 1972). This pattern is also found whenever pairs or other social units are isolated by topographic or ecological factors from their counterparts, and in populations existing at low densities. Since these areas are not known to be defended, they cannot be termed territories. One guesses that the exclusive areas of baboons and macaques might be maintained in the absence of defense by mutual or unilateral avoidance.

4–1 Territorial behavior in relation to home range in the meadow bunting (*Emberiza cioides*). The behavior of a male in relation to his nesting area is shown at left (**A, B**). A schematic model of the home range is at right (**C**). These diagrams show the transition from winning inside the territory to losing outside of it. The line from 1 to 2 (A) is a transect reproduced in B as a cross section of the home range. We see (B) that the transition may be abrupt in some parts, as in the left (uphill) part of the home range where an abrupt discontinuity in topography and vegetation occurs, or more gradual, as in the part of the home range at right (downhill), where the environmental change is not sharp. The territorial border is narrow or broad; and the gradient from win to lose is abrupt or gradual. For this male the outermost polygon is an estimate of home range. The territorial border is the zone between the next two polygon perimeters (the zone of triangles). (From Yamagishi 1971.)

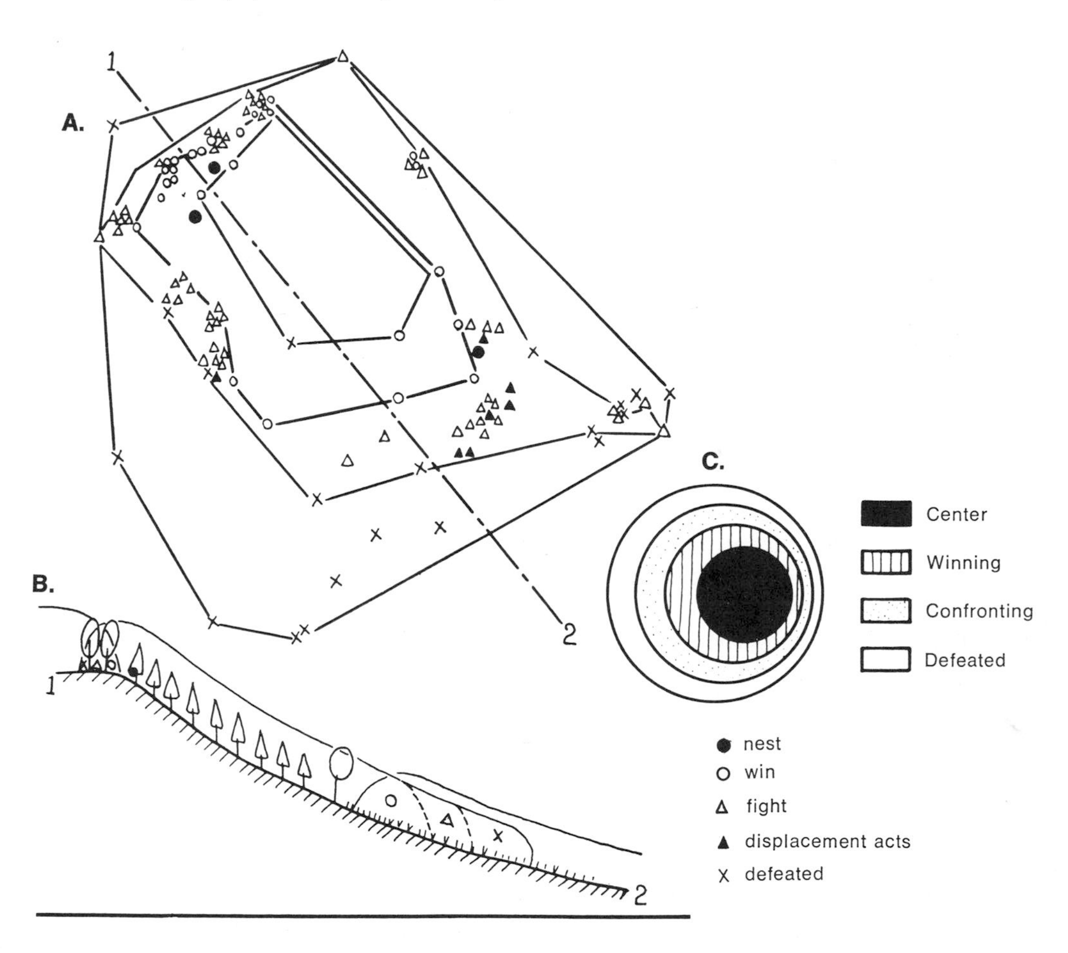

It is possible to argue that one animal does not avoid a second unless some repulsing behavior by the second has caused the avoidance either during the encounter (unperceived by the observer) or at some prior time. If one followed this line of reasoning, "mutual avoidance" would not be recognized as different from territorial defense (e.g., Krebs 1971). This argument misses the main point. It is obvious that in some species acts of "defense" directly bring about exclusive areas, while in others having exclusive areas it is nearly impossible to observe such acts. Species differ in the frequency, intensity, and ease of observation of spacing behaviors. The differences are real and can be measured; they cannot be dismissed by "logical" arguments. Avoidance can be elicited by strange, novel, or startling stimuli. It need not necessarily arise only from acts performed by a territory owner that have been specialized through natural selection for their usefulness in territory defense. There may be other ways in which mutual avoidance can arise that we do not yet know or understand. We should at least allow the possibility that such mechanisms could exist and recognize that spacing can occur without obvious "defense."

Real borderline cases are bound to occur—for example, when defense is present but relatively insignificant, when the areas are occupied almost but not quite exclusively by their owners, and when the areas tend not to have fixed boundaries. These difficulties are a reflection of the complexities of the natural world. It would be unfair to expect our compartmentalized systems of categories to fit every species perfectly.

In the past many authors have attempted to fit their observations of aggressive behavior in a particular species into a conceptual framework based on the classical picture of an all-purpose territory. In many cases this could not be done without distorting either the observations or the concept of a territory. A broader perspective is now available. The view of territoriality adopted here is that it is only one of several types of spacing mechanisms (Brown and Orians 1970). I have attempted to retain the established concept of a territory and to view it within the broader perspective of spacing behavior. In contrast, some authors unfortunately have redefined territorial behavior so as to include all kinds of behavior that space out animals, with the result that virtually any space in which an animal shows aggressive behavior has been called a territory at one time or another.

Diversity of Territories / Territories differ greatly in nearly every respect imaginable. Conspicuous differences occur in the activities performed in the territories. These activities are partially listed, together with species examples, in Table 4–1. Although there may be a *common element of competition* in the ecological selection pressures that generate territories of diverse sorts (Chapter 7), that is probably all that the territories of very different species have in common except for the incidence of aggressive behavior.

Most cases fall into one of three categories: mating, nesting, and all-purpose territories. Many species, from insects to mammals, defend an area that is used mainly for mating. Mating territories are often clumped together to form a communal mating area, or *lek*. Intraspecific defense of the nest and only a small area around it is common among colonial animals, such as seabirds that come to land mainly for nesting. Other specialized areas may also be defended. Some hummingbirds (Wolf 1970 and others) and shorebirds (Hamilton 1959; Recher and Recher 1969; Goss-Custard 1970) defend feeding areas during the nonbreeding seasons. Sleeping places may also be defended, as with some birds that sleep in holes or special crevices that may be in short supply.

The Central Role of Territory Defense / In territorial species the acquisition and retention of a territory is a sine qua non for breeding success. Consequently, selection pressures acting on all aspects of behavior related to territories are apt to be severe, and many structures and behavior patterns are likely to become specialized by selection for their role in territorial defense. The European robin (*Erithacus rubecula;* Figure 4–2), which has an all-purpose territory, illustrates this; its aggressiveness, song, and plumage are in part adaptations for defense of territory. The relationships between these factors and territoriality in the robin have been nicely summarized by Lack (1953).

The breeding cycle of the European robin is summarized in Figure 4–3. After the molt in late summer, when robins are quietest, ag-

4–2 A European robin in threat display with throat feathers erected at a stuffed robin above and out of the photo. This display plays an important role in defense of the territory. (From *Brit. Birds* 64:Plate 21. Photo courtesy of Frank V. Blackburn.)

gressiveness and song increase into the fall in both sexes. Both sexes defend territories in the fall, one bird to a territory. This activity is dampened by the winter, but in early spring (January and February) the male reaches a peak in aggressiveness and singing, while the female gives up her territory, becomes less aggressive, and seeks a mate. Upon acquiring a mate, a male typically reduces his singing somewhat but still maintains a high level of song and defends his territory. Seasonal variations in frequency of male song tend to parallel variations in aggressiveness associated with the territory.

The special role of the robin's red breast in territorial defense was

4–3 Seasonal changes in frequency of song and aggressiveness in the European robin. (After Lack 1953.)

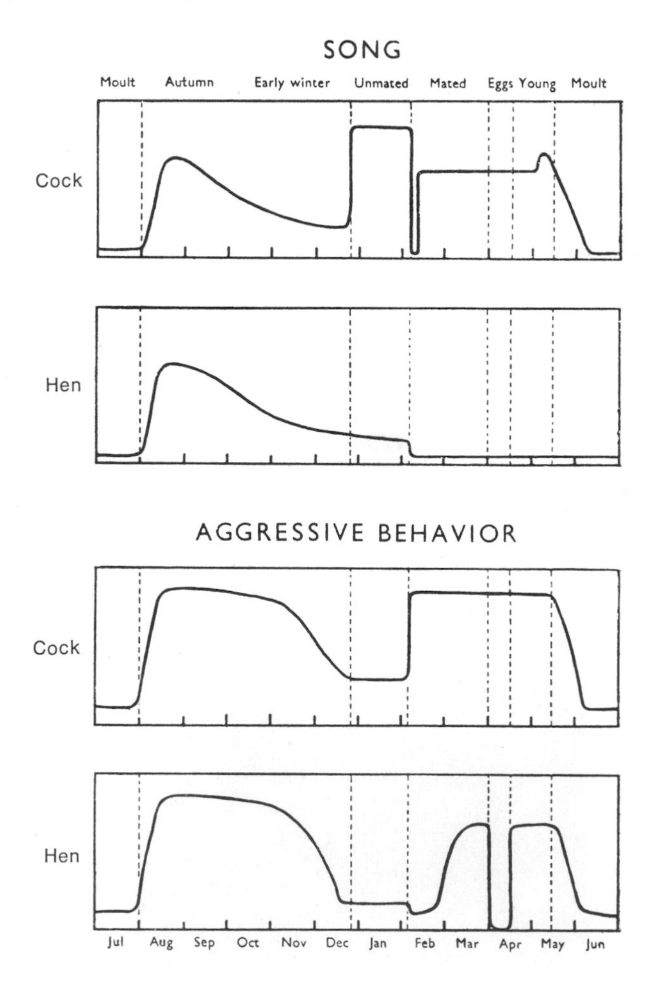

shown in the following experiments. Lack presented stuffed robins in various plumages to territorial males and observed their responses. Stuffed adults were attacked; but juveniles, which differ from adults mainly in having a dingy brownish breast instead of a red one, were ignored. A tuft of red feathers from the breast was attacked, however, even though all the other attributes of an adult robin, such as shape, size, and color, were lacking. The red breast feathers, which are fluffed out in a frontal display to make the robin appear larger and more conspicuous, are apparently more important in a territorial context than are all other attributes of a stuffed robin combined. (Experiments on the role of breast coloration in agonistic behavior in the chaffinch are described in Chapter 5.)

In summary, two of the robin's most conspicuous attributes, its song and red breast, are intimately connected with territorial defense.

The Ontogeny of Territorial Behavior / The development (ontogeny) of territorial behavior in animals that are becoming territorial for the first time in their lives is difficult to observe in most species, especially in nature. It has been observed in black-tailed prairiedogs (*Cynomys ludovicianus*) by King (1955). Prairiedogs were formerly among the most readily observed species of rodents in North America. Today, because of agriculture and often unnecessary slaughter by the federal government, few dog towns are left, and these give only a hint of the vast colonies of earlier times. The social organization, behavior, and ecology of prairiedogs have been studied in detail by King (1955) and Koford (1958), whose observations form the basis for the following description.

The town studied by King in South Dakota was rather large by present standards (75 acres). It was divided by streams, roads, and vegetation types into *wards*, each of which was divided into territories. Each of the latter was occupied by a group of prairiedogs. For six groups, or *coteries*, whose composition was accurately determined in 1950 before the young appeared, four groups had only one adult male and the other two had two. Four of the groups had from two to four adult females and two had only one. The average number of adults per group was 3.5, but there was considerable variability. Group size increased dramatically as the young emerged; one group increased from 7 to 39, although this large size did not last long because the group split up.

The behavior and importance of the prairiedog coterie is best described in King's words (King 1955:54):

> The coterie, as the basic unit of social organization in the town, consists of an integrated social group characterized by frequent friendly social contacts

within the coterie and hostile social contacts with members of different coteries. All members of a coterie are familiar with one another and are at liberty to use all of the burrows in the territory. Cohesive behavior, such as grooming, kissing, and playing, is common within the coterie. Displays of antagonism or dominance are rare among the coterie members. In contrast, the social relations between coteries are antagonistic. Members of different coteries fight with each other whenever they meet, and movements between coterie territories are undertaken only at the risk of a combat. All members of a coterie defend their territory against invasions by strange individuals. These patterns of behavior within and between coteries are characteristic of a closed society.

This pattern of organization was mediated by a variety of specialized behaviors that seemed to reduce fighting within the coterie and maintain hostility toward outsiders. A territorial call and display were conspicuous adaptations to territory defense. Kissing, playing, and

4–4 Behavior toward members and nonmembers of a coterie of black-tailed prairiedogs. *Left:* Identification kiss between young of the same coterie. *Right:* Adult female giving the territorial call at the entrance to her burrow. Notice the black mark applied by the observer for the purpose of individual recognition. (From King 1955. Reprinted with permission of the author and the Laboratory of Vertebrate Biology, University of Michigan.)

mutual grooming were conspicuous behavioral patterns that maintained peaceful relations within the coterie but were not seen between members of different coteries (Figure 4–4).

When the young first emerge from their burrows they do not, of course, know the territorial boundaries or the "rules" for their maintenance. They are treated affectionately by fellow coterie members and can roam freely in their territory. Since neighboring adults are excluded, all individuals encountered by a young animal in his territory treat him warmly, for example by grooming him. For the first week or two he may wander into an adjacent territory without resistance from its occupants. Then he begins to meet hostility from his neighbors. At first, aggression toward intruding pups is mild; but it gradually becomes severe. Soon he learns which prairiedogs greet him with a "kiss," as do all coterie members, and which do not. If his kiss is not returned by another prairiedog he barks at it, runs up to smell it, and dashes away. This sequence later develops into the tail-spreading ritual used by prairiedogs in maintaining territorial borders. The young prairiedog begins to utter the territorial calls soon after first emergence, either spontaneously or in answering others. If he does so outside his own territory he is immediately attacked. Later he limits his calling to his own territory.

In brief, the young prairiedog is born into a social system that shapes his early learning experiences. He can perform such actions as the territorial calls and the kiss on emergence, but he must learn the consequences with respect to his territory boundaries. And if he leaves his natal territory to establish a new one, he brings the experiences of his youth with him.

Artificial Territories / The development of territoriality has also been studied in laboratory experiments with mice. One of the first was done by Anderson and Hill (1965) with house mice (*Mus musculus*). They confined mice in nine arenas, each 1.8 m × 3.7 m and divided into four equal sections by partitions (Figure 4–5). Each section had food, water, a nest box, and a complexly structured central structure of concealed runways and boxes. A male and female were introduced into each section. After two days of familiarization in which all connecting holes between were closed, the holes at *a* and *c* were opened and the males allowed to establish a dominance relationship. The losers that still remained alive were removed, and then the holes at partition *b* were opened, allowing the two dominant males to make contact. The result was that 15 of 21 males established territories with boundaries coinciding with their original areas of dominance; the remaining males either became subordinate or were killed. These experiments show that

4–5 Enclosure used to study territoriality in laboratory mice. *Above:* Floor plan, showing locations of covers and passageways between compartments. Walls are of plywood 75 cm high resting on a concrete floor. *Below:* Detail of a cover in an expanded view. Stipple indicates galvanized wire screening. Mice spent 85 percent of their time within such covers when outside their nest boxes. (From Anderson and Hill 1965. Copyright 1965 by the American Association for the Advancement of Science.)

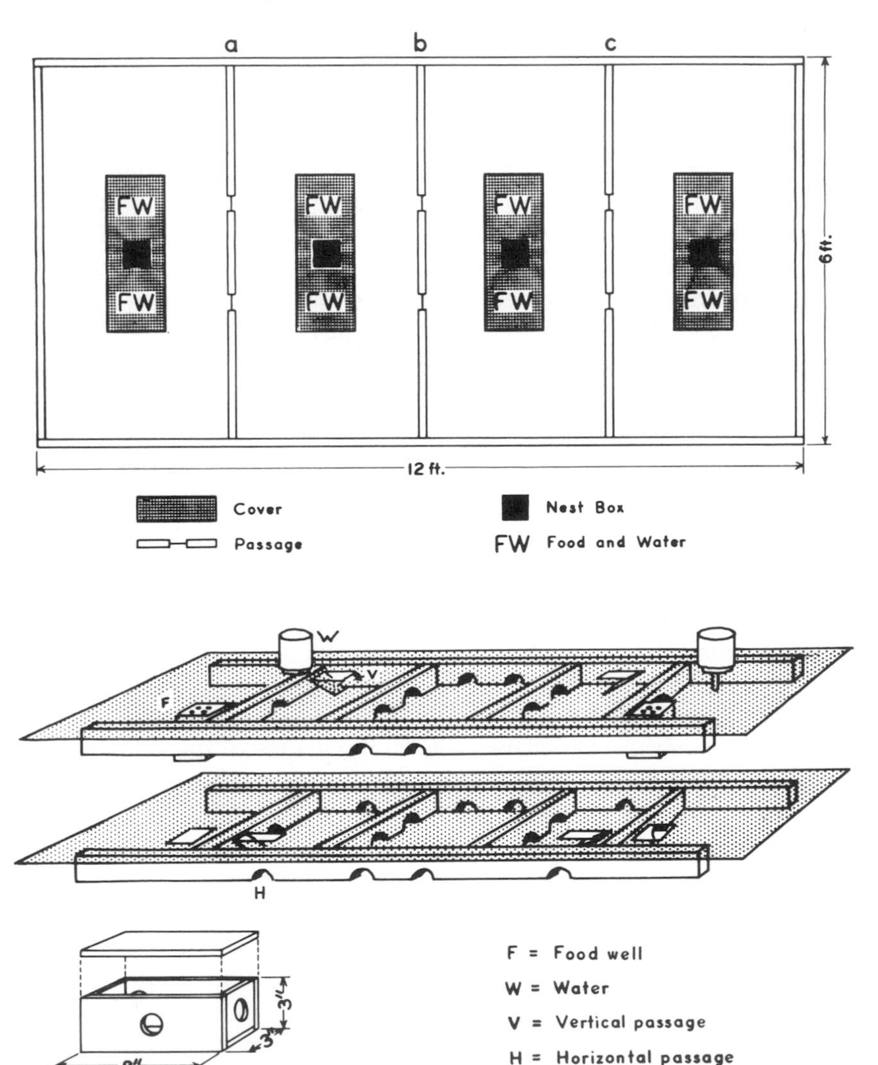

a sequence of events sufficient to establish territoriality consisted of (1) familiarization with the area followed by combat and social dominance over another male who was then removed, and then (2) defeat by a similarly experienced neighbor on the latter's home ground coupled with victory over the neighbor on familiar ground. Mackintosh (1970) obtained similar results but experimented further with different combinations of mice in a simpler environment.

CONCLUSIONS

In this chapter territorial behavior has been portrayed as the principal type of spacing mechanism among social units in species that tend toward regular dispersion patterns in some situations. However, in some species other spacing mechanisms may help bring about regular dispersion. A number of borderline cases have been described and some practical difficulties of the territory concept have been discussed. For species in which a territory is necessary for the success of individuals in breeding, adaptations, such as a song used in advertisement by a territory owner and threat display used in excluding rivals, are conspicuous features of their behavior and show the pervasive effects of territoriality on behavior and appearance. Learning plays an important role in the development of territorial behavior in the young, as shown in prairiedogs. Experiments on the development of territoriality in laboratory mice have shown that territorial social organizations may form under certain conditions. Territoriality is not an invariant species-specific trait but depends for its appearance on certain kinds of experience and environments.

5 Animals in Groups

MANY SPECIES OF animals spend part or all of their lives in groups. The mode of formation of these groups and the adaptive value of grouping for individuals vary widely from species to species. In this chapter various kinds of groups will be described briefly and some general features of social organization within groups will be considered. The emphasis will be on spacing and dominance.

TYPES OF GROUPS

Although there are certain biological relationships shared by all social groups, regardless of origin, the ecological and evolutionary factors that cause or predispose animals to form groups and to stay in them are exceedingly varied. A rough classification of these factors is given in Table 5–1 according to a simple division into major evolutionary modes.

Kin groups are groups whose members are closely related genetically because many have the same parents. This condition may come about through asexual reproduction, as in colonial hydroids and Portuguese men-of-war, but this is a special case not found in terrestrial animals. Simple groups are formed by the families of one or more females. In most species the young disperse early, but in others some, at least, persist with their parents for one or more generations as an *extended family*. Such groups are typical of the social insects and certain vertebrates, notably various primates and birds. Kin groups show some unusual kinds of social behavior and will be treated as a separate evolutionary problem in Chapters 9 and 10.

Mating groups depend ultimately on the mutual attraction of the sexes to each other for mating. This may take place in various ways. In some species a simple and long-lasting monogamous pair is formed, as in most birds. Among mammals harems of two or more females associated with one male are common. The males of some species of birds and mammals gather for courtship in communal display grounds, or leks, to which the females come only for fertilization. In many fishes and amphibians the males may also gather at localized spawning grounds, but these differ from leks because they are used for depositing

TABLE 5–1 Some grouping patterns in animal societies.

Grouping pattern	Examples
I. KIN GROUPS	
A. Clones: groups formed by asexual reproduction of sessile colonial invertebrates, typically in permanent physical contact	colonial coelenterates
B. Families: groups formed by one or two parents and their most recent offspring	goose and swan families
C. Extended families: groups formed from families by failure of many offspring to leave parents	prairiedogs (*Cynomys*) some primate groups Mexican jay (*Aphelocoma ultramarina*)
II. MATING GROUPS	
A. Pairs: monogamous groups of two	scrub jay (*Aphelocoma coerulescens*) lar gibbon (*Hylobates lar*)
B. Harems: groups in which a male attempts to keep females together and away from other males, with or without cooperation of females	red deer (*Cervus elephas*)
C. Leks: groups formed by attraction of males (and subsequently females) to a communal mating ground; eggs or young produced elsewhere	lek birds and mammals hilltopping butterflies Hawaiian *Drosophila*
D. Spawning groups: groups of both sexes formed at localized spawning grounds; no provisioning of young	many fishes and amphibians
III. COLONIAL GROUPS	
A. Groups formed by colonial nesting of pairs or one-male harems; young provisioned at nest	tricolored blackbird (*Agelaius tricolor*) many sea birds some bats and seals
IV. SURVIVAL GROUPS	
A. Groups formed by aggregation of randomly related, usually nonbreeding individuals who are mutually attracted by each other	foraging flocks night roosts of New World blackbirds ducks and geese herding mammals fish schools
V. COINCIDENTAL GROUPS	
A. Groups formed by physical factors acting on migrating or moving animals	hawks migrating along a mountain ridge land birds migrating through a mountain pass whelks on a sheltered ocean rock stream-surface insects on a calm eddy
B. Groups formed by attraction to a common resource, such as food or water	bears at a garbage dump

eggs. The evolution of various types of mating systems is a separate problem that will be treated in Chapter 8.

In a *colonial* type of social organization, pairs or females form groups where the young are to be raised; they radiate out from that location, the colony, to obtain food and other necessities. Typically the location of the colony is protected from major predators—for example, gannets on an island where no mammalian predators occur, or bats on the ceiling of a cave. Mating may or may not take place at the colony site. Colonies tend to be defended against predators but not against other members of the same species. Usually each member of the colony defends a small territory consisting mainly of its nest and the area that can be reached from it. Colonial species do not regularly defend their food supply, although they often fight over food. Colonial social systems will be considered further in connection with the evolution of territories of different sizes and uses in Chapter 7.

The category that I have designated *survival groups* includes a varied assemblage of species. Generally the individuals in such groups are not breeding, either because it is not the breeding season or because they are too young or have been somehow excluded from the breeding population. The predisposition to join large groups in these species is probably selected for on the basis of the increased survival rates of grouped versus solitary individuals. We will consider in detail in Chapter 7 the reasons why survival is higher in such groups.

Some groups seem to form mainly because local environmental factors tend to concentrate individuals either physically (Figure 5–1) or through attraction to a localized microclimate or to a resource such as food or water. Such groups are here termed *coincidental*. Terns and other seabirds come from all directions on the open ocean to gather over a surfacing school of fish. Vultures are attracted from a vast area to a carcass. Fruit bats congregate at a tree loaded with ripening fruit. At other times these animals hunt alone for food, although they also seem to watch each other from a distance. Since each coincidental group is likely to have a separate explanation and since general evolutionary principles are not apparent in their formation, they will not be considered further.

The classification in Table 5–1 is oversimplified in several ways. In any one species the selection pressures that influence grouping are numerous and complex. Although for the sake of compartmentalization certain unifying factors have been singled out, these are not the only ones important in any one species. For example, survival is important in all groups, not just survival groups; environmental factors coincidentally may predispose toward even larger groups in colonial species or in mating groups; and in the early phases of colony founding in a

5–1 A coincidental group caused by physical factors. Much of the North American population of broad-winged hawks (*Buteo platypterus*) is funneled through Panama on its migration to South America. This photo was taken on October 11, 1972 and shows some of the 41,333 broadwings that were counted passing Ancon Hill that day. The birds were concentrated by geography and their preference for following ridges, where they can glide on updrafts. (Photo by Neal G. Smith.)

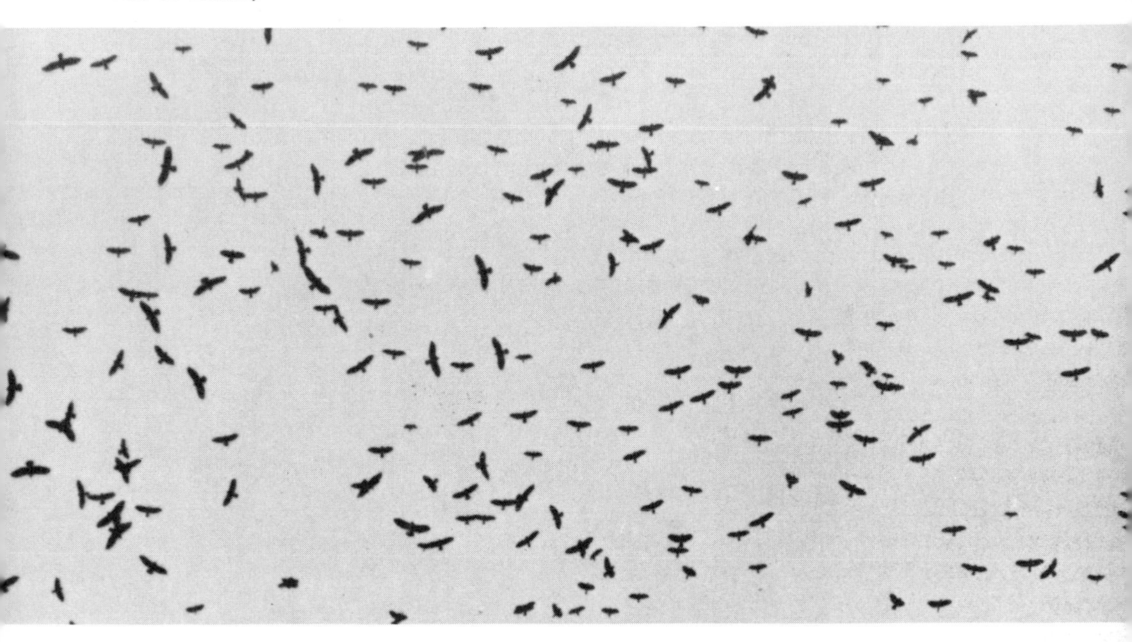

breeding season an attraction between the sexes is operating. Despite such failures, the scheme in Table 5–1 does serve to point out certain unifying biological principles common to certain social groups in diverse phylogenetic taxa. Perhaps through the recognition and study of such unifying principles in both the spacing and grouping of animals a discipline of sociobiology will take shape.

FORCES OF ATTRACTION WITHIN GROUPS

The classification in Table 5–1 is designed to emphasize the ultimate or evolutionary causes of grouping behavior. Another approach to groups is to examine the proximate behaviors that bring about the spacing pattern within a group. Aggression in the form of threat and attack is the primary repulsive force. The forces of attraction are more varied. Highly social species, such as the wolf (*Canis lupus*) and many primates, have special facial and postural positions and sounds that express *submission* and that are used in appeasement. In the more solitary foxes the ability to communicate with signs of submissiveness is not as highly developed (Fox 1970). Submissive behavior is not a positive form of attraction, but it has the result of allowing a closer peaceful approach between dominant and subordinate animals. *Sexual* attraction clearly operates in some groups, although contrary to the early theory of Zuckerman (1932; see Chapter 12), it never seems to be the primary cohesive force within a group of primates. *Soliciting* or begging behavior also brings animals together. It is used by many birds and mammals to indicate readiness for copulation or for grooming, to obtain food, and for other purposes. It is found in groups as different as social insects and primates. The attraction between parent and offspring, or the *kinship bond*, is important in many groups. Ducklings follow their mother into autumn and young Canada geese (*Branta canadensis*) may migrate and spend the winter with their parents. Young red deer (*Cervus elaphus*) remain with their mothers for a few years, and the females tend to form matrilineal herds (Darling 1964). These are the main attractive and repulsive forces in social groups, but there are others—for example, the greater ease of flying or swimming close behind a companion, or the aggressive shepherding of female hamadryas baboons (*Papio hamadryas*) and red deer by harem masters.

THE SPACE AROUND AN INDIVIDUAL

Nearest-neighbor Distance / Animals that live in groups can be roughly classified into *contact species*, in which body contact of a peaceful, nonsexual sort is normal between adults, and *distance species*, in which body contact normally occurs only in sexual or agonistic contexts (Hediger 1964). Individuals of contact species may huddle together in close groups that look like a single mass—for example, when roosting or in cold weather. They often groom or preen each other. The individuals of distance species are frequently gregarious but carefully avoid such close contact or limit it to certain contexts. A species difference of this sort in macaques is shown in Figure 5–2.

Contact species exhibit little aggressive behavior in groups, but distance species exhibit various phenomena of spacing and priority that are mediated by agonistic behavior. If the animals in a group are photographed and the distance from each individual to its nearest neighbor is measured from the photograph, a curve such as that in Figure 5–3 can be generated. The data in Figure 5–3 were taken from 29 wild flocks of sandhill cranes (*Grus canadensis*) by Miller and

Stephen (1966), and show a tendency for most individuals to be 4 to 6 feet from their nearest neighbor. The data can be evaluated with the help of the formula

$$R = \frac{\bar{r}_A}{\bar{r}_E},$$

in which the departure from randomness is expressed as the ratio, R, of the mean measured distance between nearest neighbors, $\bar{r}_A$, to the mean distance that would be expected, $\bar{r}_E$, if all individuals in the

5–2 A species difference in individual distance. The bonnet macaque (*Macaca radiata*), shown in **A,** is a contact species and spends much time in close physical contact. The pigtail macaque (*Macaca nemestrina*), shown in **B,** is a distance species and does not exhibit the huddling of the bonnet macaque. This difference between the two species was established quantitatively by Rosenblum et al. 1964. Photos from Kaufman and Rosenblum (1969) show characteristic interindividual spacing. (Photograph A reproduced by permission of the Annals of the New York Academy of Sciences.)

A.

B.

population were randomly distributed (Clark and Evans 1954; Pielou 1969). Values of R less than 1 indicate clumping; those greater than 1 indicate regular spacing (maximum $R = 2.1491$). R values for the 29 flocks from which these observations were taken were significantly greater than 1.0 in 26 cases, indicating regular spacing within most crane flocks.

Although the subject has received little attention in wild animals, R probably varies with group density. Measurements obtained from captive groups of *Drosophila paramelanica* by Sexton and Stalker (1961) showed an increase in R, hence also in regularity of spacing, as group density increased (Figure 5–4); similar observations were made of hens by Craig et al. (1969). Spacing in the flies was brought about, at least in part, by avoidance when another fly was 1 to 5 mm away and by touching other flies with extended legs. Most flies maintained a distance of at least two leg lengths from each other. The increase in R with density might be explained by the greater frequency of opportunities for spacing by leg extension at high densities.

5–3 Nearest-neighbor distances in 29 sandhill crane flocks containing 1,326 individuals. (Data from Miller and Stephen 1966.)

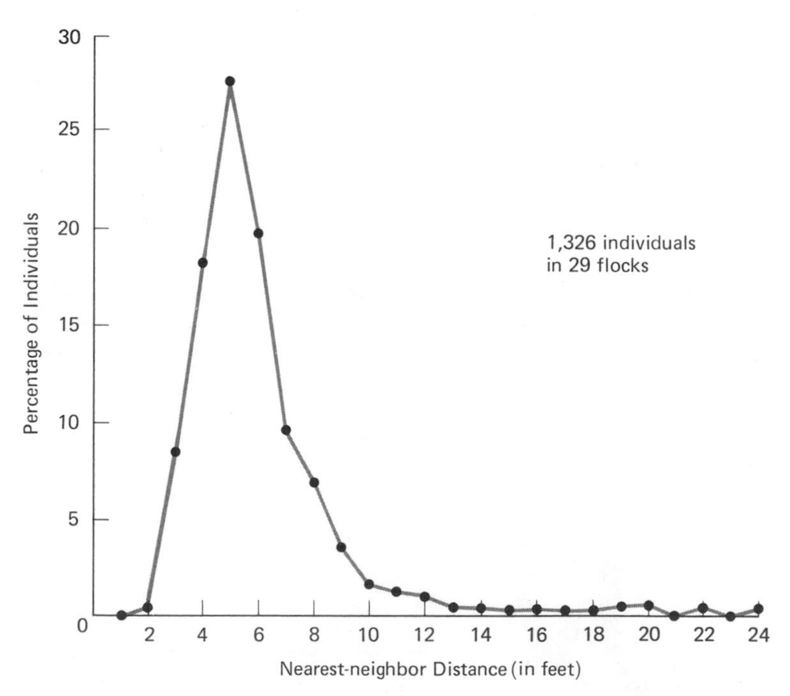

Individual Distance / The tendency for individuals of distance species to avoid proximity within a certain distance of each other was first noted by Hediger (1964). His observations led to the concept of individual distance. The concept was further applied to natural populations of birds by Burckhardt (1944) and Conder (1949). *Individual distance* may be defined as the distance from an individual at which another individual provokes aggressive or avoidance behavior. Corresponding to individual distance is the concept of *individual space* (or personal space), which may be considered to be the space, centered on an individual, that is kept free of other individuals by aggressive or avoidance behavior. Individual space is unlikely to be circular; it should be larger in front of an animal than behind.

In practice, individual distance has not yet been measured experimentally in natural populations, but it has been studied in caged birds. These studies are instructive but they should be checked for natural populations.

A series of ingenious experiments on individual distance in the chaffinch (*Fringilla coelebs*) was performed by Marler (1956). The birds were kept in groups of eight or less in aviaries 2 m × 2 m × 2 m. To study spacing behavior as a function of the distance between two individuals, a pair of movable food hoppers was used (Figure 5–5). The distance between the hoppers was varied in random order for the tests,

5–4 Increased regularity of spacing at higher densities of *Drosophila paramelanica*. *R* is the ratio of observed to expected mean values of nearest-neighbor distances. The area occupied was constant at 2,066 mm^2. Only the lowest value of *R*, 1.15, was not significantly different from 1.0, the value for populations distributed at random. The different symbols represent different days of observation. (From Sexton and Stalker 1961.)

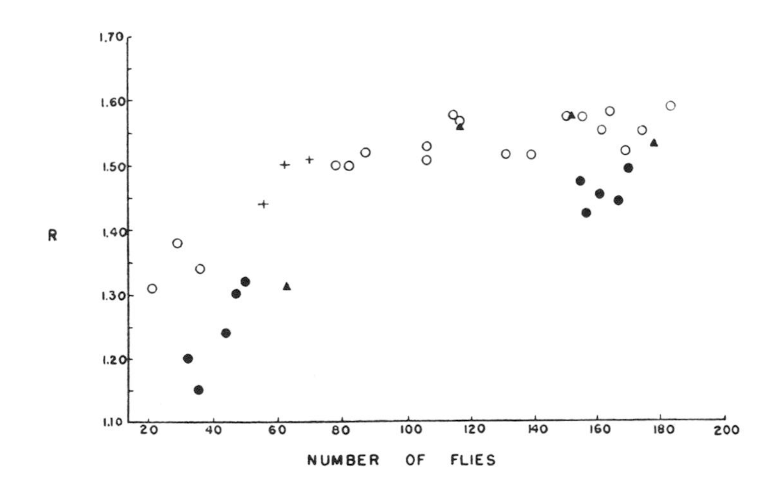

and observations were made at each distance. An encounter occurred when both hoppers were occupied simultaneously. If one bird chased the other, the encounter was classified as aggressive; if not, it was classified as nonaggressive (regardless of the possibility of other agonistic behavior). The percentages of encounters classified as nonaggressive were plotted as a function of inter-perch distance, as in Figure 5–6. From the resulting curves the distances at which 50 percent of encounters led to chases could be extrapolated and these were used for the comparisons summarized in Table 5–2.

The graph shown in Figure 5–6*A* is for a group of eight females. It shows that aggression was less frequent as the inter-perch distance was increased and suggests that the decrease is rather gradual. How typical is this graph for female chaffinches? In this graph all data for all females were lumped together. If the behavior of individual females toward all their subordinates is examined (Figure 5–6*B*), it can be seen that individuals differ in their individual distances in relation to their subordinates. And if we restrict the data to the relations between only two individuals (Figure 5–6*C*), the data suggest (though rather weakly) that the same individual can have different individual distances in

5–5 Food hoppers used in experiments on individual distance in chaffinches. *Above:* A male is about to attack another at 10 cm. *Below:* Two females feed at zero distance (From Marler 1956.)

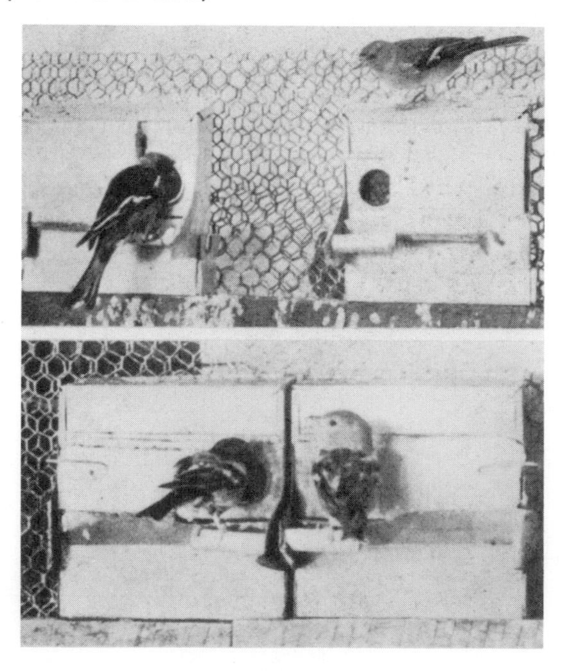

relation to each of two other individuals. These analyses revealed that not all female chaffinches have the same individual distance; variation occurs among individuals. If we compare data for a group of males (Figure 5–6*D*) with data for a group of females (Figure 5–6*A*), it is ap-

5–6 Graphs of the ratio of the number of times when two chaffinches fed from the hoppers in Figure 5–4 without fighting to the total number of times two birds fed simultaneously from the two hoppers (with or without fighting), plotted against the distance between the hoppers. **A.** Pooled observations on eight females. **B.** Results for three females taken separately toward all their subordinates. The ranks of the females are indicated by the numerals 1, 2, 3. **C.** Results for female number 2 in B against two subordinates, X and Y, taken separately. **D.** Pooled observations on eight males. (From Marler 1956.)

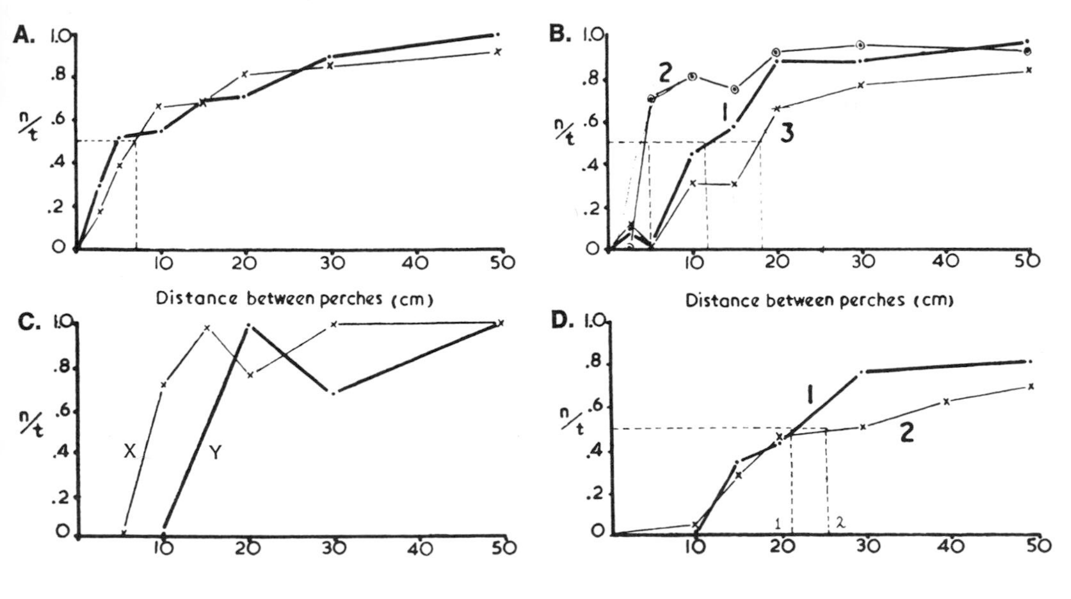

TABLE 5–2 Distances at which 50 percent of encounters were aggressive in caged groups of chaffinches. (From Marler 1956.) The multiple 50-percent-distance values represent different replicates.

Types of individuals involved	50% distances (cm)
♀ vs. ♀ in groups of 8	7, 7, 10, 12
♂ vs. ♂ in groups of 8	21, 25, 21, 18
4 ♂♂ vs. 4 ♀♀	12, 6, 9, 8
painted ♀ vs. painted ♀ in groups of 8	25, 25
4 painted ♀ vs. 4 unpainted ♀	3
♂ vs. painted ♀	21, 18

parent that male chaffinches maintain a much greater individual distance than do females. Further experiments with females dyed to resemble males show that plumage color is a major determinant of this behavioral difference. Females dyed red to look like males were treated as males by other males and females. Close inspection of the data for these experiments, summarized in Table 5–3, raises some interesting questions that only further experiments can clarify.

A closer examination of birds in confined groups reveals orderliness not only in the spacing of individuals but also in their orientation to each other (McBride et al. 1963; McBride 1964). Using photographs taken from directly above a flock of hens (*Gallus domesticus*) it was possible to measure the angle between a hen's head and the straight line to her nearest neighbor. The smaller the angle, the closer the hen was to staring directly at her neighbor. Comparing the distributions of such head angles showed that hens tend to look more directly at the nearest than second-nearest neighbor. The hens tended to face their neighbor more as the distance between them decreased, and less as the neighbor looked at them. Face-to-face situations tended not to occur except in very close proximity.

SOCIAL DOMINANCE IN GROUPS

For simplicity we may divide the effects of agonistic behavior on social organization into two main groups. The first, which was covered in Chapter 4, consists of relationships *between social units,* in which spacing behaviors in general and territoriality in particular are important. The second group, which we will consider in this chapter, consists of relationships *within social units.* In the first case agonistic behavior manifests itself in the social organization as a set, or mosaic, of areas that tend to be occupied exclusively by the social units that reside in them. Space is not shared by the competing individuals; it is *divided.* In the second case agonistic behavior manifests itself in the social organization as individual distance and in a set of *dominance relationships* among the members of a social unit, or among neighboring social units with overlapping home ranges. In this context space is *shared,* not divided, among the competing members of the group. Of course, intermediate situations occur, and social systems based on exclusive home ranges grade into those based on shared home ranges. All-purpose territories and dominance hierarchies are aspects of different social systems that intergrade.

History / The pioneering studies of dominance relationships in groups of animals were published in 1922 in German (earlier in Norwegian) by Schjelderup-Ebbe, who revised and discussed them in English in 1935. He pointed out that in flocks of hens a surprisingly highly structured social organization existed. The individual hens in his flocks were found to be members of a hierarchy, in which each individual recognized every other member and behaved toward him in either a dominant or subordinate manner, depending on the identity of the individuals concerned. Table 5–3 shows an example of the dominance hierarchy of some captive chaffinches. Since dominants tend to peck subordinates that do not get out of the way fast enough, the dominance hierarchy has been referred to as the *peck order.*

After the initial stimulus provided by Schjelderup-Ebbe, the study of dominance hierarchies attracted considerable attention, particularly in the Chicago school of animal behavior studies led by W. C. Allee. Numerous studies were done on flocks of hens (e.g., Guhl and Allee 1944; Guhl 1953, 1956; Collias 1943); the work on hens and other species has been summarized by Collias (1944) and Allee (1952). This school was interested also in the cooperative aspects of group organization, and these received emphasis in a popularized treatment by Allee (1938).

Studies of dominance relationships in captive groups have revealed important effects of hormones, population density, previous experience, hunger state, and appearance. The behavioral differences between dominant and subordinate animals have been fully described for many species. More recently the correlations between dominance status, activities of the adrenal cortex, and brain chemistry have received attention (see Chapter 6).

TABLE 5–3 A linear hierarchy in a captive group of 4 male and 4 female chaffinches. Observations were made in 10 hours distributed over 7 days in October. Since rank order was the same at all points tested in the cage it may be said to be site-independent. The individuals are designated by the initials of their color bands. (From Marler 1955.)

		LOSER									
		RWB	O	PW	RB	B	BW	R	Y	Total wins	Total losses
WINNER	♂ RWB	—	16	11	5	4	4	4	1	45	0
	♂ O	—	—	7	6	16	4	4	6	43	16
	♂ PW	—	—	—	4	7	10	10	3	34	18
	♂ RB	—	—	—	—	12	13	8	8	41	15
	♀ B	—	—	—	—	—	3	8	5	16	39
	♀ BW	—	—	—	—	—	—	5	2	7	34
	♀ R	—	—	—	—	—	—	—	11	11	39
	♀ Y	—	—	—	—	—	—	—	—	0	36
										197	197

The observation of dominance relationships within captive groups of animals is a useful technique for studies of the factors that affect aggressiveness; but because social organization is so changeable, such studies tell us very little about the structure of dominance relationships in natural populations. Progress in the latter area with regard to birds dates from the introduction by Burkitt (1924–26) of colored leg bands to enable field identification of individuals. Using color-banding with species that form groups proved that dominance hierarchies occur in natural populations as well as in confined groups. Field studies of dominance relationships among birds and other animals in natural populations have been reviewed by Collias (1944), Brown (1963*b*), and Thompson (1960).

In this section we shall be concerned primarily with groups of birds. Similar phenomena occur in fishes, lizards, wasps, lobsters, and other taxa (see Collias 1944). Primates are notable for their dominance hierarchies and will receive further attention in Chapter 12.

Methods of Observation and Dominance Criteria / Two methods of observation are commonly used in dominance studies: (1) staged paired encounters between pairs, and (2) group observation. In the pair-encounter method two individuals are placed together in a restricted space; one or the other can usually be judged dominant. By testing all possible pairs drawn from the group, a matrix of dominance relationships within the group can be constructed. This method involves removing individuals from the group and may at times yield results that differ from the actual relationships in the group context. It has the advantage of standardizing the context of the encounter and of ensuring that all possible pairs can be tested. However, it does not make possible estimates of the natural frequency of encounters between particular individuals, and it cannot be used in the field.

The second method, group observation, involves watching the normal activities of a group and recording all indications of dominance that arise normally. The frequency of encounters can be raised by providing bait, by restricting access to food or water to a single access point, or, with captives, by food deprivation. This method has the disadvantage that the observer may have to wait a long time to observe encounters between the less aggressive individuals, but this is balanced by the advantage of obtaining information on the frequency of encounters between different individuals.

In any field or laboratory study of dominance relationships within groups, the first necessity is a reliable criterion of dominance. The actual behaviors used by observers to judge dominance are as diverse as the species observed. With heifers (*Bos taurus*), butting with the head may be used; with chickens, pecking; and with primates in laboratory studies, the taking of a banana or peanut. In general, dominance implies *priority* acquired by past or present aggressive behavior, rather than by chance or some other factor. Consequently, whatever the behavior used as an indicator of dominance, it must (1) indicate priority unambiguously, and (2) indicate that the observed priority is caused, at least indirectly, by the dominance behavior of the participants rather than other factors.

A useful criterion in many field studies is the act of *supplanting*. In a typical act of supplanting the dominant approaches, with or without threat behavior, and the subordinate leaves. There is often no fighting and no obvious threat of violence. To a causal observer it may not appear an act of aggression at all. After a dominance relationship between two individuals has been established, the mere act of unhesitating, direct approach often is sufficiently threatening to cause a subordinate to give way peacefully. Criteria for ascertaining dominance in field studies have been discussed further by Brown (1963*b*).

When the observations are ready to be entered into a matrix of the form in Table 5–3, some simple rules are traditionally followed. Some of these are illustrated in Table 5–4. Various other supplemental methods of analyzing dominance have also been used with varying success, but none takes the place of that described in Table 5–4. Ranking by win/loss ratios, in particular, is not equivalent to ranking by this method.

Dominance Matrices in Natural Populations of Birds / The principal types of aggregations of birds in which dominance phenomena have been studied are summarized in Table 5–5. Since many species form aggregations only in winter and since these are usually uncomplicated by territorial behavior, the majority of studies has been done in winter. Nevertheless, in some permanent resident species such as Steller's jay (*Cyanocitta stelleri*) and probably the jackdaw (*Corvus monedula*), the dominance relationships have been shown not to vary much between breeding and nonbreeding seasons.

Dominance hierarchies have been found in flocks of sizes ranging from 2 or 3 chickadees to 500 New Zealand white-eyes (*Zosterops lateralis*). Large flocks are difficult to observe and most flocks studied range from about 5 to 30 individuals. Where it has been possible to record the composition of a flock over a period of months at a feeding station, a relatively high degree of stability or constancy of flock membership has been found in nearly all species of Section I of Table 5–5. Removal of a significant fraction from a flock of white-crowned sparrows (*Zonotrichia leucophrys*) resulted in infiltration of new members from outside (Mewaldt 1964).

Rank order in aggregations of birds at feeding stations can some-

TABLE 5–4 Steps in the construction of a dominance matrix.

(1) *Observations:* B > D, C > A, B > A, C > B, B > D, etc.*
(2) *Starting Order:* Choose an arbitrary order, *e.g.,* DEACB.
(3) *Starting Matrix:* Enter the number of wins and losses observed in the matrix:

WINNER \ LOSER	D	E	A	C	B
D		24	3	0	0
E	0		13	0	0
A	21	11		0	0
C	12	16	17		14
B	37	31	41	0	

(4) *Treatment of Reversals:* A win by one individual over another that has won the majority of encounters with the first is termed a *reversal.* Rearrange the order so that only reversals fall below the diagonal, so far as possible; that is, change the above order to CBDAE or CBEDA or CBAED.
(5) *Treatment of Nonlinearity:* An order in which an individual dominates another (wins the majority of encounters) that dominates the first is termed *nonlinear* or circular. Rearrange to minimize the inevitable ambiguity. From the circular relationship diagrammed below there are three main alternatives, as shown. In the three alternatives not shown, the departure from linearity involves two individuals rather than one.

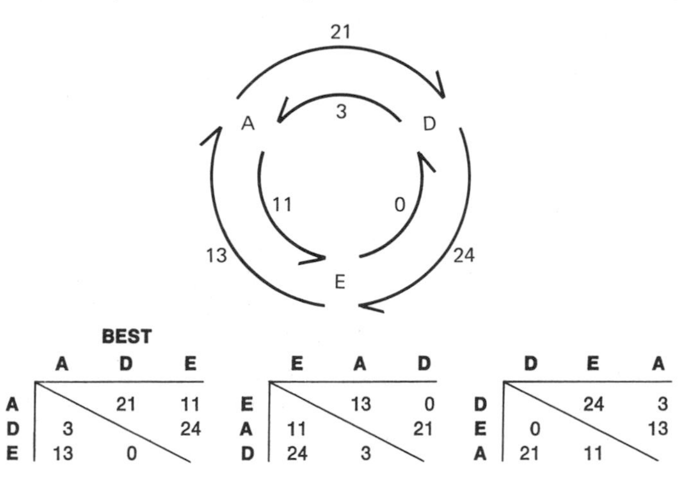

BEST

	A	D	E
A		21	11
D	3		24
E	13	0	

	E	A	D
E		13	0
A	11		21
D	24	3	

	D	E	A
D		24	3
E	0		13
A	21	11	

Place the individuals that are in the least ambiguous relationships (lowest proportion of reversals) in linear order. This procedure tends to minimize the total of encounters entered below the diagonal.
(6) *Final Matrix:* The one order that best reflects the order of dominance within the group is then CBADE. The following matrix may then be constructed:

WINNER \ LOSER	C	B	A	D	E	Wins	Losses
C		14	17	12	16	59	0
B	0		41	37	31	109	14
A	0	0		21	11	32	74
D	0	0	3		24	27	70
E	0	0	13	0		13	82

* B>D means B won an encounter with D. In most cases, these encounters take the form of supplanting rather than fighting.

times be predicted on the basis of the distance of each individual from its nesting area. Odum and Colquhoun noted that, for the black-capped chickadee (*Parus atricapillus*) and blue titmouse (*Parus caeruleus*), the dominant bird in the flock was the one nesting closest to the feeding station. Brian extended this investigation to the great titmouse (*Parus major*), showing a rough correlation between rank and distance to nest area among seven males, and finding that another male became dominant when she moved the feeding station.

The relationship between rank and distance to nest area could be studied more conveniently and intensively in Steller's jay. The population inhabited a picnic area, and the jays were accustomed to gleaning scraps from picnickers off the picnic tables. The tables were arranged in two rows and were virtually immovable. Observations of the hierarchy at each table showed that the rank of any male at any given location was correlated with the distance to his nest area at all seasons

TABLE 5–5 Species differences in the characteristics of dominance matrices in natural, nonbreeding aggregations of birds.

Characteristics of dominance matrix	Examples
I. Flock an integrated unit, with consistent composition and social structure; rank order of flock members is constant regardless of location inside flock home range; *site-independent*	
A. Composed mainly of permanent residents	3 species of chickadees (Odum 1942; Dixon 1963, 1965; Hartzler 1970; Minnock 1971) jackdaw (Lorenz 1931, 1938)
B. Composed of permanent and winter residents	juncos in Seattle, Washington (Sabine 1959)
C. Composed mainly of winter residents	tree sparrow 2 subspecies of juncos (Sabine 1949, 1959)
II. Flock of varying composition and social structure	
A. Composed mainly of permanent residents; rank order is constant at any one site but varies from place to place; *site-dependent*	2 species of titmice (Brian 1949; Colquhoun 1942) Steller's jay (Brown 1963*b*) vultures (Valverde 1959) 3 species of antbirds (Willis 1967, 1968)
B. Composed of permanent and winter residents	white-eyes (Kikkawa 1961)
C. Composed mainly of winter residents in large flocks with probably little individual recognition	white-fronted geese (Boyd 1953) sandhill cranes (Miller and Stephen 1966) Canada geese (Raveling 1970)

(Figure 5–7). This was true not just for one feeding station but for all, regardless of location. Furthermore, it was shown that as each male ranged farther from his nesting area his rank became lower. Consequently, the dominance relationships of an individual male Steller's jay may be conceived of as a series of concentric zones of diminishing rank from the center of the nesting area outward. The dominance matrix at any one location is likely to differ from those at all other locations because of the complexities in the pattern of overlap of the zones of different individuals. Somewhat similar relations were found in certain antbirds (Willis 1967, 1968), woodchucks (*Marmota monax;* Bronson 1964), chipmunks (*Tamias striata;* Dunford 1970), and *Octopus cyanea* (Yarnall 1969).

In contrast to the highly *location-dependent* rank orders in Steller's jays and great titmice, it is implicit in much of the literature on dominance hierarchies that the rank order is constant at all locations within the normal range of the flock (that is, rank order is *location-independent*). In these cases it has been shown that the flock inhabits a restricted area with little if any overlap into the areas of neighboring flocks. (How this exclusiveness comes about is unknown.) However, only in mountain chickadees (*Parus gambeli*) and black-capped chickadees (Dixon

5–7 Site-dependent dominance hierarchies in male Steller's jays. **A.** A Steller's jay. **B.** Dominance at food sources was observed at each of eight picnic tables, numbered $T_{1\text{-}16}$. Boundaries between areas of dominance or dominions are indicated by dashed lines. Nests are indicated by *N*. The observations for one winter are pooled. (From Brown 1963*b*. Photo reprinted by permission from Verna R. Johnston and *American Birds* 27:578.)

A.

B. →

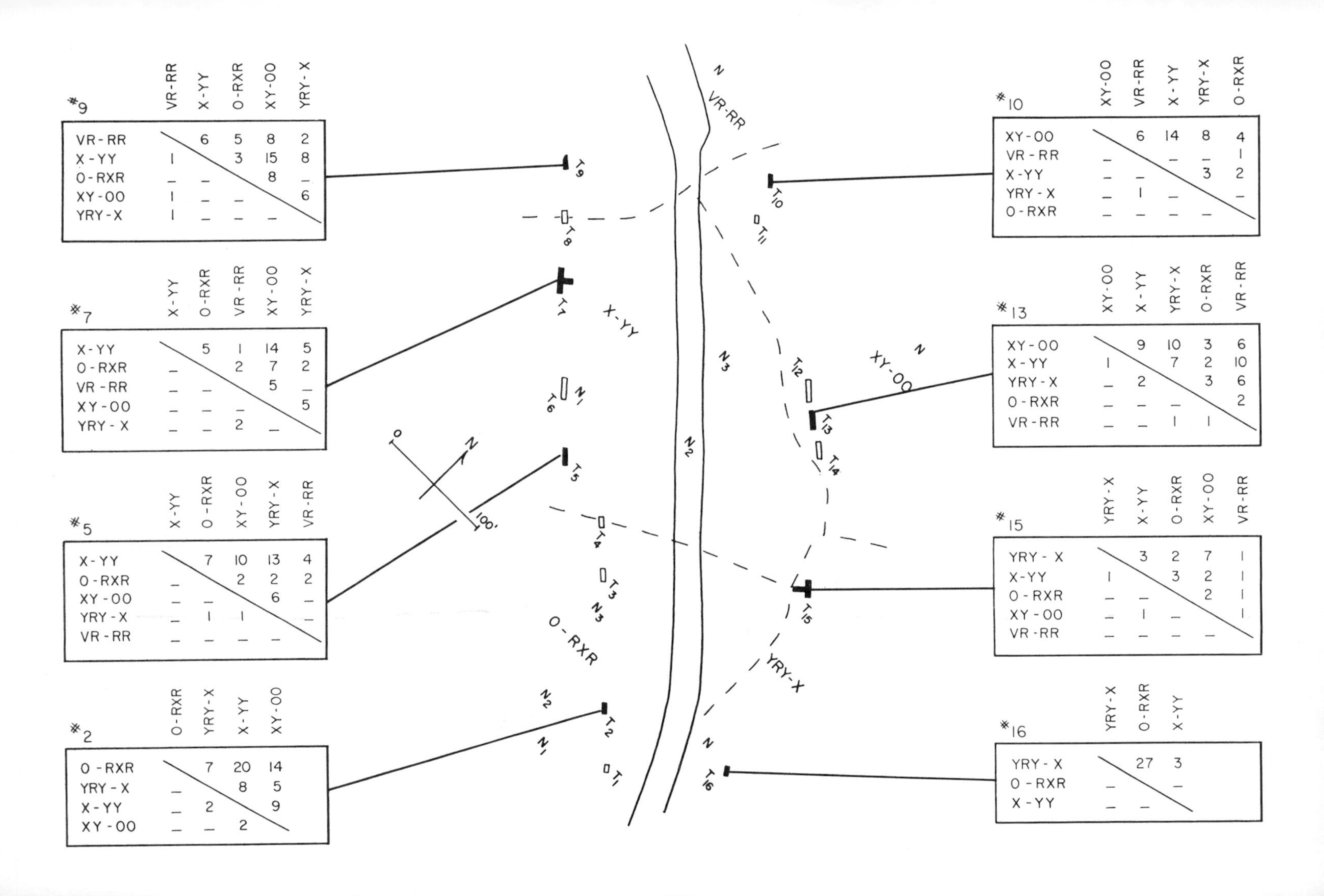

#9
VR-RR X-YY O-RXR XY-OO YRY-X
VR-RR 6 5 8 2
X-YY 1 3 15 8
O-RXR – – 8 –
XY-OO 1 – – 6
YRY-X 1 – – –
#7
X-YY O-RXR VR-RR XY-OO YRY-X
X-YY 5 1 14 5
O-RXR – 2 7 2
VR-RR – – 5 –
XY-OO – – – 5
YRY-X – – 2 –
#5
X-YY O-RXR XY-OO YRY-X VR-RR
X-YY 7 10 13 4
O-RXR – 2 2 2
XY-OO – – 6 –
YRY-X – 1 1 –
VR-RR – – – –
#2
O-RXR YRY-X X-YY XY-OO
O-RXR 7 20 14
YRY-X – 8 5
X-YY – 2 9
XY-OO – – 2
#10
XY-OO VR-RR X-YY YRY-X O-RXR
XY-OO 6 14 8 4
VR-RR – – – –
X-YY – – 3 2
YRY-X – 1 – –
O-RXR – – – –
#13
XY-OO X-YY YRY-X O-RXR VR-RR
XY-OO 9 10 3 6
X-YY 1 7 2 10
YRY-X – 2 3 6
O-RXR – – – 2
VR-RR – – 1 1
#15
YRY-X X-YY O-RXR XY-OO VR-RR
YRY-X 3 2 7 1
X-YY 1 3 2 1
O-RXR – – 2 1
XY-OO – 1 – 1
VR-RR – – – –
#16
YRY-X O-RXR X-YY
YRY-X 27 3
O-RXR – –
X-YY – –
N
VR-RR
T_9
T_8
T_7
X-YY
N_1
T_6
T_5
0
N
100'
T_4
T_3
N_3
O-RXR
N_2
T_2
N_1
T_1
N_3
N_2
T_{10}
T_{11}
T_{12}
XY-OO
N
T_{13}
T_{14}
T_{15}
YRY-X
N
T_{16}

1965; Hartzler 1970; Minock 1971) was it actually observed that rank order remained unchanged at different locations within the home range of the flock; and even for these species quantitative data are lacking except in Minock's study.

In most species there is a clear correlation between sex and rank in the hierarchy. In flocks of both sexes of Steller's jays, two species of juncos, various species of titmice and chickadees, and other species, females tend to be at the bottom of the hierarchy and males at the top. During the breeding season, however, the situation may change drastically, and females of some species become dominant over their mates, or at least are no longer subordinate. Thompson (1960) has reviewed the role of sex in dominance in various species of birds.

If the rank order of individuals has no irregularities, such as a low-ranking individual dominating a higher-ranking one, then it is said to be *linear* (e.g., Table 5–3). When the dominance relationships among many individuals are examined, most can be arranged in a linear order. It is not uncommon, however, to find a few exceptions. In the earlier literature such relationships were called *triangular,* but since more than three individuals may be involved, it seems preferable to refer to them simply as *nonlinear.*

In large flocks, the composition of which is continually changing, the probability of encountering a stranger is high and the importance of individual recognition is correspondingly reduced. Stable dominance hierarchies cannot be established under such circumstances. This situation occurs in large flocks of migrating and wintering New World blackbirds, in which flock sizes into the millions are known at winter roosts. Similar conditions obtain among nonbreeding waterfowl and large herds of mammals. Where individual recognition is lacking the outcome of aggressive encounters depends directly on threat and attack behavior. For example, in wintering flocks of white-fronted geese (*Anser albifrons;* Boyd 1953) and Canada geese (*Branta canadensis;* Raveling 1970) the outcome of most encounters can be predicted from the size of the groups participating and the age of the individuals involved. The young geese migrate and winter with their parents. Large families tend to dominate small families. The following precedence order was also observed: parents with young > paired adults > yearlings in families > single adults > unattached yearlings. Success depended most on the gander but frequently all members of a family acted as a unit. This set of "rules" rewards the genotypes that are most successful in rearing young and in keeping the family together. An exception is that a pair on the breeding grounds may capture the brood of another pair, resulting in an abnormally large family (Sherwood 1967). This superfamily then migrates as a unit. It is known that white-fronted geese separated from their families while on migration can rejoin them in some cases (Miller and Dzubin 1965).

Theoretical Considerations / Group size is a significant determinant of the complexity and type of dominance relationships that can be expected. A small increase in group size increases the complexity of social relationships disproportionately. The number of possible combinations c of two individuals in a group of size n is given by the equation

$$c = \frac{n\,(n-1)}{2}.$$

It can be seen that c increases steeply as n increases. For each individual added to the group the number of new possible pairs is increased by the number of individuals in the original group. For example, adding one individual to a group of 15 increases the number of pairs from 105 to 120. The steep rise with increasing group size in the number of possible pairs suggests that individual recognition must play a lesser role in large groups than small ones, and conversely, that overt aggressive behavior is more important than individual recognition in large groups. Similarly, larger groups are likely to show more *variability* because the stabilizing influence provided by individual recognition is lessened.

The question of linearity can also be approached mathematically. Landau (1968) developed an index of linearity. He argued mathematically that the high degree of linearity characteristic of both captive and natural hierarchies would not be likely to arise simply from differences in inherent characteristics of the individuals, such as size, strength, appearance, or displays. It should require also social relationships based on individual recognition among the individuals in the group and between a newcomer and any established flock member. Further mathematical considerations of dominance hierarchies can be found in earlier papers by Landau (1951*a*, *b*, 1965, 1968) and in a book by Rashevsky (1959).

Peck-Right and Peck-Dominance / If we were to take the data for all picnic tables shown in Figure 5–7 and lump them together, we could then construct one grand hierarchy for this jay population. If this were done we would find the X · YY dominated O · RXR 22 times but behaved submissively to him 20 times. Can one argue from such data that X · YY was dominant to O · RXR? Obviously not. Each bird had its own area of dominance with a sharp border between them. Dominance was clearcut on each side. To lump the data from both sides together and declare one bird the overall winner would not represent the real world. The result would depend mainly on the number of observations in each bird's area of dominance and would not reflect their overall relative dominance at all. The lumping together of such dominance data may seem absurd, yet it was the origin of the distinction between the terms *peck-dominance* and *peck-right*. When Masure and Allee (1934) studied dominance in confined pigeons, they failed to see that each male had its own area of dominance, and they simply lumped together all their data regardless of the area of dominance in which the observations had been made (Ritchey 1951). This procedure inevitably gave them a very confused picture, one different from that found in hens. They used the term *peck-dominance* for hierarchies of the lumped-together type, and designated the clearcut hierarchies observed in hens *peck-right*. This distinction has been perpetuated by uncritical authors. It deserves instead to be buried.

In contrast, the distinction between *site-dependent* and *site-independent* dominance is based on a useful biological distinction. It should be emphasized that *site-dependent* is not synonomous with *peck-dominance*. The criterion for peck-dominance is a muddled dominance order with many reversals; using a single hierarchy, one judges peck-dominance subjectively seeing whether there were many or few entries below the diagonal of the dominance matrix. In contrast, site-dependence cannot be established by examining a single hierarchy; to establish either site-dependence or site-independence it is necessary to compare hierarchies including the same individuals observed at two or more different localities. The two concepts are operationally different.

Natural Selection for Status-Indicating Behavior / In typical hierarchical social organizations the behavior of dominant and subordinate individuals contrasts strongly. Dominants can move around unhesitatingly within the group without care as to which individuals they might encounter. If any resistance is met, it is easily overcome by a slight gesture or display indicating the possibility of an attack. And if it comes to a fight, the dominant has the psychological advantage and nearly always wins. Subordinates behave in roughly the opposite way. They typically avoid being in the path of a dominant, and if they must come close to him they seem to be especially sensitive to slight variations in his appearance that might indicate impending aggressive behavior. In species that normally live in groups, subordinate individuals typically show what is termed *appeasement behavior* or *submissive behavior* toward dominants in situations when they are in danger of attack by the dominant. These displays seem effective in preventing the killing of one individual by another and sometimes in allowing a subordinate to approach closely a food source or other necessary resource controlled by the dominant.

The net result of these behavioral patterns is that as long as individuals continue to play the role in which they are cast and do not challenge the social order, the social organization of the group functions efficiently with little if any outward sign of aggression, and a very low frequency of actual fighting. The constructive activities of group members can proceed in such a context.

Since the status of dominance confers some advantages, we might ask why subordinates have evolved such elaborate behavior that preserves their subordinate status; and, similarly, why dominants have evolved such tolerance as to allow their rivals to remain in the same group. A part of the answer must lie in the benefits of group living. Individuals that are sufficiently tolerant of others, whether they be dominant or subordinate, will live in groups as a result and share the advantages of group living—whatever they may be. Given the desirability of group living (see Chapter 7), it is obvious that the dominant receives considerable advantage from his status. Behavior indicating a dominant status will, consequently, be selected for, but only if it is expressed by an individual that actually is dominant, or is very likely to become so. Such behavioral patterns as unhesitating, undeviating locomotion to a goal, absence of defensive or submissive postures, and general freedom of action allow a dominant animal to be easily recognized by an observer and very likely by another animal in the group thus making life even easier for him.

Submissive postures also confer advantages when expressed in the appropriate contexts. In a group-living species a subordinate probably has a better chance of surviving and ultimately reproducing by staying with a group than by setting out by himself. Since all individuals must pass through a period of subordinate status during their development and often even as adults, behavior patterns that allow them to persist in the presence of dominants without injury, without psychological damage, and without severe loss of nourishment or shelter, will be selected for, so long as they are expressed in a context in which the individual is subordinate and is unlikely to become dominant.

Since in many species most individuals are dominant to some individuals and subordinate to others, there is considerable selective value in being able to judge the context appropriately by (1) recognizing individuals and responding appropriately, (2) estimating correctly the likelihood of success or failure in encounters with strangers, and (3) choosing the most propitious time for challenging and overthrowing an aging dominant. This ability to take into consideration all the factors that might affect the outcome of an encounter, to predict the outcome, and to behave accordingly requires more intelligence than many other social processes, especially in a large group. Chance and Mead (1953) have proposed that these processes have contributed toward the evolution of intelligence in primate societies.

Survival Value of Dominance / Although many benefits of dominance in winter flocks of birds seem intuitively obvious, there is relatively little detailed evidence on its survival value. For juncos (*Junco hyemalis*) wintering in North Carolina, Fretwell (1969) showed that the survival rate of birds living in flocks was higher than those living as solitary individuals, who seemed to have been excluded somehow. High-ranking juncos and field sparrows (*Spizella pusilla*) survived better than low-ranking ones (Fretwell 1968, 1969). Murton (1967, 1968; Murton et al. 1971) observed that wood pigeons (*Columba palumbus*) low in rank could not maintain as high a feeding rate in foraging flocks as dominants and that solitary pigeons had an even lower rate due to their uneasy behavior (Figure 5–8). Solitary and subordinate pigeons had to spend a greater percentage of their foraging time looking about for dominants or predators and so had less time for feeding, ultimately losing weight and starving to death. Circumstantial evidence suggests that some types of mortality are higher in gray squirrels

5–8 Feeding rates of dominant and subordinate wood pigeons in relation to food density. The curves show the numbers of cereal grains eaten (pecks per minute) by birds in various densities of food as counted in field observations. The density of food below which an adequate intake rate cannot be maintained is given by the vertical line below *A* for a subordinate and below *C* for a dominant. (From Murton 1968.)

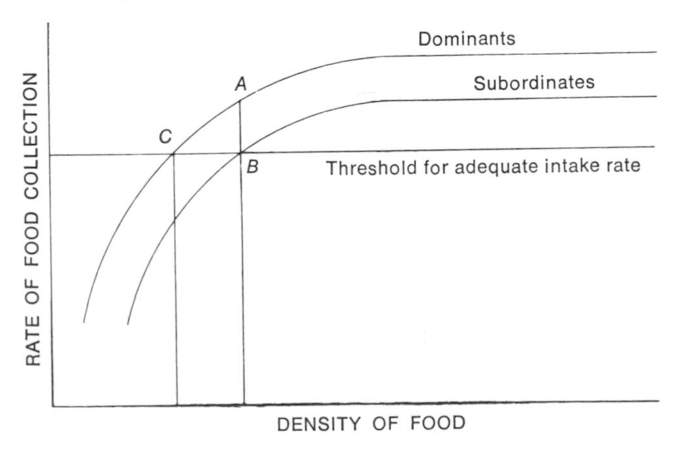

(*Sciurus carolinensis*) among low-ranking individuals (Pack et al. 1967).

Social Roles / In many groups of animals the individuals comprising the group are virtually indistinguishable from each other in appearance or behavior. This is especially true in nonbreeding aggregations. In other groups such factors as sex, age, and family relationship are correlated with detectable differences in social roles. These roles are most strikingly differentiated in the bees, ants, wasps, and termites, where a queen, drones, and one or more classes of workers can be recognized behaviorally and morphologically (see Chapter 10). Baboons and monkeys also show role differentiation with respect to sex, age, and other factors (see Chapter 12). The importance of these different social roles in the social organization of primates and social insects will emerge in the chapters devoted to those groups. They are mentioned here to emphasize that dominance is not the only important aspect of social organization in groups. For many species of primates these roles have not yet been adequately described, and their investigation constitutes an active field of research.

CONCLUSIONS

Many types of animal groups occur in nature but most of them can be associated with a comfortably small number of evolutionary problems. These deal with sociobiological principles that range widely across various phylogenetic groups or taxa. Within a group social forces of attraction and repulsion bring about a spacing pattern that is characteristic of the species. In distance species the spacing between individuals has been studied by the nearest-neighbor method in cranes, crickets, and flies, and in chaffinches by experimental manipulation of inter-individual distances.

The agonistic behavior that mediates individual spacing is typically associated with dominance phenomena. These have been studied in a variety of birds, where they were first discovered, and more recently in wild populations of primates. In groups of relatively stable composition that travel as a unit the dominance hierarchy tends to be linear and constant regardless of the location of the group. Such hierarchies are *site-independent*. In groups whose composition shifts frequently and in which individuals have separate areas of dominance (the core of their home range), the rank order of dominance tends to differ greatly from place to place. Such hierarchies are *site-dependent*. In large groups of changing composition, as with wild geese, threat behavior seems to be more important than individual recognition, and clear-cut linear dominance hierarchies tend not to form. In the groups of many species individuals are indistinguishable behaviorally and morphologically from each other and play similar roles; but in the social insects and some primates individuals differing in sex, age, and other factors may play highly specialized social roles in the group.

6 Behavior and Populations

THE STUDY OF fluctuations from year to year in the density of a population has traditionally been the province of ecology. Ecologists at first thought that the causes of such fluctuations could be traced almost exclusively to such factors as the amount of food available, inclement weather, predation, and disease. Although a few authors (e.g., Moffat 1903; Howard 1964; Huxley 1934; Kalela 1954) had noticed that in some species behavior participates in determining the number of individuals living in a particular area, relatively little attention was paid to behavior as a factor in population ecology until the 1960s (e.g., Wynne-Edwards 1962).

In most animals behavior is a critical factor in bringing the sexes together for reproduction, and in many species it is also critical in parental care. It is clearly a factor of importance in population growth. However, the interest of ecologists in behavior has centered on its role in curtailing increases and initiating decreases. Behavior may dampen population fluctuations in some species. The "limitation," "control," or "regulation" of population numbers by behavior is often discussed as if behavior were the only important factor. Unfortunately, the champions of the ecological importance of behavior have frequently oversimplified their arguments and neglected other important aspects of population dynamics, particularly food supply. Since no population can grow beyond its food supply, food is ultimately a limiting factor for all populations, and consequently one that can never be disregarded. Although this chapter will necessarily emphasize behavior, the reader should attempt to view behavior as only one of several important factors that influence population numbers.

The dampening effects of behavior on population fluctuations are mediated primarily by agonistic behavior. Two classes of effects can be recognized: *exclusion effects* and *subordination effects*. Defense of an entire home range by the male in a monogamous species excludes other males in two ways. It excludes one individual from using the home range of another, and so tends to bring about *regular dispersion*. And it can exclude some individuals from an entire area or habitat if all available space has been taken. These exclusion effects will be the subject of the first section of this chapter. If individuals are not excluded, they may remain in a subordinate status. Subordination effects may include increased mortality rate, lowered reproductive rate, and various behavioral and physiological correlates of subordinate status. The effects of subordination on populations will be considered in the second section of this chapter.

EXCLUSION EFFECTS

The most common cause of exclusion is territorial behavior; but subordinates may leave an area of dominance of another individual to find an area of their own, and some individuals may simply avoid others without territorial defense being involved. Exclusion affects populations in various ways. We may ask, "What are the population effects of exclusion on density and dispersion? On reproduction? On survival?"

Effects on Breeding Density / In a monogamous population that breeds once a year the number of mature male adults, N_T, at the start of a breeding season will depend on reproductive success and survival from previous breeding seasons, and so will vary from year to year. Given that the entire home range is defended by the adult male, two extreme hypothetical cases can be recognized. If territory size, T, is a simple function of N_T, then in a fixed area, A, if all the available space is divided equally among the number of adult males in that year,

$$T = \frac{A}{N_T} \cdot$$

This relationship roughly describes the mean territory size in a given area if the number of territory holders is known, as shown in Figure 6–1. If all males receive their share of the area and none is excluded from breeding regardless of how high the density becomes, the number of breeders, N_B, will equal the total number of males, regardless of population density, as in Figure 6–2*A*: $N_B = N_T$.

At the other extreme (Figure 6–2*B*), there may be a rigidly fixed minimum territory size, T_{min}. The maximum number of territory-owning males that the area can accommodate, N_{max}, would then be as follows:

$$N_{max} = \frac{A}{T_{min}}$$

If the number of adult males exceeded N_{max}, the remainder would be excluded from holding territories and from breeding. These nonbreeding adults, N_F, are referred to as the *surplus* or *floating* population. The total adult population (of males, in this example) can then be divided into a fraction that is breeding, N_B, and a fraction, N_F, that is excluded from breeding by the territorial behavior of the breeders:

$$N_T = N_B + N_F$$

The nonbreeding fraction, or floaters, either are widely dispersed wanderers moving from territory to territory, always looking for a vacancy

6–1 Territory size as a function of breeding density in male dickcissels (*Spiza americana*). The upper curve gives the theoretical curve discussed in the text. The lower curve is plotted from observations. The actual values of territory size depart from the hypothetical ones at lower densities because at low densities it is unnecessary and uneconomical in terms of the individual's energy budget to defend such a large area (that is, not all the study area is actually used and defended). At high densities the observations approach the expected values. (From Zimmerman 1971. Courtesy of American Ornithologists' Union.)

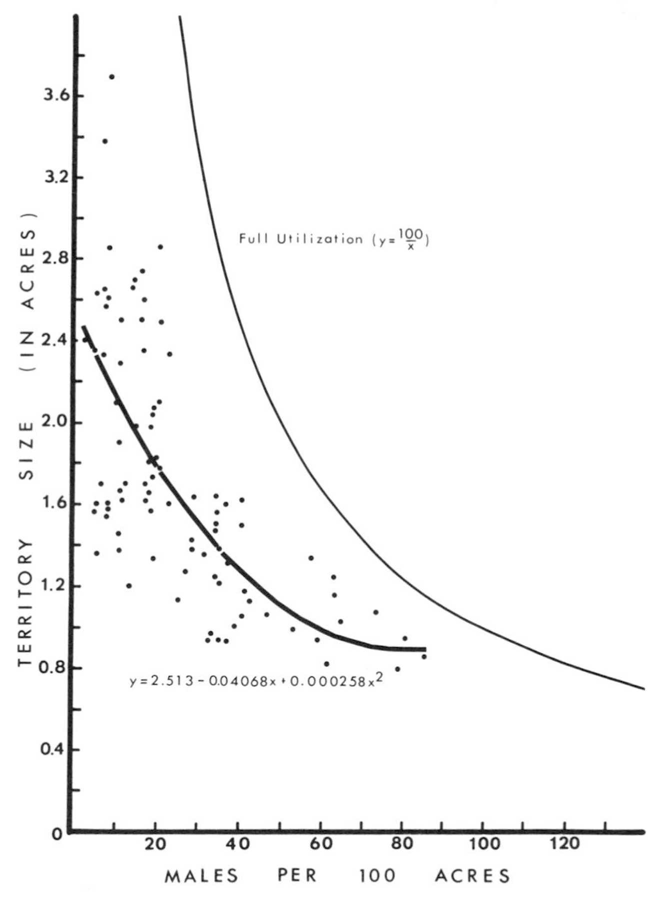

and usually being chased out, or live as a group in some neutral area too poor for breeding.

It is unlikely that either of these extreme hypothetical cases occurs in nature. More likely is that an intermediate situation exists, such as that described by Julian Huxley (1934:277) in the following words:

There appears to be a minimum size of territory, any encroachment on which is bitterly resisted. Above this size, resistance to encroachment is less whole-hearted, and compressibility therefore greater. . . . Territories are thus par-

6–2 Hypothetical patterns of population fluctuations as a function of behavior. Each point represents the density of animals of the indicated status in a particular breeding season or year. The lower graphs show population changes over time. The upper graphs show the number of territory owners and floaters in relation to the total number of competitors, N_T. N_B = number of individuals holding breeding territories. N_F = number of individuals (floaters) not holding territories because all available space has been monopolized by the territory holders. **A.** Fully compressible territories; density independence; no effect of territorial behavior on density of territory holders. **B.** Incompressible territories; upper limit on density of territory holders is set by a rigidly maintained minimum territory size; excess individuals are floaters. **C.** Partially compressible territories, as in Huxley's elastic-disc analogy.

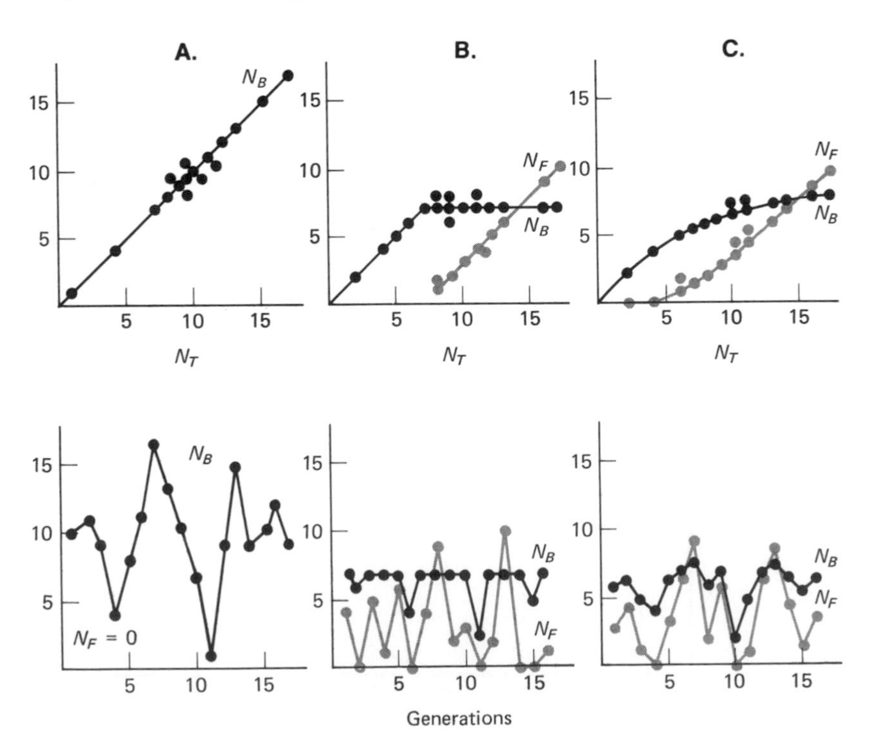

tially compressible, but their compressibility is not complete. They are like elastic discs, of which there is a lower as well as a higher number which can be placed together to cover a given area. If this view is correct, territorial instinct (i.e., male pugnacity while in possession of a territory) *will* be one of the more important of the factors determining the population of breeding pairs in a given area.

The elastic-disc analogy is represented in Figure 6–2*C*.

Many species inhabit not just one habitat, as the simple models of Figure 6–2 imply, but a range of habitats varying in their desirability for breeding individuals and in their productivity in terms of young produced per pair. The model in Figure 6–2 can be made more realistic by adding a second habitat of lower attractiveness to the species and lower productivity. These are designated *rich* and *poor*. The revised model is shown in Figure 6–3. Variability in T_{min}, shown in Figure 6–2*C*, is omitted in the revised model for simplicity.

According to the revised model, individual males failing to obtain a territory in the rich habitat (because it is saturated with territories of the minimal size for that habitat) can establish territories in the poor habitat. A class of floaters is created only at densities so high that the poor habitat has become saturated with territories of the minimum size for that habitat.

We shall use Figure 6–3 as a model from which to make some predictions that can be tested by examining natural populations or in the

6–3 Hypothetical effects of territoriality on breeding densities in two habitats, assuming fixed minimum territory sizes differing in the two habitats. N_F = number of floaters; N_{Br} and N_{Bp} = number of breeders in rich and poor habitats respectively; N_{Fr} and N_{Fp} = number of floaters in rich and poor habitats, respectively. In Level 1, neither the rich nor the poor habitat is saturated; in Level 2, the rich habitat is saturated; and in Level 3, both habitats are saturated. (From J. L. Brown, *American Naturalist,* University of Chicago Press, Copyright 1969 by the University of Chicago.)

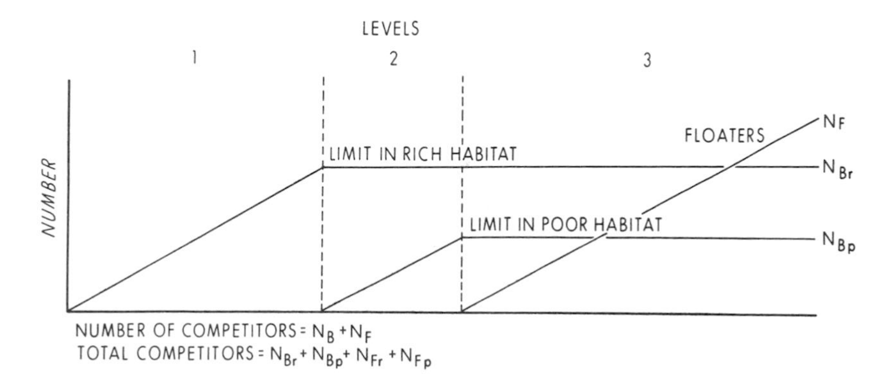

laboratory. The model incorporates the limiting effects of territorial behavior on populations in extreme form. To what extent, in what species, and under what circumstances does it accurately reflect the real world? It is clear that the predictions of the model vary according to the number of competitors and according to whether neither, one, or both habitats are saturated. These levels of saturation will be referred to as Levels 1, 2, and 3, respectively, as shown in the model.

At Level 1 an optimal habitat for a species is not fully utilized, so that suboptimal habitats which could be occupied are not. It requires that the population fluctuate below the level at which another habitat would be occupied. Such conditions are found in various birds and mammals.

At Level 2 the model predicts that at certain densities in a preferred habitat individuals will begin to move into another habitat. Glas (1960) obtained data on chaffinches (*Fringilla coelebs*) that agree with this prediction. He counted the number of territory holders in two habitats, a mixed wood and a pine wood, as the birds were establishing territories in the spring. Figure 6–4 shows that the first settlers chose only the mixed wood and that the density of breeders in the pine wood tended to lag behind that in the mixed wood. This is consistent with the hypothesis that territorial defense in the mixed wood led some chaffinches to nest in the pine wood after competition for space in the mixed wood built up to a critical level. (Unfortunately it is also consistent with the hypothesis that the food supply developed later in the spring in the pine wood than in the mixed wood—perhaps because of the timing of insect emergence—and that this factor determined the temporal pattern of attractiveness of the two habitats.)

A second prediction of the model in Figure 6–3 for Level 2 conditions is that when the poorer habitat is occupied incompletely, the yearly variation in density of breeding males should be greater in the poor habitat than the rich one. This proved true in long-term studies of the great tit (*Parus major*), the coal tit (*Parus ater;* Kluyver and Tinbergen 1953), and the chaffinch (Glas 1960). These results agree with the hypothesis that territorial behavior in the mixed wood combined with preference for the mixed wood stabilized the breeding densities there. (The results also agree with another hypothesis. During winter great tits from both woods forage mostly in the mixed wood. Since great tits are dominant in winter in their own nesting areas (Brian 1949), mixed-wood tits are less likely to die or emigrate because of food shortage than are the pine-wood tits. This should cause their numbers to vary less than those of pine-wood tits.)

At Level 3 the model predicts the existence of a surplus population that is prevented from breeding by the behavior of the territory owners.

These surplus individuals should be able to breed if a territory becomes available to them. The existence of a surplus population in many species is difficult to establish. However, in studies of the arctic ground squirrel (*Spermophilus undulatus*) by Carl (1971) the status of individuals could be determined by marking and visual observation. Males of this species defend territories covering one-quarter to two-thirds of their home range and including burrow systems in which one to four females nest. All females are allowed to nest; the surplus is composed entirely of males. Figure 6–5 shows the fate of the floaters and territory owners over two breeding seasons. The number of territorial males remained relatively constant in a given year because losses were replaced from the surplus population. In contrast, the number of floaters decreased drastically over the same period. This example shows the role of territorial behavior and the surplus population in stabilizing the number of breeding males.

The existence of a surplus population is less easily demonstrated

6–4 Number of chaffinch territory owners in two habitats as the territories were being established in the spring. The mixed wood was settled first and supported a higher density after settlement was complete for the season. Note that the difference in density of the populations in the two areas is greater than indicated on the ordinate of the graph because the two areas were of unequal size. (From Glas 1960.)

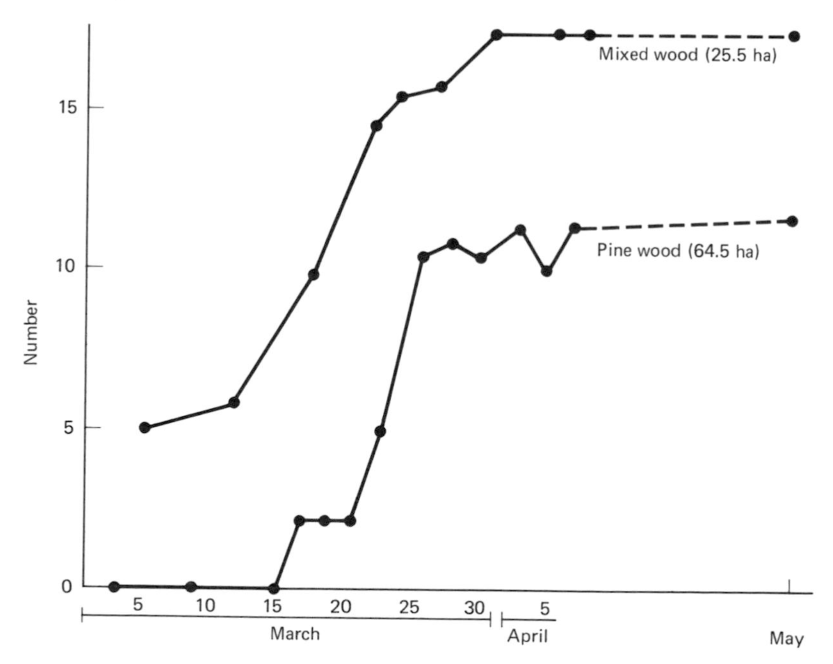

but is known in a variety of bird species (Brown 1969*a*; Watson and Moss 1970). In the red-winged blackbird (*Agelaius phoeniceus*) a surplus of males was demonstrated by removing territorial males from their territories and observing their replacement by other males (Orians 1961). The effect of male territorial behavior on the stability of the population density of breeding males in a marsh was demonstrated by Brenner (1966) and Davis and Peek (1972). In this species the males

6–5 Role of territorial behavior in maintaining stability of a population of arctic ground squirrels. The number of territorial males is maintained by additions from the floaters when a territorial male disappears. The number of territorial males remains about the same through a season, while the number of floaters diminishes drastically. Arrows to the left indicate recruitment, and arrows to the right show losses from the population. (Photo by E. Carl. Graphs from E. Carl, *Ecology.* Reprinted by permission of the Publisher. Copyright 1971, Ecological Society of America.)

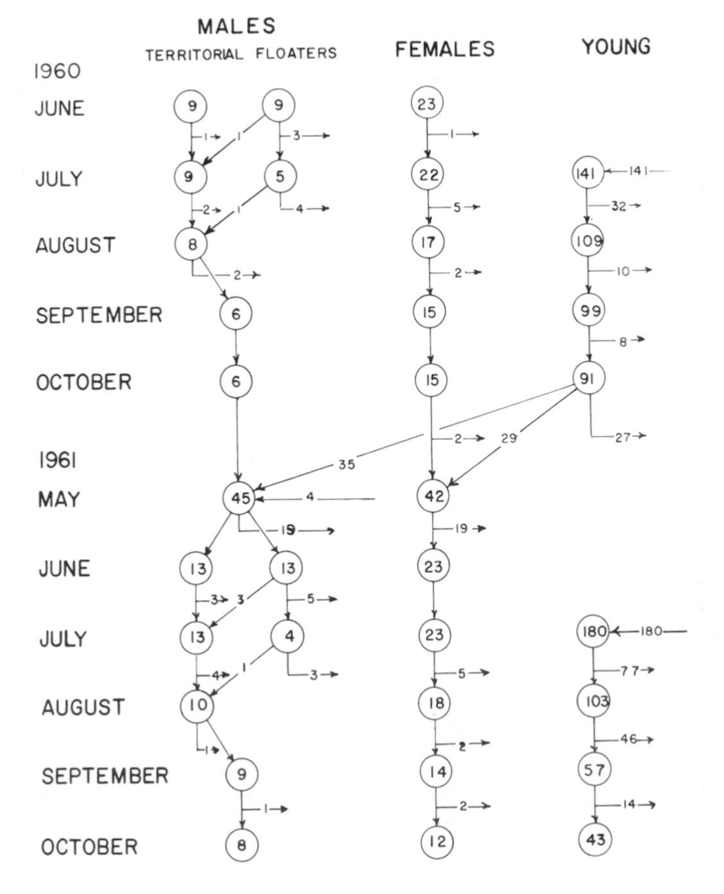

have usually one to three females each. Therefore, all females can breed while some males are prevented from doing so. In some years the number of females in a marsh may reflect the total number of males in the population. Assuming an equal number of females and males, which is not necessarily the case, the number of floaters can be estimated crudely by subtracting the number of breeding males from the number of breeding females. Figure 6–6 shows that a surplus probably existed in 1960 and 1961; but more importantly it illustrates the stabilizing effects of territorial behavior on the number of males holding breeding territories from year to year, despite considerable yearly variation in the number of females (and males) in the population.

This is a good example of a natural experiment with a built-in control. In most species the total breeding population fluctuates from year to year, but since the breeding density in a given year is influenced by many different factors, no one of them can be singled out as "determining" the number that breed. In this case, if it is assumed that the number of females reflects the total of males in the population, then the difference between male and female representation in the breeding

6–6 Number of red-winged blackbirds on a small marsh. The density of breeding males was more stable than that of females. Variations in the quality of the habitat due mainly to rainfall differences were a major influence on the number of females. (Data from Brenner 1966; Davis and Peek 1972.)

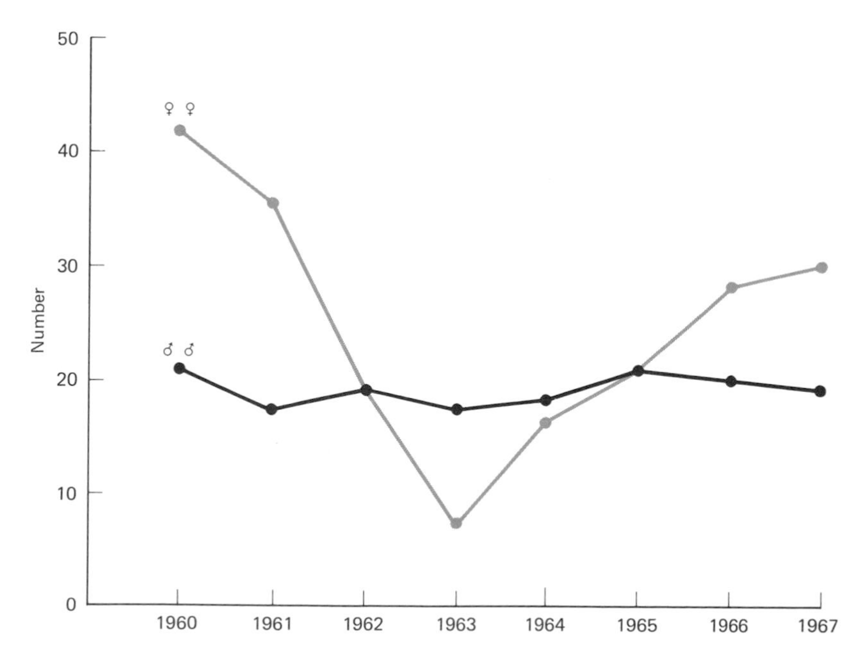

populations should be due almost entirely to the differences in territorial behavior between the sexes: female redwings are not as territorial as males. Unfortunately, the excess of males over females in 1963 and 1964 suggests that other variables are also involved (for example, habitat preference).

Experiments on the consequences of territorial behavior in captive populations have been rare, though Bustard (1970) has carried out one on the Australian lizard *Gehyra variegata*. He first observed that the population in nature was rather stable (1969) and that the males were strongly territorial all year (1968). Each male defended a tree stump together with two or three females. Using artificial tree stumps and supplying food in excess, he added females at regular intervals to a confined population. The number of females per stump remained stable despite the introduction of extra females in the presence of more than enough food. This suggested that territorial behavior, rather than food supply, limited the number of females per stump. The potential of experiments on confined populations to assess quantitative aspects of the impact of behavior on populations has hardly been tapped.

Effects of Territoriality on Reproduction / The evidence we have explored indicates that territorial behavior spaces out breeders and sometimes prevents mature adults from breeding. This is likely to be true in any territorial species, regardless of its phylogenetic position. We may now consider the consequences of these effects for reproduction. Let us consider first the studies that have demonstrated a surplus population. These are summarized in Table 6–1.

In many species with a sizable surplus population there is no effect on reproduction simply because no females are prevented from breeding. This category includes polygynous species of mammals and birds. It also includes monogamous species having an unbalanced sex ratio with extra males in the breeding population, usually for unknown reasons. Judging from experiments on birds that repopulated a depopulated area of Maine spruce forest, this category includes a surprisingly large number of species (Stewart and Aldrich 1951; Hensley and Cope 1951).

Species having specialized nesting requirements are sometimes limited by a shortage of suitable nest sites. This factor alone might limit the breeding population in species of birds that depend entirely on old woodpecker holes for nest sites. In these species only one female can make use of any one hole for physical reasons, and the number of eggs she can effectively incubate is close to the number she lays herself. These holes are vigorously defended; but even if they were not, the number of females breeding in an area would still be limited by the number of holes if there were fewer holes than adult females. For the lizard *Gehyra variegata*, stumps suitable for "nesting" are limited in number, and so in conjunction with territorial behavior, the number of stumps tends to limit the density of breeders. However, there is no physical limitation on the number of lizards that can inhabit a stump; the limit is set by agonistic behavior.

Some species of birds have been unusually successful in cities and towns and have reached population densities unknown in their original natural habitats. The American robin (*Turdus migratorius*) is one such species, but it has been less thoroughly investigated than its counterpart, the European blackbird (*Turdus merula*), whose behavior and ecology were studied by Snow (1958; see also Lack 1966). This blackbird may have a surplus population in "artificial" habitats like cities, towns, and botanical gardens, where its densities may reach ten times those recorded in its more natural woodland habitats. Other examples of species with surpluses due to unusually high densities in artificial habitats are discussed by Brown (1969*a*). These studies are useful in showing that territorial behavior can limit breeding density under abnormal conditions, but they do not provide evidence that territorial behavior is an important contributory factor in determining breeding density under more natural conditions. On the contrary, if it is

TABLE 6–1 Summary of effects of population surplus on reproduction in species defending large fractions of their home ranges. (After Brown 1969*a*.)

Species characteristics	Surplus Males	Surplus Females	Reproduction limited	Reproduction limited by territoriality	Reproduction limited by territoriality under natural conditions	Example
Polygynous species	+	–	–	–	–	arctic ground squirrel (Carl 1971)
Monogamous species with excess males	+	–	–	–	–	many songbirds (Hensley and Cope 1951)
Species dependent on nest sites that are limiting	+	+	+	–	–	hole-nesting birds
Species reaching abnormal densities in unnatural habitats	+	+	+	+	–	European blackbirds in Oxford Botanical Garden (Snow 1958)
Species at normal densities in natural habitats	+	+	+	+	+	rock ptarmigan (Watson 1965)

true that a surplus does not become evident until abnormal densities are reached, this would suggest that under normal conditions territorial behavior is insignificant in limiting breeding density in these species.

After the first four categories listed in Table 6–1 have been considered, there is a residuum of cases where territorial behavior has been shown to play a role in determining both breeding density and number of young produced in relatively natural habitats. These include the rock ptarmigan (*Lagopus mutus*, Watson 1965), which nests in Alpine tundra that has resisted agriculturalization so far; the great tit (Krebs 1970, 1971), studied in semi-natural woodland; and the red grouse (Watson and Jenkins 1968), investigated under relatively natural conditions.

Another factor that interacts with territorial behavior in influencing reproduction is the density of breeders. If each breeder produced a fixed number of young each breeding season, the number of young produced would be a simple linear function of the number of breeders. Actually this is usually not the case in natural populations. The great tit is one of the most intensively studied species with an all-purpose territory, and it will be used as an example. Its population biology has been studied for many years, mainly in the Netherlands and Britain (reviewed in Lack 1966). As breeding density increases in this species, the production of young per pair is reduced. For example, in a study by Perrins (1965) the number of young produced per pair (estimated from those surviving in autumn) was greater in a sparse population than in a dense one, as shown in Figure 6–7.

The difference in production of young between dense and sparse populations could theoretically be as great as that between populations living in rich and poor habitats. When both these effects are considered together, as in Figure 6–7, it becomes clear that the effects of territorial behavior on the reproduction of a population living in two or more habitats cannot be easily predicted. For example, it would not necessarily be true that tits forced to breed in a poor habitat would produce fewer young per pair than those breeding in a rich habitat, since the penalty for breeding in a densely populated rich habitat might equal (as in the "ideal free distribution" of Fretwell and Lucas [1969]) or exceed that for breeding in a sparsely populated poor habitat.

When Krebs (1970, 1971) compared the effects of territorial behavior on populations of the great tit with the effects of other factors, he concluded that "territory has at the most only a weak density-dependent effect," because territory size varied so from year to year and seemed to depend mainly on the number of great tits surviving from the previous breeding season. The moral of the great tit story is that it is possible for territorial behavior to have a conspicuous effect in determining breeding density but at the same time not be very important as a regulative factor. In other words, territoriality apparently acts only a little more strongly on the population in years of high density than in years of low density. It remains an open question how far this conclusion is applicable to other territorial species, whether birds, mammals, lizards, or other groups.

6–7 Effects of territorial behavior on production of young as a function of habitat and population density in the great tit. The data are from Perrins (1965). He used the number of young per brood surviving to autumn as an index of production of young. **A.** Production of young *per pair* was greater in a sparse population in Great Wood, than in a dense population inhabiting a comparable habitat in Marley Wood. **B.** The number of young produced *per acre* was about the same in the two areas despite differences in the numbers of pairs. **C.** A graphic method for estimating the disposition of the population into two habitats that would maximize the number of young produced per acre. It is assumed in C that the number of pairs present in both habitats combined is such that there would be a density of 1.9 pairs per acre if all birds settled in the rich habitat and none in the poor habitat. (Data from Perrins 1965; graphs from J. L. Brown, *American Naturalist,* University of Chicago Press. Copyright 1969 by the University of Chicago. Drawing of great tits from D. Lack, 1966, *Population Studies of Birds,* The Clarendon Press, Oxford.)

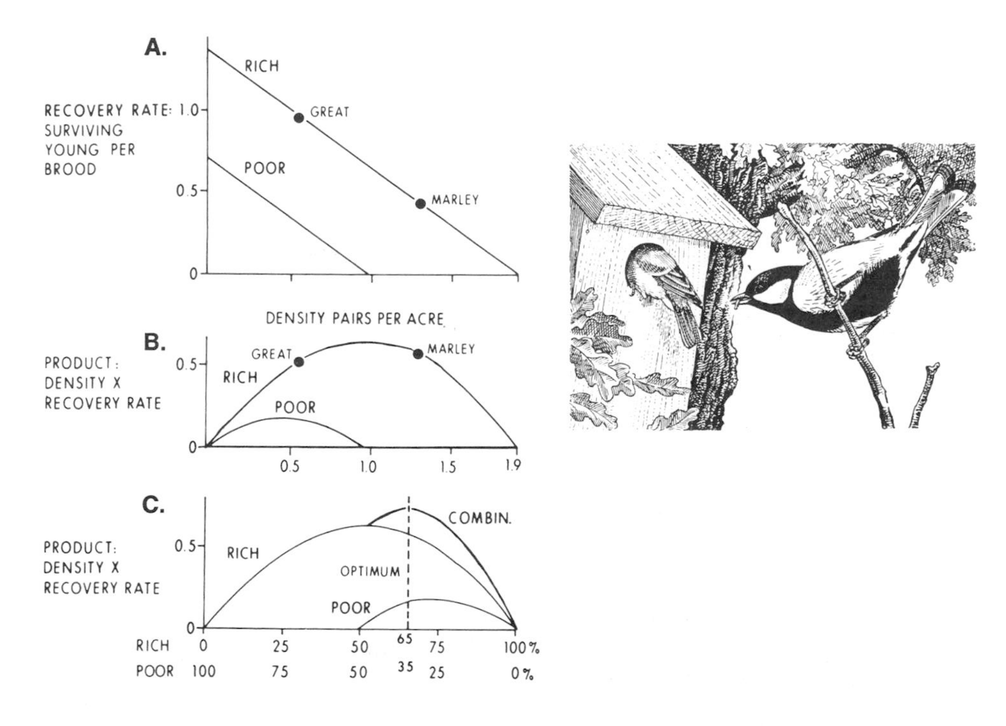

Effects on Survival / Territorial behavior occasionally causes the death of an individual by injury while fighting, but its main effects on survival are more subtle. Individuals excluded from an area have no "home" of their own. They may be driven out of desirable feeding areas and are denied the security that comes from familiarity with a particular area. Consequently, their mortality rate is often higher than that of territory owners. Death may come through predation, which may be facilitated by the prey's undernourishment and lack of familiarity with safe places for escaping from predators. These effects on survival are not restricted to the breeding season, as are the main effects of territoriality on reproduction; they may occur in winter in species that maintain a territory throughout the year, or maintain a territory on the wintering grounds. For details of the effects of territorial behavior on mortality and a discussion of examples, see Brown (1969*a*) and Watson and Moss (1970).

The Regulation of Density by Aggressive Behavior / Of the many processes that influence numerical fluctuations in a population, some occur independently of the density of individuals in the population (*density-independent processes*) and some occur as a function of population density (*density-dependent processes*). For example, changes in the weather occur independently of density and can cause a change in population density. A cold snap will kill all cold-susceptible individuals regardless of their density. Starvation due to a food shortage may be density-dependent, since the shortage may be aggravated by high population densities and since the fraction of a population that dies through starvation when food is scarce is typically dependent on the size of the population. Ecologists favor the view that density-dependent processes stabilize the size of a population more efficiently than density-independent processes.

Some features of territorial behavior suggest that its action on populations is density-dependent. As shown in Figures 6–2 through 6–6, the exclusion of individuals from preferred habitats or their exclusion from breeding is density-dependent. The depressing effects of territorial behavior on breeding density, reproduction, and survival tend to inhibit population growth beyond a certain level. For example, the larger the population, the greater the fraction forced into a surplus status and deprived of a chance to breed (at least, this is true in the models).

An unusual example of the regulation of density among breeding males through aggressive behavior was described for the dragonfly *Aeschna cyanea* by Kaiser (1968, 1969). Both sexes come to ponds only for copulation and egg-laying; they forage for food away from the mating grounds. A male in the mating ground chases other males aggressively, making them spread out and keeping the density of breeding males in any one spot low. This aggressive behavior lasts only brief periods (up to 40 min.) and alternates with foraging away from the breeding area. How long a male stays in the breeding area depends on internal factors and on the behavior of other males that he meets. Even if he meets no other males, his level of aggressiveness, as indicated by the frequency of buzzing-flights, declines to zero during his stay. If he

6–8 Regulation of breeding density by aggressive behavior in male dragonflies. **A.** The number of males present is relatively independent of their rate of arrival. Each point represents the mean for an hour; the hypothetical density was calculated using a mean duration of 31 min. **B.** The average duration of a stay decreases at higher arrival rates. **C.** Feedback model for the regulation of breeding density; the circle represents a multiplicative function. (After Kaiser 1969.)

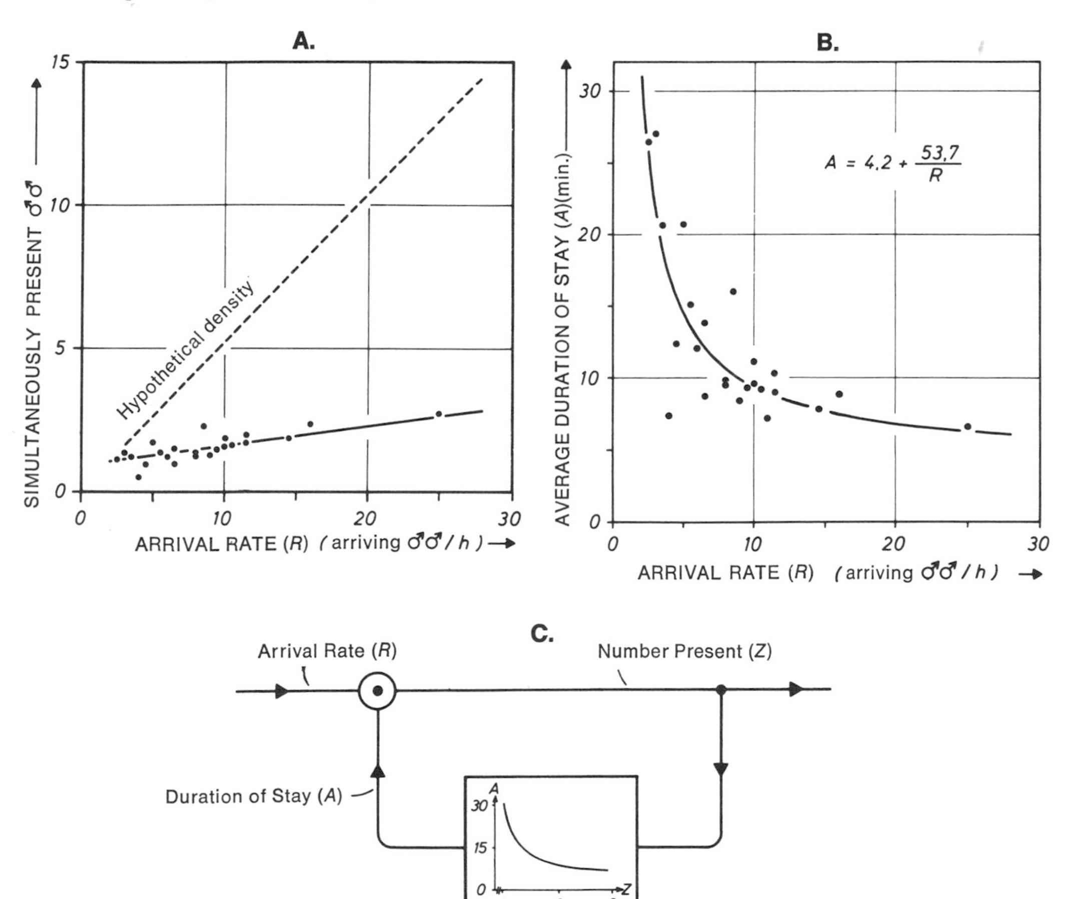

meets more resistance, and especially if he loses a fight, he is likely to leave sooner. A higher arrival rate of males results in more frequent fights and a lower average duration of stay by each male. A male leaves after losing a fight, and he loses sooner when the arrival rate is higher.

Figure 6–8*A* shows how the density of males at one breeding area depends on the arrival rate. The number of males present is remarkably invariant over a wide range of arrival rates. The curve shown in Figure 6–8*B* was calculated from the field data presented in Figure 6–8*A*. It shows, for males, the average duration of stay as a function of the arrival rate. This information allowed Kaiser to construct the simple feedback model shown in Figure 6–7*C*. According to the model, the density of breeding males is determined by the average duration of stay, which is regulated by aggressive behavior according to the illustrated relationship between stay-duration and arrival rate. The relationship is temperature-dependent, and the curve would be different at a different temperature. The consequence of aggressive behavior in this case is a relatively stable regulation of the density of males at the breeding area.

SUBORDINATION EFFECTS

In some species aggressive behavior may space out individuals without necessarily producing exclusive areas and recognized boundaries. Such behavior may lead to areas of dominance within which other individuals are tolerated if they behave submissively. At low population densities there may be few social encounters, these being restricted to small areas of overlap in the home ranges of neighbors. However, as population density grows, areas of dominance shrink, overlap in the home ranges of individuals increases, and individuals must spend more of their time outside their areas of dominance. A point is reached where many individuals are subordinate everywhere; they have no area of dominance. Increments to growing populations then appear most prominently in the class of subordinates, mainly the young. If the number of dominants remains the same, the entire increase in population will be in the class of subordinates. In species in which the breeding density is not rigidly fixed, but varies widely and with much social friction, the role of subordinate individuals is of central importance, since these animals compose the majority of the population—as in man.

Fluctuations in population density, and consequently in the importance of subordinates, are particularly dramatic among rodents. For convenience, further discussion of subordination effects will be limited mainly to rodents, but these effects are well known in other groups too.

The density of collared lemmings (*Dicrostonyx groenlandicus*) near Churchill, Canada was reported to vary by a factor of roughly 600 over several years (Shelford 1943). The density during a low was estimated at six to seven lemmings per 100 hectares, and the animals were exceedingly difficult to find. During a high the density was estimated at 4,000/ha and the lemmings were conspicuous everywhere. Variation in population density in the brown lemming (*Lemmus trimucronatus*) in Alaska is regarded as comparable in magnitude, or possibly greater in certain local areas (Pitelka 1957). The brown lemming population at Point Barrow, Alaska fluctuates from boom to crash in cycles of three to four years. Population increases by a factor of 10 or more are routine for many temperate-zone rodent species, but the cycles are less regular than in the arctic. Mouse "plagues" are fairly regular events in many areas.

Species differ in the extremes of their density fluctuations. Terman (1966, 1968) has described these differences for various rodent species (Figure 6–9). The differences are related partly to the individuals' ability to produce a large number of young in a short period of time. The genus *Peromyscus* provides a good example of closely related species that differ in population variability (McCabe and Blanchard 1950), although they are relatively stable compared to other mice and voles. As shown in Figure 6–10, the strongly territorial *P. californicus* in a California study area reached a seasonal peak number of mice, 14, that was not much greater than the minimum number for that year, 10. The somewhat less aggressive *P. maniculatus* in the same area reached a peak, 75, that was proportionately much higher than the minimum for the year, 30. *P. truei* was intermediate. These differences in seasonal population fluctuation among the three species are partly due to differences in the birth rates of the species involved, as shown in Table 6–2. Species differences in population stability are also related to a host of other factors, including community complexity, geographical latitude, survival rates, and aggressive behavior.

Responses to Population Increases: Emigration / A natural response of an animal living in a state of forced subordination is to try to find an area where he can be dominant. This means leaving the area where he is subordinate and going elsewhere. Emigration of animals is a typical response to the effects of crowding. It is seen in natural populations and laboratory experiments. Probably the most famous exodus is that of the lemmings (*Lemmus lemmus*) of Scandinavia, but mass

6–9 Variability in populations of species of small mammals. Data are from the North American Census of Small Mammals. *P#* = the number of different populations in which a mean of at least four animals was trapped per trapping period; $\bar{X}N$ = the mean number of animals trapped per trapping period. For details see Terman (1966, 1968). (From Terman 1968. Reproduced by permission of the American Society of Mammalogists.)

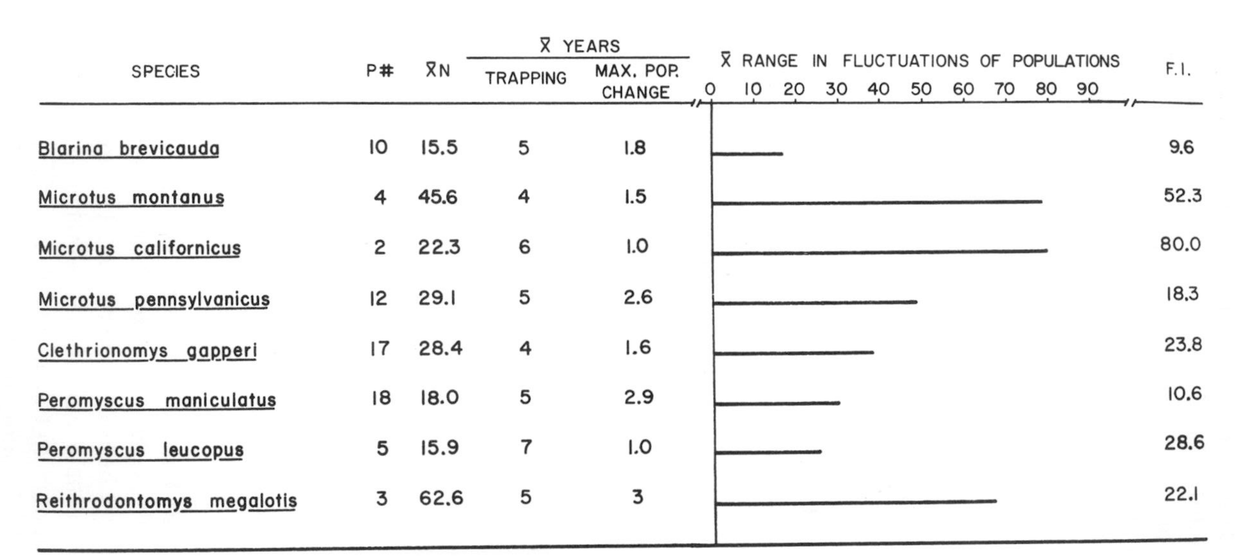

SPECIES	P#	X̄N	X̄ YEARS TRAPPING	X̄ YEARS MAX. POP. CHANGE	F.I.
Blarina brevicauda	10	15.5	5	1.8	9.6
Microtus montanus	4	45.6	4	1.5	52.3
Microtus californicus	2	22.3	6	1.0	80.0
Microtus pennsylvanicus	12	29.1	5	2.6	18.3
Clethrionomys gapperi	17	28.4	4	1.6	23.8
Peromyscus maniculatus	18	18.0	5	2.9	10.6
Peromyscus leucopus	5	15.9	7	1.0	28.6
Reithrodontomys megalotis	3	62.6	5	3	22.1

$$\text{F.I.} = \text{Fluctuation Index} = \frac{\text{Mean range of population fluctuations}}{\text{Mean years for range to occur}}$$

emigrations from regions of high population density are well known in many birds and mammals. The nature of the mass movements of Norwegian lemmings appears to differ from place to place and time to time. In some areas a two-way migration is the rule: the lemmings leave the higher areas of harsher climate for the winter and return in the summer, regardless of population density (Aho and Kalela 1966; Archer 1970). In other areas mortality among emigrants may be so high that virtually none return. The report of Myllymaki et al. (1962) suggests that there is some regularity in the direction of the movements, a regularity fixed at their inception. Most other species of rodents do not migrate regularly, emigrating only at times of high density and apparently without predetermined direction.

Detecting the emigration of individuals in natural populations

6–10 Seasonal changes in population density in three species of *Peromyscus.* (Data from McCabe and Blanchard 1950.)

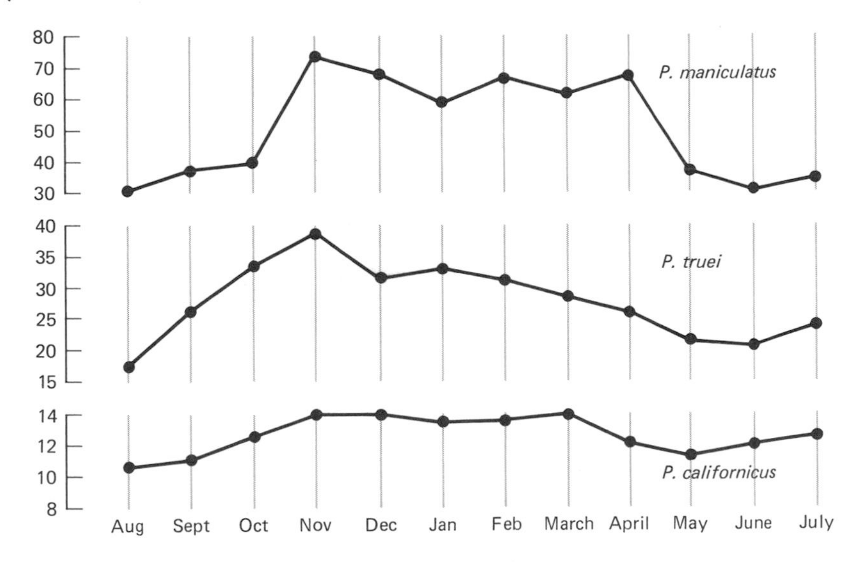

TABLE 6–2 Comparison of reproductive rates in three species of *Peromyscus* in a study area in California. (McCabe and Blanchard 1950.)

	P. maniculatus	*P. truei*	*P. californicus*
Average number of litters per season	4.00	3.40	3.25
Average number of young per litter	5.00	3.43	1.91
Average offspring per breeding female per season	20.00	11.66	6.21

under normal conditions may require experimental techniques. Two methods have been used. In the first, newcomers to an experimentally depopulated area are recorded (as reviewed by Stickel 1968). In the second, overpopulation is created artificially by the release of additional individuals into the population while movements of all animals in the study area are monitored.

Movements into a depopulated area have been observed in natural populations of *Microtus pennsylvanicus* (Van Vleck 1968), woodchucks (*Marmota monax*, Lloyd et al. 1964), prairie deermice (*Peromyscus maniculatus bairdi*, Blair 1940), Norway rats (Orgain and Schein 1953), nutria (Ryszkowski 1966), *Peromyscus leucopus* (Stickel 1946), and other species (Archer 1970). In *P. leucopus* inhabiting a 17-acre plot in a Maryland lowland forest, Stickel showed that (1) individuals entering the depopulated acre came from all directions, (2) most came from nearby established home ranges, (3) those individuals nearest the depopulated acre were the first to be caught, and (4) most of the ingressants were adults (54 adults, 25 juveniles). Their dominance status was unknown. The locations of the home ranges of the ingressants in early and late portions of the study are shown in Figure 6–11. Species probably differ in the proportion of

6–11 Original home ranges of ingressant male *Peromyscus leucopus.* The area of the large circle is 17 acres, and of the small inner circle, 1 acre. In **A** the home ranges of adult males caught on the first day of trapping in the central area are shown in black. In **B** the home ranges of adult males caught from the 18th through 35th days of trapping are shown in black. (From Stickel 1946. Lucille F. Stickel, Bureau of Sport Fisheries and Wildlife, Patuxent Wildlife Research Center, Laurel, Maryland. *Journal of Mammalogy.*)

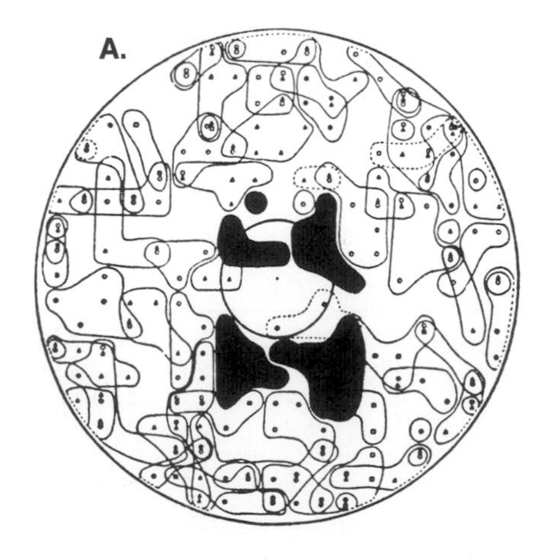

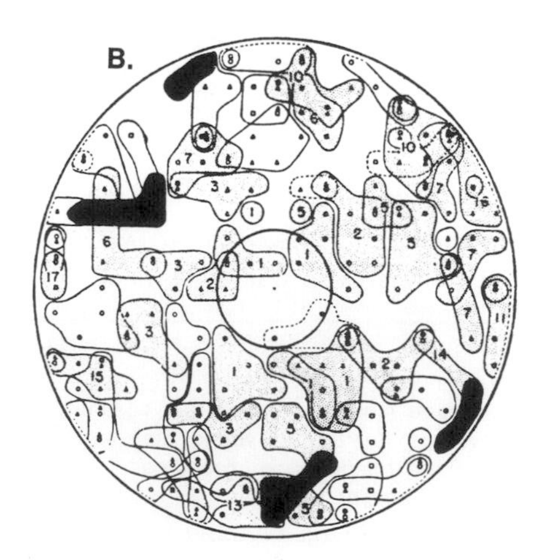

transients in their populations (Stickel 1968). Seasonal peaks occur in the frequency of transients, as suggested in Figure 6–10.

Survival of Emigrants and Transients / Mice that leave their home ranges as a result of "population pressures" probably tend to wander until they find a relatively suitable and less crowded area for a home range. At times of high densities the chances of these individuals surviving and reproducing are probably poor. Mice attempting to enter established territories or dominance realms are frequently killed or excluded in laboratory populations and apparently also in natural populations. In populations of house mice (*Mus musculus*), subordinate emigrants probably are successful in entering other territories only rarely, judging by the distinct genetic differences found among local populations (Anderson 1964; Selander 1970; Klein 1970). Experiments showed that artificial introductions of strange mice into established populations of house mice were unsuccessful unless at least four individuals were added simultaneously (Reimer and Petras 1967).

The transient not only suffers the risks of trying to enter an established group, but probably is also more susceptible to predation because he lacks familiarity with his surroundings. In a laboratory situation, white-footed mice (*Peromyscus leucopus*) exposed to limited predation by a screech owl (*Otus asio*) survived better after several days of prior familiarization with the test room than did naive mice (Metzgar 1967).

Flexibility in Social Organization in Response to High Densities / Defense of an exclusive area of fixed size by a male or a mated pair is a relatively inflexible system. If space is limited when population density increases drastically, the entire surplus population is sentenced to a nonbreeding, nonterritorial existence, or to death. Such a system should effectively prevent breeding density from rising. If the exclusive areas are not of fixed size, the presence of more competitors for them may force the average territory size to become smaller. At this point species that are more sensitive to crowding would stop breeding, or at least breed unsuccessfully, thereby preventing further increases. Species less sensitive to crowding would continue breeding. Therefore, in theory we can recognize a continuum ranging from species that are highly sensitive to and intolerant of crowding to those that are highly insensitive and tolerant of crowding. Among rodents, species of the genus *Peromyscus*, particularly *P. californicus*, are at the sensitive end of the spectrum; the brown and collared lemmings must be counted as relatively insensitive, since breeding continues at extremely high densities. Eisenberg (1967) has investigated species differences in sensitivity to crowding under standardized laboratory conditions in various rodents. Among the species studied, *Dipodomys nitratoides* was found to be highly sensitive to crowding; *P. californicus*, intermediate; and *P. maniculatus*, relatively insensitive (using as criterion the density at which reproduction ceased).

In nature, when rising population density forces animals to have smaller exclusive areas and more overlap in their home ranges, various accommodations to increased density may emerge. There is no general rule to predict the nature of these accommodations; however, they may involve communal groups helping or replacing the pair or dominant male.

Experiments with house mice in population cages by various authors indicate that subordinate males may be present in the artificial territories; however, when ample space is provided, the subordinates emigrate from the territories of the dominant males to unclaimed areas even when food is not scarce (Strecker 1954; Archer 1970). The situation in natural populations is not clear. Territories containing an adult male and one or more adult females appear to be the rule, but whether subordinate adult males are a regular feature of these territories, and under what conditions, is difficult to say. Eibl-Eibesfeldt (1950) reported group territories with peaceful relations among group members but the conditions of density, environment, and past history necessary for such a social organization to emerge require further elucidation. It seems clear that the dominance hierarchies so often studied in laboratory mice are not necessarily a regular feature of natural populations of the same species.

Tolerance of subordinate males in groups of Norway rats (*Rattus norvegicus*) appears to be greater than in house mice in wild and captive populations. In European species of *Microtus*, females normally defend exclusive areas around their nests; at high densities these areas shrink and two or more females may share a territory by remaining in their natal area until mature. Judging by the population-pen experiments of Houlihan (1963), *M. californicus* has a different social structure; some spacing of individuals occurred, but defense of home range was not observed. At higher densities aggression and home-range overlap became extreme. In both European and Californian *Microtus* there was some evidence of a relatively peaceful arrangement, or "organization," among members of a local population, which might break down at very high densities. The presence of such unspecified orderly relationships in a natural population of cotton rats (*Sigmodon hispidus*) was indicated by an experiment of Wolfe and Summerlin (1968);

in staged encounters of four animals at a time, animals captured from the same small area, where they presumably had previously encountered each other in their extensively overlapping home ranges, fought and chased each other significantly less than did animals from widely separated areas who were strangers to each other. These results indicate that an individual familiar under natural conditions may provoke less aggression than a strange one (when tested in the laboratory), a conclusion that is certainly consistent with pure laboratory studies.

Effects of High Densities on Recruitment / *Recruitment* refers to the processes by which animals are added to the breeding population. It includes not only their birth, but also their subsequent survival to maturity. It is a composite of birth and death processes distributed over age classes. High population densities may reduce recruitment through effects on survival or reproduction. Reports of the adverse effects of crowding on survival and reproduction in penned populations or caged groups of rodents are numerous (Snyder 1968; Archer 1970; Thiessen 1964). Typically the subordinate rodents are the first to be affected, but dominants also may suffer by fighting each other.

Reports regarding natural populations are fewer. Chitty (1952) found high juvenile mortality at peak population densities of voles. Various reproductive functions were adversely affected by high densities in populations of *Mus musculus* (DeLong 1967), *Microtus californicus, M. montanus* (Hoffman 1958), *M. pennsylvanicus* (Hamilton 1937), *Peromyscus maniculatus bairdii* (Helmreich 1960), and *Clethrionomys rufocanus* (Kalela 1957). There are reasons for doubting that these deficiencies were due simply to shortage of food or shelter; the effects of repeated aggressive encounters seem to be primarily responsible.

Physiology as a Function of Population Density / One explanation of population declines in mammals that has attracted much attention invokes physiological changes believed to result from the increased frequency of social contacts among individuals. Reviews by Christian (1963), Christian and Davis (1964), Christian et al. (1965), and Myers (1966) have presented this point of view, with special reference to the effects of population density on the adrenal cortex (see also the review of Brain 1971). The primary physiological changes responsible for the ultimate losses to the population at high densities have been attributed to adrenal hypertrophy. "It was suggested . . . that increased adrenocortical secretion would increase mortality indirectly through lowering the resistance to disease, through parasitism or adverse environmental conditions" (Christian and Davis 1964:1551). To test this hypothesis many laboratory experiments have been carried out.

The physiological effects of crowding mice into a small cage are diverse. The list of reported findings includes increased weight of spleen, increased hematopoiesis, thymic involution, suppression of somatic growth, increased plasma corticosterone, delay or prevention of sexual maturation, delay of spermatogenesis, decline of weight of accessory sex organs, diminished ovulation and implantation, increased intrauterine mortality of fetuses, inadequate lactation to the point of stunting size at weaning, and in some circumstances, complete curtailment of reproduction. Growth in laboratory populations has been halted by various of these factors, such as decline in birth rate, infant mortality, and combinations of these. Disease may also decimate a population as a secondary consequence of such conditions.

Neuroendocrine Physiology as a Function of Defeat / Early studies of the physiological consequences of crowding in mice revealed considerable variability in the responses of individuals in the same group. Further observation revealed that the dominant animal in a group tended to be nearly immune to the typical changes induced by crowding, while low-ranking subordinates were most affected (Davis and Christian 1957; Louch and Higginbotham 1967). As a result, the focus of many experiments shifted from the effects of crowding to the more specific question of the consequences of defeat. These revealed significant correlations between social rank and adrenal physiology; lower-ranking mammals could be differentiated from dominants by eosinophil counts, adrenal lipid and cholesterol concentrations, hydrocortisone secretion, adrenal ascorbic acid content, *in vitro* corticosteroid production, and adrenal weight.

Critics have argued that these changes should not be expected in natural populations, since the densities in laboratory experiments of this sort are usually far in excess of natural conditions (Negus and Gould 1965, and others). There is not complete agreement that the phenomena found in crowded laboratory mice actually are found in nature (see references in Archer 1970; Thiessen 1964; Terman 1968), although comparable endocrine changes have been reported in natural populations of Japanese deer (sika), woodchucks, rabbits, voles, rats (Christian and Davis 1964), and lemmings (Andrews 1968). Furthermore, artificially reducing the density of a dense natural population reduced adrenal weight in rats, deer, and woodchucks.

In mammals, emotions are mediated by brain systems involving various synaptic transmitter substances and related precursors and enzymes that are susceptible to manipulation both by behavioral experience and by drugs (Schildkraut and Kety 1967). It has been shown in laboratory experiments that levels of the relevant chemicals in the brain stems of mice are susceptible to modification by exposure to fighting and defeat (Eleftheriou and Boehlke 1967; Boehlke and Eleftheriou 1967; Eleftheriou and Church 1968*a*, *b*) or simply by watching vigorous fighting (Welch and Welch 1968). Fearful behavior is also thought to be influenced by ACTH (adrenocorticotrophic hormone) and by adrenal steroids (Weiss et al. 1969). These data suggest that the function of brain structures involved in agonistic behavior (reviewed by Kaada 1967) may be modified chemically by agonistic experiences at high population densities. A comprehensive theory integrating the roles of environmental stimulation and nervous and endocrine systems was developed by Welch (1965). The role of neurochemical changes in the dynamics of natural populations is unknown.

Evolutionary Fate of Emigrants / A prerequisite for genetic adaptation to new ecological situations is dispersal from established populations. Social processes that are conducive to emigration are important in range extensions and occupation of new environments. The individuals who find themselves in a new environment previously unoccupied by their species are typically not genetically adapted for it. Consequently, natural selection acts strongly on them. The amount of genetic change that takes place in the new population is to a large extent determined by how much the new environment differs from the old (and by the genetic variability among the founders). Since stabilizing selection probably predominates in well-adapted populations, relatively large evolutionary change in established populations should be slow. In a new environment, however, directional selection should be more important, and larger evolutionary changes may be predicted.

The movement of animals into unoccupied habitats comes about through the processes of dispersal (longer movements) and spacing (smaller movements). Both of these occur regularly regardless of population density, but they are intensified by high population densities. At such times individuals have great difficulty finding unoccupied space and are forced to look farther afield for it. At high densities, what started out to be a short-distance spacing movement might end up as a long-distance dispersal movement.

Therefore, two major factors may be identified in the colonization of new environments: (1) population density, and (2) the combined effects of spacing and dispersal behavior. Both of these are strongly influenced by social organization. Somewhat similar ideas were expressed by Christian (1970).

Two intergrading patterns of response to increasing population density may be recognized in rodents and other animals, building upon a distinction introduced by Eisenberg (1967). *Density-tolerant species* generally have high reproductive rates; they keep on reproducing at relatively high densities because their social behavior is flexible enough to allow them to tolerate considerable crowding. Their dispersal movements are likely to be concentrated in years of high density and to vary greatly from year to year. When high densities come, large numbers of animals are forced to emigrate. Many of these dispersing animals would normally have established dominance somewhere if space had been available; they are not necessarily cast in the role of subordinates because of genetic inferiority but because they had the misfortune to be born at a time of high density.

Density-intolerant species generally have low reproductive rates; they cease reproduction when forced to live together at high densities because they typically lack social mechanisms that would allow them to live at close quarters without fighting. They are likely to be solitary or to live in pairs and defend a territory strongly. For these reasons these species do not fluctuate much in density from year to year. Seasonal changes in density may, however, be marked (Figure 6–10). Dispersal is more likely to occur in these species at a particular season, when density is high, and to be similar in magnitude from year to year. Because of the low reproductive rates and small numbers of emigrants, these species are not adapted for colonizing transient habitats, as density-tolerant species are. The emigrants from populations of density-intolerant species are more likely to have had an even chance to establish a territory in their original habitat; that is, the number of competitors for available space varies little from year to year. If an individual fails to win out in the struggle for territory, it is not because he was born in a year of high density but more likely because he was inferior genetically or because of some acquired inferior characteristics.

CONCLUSIONS

Aggressive behavior tends to space out individuals or pairs in the available habitat. Territorial behavior (aggressive behavior resulting in an exclusive area) has, in theory at least, the potential for setting an upper limit to breeding density, provided there is a fixed minimum territory size. A simple model based on this latter assumption is used to generate predictions concerning breeding density, movements between habitats, and stability of populations. These predictions are compared with some findings from experimental and natural populations of lizards and birds. Some support is found for the idea that territorial behavior sets an upper limit to population density in natural populations, but several complications make interpretation difficult.

Aggressive behavior may affect populations through *exclusion* of individuals from key areas, as in territoriality and other forms of aggressive spacing. It may also affect populations through *subordination* of individuals; subordinate animals remain in the presence of dominant animals (they are not excluded), but typically at the cost of restricted access to the essentials for reproduction and survival.

The effects of subordination have been studied mainly in rodents. Species differ in the degree of fluctuation in their populations. Those with drastic fluctuations tend to have higher birth rates and to be more tolerant of crowding. When crowding occurs in the laboratory, a variety of physiological effects involving the adrenal glands and other endocrine structures can be demonstrated. Animals that have suffered defeats are prone to such changes, but winners are more resistant. The significance of these phenomena in natural populations is a controversial subject. For various reasons survival and reproduction in natural populations tend to be lower at high densities.

Density-tolerant species tend to fluctuate widely in numbers. At peak densities they produce many animals that disperse widely and colonize new or temporarily vacant localities. Populations established in new habitats by such dispersing individuals are subject to new selection pressures and possibly rapid evolution.

7 The Evolution of Diversity in Spacing Systems

THE DISPERSION PATTERN of a species reflects the norms of social behavior among individuals of that species. The individual's role in the dispersion pattern is the result of a variety of behavioral mechanisms that promote the opposing extremes of spacing-out and grouping. The compromise between these opposing factors that is most successful for the individual becomes the norm favored by natural selection (in this case, by individual selection). Individual genotypes near the norm leave more descendants in successive generations than genotypes divergent from the norm (stabilizing selection). By this means the dispersion pattern of a species, although it is a population characteristic, can be easily explained by theories employing individual selection.

In many cases alternate theories based on population selection ("group selection") have been advanced (e.g., Wynne-Edwards 1962, 1963); but in all cases known to me the dispersion pattern and its underlying social system can be adequately explained by theories based on individual selection without recourse to population selection. Regardless of the final judgment on the importance of population selection, the ecological ramifications of individual spacing and grouping behavior must be studied in order to detect and estimate the selection pressures influencing individuals *within* a population.

In the comparative study of social systems one commonly finds the same system occurring in unrelated taxonomic groups (convergent evolution) and different systems in the same taxonomic group. The wide generality of evolutionary convergence in social systems suggests that the individual behaviors composing a social system are quite susceptible to selection pressures. Furthermore, a correlation often exists between the presence of a particular social system and some major feature of the environment (e.g., Crook 1965). All these findings indicate that the social system of a species is a function of, and an evolutionary response to, its ecological situation. This means that social systems cannot usually be used to study phylogenetic relationships. Nor can evolutionary trends toward "better" social systems be identified. A social system is better or worse for the individuals of a species only in relationship to their environment, and not usually for any intrinsic reason.

In this chapter spacing behavior is emphasized in the first section and grouping behavior in the last. Particular emphasis is given here to agonistic behavior and ecological relationships. The special evolutionary and ecological aspects of sexual and aid-related behavior, which also affect spacing and grouping, are treated in later chapters.

THE EVOLUTION OF TERRITORIALITY

We shall deal mainly with animal species and groups that have been intensively studied ecologically—many of these are birds. The principles with which we shall be concerned should apply to man as well as to all other social animals. Since territoriality is the most extreme and effective kind of spacing behavior, the principal concepts will be developed in a discussion of territoriality. They may also be applied to other spacing systems.

Two related problems will be considered: (1) How did *territoriality* evolve? (2) How did the *diversity* in territorial systems evolve? These must be considered together.

Early Hypotheses / Since there is such diversity in the size and type of areas animals defend, it is difficult to find anything in common about them or the behaviors performed in them. Defended areas include areas used for mating, feeding, or nesting, and for combinations of these activities. Opinions since Howard's day (1920) regarding the evolutionary origins of territorial behavior have been divided. At first, observers sought to find a "function" for territoriality, and in the process suggested many conceivable, but sometimes far-fetched, consequences of territorial behavior. These have been reviewed in detail

TABLE 7–1 Some hypothesized functions of territorial behavior. (Modified from Hinde 1956.)

1. Pair formation	6. Reduction of time devoted to aggression
2. Defense of nest and young	7. Prevention of epidemics
3. Guaranteeing a food supply	8. Limitation of population density
4. Reduction of losses to predators	9. Reduction in interference with copulation and reproduction
5. Maintenance of the pair bond	

by Nice (1941) and Hinde (1956). Some of them are listed in Table 7–1. Only the more important can be discussed here.

The Mating-Territory Hypothesis / Some workers have felt that there is no general explanation of the evolution of territorial behavior. It was thought that each type of territory evolved under its own set of selection pressures and that the various types have essentially nothing in common with each other. Certainly a territory used only for feeding has evolved under different selection pressures than one used only for mating. There can be no doubt that in lek types of social organization, in which males hold territories that are essentially only mating stations, success in mating is an important selection pressure in maintaining the system. Males without mating territories in the lek usually do not mate. O'Donald (1963*b*) has argued that even all-purpose (mating-nesting-feeding) territories can be explained as adaptations for mating. This does not, however, tell us why some species have evolved mating territories while others have not. Consequently, it is not a sufficient answer even for mating territories, let alone for more complex types.

The Food-Territory Hypothesis / A popular theory that can be traced at least to Altum (1868) and Howard (1920) and has many adherents today is that the primary advantage of defending an area, in those species that feed in their territories, is to reserve a supply of food or a foraging area for feeding the young when they hatch. That species which typically hold all-purpose territories do all or most of their foraging there favors this view. And some species, the pectoral sandpiper (*Erolia melanotos*), for instance, hold territories in fall and winter ranges that are clearly feeding territories (Hamilton 1959).

Although it is difficult to prove or disprove this hypothesis, some interesting correlations can be shown between territory size and food supply. Ornithologists have long felt that the breeding densities of birds are typically higher in habitats having a richer food supply for the species concerned, and a number of workers have stated this in a general way making use of census data. Since in these species territories are mutually exclusive, the average territory size must decrease as the density of territorial individuals increases.

A more thorough study of this general rule in a single species was done by Stenger (1958) with the ovenbird (*Seiurus aurocapillus*) in Ontario. The ovenbird's territory is large and essentially equal to its home range. Stenger determined what foods were being eaten through stomach analysis, and by collecting many samples of litter from the forest floor, where most feeding was done, she discovered what foods were available and in what amounts. Since the major items in both diet and litter occurred in similar proportions the assumption was made that food items were eaten in numbers "approximately proportional to their availability." Analysis revealed a *negative correlation* between territory size and food density: food-rich territories were smaller.

Two hypotheses seem possible at this point: (1) Areas with richer food supplies attract a greater density of birds, who then defend smaller exclusive areas because of stronger pressure from their more numerous neighbors. The available space is divided up among the number of territory aspirants, and there is no minimum territory size. (2) In areas with a richer food supply, individuals can satisfy their food needs in a smaller area, thereby leaving more space for other individuals. Consequently, the minimum territory size is smaller. They do not do this altruistically or for the good of the species; they do it because it would be wasteful of their time and energy to defend an area larger than needed. Proponents of the food-territory theory favor the second interpretation; opponents favor the first. The two hypotheses are not mutually exclusive.

A second study in which food supply was carefully measured and correlated with size of territory was made by Smith (1968) with two species of tree squirrels (*Tamiasciurus*). He found that in hard years squirrels tended to be lost from energy-poor territories but survived in energy-rich ones. In two years of food shortage the ratio of energy available to energy required per year varied from 0.1 to 2.8 in 19 territories. Among 12 squirrels that definitely disappeared over the winter the ratios varied from 0.1 to 1.8, while the 3 that definitely remained had ratios of 1.3 to 2.8.

In certain large predators the relations between breeding density and the richness of the food supply can be shown clearly. Taking advantage of a relatively simple arctic community, Maher (1970) showed not only that territory size in the pomarine jaeger (*Stercorarius pomarinus*) was smaller when its major food supply (the brown lemming) was abundant, but also that beyond certain prey densities, the predators ceased reducing their territory sizes (Figure 7–1). Here again cause and effect are not completely separable, but the results clearly indicate that territory size was not determined by food alone. If it had been, territory size would have been a linear function of prey density instead of reaching a minimum size.

The studies just described show a close relationship between food supply and home-range size, but only Smith (1968) has obtained data directly relevant to the need for defense, and more work is needed on this problem.

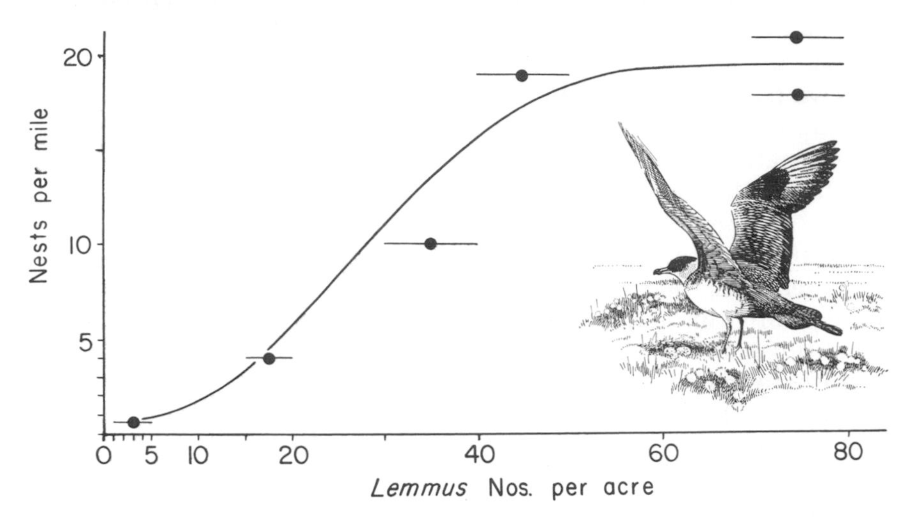

7–1 Relationship between nesting density (reciprocal of territory size) and food density for pomarine jaegers. Note that the relationship is not linear. (From Maher 1970. Drawing of jaeger from D. Lack, 1966, *Population Studies of Birds,* The Clarendon Press, Oxford.)

A General Theory / An inadequacy of the food- and mating-territory hypotheses is that they do not suggest why some species have one type of territory and other species another. Why is food the critical selection pressure in one species and mating in another? Furthermore, how the selection pressures involved act is not clearly specified.

It is profitable to rephrase the question of the evolution of territoriality in such a way as to eliminate the extraneous notion of "function." We wish to know the environmental conditions that facilitate or hinder the evolution of particular types of territoriality. More precisely, what are the relative fitnesses of different types of individuals with greater or lesser predisposition toward defending various types of territory in a series of specified environments? Only by emphasizing the behavioral relationships between the fitness of individuals and their environment can the selection pressures affecting behavior be fully understood.

Aggression for its own sake is not likely to benefit an individual. It exposes him to risks of death or injury in battle, and it makes him more vulnerable to predators. Furthermore, it may take time and energy away from reproductive and parental activities. The latter phenomenon has been termed *aggressive neglect* and has been discussed by Hutchinson and MacArthur (1959) and Ripley (1961) in relation to interspecific aggression. The concept is also applicable to intraspecific aggression. Field observations of young that were starving in the nest while their parents defended their territory have been reported in the red-winged starling (*Onychognathus morio;* Rowan 1966).

The absence of any intrinsic benefit in aggressive behavior (in contrast to feeding, mating, or nesting) suggests that when it does occur conspicuously and consistently in a species it must be providing something worth its risks. In order to illustrate this, let us consider two alternatives.

1) *Competition lacking:* Assuming monogamy, which in birds is more common than polygamy, and assuming an optimal supply of nest sites, food for raising young, and all the other requisites for reproduction and survival, would the fitness of a territorial individual be greater than that of a nonterritorial one? Since with a little persistence, according to our assumptions, either individual can obtain a mate, an adequate food supply, and an adequate nest site without defending a territory, aggressive behavior would only expose an individual to risks without gaining for him anything he did not already have. Under such conditions territorial behavior would be disadvantageous to an individual. Territorial individuals would be less fit than nonterritorial ones, and their contribution to the next generation would be less. Territoriality could not evolve through individual selection under such conditions and would be selected against if it were present. This suggests that competition is present in territorial systems today and was present during the evolution of territoriality.

2) *Competition present:* If we alter our original set of assumptions so that the situation is no longer ideal, but closer to reality, the results differ. If instead of specifying optimal conditions for all individuals we say that there is less than enough for all of some requisite for reproduction—food, cover, mates, or nest sites—some individuals will probably receive less than others of the resource in short supply. The "haves" would then leave more offspring than the "have nots," other things being equal.

In some species the difference in the number of young raised by "haves" and by "have nots" can be quantified. Such long-term studies are difficult to perform and require unusual circumstances; consequently, few have been made, and the data are, for the most part, indirect. Working with the side-blotched lizard (*Uta stansburiana*), Tinkle (1969) was able to show a direct correlation between territory size and number of offspring surviving to maturity (Figure 7–2). Female lizards having larger territories tended to be more fit. Since the habitat in the various territories was similar in quality, area (quantity) was apparently the critical factor. This suggests that the quantity of food in each territory was the main reason for the correlation, but further research would be necessary to establish this. Assuming that survival rate was not negatively correlated with territory size, we may conclude that a correlation between territory size and individual fitness has been demonstrated.

Correlations between indices of fitness and territory size have been demonstrated in other species (e.g., Robel [1966], who used number of copulations as the fitness index), and the argument does not rest on the Tinkle study alone. The unique and valuable aspect of Tinkle's investigation is that he used realized fitness (that is, actual numbers of *mature* progeny).

7–2 Territory size as a determinant of fitness in female side-blotched lizards. **A.** Mean recapture radius represents the distance from the center of the territory to all recapture points and is proportional to territory size in this species. The abscissa is the number of young surviving to maturity per female territory owner over a period of three generations. The correlation is significant ($p \sim 0.001$). **B.** Spatial disposition of home ranges plotted with the minimum polygon method during a year of high density of adults. In this population the home ranges are territories; the owner excludes rivals by its aggressive behavior. **C.** Photos of male lizard (*below*) and female (*above*). (A and B from Tinkle 1969. Photos by Isabelle Hunt Conant, from Tinkle 1967.)

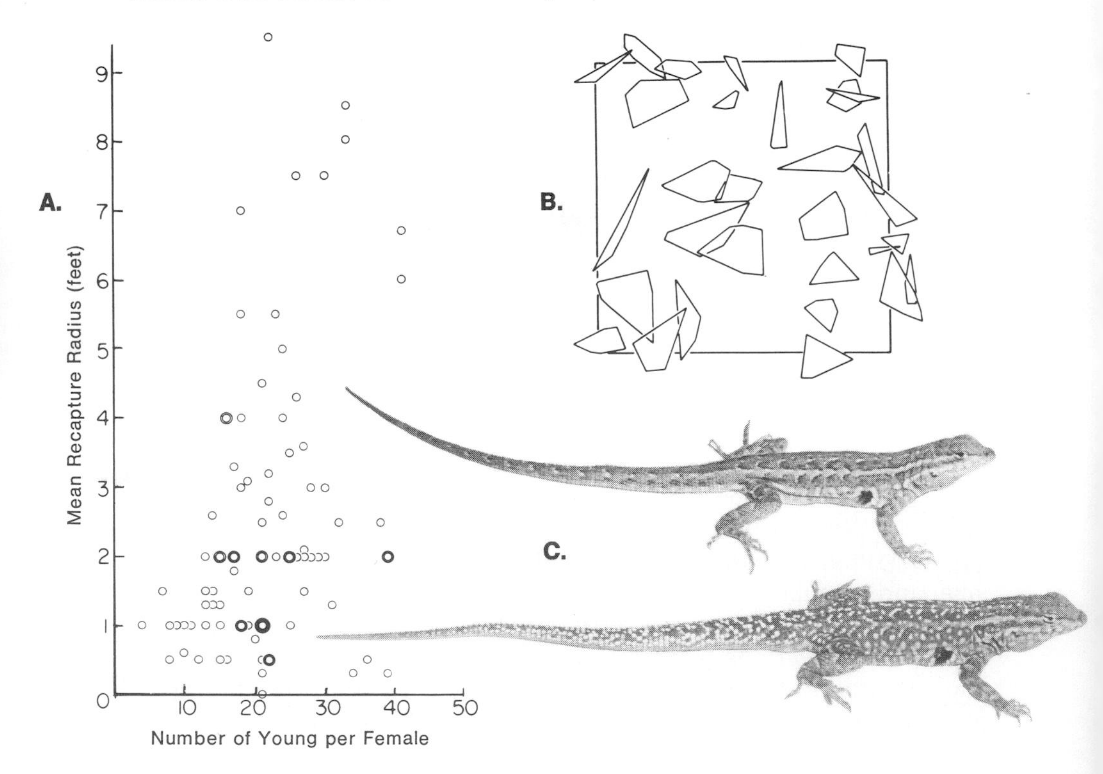

The rewards of aggression in terms of fitness depend on the stakes. If there is little to be gained by aggression and much to be lost by it, territorial behavior will be selected against. If there is much to be gained or guaranteed by aggression and little to be lost by it, territorial behavior will be selected for. Under steady-state conditions of competition, a norm for intensity of territorial behavior will most likely be established, with extremes in both directions selected against.

The intensity of the territoriality that evolves in a species should bear some relationship to the prevailing degree of competition. If the species can barely reproduce itself sufficiently to replace losses to predation and disease, there should be little competition during the breeding season and little need for territorial defense. But if the species is so prolific that there is much competition for the requisites for reproduction, territorial behavior should be intense, and advertisement and threat behavior should become highly specialized.

The general theory described here is summarized diagrammatically in Figure 7–3. It may be distinguished from the preceding separate theories by its emphasis on the *balance* between advantages and disadvantages of territorial behavior, and by its emphasis on the common denominator of *ecological competition,* which itself depends on the availability of resources relative to population density and needs. The important difference from other theories is the emphasis on *crucial resources.*

7–3 Outline of a general theory of the evolution of diversity in intraspecific territorial systems. This is a resource-centered theory. The value of territorial behavior for a species is determined by the supply of a resource that may be critical for the species and by the energy costs and gains associated with fighting for supplies of the resource. (From Brown 1964.)

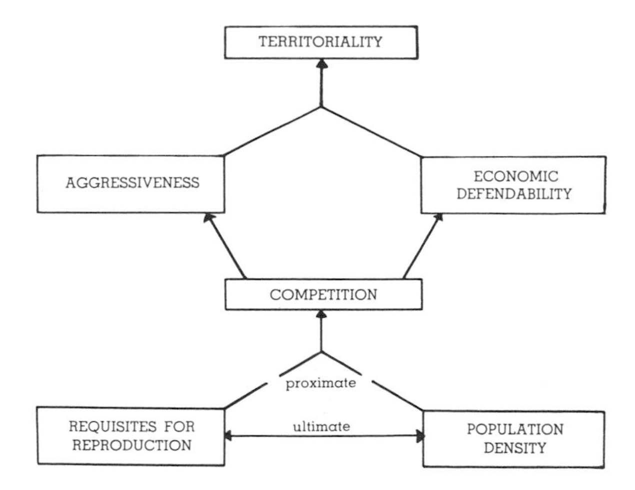

One advantage of this resource-oriented general theory is that it specifies the type of behavior likely to characterize a species according to (1) the nature of the resource in short supply, and (2) the cost of defending that resource. If we know these facts, we can predict the kind of territory a species has. For example, aerial or oceanic food resources could not be feasibly defended even if they were in short supply; the general theory predicts that species utilizing such food resources lack feeding territories. When in these species the nest site is relatively safe from predators it may be defended, as in the typical social organization of island- and cliff-nesting colonial sea birds. When neither defense of the nest nor feeding area by the male is required, the lek type of social organization may result, since the male is free to devote nearly all his energies to competing for females (see Chapter 8). In lek species aggressive neglect seems to be insignificant because the young are precocial and need not be fed by the parents. Further discussion may be found in Brown (1964).

Population Selection and Territorial Behavior / The idea that the evolution of a species characteristic could be explained by saying that it was "for the good of the species" is a favorite of the popularizers of natural history (e.g., Ardrey 1966; Lorenz 1963). The idea continues to be attractive, even though it was effectively debunked in 1930 by Fisher (1958). Perhaps the resemblance of this concept to modern moral ideas that codify the individual's subordination to society predisposes people who have not taken the time to think it through to accept it.

The most recent version is that a behavioral trait can evolve merely because it is "good for the population." The main points of this version were first fully stated in modern form by Kalela (1954), and restated, together with voluminous interpretations of many specific cases, by Wynne-Edwards (1962, 1963). Territorial behavior illustrates this viewpoint nicely and can be used as a test case with which to examine its validity. The argument may be summarized as follows:

1) Territorial behavior limits population density.

2) Limitation of population density curtails reproduction (prevents overpopulation).

3) Curtailment of reproduction prevents permanent reduction or extinction of the food supply.

4) Populations having such "self-regulatory" properties survive longer than those lacking them (through population selection), causing the evolutionary fixation of behavioral traits that benefit the population by limiting it at safe levels.

As we progress from arguments 1 through 4, these assertions become more and more difficult to verify. As we saw in the preceding chapter, territorial behavior does tend to limit breeding densities in some species, but often these effects are confined to the males or to unnatural environments. Consequently, the existence of territorial behavior does not necessarily mean that growth of the population is being prevented. Too much demand on food resources because of overpopulation has been observed in some simple island communities, but most mainland communities of territorial species are sufficiently complex, with alternate predator and prey species, that the long-term damage of high population levels would be minimal.

The most difficult argument to accept is that territorial behavior evolves through population selection (sometimes imprecisely termed *group selection*). This was much debated in the 1960s. Arguments against population selection specifically in regard to territorial behavior were advanced and summarized by Brown (1964, 1969*a*). With respect to the evolution of social systems generally, the population-selection hypothesis was rejected by Crook (1965) and Lack (1968). Both sides of the argument were concisely summarized by Wiens (1966), who favored individual selection. Williams (1966) considered the general nature of population selection at length, and rejected it as far as applications to behavior were concerned. Most evolutionary biologists have rejected population selection as the main mechanism responsible for social systems, at least with the currently available evidence. Most also qualify their rejection by pointing out that it is a mathematically acceptable theory. The principal objections are (1) that population selection is inherently slower and weaker than individual selection (or kin selection), and (2) that the required subdivision of populations into isolates that are neither too close together nor too far apart is rare. The component processes of individual selection (differential birth and death of genotypically different individuals) are intrinsically faster and more frequent than those of population selection (differential colonization and extinction of genotypically different populations).

The reader at this point cannot be expected to master the mathematical details of population selection theory (e.g., Eshel 1972), but should be able to see that a trait "beneficial to the population or species" cannot evolve *within* a single large population unless it is also advantageous to the individuals (or genotypes) carrying it. By definition, only genotypes with above-average fitness increase in frequency. If the behavior of a genetically "altruistic" individual benefitted all members of a population equally, regardless of the genotype of the recipient, selection would not favor the "altruist," and no change in gene frequency would occur. Only when the net benefit is applied selectively to the genotype of the bearer (usually through the bearer's own offspring) can the gene increase in frequency within a population. Consequently, no behavioral trait that benefits other genotypes more than its own (or equally with its own) can increase in frequency by natural selection *within* a population.

THE EVOLUTION OF COLONIALITY

Clumped versus Regular Dispersion in Blackbirds / What correlations exist between environment and social system? Can we identify patterns in the spatiotemporal distribution of resources that are associated with certain social systems? Comparative ecological studies of natural populations are necessary to answer these questions.

The work of Orians and his students on the ecology of blackbird social systems exemplifies this approach to social systems (Orians 1961, 1966; Willson 1966; Horn 1968). The red-winged (Figure 15–9) and tricolored blackbirds (*Agelaius phoeniceus, A. tricolor*), although very similar in appearance, are dramatically different in social organization.

This situation offers a natural control, for we can compare the social systems of these two species without any major complications due to morphological differences of the sort that commonly accompany large species differences in social organization. The case is of special interest because it shows that major changes in social organization can evolve in a relatively short time, geologically speaking (probably since about Pleistocene time), and that social organization is relatively sensitive to natural selection.

The principal differences between the two species are summarized in Table 7–2. The social system of the redwing is adapted for a dependable food supply that appears every year predictably at about the same time and persists over about ten weeks. The males, which are polygynous, devote considerable time to attracting females and keeping other males out but usually none to feeding the young of their females. The females are also territorial, though to a lesser degree. Both males and females must stay on or near their territories to defend them successfully.

The social system of the tricolor is apparently adapted for a food supply that appears unpredictably at various times, and varies in amount and duration. All individuals begin breeding simultaneously; the adult males, polygynous like the redwings, all begin territorial defense at the same time, and are joined by all the females. Consequently, there is little advantage to prolonging the period in attempts to acquire more females. Also, the males are free to feed their young and do so as much as the females. Freed from the necessity of prolonged territorial defense, the males and females can forage long distances from the colony. This explains the females' greater expenditure of energy in finding food for their young.

The critical environmental difference in the geographical range of the tricolor that selected for a nomadic, highly gregarious, synchronous social system was thought by Orians to be the unpredictable nature of the food supply in the main river valleys of central California, due to the variability in time and extent of spring flooding (before flood control) and perhaps also to locust plagues. Rich sources of food were available, but the times and places of their occurrence could not be predicted in advance. Nomadic flocks that traveled widely and could put off breeding until a rich food supply was discovered were better suited to such an environment. The breeding schedule of the tricolor is so sensitive to temporary abundance of food at irregular times that the species may even attempt breeding on a large scale in the fall (Orians 1960; Payne 1969*a*).

That the colonial type of social organization might be inherently more efficient for the harvesting of food supplies that are highly

TABLE 7–2 Comparison of social system and environment in two closely related species of blackbirds, *Agelaius* spp., in California. (Data from Orians 1961.)

	A. phoeniceus	***A. tricolor***
Geographic range	North America	California
Dispersion in marsh	regular	clumped, highly gregarious
Territory size of male	2,500–32,300 ft^2	35 ft^2
Massed feeding flights	absent	characteristic
Energy used in territorial defense*		
♂, per year	2.2	0.5
♂, per day	10.7	13.0
Period of territorial defense	10+ weeks	1 week
Energy used in feeding young*		
♀, per day	157.5	317
♂, per day	0	317
Feeding of young by male	unusual	important
Foraging area	in and near territory	30 sq. miles
Breeding schedule	regular, protracted, asynchronous	opportunistic, synchronous
Food availability (before civilization)	regular, predictable	irregular, unpredictable

* Energy estimate is expressed as the percentage above a value for resting.

clumped in time and space was demonstrated mathematically by Horn (1968; see also below). Unpredictably located sources of rich food would give an even greater advantage to coloniality. Horn concluded that the colonial dispersion pattern of Brewer's blackbird (*Euphagus cyanocephalus*) is primarily adaptive to the variable nature of its food supply.

An Energy-Budget Approach / In general, a regular dispersion of nests in all-purpose territories seems characteristic of species whose food is rather uniformly distributed in time and space. Conversely, a clumped dispersion, as in colonial breeding, seems characteristic of species or populations whose food is nonuniformly distributed spatially—food may be abundant in one place at one time and in another place an hour, day, or week later. Such a food supply may be predictable in amount but not in location. Not being required to defend a large foraging area, colonial individuals are free to go wherever food can be found.

Other things being equal, it can be expected that an individual will attempt to minimize its expenditure of energy in foraging and to maximize the amount of food gained per unit of energy invested in foraging (including travel time to and from the food). Mathematical considerations (Figure 7–4) show that it is more efficient (requires less travel time) for a fixed number of individuals to space themselves out and for each to forage within a small radius of its nest than it is for them to form clumps of two (or more) within a foraging area of wider radius but equivalent area and resources (Smith 1968). This holds only if the food is uniformly distributed. If food is concentrated unpredictably at various places and times, a clumped dispersion pattern, in which the colony is located centrally, becomes more efficient (Horn 1968), as shown in Figure 7–5. And Horn has shown that the main food of colonial Brewer's blackbirds (*Euphagus cyanocephalus*) was in fact distributed irregularly.

An advantage of the energy-budget approach to the evolution of social systems as expressed in the general theory is that the energy gained and lost because of various actions can be estimated quantitatively. It is possible in favorable cases to estimate how many calories a wild animal expends in each of a day's activities, by determining how much time he spends in each of his different activities and then multiplying each of these totals by a calorie requirement that has been determined experimentally, usually under laboratory conditions (Orians 1961; Verbeek 1964, 1972; Verner 1965*b*; Schwartz and Zimmerman 1971; Wolf and Hainsworth 1971; Stiles 1971; Collias and Collias 1971; Collias et al. 1971). Naturally, more calories per unit of time are required for some activities, such as flying, than for resting. Calorie income and foraging efficiency can also be estimated if the main foods and the feeding schedules are precisely known (e.g., Smith 1968; Wolf et al. 1972). It should eventually be possible to estimate the energy cost of territorial behavior and the energy gain in holding a territory, and to evaluate the cost/benefit ratio in terms of competition among individuals (as in Figure 7–3). Although calories are not easily converted to fitness, they do provide a useful common currency for comparing many costs and benefits of behavior. For a detailed review of energy-budget studies, Schoener (1971) is valuable.

In general, natural populations usually have dispersion patterns that accord with the above mathematical models of foraging efficiency: that is, dispersion correlates well with energy-budget considerations. But is this the whole explanation?

7–4 Energetics of sharing a territory in food-storing squirrels of the genus *Tamiasciurus.* A squirrel harvests the cone crop in its territory and stores it near its nest, which is centrally located. Assume that the cones before harvesting are uniformly dispersed in the territory and that a squirrel must make one round trip ($2r$) from the center of its territory to each point in its territory in order to harvest the crop. If two squirrels share a territory and share the work, the total distance traveled per squirrel—and hence the energy cost—is more by a factor of $\sqrt{2}$ than if each squirrel lived alone in its own territory half the size of the shared territory. (After C. C. Smith 1968.)

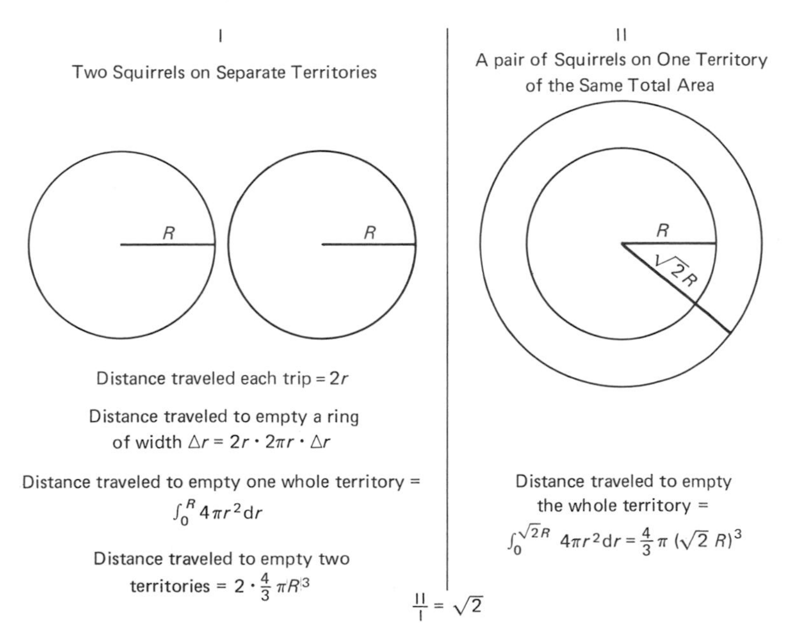

A Survival Approach / Another important factor in the reproductive success of individuals is the safety of the nest site, which is also correlated with dispersion pattern. Clumped nests are usually in safe places and not well hidden; widely dispersed nests are hard to find simply because of their dispersion, and they are often well hidden (*cryptic*) too. According to an alternate hypothesis, the clumping of

7–5 Energetics of foraging trips with two distributions of food (circles) and two distributions of nests (triangles). In a *stable food distribution,* food is equally divided among the 16 solid circles and is equally abundant at each site all the time. In a *moving food distribution,* the same total amount of food is present, but its location varies among the 16 open circles such that food is present at only one place at a time and is present at each open circle equally often. Each grid contains just enough food for the four nests. The shortest distance between two circles equals one unit of distance; *d* is the average length of trips from nest to food.

A. Food stable; nests spaced. Adults at each nest must fly only to the 4 nearest circles. Average round trip between nest and food is 1.42 units.

B. Food stable; nests clumped. Average round trip is longer than in A because adults must utilize all 16 food sites from one central location.

C. Food clumped; nests clumped. Same as B, but temporal order of utilization is different.

D. Food clumped; nests spaced. Average round trip is longer than in C because nests are not centrally located.

(From Horn 1968.)

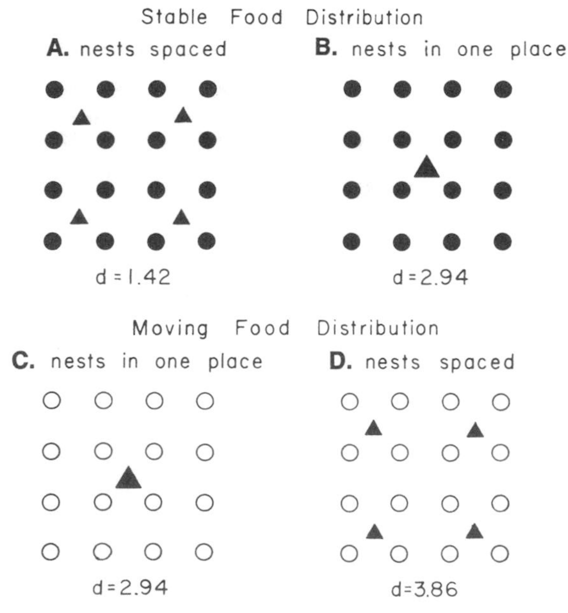

nests in some species is better explained in terms of the greater survival rate of eggs and nestlings raised in dense colonies than those raised in nests having a regular dispersion pattern. Survival might be higher because of two factors: (1) the inherent safety of the nest site (on an island rather than the mainland; in a tree rather than on the ground; on a cliff face rather than on flat land), and (2) massive common defense against predators by most colony members.

It has been argued (Lack 1966:279; Crook 1965) that in noncolonial species predation (rather than energy considerations) provides the principal selective agent favoring regular dispersion. That predation does select against clumping has been demonstrated experimentally using wild predators (Tinbergen et al. 1967). If ordinary chicken eggs are laid out in the grassy coastal sand dunes of England, crows (*Corvus corone*) are likely to see and eat them, even if they are painted to resemble gull eggs, which are cryptically colored. By laying the eggs out in various dispersion patterns, Tinbergen and his coworkers found that regular dispersion provided better protection against predators than did clumping. Although most of the bird species in the same study area have regular dispersion patterns, the most abundant species, the black-headed gull (*Larus ridibundus*), nests in dense colonies in spite of the depredations of crows, foxes, and other predators on adults, eggs, and young (Kruuk 1964). The species having regular dispersion patterns (e.g., skylark; Delius 1965) have widely dispersed terrestrial food supplies; the gull usually eats marine animals in local and transient concentrations.

Social Stimulation / Among colonial birds it has been observed in several species that reproductive success is higher in large colonies than in small ones (references in Tenaza 1971). Large colonies also tend to breed earlier in the spring than smaller ones. It was proposed by Darling (1938) that these differences were the result of greater social stimulation in larger colonies than smaller ones. However, it was found that smaller colonies often had a higher percentage of birds breeding for the first time than did large colonies. Since younger breeders generally breed later in the season than older ones and are typically less successful, the differences between large and small colonies have been attributed to the age factor, and social stimulation is not considered proven as an advantage of colonial nesting.

A third factor that may be responsible for the difference in breeding success between large and small colonies was identified by Tenaza (1971), as shown in Figure 7–6. Nests at the periphery of a colony should be more susceptible to predation than those in the center, because most predators enter the colony from the edge and encounter

7–6 Perimeter effect in a breeding colony. **A.** Each concentric hexagon represents the perimeter of a hypothetical colony. The numbers show the percentage of the total number of nests in each colony that occur on the perimeter. They show that the proportion of perimeter nests increases with decreasing colony size. **B.** Hypothetical curve based on observed differences in clutch size among colonies of Adelie penguins. Clutch size is reduced in smaller colonies, according to the hypothesis, because of the greater dangers of nests on the periphery. **C.** An adult Adelie penguin rolls a displaced egg back into its nest. **D.** A large Adelie penguin colony with a relatively small periphery. (From Tenaza 1971.)

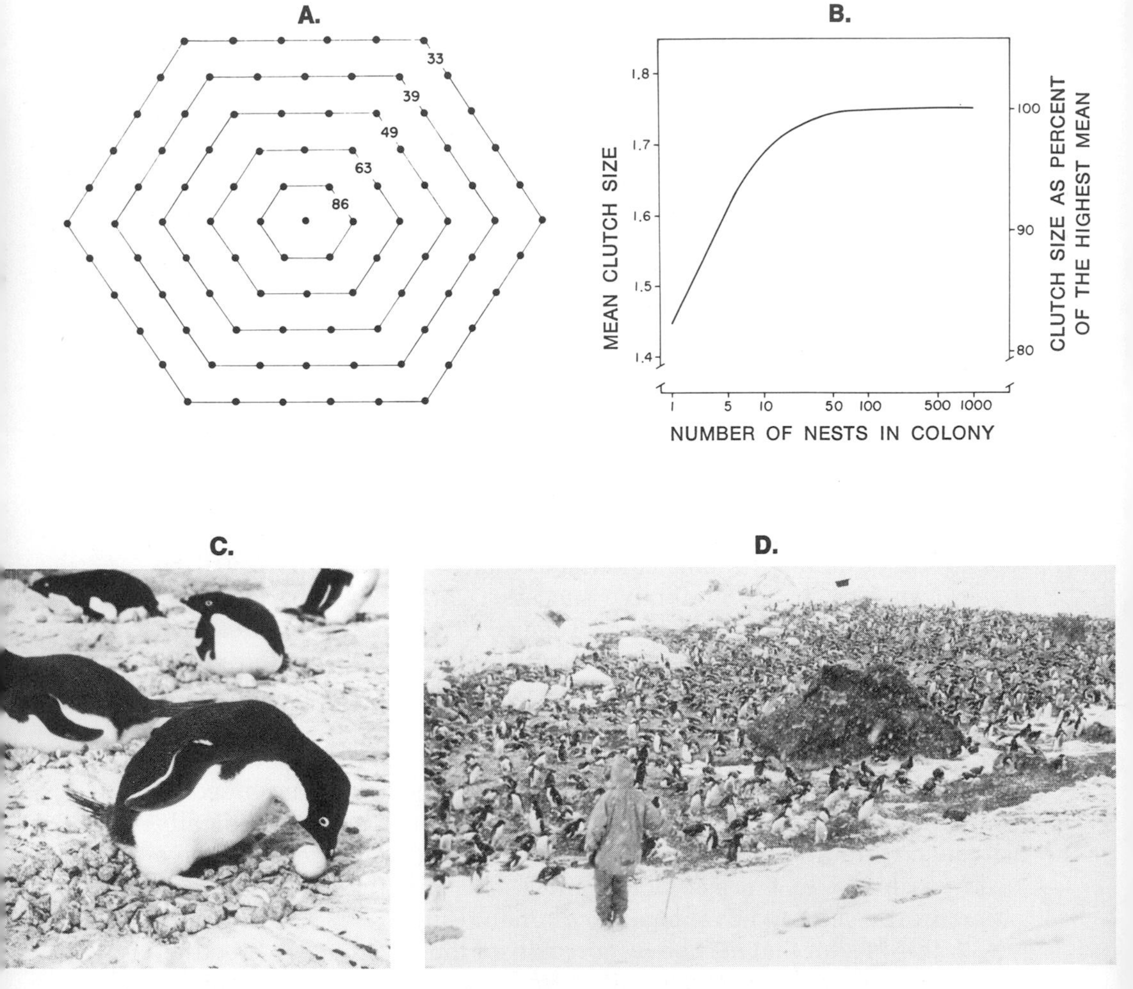

peripheral nests first. Since in small colonies the percentage of nests in the periphery is larger than it is in large colonies, this factor alone could cause a lower nesting success of small than large colonies. This is a variant of the "you-first" principle discussed below.

THE EVOLUTION OF SURVIVAL GROUPS

Some Benefits of Survival Groups / Some animals in a nonreproductive state react to others of their own kind by joining together to form a group, while in other species the response is avoidance or aggressive repulsion. As indicated in Chapter 5 (Table 5–1), the selection pressures favoring group-joining behavior in individuals are varied. In this section we will consider some of the ecological selection pressures that tend to make nonbreeding individuals clump together to form large and cohesive schools, herds, or flocks (Type IV in Table 5–1); the other types of groups in Table 5–1 are omitted here, but some of the principles discussed may apply to more than one type of group.

Since reproduction is explicitly excluded in the groups under discussion, survival is the key to fitness. Survival will depend in many species on the impact of predators and on the food supply. These two factors are reflected in the hypothesized selection pressures favoring group formation listed in Table 7–3.

A factor of probably universal significance in groups is the increased efficiency of *predator detection.* This was expressed colorfully by Galton (in Hamilton 1971*b*) in regard to African cattle:

"To live gregariously is to become a fibre in a vast sentient web overspreading many acres; it is to become the possessor of faculties always awake, of eyes that see in all directions, of ears and nostrils that explore a broad belt of air; it is to become the occupier of every bit of vantage ground whence the approach of a lurking enemy might be overlooked."

To be fair, one should admit that the group might also interfere with some aspects of predator detection; for example, the noise of the group might mask the sounds of an approaching predator even for the prey closest to the predator. An example of the importance of predator detection in the energy economy of the wood pigeon (*Columba palumbus*) was described by Murton (1968). In flocks of wood pigeons the birds can devote a relatively large proportion of their time to feeding, because the time each individual devotes to keeping an eye out for predators can be reduced. Isolated wood pigeons were observed to have such a low feeding rate due to the higher frequency of looking for predators that they were judged likely to starve in the midst of a food supply ample for flocked wood pigeons (see Figure 5–8).

Cooperative defense against predators is common in colonial breeding sea birds but rather uncommon in nonbreeding animals. The musk ox is a famous example: adult oxen form a ring with their horns facing outward against the predator, a wolf for example; the young stay within the ring. The mobbing of hawks and owls by small birds is a puzzling case of cooperative harassment, but whether its selective advantage lies more in defense or in aiding future detection or avoidance of the predator is debatable.

Confusion of a predator because there are so many prey individuals to choose from might occasionally occur, but this seems unlikely as a major advantage of clumping. The erratic behavior of individuals or flocks is no doubt important in confusing predators ("protean" behavior; Humphries and Driver 1967), but this is not an advantage unique to groups.

Facilitation of escape may occur because of the aerodynamic and

TABLE 7–3 Survival groups: some proposed selection pressures favoring their formation.

Selection pressures	Examples
I. Predator-related	
A. *Detection of predator* is quicker and more reliable by a group than by an isolated animal	
B. *Cooperative defense against a predator*	musk oxen forming a circle with heads out in defense against wolves
C. *Confusion of predator*	falcon attacking a flock of birds who hesitates while deciding which individual to attack
D. *Facilitation of escape from predator* by increasing speed and reducing energy expense of swimming or flying in close formation	
E. *Collective mimicry*	dense school of fishes might resemble a single large fish
F. The *cicada principle:* numbers of prey present are beyond the capacity of predators to consume even though prey are easily caught	17-year and 13-year cicadas
G. The *you-first principle*	schooling fish
II. Food-related	
A. *Cooperative foraging*	teamwork in running down caribou by wolves or wildebeests by wild dogs; cooperative fishing by cormorants
B. *Food detection.* Ecological enhancement of foraging efficiency: to find food, go where other group members go	

hydrodynamic advantages of certain positions in a group relative to others. Speed may be increased and energy conserved in this way (Lissaman and Schollenberger 1970).

It has been suggested that large, dense schools of fish might gain an advantage by their resemblance to a single large fish—a form of *mimicry*—thus scaring off some potential predators (Breder 1959). This fanciful idea has so far received little serious consideration, perhaps because field observers find no support for it.

Living in a large group may aid individuals of some species in *obtaining food cooperatively*. Wolves and African hunting dogs that run down their prey probably benefit from teamwork in tiring and attacking large mammals (Mech 1970; Estes and Goddard 1967; Van Lawick-Goodall 1971). Cormorants have been described as herding fish schools in such a way as to concentrate them in shallow water where they have less chance of escaping (Bartholomew 1942). Further examples are mentioned by Schoener (1971).

Food detection may also be more efficient in a group. Where food supplies are patchy and transient but rich, individuals may locate and utilize them more efficiently in a group. Individuals of many species apparently communicate with each other about the location of good food sources, probably in the main through mutual observation. Individuals in roosts, and large flocks of many species of birds in the nonbreeding season, probably benefit from grouping principally in this way (Hamilton and Watt 1970; Ward 1965; Zahavi 1971; Murton 1971).

The benefits of survival groups vary with the species under consideration, as suggested in the previous section. But certain properties of groups are quite general and probably benefit all or most nonbreeding groups not composed of close relatives. Two of these are the gluttony principle and the you-first principle.

Where the density of prey relative to the density of predators is very high, forming large groups tends to maximize the safety of individuals. Predators can harvest only as much as their stomachs can hold. Any remaining prey are safe. For prey it is better to live in an area where predators have full stomachs than where they are empty. I have called this the *gluttony principle* because it caters to the most gluttonous predators. Since predators (of the same species, at least) tend to space themselves out, prey that are highly clumped probably are at an advantage.

A corollary might be called the *cicada principle*. This was discussed from an evolutionary viewpoint by Lloyd and Dybas (1966). The 17- and 13-year cicadas in any one area appear in large numbers at intervals of 17 and 13 years, respectively, and for relatively brief periods in the season. There are virtually no adults in the intervening years. They emerge in such large numbers and so synchronously that the available predators are far too few to make much impact on the numbers of cicadas. The increase in the predator population due to a cicada year would be lost in subsequent years because of the absence of cicadas in those years. The cicada life cycle is summarized in Figure 7–7.

The *you-first principle* may be illustrated with the parable of the frogs and the snake (Hamilton 1971*b*). Imagine a small lily pond containing one water snake and a population of frogs perched around the rim on the shore at random intervals. Once a day the snake appears at some random point in the pond and eats the *nearest* frog. What dispersion pattern will result if each frog spaces himself so as to minimize his chances of being the frog nearest to the snake? The risk of a frog F being taken by the snake on a single day is given by the proportion of the circumference of the circle that is occupied by a line running from the point halfway between F and his right-hand neighbor to the point halfway between F and his left-hand neighbor. To minimize his chances of being selected by the snake, F has only to minimize the distances between himself and his neighbors. Thus with a great show of love for his fellow frogs, he moves between two of them that are closer together than his own neighbors—thereby increasing the chance that some other frog will be taken by the snake.

Even in two- or three-dimensional space it is intuitively clear that it is safer to have one of your fellows between you and the predator. Field observers of birds and mammals suggest that individuals prefer the center of a group. In wood pigeons the dominant individuals occupy the central position in a feeding flock (Murton 1968). In colonial nesting birds older individuals get the nest sites near the colony center where predation on nests is less than at the periphery (Ahlen and Andersson 1970; Tenaza 1971; Patterson 1965; Coulson 1968).

For the you-first principle to operate, the nature of the sensory cue that triggers approach is irrelevant. Any effective cue will do. It has been argued that in fishes the stimulus to school is basically that of seeking *cover* (Williams 1964). It is noteworthy that schooling is most conspicuous in species of fish that live in habitats lacking in cover. Bottom fishes typically hide in crevices rather than behind each other. A similar generalization can be made for mammals and birds: large, dense groups are characteristic of flat, open country where visibility is not broken by many trees or elevations, and where there are no good hiding places. In such habitats concealment is of less importance and other factors loom larger for survival.

Another factor relevant to the you-first principle is the ease with which a predator can capture its prey. A falcon or goshawk pursuing a bird can usually outfly it. A predatory fish feeding on a school can usually outswim it. Consequently, these prey species are more vulner-

7–7 Diagrammatic representation of ecological relations between 17-year periodical cicadas and their natural enemies. Above ground, significant numbers of adult cicadas appear and lay eggs only once every 17 years. Birds use the abundant supply of cicadas to feed their nestlings, which could conceivably reduce nestling mortality and hence increase the population of young birds in the autumn of that year, but this effect would not be expected to persist for 17 years. Similarly, the parasitoids, which must reproduce unusually well every seventeenth year at the expense of adults and eggs of periodical cicadas, presumably have life cycles of one year and depend on alternative hosts during the intervening years. One would expect a temporary buildup in the numbers of parasitoids, perhaps leading to a depression in the numbers of alternative hosts, with a return to some kind of equilibrium by the time of the next 17-year cicada emergence. Underground, we postulate that moles may gradually decrease their territory sizes (hence, increase their numbers) over a 17-year period, as the biomass of periodical cicada nymphs increases. With the emergence of the nymphs, this extra food supply suddenly vanishes for two months and is then replaced by myriads of tiny nymphs which, at that size, may be impractical as food for moles. Faced with this sudden crisis, we suppose, the moles may increase their territory sizes once more, whereupon the "surplus" moles would be driven into less suitable habitats. None of these hypothetical population changes in predatory species have been documented with numerical evidence. (Drawing by Marion Pahl. Legend and drawing from Lloyd and Dybas 1966.)

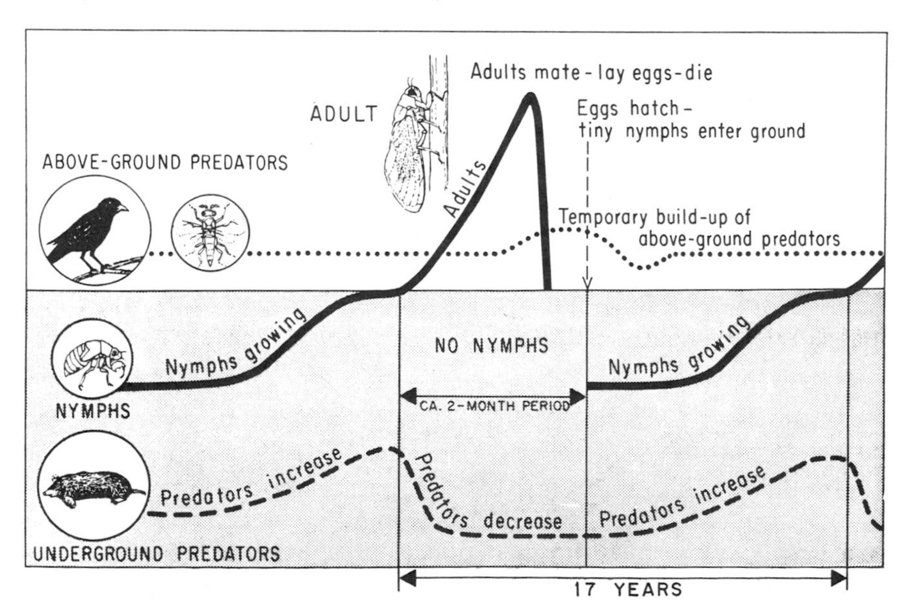

able and depend more on statistics than stealth to safeguard their survival. The group is usually conspicuous, so finding the prey poses no great difficulty for the predator. Certainly the sick and weak tend to be caught first, but this does not eliminate the role of chance.

If predators were more attracted to groups than to individuals, which seems likely, then part of the advantage of being in a group would be offset. But how much? It is known that wide spacing of stationary prey items results in a higher rate of predation than does clumping, when the predators are crows (*Corvus corone*) and the prey items are eggs (Tinbergen et al. 1967). But relatively little attention has been given to the general problem of optimal group size, or specifically, the ease with which predators detect groups of different sizes.

Perhaps the most interesting aspect of the frog-and-snake fable is the idea that group formation might conceivably be detrimental to the species while benefiting the individual. The you-first principle benefits an individual mainly at the expense of other individuals. Its net effect on the species might be to make the prey more conspicuous to the predator. This might be expected to lower the biomass, carrying capacity, or some other index of the probability of nonextinction. Could group formation be considered spiteful behavior (see Table 9–3) in the sense of Hamilton (1970, 1971*a*)?

THE EVOLUTION OF OVERLAPPING HOME RANGES IN RODENTS

Paradoxically, interest in behavioral aspects of population "regulation" in rodents has centered on those species that are among the least successfully regulated—that is, most variable—such as lemmings, voles, and others susceptible to a boom-and-crash type of population fluctuation. It is for these species that theories based on social stress have been most popular. Representative of this school is the position of Christian and Davis (1964:1560, 1550). They believed that "the behavioral-endocrine feedback system is important in the regulation of populations of rodents, lagomorphs, deer and possibly other mammals" and that the system "acts as a safety device, preventing utter destruction of the environment and consequent extinction." More generally, their position was that "within broad limits set by the environment, density-dependent mechanisms have evolved within the animals themselves to regulate population growth and curtail it short of the point of suicidal destruction of the environment." If true, this thesis would have fundamental implications for theories of the evolution of behavior. It would strongly imply that many of the characteristics of mammalian agonistic behavior have evolved through population ("group") selection, rather than conventional individual selection, a position also held by Wynne-Edwards (1962, 1963) but by few evolutionary biologists.

The validity of evolutionary hypotheses based on population selection is highly questionable, as discussed above; however, certain aspects deserve consideration here. The problem is that the adrenal-hypertrophy hypothesis postulates a physiological mechanism believed to be *harmful to the individual* at high population densities (because it hastens his death), but supposedly *beneficial to the population.* If adrenal hypertrophy is really harmful to the individual, it would be difficult to conceive how such a trait might have evolved through individual selection; consequently, because of the density-limiting effects of this syndrome, this hypothesis implies that population selection is the principal evolutionary mechanism.

Because the social environment of lemmings and other widely fluctuating species changes drastically with population density, it has been proposed by Chitty (1967) and others that individual selection brings about genetic changes in the populations that are associated with and influence the pattern of density fluctuation. Although genetic changes correlated with population density are known (Tamarin and Krebs 1969), there are still too few data with which to assess Chitty's hypothesis.

If individual selection were sufficient to explain these phenomena, the disadvantages suffered by genotypes predisposed toward the boom-and-crash pattern would have to be more than counterbalanced by certain as yet unknown advantages. Little thought or research seems to have been devoted to discovering such advantages. The following ideas are offered in hopes of providing some testable hypotheses based on individual selection.

According to the present hypothesis, the crucial difference in social organization between rodent species with stable and fluctuating densities is in the intermediate and low stages of population density. In the stable species, a normal survival rate of young in a given year provides more than enough individuals to saturate the available territories, thus giving rise in a single season to severe competition for space and dominance. In fluctuating, density-tolerant species, on the other hand, several years and several generations may be required to reach limiting conditions. Then, after it is too late, social restraints are insufficient to stabilize the population and it crashes.

The critical difference may be in the number of generations that occur during periods when food and space are plentiful, and in the number of young born then. The more generations raised under these conditions, the greater will be the selection for rapid reproduction and against density intolerance *during this period.* This amounts to selection for a "loose communal type" of social organization (Eisenberg 1963, 1965) that, because of its weak restrictions on new individuals entering the social order, is prone to boom-and-crash population fluctuations. Selection against high reproductive rates would occur only at peak densities, when little reproduction occurs.

A mathematical model for selection of similar life-table phenomena in a periodic environment has been given by MacArthur (1968). Figure 7–8 plots a fitness function, r_{max}, against environmental quality. This model can be adapted to the present context by assuming high population density to represent a low level of environmental quality, and low density, a high level of environmental quality. The figure shows that a genotype that is less fit under average, steady-state conditions may be more fit in a fluctuating population over an extended number of generations, by virtue of its superiority during times of low density and rapid population increase. A density-tolerant, nonterritorial genotype might achieve a net advantage under conditions of cyclic variations in population density. Therefore, it is hypothesized that the advantage of this genotype at low densities outweighs its

7–8 A model for the evolution of density tolerance and high r_{max} in a population of rodents subject to recurring booms and crashes. The density-tolerant genotype *T* has a higher birth rate (*b*) under conditions of low population density than does the density-intolerant (more territorial) genotype *I.* But at intermediate and high densities, *I* is superior. Despite the advantage of density intolerance at medium and high densities, if the population fluctuates between conditions of high and low density *T* can have the overall net advantage, as indicated by the chords of the respective curves. (From MacArthur 1968.)

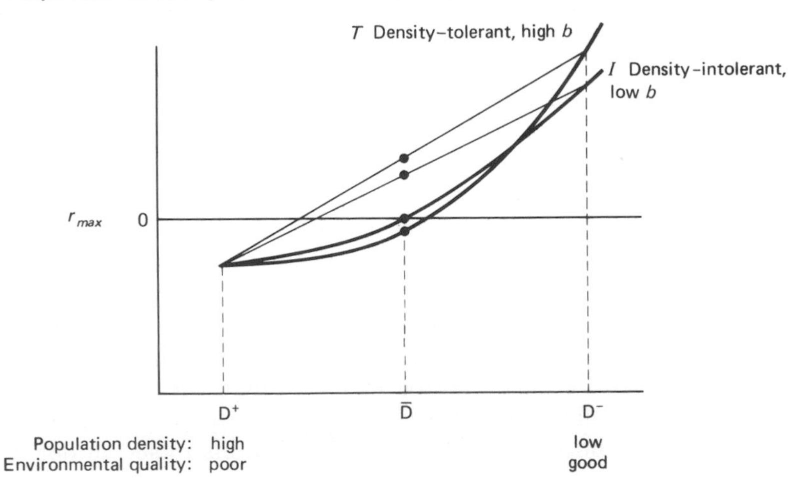

disadvantage at higher densities if the population undergoes cyclic fluctuations. The susceptibility of these individuals to various causes of death at high population densities can be interpreted as the price they have to pay for superior reproductive rates at low densities.

CONCLUSIONS

The evolution of diversity in vertebrate social systems can be explained adequately using concepts based on individual selection and kin selection. Population selection (group selection) cannot be eliminated as a possibility in at least some cases, since it is mathematically possible; but there is no compelling evidence from real populations implicating it in the evolution of behavior. The general problem of the evolution of diversity in vertebrate social systems can be divided into a set of problems concerned with various patterns of spacing out individuals and another set concerned with grouping of individuals.

Spacing is usually mediated by agonistic behavior and is generally concerned with competition for a necessary resource that is in short supply, such as food, females, or cover. Grouping has a variety of advantages, but in nonbreeding seasons it seems to be mainly concerned with advantages in survival (rather than in competing for a resource to be used in reproduction). The most conspicuous manifestation of spacing behavior is territoriality. This risky, energy-burning behavior benefits individuals in various ways, depending on which resources are in short supply. If a critical resource is not in short supply or is not defendable economically (in terms of time-and-energy budgets), it should not be defended. The diversity of territorial types (e.g., feeding, mating, nesting, all-purpose) is explainable in terms of which resources are critical for a given species and which are economically defendable.

The forming of groups by individuals in nonbreeding seasons has a variety of consequences that are beneficial in terms of individual fitness. Two of these that may be of general importance for many species can be designated the cicada principle and the you-first principle. The number of individual cicadas appearing all at once is much too large for predators to consume. Some prey are taken by predators, but most escape being captured because the predators soon reach the upper limit of the number of prey they can put in their stomachs. The you-first principle implies that an individual's chance of being singled out for capture is less if it is a member of a large group than if it is alone or in a small group. This effect depends on the ability of prey to reproduce faster than their predators.

The populations of certain species of rodents, especially arctic populations of lemmings, exhibit a curious instability in which large numbers of individuals die in cycles of population growth and decline. The persistence in these populations of genotypes that are prone to death at certain phases of the population cycle is difficult to explain in evolutionary terms. It is hypothesized that the reproductive advantage of density-tolerant genotypes at low and medium population densities compensates for their susceptibility to death at high densities.

8 Mating Systems and Sexual Selection

This form of selection depends, not on a struggle for existence in relation to other organic beings or to external conditions, but on a struggle between the individuals of one sex, generally the males, for the possession of the other sex. The result is not death to the unsuccessful competitor, but few or no offspring.

Charles Darwin, 1859

THE COURTSHIP OF animals has always fascinated man, and many questions have been raised about animal mating behavior. Why are some species monogamous and others polygamous? Why do males fight over females in some species but not in others? Why are the courtship rituals of some species so bizarre and conspicuous, while other species accomplish the same job in a much more prosaic way? And what factors influence the choice of a mate?

The major trends in the study of mating systems and sexual selection were begun with a brief mention by Darwin in 1859 (see Darwin 1928) in *The Origin of Species*. There, he wrote about what he termed *sexual selection*. He developed this theme fully with respect to the animal kingdom in *The Descent of Man and Selection in Relation to Sex*, published in 1871. Nearly three-quarters of this lengthy book was devoted to the courtship of animals. Progress during the hundred years since Darwin is summarized in Campbell (1972).

TYPES OF MATING SYSTEMS

Those aspects of a species' social organization that determine the ways in which males and females come together for breeding are referred to as its *mating system*. The principal types of mating systems are shown in Figure 8–1.

In a *monogamous system* each breeding adult is mated to only one member of the opposite sex. Monogamy is rare among mammals and most other animals, but over 90 percent of the species of birds in the world are monogamous (Lack 1968:149). *Perennial monogamy* is a system in which individuals mate for life, or at least for several years, with the same partner, and in which at least some sort of loose association between the sexes in the pair is maintained even in the nonbreeding season. Swans, certain geese, certain cranes (*Grus*), kittiwakes, probably certain gibbons (*Hylobates* spp.), and a few other species are of this type. *Seasonal monogamy* characterizes species that are monogamous during a breeding season, but in which males and females separate during the nonbreeding season. Most small migrant birds are of this type, since it is presumably difficult and of doubtful advantage for two small birds to stay together throughout a long migration of hundreds or thousands of miles. Such species have a strong predisposition to return to the same breeding place in successive years, and this may result in the pairing of the same individuals in successive years. Since many small birds have a life expectancy of only two or three years, it would be difficult to distinguish perennial from seasonal monogamy in these species.

Polygamy is a system in which an individual has two or more mates, none of which are mated to other individuals. In the strict sense it implies some sort of a bond, whether by dominance or attraction, between the mates. The bonds may be successive, as in serial

8–1 Some types of mating systems in vertebrates. Length of bar indicates duration of pair bond. Each bar represents a female.

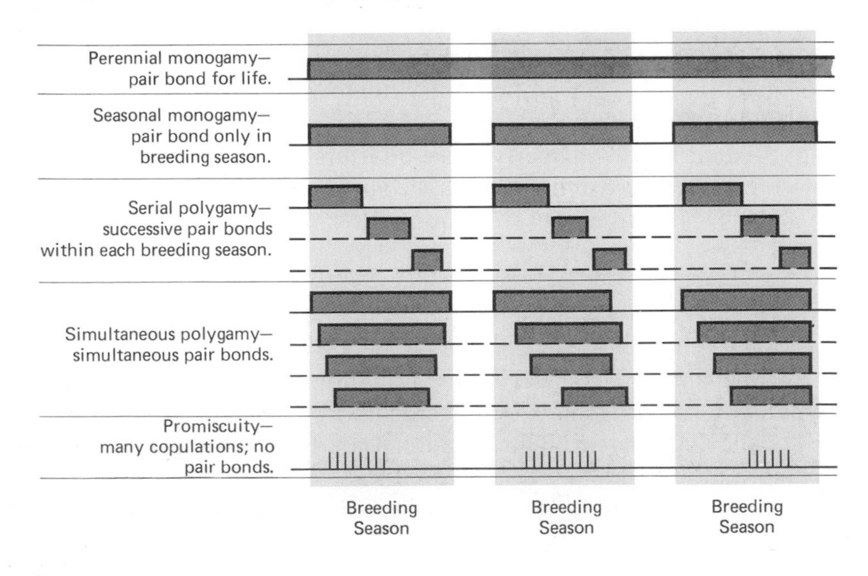

polygamy, or simultaneous, as in simultaneous or harem polygamy. *Serial polygamy* (or serial monogamy) implies that a male can have several mates in succession in a given breeding season, but no more than one at a time. It occurs in the pied flycatcher (*Ficedula hypoleuca*). A pair bond is formed and then broken with each successive female.

Polygamy can be divided into two types. In *polygyny*, the most common, one male mates with two or more females. This is the case in most mammals and in about 2 percent of the world's bird species (Lack 1968:150). In *polyandry*, one female mates with two or more males, but not vice versa. It is found spectacularly in the jaçanas (Jenni and Collier 1972). In these species the female is the colorful, dominant, and more territorial sex. After a male is courted and won by the female, he incubates the eggs while the female attempts to win more males for successive clutches.

In *promiscuity* there are no pair bonds; males and females copulate with from one to many of the opposite sex. This system differs from polygamy in that no one individual has exclusive rights over any individuals of the opposite sex (Figure 8–2). It is found in 6 percent of the world's bird species (Lack 1968).

The above types of mating systems have been described in simple terms. In fact, a species may have a *mixture* of mating systems. Different individuals or different populations within a single species may have different mating patterns. For example, a worldwide survey of mating systems in human populations reveals perennial monogamy, serial monogamy, polygyny, polyandry, and promiscuity. To categorize the human species into any one of these types would be a gross oversimplification.

8–2 Types of polygamy and promiscuity. Lines indicate mating bonds (polygyny, polyandry, monogamy) or copulations (promiscuity). In some systems some males may be denied females because the other males have taken all the females (in polygyny) or because of an unbalanced sex ratio (in monogamy); excluded males (or females, in polyandry) are designated with a dashed line.

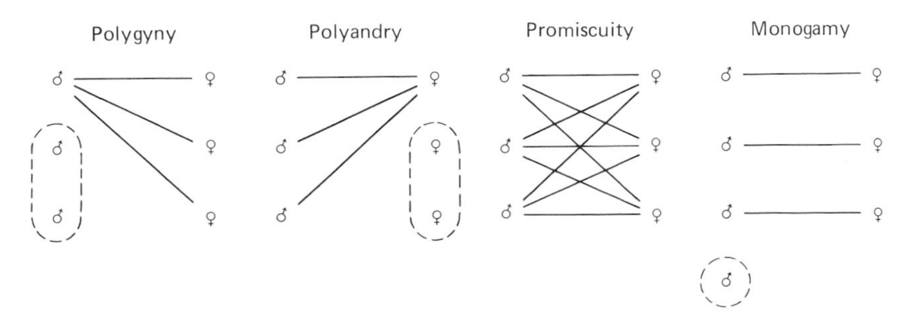

The mating systems defined above cover the familiar cases among normal, bisexual animals. Various other specialized mating systems exist, and a few should be mentioned because they pose some interesting problems. Individuals of some hermaphroditic species may simultaneously carry functional male and female organs and may be capable of self-fertilization, as in the fish *Rivulus marmoratus* (Harrington 1961). The related case in which an individual is first a female and then a male, both structurally and behaviorally, is known as *protogynous hermaphroditism.* Its behavioral features have been described for the cleaner fish, *Labroides dimidiatus* (Robertson 1972). In this species the male dominates a harem of three to six mature females, who maintain a linear dominance hierarchy. When the male dies or is removed, the largest and most dominant female begins to transform into a male behaviorally after a few hours and may complete the transition in several hours. Male spawning behavior then appears in 2 to 4 days and sperm can be released within 14 to 18 days. In this system, new males are produced only when needed. Still other "mating" systems occur in unisexual species. Among certain fishes (e.g., *Poeciliopsis;* Schultz 1961) and salamanders (e.g., *Ambystoma;* Wilbur 1971), males are needed only to supply sperm to initiate development; they contribute nothing to the genome of the offspring and must belong to a related species (a form of social parasitism).

SEXUAL SELECTION

These various mating systems differ in their potential for rapid evolution of mating behavior. Their differences are mainly (but not only) a function of the proportion of males in the population failing to find a mate, or to be more to the point, prevented from mating. In a monogamous system nearly all males find females if they look long enough and if the sex ratio is even. In a polygamous system with an even sex ratio, for every extra female that one male gets, another male must be deprived. The larger the harems of individual males, the greater the fraction of males in the population that cannot mate at all. Promiscuity has similar but more extreme effects. Since males deprived of mates leave no offspring, it is clear that selection works against them in the most extreme fashion. The differential production of progeny by different genotypes as a result of competitive mating is known as *sexual selection.*

Darwin distinguished between "sexual" and "natural" selection. This has caused some confusion because sexual selection is certainly natural, as opposed to being artificial, and because Darwin's sexual selection and natural selection are both forms of individual selection.

(Some authors have tended to equate Darwin's natural selection with modern individual selection, but this practice is unsatisfactory because Darwin's sexual selection is also individual selection.) The dichotomy recognized by Darwin seems to be real, but his choice of terms to express it is misleading in a modern context. His dichotomy can be expressed by the terms *sexual selection* and *nonsexual selection*—either of which can be natural, if occurring in nature, or artificial, if arranged by man.

To summarize: in the modern context sexual selection is a special case of individual selection that deals with the evolutionary consequences of competitive mating; it can be either natural or artificial.

The behavioral and morphological traits (e.g., courtship behavior and structures) that sexual selection is invoked to explain have other consequences in addition to influencing competitive mating. Consequently, one must be cautious in ascribing aspects of courtship to sexual selection. For example, sexual displays may bring individuals together that have been dispersed over large areas; or cause mating to occur in or near a nest; or synchronize the physiological and behavioral events between members of a pair *after* pair formation; or synchronize the breeding activities of a colony that depends for protection against predators on simultaneous defense by the colony members. Similarly, structures and color patterns used in courtship displays may also be used in threat displays between males. These observations are simply manifestations of the fact that, as is usual in genetics, a single gene affects more than one character or trait, and all characters are likely to be affected by more than one gene. Nevertheless, this does not prevent us from analyzing the relationships between certain genes and one type of effect, namely, influence on competitive mating.

TYPES OF SEXUAL SELECTION

Darwin (1871) recognized two main processes involving competitive mating: (1) the influence of male *dominance* and (2) the influence of female choice, or "sexual *preference*." This division has been recognized by most subsequent authors, but it should be pointed out that the categories are not mutually exclusive and that combinations of the two often occur.

MALE DOMINANCE AND COMPETITIVE MATING

Some species utilize agonistic behavior in competing for mates. Such species are characterized by conspicuous fighting among males for females. Darwin believed that the evolution of special weapons used in fighting could be explained by their use in competing for mates. Not all competition involves overt fighting; threat display may also be important. The antlers and horns of deer, sheep, goats, and other animals were thought by Darwin to be explainable on the basis of their usefulness in fighting. But Geist (1966*a*, *b*) has argued, rightly I believe, that their main value for many species is in their psychological effect on opponents—as threat or intimidation—by indicating the status of their bearers.

The Fur Seal / Perhaps the most conspicuously polygamous mammal on earth is the fur seal (*Callorhinus ursinus*). Since it is such a good example of the male-dominance type of sexual selection, we shall look into its mating system in some detail (Bartholomew 1970). Fur seals range over the northern Pacific Ocean in the winter and breed mainly on the Pribilof Islands in the Bering Sea and the Kurile, Robben, and Commander Islands. As the breeding season approaches in late spring the males come ashore for the first time, and establish territories along the edge of the water in places where it is easiest for seals to pull out of the water and rest (Figure 8–3). Later, in June and July, the females come ashore in the territories of the males to have their pups and mate. About 90 percent of females six to ten years old found on the open sea are pregnant; slightly fewer in the four-to-six-year-old group. The pups are born after the female has been ashore from zero to three days. From four to seven days later the mothers come into estrus, which lasts about one day. After copulation they alternate five-to-ten-day trips to sea with one-day visits to nurse their young ashore.

The males lead a more strenuous life. The holding of a mating territory by a male is something of a marathon feat. The strongest and most mature males hold the best territories near the water's edge where the females prefer to rest. They defend their territories day and night for up to two months without food. They fight fiercely and frequently die; the mortality rate of territorial males is three times that of females. Males maintain harems within their territories. When a female attempts to leave the territory of one male and enter another's, she may be picked up in the teeth of the male and forcibly returned to her harem. Males may try to steal females from neighboring territories, and sometimes there is a tug-of-war between two males with a female in the middle.

The harems of male fur seals vary in size up to a maximum observed of 153, with a mean of 16 ($n = 19$). Mating activities within the harem of one male last from 18 to 41 days, with a mean of 31 days. In the territories most populated with females there is a high turnover of males.

8–3 A fur seal mating ground. (Photo by Robert T. Orr. From Orr 1970. *Animals in Migration.* Macmillan.)

As with many polygamous species, *maturity is delayed in the males;* male fur seals reach full maturity only at seven years, while the females are mature at the age of three to four years. *Sexual dimorphism* is also pronounced, with the males being more than two times larger than the females. Males play no part in parental care and even show callous disregard of their own pups.

Sexual dimorphism is even more pronounced in the elephant seal (Figure 8–4), which was studied by Bartholomew (1952). In this species, place-independent dominance hierarchies are established where the females come ashore to give birth and to mate, but the males do not forcibly prevent females from leaving their harems. Here again the *dominant bulls perform most of the matings;* for example, in an extensive observation period on an island off California 6 percent of the 71 males inseminated 88 percent of the 120 females (LeBoeuf and Peterson 1969).

8–4 Elephant seals. (Photo courtesy of B. J. LeBoeuf.)

Red and Roe Deer / The social organization of the fur seal outside of the breeding season is rather simple, with males and females leading a solitary life on the open sea. In some other species, however, a polygamous mating system may also influence social organization outside of the breeding time. The red deer (*Cervus elephas*), which was the subject of a classic study in 1937 of behavior and ecology by Darling (1964), illustrates this well, especially when contrasted with the roe deer (*Capreolus capreolus*), which tends to be monogamous (Table 8–1). The red deer is a large species that inhabits northern Europe and Asia and resembles the North American wapiti or elk (*Cervus canadensis*) in its large antlers (Figure 8–5). As the autumn rut, or mating season, approaches, the males begin to round up herds of females that become their harems. Males compete among themselves for the females with much fighting and threatening, making use of their large body size and impressive antlers. The strongest and most dominant males reserve for themselves the most females. Males from which antlers were removed experimentally were less successful in obtaining females (Lincoln 1972). After the rut is over, males and

TABLE 8–1 Comparison of social systems of red and roe deer in western Europe. (After Darling 1964.)

	Red deer	**Roe deer**
Mating system	polygamous	monogamous
Habitat	open country	thick brush
Sexes	separate	together all year
Nonbreeding male groups	aggregations in favorable areas may be large	rare; small
Nonbreeding female groups	socially integrated; group size to over 100	absent
Harems	large	absent or small
Maternal care	to 3rd year	to first year
Male aggression	only in rut	all year
Size	large (95–220kg)	small (15–27kg)
Antlers	large	small

8–5 Red and roe deer.

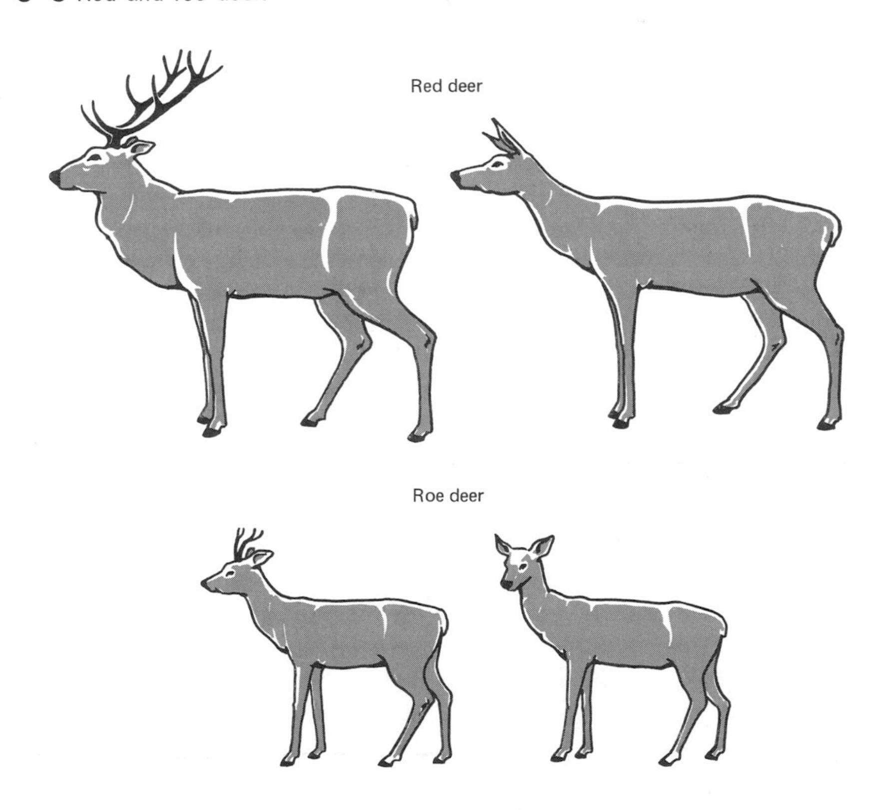

females retire to largely separate home ranges. Males lead an independent life and do not form stable, integrated groups. Females, on the other hand, form family groups that are well integrated and organized, with a "benevolent" leader and a dominance hierarchy maintained with slaps of the forefeet at those rare times when it is necessary. The young do not mature for three years, and they stay with their mother for this time. The polygamous system and the absence of a pair bond between male and female red deer allow unisexual groups to form. In the males these have no unifying force and amount to mere coincidental aggregations, but in the females the bond of family (and the absence of disruptive male aggression) keeps individuals together over a protracted period and allows leadership and cooperation to become important.

The roe deer (Figure 8–5) is mainly monogamous and does not usually form harems. Sexual selection is less intense, body weight and antlers are small. The male and female are together throughout the year, and the male jealously dominates the female and drives off rival males. Parents care for their young only one year.

The pair bond in this monogamous system persists throughout the year, preventing the formation of large unisexual groups. The short period of parental care is not conducive to formation of large family groups.

Other Mammals / Among mammals the role of aggressive male behavior tends to be more important than that of female preference. Mating systems and social organization in mammals have been reviewed by Eisenberg (1966) and Ewer (1968). In solitary species males leave their territories to find females in heat, and the dominant males probably achieve the most copulations, as in domestic cats and squirrels. In the large grazing mammals, harems are common (Ewer 1968). Individual males may either defend a (stationary) territory containing a group of females, as in the vicuna, Grant gazelle, or waterbuck (*Vicugna vicugna, Gazella granti, Kobus ellipsiprimnus*), or they may defend a group of females that they herd together but that is not restricted to a fixed territory, as in the impala, blue wildebeeste, and horse (*Aepyceros melampus, Connochaetes taurinus, Equus caballos*). In other species a group of females may be shared by a group of males, with mating access to the females determined mainly by dominance relations among the males, as in chacma baboons, prairiedogs, and perhaps wolves (*Papio ursinus, Cynomys ludovicianus, Canis lupus*).

General Characteristics / Species in which sexual selection of the male-dominance type is believed to be important share several characteristics that may be summarized as follows. (1) Aggressive conflict among males is unusually frequent and conspicuous during the mating season. (2) Females are clearly the object of aggressive competition in most cases, especially when food and cover can be eliminated from consideration because of the nature of the area defended. (3) The variance in number of females per male is unusually high; in other words, a few males win many females while many males win none or few. (4) Success in acquiring females is correlated with dominance over males, and female preferences for individual males are either absent or difficult to detect. (5) Pair bonds tend to be absent or brief and weak. (6) Sexual dimorphism in structure and behavior is extreme, especially in features of the male which help him win contests with other males. (7) Full physical maturation comes a year or more later in males than females. (8) Mortality rates in males are likely to be higher than in females, at least during the mating season. (9) There are likely to be more females than males among adults. This may be caused by the delay in maturation of males relative to females and by the greater mortality rate of males. (10) There is little or (usually) no paternal care of offspring.

MALE ADORNMENTS AND FEMALE CHOICE

Darwin probably overemphasized the role of female choice in the evolution of bizarre plumages in birds and other attractions serving in male courtship behavior. Because of his impression that many male and female birds remained unpaired despite the availability of mates, he concluded that these unpaired individuals were simply too fussy; he apparently believed that they were not pleased with the appearance of the available potential mates so they did not mate at all (a trait unlikely to evolve through individual selection). With the benefit of many more years of study we now know that whatever surplus might exist probably is the result of territoriality, an unequal sex ratio, or a shortage of nest sites, rather than fussiness (see Chapter 6). This does not mean, however, that the idea of genetically correlated sexual preferences need be rejected. Nevertheless, the idea that males have evolved "ornaments" through their effects on females has aroused much skepticism, and caution is in order (cf. Bastock 1967; Faugères et al. 1971).

The cases that have been interpreted as resulting from sexual selection due to female preference may be divided into three types, differing in the types of territories defended and the relative importance of dominance in the mating system. These are (1) isolated mating territories, (2) communal mating grounds, and (3) all-purpose territories.

Isolated Mating Territories / The argument for the role of female choice in the evolution of unusual displays and plumages is perhaps nowhere stronger than in regard to some of the birds of paradise, bowerbirds (Gilliard 1969), hummingbirds, and other promiscuous species that station themselves solitarily in the forest. Since the mating stations are isolated, inter-male aggression does not prevent females from exer-

8–6 Displays of males of various species of birds of paradise. **A.** Twelve-wired bird of paradise (*Seleucidis melanoleuca*). **B.** Lesser bird of paradise (*Paradisaea minor*). **C.** King bird of paradise (*Cicinnurus regius*). **D.** Magnificent rifle-bird (*Ptiloris magnificus*). **E.** Superb bird of paradise (*Lophorina superba*). **F.** Magnificent bird of paradise (*Diphyllodes magnificus*). Lines indicate reported hybrid combinations. (From Johnsgard 1967.)

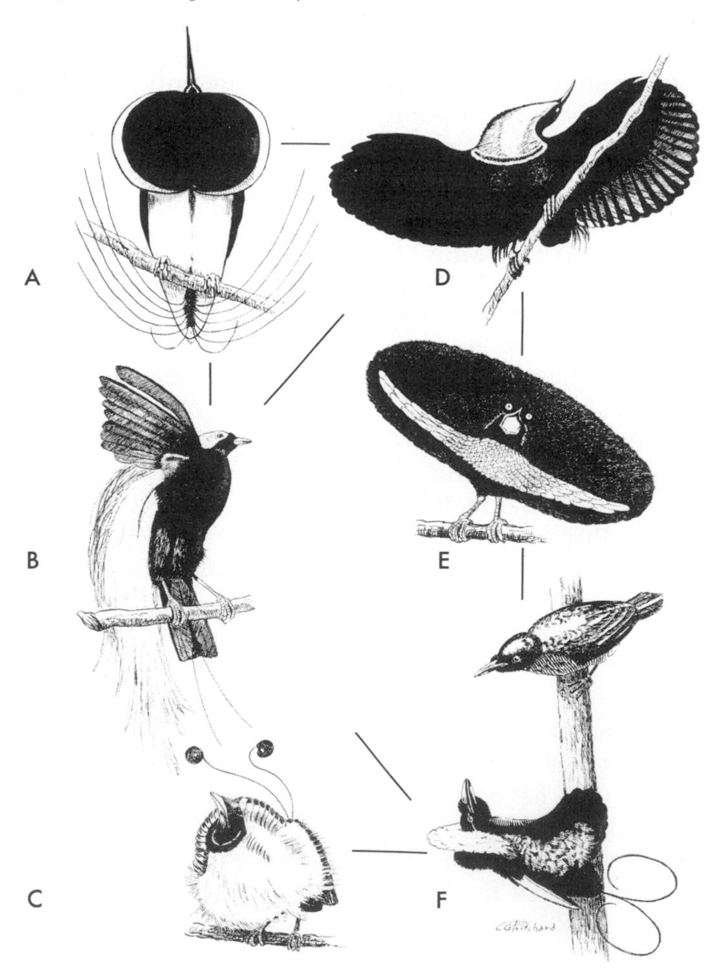

cising their choices. Simply to see the males of these species displaying is to be convinced that some unusual form of selection must be operating (Figure 8–6). Under these conditions the decision of a female to copulate with a particular male may depend entirely on his appearance and behavior toward females. If females visit several males while coming into breeding condition, the competition to impress the female may theoretically be quite intense. In the ruffed grouse (*Bonasa umbellus*), where the displaying males are dispersed widely in woodland and a female cannot survey all the local males at once, it has been shown by attaching radio transmitters to females that they visit most of the local males at their display grounds (Brander 1967). Similarly, telemetered observations have shown that female black grouse (*Lyrurus tetrix*) visit several leks. These grouse are promiscuous rather than polygamous, in the strict sense of the terms. Even when mating territories are isolated and solitary, the role of male aggression is not completely eliminated. The number of optimal display sites may be limited, and competition for them might be important.

Communal Mating Grounds / Communal mating grounds are traditional gathering spots where males come in the breeding season to establish small mating stations next to each other and where females come for copulation. These gathering spots attract males and females from a wide area and are often situated on particularly prominent sites where the groups have a better chance to attract females. At most communal mating grounds the combined effect of all the males displaying or calling there can be seen or heard from a much greater distance than would be the case for a single individual (e.g., grouse; Hjorth 1970); this seems to be part of the reason for the widespread occurrence in various groups of birds and mammals of this type of mating system. Attracting females is most important in nonmonogamous species, and a large group of displaying males is probably more attractive than a small group or a lone male. In agreement with this hypothesis, it has been observed that more females per male nest in large colonies of the village weaver (*Ploceus cucullatus*) than in small ones (Collias and Collias 1969).

The term *lek* is often applied to a communal mating ground. It was first used for the communal mating ground of the ruff (*Philomachus pugnax*), a species of sandpiper found in Eurasia. The term has come to be associated with this particular mating system, and we now use it for other species having the same type of system. These include many members of the grouse family (Tetraonidae; Hjorth 1970), at least one cotinga, some pheasants, some hummingbirds, some birds of paradise, some bowerbirds, some weaverbirds, and some manakins

(Lack 1968). Among mammals with communal mating grounds the lek type with an element of female choice is rare, but has been observed in the Uganda kob (*Adenota kob;* Buechner and Schloeth 1965; Leuthold 1966). Leks are sometimes known as arenas, and communal display is then called arena display. These species have some of the most highly specialized displays and courtship devices to be found in the animal world. They even include cases of apparent cooperation between males to attract females, such as the associations between the independent and satellite males of the ruff (Hogan-Warburg 1966; discussed below), and the cooperative jumping displays of pairs of male blue-backed manakins (*Chiroxiphia pareola;* Snow 1963).

Sage Grouse and Other Lek Species / A North American species having a lek type of mating system in an extreme degree is the sage grouse (*Centrocercus urophasianus*), which has been studied by Scott (1942) and Patterson (1952). This species inhabits the high sagebrush plains of western North America. It is highly dimorphic sexually; the female is brownish and rather nondescript, but the fully adult male is striking because of his much larger size, his yellow air sacs that make a popping noise when he is displaying, his white breast, and his spiky tail feathers. The leks vary in the number of males taking up mating territories there. In one sample of 28 leks the number of males varied from 14 to 400 or (rarely) more, with a mean of 78 males per lek. They are distributed over the plains at a density of about 1 lek per 6 square miles. A large lek may be divided into as many as five centers of mating activity. At each of these there is a dominance relationship among the males, with the dominant male performing the large majority of matings; for example, in 154 observed copulations 114 were performed by a dominant cock and only 40 by the several subordinate cocks. As is typical of highly polygamous and promiscuous species, the maturity of the male sage grouse is delayed, and males do not care for the young, as the males and females live apart for most of the year.

On a lek a mixture of territorial and dominance effects contributes to the social organization of the males. Robel (1966) showed that territory size was correlated with success in mating; by observing six marked males in one year he determined that 70 percent of the 21 observed matings were performed by the male with the largest territory. That the females of lek species take the occasion of the gathering together of the males to look them all over is suggested by observations of females moving around the lek and between leks.

All-purpose Territories / In migratory birds with all-purpose territories males typically return to the breeding grounds earlier than females; this gives the arriving females their choice of males and of territories. This group of species includes both polygynous and monogamous types. In the former category are the red-winged blackbird (*Agelaius phoeniceus*), yellow-headed blackbird (*X. xanthocephalus*), dickcissel (*Spiza americana*), long-billed marsh wren (*Telmatodytes palustris*), and many others. The number of females mated to a male in these species tends to be limited by the space available for nesting within the male's territory or by the male's nest-building activities. These depend heavily on the male's ability to defend suitable areas. Sexual dimorphism may be present or absent in these species, but it is not usually extreme. Maturation may be delayed in males or not. And some populations of a species may be polygamous, some monogamous, as with the long-billed marsh wren (Verner 1964). Because these species have smaller harems, the intensity of sexual selection should be much less than in the preceding two groups; this probably helps explain the reduced frequency and degree of sexual dimorphism and delayed male maturity.

In monogamous species females may have a choice of male in much the same way. Females in search of a mate are mobile, while unmated males are restricted to their territories. In these species fewer males are prevented from breeding by territorial exclusion than in polygamous species, so the action of sexual selection is probably less intense. Darwin felt that early breeding was a component of mating fitness in monogamous birds, but it is difficult to demonstrate a behavioral effect that might be due to the role of the female in choosing early breeding males (see, however, O'Donald 1972).

General Characteristics / Species in which sexual selection of the female-choice type is believed to be important share many general characteristics with those in which sexual selection of the male-dominance type predominates. The two types differ mainly in the frequency and violence of male-to-male encounters and in the types of adaptations that have evolved for obtaining females. In the male-dominance type, signals and structures useful in fighting against males have been favored, while in the female-choice type signals and structures are specialized mainly to impress females. In many species of birds anti-male and pro-female functions seem to be served by the same structures, for example, in the red epaulets of red-winged blackbirds. In these cases the relative weighting of these two functions can sometimes be determined by experiment. In cases where experiments are lacking, such as the bowers of bowerbirds and the plumes and displays of birds of paradise, the displays and structures are so extreme that it is difficult to avoid the conclusion that their main function is to attract females. Experiments on the factors determining female choice are discussed below.

ECOLOGY OF MATING SYSTEMS

Ecological Models of Sexual Selection / Although a mating system is imposed on a species partly as a secondary consequence of the spacing behavior of the species, how many females a male typically has is to some extent a separate problem. Since it is generally to the advantage of the male to have more than one mate, the selection pressures acting on the females for or against polygamy are critical (Verner 1964). The nature of the environment in which the species lives may be an important influence on these selection pressures. A generalized model that takes into account various environmental features and other factors in the evolution of mating systems has been proposed by Orians (1969) and has given direction to research on this problem. This model is presented in modified form in several variations in Figure 8–7. It applies strictly to females.

The role of *parental care* is critical in the evolution of mating systems. When the male plays no role in feeding or protecting the young, there may be no advantage to the female in remaining with

8–7 Models for the evolution of polygyny or monogamy adapted from models of Orians (1969). The fitness values of all female members of the population are shown as a function of the territories occupied. Different curves on a single graph represent different numbers of females per male. Territories are ranked from right to left in the order in which they become occupied (in a migratory species), and with the assumption that this order corresponds to the relative fitness values of the combination of one male and one female on a territory. It is assumed that the attractiveness of all males to females is equal, that the contribution to feeding the young is the same for all males unless specifically designated otherwise, that the intrinsic fitness values of all males are equal except for the abilities to choose and defend territories, and that females make their choice of mate on the basis of territory quality. The curves for one female per male show the range of territory qualities held by males with one female. Dotted extensions of these curves are shown where the addition of more females has eliminated that part of the curve. Bachelor territories are not shown but would be of lower rank than for the lowest-ranked monogamous males. The curves are drawn so that average fitness for all members of the population approximates 1.0 (population neither increasing nor decreasing) and to show a situation in which directional selection will act for or against genetic factors that facilitate polygyny.

A. Selection for monogamy in a uniform environment. Territories (all-purpose or mating-nesting) vary little in productivity; fitness of females attempting or tolerating bigamy is below the lowest fitness of monogamous females.

B. Selection for polygyny in a diverse environment. Territories (same types as in A) vary greatly in productivity. Even when male care of young is diluted when the addition of a third ♀ brings additional young to feed, polygynous females still have above-average fitness if they choose rich territories.

C. Critical role of male care of young. In 2♀♀*a* the role of the male in caring for young is minor, and dilution of his aid is of little consequence: polygyny is selected. In 2♀♀*b* the role of male care is important, and dilution of his aid results in below average nesting success.

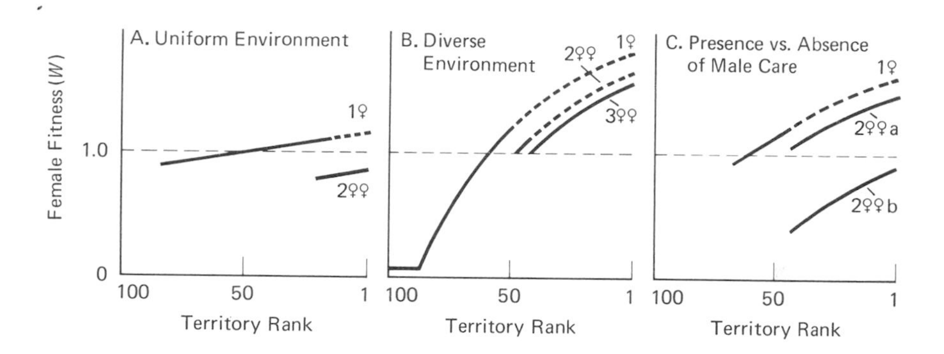

D. Selection for female choice in a promiscuous species not having male parental care, such as a lek species. Following Fisher (1930), those females that mate with the most attractive males have the most attractive sons and, consequently, more offspring in the second generation (but not the first). In this case territories are all of equal rank, and are not used for feeding or nesting.

E. Selection for monogamy in a sparse population. Because the number of individuals is small, none is forced to breed in unproductive territories. Bigamous females are less productive than average because the male's aid is diluted.

F. Selection for polygyny in a dense population in the same area as in E. Bigamous females are no more productive than in E in absolute terms, but they are relatively more productive than monogamous females in much worse territories.

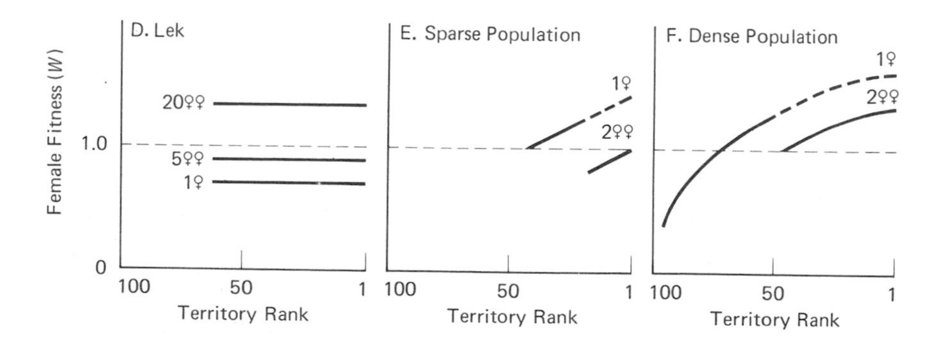

the father. Such conditions are common in mammals. Because in solitary mammals the males are physically incapable of nursing their young and are of negligible use in protecting them, the balance of selective forces tends to favor polygyny in those species. Monogamy occurs mainly in the few species in which the male brings food to the female (who converts it to milk) and to the young (in later stages of growth), as in the red fox (*Vulpes fulva*), and some other carnivores. Even if the males of grazing or browsing species could accomplish this, it would be a dubious advantage, since the females are usually surrounded by adequate food. Similar arguments apply to the remainder of the animal kingdom; since parental care of the young by the male is rare in most invertebrates and lower vertebrates (e.g., mouthbreeding fishes), monogamy is correspondingly rare. In birds, the male may be important in the care and feeding of the young, and monogamy is the general rule. Exceptions are numerous, however, and the illustrations in Figure 8–7 apply primarily to birds and to a few other species in which both sexes feed the young.

The models in Figure 8–7 formulate certain hypotheses. The effects of a uniform as opposed to a diverse environment are shown by *A* and *B*. In a nearly uniform environment there is only a small differential in territory quality from the best to the worst; the number of young raised by a monogamous female in the worst territory might be only slightly less than in the best. In a uniform environment the loss of half of the male's parental care might result in a relatively large loss in production of young. However, in a diverse environment the differences between the best and the worst territories occupied are large in comparison with the loss due to reduced participation of the male in the feeding of the young. If the best of these territories have enough food so that the male's help is negligible or unnecessary and sufficiently more food than the worst territories, females may raise more young by sharing a male in a rich territory than by breeding monogamously in a poor territory.

The difference male care can make under certain conditions is shown in *C*. As the contributions of the male to the young become less important, the curve for bigamy approaches that for monogamy; the sacrifice in fitness for a female breeding bigamously diminishes toward zero. The illustration shows a situation in which it is advantageous for the females to breed bigamously if the contribution of the male is small (2♀♀*a* in *C*), but not if it is large (2♀♀*b* in *C*). The extreme case is with communal mating grounds, leks, in species where the male plays no role at all in care of the young. There is then no disadvantage to the females in being polygynous and there may be an advantage in mating with the most polygynous of the available males (see below).

The subtlety of selection for or against monogamy is illustrated by a comparison of the effects of sparse and dense populations in *E* and *F*. The situation in *E* is much like that in *A*: there is more than enough space for all individuals to breed monogamously in a good habitat. Selection would favor monogamy for the same reasons as in *A*. But in a dense population in the same breeding area, competition for territories among the males would be much more intense, and some of the males would have to settle for territories of relatively inferior quality. Under such conditions, it would be advantageous for some females to mate bigamously rather than monogamously in poor habitats.

Ecological and Behavioral Factors Influencing Female Choice / Because the above hypotheses concerning environment and the evolution of polygamy in species with male parental care are relatively recent, there have been few direct attempts to test them in field studies. Both field and laboratory studies usually fall into one of two categories: studies of the properties of males, such as appearance and sounds, or studies of the properties of the territories held by the males. Further material on the devices employed by males to attract females and on the responsiveness of females will be found in Part IV on animal communication and in Part V on stimulus recognition mechanisms. Here, attention will be restricted to intraspecific variations in the fitnesses of males as a result of female mating choices.

• Vocalizations and Other Sounds: Males of some promiscuous or polygamous species may produce songs, mating calls, or wing and tail sounds as a part of premating courtship. In some monogamous species, such as the European robin (*Erithacus rubecula*) or certain duetting species, the female may sing much like the male. But in polygamous species females tend to lack a song or to sing quite differently from the male (as in the red-winged blackbird). Singing is characteristic of unmated males in many species and may either diminish (European robin) or disappear (brown towhee, *Pipilo fuscus*) after pairing in monogamous species. In polygamous species whose males pair with several females successively, singing may diminish after each mate is acquired and increase again before the next (Verner 1963); or it may continue at a high rate throughout the breeding season. Songs are known to be important in species recognition by female frogs and male birds (Chapter 18), but their role as a component of male fitness in nonmonogamous species remains unclear. Further studies on the relationships between song characteristics and male fitness like that of Kok (1972) are needed.

• Display and Plumage: Darwin believed that the conspicuous male plumages and displays of sexually dimorphic species evolved through sexual selection. To support this we may cite the generalization that the most promiscuous species with the largest number of females per attractive male (and hence, the strongest sexual selection) tend to have more conspicuous male plumage and displays than polygamous species with few females per male; and these, in turn, tend to be more conspicuous than males of monogamous species. Thus lek species like certain grouse (Tetraonidae), the cock of the rock (*R. rupicola*), manakins (*Pipridae*), and some birds of paradise (Figure 8–6) are in general more specialized in male courtship display than are less polygamous species like red-winged blackbirds, long-billed marsh wrens, and other species that defend all-purpose territories. In a detailed survey of North American passerine birds, Verner and Willson (1969) established that the frequency of sexual dimorphism (mostly in plumage) was significantly higher among promiscuous and polygamous species than among monogamous species. The degree of sexual dimorphism is also correlated with the mating system. This relationship is illustrated in various species of the New World family Icteridae (orioles, American blackbirds, meadowlarks, bobolinks) in Table 8–2, which shows that in this family promiscuous species tend to be most dimorphic, and monogamous species, least.

A word of caution is needed: sexual dimorphism may arise or be maintained by selection pressures in contexts other than sexual selection (Selander 1972). Dimorphism in body size and bill size in some species appears to be related to the range in size of food items that can be utilized by a pair (Storer 1966; Schoener 1967, 1969; Frochot 1967). This range can be reduced by interspecific competiton. Sexual dimorphism in hole-nesting ducks may be an adaptation to allow the female a greater choice of nest holes (Bergman 1965). A comparison of 129 species of ducks revealed that sexual dimorphism is considerably greater in hole-nesting than in ground-nesting species.

Some plumage specializations in birds (Fisher 1958) and most cases of sexual dimorphism in the horns and antlers of mammals (Geist 1966*b;* Lincoln 1972) are useful primarily in threat displays against males, or as indicators of adult status. Male red-winged blackbirds whose red epaulets were dyed black soon lost their territories to other males, while similarly handled, undyed controls kept their territories (Peek 1972; Smith 1972).

• Courtship Nests and Bowers: In some species the male builds a nest, or constructs a bower, or clears a space that is an integral part of his display to females. In the long-billed marsh wren it has been shown that males whose courtship nests are destroyed experimentally are unsuccessful in attracting females (Verner 1965*a*). The most famous species in this category are the bowerbirds (Ptilonorhynchidae) of New Guinea. In many of these species the males clear spaces on the ground and decorate them with overturned leaves or colored stones (Figure 15–11); the females then come to the bowers for copulation (Marshall 1953; Gilliard 1969). One of the most famous of these, the male satin bowerbird (*Ptilonorhynchus violaceus*), uses colored fruits to paint his bower in a primitive form of tool-using.

Summarizing the role of male courtship in female choice, there is highly suggestive evidence arising from general correlations that male plumage color and structure, displays, and vocalizations are important in the female's choice of a mate in some species. Experiments can be performed in the field to test some of these ideas by altering males of promiscuous species in various ways. Considerable experimentation has been done on the performance of females in choosing between mates of two different species, and this will be reviewed later in the section on isolating mechanisms. Alterations in appearance are also known to affect recognition within a pair in monogamous species like flickers (*Colaptes auratus;* Noble 1936), mourning doves (*Zenaidura*

TABLE 8–2 Sexual dimorphism and mating system in various icterids. The mean combines the data for wing, tail, tarsus, and bill length. Figures are percentage differences between the sexes. (After Selander, 1958.)

Species and race	Wing length	Tail length	Mean	Mating system
Zarhynchus wagleri wagleri	28.6	22.8	23.3	promiscuous
Gymnostinops montezuma	24.8	22.8	22.8	promiscuous
Cassidix mexicanus prosopodicola	22.2	29.1	22.3	promiscuous
C. m. major	22.1	26.8	21.3	promiscuous
C. m. mexicanus	20.2	27.4	20.5	promiscuous
Agelaius phoeniceus	18.2	19.1	17.4	mainly polygynous
Scaphidura oryzivora	21.7	23.0	16.5	probably promiscuous
Holoquiscalus niger	15.7	14.8	13.5	monogamous and probably also polygynous
Quiscalus quiscula	11.1	17.4	12.1	monogamous
Molothrus ater	8.5	11.2	8.9	monogamous or promiscuous
Euphagus cyanocephalus	8.5	10.0	8.8	monogamous, though polygyny common
Dives dives	11.2	8.5	8.2	monogamous
Amblycercus holosericeus	9.6	7.1	6.7	monogamous
Icterus galbula	8.4	9.6	6.4	monogamous
Euphagus carolinus	5.7	9.3	4.4	monogamous
Icterus spurius	6.4	3.8	3.5	usually monogamous

macroura; Goforth and Baskett 1965), and gulls (*Larus;* Smith 1966), particularly when the appearance of the facial region is affected.

● Territory Size: Tests of the hypothesis that males with larger territories acquire more females were made by Verner (1964), Verner and Engelson (1970), Zimmerman (1966), Willson (1966), Watson and Miller (1971), and Kok (1972). They found that polygamous males had the largest territories, and bachelor males the smallest, in the long-billed marsh wren, dickcissel, and red grouse, while in yellow-headed blackbirds no significant differences in territory size could be found among polygamous, monogamous, and bachelor males. In cases where territory size was not correlated with number of females, territory quality seemed important. In the prairie chicken, Robel (1966) showed that the male with the largest mating territory performed most of the copulations; however, this is a lek species and the territory is used only for mating.

● Territory Quality: In a species with all-purpose territories it is important that the territory supply adequate nesting habitats and food. More evidence is available concerning the importance of nesting habitat than food availability. That this can be of crucial significance in the production of young by females was shown by Kluyver's (1955) study of the great reed warbler, an Old World species. These birds nest mainly in reed beds, but some males also hold territories in beds of *Scirpus* or *Typha*. Some females try to breed with these males, but none succeed in raising young because the heavy nest, which is built of wet material, is not adequately supported by *Scirpus* or *Typha* vegetation and falls over into the water.

In certain North American species the area of suitable nesting habitat within the territory of a male appears to be correlated with the number of females nesting in his territory. Zimmerman (1966) found that for the dickcissel the area and height of shrubby herbaceous plants suitable for nest sites were greatest in the territories of males who were mated and least in territories of bachelor males. As shown in Figure 8–8, a correlation was found between the number of females nesting in a territory and the volume of vegetation. In hole-nesting birds that do not make their own holes, polygamy may arise when some territories have several suitable holes while others have none (Haartman 1969).

Another important aspect of the habitat is its food supply. Lack's (1954) conclusion that the number of young a pair can raise tends to be limited by the food the parents supply suggests that food availability is an important factor in the female's choice of a territory in which to nest. Measurements of food availability in blackbird territories have shown that the number of females per male is correlated with the richness of the food supply in the male's territory, in those cases where the females use the male's territory for foraging (Orians, in press). Other data summarized by Orians also indicate that food is an important aspect of territory quality.

Assessing the effects of territory quality on a female's breeding success is easier if the contributions of food and nesting habitat are combined. According to the Orians model it should be more advantageous for females to breed polygamously in good territories than monogamously in poor ones, or at least polygamy should not mean a lower production of young. Several studies on red-winged blackbirds confirm this prediction (summarized in Orians [in press]). Table 8–3 illustrates the point. It shows that the number of young fledged per nest was more than twice as high for females nesting at the ratio of six to seven per male than for females at one to five or more than seven per male. Since male redwings rarely feed their nestlings, there is no loss in parental care for nestlings when females nest polygamously. The main losses should be in the feeding of fledglings (where males help) and in female foraging efficiency, which is reduced because the density of females is greater. Apparently these losses do not come

8–8 Correlation of territory of male with number of females nesting on it in the dickcissel. The vegetation index reflects the amount of tall and dense vegetation suitable for nest sites in the territory. Each point represents one territory. The number of females per single male is the average over 4 weeks in June; since changes occurred, the average may not be a whole number. This accounts for the "fractions of a female" indicated on the graph. $t_b = 2.52$, $df = 31$, $p < 0.02$. (From Zimmerman 1971. Courtesy of American Ornithologists' Union.)

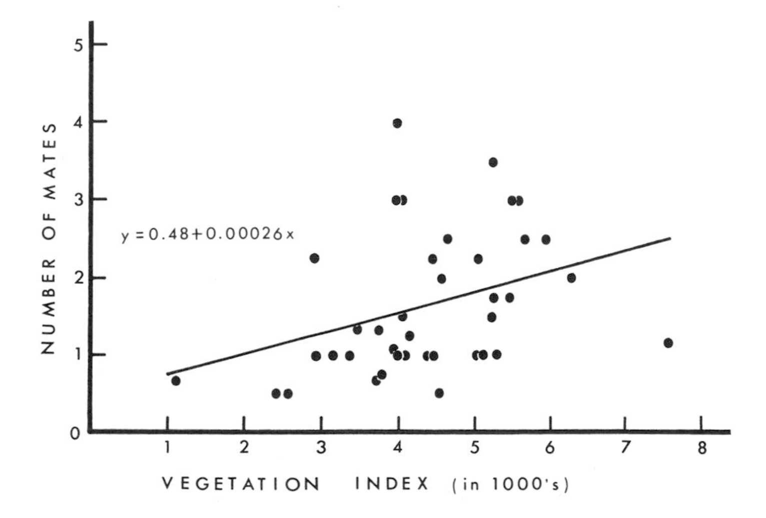

TABLE 8–3 Relationship between degree of polygamy and nesting success for red-winged blackbirds in Oklahoma. (Data from Goddard and Board rearranged by Orians [in press].)

Number of females per territory	Number of nests	Number of young fledged	Number fledged per nest
1–3	29	16	0.55
4–5	54	29	0.54
6–7	55	72	1.32
>7	105	79	0.79
Totals	*243*	*196*	*0.81*

close to counterbalancing the advantage of choosing a high-quality territory.

Habitat and Diet / Crook (1964) argued, following the lead of Armstrong (1955), that polygamy in birds was an adaptation to environments with a marked seasonal peak in food availability. However, this theory fails to specify the advantage of polygyny to the female, who in any case should be able to raise more young with a male's aid than without. A major improvement of this theory was made by Verner (1964), who suggested that a critical factor should be the diversity of qualities of territory available to a female. The models in Figure 8–7 show how diversity in habitat quality could be an important factor in the evolution of monogamy or polygamy. A wide range of environments provides some that are so rich that the penalty for a female mating polygynously in them is relatively small compared to the penalty for mating monogamously in a poor habitat. A correlation between the prevalence of polygyny and habitat was found by Verner and Willson (1966, 1969) in a survey of the mating systems of North American passerine birds. Of the 291 species, 14 are polygynous; and of these, 13 breed in marshes, prairies, or savannah-like habitats. None are polyandrous, and brood parasites (cowbirds) were excluded from the sample. In a survey of African weaverbirds, Crook (1965) found a correlation between polygyny and a grassland-savannah habitat on the one hand and monogamy and a forest habitat on the other. Haartman (1969) did not find similar correlations in a study of European passerine birds, but this might be due in part to the relatively greater extent of woodland habitats there.

Verner and Willson argued that greater richness and diversity of food supplies are important selection differentials that favor polygyny. For polygyny to evolve there, the food supply in a male's territory must be rich enough to provide for the young of two or more females (in a space small enough to be economically defendable by the male), and the differences in territory quality must be great enough to make it more advantageous for females to mate polygynously in a good territory than monogamously in a poor one.

According to Lack (1968:150), unusual pairing habits in birds are linked with diet. Nearly all species utilizing animal food are monogamous. Most nonmonogamous species eat plant food, with a preference for fruit and nectar among promiscuous species (e.g., hummingbirds) and for seeds among polygynous species. The reasons for these correlations are unclear and the causal relationships are probably quite indirect.

GENETICS AND SEXUAL SELECTION

Fisher's Concept of Sexual Preference / Darwin's original statement of the concept of sexual selection (1859, 1871) was reaffirmed in its principal parts and elaborated in 1930 by Fisher (1958), who introduced the idea of selection acting also on the female's ability to differentiate between males on the basis of their appearance or other individual characteristics. Fisher recognized two classes of effects that might be expected on theoretical grounds to be due to the action of selection on females. In Type 1, females prefer males having traits that are generally advantageous, such as general health, vigor, stamina, absence of debilitating deformities, and any beneficial trait not included in Type 2. Fisher reasoned that if females could discriminate between males on the basis of characters affecting their relative fitnesses, selection would favor females who chose males of superior fitness. Any genetic predisposition that improved this female ability would be selected for. Genetic effects of this type would not be easily detectable in practice, since their action would be mainly that of speeding up selection for traits already known to be beneficial for other reasons.

In Type 2, females prefer males having traits that are advantageous *only* for their value in attracting females and inducing them to mate. Clearly, every trait must be considered as the result of a balance of opposing and complexly interacting selection pressures. But Fisher's point is that some traits might be eliminated by selection were it not for their advantage in attracting females. The bizarre plumages and displays of birds of paradise, manakins, bowerbirds, and other birds that draw attention to themselves by their displays and vocalizations must surely make them more noticeable to predators as well as to females of their own species. Some evidence that highly adorned males of sexually dimorphic species do in fact suffer a higher mortality rate than females was obtained for the great-tailed grackle (*Quiscalus mexicanus*) by Selander (1965). Females outnumbered males by a factor of 1.34 in October and by 2.42 in March (a month before the breeding season). This change in the adult sex ratio would require a male mortality rate roughly twice that of females. (The sex ratio of nestlings is 1:1 in this species, as in other nonmonogamous species of birds in which it has been checked.)

Genetic Models of Sexual Selection / The genetic implications of Fisher's concept of sexual selection were worked out by O'Donald (1962; 1963*a*, *b*; 1967) with the use of mathematical models and computer simulations. Although these exercises are strictly theoretical, they constitute a set of concrete hypotheses describing the evolutionary relationships between the relatively fixed patterns of courtship in males and the genetic factors of the responsive mechanisms in the female, mechanisms of approach to and mating with males of the same species. They are outlined here because of their considerable theoretical significance for the evolution of behavior and in the hope that ways will be found to test them.

Assume the following: (1) a polygamous species with an equal number of sexually mature males and females, and (2) that females mate only once while males can mate many times. For every female beyond the first, fertilized by a given attractive male, a male designated as unattractive must be deprived of mating at all. Selection in this context should favor attractive males.

Attractiveness in the model is determined by a single locus; M^+M^+ males are attractive and M^-M^- males are not; heterozygotes are intermediate. Preference in females is determined by a different locus; F^+F^+ individuals are discriminating and mate preferentially with M^+M^+ males, up to an arbitrary upper limit of harem size for a given male, if a choice is available, but will mate with M^-M^- males if it is not. F^-F^- females mate at random with respect to male genotype. Heterozygous females are intermediate.

Starting with a population in which M^+ and F^+ are rare, attractive males have an advantage only in the presence of discriminating females, and will increase in frequency only as discriminating females increase. If there are any discriminating females, attractive males will have a clear advantage and their number will increase *exponentially* as the proportions of attractive males and discriminating females increase. In computer simulations the increase in these alleles follows an

S-shaped curve, being slow at first, rapid in the middle, and slow at the end.

Discriminating females have no intrinsic selective advantage. They leave no more offspring than nondiscriminating females. But if they mate with an attractive male, and especially if the male is homozygous for M^+, then their sons will attract more females than will the sons of M^-M^- males. Consequently, they will leave more grandsons and granddaughters by mating with an attractive male than with an unattractive one. Since the advantage of discrimination is not shown in the first generation, there is in a sense a *delay* in the manifestation of superior fitness for such traits. The selective advantage of F^+F^+ females lies wholly in that the F^+ genes come to be associated in the same individuals with M^+ genes more frequently than do F^- genes. The closer the association between the M^+ locus and the F^+ locus, the faster sexual selection will proceed. Consequently, any factor facilitating this association, such as close *linkage* between M^+ and F^+ loci, will be selected for. And any factor slowing down this association, such as dominance which masks the presence of M^- or F^-, should be selected against. By this argument, *recessiveness* should actually lead to faster sexual selection than dominance (O'Donald 1963*a*). Thus the model predicts that genes determining structures and behaviors subject to sexual selection are more often recessive than dominant, and more often linked than not.

Another prediction of the model is that the linkage should not be expected to occur on the Y chromosome, since both sexes are affected by one of the two loci (O'Donald 1962) and the Y chromosome is found only in female birds and male mammals.

Frequency-dependent Sexual Selection / Courting behavior appears to have been highly resistant to evolutionary change in some cases. For example, the mating song of a cricket on the Bermuda Islands in the Atlantic closely resembles that of its presumed mainland progenitor (*Gryllus firmus*), even though the two species have been isolated by many thousands of generations (Alexander 1962). In the purple martin (*Progne subis*), geographic variation in plumage is conspicuous in the females, in which it is interpreted as reflecting adaptation to different regional conditions, but absent in the males (Johnston 1966). This might be due to the role of the male's plumage in mate selection by females. In such cases stabilizing selection is invoked as an explanation; deviant songs and plumages in such species are presumably less successful in attracting females and so are continually eliminated from the population, thus preserving the norm.

More direct evidence of stabilizing selection in mating has been found by comparing samples of insects caught during copulation with samples caught in the same season but not copulating. By this means Mason (1964) showed that the variances of structural characters in wild populations of a cerambycid beetle (*Tetraopes tetropthalmus*) were less in copulating than noncopulating individuals. Similarly, in the California oak moth (*Phryganidia californica*) variances of structural characters in males were less in copulating than noncopulating individuals (Mason 1969). In the latter species, however, copulating males averaged longer wings than noncopulating males. There were no differences among the females. These cases provide evidence of selection against the unusual.

In contrast, the outlandish visual appearances and courting actions of promiscuous species seem to require selection for the unusual and striking for their explanation. For example, if peahens (genus *Pavo*) were more stimulated or attracted to unusually bizarre peacocks, directional selection would favor such males. When this male type became the norm for the population, the tastes of the peahens might be even better satisfied by even more bizarre peacock display, and so again the unusually effective male displays would be selected for. In this case selection is not, strictly speaking, frequency-dependent; the selected trait is both rare and striking at first, but common after selection. By endless repetition of this process (selection of the rare and striking), the displays of the cocks would become extremely specialized. Presumably the process would come to a halt only when the advantages of attractiveness to females were counterbalanced by the disadvantages of susceptibility to predation and awkwardness in fighting or foraging. If such a balance were present, it should be reflected in sexual differences in mortality.

Selection by females for rarity of male types was first found in laboratory experiments on *Drosophila* by Petit (1951). For example, in a group of 25 pairs of *D. pseudoobscura* from Texas or California, when 5 of the 25 males were of one type and 20 of the other, the minority males achieved significantly more than 20 percent of the copulations (Ehrman 1966). When 12 males were Texan and 12 were Californian, no difference in success was observed. Similar rarity effects were found for strains with different chromosomes, for flies differing by one mutant gene, and for flies of the same strain raised at different temperatures. These effects are known for at least six species of *Drosophila* (Ehrman 1969). Exceptions to the pattern described above have been found, especially, as expected, between populations that have evolved incipient behavioral isolating mechanisms (see Chapter 18), as in *D. paulistorum*. The rarity effect is found commonly among males (discrimination by females) but rarely among females (discrimination by males). The sensory basis of the female's ability to make such apparently subtle distinctions seems to be primarily olfactory (Ehrman 1969), but in theory the rarity effect could be mediated by any sensory mode, depending on the species.

In the context of sexual selection these findings are important because they show that females are in fact capable of making extremely fine distinctions among males, and that genetic factors can be correlated with the preferred phenotype. Moreover, some of the discriminations tested—for example, between karyotypes (Spiess 1968)—are of a sort that might occur in natural populations. The results suggest that sexual selection through female preference does exist in natural populations of some *Drosophila* species, and that it has significant genetic consequences quite apart from the selection pressures generated by the risk of hybridization with related species (Chapter 18).

Female preference for minority genotypes may be expected to preserve genetic variability in a population. There exists among birds one striking case where such an explanation might be applicable, namely the preservation of genetic diversity responsible for the various color morphs of the ruff. In this lek species two types of male occur. *Independent* males are aggressive and defend a small mating territory about 30 cm in diameter. *Satellite* males (38 percent of the male population) are conspicuously nonaggressive and do not defend a mating territory; they achieve copulation by associating with a particular independent male and "stealing" copulations while the independent is busy defending his territory (Hogan-Warburg 1966). Two types of conspicuous plumage differences exist among male ruffs. Satellites differ from independents by their white ruff; independents have dark ruffs, but are much more variable in appearance than are satellites—to the point where no two independent males on a lek are identical. The presence of the white morph can be correlated with its greater visibility at a distance and effectiveness in attracting females. It is rare at large leks but more common at small ones. Since small leks would otherwise be less likely to be visited by females, the presence of some satellite males is a small price for the dominant independent males to pay for luring more females to the lek. The presence of conspicuous variation among the independent males requires a separate explanation. One possible hypothesis is that females are stimulated by novel plumage colors.

Genetics of Male Courtship / Although there are not many genetic experiments directed specifically to the models of sexual selection discussed above, some information is available in studies devoted to other topics, especially the genetics of isolating mechanisms and species formation. Genetic effects on male courtship will be considered first.

The display structures and colors specialized for male courtship are integral parts of the male's appearance and can be expected to be significant in influencing the female. Since they are easier to study than behavioral characteristics, there is more information available about them and they offer many opportunities for experimental study of sexual selection. Abundant evidence exists that variations in many structures important in courtship display and behavior of males are genetically determined. Examples include the antlers and horns of mammals; the color patterns and feather structure of the varieties of pigeons, ducks, domestic fowl, and domesticated finches; the colors of domesticated fishes like guppies, swordtails, and mollies; the color phases of morphs of many species of birds; wing mutants of *Drosophila melanogaster;* and many others.

Evidence for genetic effects on the courtship of males is also available from many sources. Sex linkage of genes controlling cricket song was found by Bigelow (1960) and Bentley (1971). Ewing (1969) analyzed the genetic basis of sound production in the courtship song of male *Drosophila pseudoobscura* and *D. persimilis* by examining F_1 hybrids and backcrosses. Some genes affecting song repetition rate and frequency within a sound pulse were located on the X chromosome;

8–9 Comparison of courtship of male *Drosophila melanogaster* in lines selected for large and small body size with controls. Courtship is divided into three phases: *O,* orientation, *V,* vibration, and *L,* licking (see Figure 8–10). Males were tested against females of the control stock. Males selected for large body size spent proportionately more time in orientation and less in licking than did controls. Males from the small line spent proportionately less time in orientation and more in vibration than did controls. There were two large lines, L_1, L_2; two small lines, S_1, S_2; and controls, *C.* (After Ewing 1961.)

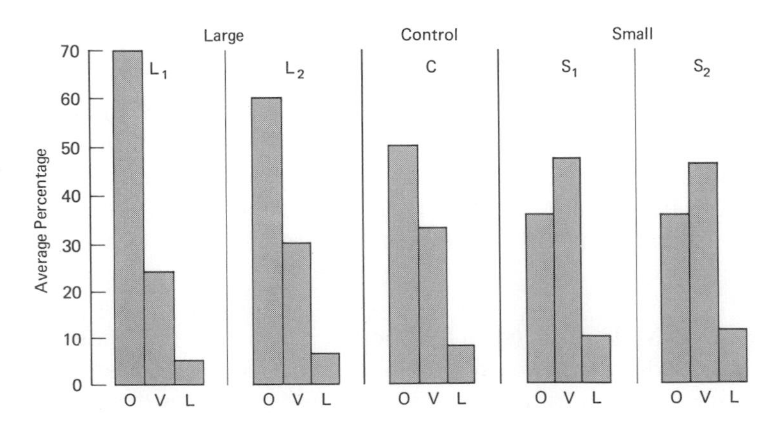

other genes affecting the song were autosomal. It would be interesting to know the location of genes affecting female response to the song. Some mutants of *D. melanogaster* are deficient in courtship display (Spiess 1970). For example, *yellow* males have shorter vibration-bout lengths than do wild type males (see Chapter 1; Bastock 1956), and *ebony* males have difficulty orienting toward the female in courtship because they are blind. Color morphs of the ruff differ also in display posture to females, and the light morph tends to be subordinate to the independent males (Hogan-Warburg 1966). Breeds of chickens differ in the crows of the cocks (Siegel, Phillips, and Folsom 1965). Further information about genetic effects on insect courtship is available in the review of Ewing and Manning (1967).

An unexpected case of sexual selection acting on males was encountered by Ewing (1961) while selecting for body size in *D. melanogaster*. Starting from one population, separate lines were established in which flies were selected for large and small body size. Flies from generations 8 to 13 showed changes in thorax length of 6.3 to 8.9 percent in the selected direction from the mean of the starting population. Comparing the courtship of males of large and small lines when exposed to females in the control population revealed the differences shown in Figure 8–9.

Three stages of courtship in *D. melanogaster* were described by Bastock (1956), as shown in Figure 8–10. In the first, *orientation*, the

8–10 Courtship and copulation in *Drosophila melanogaster.* Courtship has three phases: **A,** orientation, in which the male faces and approaches the female; **B**, vibration, in which the male extends a wing and vibrates it; and **C**, licking, in which the male stimulates the genitalia of the female with his mouthparts. **D** and **E** show attempted and actual copulation. (From A. Manning, in *Viewpoints in Biology,* eds. J. D. Carthy and C. L. Duddington. Butterworths and Co., Ltd.)

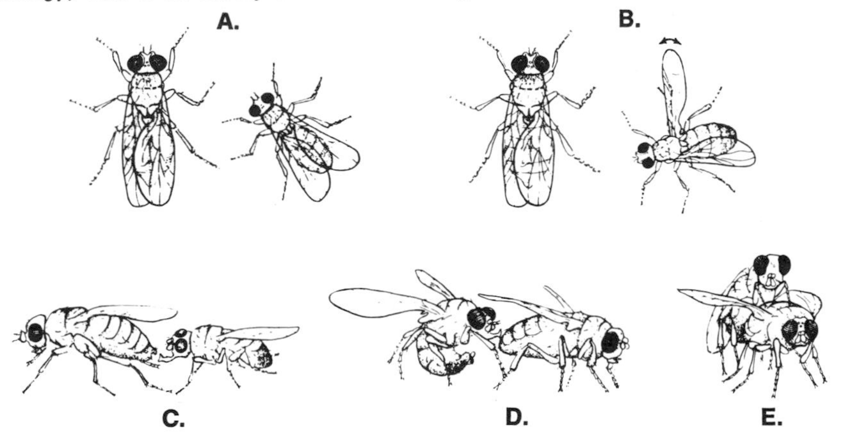

male faces and approaches a female. In the second, *vibration*, the male raises and vibrates his wings. In the third, *licking*, the male stimulates the genitalia of the female with his mouthparts and then copulates. Males from the large line spent proportionately more time in the orientation phase and less in vibration than did controls; males from the small line spent less time in orientation and more in vibration than control males.

The question then was how this difference had arisen. Changes in general activity were ruled out as a cause, since both large and small lines were less active than controls. Another hypothesis was that the differences in courtship could have arisen through sexual selection. To test this possibility, the selection regime was changed from putting the ten selected pairs together in a bottle to one in which each selected pair was isolated in a vial. This eliminated competitive mating. The results of this experiment are shown in Figure 8–11. In the small line, housing flies in pairs eliminated the evolutionary change in vibration. Housing

8–11 Sexual selection caused by artificial selection for body size in *Drosophila melanogaster.* The *S* and *L* lines were selected by the experimenter for small and large body size, respectively, and housed one pair to a vial for the first nine generations. At the ninth generation the *L* line and some (*Sc*) of the *S* line were housed ten pairs/bottle, while the rest of the *S* line was maintained at one pair/vial, as earlier. The presence of ten pairs/bottle is thought to have allowed competitive mating such that small males that were more active in vibration were more successful in copulation, thus explaining the increase in vibration. The reduction of vibration in the large line is unexplained, but the increase after introduction of competitive mating is suggestive of sexual selection for vibration. (After Ewing 1961.)

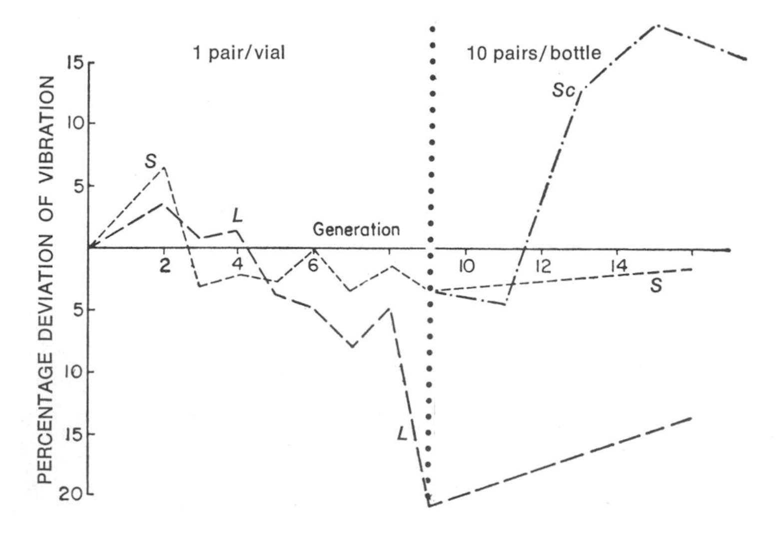

flies in groups of 20 led to evolutionary changes in vibration. These changes were not dependent on small body size *per se* but on the combination of small body size and sexual selection. Wing area in males influences their effectiveness in courtship. A probable explanation of the results is that smaller males varied in their capacity to stimulate females and that those who vibrated more were more effective in competing for females (that is, were "selected" by the females). The smaller wing area in small males was compensated for by an increase in wing vibration. (The explanation for the large line must be different, since the decrease in vibration occurred even when flies were housed in pairs and since switching to mass mating brought an increase in vibration.)

Genetics of Female Selectivity / The genetic effects of sexual selection in females predicted by the above models are more difficult to find. However, several workers have performed selection experiments on *Drosophila* which place an advantage on the ability of the female to discriminate, either between males of her own and another species or between males of different types in her own species. These experiments are reviewed in the chapter on isolating mechanisms (Chapter 18).

CONCLUSIONS

In this chapter an attempt has been made to bring together genetic and ecological theories and observations about the evolution of mating behavior and mating systems.

Many animals are sexually dimorphic in the structures and behaviors used in courtship. Darwin (1871) attributed these features in the males to an evolutionary process that he called sexual selection. Although Darwin distinguished between "sexual" and "natural" selection, in a modern context sexual selection is a type of natural selection and of individual selection—a type based on competitive mating. The type of mating system a species has strongly influences the potential for, and forms of, competitive mating. In polygamous and promiscuous species the potential for competitive mating is high, while in monogamous species it is relatively low (since a male gets no more than one female).

Darwin thought that when males have opportunities to win many females by fighting for them, structures and behaviors used in threat and fighting evolve—for example, the antlers of male deer and the large size of male fur seals. When males have less opportunity to fight over females because of greater distances between males, females are more able to choose a mate of their preference. Darwin thought that the bright colors and ornaments of many male birds evolved because of the advantage they give males in attracting females, but this has been partly refuted. We know through experiment and observation that the bright colors of males of certain species of birds are useful mainly in threats against other males. There remain, however, several groups of birds where, because of their extreme ornamentation or bizarre behavior, the hypothesis of female preference remains in strong contention (e.g., birds of paradise). It seems likely that both types of sexual selection operate together in many promiscuous and polygamous species.

The following characteristics are typical of species in which competitive mating is severe: (1) extreme sexual dimorphism, (2) many females fertilized by relatively few males, (3) a surplus of males who achieve no copulations, (4) weak or nonexistent pair bonds, (5) delayed maturity in males relative to females, (6) little or no paternal care, and (7) mortality rates that are higher among males than females. Species in which males compete for females mainly by inter-male aggression are noted for males with traits that are useful in fighting and threatening (e.g., antlers, large body size). Species in which inter-male aggression is less conspicuous and in which females appear to have greater freedom of choice of males are noted for bizarre colors, plumes, displays, and ornaments that seem to be of little use in fighting but useful in courting females.

The variations in mating systems among species are due in part to phylogeny and in part to ecology. Mammals exhibit the influence of phylogeny. Males, lacking functional mammary glands, are unable to nurse their young and so in many species can contribute little or nothing to the care of their offspring that cannot be satisfactorily provided by the mother. As a result there is often no advantage to a female in retaining the male's aid by means of a social bond, especially when he may be competing with the female and young for food. In short, monogamy would be disadvantageous to such a female. The rarity of monogamy in mammals may thus be partially explained. In birds, on the other hand, both sexes are usually equally capable of nourishing the young, and monogamy is the rule.

In the evolutionary transitions between monogamy and polygamy in birds the role of parental care is felt to be significant. When forsaking monogamy for polygamy, a female must accept a reduced role of the male in feeding her offspring. If this loss is small relative to the gain achieved by mating with a male in a rich territory, then polygamy will be beneficial for her. The richness and range of territory qualities in a given area may therefore be expected to be important. Field studies of polygamous birds have borne out this prediction to some extent.

The factors that influence a female's choice of a male under natural conditions are critical for testing Darwin's ideas about sexual selection. To some extent they can be tested by correlation and experiment. These factors are covered here in relation to sexual selection and in Chapter 18 in relation to species recognition.

In birds and mammals the relevant factors can be divided into two categories, (1) the appearance and behavior of the male, and (2) the territory of the male. The relative importance in female choice of natural variations in male song and plumage has not yet been adequately studied. Correlations suggest that song is necessary to obtain a female in some species, as well as to defend a territory. There is a rough correlation in birds between conspicuousness of sexual dimorphism of plumage and promiscuity. The importance of a male's territory has been indicated in various studies. Territory size is important in some species; in others, the quality and quantity of the territory's food or nesting habitat may correlate well with the number of females a male acquires. In short, we still know relatively little about how female birds and mammals choose a male from the available candidates.

Genetic models of the consequences of sexual selection predict the presence of genes that help females choose the most attractive males, and predict that these genes will be closely linked with genes influencing male attractiveness. It is also expected on theoretical grounds that some genes and structures associated with courtship in males would be eliminated by nonsexual selection if they were not retained in the population by sexual selection. These predictions need further testing.

Frequency-dependent selection for or against rare types is suspected in several species. It has been demonstrated experimentally in *Drosophila,* in which case minority phenotypes among males have an advantage in competitive mating. The significance of novelty as a factor in competitive mating remains an unsolved problem.

9 Aid-giving Behavior: Diversity, Ecology, and Evolution

It would be instructive to know not only by what physiological mechanism a just apportionment is made between the nutriment devoted to the gonads and that devoted to the rest of the parental organism, but also what circumstances in the life history and environment would render profitable the diversion of a greater or lesser share of the available resources towards reproduction.

R. A. Fisher, 1930

THE THIRD major behavioral component of social organization, after agonistic and sexual behavior, is aid-giving behavior. It is the primary part of any social relationship involving the giving and receiving of benefits, whether the latter be goods (e.g., food) or services (e.g., status, reduction of risk). Under the heading of aid-giving may be included all types of behavior in which one or more individuals enhance the relative fitness of one or more other individuals. Aid-soliciting behavior is also an important part of this social context in many cases, but it will be neglected for the sake of brevity.

Examining aid-giving behavior from an evolutionary perspective not only helps us understand the evolution of social organization in general; it is critical for understanding the evolutionary antecedents of human behavior and social organization. Man differs from other animales far more importantly in the development and complexity of aid-giving behavior than in those aspects of social organization deriving from agonistic or sexual behavior. Aid-giving in most species is founded on a social bond, such as occurs between parent and offspring, between the members of a monogamous pair, or among members of an expanded family group. The capacity to form the bonds that characteristically accompany aid-giving has evolved in man to include relationships based upon abstractions far more complex than those found in any other species. For example, man develops loyalties to his peer group, his village, his ethnic or tribal group, his nation, his labor union, and his political ideology. These abstractions are the beneficiaries of aid given by individuals often without monetary compensation. Just as important are mutual aid agreements between employer and employee, and others involving a monetary adjustment of benefits acceptable to both parties. Although the term *civilization* connotes different things to different people, high on the list of attributes of civilization must be the ability of unrelated individuals to live in harmony through the establishment of mutually beneficial aid-giving relationships.

Diversity and Phylogenetic Incidence / Simple forms of aid-giving behavior, such as the carrying or protecting of eggs or young, can be recognized even in certain coelenterates. The more complex forms of aid-giving are found mainly among the more complex phyla. This is not a very useful generalization, however, since the aid relationships among social insects tend to be more complex than in many mammals. Aid-giving behavior is conspicuously well developed in two groups of animals, the vertebrates and the social insects. Although the physiological basis of such behavior in the two groups is, of course, completely different, ecologically there might be some common ground.

The major categories of aid-giving behavior are listed in Table 9–1. Since many of these are well known, it is not necessary in the present context to describe them. Certain generalizations can be stated, however; and some of the evolutionary problems related to them will be discussed.

It is apparent that considerable diversity exists in aid-giving behavior. Some species exhibit little or no parental behavior, while the parental or communal behavior of others is long-lasting and complex. Furthermore, there is no detailed correlation of the complexity of aid-giving behavior with phylogenetic groups. Some of the more highly developed forms of parental care among vertebrates occur in fishes and birds, while certain mammals provide little more than a nest and milk. Although the most complex kinds of aid-giving, involving leadership, alliances, and care of a group's injured members, are virtually restricted to mammals, it could not be argued that motivation to aid others has been dependent on the evolution of intelligence. Ecological factors are probably more important.

The basic question of what factors determine the apportionment of energy in a species toward (1) egg production for the present reproductive effort, (2) parental care for the present, and (3) survival of the parents for future reproductive efforts was raised by Fisher (1930) in the quotation at the head of this chapter. The first part of this chapter provides an introduction to some of the ecological concepts relevant to this problem.

TABLE 9–1 Some types of aid-giving behavior found in animals, with examples of representative species

Behavior type	Examples
I. Care of eggs by parents	
A. None	some pelagic fishes; some starfishes
B. Choice of site	most frogs and salamanders
C. Construction of nest	fishes; reptiles; birds; platypus
D. Defense of nest or eggs	fishes and others
E. Carrying of eggs	echidna; mouth-breeders; South American pipid toads; pipe fishes
F. Aeration	sticklebacks
G. Incubation	birds
II. Care of young by parents	
A. None	most amphibians; reptiles; some birds (Megapodidae)
B. Provisioning of young before hatching	some wasps
C. Feeding or guiding of young	birds; mammals; social insects
D. Protection of young—by attack, threat, and distraction displays	birds; mammals
E. Leadership	red deer hinds; sheep
F. Enhanced dominance	monkeys
G. Carrying of young	marsupials; primates; grebes
III. Care of relatives by nonparents	
A. Feeding of young by young of earlier brood	bird helpers; social insects
B. Feeding of young by adults in same family group	Mexican jays
IV. Communal and cooperative behavior	
A. Communal display	lek species
B. Communal territory defense	Mexican jays
C. Communal colony defense	terns; gulls; and others
D. Communal foraging and hunting	cormorants; pelicans; hyaenas; hunting dogs
E. Cooperative display	ruffs; manakins
F. Mobbing and alarm-calling	monkeys; birds
G. Schooling	fishes
H. Communal herd defense	musk oxen; baboons
I. Grooming of others; hetero- and allo-grooming	primates; birds
V. Symbioses	cleaning fishes; mixed-species flocks of birds
VI. Social parasitism	
A. Slavery: feeding of young by slaves captured from another species	certain ants
B. Brood parasitism: feeding of young of another species that has laid its eggs in the nest of the foster parents	cuckoo-bees; cuckoos

THE EVOLUTION OF REPRODUCTIVE RATES AND PARENTAL CARE

The evolution of reproductive rates is a problem bearing on parental care that is of general interest to ecologists. When parental care is significant in the rearing of young, a change in the amount of care given to an individual can have a large effect on its survival. If there is an upper limit to the total amount of parental care that a parent can give to a brood of young, for instance in the food it can provide, then an increase in the size of the brood must sometimes cause a corresponding decrease in the maximum amount of parental care given to each brood member. Consequently, some knowledge of the evolution of birth rates and brood size is essential for understanding the evolution of parental care.

Clutch Size / The factors that influence clutch size (the number of eggs laid in a set) in birds have been analyzed by Lack (1954*a*, *b;* 1966; 1968) and others. His thesis (Lack 1954*a*:173) was that "clutch size has been adapted through natural selection to correspond with the maximum number of offspring for which the parents can, on the average, find enough food without seriously harming themselves." The opposing argument is that animals raise fewer young than they are capable of rearing and so prevent overpopulation (Wynne-Edwards 1962). Lack made a classic study of this problem, observing the success of clutches of different sizes in the European swift (*Apus apus*). This species feeds its young on insects that it catches in midair like a swallow. Lack showed for the period 1946–50 that in broods starting with 2 young (the most common number) 82 percent safely left the nest, a rate of 1.6 young per nest. In contrast, broods starting with 3 young fledged only 45 percent, or 1.4 young per nest. The losses were due to starvation. Consequently, 2 was a slightly more productive clutch size than 3 during 1946–50.

A similar study of starlings in Switzerland showed no significant differences in pre-fledging success among broods of five, six, seven, and eight nestlings; but the average number of survivors of the post-fledging period of dependence on the parents was about equal for these brood sizes. This result reflects an above-average mortality rate for young from larger broods. Since the young from large clutches tended to weigh less, their higher mortality is understandable. Similar effects have been shown in other species by experimentally raising clutch sizes (reviews in Lack 1966, 1968; Klomp 1970). The argument that animals avoid raising a large number of young to prevent overpopulation has been largely discredited. The factors limiting clutch size in most of these cases seem to be both the availability of food near the nest and the activity of the parents in finding food and bringing it to the young. Parental care is clearly critical.

Evolution and Environment / With the three-factor relationship among parental care, clutch size, and environment in mind, some simple and basic predictions about the evolution of parental care can be made. For example, the amount of food parents bring to the nest could be increased by a rise in behavioral efficiency in finding and transporting food, or by an increase in the food supply. Genes causing increased efficiency are likely to be selected for. If the environment becomes richer, selection will favor genes causing higher clutch size, causing an increase in the birth rate, *b*. If the environment worsens, selection will favor genes causing a decrease in clutch size, resulting in a decrease in *b*. In the latter situation individuals who produce fewer eggs actually would produce more mature offspring than those with higher clutch sizes, because parents with small clutches can provide more care per offspring than those with large clutches. In all of these situations the quality of the environment is critical.

Uncrowded Environments / A major determinant of environmental quality for most individuals is the density of their major competitors, namely, other members of their own species. If in theory an ideal environment were provided without predators and without competition for resources from one's own species, the population would grow exponentially as a function of birth rate, *b*, and death rate, *d*, as shown in Figure 9–1. The growth of a population over generations in an ideal environment can be described by the equation

$$N_t = N_0 e^{r_{\max} t},$$

where the number at a given time, N_t, is a function of the starting number, N_0, and the rate of increase in numbers, r. In an ideal environment the population increases at its maximum rate, and r is termed r_{max}. If the population were neither increasing nor decreasing, r would be zero.

The value of r is determined by the overall birth rate, *b*, and death rate, *d*.

$$r = b - d$$

Thus r_{max} is obtained in theory by removing all causes of death other than "old age"; by minimizing *d*, the maximum value of $(b - d)$ is obtained. Species with high death rates from factors other than old age have high birth rates and correspondingly high values for r_{max}. Species with low death rates have low birth rates and correspondingly low values for r_{max}.

9–1 Exponential population growth in an ideal, uncrowded environment (*left*). More rapid population growth is achieved by raising r_{max} in a process called r_{max}-selection (*right*).

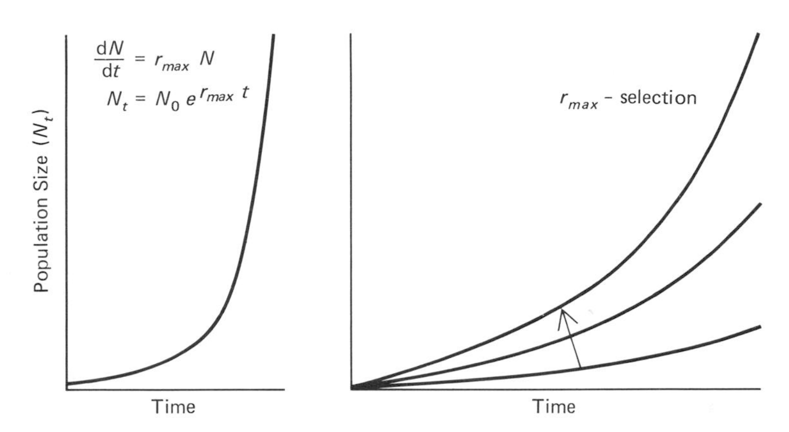

In natural individual selection, genotypes with the highest values of r are selected (sometimes designated *m*, when applied to genotypes; Fisher 1930). The value of r for a genotype can be raised by increasing *b* or decreasing *d*, or both. Whether r increases by increasing *b* or decreasing *d* seems to vary with the environment. Two extreme types of environment will be considered, the first nearly ideal, the second, highly competitive and rigorous.

r_{max}-selection / If a species in which r_{max} is adapted to a realistic environment is introduced into a theoretically ideal environment, what types of genetic changes in the factors underlying r_{max} can be expected? Changes in both *b* and *d* will be considered. An increase in r_{max} can be achieved by decreasing *d*, but long life by itself does not add individuals to the gene pool. Long life is selected for mainly when it promotes repetitive (*iteroparous*) reproduction, thus allowing longer-lived individuals to contribute more offspring to the gene pool than short-lived ones. The extreme opposite is to reproduce only once (*semelparous* reproduction) and then die.

Our question can now be seen to be better put as follows: Will a species introduced from a realistic environment into an ideal environment (with mortality minimized) evolve toward semelparity and an increased clutch or litter size, or toward greater iteroparity and more frequent production of a standard-sized clutch? It has been shown mathematically by Cole (1954) that under ideal conditions (i.e., $d = 0$), adding just one more egg to a clutch in a semelparous species is equivalent to living and breeding forever in an iteroparous species. In Cole's words (p. 118), "For an annual species, the absolute gain in intrinsic population growth which could be achieved by changing to the perennial reproduction habit would be exactly equivalent to adding one individual to the average litter size." Our answer is that in near-ideal environments a high initial clutch size is favored by natural selection, with the result that r_{max} is probably increased mainly by a rise in b rather than through longer life (lowered d) and greater iteroparity (Figure 9–1). Because uncrowded and near-ideal environments tend to favor increases in b, and thus in r_{max}, selection acting to raise r_{max} has been termed r-selection (MacArthur and Wilson 1967). To be precise it might better be called r_{max}-*selection*, since r for genotypes is also increased during the process of K-selection, in which case r_{max} becomes reduced (see p. 194). For further clarification of this point see Hairston et al. (1970).

Colonizing Species / Ideal conditions are rare in the real world, but when individuals discover a rich environment that has not previously been colonized by their species, conditions are favorable for those individuals and for rapid population growth. To be the first to colonize such habitats, genotypes must produce many individuals with good dispersal abilities. Once an unexploited environment has been found, successful genotypes must be able to grow rapidly in numbers so as to be able to colonize other such areas before the population of the colonizing species is reduced by the arrival of other species. Species that are adapted for such a way of life are called *colonizing species* or *opportunistic species* (Baker and Stebbins 1965). Good examples are weeds and parasites. Many habitats if disturbed or destroyed go through a series of natural changes in vegetation and fauna. For example, when the pastures of eastern North America are abandoned, the grassland is soon invaded by weeds, then by shrubs, then by rapidly growing trees such as poplars (*Populus*) and finally by slower-growing but better-competing trees such as oaks (*Quercus*), which form the climax vegetation. Such a transition is called a *sere* or plant succession, and its various stages are termed *seral stages,* the final stage being termed the *climax*. Along with the vegetational changes go corresponding changes in faunal composition. Species that are adapted to transient habitats undergo many episodes of population growth and decline as their habitat changes. Species with strong cyclic fluctuations in population density, such as voles and lemmings, also undergo repeated episodes of population growth. Other correlates of r_{max}-selection are listed in Table 9–2.

TABLE 9–2 Simplified summary of some aspects of r_{max}-selection and K-selection. (Adapted from MacArthur and Wilson 1967:189–190, Pianka 1970, and Hairston et al. 1970.)

	r_{max}-selection	K-selection
Defining characteristics	favoring a higher population growth rate and higher productivity	favoring more efficient utilization of resources, such as closer cropping of the food supply
Conditions of occurrence	1. far below K	1. at K
	2. little competition among and within species	2. much competition
	3. transient habitats	3. long-lasting habitats; climax
	4. frequent and large population fluctuations	4. infrequent and small population fluctuations
Consequences	1. r_{max} higher	1. K raised
	2. K rarely approached	2. r_{max} often lower
	3. b and d higher	3. b and d lower
	4. rapid development	4. slower development
	5. earlier reproduction	5. later reproduction
	6. shorter life	6. longer life
	7. less parental care	7. more parental care
	8. good dispersal	8. poor dispersal

Crowded Environments / In climax habitats the "balance of nature" is more stable. Many species live in approximate equilibrium with each other, and their densities do not fluctuate much. The number of individuals of one species that an environment can support at equilibrium is known as the *carrying capacity*, K, of that environment for that species (see Figure 9–2, Table 9–2). The effects of crowding on population growth can be described by the following equation:

$$\frac{dN}{dt} = r_{max} \left(\frac{K - N}{K}\right) N$$

9–2 Density-dependent population growth (*left*). Selection resulting in an increase in the carrying capacity, K, is known as K-selection (*right*).

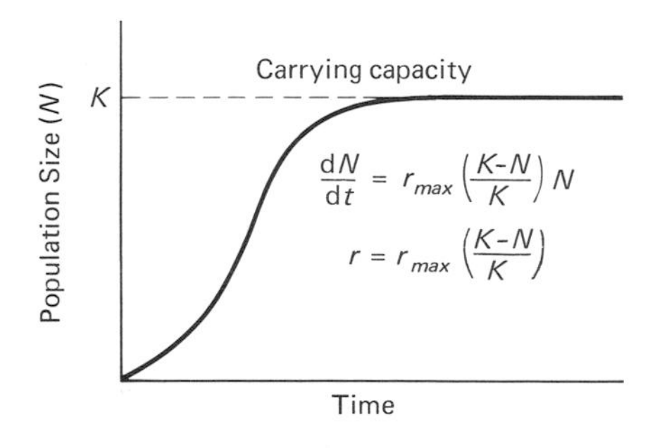

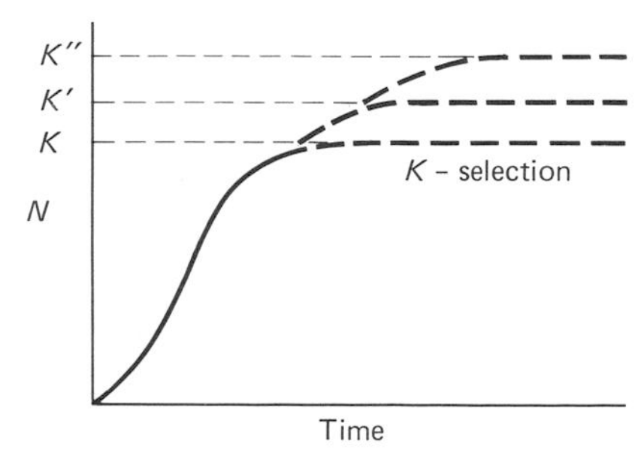

Departures from carrying capacity tend to be corrected by population growth or decline in proportion to the deviation from K. The rate of growth of a population will be positive if N is below K and negative if above. The rate of growth will be high if N is far below K, and low if only a little below K. At K, the rate of growth is zero. Since the processes that raise and lower N under such conditions are dependent on (K − N), they are said to be *density-dependent*. Examples of these processes are aggressive behavior, predation, and disease. In contrast, processes such as the action of weather, occurring independently of K, are termed *density-independent*.

K-selection / It has been argued that increases in the ability of individuals to survive and reproduce under stable environmental conditions lead to an increase in the number that can live in a given area, and hence to increased carrying capacity (MacArthur 1962; MacArthur and Wilson 1957). An increase in K can only come about through a period of population growth during which r increases and becomes temporarily positive due to an excess of *b* over *d*. This can in theory be achieved by a decrease in *d* (leaving r_{max} unchanged or lower) or an increase in *b* (raising r_{max}), or both. What actually happens?

We cannot observe all these changes in nature, but a hypothetical example may serve to illustrate the concepts. In the case of Lack's swifts increasing *b* by laying more eggs in a clutch would be selected against (depending on clutch size); but reducing *d* by increasing adult survival, thus allowing more years of breeding, could be selected for. In a species living in a climax habitat there are no uncrowded habitats to find and exploit, so there would be little advantage in producing large numbers of young. That avenue of adaptation is mainly closed. By contrast, there is a premium on simply being able to survive, especially for a long period. As a result, species living under climax conditions tend to have higher survival rates and lower reproductive rates than related species living in more temporary habitats. Further correlates of K-selection are listed in Table 9–2.

The changes that occur in r, r_{max}, *b*, and *d* during selection can be confusing. A brief summary may help. In r_{max}-selection, genotypes that achieve a higher r by virtue of a higher *b* are more successful. Higher values of r_{max} and *d* result. In K-selection, genotypes that achieve a higher r by virtue of a lower *d* are more successful. The population grows from an older to a newer and higher K. During growth, r is positive, but it returns to zero when the new K is reached. Afterward K is higher; *b* and r_{max} are likely to be lower than before. For a more detailed treatment of these matters see an ecology textbook (e.g., Emlen 1973).

Parental Care and K-Selection / The above introduction to *K*- and r_{max}-selection has focused on only one of many parameters influencing the evolution of parental care, namely the reproductive rate. The general thesis is that *K*- selection is conducive to the evolution of care-giving strategies, while r_{max}-selection is antithetical. Basically, all adaptations for increased or specialized parental care tend to increase the survival of the young by making the young more dependent on the parents. In r_{max}-selection, the young become less dependent.

The concepts of *K*- and r_{max}-selected species are relative ones, and so are best demonstrated by comparisons. The correlations can be seen in a wide variety of species. On the level of major groups the taxa in which parental care has evolved furthest, mainly terrestrial vertebrates, seem to be relatively *K*-selected, while most invertebrates tend to have characteristics correlated with r_{max}-selection (Pianka 1970). Among insects, the highly social forms with the most complex kinds of parental care also have the population characteristics of *K*-selected species, though exceptions occur (e.g., long-lived cicadas lack specialized parental care).

The influence of ecological conditions conducive to *K*-selection might explain the parallel evolution of similar adaptations for larval survival in unrelated groups, such as the adaptations for larval survival in leptodactylid frogs of South America and Australia (Martin 1970). On both continents there have been independent trends toward fewer and larger eggs, and a change from aquatic oviposition sites to terrestrial burrows or incubation chambers.

An impression of the effects of different weightings of fecundity and survival in various species can be gained by examining different species in the same genus. A comparison was made in Chapter 6 between the slowly reproducing *Peromyscus californicus,* with its well-constructed nest, long pair bond, and stable population, and the more rapidly reproducing *P. maniculatus,* with its relatively insignificant nest, brief pair bond, and relatively large numbers of young (Table 6–2 and Figure 6–10).

Extreme differences in fecundity rates are found among fishes. Oceanic species may spawn many thousands of small eggs and leave them to develop with no parental care. The Pacific species of salmon (*Onchorhyncus* spp.) leave the ocean at sexual maturity to spawn in streams, depositing many thousands of eggs per female and then dying. At the other extreme associated with specialized parental care, relatively few eggs are laid in the nest of a male stickleback (*Gasterosteus aculeatus*) or in the pouch of a male seahorse (*Hippocampus* spp.). As attractive as such comparisons of parental care are in fishes, the matter is complicated by the roles of body size, longevity, and egg size (Williams 1966).

In birds, species with precocial young (capable of locomotion and foraging at hatching) tend to have larger clutch sizes than species with altricial young (not capable of locomotion or foraging at hatching), despite the ability of altricial species to lay more eggs if eggs are removed before a clutch of eggs is complete (Kendeigh 1952). In altricial species newly-hatched young are in a less advanced state and require considerable more parental care, such as feeding and temperature regulation.

THE EVOLUTION OF ALTRUISTIC BEHAVIOR IN VERTEBRATES

What Is Altruism? / The evolution of parental care raises no special problems, since it can be explained according to the classical theory of individual selection. Clearly, when fitness is measured in terms of offspring surviving to the next generation, a certain amount of aid given to one's own offspring raises one's own fitness. Such aid is normally no sacrifice (contrary to what some human parents like to believe), since it benefits the donor directly by increasing the number of his offspring. Consequently it is, up to a point, normally not altruism (for discussion see Trivers 1974). In contrast, many cases exist among animals in which aid is given to individuals of the same species who are *not offspring* of the donor. For example, some insects, primates, and birds may regularly feed young that are not their own. Some may even spend most of their life helping others and may forgo entirely the production of their own young. When fitness is reckoned only in numbers of offspring, such behavior should be selected against. How could such behavior possibly evolve? This is the major problem that we shall examine in the remainder of this and in the following chapter.

Altruistic behavior can *in theory* be distinguished from other types of intraspecific social interactions according to the way in which benefits are apportioned. In a social interaction between two individuals, the behavior of one individual toward another can alter the fitness of both parties in four ways. A simple framework for the consideration of these effects has been provided by Hamilton (1964) and is shown in Table 9–3. The currency of comparison in this classification is the number of offspring, or what might be called *individual fitness* (since it is measured in terms of offspring produced by the individuals con-

TABLE 9–3 Classification of intraspecific social behavior according to effects on fitness. Neutral effects (no gain or loss) are omitted. (After Hamilton 1964.)

		Individual reproductive success of recipient	
		Gains	**Loses**
Individual reproductive success of donor	**Gains**	cooperative	selfish
	Loses	altruistic	spiteful

cerned). Behavior of one individual (the donor) toward another (the recipient) that raises the number of offspring of both parties (relative to nonparticipants) is termed *cooperative*. In *selfish* behavior the donor gains relative to the recipient, who loses. In *spiteful* behavior they both lose (relative to nonparticipants). In *altruistic* behavior the donor loses relative to the recipient, who gains. In this simplified scheme, cases in which a participant neither gains nor loses are ignored. Since fitness is relative, it is not possible for one participant to gain or lose without a change in the fitness of another.

Contrary to their customary usage in describing human behavior, the terms *selfish, cooperative, spiteful,* and *altruistic* in Table 9–4 are not intended to specify the emotional state of the performers nor to imply any commonality in the physiological or motivational mechanisms underlying the behaviors in different taxa of animals. Some readers may balk at the use of these terms for animal behavior. But they are already in common use, and it is useful, in my opinion, to employ these terms derived from human experience to emphasize that these situations and relationships are common to animals generally and are not uniquely human. Recognition of this fact should encourage the practice of viewing the human species in biological perspective. There might actually be more similarity in the underlying mechanisms than we presently believe. Certainly we are in no position at present to conclude that there is no biological basis for the similarities in such behavior among the various groups of vertebrates. Finally, the reader who objects to such terms as *altruistic behavior* might reflect on the fact that nearly *all* our terms used to describe the behavior of animals are derived in one way or another from terms used to describe human behavior.

The definition of altruism given above is based on full knowledge of the effects of the behavior on the fitnesses of the participants. It is a useful definition for theorizing in evolutionary biology (see, e.g., Hamilton 1963, 1964, 1972). However, such full knowledge of the effects of behavior on individual fitness can virtually never be achieved in practice, so the definition is a theoretical rather than operational one. Altruism can be defined operationally as the *giving of aid in the form of arbitrarily defined goods or services to individuals of the same species who are not offspring or direct descendents of the donor and without direct benefit to the donor or its mate.* At the risk of creating still another bit of jargon, this kind of aid-giving will be referred to as *operational altruism* because it is defined by specified operations that can be performed (giving and parenthood). When not otherwise specified, *altruism* in the following account will mean *operational altruism.*

Where Does Operational Altruism Occur? / Operational altruism has an interesting distribution in the animal world. It is found conspicuously among social insects (Hymenoptera, Isoptera), but it is apparently lacking in all vertebrate groups except birds and mammals. Among the vertebrates the clearest examples occur in communal forms of social organization.

A distinction is commonly drawn between communal and colonial or other gregarious types of social organization. A *communal* group tends to have constant membership over a relatively long period of time; the individuals in the group tend to have established roles and relationships with each other; the home range of the group is often maintained as a territory or dominion; and outsiders are usually excluded from the group. The members of the communal group are often close relatives. The group is kept together by social bonds and may often exhibit a high degree of cooperative or altruistic behavior. Examples of mammalian communal groups include savannah baboons (of the *Papio cynocephalus* superspecies), monkeys of several genera, wild dogs (*Lycaon pictus*), African lions (*Panthera leo*), and spotted hyenas (*Crocuta crocuta*). Communal birds include many species of New World jays (e.g., *Aphelocoma ultramarina*), anis (e.g., *Crotophaga ani*), acorn woodpeckers (*Melanerpes formicivorus*), Australian white-winged choughs (*Corcorax melanorhamphus*), and many others (references below).

Noncommunal groups, such as schooling fish, colonial birds, and herding mammals, tend to have a shifting composition. As a result new relationships are continually being established and old ones dropped. Because of the shifting membership of the group, it may not even be possible to define its home range. When home ranges can be recognized for such groups, they are usually not defended against other such groups. For example, a colony of sea birds does not defend its foraging area against members of other colonies. The forces that keep noncommunal groups together center on attraction to a common resource, such as a colony's nesting area, and an attraction to the same or similar species, rather than on social bonds to particular individuals. Where cooperation occurs in such groups, it seems to be fortuitous rather than coordinated, and it is typically of a lower grade than in communal groups. These differences are summarized in Table 9–4.

The distinctions between communal and noncommunal groups are not sharply dichotomous, but rather, are mainly matters of degree. Intermediate species and situations occur. Since extremes are easily recognizable and apparently more common than intermediates, the distinction is useful.

Operational altruism is rare among animals when the million or more animal species are considered, but it can be found in hundreds of species in a variety of contexts. Of these contexts the one most studied is *food sharing.* The sharing of food is the basis of the energy economy of social insects, and is conspicuous in African lions (Schaller 1972), wild dogs (Kühme 1964, 1965), spotted hyenas (Kruuk 1966, 1972), and many birds. Among nonhuman primates food sharing is not conspicuous, but has been observed after cooperative hunts (e.g., in chimpanzees, *Pan troglodytes;* Teleki 1973).

Individuals other than the parents who aid in reproduction (for example, by bringing food to the young) or who otherwise aid individuals who are not descendents have been termed *helpers* (Skutch 1935), a term that should be useful in regard to both birds and mammals. Helping behavior includes much or all that has been termed *aunt* behavior in studies of primates (Rowell et al. 1964; Rowell 1972). Helping

TABLE 9–4 Summary of differences between communal and noncommunal groups. Only extremes are listed here, but intermediate states and species occur in nature.

	Communal group	Noncommunal group
Group size	relatively small and constant	more variable; can be very large
Group composition	relatively constant	more variable
Aggressive repulsion of other groups from home range	usually present	typically absent
Social relationships among members	established and relatively constant	often in flux because of shifting group membership
Attractive force	social bonds to recognized individuals	common resource, e.g., colony site; any member of same species

is equivalent to operational altruism. The commonest context of helping is sharing food with young. The occurrence of helpers in birds has been reviewed by Skutch (1961). The occurrence in birds of communal breeding systems, which are characterized by a conspicuous role of helpers, has been reviewed by Lack (1968), Harrison (1969), Rowley (1968), and Brown (1974). Similar reviews for mammals are unavailable, but much relevant information is available in the surveys by Ewer (1968), Eisenberg (1966), Fisler (1969), Schaller (1972), and Jolly (1972).

Another frequently discussed (but little investigated) example of operational altruism is *alarm-calling* (e.g., Smith 1964; Trivers 1971). The assumptions are (1) that alarm-calling benefits the hearers by alerting them to dangers, (2) that alarm-calling risks the life of the caller by attracting the attention of the predator to the location of the caller, and (3) that the hearers are frequently not descendents of the caller and perhaps not even related. The first assumption can be verified by any field naturalist; alarm calls do cause alarm. The second is tenuous. There seem to be no naturalistic observations published which report that an alarm-caller lost its life by attracting the attention of a predator while saving its fellow prey. It has even been argued that alarm-calling benefits the caller by confusing the predator (Perrins 1968); alarm-calling might also bring to the attention of the predator the individuals who are stimulated by the call to move. Moreover, the first animals of a group to call are usually the first to reach protective cover, which makes them less likely to be caught. The third assumption seems to be frequently false with respect to communal groups; the benefited receivers often are closely related. Its validity for noncommunal groups depends on the frequency with which young stay with their parents. During the nonbreeding season, the young of certain large waterfowl and cranes are thought to stay with their parents, but comparable information for small migrant birds is unavailable and the popular assumption has been that young are most likely to become separated from their parents on migrations. (In this connection it is worth asking the following question: Do the calls given by migrating small birds at night serve primarily to keep members of a family or pair together?)

Other examples of operational altruism are known. *Mutual grooming* in some primates and mutual preening in birds may involve aiding nondescendents. *Anti-predator behavior* by members of a group may have similar consequences. Other supposed examples of altruism, such as submissive or appeasement behavior, or losing in a contest for territory or status (e.g., Darlington 1972), can be explained in a straightforward manner by means of models based on individual selection and on the direct effects on the number of the individual's own offspring. In these cases the loser (the supposed altruist) either lost in an outright bid for victory, or forfeited the contest because it had little or no chance of winning. Deciding not to fight because the odds for winning are unfavorable is self-preservation, not altruism.

Sketches of Some Communal Breeders / Birds have certain advantages over mammals for the study of altruistic behavior. It is possible for one individual to incubate another's eggs in birds, but not in mammals; and the absence of nursing in birds greatly facilitates the feeding of young by nonparents of both sexes. For these reasons the effort devoted to care of young by nonparents can be more easily quantified in birds than in mammals by recording the time spent on the nest and the number of feedings.

One communally breeding species that has been intensively studied (Brown 1963, 1970, 1972, 1974) is the Mexican jay (*Aphelocoma ultramarina*). These jays are nonmigratory and live in flocks of 5 to 15 in pine and oak woodlands. Each flock defends a territory communally all year and usually excludes members of other flocks from most or all of its home range. Nests are built by pairs, and the eggs in each nest are laid usually by only one female. There may be one to four active nests in a flock, but two is a common number. At each nest most or all of the members of the flock help bring food to the nestlings. All but two of the feeders at a given nest are, therefore, altruistic helpers. The contributions of each member of one flock at two nests in successive years are shown in Figure 9–3. The parents tend to be the most active in feeding their young, but their contribution typically amounts to only about half the total food. The other half is brought mainly by nonbreeding adult and immature helpers.

The population ecology of communally breeding birds is often characterized by a relatively low reproductive rate and high survival rate. The low reproductive rate is caused by a delay in the age at which individuals first breed. Mexican jays do not begin effective reproductive behavior until they are three years old or more, while their close relative, the scrub jay (*A. coerulescens*), begins at one year. In most communal breeders, some or all of the helpers are immatures of a previous brood, or adults or subadults from previous broods, who failed to find mates of their own. Although the helpers contribute to reproductive success, their individual activities are usually less than those of breeding adults. Another feature of the ecology of many communal breeders that is present in the Mexican jay is reduced dispersal. As shown in Figure 9–3, some of the young produced in one year were still in the flock a year later, and only one individual was known to have moved to another flock.

A more advanced stage of altruistic behavior is found in the group

9–3 Communal feeding of nestlings in the Mexican jay. The percentage contribution of each member of one flock is shown at both active nests in two successive years. An asterisk designates a statistically significant ($p < 0.01$) preference for delivering food to one of the nests. Year of hatching is indicated where known; jays in class of 1967⁻ were hatched in 1967 or an earlier year and are of uncertain ages. The sample size, *n*, refers to the total number of deliveries of food. (From Brown 1972.)

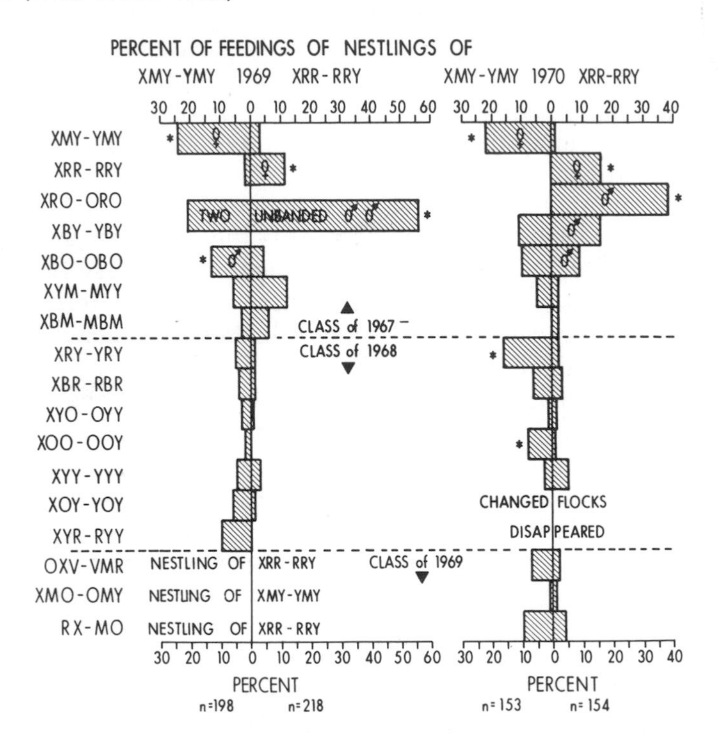

of four species of anis which comprise the subfamily, Crotophaginae, of the cuckoos (Cuculidae). Their behavior and social organization were studied by Davis (1942). These species live in a savannah habitat with much open country and scattered trees; they range over much of South and Central America, are quite gregarious, and normally occur in flocks that defend flock territories. In the least social of the four species (*G. guira*), pairs segregate out from the flock to defend pair territories and to breed. They defend the flock territory only weakly. In the most social species (*Crotophaga sulcirostris, C. ani*), the pairs may unite to build a communal nest in which several females may lay their eggs and take turns incubating.

Similar communal relations occur to a greater or lesser extent in a number of other species of birds (Lack 1968; Rowley 1968; Harrison

1969). Many of these are found in Australia; they include the superb blue wren (*Malurus cyaneus*), the white-winged chough, and the apostle bird (*Struthidea cinerea*, so named because it tends to occur in groups of 12). Sporadic examples of self-sacrifice involving the feeding of young not an individual's own are more common. Their occurrence has been reviewed by Skutch (1935, 1961).

Is Kin Selection Involved? / Various solutions to the problem of the evolution of operational altruistic behavior have been proposed and debated (see Haldane 1932, 1955; Wright 1945; Williams 1966; Trivers 1971; Darlington 1972; Eshel 1972), but only that of Hamilton (1963, 1964, 1971) has won general acceptance among evolutionary biologists. Hamilton recognized that aid to offspring was a special case of the more general category of aid to relatives. If selection could favor aid given to one's offspring, it could also favor, to various degrees, aid given to siblings and more distant relatives. Hamilton argued that aid to relatives would be favored when the ratio of gain in fitness by the recipient to loss in fitness by the donor was greater than the reciprocal of a measure of the degree of genetic relationship between the donor and recipient, Wright's coefficient of genetic relationship, r:

$$\frac{\text{gain in fitness to recipient}}{\text{loss in fitness to altruist}} = k > \frac{1}{r}$$

For simple relationships in diploid species without inbreeding,

$$r = \Sigma \left(\tfrac{1}{2}\right)^L,$$

where L is the number of generation links in a lineage between the two individuals (Li 1955). Consequently, for a sib-sib relationship, r equals $\frac{1}{2}$; for parent-offspring, $\frac{1}{2}$; grandparent-grandson, $\frac{1}{4}$; uncle-nephew, $\frac{1}{4}$; first cousins, $\frac{1}{8}$. Examples are shown in Figure 9–4. When full sibs (brothers and sisters) are involved, the gain in fitness must be more than twice the loss if selection is to favor altruism. To take an extreme example, it would be beneficial for a surviving sib to sacrifice his life to save his drowning brothers and sisters ($r = \frac{1}{2}$) only if by doing so he could save more than two of them ($k > 1/\frac{1}{2}$). The sacrifices need not be so great if the gains in fitness are smaller. Selection acting in the above manner on groups of relatives has been aptly termed *kin selection* by Smith (1964; see also Brown 1966).

The basic concept of Hamilton's theory is that genotypes causing individuals to aid others who are likely to carry the same genes may increase in frequency in a population under certain conditions. These occur when a positive *net gain* to the genotype occurs as a result of certain types of aid-giving behavior that require a loss in the donor's

9–4 Calculation of the coefficient of genetic relationship for various pairs of relatives (black dots) in a large, panmictic population of diploid animals according to the formula $r = \Sigma(\frac{1}{2})^L$. L is the number of links in a path between the pair. Where more than one path exists, the values of $(\frac{1}{2})^L$ for all possible paths are summed.

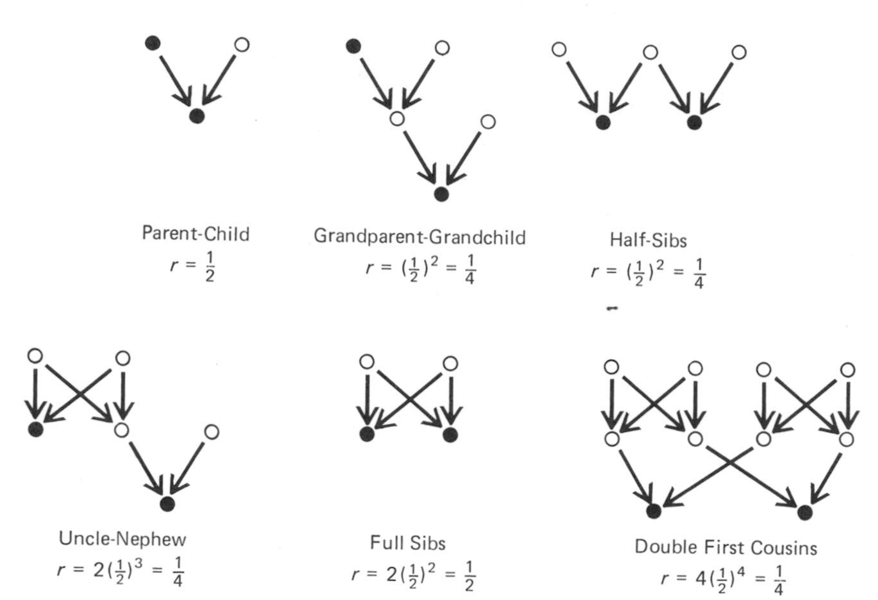

individual fitness: although the altruist may suffer a loss in individual fitness (has fewer offspring than nonaltruists), the genotype of the altruist receives a net gain compared to genotypes of nonaltruists. In such cases the altruistic genotype is said to have a higher *inclusive fitness* than the nonaltruistic genotype. Inclusive fitness is based on gene frequency rather than on numbers of offspring.

Individual Mexican jays whose nests have failed or who are immature can raise virtually no young of their own at that time. If they were to act altruistically, k for them would be very high, because they have already lost their chance to breed or have never had a good chance. Therefore, they have little immediate reproductive potential of their own to sacrifice at that time. In addition, since the young of the recipient are likely to be full sibs (or even closer because of inbreeding), they are probably at least as likely (or more so) to carry the genes of the altruist as would its own young if the altruist were to mate with a stranger outside the flock. For these individuals, who are the main helpers, the situation is favorable for altruism and kin selection both in terms of k and r. Selection in this situation should favor the ability to act altruistically in circumstances when an individual has little chance to raise its own young, but should also favor the genotype of individuals who would act selfishly when the chances of raising their own young were good. The ability to act altruistically involves both being near the parents and responding appropriately to the social environment.

It should be possible to test some of the predictions of Hamilton's theory in field observations. Three predictions will be considered. (1) Since the strongest selection pressures for altruistic behavior are among near relatives, we should expect to find altruism occurring *more often among closely related individuals than distantly related ones.* (2) Since dispersal tends to break up family groups, altruism should be more common in species with relatively *poor dispersal mechanisms* than in those with efficient dispersal. (3) Any mechanism increasing the frequency of association among altruists should be favored by selection in cases of existing altruism; consequently, altruists should possess the capacity for recognizing each other, including *individual recognition.* We will now discuss these hypotheses briefly.

1) Does altruism occur more commonly among closely related individuals than among more distantly related ones? Studies of communal groups have revealed that many members of the group are descendents of other group members, for example, avian helpers in many cases (Lack 1968; Harrison 1969; Rowley 1968; Brown 1974). There is some movement of individuals from one group to another, but in mammals and birds members of communal groups are likely to be more closely related to others within their group than to others outside of it. In such groups altruistic and cooperative behaviors, such as food-sharing, predator defense, and grooming of others, benefit one's own group almost exclusively. In short, a positive correlation between altruistic aid-giving and genetic relationship probably exists, but it remains to be confirmed generally by quantitative studies.

2) Is altruism more common among species with relatively little dispersal? Because individuals of communal species rarely leave their home group, their dispersal must be very low. When they do leave, long-distance movements appear rare. Among mammals and birds, communal breeders tend to be nonmigratory, although some wandering occurs (birds, Brown 1974; hyenas, Kruuk 1972; lions, Schaller 1972). In the few communal species that do leave their territories during unfavorable conditions, the group remains an integral social unit and returns to its territory when conditions improve (e.g., white-winged choughs; Rowley 1965*b*). It is notable that in avifaunas where migratory species are numerous, as in the temperate and colder parts of North America and Eurasia, communal breeding (as defined above) is virtu-

ally nonexistent. Further, no conspicuously migratory species has a communal social system. In contrast, in tropical and Australian avifaunas containing few migratory breeding species, communal breeding is relatively common (Fry 1972). In brief, altruistic behavior as found in communally breeding species is associated with relatively low rates of dispersal in comparison to noncommunal breeders.

3) Are there mechanisms that keep individuals of a group together, and do they tend to keep relatives together more than nonrelatives? Is individual recognition more highly developed in altruistic species? The question concerning individual recognition cannot be considered a test of Hamilton's theory because the need for individual recognition among birds and mammals is also strong in many types of nonaltruistic social systems. The problem of recognition is especially great when large numbers of pairs nest together or winter together. Individual recognition has been shown to be present both in colonially nesting birds (Chapter 16) and in wintering, nonbreeding flocks of birds in dominance hierarchies (Chapter 5). Individual recognition is clearly present also in communal groups of monkeys, baboons, wild dogs, hyenas, and various other species that live permanently in small, close-knit groups, as can be inferred from studies of dominance and other field observations. More important than individual recognition for the evolution of altruism in such cases is recognition and repulsion of outsiders, since a single outsider entering a communal group can have a large and enduring genetic influence if he is allowed to breed. Few field studies have been carried on long enough to evaluate the frequency of genetic infiltration by outsiders, but it is clear from numerous observations of antagonism to outsiders that they are immediately recognized and usually repulsed in communal species (e.g., Mexican jays; Brown 1970) and typically ignored in colonial species.

A small amount of infiltration into communal groups by outsiders is to be expected. Successful groups are those that maintain their own genetic superiority in fitness while exporting their genes to less successful groups. This means that they should be simultaneously selected for, first, staying in the home group and repelling outsiders, and, secondly, producing in appropriate circumstances individuals that will leave the home group and successfully penetrate other groups or found new ones. The result in communal breeders is a balance in which relatively many stay and few leave. To return to the initial question in (3) above, the mechanisms that keep potential altruists together are social bonds formed in part as a result of growing up in a communal and often territorial group. Since such individuals tend to be more closely related to each other than to others in the population, it can be concluded that the features of communal social systems do work to keep related altruists together and to prevent them (most of the time) from aiding individuals who are genetically relatively distant.

Summarizing the field observations relative to these three predictions from Hamilton's (1964) theory, it appears that the predictions have all been affirmed, at least qualitatively. The observed facts agree well with requirements of the theory and lend weight to its plausibility. Consequently, it seems likely that genetic relationship is one of the factors that have been important in furthering the evolution of operational altruistic behavior. Alternative theories will be considered below.

Are Altruists Basically Selfish? / Zahavi (1974) suggested that helpers in flocks of Arabian babblers (*Turdoides squamiceps*) were no help at all, and actually did more harm than good to their parents by getting in the way, attracting predators to the nest, and competing for food. He felt that the helpers were acting to maximize their own individual fitness and were not really helping others to produce more young. If so, then it would be possible to explain the "helping" by means of individual selection *without the need to invoke kin selection.* Although he produced no convincing evidence to bolster this hypothesis (Brown 1975), the idea must be taken seriously. For individuals raised in a communal flock, the strategy of staying in their home flock as nonbreeders may be more likely to result in their ultimately acquiring an opportunity to breed than if they were to disperse far and wide immediately on reaching independence (Brown 1974). Because all possible territories are likely to be filled and difficult to enter, a young bird may have little chance to establish itself or enter a group. Since it would be advantageous to parents and offspring alike to allow the young to remain as long as optimally necessary, it is not hard to see many advantages to individuals in such a system, advantages that do not require kin selection as an explanation. The food given by helpers to younger sibs might even be interpreted as "practice nesting" by the helpers rather than wasted energy. To explain such operational altruism completely on the basis of selfish advantage requires that all aid given by nonbreeders to breeders, including all aid by young to their parents, be interpreted as aiding the nonbreeders and not the recipients. It is difficult to see how this could be true in Mexican jays; in that species, adults who have lost their own eggs or young may even contribute more aid to the young of other nests than do the actual parents (Brown 1972). At present, however, it seems too early to completely rule out individual selection as a sufficient explanation of altruism in all communal birds.

The critical evidence that is needed in order to require kin selection as part of the evolutionary explanation of operational altruism includes a demonstration that more young survive to maturity in groups with helpers than without. Such data are difficult to obtain. The closest approach so far has been in the study of the superb blue wren in Australia (Rowley 1965*a*). This species breeds in pairs, with or without one male helper. Rowley found that the number of young raised by threesomes was greater than that raised by pairs.* If it could be shown that the difference was caused by the helper, the case for kin selection would be strong. Alternatively, the higher production of young by threesomes might be at the cost of lower weight and/or survival rate of the young. Or the difference might be due to the quality of the territories rather than to the behavior of the helpers. Further work along these lines will be required before the case for or against kin selection will be fully convincing.

Reciprocal Altruism / Not all cases of operational altruism involve individuals suspected of being close relatives. For these exceptions it would seem at first sight that kin selection cannot be invoked. Individual human acts of altruism or heroism often involve unrelated strangers. However, it seems likely that for most of his evolutionary history man lived in extended family groups. The breakdown of the family and the tribe in man is a very recent phenomenon. Therefore, the possibility of kin selection being a critical factor in man's evolution cannot be eliminated. Nevertheless, the human capacity for cooperating and exchanging favors is unique among animals and might require additional explanation. For the human species the theory of reciprocal altruism, as elaborated by Trivers (1971), seems appropriate. If an altruist, by doing a large favor for a recipient at little cost to himself, could on the average ensure the return of the favor with great benefit to himself, his individual fitness would be enhanced. Individuals who were most successful at winning the favors of others at small cost to themselves would tend to be more successful and would then pass on to the next generation an increased number of genes conferring the ability for reciprocal "altruism." For selection to further this kind of "altruism," individuals must (1) live together for long periods of time so as to have sufficient opportunities for reciprocation, (2) be able to recognize each other individually, (3) be able to remember their indebtedness, and (4) be motivated to reciprocate. These are stiff requirements. Consequently, it seems likely that if selection has operated in this way it has done so only in the immediate progenitors of the human species.

In Mexican jays reciprocity is brought about in a different way,

* For further discussion of the fit of these data to the predictions of kinship theory, see Appendix.

which is not, strictly speaking, what Trivers meant by reciprocal altruism. Altruists are mainly adults who have not found a mate or whose nests have failed, but they also include immature birds (one and two years old). Successful parent-recipients are usually four years old or older. Each adult has passed through an ontogenetic sequence in which it started out as an inefficient helper (one to two years), improved, and perhaps bred unsuccessfully (three years), and finally bred successfully (five years and up). The help that an individual, X, gave to its elders while it was young is in turn given to X by younger jays when X matures. Because most birds die or disappear before becoming successful breeders, the successful breeders end up receiving more aid from the flock than they as younger individuals gave to the flock.

Other Theories / A full discussion of various other theories cannot be given here, but the following alternative or supplementary theories deserve mention. The idea of *population selection*, for instance, has been invoked to explain the evolution of altruism (Eshel 1972; Darlington 1972). The idea is that groups containing altruists might survive better or colonize better than groups without altruists. Although these theories are probably formally and mathematically correct, there seem to be no biological facts about altruism that cannot be explained more simply by combinations of kin selection and individual selection. A supplementary model suitable for cooperative hunting (e.g., in wild dogs) has been suggested, in which cooperating hunters need not be kin and in which the advantage of cooperative hunting is frequency-dependent (Boorman and Levitt 1973).

Symbiosis or Mutualism / When different species live together so as to benefit each other, the relationship is known as symbiosis or mutualism. Examples include (1) the crabs that carry small sea anemones on their shells, providing protection for the crab and food for the anemone (Dales 1966); (2) the small fishes that live unharmed among the poisonous tentacles of large sea anemones, luring food to the anemone and enjoying the protection of its tentacles (Schlichter 1968; Allen 1972), and (3) the small cleaning shrimp and fishes that remove ectoparasites from larger fishes (Feder 1966; Losey 1972). Since both species benefit from the association, these cases have not traditionally been regarded as altruism. Yet as Trivers (1971) pointed out, the cleaning symbiosis presents an interspecific parallel to intraspecific reciprocal altruism. Both cleaners and cleaned benefit, and both show restraint. The benefits are food for the cleaner and removal of ectoparasites for the cleaned. The restraints are reluctance of the cleaned to swallow the cleaner (Figure 9–5), and reluctance of the cleaner to take a bite out of the flesh of the cleaned. Some "cheating" seems to occur. Some

9–5 A cleaner fish (wrasse, *Labroides dimidiatus*) at work inside the mouth of a host (coral trout, *Plectropomus maculatus*).

cleaners are swallowed. And certain small fishes gain access to the cleaned fish by mimicking a cleaner fish, only to take a bite of flesh from the cleaned fish (Wickler 1963; Springer and Smith-Vanig 1972).

It is easy to imagine a plausible origin of the cleaning symbiosis. The large fish, when not hungry, rested in an area where small fish that feed on organisms growing on rocks might also remove such organisms from the large fish. Large fish that consumed their cleaners would suffer more from ectoparasites than those who did not. If the benefits were strong, selection might then favor the evolution of genetic mechanisms to facilitate the ontogenetic development of the cleaner-cleaned relationship. This evolutionary explanation is far from complete, however, and many relevant hypotheses remain to be tested.

Types of Altruism / Definitions of altruism and evolutionary theories about it have proliferated since Hamilton's (1963) stimulating rediscovery of the subject. Different authors have defined altruism in fundamentally different ways. These have been discussed above, but to facilitate direct comparison they are summarized in Table 9–5. The evolutionary theory that an author invokes is directly dependent on his definition, which is based on the "currency" in which gains and losses are measured. The various currencies include individual fitness (reckoned by counting offspring), inclusive fitness (reckoned by examining changes in frequency of genes or genotypes), and certain goods

TABLE 9–5 Types of altruism discussed in evolutionary biology. In all cases aid is given by one individual, the altruist, to recipients who are not offspring of the altruist, without immediate and direct benefit to the altruist.

Name	Hypothetical sacrifice to altruist	Hypothetical gain to recipient	Postulated gain to genotype of altruist	Type of natural selection invoked	References
Operational altruism	arbitrary, e.g., food	arbitrary, e.g., food	not specified	not specified	this chapter
Individual fitness altruism	individual fitness as reckoned in offspring	offspring of recipient	offspring of close relatives; inclusive fitness	kin selection	Hamilton 1964
Intraspecific reciprocal altruism	offspring; inclusive fitness	offspring; inclusive fitness	offspring of altruist; individual and inclusive fitness; gain exceeds loss	individual selection	Trivers 1971
Interspecific reciprocal altruism	individual and inclusive fitness; e.g., risk of injury	individual and inclusive fitness; e.g., removal of parasites	individual and inclusive fitness	individual selection	Trivers 1971
Inclusive fitness altruism	inclusive fitness of genotype of altruist	inclusive fitness of genotypes of recipients	population survival	population selection	Eshel 1972

or services that are normally associated with fitness concepts (e.g., food, grooming, protection).

CONCLUSIONS

The giving of aid is an important aspect of social organization, especially in man. The phylogenetic antecedents of most types of aiding behavior are found in the behavior of parents. Parental aid varies from the simple choice of a site for oviposition to enduring social bonds like those occurring in some monkeys and man. From an evolutionary perspective, aiding is independent of intelligence but dependent on ecological relationships.

Brood size is a fundamental determinant of parental care. Lack maintained that animals produce as many offspring as they can successfully rear, rather than holding back reproduction for population control. Observations and experiments have shown that members of broods that are larger than normal tend to weigh less and suffer greater mortality than those in smaller broods. Many other factors also influence brood size.

Selection favoring more rapid population growth through more rapid reproduction by individuals (e.g., by producing more eggs or babies) is known as *r*-selection. It tends to occur in new or transient ecological situations and to reduce parental care.

Selection favoring greater efficiency and effectiveness under conditions of stable population density at the carrying capacity of the environment for the species is known as *K*-selection. It tends to increase parental care and other mechanisms that enhance survival. The birth rate and maximum population growth rate (r_{max}) of *K*-selected species tend to be lower than in *r*-selected species.

For the purposes of evolutionary theory, behavior between two individuals can be classified according to effects on the production of offspring (i.e., according to individual fitness) into four types: cooperative, selfish, altruistic, and spiteful. These names refer to effects on fitness only and not to emotional states.

Since full knowledge of the effects of behavior on fitness is difficult to obtain, an operational definition of altruism is needed. Operational altruism is defined as the giving of aid in the form of arbitrarily designated goods or services to individuals of the same species who are not offspring or descendents of the altruist, without direct benefit to the altruist or its mate. So defined, altruism is found conspicuously in the social insects and among communally living vertebrates. The latter include wild dogs, certain primates, and various birds. These groups often share food, defend territories, and form strong social bonds among members. They are to be distinguished from colonial and other noncommunal groups.

Avian operational altruists have been called *helpers*, a term that should also be useful in dealing with mammals. Aid to nonoffspring may take the form of food-sharing, predator detection, alarm-calling, mutual grooming, or territorial defense.

The most popular theory (Hamilton 1964) for the evolution of operational altruism is based on kin selection. If altruists aid mainly their siblings and other close relatives, the benefits will be received by other individuals who are likely to carry the same genes as the altruists. In this way the genotype of an altruist might increase in frequency, even in cases involving lower-than-average reproductive success by altruists with respect to their own young. In the vertebrates, selection should favor altruism toward relatives by individuals whose chances of breeding are low at the time, but should favor selfish behavior in these same individuals when chances of successful breeding are high. The ontogenetic history of altruistic behavior should in these species come to be sensitive to the social environment.

Other theories about the evolution of operational altruism have also been considered. In some species it has been suspected that "helpers" are acting purely selfishly by increasing their own later reproductive success while not significantly aiding their hosts. In other cases it has been argued that reciprocity (a form of cooperation) might be involved. Even population selection has been invoked. The cleaning symbiosis has also been interpreted as reciprocal altruism, but it is fundamentally different since the relationship is interspecific. The various types of altruism and associated theories are summarized in Table 9–5.

10 The Evolution of Sociality in Bees

. . . With social insects, selection has been applied to the family, and not to the individual. . . .

Charles Darwin, 1859

THE EVOLUTION OF aid-giving behavior in insects has followed different paths than in the vertebrates, but has reached an amazing level of complexity, especially in its crowning achievement, the honeybee (*Apis mellifera*). This species has so fascinated and nourished man from the time of Aristotle and Pliny to the present that many books, and no fewer than 20 periodicals, have been devoted exclusively to bee lore. A comprehensive guide to this wealth of literature is fortunately available in the works of Chauvin (1968), von Frisch (1967), and Michener (1974).

The honeybee is at the apex of only one of at least 11 independent evolutionary origins of sociality in insects (Wilson 1966), where *sociality* implies the cooperation of offspring in rearing their siblings. Sociality in its highest form, eusociality, occurs in only two orders of insects, the Isoptera (termites) and the Hymenoptera (bees, ants, wasps). All modern termite species are eusocial and are thought to have descended from a single, original social species. In the Hymenoptera all degrees of sociality occur, from strictly solitary species through semisocial and subsocial to fully eusocial species. Since sociality is thought to have evolved independently in at least ten different branches of the hymenopteran phyletic tree, we may ask, why has sociality evolved so often in the Hymenoptera and only once (in termites) in all other insects? This problem will be approached in two ways: first, by examining eusociality in the honeybee and in other species that exhibit varying degrees of sociality, and second, by considering the application of Hamilton's (1964) theory for the evolution of altruism to the social insects.

Although the bees are best suited for a consideration of the evolution of sociality among insects because there exists a spectrum of intermediate semisocial and subsocial species and because a wealth of information is available about them, communal living is as much or more specialized in some of the ants, wasps, and termites. For information about these other equally fascinating social insects the reader is referred to the works of Wilson (1971), Evans and Eberhard (1971), Sudd (1967), Krishna and Weesner (1969–70), and Brian (1965).

The Spectrum of Bee Societies / Any major evolutionary achievement, such as sociality in the honeybee, has come about through a long series of small genetic changes. Often adaptation to one set of environmental conditions may be a favorable starting point, or *preadaptation,* to quite a different adaptation. In other words, the effect of a previous selective process may be a partial cause of the next selective process. Although the precise sequence of changes that led to the present state of *Apis mellifera* can never be recreated or known with certainty, it is possible, by examining other species of bees, to construct logical evolutionary hypotheses concerning the general nature of the evolutionary changes that have occurred. As background for the following discussion, Figure 10–1 shows a partial classification of the Hymenoptera.

Among nonsocial bees the simplest social organizations are found in the solitary species, in which the only social acts are courtship and copulation. In these species it is common for the mothers to die before their young emerge. Consequently, there is essentially no opportunity for the young to aid her. The poppy bee, *Osmia papaveris* (Figure 10–2), may be chosen from many as an example (Lindauer 1961). Fertilized females of this species dig a hole in the ground, line it with poppy petals, place a food ball of nectar and pollen in it, lay eggs inside it, cover it, and never return. The young emerge, mate, and repeat the pattern unaided. In other solitary bees, such as *Megachile,* the holes may be placed together rather than separately, but the basic pattern is similar.

The first important step toward sociality is found in those species in which the females are alive and working on the nest when the young emerge. This superficially insignificant fact is a critical preadaptation, because it makes possible social contact between the emerging daughters and their mother, thus allowing selection to operate to elaborate on this crucial relationship. This may lead to a condition in which emerging females may (1) found new solitary nests, (2) found new colony nests jointly (sib-foundress associations), or (3) remain at their native nest and contribute to its economy by working and laying eggs.

In some species, such as *Lasioglossum marginatum,* emerging females have this sort of "choice." There is little or no division of labor; the founding female of a colony lays eggs and cares for them; early-hatched females remain, lay eggs, and care for their own eggs and offspring. The nest is defended jointly by all females.

Division of labor between females of different types probably ap-

10–1 Partial classification of bees and other Hymenoptera. (From Rothenbuhler 1967.)

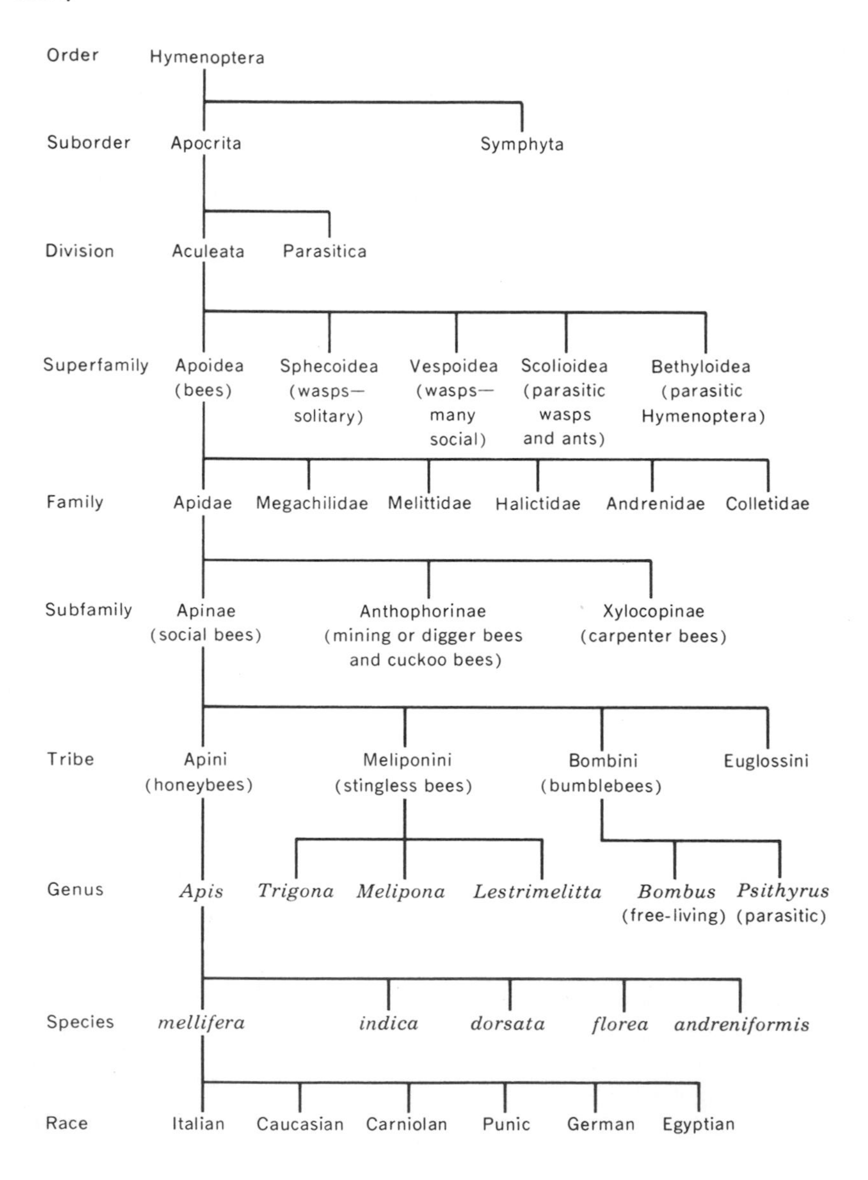

pears first in species such as *Augochloropsis sparsilis,* which have been described as semisocial (Michener and Lange 1958). Examining the nests of this species revealed two kinds of females. Some foraged outside the colony, fed the young, and lacked enlarged ovaries; these bees functioned as workers even though some had mated (sperm was found in the spermatheca). Others laid eggs but were not involved in foraging. Observations of marked individuals suggested that mated

10–2 The poppy bee is a solitary species. Fertilized females dig a hole in the ground, line it with pieces of petals from a poppy flower, place a food ball of pollen and nectar in it, cover it and never return. The young emerge and repeat the sequence unaided. (Reprinted by permission of the publishers from Martin Lindauer, *Communication among Social Bees.* Cambridge, Mass.: Harvard University Press, Copyright 1961, by the President and Fellows of Harvard College.)

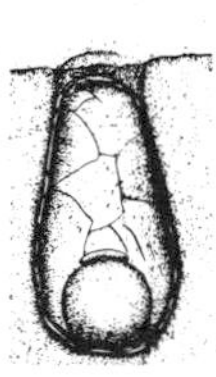

females tended to develop in one of two directions; they either became predominantly pollen-gatherers and workers in the colony, or they became predominantly egg-layers. Most unmated females probably became workers. These two types were not significantly different in structure upon emergence, but those specializing in foraging and handling food tended to develop ragged wings and worn mandibles, while those involved in egg-laying remained relatively fresh in appearance. It is in such cases that Hamilton's (1963, 1964) theory for the evolution of altruism may have the greatest bearing (see below).

Eusociality is reached in those species in which the caste assignment of a female is determined before she leaves the larval stage, and in which there is a morphological difference between the castes (queens and workers). In the bumblebees, for example, large overwintering queens found new colonies annually each spring. The queens care for their first brood; but these young are smaller than the queen and are not able to found new nests; they become workers for the succeeding broods. The queen then leaves the work of the colony to her first and subsequent broods, loses her wings, and devotes herself to laying eggs.

It seems likely that stages similar to those just described were passed through in the evolution of sociality in bees. The species mentioned above do not constitute an evolutionary series themselves; they only represent a few of the types that could have been involved.

Only a few of the factors involved in the evolution of sociality were mentioned in the above series, mainly maternal survival and division of labor. A more complete, but diagrammatic, picture of the factors involved is shown in Figure 10–3. *Storage* of honey for use by the colony in times of food scarcity is found mainly in the highly social species, and is an important factor in the survival of colonies, especially in perennial colonies like those of the honeybee. *Communication* is highly developed in the honeybee, but also occurs, in various forms, in the more primitive species (see Chapter 19 on communication in honeybees). *Predigested* or secreted food is an important factor in communal life because it enables the larvae to grow at a much faster rate than they would otherwise; this leads to a faster increase in the number of workers and in the ability of the colony to grow and produce successful queens. The *passing of food* from one individual to another (*trophallaxis*) is also an important form of communication in social insects. Maternal care in the simpler species consists of providing a fixed amount of food for the larvae at the time of oviposition, then sealing the cell (*mass provisioning*). In the advanced species the cells of developing larvae are opened and food is supplied as the larva grows (*progressive provisioning*). This also speeds development, and leads

10–3 Grades of evolutionary specialization in different aspects of social life in various bees. (From Rothenbuhler 1967.)

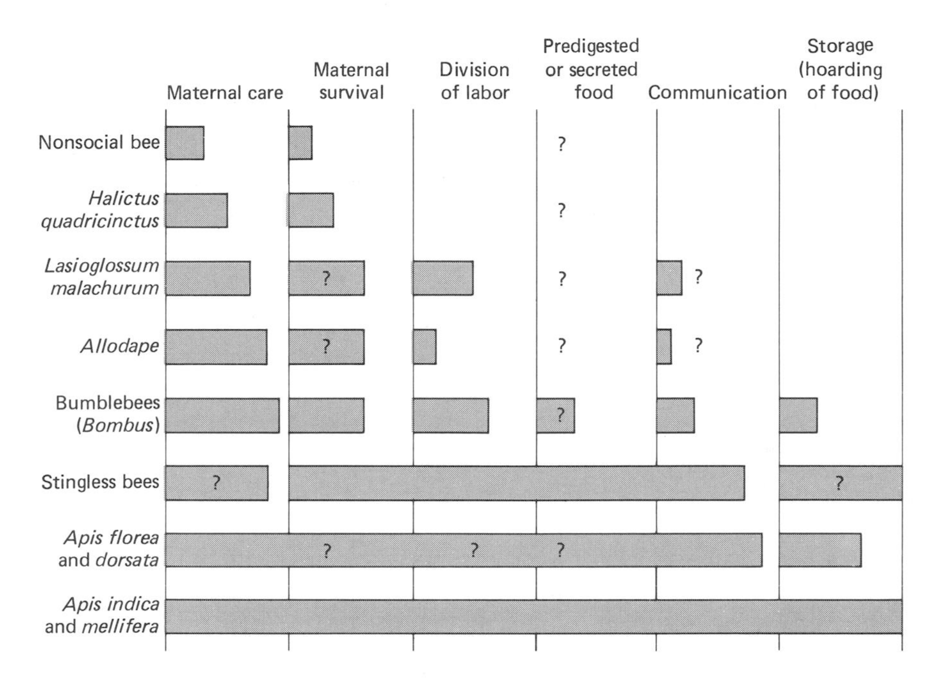

to larger colony size and greater success in production of queens and drones. In the most advanced species all or most of these factors are highly developed.

The Honeybee / In the Hymenoptera, sociality is typically accompanied by the existence of two female castes, whereas there is only one type of male (Figure 10–4). In termites, by contrast, castes are also found in males. In the honeybee the two female castes are determined by the treatment given them before they emerge from their larval cells. Queens are reared in special enlarged cells; they are fed a secretion of the mandibular glands ("bee milk") for their entire larval life, 21 days. When they mature they have small brains and a large abdomen. They are adapted primarily for producing eggs rapidly, and are able to do this because they are fed mandibular secretions throughout their egg-laying life. A queen may produce up to 2,000 eggs per day, or up to half her own weight in eggs daily. Except when new queens are being produced, there is only one to a colony.

Workers do not differ from queens genetically, but they are raised differently. The eggs are laid in regular cells, and the larvae are fed

10–4 Drone, queen, and worker honeybee (*from left to right*). Note the long wings, large eyes, and wide abdomen of the drone, and the long abdomen of the queen.

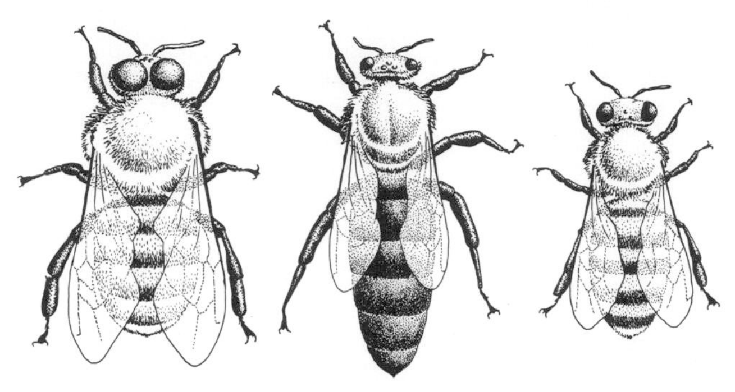

mandibular secretions only for two days, beyond which they are fed a mixture of pollen and nectar ("bee bread"), a less nutritious food. Although workers do not lay eggs they perform all the constructive activities in the hive. Up to 60,000 in number in a single colony, they participate in a succession of essential activities over their 30-day summer life-span. This sequence is very flexible and the different activities wax and wane with much overlap. Each worker passes through the following successive and overlapping phases: cleaner (day 0–day 3), brood nurse (days 4–9), builder (days 10–16), receiver and storer of pollen (days 16–18), guard (days 19–20), and forager (days 21–30). These are illustrated in Figure 10–5.

Paralleling these behavioral activities, nursing glands are prominent from day 5 to day 10 and wax glands from day 10 to day 18; both are reduced from day 18 until death. The behavior and the condition of the glands can be modified to some extent according to the stimuli indicating the need for them. An individual worker commonly engages in more than one of these chores on a given day, and actually spends most of her time just loafing (Lindauer 1961).

The drones, up to 400 of which may be produced in a season, do not help care for the young. Their function is to carry the genes of the colony to the virgin queens of other colonies. Their long wings and large eyes are presumably adaptations that help them find the females. Mating takes place in midair at least 20 feet above the ground. The drones are presumed to range widely (up to 5 miles) in search of queens on their mating flights.

Haplodiploidy and Altruism / The worker in a social insect species is probably the best example of altruism among animals; her reproductive

10–5 Schematic (*left*) and actual (*right*) schedule of a worker honeybee according to her age. Conditions of mandibular or nurse glands (in head) and wax glands (in abdomen) shown at left are correlated with behavior. The idealized developmental sequence for a worker is as follows: cell-cleaning, brood care, building, guarding, foraging; however, as shown at right these activities overlap greatly. The figure at right shows the activities of a single honeybee painstakingly observed during her entire life by Lindauer. (Reprinted by permission of the publishers from Martin Lindauer, *Communication among Social Bees*, Cambridge, Mass.: Harvard University Press. Copyright 1961, by the President and Fellows of Harvard College.)

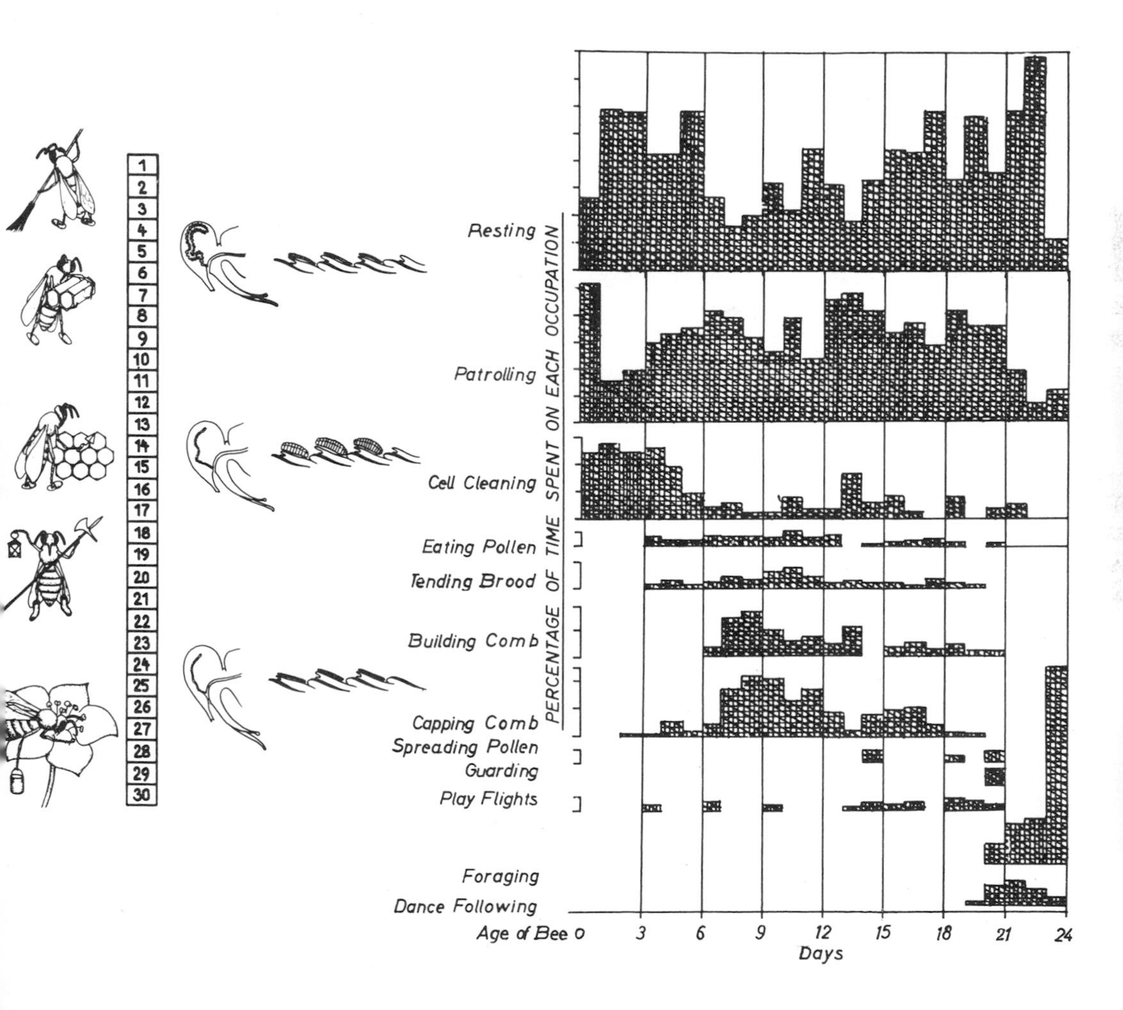

life is sacrificed for the "state." In the highly social species the worker forfeits any egg production of her own and devotes her energies to raising the young of one or more other individuals. Such behavior clearly satisfies the conditions for operational altruism given in the preceding chapter, even though individual workers have no "choice" in the matter. In this extreme case the operation of kin selection is most understandable, since in the social insects the colony is a reproductive unit. Workers, drones, and queens all benefit the colony by helping it to survive and to increase the frequency at which its genes are represented in the colonies of the region. It is the genetic constitution of the colony that is critical, rather than that of the individual, and genetic theories of evolution in social insects are based on this understanding (Wilson 1963, 1966).

One of the questions in the evolution of sociality is to explain its independent origin at least ten times in the Hymenoptera and only once (in termites) in the other two dozen orders of insects. When Hamilton's (1963, 1964) theory for the evolution of altruism is applied to this group, a persuasive hypothesis results. The key may lie in the fact that male hymenopterans are haploid and females diploid, a condition known as haplodiploidy. Males develop parthenogenetically from unfertilized eggs. Consequently, for a worker bee the proportion of her genome identical (by descent) with that of her mother is $\frac{1}{2}$, while the proportion with a sister by the same father is $\frac{3}{4}$.

If you have trouble seeing this, recall that all sperm from a single haploid father must be genetically identical, thus making at least half the genome in each of his daughters identical, and that the chance of receiving either of the two types of haploid gametes from the queen is $\frac{1}{2}$. Put in more familiar language, a daughter is more closely related to her full sisters than to her mother in a haplodiploid species, but in a normal diploid species she is on the average equally close to both. For sons, the situation is different; two sons of the same mother (there is no father) may be genetically identical or completely different, but on the average they share $\frac{1}{2}$ of their genome. Hence, sisters are more closely related to each other than brothers are to each other.

When these data are substituted in the formula $k > 1/r$ (see preceding chapter), we see that for selection to favor altruism, the critical ratio of gain-to-loss in fitness must be 2.0 between daughter and mother or between two sons, but that it is lower, $1\frac{1}{3}$, for two daughters of the same parents. Consequently, selection for altruism, where possible, will be favored more in haplodiploid organisms than in diploid organisms, thus seeming to explain in part the high incidence of altruism in Hymenoptera. It will also be favored more among daughters than among sons, thus explaining in part why there are two types of females in Hymenoptera but only one of males, rather than some other arrangement. Exceptions to these explanations (such as the termites, which are diploid) have been discussed by Wilson (1966).

To imagine how kin selection facilitated by haplodiploidy might lead to sociality, consider the hypothetical case of a species, such as *Augochloropsis sparsilis*, in which the determination of role as worker or queen is apparently not made until the female has become free-flying. Such a female has a limited amount of time and energy after fertilization; she can use it to found her own colony and rear her own young, or she can use it to help rear her sisters by taking care of her mother's larvae. Assuming she can rear the same number of her own young or of her mother's young with this energy budget, the proportion of her genes in the young so reared would be 75 percent if she chose to rear her mother's young (full sisters) but only 50 percent if she chose to rear her own young—all other factors being equal. Under such conditions, genes favoring the care of sibs would increase in frequency, and genes favoring the care of daughters would decrease. What actually happened would be determined by a host of ecological factors and by the proportion of the progeny that became queens as opposed to workers.

For a full discussion of bee societies, see Michener (1974). The origin of sociality in insects is discussed in detail by Lin and Michener (1972) and Alexander (1974).

CONCLUSIONS

The evolution of aid-giving behavior has been especially prominent among social insects. These species have a high degree of *altruism:* many individuals contribute to the inclusive fitness of their genotype not by raising their own young but by raising young from eggs produced by the colony queens. A precondition for the evolution of sociality in insects seems to have been provisioning of the eggs. Increased longevity of the females to the point at which social interactions with their daughters became possible may have provided conditions under which kin selection could operate to elaborate on this relationship, making the young more subservient to and dependent on the colony. The suggestion that haplodiploidy facilitated the evolution of social behavior and altruism in the Hymenoptera is attractive.

11 Social Parasitism

IN MANY unrelated groups of animals the young of one species are raised by another. This *social parasitism* is accomplished by laying eggs in the nest of another species or by capturing slaves of another species. Since these practices are detrimental to the victims in most (but not all) cases, their evolution poses a variety of problems. How have such social systems evolved? Does the victim evolve protective devices against the parasite or enslaver? What are the ramifications of social parasitism in problems of early learning (imprinting), species recognition, the origin of species, and the physiology of reproduction? Social parasitism is of interest not only because of its bizarre nature but also because of the unique insights it offers into basic problems in the evolution of behavior and the opportunities for experimental studies on species with unusual reproductive adaptations.

Polyphyletic Origins / Social parasitism is outstanding as a type of social organization that occurs in widely divergent animal groups, including birds, bees, ants, and wasps. In each of these groups, social parasitism has evidently evolved independently (convergently) from separate origins. This has happened many times in the Hymenoptera (bees, ants, wasps) in several different ways, and at least six times in birds (ducks, cuckoos, honeyguides, cowbirds, weaverbirds). Could there be any common denominator in the evolutionary histories of social parasitism in these groups? Perhaps there are some simple situations and life history patterns that facilitate socially parasitic behavior in all these groups. The behavior appears to be restricted to species that require *parental care*, that utilize *nests*, and that *lay eggs*. Such species offer greater opportunities for social parasitism than do species that give birth to living young.

SOCIAL PARASITISM IN INSECTS

Ants / Probably a greater diversity of forms of social parasitism exists in the ants than in any other group of animals. Wilson (1971) listed over 160 species of socially parasitic ants, and added that no two species are exactly alike in their adaptations for parasitism. The most extreme form of social parasitism in insects is known as *inquilinism*. This is a form of permanent and completely dependent parasitism in which the entire life cycle of the parasite is carried out in the nest of the host. If the inquiline is an ant, workers are typically nonexistent or at least conspicuously degenerate. The following account of social parasitism in ants is drawn mainly from the works of Wilson (1971) and Sudd (1967).

Teleutomyrmex schneideri is an inquiline of a single species of ant in a small region of Switzerland. Extreme inquilines typically are rare and have small geographic ranges. *T. schneideri* is found only in the nests of its host species and lacks a worker caste. The tiny queens ride on the backs of their hosts and probably obtain food by soliciting from the host workers. The eggs, larvae, and pupae of the parasite are mixed with those of the host. Mating occurs within the host nest. The mated queens fly out to find new host nests or remain at home.

Comparing the behavior of different ant species that are social parasites reveals three main series of intermediate forms leading to inquilinism (Wilson 1971). This indicates that the extreme forms of social parasitism in ants have been arrived at by convergent evolution via three quite different routes. These routes—slavery, temporary parasitism, and food parasitism—will now be described. Figure 11–1 shows the inferred pattern of convergent evolution.

Inquilinism via Slavery / Some species of ants are quite aggressive and territorial among their colonies. Predation by one species of ant on others is also well known, and is characteristic of certain groups, such as the army ants, that specialize in plundering nests of other species of ants. If some of the captured pupae were brought home by the raiders and not eaten, they would develop into workers in the nest of the raiding species. This is thought to be the origin of slavery (*dulosis*) in ants. Such acquired workers might be more efficient in raising the young of the slave-owning species than its own workers. If so, selection would favor the bringing home of pupae after a raid, as well as adaptations making the raiders more successful. A well-known slave-keeping species, *Formica sanguinea*, is sometimes found in colonies without slaves and does not seem to be dependent on them. It represents an early stage in this type of evolutionary series.

In other species, such as the famous *Polyergus rufescens*, or amazon ants, slavery is carried to an extreme. It would be difficult to improve on the following description of *P. rufescens* by the great student of social behavior, William Morton Wheeler (1910:472–473).

The worker is extremely pugnacious, and, like the female, may be readily distinguished from the other . . . ants by its sickle-shaped, toothless, but very minutely denticulate mandibles. Such mandibles are not adapted for digging in the earth or for handling thin-skinned larvae or pupae and moving them about in the narrow chambers of the nest, but are admirably fitted for piercing the armor of adult ants. We find therefore that the amazons never excavate nests nor care for their own young. They are even incapable of obtaining their own food, although they may lap up water or liquid food when this happens to come in contact with their short tongues. For the essentials of food, lodging and education they are wholly dependent on the slaves hatched from the worker cocoons that they have pillaged from alien colonies. Apart from these slaves they are quite unable to live, and hence are always found in mixed colonies inhabiting nests whose architecture throughout is that of the slave species. Thus the amazons display two contrasting sets of instincts. While in the home nest they sit about in stolid idleness or pass the long hours begging the slaves for food or cleaning themselves and burnishing their ruddy armor, but when outside the nest on one of their predatory expeditions they display a dazzling courage and capacity for concerted action compared with which the raids of *sanguinea* resemble the clumsy efforts of a lot of untrained militia. The amazons may, therefore, be said to represent a more specialized and perfected stage of dulosis than that of the sanguinary ants. In attaining to this stage, however, they have become irrevocably dependent and parasitic. . . .

The tactics of *Polyergus*, as I have said, are very different from those of *sanguinea*. The ants leave the nest very suddenly and assemble about the entrance if they are not, as sometimes happens, pulled back and restrained by their slaves. Then they move out in a compact column with feverish haste, sometimes, according to Forel, at the rate of a meter in $33\frac{1}{3}$ seconds or 3 cm. per second. On reaching the nest to be pillaged, they do not hesitate like *sanguinea* but pour into it at once in a body, seize the brood, rush out again and make for home. When attacked by the slave species they pierce the heads or thoraces of their opponents and often kill them in considerable numbers. The return to the nest with the booty is usually made more leisurely and in less serried ranks.

New *Polyergus* colonies are apparently founded when a queen takes over a raided nest, or when she enters alone into a host nest and subsequently kills the host queen.

Adaptations for social parasitism through slave raids are conspicuous in the workers of slave-making species. As Wheeler mentioned, these workers are specialized for fighting. The morphological adaptations of the head and mouthparts for fighting are shown in Figure 11–2 and 11–3. These facilitate the success of the behavioral traits of the raiders.

In some inquilines (e.g., *Strongylognathus testaceus*) the mouthparts of the few, vestigial workers show that the species has evolved

11–1 Convergent evolution of social parasitism in ants. (Inspired by Wilson 1971.)

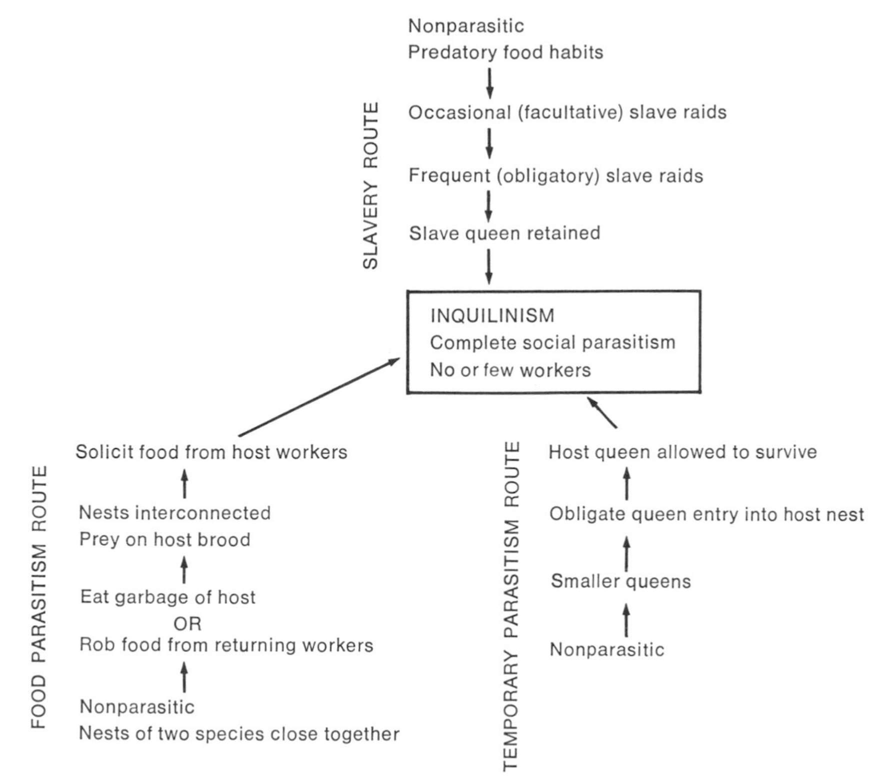

11–2 Adaptations for fighting in a slave-making species (*Harpagoxenus americanus*), **A,** compared to one of its slave species (*Leptothorax curvispinosus*), **B.** Note the enlarged head and mandibles. (From Wheeler 1910.)

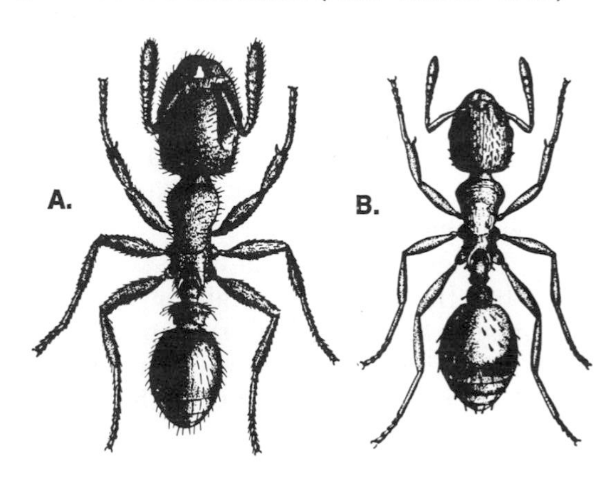

11–3 Adaptations for fighting in slave-making ants, showing various degrees of modification of the mouthparts. **A.** *Polyergus rufescens.* **B.** *Strongylognathus alpinus.* **C.** *Strongylognathus testaceus.* **D.** *Formica sanguinea.* **E.** *Harpagoxenus sublaevis.* In the first three the mandibles are sharply pointed and used to pierce the exoskeletons of the victims. *Formica sanguinea* is a facultative slave-maker that must also use its mandibles for normal work; consequently, its mandibles are unmodified. The tactics of *Harpagoxenus* are different; it has clipper-shaped mandibles used to nip and cut the appendages of opponents. (From Kutter 1969.)

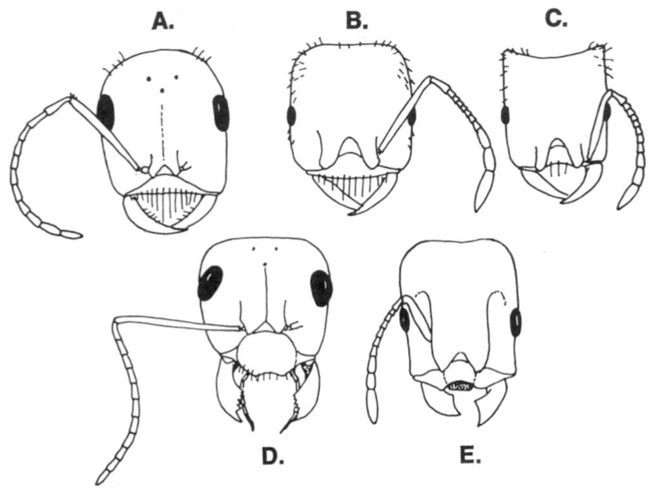

to inquilinism from a slave-keeping state. Other species bridge the gap between slavery and inquilinism more completely.

Inquilinism via Temporary Parasitism / A quite different form of social parasitism that is also a route to inquilinism is what is known in entomology as *temporary parasitism.* The early stages of this specialization are shown by *Formica lugubris.* After their mating flights the queens of this species normally return to their home colony and later leave with a group of workers to found new colonies. But some mated queens fail to find their own colony and attempt to enter colonies of other species. They may then succeed in somehow eliminating the host queen and in having the foreign workers care for the usurper's brood. Thus for a brief period the nest consists of workers of both species, but eventually the *F. lugubris* workers become numerous and those of the former owner die out. This occasional manner of colony foundation is probably the principal origin of temporary parasitism.

In the more parasitic species like *F. rufa,* queens depend on temporary parasitism to found new colonies. The queens of parasitic species employ various strategies for entering the host nest. They may enter by stealth, by aggression, or by chemical deception. They may even be carried in by the host workers if they first "play dead."

Nonparasitic queens that found new nests on their own are frequently much larger than their workers so they can carry enough nourishment and eggs to initiate a colony. For a parasitic queen it is unnecessary to carry such a large energy store, since soon after entry she will be fed by her adopted workers. Consequently, there is typically a reduction of the size difference between queen and workers in parasitic species, as shown in Figure 11–4.

The means by which parasitic queens dispose of host queens are difficult to discover. The queen of *Lasius reginae* rolls over the host queen and strangles her. The temporarily parasitic queen of *Bothriomyrmex decapitans* climbs onto the host queen and slowly gnaws her head off (Figure 11–5). Assassination of the host queen deprives the

11–4 Relative sizes of queens (*Q*) and workers (*W*) of parasitic and nonparasitic ants. *Lasius niger* and *L. flavus* found new colonies independently; *L. umbratus* and *L. fuliginosus* are temporary social parasites and found their new colonies by entry of a queen into a nest of another species. (From Eidmann 1926.)

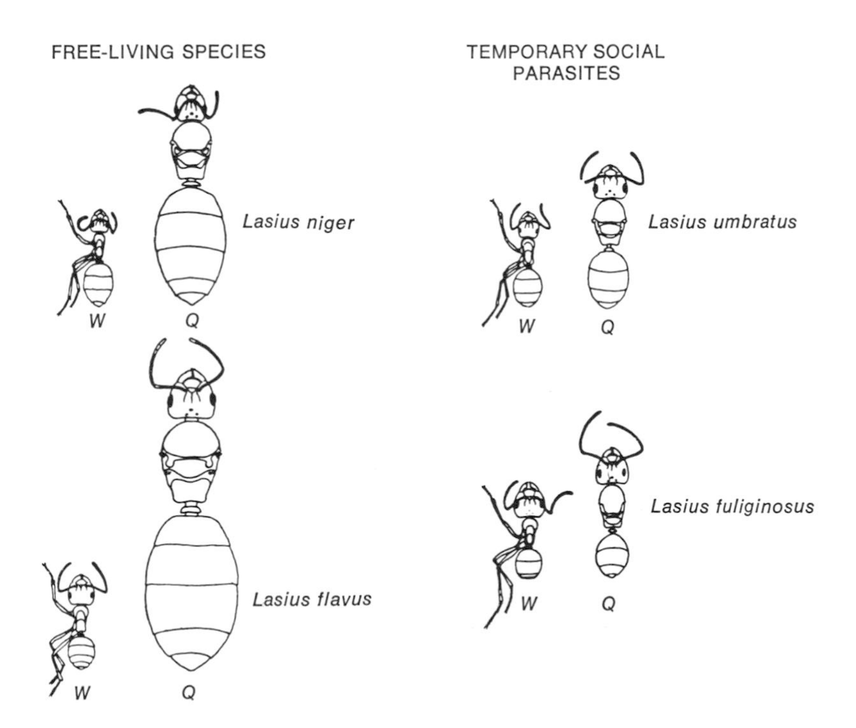

11–5 Queen assassination by a temporary social parasite, *Lasius reginae;* the victim belongs to *L. alienus.* (From Faber 1967.)

new queen of a continuing supply of host workers. So it is not surprising that in some cases the host queen is allowed to survive. The workers of the parasitic species may then become unnecessary or even useless, in which case inquilinism is achieved.

Inquilinism via Food Parasitism / In a given region, several species of free-living ants are likely to coexist. Closely related species tend to be aggressive and so to space themselves out. Dissimilar species may develop close associations. Some small species nest near large species and feed on the uneaten detritus collected by the large species. Alternatively they may rob homecoming ants laden with food (*clepto-parasitism*). The nests of such pairs of species may be in close contact, which makes possible further exploitation. "Thief ants" (*Solenopsis fugax* and *S. molesta*) enter the nests of their close neighbor and prey on the brood. Some ants exploit termite nests in the same way. Other neighboring ants are able to obtain food from host workers by soliciting regurgitation. Beginning with a casual food-sharing situation, behavior patterns more and more like inquilinism may evolve.

Evolution of Inquilinism / Three evolutionary pathways to inquilinism have just been described. No matter which route was involved, inquilines tend to share many differences from free-living species. These include the following: (1) loss of the worker caste, (2) reduction of the size of the queen, (3) reduction or loss of wings in the male, (4) loss of nuptial flights and replacement by mating within the host nest, (5) reduction of the mouthparts to nonfunctional state useful only in soliciting food, (6) reduction and fusion of the central nervous system, and (7) increased chemical attraction of the queens to the host workers, which lick the queens frequently.

Except where the food-parasitism route has been involved, social parasites among ants and other Hymenoptera tend to be *closely related* (Emery's rule). The closest relative of a social parasite is often a free-living species of the same or a closely related genus. Perhaps this is true because association between two species is facilitated if they share common methods of grooming, food exchange, trail-sharing, or other behaviors. This is likely to be crucial for temporary parasites especially, but also valuable for slave-keeping species that depend on harmony between the two species within their colonies.

I have chosen to limit the discussion of social parasitism in insects mainly to ants because the subject has been extensively analyzed in the works of Wheeler, Wilson, and others, but it should be mentioned that inquilinism is extremely well developed in many other groups that utilize the nests of social insects (see the summary provided in Wilson [1971]).

Nest Parasites of Bumblebees / The nests of bumblebees (*Bombus*), of which about 300 species are known, are subject to attack by bees of the genus *Psithyrus* (Rothenbuhler 1967). Late-emerging queen bumblebees sometimes attempt to take over active nests of their own species, thus lending weight to the theory that the genus *Psithyrus* evolved from *Bombus*. The various species of *Psithyrus* enter the nests in different ways. Some are aggressive and kill the host queen; others are submissive and succeed by avoiding fights. Some are adapted for forcible entry by having unusually thick skins, pointed mouthparts, strong mandibular muscles, and strong stings. Once inside, the parasite lays eggs that are then cared for by the *Bombus* workers. *Psithyrus* queens produce no workers of their own. Having no need of them, females of *Psithyrus* have lost the pollen-collecting baskets (Figure 11–6) and the pollen-collecting, building, and brood-nursing behaviors of other bees. The few young produced by *Psithyrus* are all reproductives and are sufficient to perpetuate the species. Other types of social parasitism in bees, especially pillaging or nest robbing, have been discussed by Nogueira-Neto (1970), Wilson (1971), and Michener (1974).

Social Parasitism of Solitary Wasps / More similar to the kind of social parasitism found in birds is that found in the wasp genus *Evagetes*. Some species of solitary wasps have lost the behaviors of nest-building and provisioning. Their way of avoiding these steps is to utilize the nests and labors of other solitary species. After the rightful owner has prepared the hole, deposited the paralyzed larvae as food for her young, and laid an egg, *Evagetes* enters. The parasite then may devour the host egg and substitute one of her own. Some species of *Evagetes*

11–6 The fringe of hairs (corbicula) on the hind tibia of *Bombus lapidarius* (*above*), used in collecting pollen, is lacking in the socially parasitic queen of *Psithyrus rupestris* (*below*). (After Sladen 1912.)

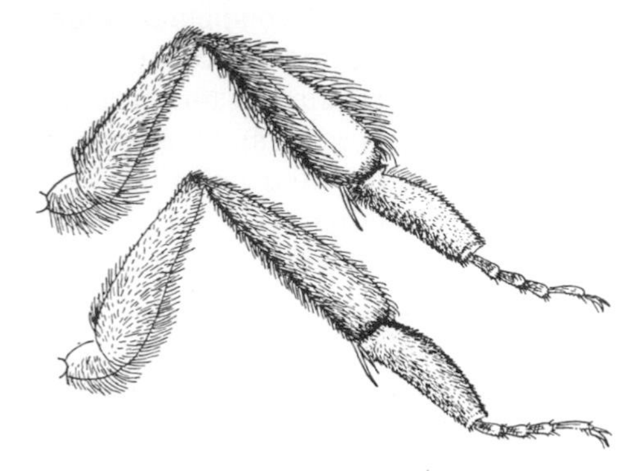

leave the host egg in place; in this case the young *Evagetes*, on hatching, seeks out the host egg and eats it, thus getting all the food for itself plus the host egg. The behaviors of these parasites and their hosts have been described in a delightful book of natural history writing at its best by Evans (1963), who also gives references to the original articles.

SOCIAL PARASITISM IN BIRDS

The best-known form of avian social parasitism is termed *brood parasitism*. The female brood parasite lays an egg in the nest of another species and leaves it for the host to rear, usually at the expense of the host's young.

Brood Parasitism and Egg Mimicry in Cuckoos / The most famous example of brood parasitism in birds is the European cuckoo (*Cuculus canorus*), a member of the Old World subfamily Cuculinae, containing some 50 species, all of which are brood parasites. The female European cuckoo, as is typical of avian brood parasites, selects a nest in which the eggs of the host are currently being laid. Most small birds likely to be hosts for the cuckoo lay one egg each day until the clutch is complete. The host female is away from her nest much of the time during laying; at about the time the clutch is completed, incubation begins, and one or both members of the pair incubate the eggs 90 to 100 percent of the time. The cuckoo must deposit her eggs before incubation begins, so as to have easy access to the nest and to ensure that incubation of her egg begins at the same time as those of her host. A cuckoo lays only one egg in each host nest. When she deposits it, she first removes one egg of the host, and she may hold it in her bill while laying.

Having an incubation period that is a little shorter than that of its hosts, the young cuckoo hatches ahead of its nestmates. It then proceeds to eject the eggs and newly-hatched young of its host from the nest by rolling them up to the edge of the nest and out (Figure 11–7). Young cuckoos grow rapidly and soon become considerably larger than their foster parents.

Since cuckoos tend to specialize on one host species in a given region, the breeding population of the chosen host may suffer accord-

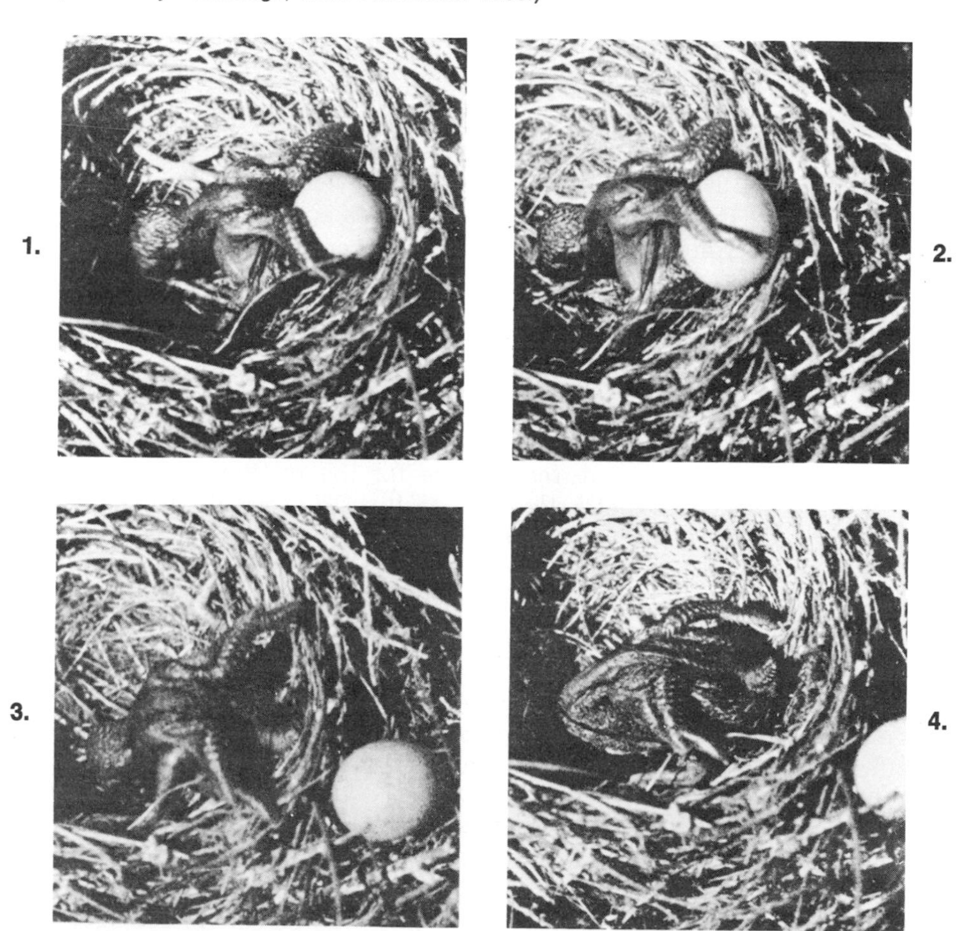

11–7 Nestling of red-chested cuckoo (*Cuculus solitarius*) ejecting host's egg by rolling it over the edge of the nest with its back and wings. Sequence: 1, 2, 3, 4. (Photos by Liversidge, from Friedmann 1956.)

ingly. Table 11–1 shows that over a period of years in an area in Germany the number of nests of the reed bunting (*Emberiza schoeniclus*), the main host of the cuckoo in that region, declined progressively as the proportion of nests parasitized by cuckoos increased. This evidence alone does not prove that cuckoos caused the decline, but the idea is at least plausible.

TABLE 11–1 Correlation of breeding density of parasite, European cuckoo, and host, reed bunting, in a marsh in Germany. (After Schierman 1926.)

Year	1919	1920	1921	1922	1923	1924	1925
Number of host nests	14	15	12	11	9	9	8
Percentage parasitized	29	40	50	73	67	67	88

The data in Table 11–1 suggest that buntings able to recognize and remove cuckoo eggs would have a considerable advantage over those not so able. Experiments by Rensch (1925) and Rothstein (1971) showed that some hosts can recognize an egg that is different from the others. Species that were regularly parasitized were better at removing strange eggs than those not parasitized. Certain species are apparently unable to discriminate their own from the cuckoo's eggs: if three cuckoo eggs were placed with one host egg, the host removed its own egg. An alternative, but less effective, response to detection of a strange egg is to desert the nest, build another, and lay another clutch of eggs.

As a consequence of the ability of hosts to recognize strange eggs, cuckoo eggs resembling those of the host should be more likely to escape detection than those not so protected. This is thought to have been the selection pressure responsible for the phenomenon of egg *mimicry* in cuckoos and other brood parasites. The eggs of the cuckoos breeding in a given region resemble very closely the eggs of the principal host species in that region. The European cuckoos living in different habitat regions of Europe have different hosts, and their eggs mimic those of the host even when the various hosts are quite different in egg coloration. With the breakdown of the former major habitat divisions in Europe, these egg races interbred widely so that several egg types may now be found in one region (Southern 1954). In Assam, India, as many as six egg races have been found in the Khasia Hills district; each was thought to be associated with a different host in its appropriate habitat. The success of egg mimicry for the cuckoo may be judged by Baker's finding (in Southern 1954) that eggs laid in nests of the mimicked host were less often deserted (8 percent of 1,642 eggs) than eggs laid in nests of nonmimicked hosts (24 percent of 298 eggs).

The failure of these egg races to evolve into full species is probably due to the ability of cuckoos to recognize their own species despite being raised in the nest of another species. The early experience of cuckoos with their hosts might, however, be responsible for their strong host preference as adults. An individual female cuckoo is thought to pay attention to only one host species. Chance (1922) observed over several years that one cuckoo laid her eggs in the nests of meadow pipits (*Anthus pratensis*) 58 out of 61 times. Other individual cuckoos laid exclusively (so far as is known) in the nests of other species of host.

Cuckoos and their hosts illustrate nicely the interactions of selection pressures that are important in the evolution of a complex behavioral trait like brood parasitism.

Nestling Mimicry in the Viduines / The viduines are an African subfamily of the family of weaver finches (Ploceidae). They are apparently all brood parasites of various species of the family Estrildidae, and are highly host-specific; although they may occasionally lay in the nest of an abnormal host, successful reproduction is thought to occur only in nests of the appropriate hosts. Viduine parasites do not eject or kill the eggs or young of the hosts. The eggs of both host and parasite are white, but since white eggs are common in birds that utilize covered nests, their resemblance cannot be attributed to mimicry. The success of viduines as brood parasites can only be appreciated through familiarity with their hosts. Estrildids have evolved a highly specialized method of feeding their young. Unlike most passerine birds, which feed insects to their young, the estrildids feed their young food that has been predigested in the crop. The feeding is done in a manner which is ordinarily impossible for the nestlings of nonestrildids to handle. The parent puts its bill deep in the nestling gape and pumps the food out of its crop into the nestling. A special tongue position, neck position, and breathing technique are required by the nestling. In addition, nestling estrildids have highly distinctive color patterns on the insides of their mouths, which are shown to the parent when the young open their mouth to beg for food. These are so species-specific that it is generally impossible for the young of one species of estrildid to be reared by adults of another species because the parents will not feed them (Figure 11–8). These are formidable problems for a brood parasite, but the viduines have overcome them. The nestlings of each viduine parasite almost exactly match the feeding technique and gape pattern of their foster species.

The problem of species recognition and mating is solved quite differently in the viduines than in the cuckoos. Instead of relying on species differences in songs and calls that are genetically determined,

11–8 Nestling gape mimicry. The pictures show the markings on the inside of the mouth of a nestling as it would be seen by a parent faced with the begging young with mouth wide open. The diagnostic spots and lines may be of various colors and are characteristic of the species. The markings are located inside the mouth, on the tongue, and at the corners of the mouth at the sides. In **1–4** the name of the host is given above the name of the parasite. *Lagonosticta senegala* (**5**) is thought to be the host of *Vidua chalybeata* (**6**). From Friedmann 1960.)

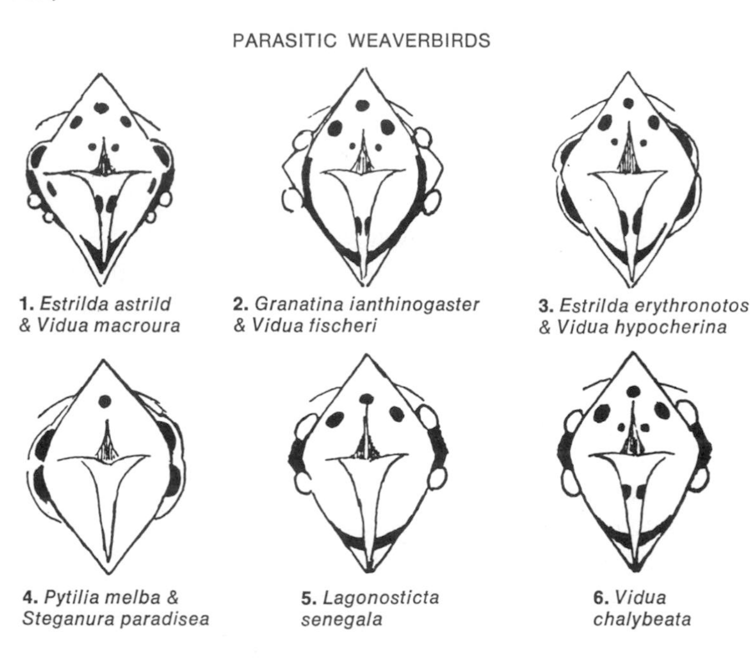

the viduines learn the characteristic calls and song of their hosts and reproduce them during courtship and at other times. The appearance of male viduines is very different from that of their hosts, however, and this is probably responsible for preventing confusion between the viduine species and their hosts. On the other hand, in some cases the different species of viduines look very much alike, so that confusion between two species of viduines might be prevented primarily by the vocalizations learned from the foster parents (Payne 1968).

As might be expected from the above description, the geographic range of a species of viduine does not extend outside the range of its host. It has been found through close study of viduines with different hosts that some species had not previously been recognized because of their close morphological resemblance to other viduines. Unlike the European cuckoo, the origin of new viduine species seems to have followed the adoption of new host species. Since viduines learn their courtship songs from their foster parents and cuckoos do not, it is tempting to think that this may partly explain the proliferation of species among viduines. Before conclusions on this point can be reached, the role of geographical isolation must be carefully considered. The viduines and other brood parasites pose interesting problems in the origin of species and of species-specific behavior.

This discussion of viduine behavior is drawn primarily from the works of Nicolai (1964, 1967), whose discoveries were instrumental in reawakening interest in this problem. Subsequently, Payne (1968, 1973) has contributed new field data. Summaries of the present state of knowledge about this group may be found in the works of Wickler (1968) and Lack (1968).

Some Intermediate Stages in the Evolution of Brood Parasitism / The New World family Icteridae includes several species of brood parasites, whose behavior patterns were studied in a classic monograph by Friedmann (1929). Four of these species now placed in the genus *Molothrus* provide examples of various kinds and degrees of evolutionary specialization for brood parasitism (Table 11–2). One of these, *M. badius*, is not a brood parasite. It usually obtains a nest by capturing one from another species, although some individuals have been seen building their own. It is monogamous and rears its own young. Nest piracy of this sort is also known in the piratic flycatcher (*Legatus leucophaius;* Skutch 1969), the chestnut sparrow (*Passer eminibey*), and certain members of the Estrildidae (Payne 1969*b*). The habit of using *old* nests of other species is widespread in birds, so it is not hard to imagine that nest piracy evolved from this condition. Many species use old woodpecker holes (e.g., various flycatchers, titmice, warblers, barbets, and wrens). Some sparrows use old weaver nests, which they reline with feathers. Nest piracy is not regarded as a nec-

TABLE 11–2 Comparison of four species of New World Icteridae tentatively placed in the genus *Molothrus*. (After Friedmann 1929.)

	M. badius	*M. rufo-axillaris*	*M. bonariensis*	*M. ater*
Range	pampas of S. Am.	pampas of S. Am.	widespread in S. Am.	widespread in N. Am.
Nest	few build, most capture	none	none	none
Hosts	many species	only *M. badius*	98+	195+
Mating system	monogamous; helpers	monogamous	promiscuous	promiscuous
Defended area	nest and surroundings	song tree	song tree	song tree
Display of male	inconspicuous	inconspicuous	conspicuous	conspicuous
Plumage	monomorphic; dull brownish	monomorphic	sexually dimorphic	sexually dimorphic

essary precondition for the evolution of brood parasitism, but the possibility of this route to parasitism exists.

It is ironic, perhaps, that *M. badius*, having pirated a nest from another species, is then subject to brood parasitism by another species of the same genus, *M. rufo-axillaris*, which lays its eggs nearly exclusively in the nests of *badius*. The geographic range of *rufo-axillaris* does not extend beyond that of *badius*.

The remaining two species of small cowbirds (*Molothrus ater*, *M. bonariensis*) are widespread in North and South America respectively. The list of species in whose nest an egg of *M. ater* or *M. bonariensis* has been laid extends to nearly 100 or more for each species. The length of this list is due to the wide range of habitats and geographic regions inhabited by these species. In any one district the burden of raising cowbirds falls on only a few species. Nevertheless, it is clear that these cowbirds do not approach the degree of host-specificity found in many other brood parasites (Mayfield 1965; but see McGeen 1972 and McGeen and McGeen 1968). Their strategy seems to be just the opposite, namely, to spread the burden among many species, thus reducing selection pressures that favor the development of discriminatory responses in the hosts against the parasites, and making a greater number of foster parents available to the cowbird. This allows a greater number of eggs to be laid (Payne 1965). Egg mimicry apparently does not occur in these species.

The two most widespread icterid brood parasites, *M. ater* and *M. bonariensis*, are promiscuous, with conspicuous male display and sexual dimorphism. This might be expected because the males provide no parental care, so the advantage to the female of monogamy is reduced or eliminated. On the other hand, *M. rufo-axillaris* is monogamous, and both sexes have a dull plumage. The interesting question about mating systems is this: Considering the absence of parental care in all brood parasites, why are any species monogamous, as the European cuckoo (Lack 1968) and *M. rufo-axillaris* (Friedmann 1929) seem to be? The literature does not answer this question for us.

One possible origin of brood parasitism that has received little attention is communal breeding. Species of communal breeders often have reduced reproductive rates, and so might be more likely to be able to raise their reproductive rate through brood parasitism. They are also accustomed to having their young fed by other birds and to laying eggs in nests over which they lack full ownership. It is perhaps more than coincidence that, among the cowbirds, *Molothrus badius* is a communal breeder (Fraga 1972), and that one subfamily of New World cuckoos (Crotophaginae) is composed mainly of communal breeders (Davis 1942), while most of the remaining New World cuckoos are nonparasitic (Lack 1968). Furthermore, both *M. badius* and *Cro-*

11–9 Foster homes of the giant cowbird. This tree contains a large nest of stingless bees (*Trigona*) near the center on one of the main trunks, and many woven, pendant nests of caciques and oropendolas further out on the branches. In this colony the host species, which are protected against botflies by the bees, discriminate against cowbird eggs. (Original photo by N. G. Smith.)

tophaga sulcirostris use old nests of other species at least some of the time (in *C. sulcirostris*, 13 of 22 nests in Ecuador were built by *Mimus longicaudatus* [Marchant 1960]).

The Advantage of Being Parasitized / Parasite-host systems may be said to exist in a state of balance. If the parasite becomes too successful in destroying the host, both are likely to become extinct. On the other hand, the parasite may also become extinct if the host develops resistance. The most successful parasites reach a point at which they do not harm the host and may even benefit it, in which case natural selection on the host no longer works to resist the parasite. Among brood parasites such a condition has been reported only once. This case was described by Smith (1968).

The giant cowbird (*Scaphidura oryzivora*) lays its eggs in the large pendant nests of several colonial oropendolas and caciques. These tropical species were studied in Panama, but the range of the cowbird extends from southern Mexico to northern Argentina. The hosts suffer from predation by vertebrates and from parasitism by botflies (*Philornis*). Some protection from both these agents is obtained by locating the nests in a tree that contains the nests of bees (*Trigona*) or wasps (*Protopolybia, Stelopolybia*). In colonies not protected by bees or wasps, the botflies deposit eggs or larvae on the nestlings. The larvae then burrow into the nestlings. If more than seven enter the nestling, he is likely to die. The hosts do not remove the botfly eggs and larvae from their young, and mortality from botfly larvae is probably the greatest single source of mortality for the host. The advantage of being parasitized by cowbirds is that the young cowbirds remove the eggs and young of the botfly from the nestlings of the host (and their cowbird nestmates). The average number of host fledglings produced per nest in a colony unprotected by bees or wasps is nearly three times greater from nests containing a nestling cowbird (0.52, 0.53) than from nests containing only host young (0.17, 0.19).

In colonies protected by wasps or bees, the fledging rate (0.53, 0.55) of host young without cowbirds was greater than that of nests containing cowbirds (0.28). Strange eggs are removed from the nest in protected colonies, but not in unprotected ones, thus generating strong selection pressures on the cowbirds to produce mimetic eggs. In a given colony, the cowbirds laid three types of eggs: a cacique egg mimic, an oropendola egg mimic, and a nonmimetic egg. Females laying mimetic eggs were shy and skulking in their behavior; they typically would lay only one egg in a nest and only in a nest already containing one host egg (the normal clutch of hosts is two). Females laying nonmimetic eggs were gregarious and aggressive in entering nests, and they laid from one to five (usually two or three) eggs in a nest. This earned them the name *dumpers*. The dumpers laid in empty nests, nests with full clutches, and all stages in between. They were more productive in unprotected colonies than in protected ones. In protected colonies females laying mimetic eggs were apparently more successful than dumpers. This case illustrates nicely (1) the complexity of the selection pressures that may be involved in the evolution of brood parasitic behavior, and (2) the balance between opposing pressures.

Other Brood Parasites / Space does not permit a full description of the other groups of brood parasites. Much diversity in the brood parasitic pattern is found in the cuckoos (e.g., Friedmann 1967, 1968*a*; Reed 1968). All known honeyguides (Indicatoridae) are brood parasites (Friedmann 1955, 1968*b*). In some of these species the nestling uses a razor-sharp structure on its bill to kill or injure the host nestlings. Among ducks, brood parasitism is partially developed in the redhead duck, *Aythya americana* (Weller 1959), and the tufted duck, *Aythya fuligula* (Fredrikson 1968). Obligate brood parasitism occurs in one duck species (*Heteronetta atricapilla;* Weller 1968).

The Evolution of Brood Parasitism / This subject has been reviewed by Hamilton and Orians (1965). They pointed out that the theory of natural selection does not allow evolutionary interpretations based on "degeneration." Early brood parasites must have had superior fitness for brood parasitism to have evolved. Explaining why the behavior arose in some groups and not others requires a separate and detailed examination in each case. For the present purposes, some of the preconditions favorable to brood parasitism are listed below. These are found in varying degrees and combinations in the various groups thought to be ancestral to brood parasites. They include (1) incubation period slightly shorter than the host's, (2) nest built by male instead of female, (3) old nest of another species typically used, (4) rapid nestling growth, (5) hole or open nesters, as the case may be, (6) low fledging rates when fed by own parents, (7) tolerance of host's diet, (8) incubation by male, (9) rarity relative to the hosts, and (10) range and habitat matched with those of host. Further evolution included (1) loss of nest building, (2) loss of incubation behavior and incubation patch, (3) loss of feeding of young, (4) removal or destruction of host eggs and young, (5) mimetic eggs, (6) mimetic nestlings, (7) possible hawk-mimicry (Kuroda 1966; Lack 1968), (8) possibly quickened nestling growth, (9) behavior facilitating approach to the host's nest (e.g., Payne 1967), (10) laying only one egg per nest, (11) laying only when host is laying, (12) modification of the endocrine physiology of reproduction. The secondary consequences of brood parasitism probably include a shift toward a promiscuous mating system, a shift away from all-purpose territories, and intensified sexual selection with the resultant elaboration of sexual display and dimorphism. Counteradaptations by hosts include a greater ability to recognize and remove strange eggs or to desert a parasitized nest, and a greater ability to refuse food to strange nestlings.

CONCLUSIONS

The phenomenon of social parasitism illustrates how the social organization of a species may be drastically reordered through its relations with other species. It also shows that behavior patterns as conservative and deep-seated physiologically as nest building and the care and feeding of an individual's own young can be selected against when increased fitness results.

12 Diversity in Primate Social Organizations

THE SOCIAL ORGANIZATIONS OF primates are best treated separately from those of other species for various reasons. First of all, they have an interest all their own because of man's close phylogenetic affiliation with other primates. It is as legitimate to make comparisons between man and other primates as between any two species of animals, and it should be clear that the comparative method is in principle as capable of shedding light on man's behavior as on the behavior of any other species. This approach must be used with caution, however, because of the famous plasticity of primates in behavior. Indeed, the *variability* in primate social organizations is one of their most interesting aspects. However, it would be a great mistake to dismiss all interspecific differences in social organization among primates as being due to culture, conditioning, learning, tradition, or other manifestations of the plasticity of behavior.

A more biological reason for treating primates separately derives from the greater *complexity* of their social organizations, at least in some species. The three main determinants of social organization that have been discussed earlier (agonistic, sexual, and aid-giving behavior) interact in primates in ways that make it difficult to consider the action of one except in relation to the other two—probably to a greater extent than in other groups. This complexity, combined with the demonstrably greater capacity of individual primates for intelligent manipulation of social relationships to benefit themselves, has given rise to phenomena of social organizations (e.g., coalitions, genealogical effects) that are unknown or insignificant in other groups.

A further advantage of treating the primates at this point is that the principles developed through analysis of other groups can now be seen to apply to primates as well, and their significance for understanding man's own behavior and social organization can be more fully appreciated.

The order Primates is moderately large and diverse. It has 11 living families (plus 7 extinct ones), 50 living genera, about 200 species, and roughly 600 described subspecies. An outline of their classification, including the species mentioned in the following account, is given in Table 12–1. Napier and Napier (1967) have written a book-length review of primate biology. Jolly (1972) presents a wide-ranging introduction to primate behavior. Primate social organizations have been reviewed by Hall (1964, 1965*a*), Eisenberg (1966), Crook (1970), Chance and Jolly (1970), Kummer (1971), and Eisenberg et al. (1972). Much material on primate social organization is found in the volumes edited by DeVore (1965) and Jay (1968) and in the journal *Primates.*

MATING SYSTEMS

A description of the types of groups in which a primate species lives is a good indication of its mating system. Since intra-group and inter-group relations define much of primate social organization, it is convenient to begin by recognizing a few common types of groups, describing them briefly, and providing examples. The following types of social units characterize the social systems of most primates and serve as a framework for discussion:

1) Solitary individuals, coming together mainly for mating.
2) Solitary pairs of adults.
3) One-male harems.
4) Age-graded-male troops.
5) Multi-male troops.

It is assumed that adult females are accompanied by their dependent offspring. The term *harem* is used only to indicate the usual preponderence of females in the primary social units; the excluded males in such species either form all-male groups or live alone. They are probably subject to heavier mortality than the females. Examples of species in each of the principal types are listed in Table 12–2.

The importance of sexual bonds in the social organizations of primates was recognized early by Zuckerman (1932). Not all subsequent workers agreed with Zuckerman's interpretation that sexual attraction is the primary cohesive force in primate societies. A realistic assessment of its importance had to await the time when a fair number of primate species had been studied under natural and laboratory conditions. The post–World War II surge in field and laboratory studies of primates has made possible the beginning of a comparative approach to social organizations in primates. As a result, the role of mating sys-

TABLE 12–1 A classification of the primates. The Tupaioidea (tree shrews) are not considered to be primates by some taxonomists, but they are probably closely related. (From Melnechuk and Ploog 1969.)

Order	Suborder	Infraorder	Superfamily	Family	Subfamily	Genus	Common name
PRIMATES	PROSIMII	Lemuriformes	Tupaioidea	Tupaiidae	Tupaiinae	*Tupaia*	common tree shrew
						Dendrogale	smooth-tailed tree shrew
						Urogale	Philippine tree shrew
					Ptilocercinae	*Ptilocercus*	pen-tailed tree shrew
			Lemuroidea	Lemuridae	Lemurinae	*Lemur*	common lemur
						Hapalemur	gentle lemur
						Lepilemur	sportive lemur
					Cheirogaleinae	*Cheirogaleus*	mouse lemur
						Microcebus	dwarf lemur
				Indridae		*Indri*	indris
						Lichanotus	avahi
						Propithecus	sifaka
				Daubentoniidae		*Daubentonia*	aye-aye
		Lorisiformes	Lorisoidea	Lorisidae		*Loris*	slender loris
						Nycticebus	slow loris
						Artocebus	angwantibo
						Perodicticus	potto
				Galagidae		*Galago*	bush baby
		Tarsiiformes	Tarsioidea	Tarsiidae		*Tarsius*	tarsier
	ANTHROPOIDEA	Platyrrhini	Ceboidea	Callithricidae		*Callithrix*	plumed and pygmy marmosets
						Leontocebus	tamarin
				Cebidae	Callimiconinae	*Callimico*	Goeldi's marmoset
					Aotinae	*Aotes*	douroucouli
						Callicebus	titi
					Pithecinae	*Pithecia*	saki
						Chiropotes	saki
						Cacajao	uakari
					Alouattinae	*Alouatta*	howler
					Cebinae	*Cebus*	capuchin
						Saimiri	squirrel monkey
					Atelinae	*Ateles*	spider monkey
						Brachyteles	woolly spider monkey
						Lagothrix	woolly monkey
		Catarrhini	Cercopithecoidea	Cercopithecidae	Cercopithecinae	*Macaca*	macaque
						Cynopithecus	black ape
						Cercocebus	mangabey
						Papio	baboon, drill
						Theropithecus	gelada
						Cercopithecus	guenon, vervet
						Erythrocebus	patas monkey
					Colobinae	*Presbytis*	common langur
						Pygathrix	douc langur
						Rhinopithecus	snub-nosed langur
						Simias	Pagi Island langur
						Nasalis	proboscis monkey
						Colobus	gueraza
			Hominoidea	Hylobatidae		*Hylobates*	gibbon
						Symphalangus	siamang
				Pongidae		*Pongo*	orangutan
						Pan	chimpanzee
						Gorilla	gorilla
				Hominidae		*Homo*	man

TABLE 12–2 Range of social organization and feeding ecology for selected primate species (see text for descriptions). (From Eisenberg et al. 1972. Copyright 1972 by the American Association for the Advancement of Science.)

Solitary species	*Parental family*	Minimal adult ♂ tolerance* (uni-male troop)†	Intermediate ♂ tolerance ‡ (age-graded-male troop)†	Highest ♂ tolerance § (multi-male troop)†
A. Insectivore-frugivore	A. Frugivore-insectivore	A. Arboreal folivore	A. Arboreal folivore	A. Arboreal frugivore
Lemuridae	Callithricidae (Hapalidae)	Cebidae	Colobinae	Indriidae
Microcebus murinus	*Saguinus oedipus*	*Alouatta palliata*	*Presbytis cristatus*	*Propithecus verreauxi*
Cheirogaleus major	*Cebuella pygmaeus*	Colobinae	*Presbytis entellus*	Lemuridae
Daubentoniidae	*Callithrix jacchus*	*Colobus guereza*	Cebidae	*Lemur fulvus*
Daubentonia	Cebidae	*Presbytis senex*	*Alouatta palliata*	
madagascarensis	*Callicebus moloch*	*Presbytis johni*		
Lorisidae	*Aotus trivirgatus*	*Presbytis entellus*		
Loris tardigradus				
Perodicticus potto				
B. Folivore	B. Folivore-frugivore	B. Arboreal frugivore	B. Arboreal frugivore	B. Semiterrestrial frugivore-omnivore
Lemuridae	Indriidae	Cebidae	Cebidae	Cercopithecidae
Lepilemur mustelinus	*Indri indri*	*Cebus capucinus*	*Ateles geoffroyi*	*Cercopithecus aethiops*
	Hylobatidae	Cercopithecidae	*Saimiri sciureus*	*Macaca fuscata*
	Hylobates lar	*Cercopithecus mitis*	Cercopithecidae	*Macaca mulatta*
	Symphalangus	*Cercopithecus*	*Miopithecus talapoin*	*Macaca radiata*
	syndactylus	*campbelli*		*Papio cynocephalus*
		Cercocebus albigena		*Papio ursinus*
				Papio anubis
				Macaca sinica
				Pongidae
				Pan satyrus
		C. Semiterrestrial frugivore	C. Semiterrestrial frugivore-omnivore	
		Cercopithecidae	Cercopithecidae	
		Erythrocebus patas	*Cercopithecus aethiops*	
		Theropithecus gelada	*Cercocebus torquatus*	
		Mandrillus	*Macaca sinica*	
		leucocephalus		
		Papio hamadryas		
			D. Terrestrial folivore-frugivore	
			Pongidae	
			Gorilla gorilla	

* Troop with one adult male and strong intolerance to maturing males.
† "Troop" refers to the basic social grouping of adult females and their dependent or semidependent offspring.
‡ Troop typically showing age-graded-male series.
§ Troop with several mature, adult males and age-graded series of males.

tems can now be seen in better perspective, and sexual attraction is no longer considered the primary cohesive force in primate groups. In fact, there are many cohesive forces involved.

Solitary Species / Social systems based on widely spaced individuals who have social contact only for courtship and copulation and at the edges of their home ranges are found in several species in the suborder Prosimii (Charles-Dominique 1972; Jolly 1972). Most of these species are small-sized, nocturnal, arboreal forest-dwellers, and are consequently not as well known as the large, more conspicuous diurnal species. Several have a diet that is heavily insectivorous.

Monogamous Pairs / One monogamous species, the lar gibbon (*Hylobates lar*), is fairly well known (Carpenter 1940; Ellefson 1968). Mated pairs appear to live together for life. The young stay with the parents while they are sexually immature, but they leave the parents as they reach breeding condition, and single individuals of both sexes are found. A similar situation occurs in *H. syndactylus* (Kawabe 1970). A New World species, the titi monkey (*Callicebus moloch*), has a similar social organization. Defended and largely exclusive areas are inhabited by pairs of adults and their young in all these species (see Figure 12–2).

Single-Male Harems / The mating systems of many primate species are based on groups composed of one adult male and several adult females plus their young. The gelada baboon *Theropithecus gelada* of the montane grasslands of Ethiopia has been studied by Crook (1966*a*; Crook and Aldrich-Blake 1968). It is a vegetarian species that rarely strays far from the safety of steep precipices and gorges. There it is found in aggregations up to 400. Under poor feeding conditions these herds break up into one-male groups, all-male groups, parties of young juveniles, and isolated single males. The one-male harems maintain their integrity even when they come together in large groups. Males are aggressively possessive about their harems but show no loyalties to a particular herd.

The high number of adult females per adult male in a one-male group (Table 12–3) indicates a relatively high degree of sexual selection in this species. Correlated with this (but not necessarily caused entirely by it) is the conspicuous sexual dimorphism of this species.

The hamadryas baboon (Figure 12–1*A*) occurs in the arid brush lands of northeastern Africa and southwestern Arabia. It has been the subject of extensive field studies (Kummer and Kurt 1963; Kummer 1967; 1968*a*, *b*). Like the gelada, it aggregates in large numbers at

TABLE 12–3 Correlates of sexual dimorphism in selected primates. (Data are from Hall 1965*a*, Napier and Napier 1967, and additional sources listed below. Data from different authors are not fully comparable.)

Species	Adult body weight: Female as percentage of male	Adult body weight: Male (kg)	Age at physical maturity: Male (yrs)	Age at physical maturity: Female (yrs)	Reproductive social unit	Approximate mean number of adult females per adult male	Additional sources
Hylobates lar	93	5.7	5–8	5–8	pair	1	Reynolds 1967
Pan satyrus	89	45	7–8	7–10	loose multimale band	1	Reynolds 1967
Allouatta palliata	77	7.4	?	?	one-male and age-graded-male groups	2.3	Carpenter 1934
Macaca mulatta (forest)	70	8	4.5	3.5	multi-male group	3	Southwick et al. 1965
Presbytis entellus	67	15	6–7	3–4	one-male and age-graded-male groups	1.6–7.5	Yoshiba 1968
Erythrocebus patas	54	10–12.7	?	?	one-male group	6.3	Hall 1965*b*
Papio hamadryas	53	18	7	5	one-male group	2.3	Kummer 1968*b*
Gorilla gorilla	53	136–204	9	7	age-graded male group	1.8	Schaller 1965
Papio anubis	50	22–30	8	4	multi-male group	2.0	Hall and DeVore 1965
Theropithecus gelada	49	20	?	?	one-male group	3.9–4.6	Crook 1966*a*

sleeping sites and radiates out to forage during the day in one-male harems, all-male groups, groups of young, and isolated males. Males are highly oppressive in preventing escapes of their females, and punish strays by biting them on the neck. Kummer (1968*b*:41) wrote of the mating system, "Adult males copulate only with females belonging to their own unit. We have never found any exception to this rule. At most, a male would follow the swellings of a passing female with his eyes. This means that males with only one female may sometimes not arrive at copulation for months at a time. Those having no females at all may remain without copulating for years." About 20 percent of adult males had no females. Occasionally males tried to acquire anestrous (nonreceptive) females from other units; on the other hand, they also were seen to ignore solicitations by estrous females of another unit. Adult males seemed more concerned with keeping a set of females around them than with chance copulations with strangers. Females, by contrast, were commonly seen copulating with young males when not under the surveillance of their owner. Extreme sexual dimorphism and delayed male sexual maturity are characteristic of this species also.

The patas monkey (*Erythrocebus patas*) was studied in the grassland-savannah of Uganda by Hall (1965*b*). In this habitat there are no cliffs, and the safest night roosting sites are the scattered trees. The species is slim and is probably the fastest terrestrial primate. The typical social unit is a one-male group composed of one large adult male and up to 22 females and young. In contrast to other species with one-male groups, the male patas monkey is more benevolent. He acts as sentinel and allows the females to lead. Groups tend to avoid each other, but chases of one group by another do occur. There is a relatively high number (6.3) of females per adult male in the one-male groups, and consequently a high degree of sexual selection of the male-dominance type. Sexual dimorphism in body size is extreme, adult females being only half the size of adult males (Hall 1964, 1967).

Age-Graded-Male Troops / In species characterized by single-male groups, the young males are expelled from the group early after they become independent of their mother. If the dominant male exhibits more tolerance of maturing, but still subadult, males, the troop is often larger and contains males of various ages. The males in such cases are said to be age-graded (Eisenberg et al. 1972). These additional males are prevented by the dominant from assuming the roles of a full adult. This type of social group is found in a variety of primate species inhabiting a wide range of environments (Table 12–2). The vervet

(*Cercopithecus aethiops*) and gorilla (*Gorilla gorilla*) are among the best known. The age-graded-male type of group can be easily derived from the single-male type, and is probably an intermediate stage in the evolution of multi-male troops.

Multi-Male Troops / Multi-male troops are distinguished from the age-graded-male troops by the presence typically of two or more males who are full adults, physically and behaviorally. This type of social group is best known among the more-or-less terrestrial macaques and the savannah-living baboons. Some representative species are listed in Table 12–2.

The mating systems and social organizations of baboons (Figure 12–1) have been studied intensively in various habitats and parts of Africa (Hall and DeVore 1965; DeVore and Hall 1965; Rowell 1966, 1967*a*, *b*, 1972; Altmann and Altmann 1970). The composition of some troops in which every member was individually recognizable is shown in Table 12–4. Troops of savannah-living baboons vary greatly in size and composition, but they typically contain 2 to 8 adult males and 6 to 18 adult females. As Table 12–4 shows, the number of immatures per female is lowest in one-male groups and highest in groups with two to eight adult males; it is also low in groups with high female:male ratios. These facts suggest (but do not prove) that troops differing greatly from middle values in composition and size are the least

TABLE 12–4 Composition of baboon troops. (Data from DeVore and Hall 1965; Rowell 1966, 1967*a*.)

Group	Adults ♂	Adults ♀	Adult females per adult male	Total immatures (both sexes) per adult female
PAPIO URSINUS				
Rhodesia	13	31	2.4	1.9
Cape C	8	18	2.2	3.0
Cape TM	3	12	4.0	1.1
Cape N	1	10	10.0	0.9
PAPIO ANUBIS				
Kenya SV	5	12	2.4	2.0
Kenya SR	6	7	1.2	2.6
Kenya AR	1	9	9.0	2.0
Kenya LT	2	3	1.5	4.0
Kenya MR	1	6	6.0	0.8
Uganda S	5	5	1.0	3.2
Uganda V	17	16	0.9	0.8

productive. It would be reasonable to suppose that this reflects stabilizing selection for individual properties that determine troop size and organization.

Troop Size and Sexual Selection in Baboons / The size and sex ratio of baboon troops seem to be correlated with the appearance of the males. As shown in Figure 12–1, the hamadryas males have a conspicuous cape and average a relatively high number of females per male (2.3); there is one male in each primary social unit, which is relatively small (3.3 adults). At the other extreme, males of *P. cynocephalus* have only a suggestion of a cape and a smaller number of females per male (1.3); there are many males in a primary social unit, which is relatively large (28.8). Anubis and chacma baboons are intermediate.

Two principal evolutionary explanations for these correlations will be considered. Probably both are important. The first is based on anti-predator strategy; the second, on sexual selection.

In rich habitats, where large predators such as leopards, lions, and cheetahs are relatively abundant and baboons more numerous, the value to a troop of having many males, who cooperate in defending the troop from predators (Hall 1964; DeVore and Hall 1965), is high. If a male is lost while defending the troop, other males will continue to protect the troop. On the other hand, baboons living in food-poor regions, such as the semi-desert inhabited by *P. hamadryas*, might well be subject to far less pressure from large predators. Under such conditions one adult male might usually suffice for protection of the (smaller) group.

Two other variables that promote larger troop sizes should be mentioned. First, in richer habitats the troops could be larger than troops with equal home ranges in poorer habitats, simply because reproductive success would be greater. Second, larger troops tend to dominate smaller ones and so may gain priority to favored foraging areas during times of food scarcity.

The selection pressure favoring numerous males in a troop tends to be counteracted by sexual selection. A male that succeeds in driving rival adult males out of his troop should father more young than one who is less aggressive toward rivals; such aggressive males are consequently favored by sexual selection. If selection for acquiring females is opposed to selection for predator defense, then we should expect both mantle development and troop size to be intermediate in *P. anubis*—between *P. hamadryas* at one extreme (in a predator-poor environment) and *P. ursinus* and *P. cynocephalus* at the other (in a predator-rich environment). The facts agree in these respects; but data on adult sex ratios in *P. anubis*, which should also be intermediate, are

12–1 Correlation of sexual dimorphism with adult sex ratio in baboon species. The pictures are of male baboons only. They illustrate differences in the cape or mantle of the male. Females, in all baboons, lack the cape entirely. (Data from Altmann and Altmann 1970 and Kummer 1968; pictures and maps after Dorst and Dandelot 1969.)

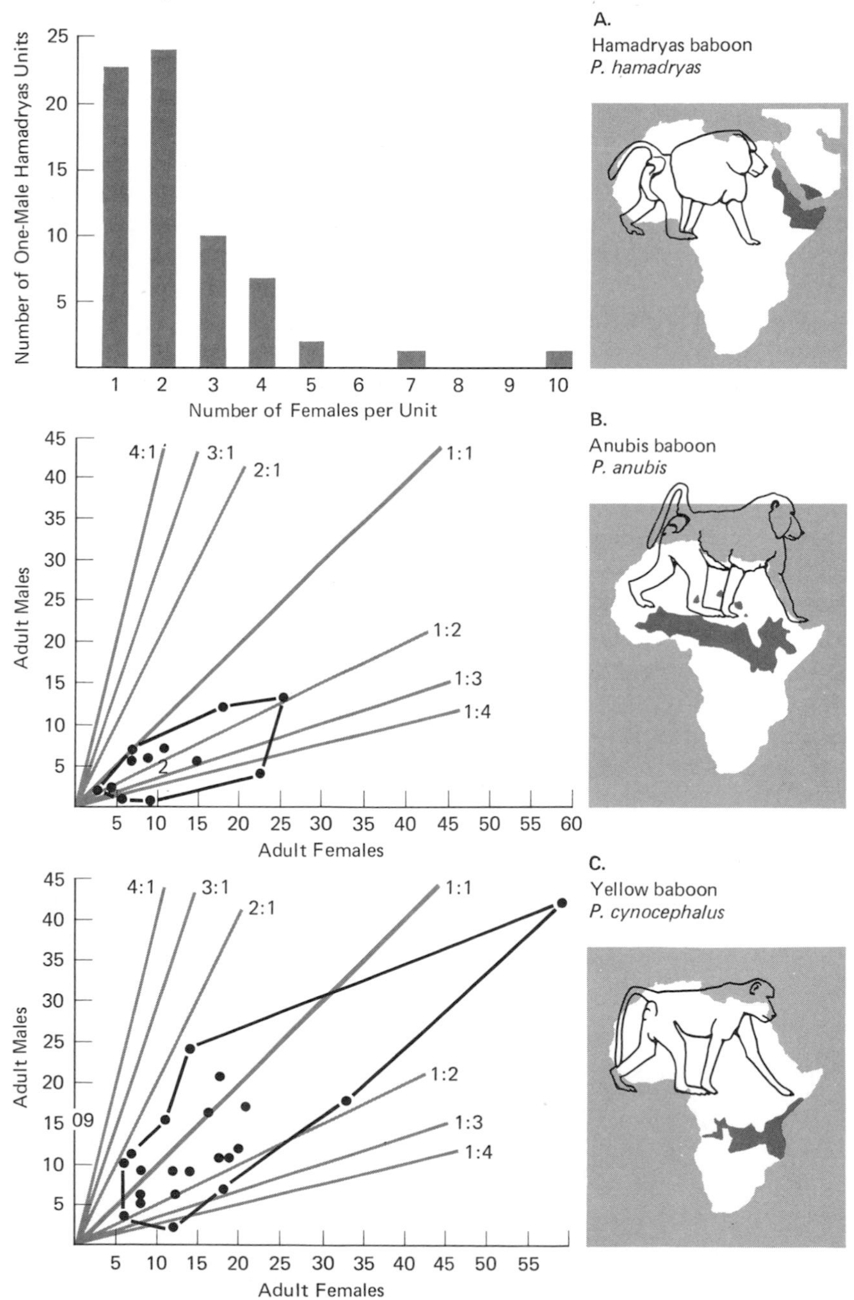

TABLE 12–5 Sex ratio and group size in baboons. The mean is followed by the standard deviation in parentheses. (Data are from the literature and observations summarized in Altmann and Altmann [1970].)

	Number of adult males	Number of adult females	Number of females per male	Total number of adults	Number of groups
P. hamadryas	1	2.3 (1.6)	2.3	3.3	68
P. anubis	5.94 (3.86)[ab]	8.5 (6.5)	1.4	14.4	18
P. ursinus	10.80 (6.17)[b]	19.1 (7.3)	1.8	29.9	20
P. cynocephelus	12.62 (8.82)[a]	16.2 (12.6)	1.3	28.8	21

[a] Probability that both means marked a are from same population <0.01.
[b] Probability that both means marked b are from same population <0.01.

apparently inconsistent (see Figure 12–1 and Table 12–5) and difficult to evaluate.

The accuracy with which the intensity of sexual selection can be estimated from such data may be low. A complicating factor in multimale groups is that dominant males tend to get more copulations in the peak of estrus than do subordinates, thus causing the adult sex ratio to underestimate the intensity of sexual selection. Another complication arises from uncertainty about the variability in the length of time that each male is dominant; each male tends to rise in rank and success in copulation as he matures. Those that are subordinate one year might be dominant the next. The whole problem of dominance is complicated by probable genealogical effects (as in macaques; see below). The length of time that each male is dominant and the total number of young fathered by him are usually unknown. What is needed, ideally, is an estimate of the variance of the number of young fathered per male in his total lifetime in various baboon populations.

Sexual Dimorphism and Sexual Selection / The large size and the adaptations for threat behavior, status indication, and fighting of male primates of certain species have been interpreted as adaptations for defense from predators (Hall 1964; Devore and Hall 1965). But other explanations also deserve consideration, especially sexual selection (Struhsaker 1969; Crook 1972). Sexual dimorphism in baboons is extreme for primates, the males being roughly twice as heavy as the females as well as having conspicuous manes and exaggerated canine teeth. These features are probably useful in defense against predators for baboons, but not apparently for such sexually dimorphic species as patas monkeys, langurs, or other smaller species, which flee from large predators rather than challenging them. Although sexual selection cannot be the whole explanation for sexual dimorphism, the degree of dimorphism appears to correlate fairly well with the potential number of females per dominant male (Tables 12–3, 12–5), being extreme in patas monkeys and baboons and slight in gibbons; the chimpanzee appears to be an exception, but the mating system of chimps is still unclear. Where harems are not maintained, as in tree shrews, mouse lemurs, chimpanzees, and gibbons, sexual dimorphism and delayed male maturity are less extreme. This is true in chimpanzees despite a possible need for protection against leopards (Kortland 1962).

The sexual selection hypothesis has several advantages. It helps us understand not only sexual dimorphism, but also delayed male maturity, species differences in harem size, the existence of all-male groups or solitary males in certain species, and priority of access to estrous females, all of which seem to be related (see Chapter 8).

THE ROLE OF AGONISTIC BEHAVIOR IN PRIMATE SOCIAL ORGANIZATION

The use of space by individuals is reflected in their dispersion pattern. The dispersion pattern is typically the result of cohesive behavior, tending to keep individuals together, and spacing behavior, tending to keep them apart. In primates, groups of various sizes are achieved and maintained primarily through family cohesion; offspring tend to remain with their parents even as adults in some species. For adult males, sexual attraction is important. Chief among the factors that oppose cohesion is aggressiveness. This is manifested in terms of the social organization in two ways, as aggressive behavior *between* groups and *within* groups. The former may give rise to regularity in dispersion and minimize the overlap of the home ranges of groups. The latter is reflected in dominance phenomena, which probably tend to limit group size, particularly by causing males to leave. It appears that aggressive behavior is the principal social agent determining species differences in group size and composition among primates. In some species aggressive behavior is so extreme that another adult, sexually active male is not tolerated anywhere near by a dominant male. In others, two or more such males may live together with little sign of aggression among them except for occasional manifestations of an established dominance relationship. Dominance relations within groups will be considered first.

Dominance Phenomena within Groups / Dominance phenomena have attracted considerable attention in those species in which the primary social unit is the multi-male troop, since it is in these species that dominance relationships are the most complex. An exemplary study of dominance in a wild population of primates is that of Struhsaker (1967*a*, *b*) on the vervet monkey in the Masai-Amboseli Game Reserve, Kenya. The groups he studied were of relatively stable composition, with a mean size of 24.1 individuals and an adult sex ratio of 2.1 females per male. A *linear dominance order* with very few reversals or triangular relationships characterized each group, the males generally dominating the females. Selected data for four of the males in one troop are shown in Table 12–6.

The data in Table 12–6 were obtained by observing which individuals supplanted others for access to space and preferred food items. Dominance was also reflected in the red-white-and-blue display, in which a male would display his highly colorful genital area to a subordinate. Dominant males displayed more often than subordinates. Dominants also achieved more copulations, the first-ranking male achieving 26 during the observation period, the second-ranking male, 3.

A unique feature of primate dominance orders is the occurrence

TABLE 12–6 Dominance and display among the four top-ranking males in a wild troop of vervet monkeys studied for one year by Struhsaker (1967*a*).

Dominance in natural dyadic encounters over access to food and space

Aggressor and/or supplanter	Agressee and/or supplantee: SG	SA	LP	LY
SG	—	5	25	22
SA	0	—	5	1
LP	0	0	—	2
LY	0	0	0	—

Red-white-and-blue displays

Displayer	Individual at whom display directed: SG	SA	LP	LY
SG	—	18	66	14
SA	0	—	6	0
LP	0	0	—	0
LY	0	0	0	—

Number of copulations observed

	SG	SA	LP	LY
	26	3	0	0

of *coalitions*, in which two or three animals join forces against another. A typical vervet coalition is a transient momentary affair in which one individual of lower rank solicits aid from another in a contest that he would not win alone. In some cases two or more subordinates combine to neutralize or reverse the actions of a dominant.

Similar transient coalitions were observed by Rowell (1966) among forest-edge *anubis* baboons in Uganda and by Koyama (1967) in Japanese monkeys. In Kenyan *anubis* baboons, longer-lasting coalitions were observed in rare instances (Hall and DeVore 1965). In hamadryas baboons coalitions take the form of "protected threats" (Kummer 1967): a female positions herself in front of her male in order to threaten a rival female from a safe position. Similar tripartite relations were observed by Altmann (1962) in rhesus monkeys (*Macaca mulatta*) on Cayo Santiago off Puerto Rico.

The importance of dominance phenomena in primate groups seems to vary from species to species and may also vary among different populations of the same species or superspecies. For example, Hall and DeVore (1965) stressed the importance of dominance relations among baboons in South Africa and Kenya, but Rowell (1966) found that in Uganda aggressive encounters were rare, the composition of troops was more variable, home ranges were smaller, and the numbers of adult females per adult male were much reduced. Gartlan and Brain (1968) stressed the transitory relationships among male vervets studied on Lolui Island; these relationships tended to reduce the emergence of stable hierarchies such as Struhsaker observed at Amboseli. Social structure also differed in other ways from that at Amboseli (smaller group size, smaller territories, fewer adult females per adult male). Intraspecific variations in social structure from place to place in various other primates were summarized by Rowell (1967*b*, 1972), Struhsaker (1969), and Crook (1970). Because much intraspecific variation in social organization seems to exist in most primate species, the stable and linear hierarchy found in Amboseli vervets should probably be viewed as typifying one end of a continuum toward which multi-male troops of primates tend under various conditions.

Perhaps the most important statement that can be made about dominance relations in multi-male groups in all species studied is that the boss typically performs more copulations with females at their peak of estrous receptivity than does any other individual. Consequently, he fathers the most young and leaves more of his genes to future generations than do his less-powerful rivals. Although it is true that the boss receives other minor benefits (easier access to food, grooming, and space), the frequency of fathering appears to be a much more powerful selection pressure. Consequently, it seems fair to state that the primary effect of male dominance on male primates in an evolutionary sense is in relation to the fathering of offspring (sexual selection, male-dominance type), even if the males do not appear to the observer in the field to be competing directly for females.

Females also have dominance orders in wild baboons, vervets, macaques, and other primates. For females, dominance confers easier access to food, space, water, and grooming partners. More important, it has been shown in *Macaca fuscata* in Japan (Koyama 1967, and earlier Japanese writers) and in *M. mulatta* on Cayo Santiago (Koford 1963; Sade 1967; Missakian 1972) that male and female offspring of dominant females have an early advantage over the offspring of subordinate females, and tend to rise sooner and higher in the hierarchy as they mature. Consequently, aggressiveness and dominance among female primates is of considerable value as a component of fitness and may perhaps affect the success of their sons in sexual selection (male-dominance type).

A feature of the dominance relations in some primate groups is the role of the dominant male in breaking up fights, chases, or other disputes between subordinates. This has been commonly observed in troops of baboons and macaques. It has been shown experimentally in captive macaques that the frequency of disputes rises when the dominant male is removed (Tokuda and Jensen 1968). This role of the dominant is consistent with Hamilton's (1964) genetical theory for the evolution of social behavior; a long-dominant male is likely to be closely related with more other troop members than is any other individual, and it is to his advantage genetically to minimize mutual damage among his close relatives. The *peace-making role* of the dominant and the occurrence of coalitions are apparently unique to primate social systems. They have not yet been observed in the dominance relations of birds or other mammals.

Agonistic Behavior between Groups / In examining the relations between groups a number of related questions arise. Is all or any part of the home range defended? How much and which parts of the home range overlap with the home ranges of neighboring groups? If certain parts of the home range do not overlap (are exclusive), does this arise mainly (1) because of low population density in relation to home range size, (2) because of geographic barriers, (3) because of mutual avoidance of groups, (4) because of strong defense of core areas but weaker defense of non-core parts of the home range, (5) because of strong-to-moderate defense of the entire home range, or (6) because of other factors or combinations of the above?

Primates as a group show much variability in how much they tolerate neighboring groups or individuals from outside the social unit. Thus in some species, such as the lar gibbon (Carpenter 1940), vervet (Struhsaker 1967*a*), black and white colobus (Marler 1969*a*), the lutong, *Presbytis cristatus* (Bernstein 1968), and the titi, *Callicebus moloch* (Mason 1968), intolerance of neighboring groups is intense, and classical all-purpose territories can be recognized (see Figure 12–2). In others, such as chimpanzees and baboons, little if any intolerance or spatial exclusiveness of groups can be detected. Many species fall between these extremes.

In some of the *intermediate* species, the nature of agonistic encounters between troops varies considerably from occasion to occasion, region to region. With such species, at least in the current state of our knowledge, using the terminology of territoriality seems to be oversimplification. By suggesting a concept of fixed, exclusive areas that are rigidly defended with specialized behavioral patterns, the subtleties of social organization, which are of great interest in primates, might not be fully appreciated. Even though spacing behavior occurs in baboons and macaques, it would be stretching a point to term it territorial behavior (see Chapter 4). Spacing usually occurs through mutual

12–2 Composition of groups of titi monkeys (*Callicebus moloch*). Home ranges of three of the groups are shown. Territorial defense occurred at sites of confrontation. *A* = Adult. *I* = Immature. *J* = Juvenile. (From Mason 1968.)

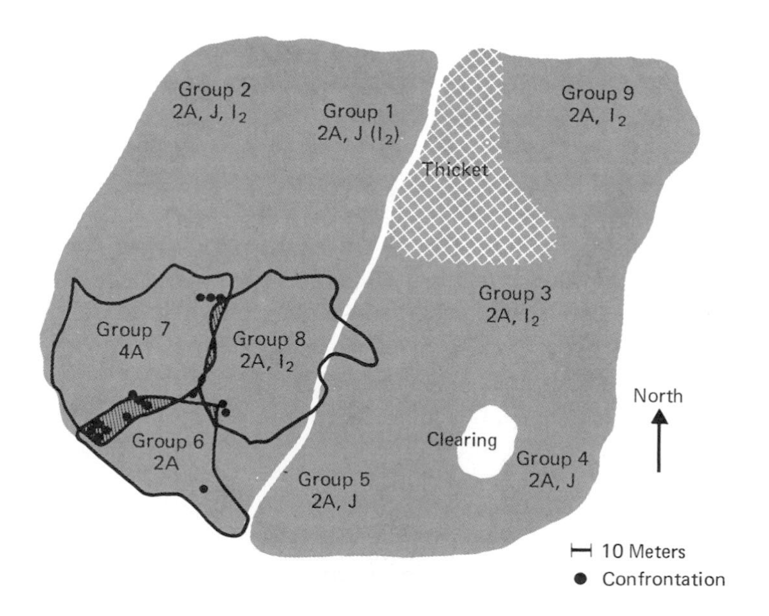

avoidance in these species or through dominance of one troop over another; occasionally one troop is seen chasing another. Since new troops may arise by fission of an old troop in macaques (Furuya 1968, 1969) and since the members of neighboring troops may consequently be close relatives, selection for extreme aggressiveness between troops might be expected to be reduced.

The chacma and olive baboons illustrate the state of at least several species of monkeys. In over 2,000 hours of observation by Hall and DeVore, "no aggressive interactions between groups have ever been recorded" (Hall 1964:56). Relationships between troops are characterized by mutual avoidance in most situations and temporary mutual tolerance in special circumstances, such as at water holes in arid regions. Typically, a troop forages and spends most of its time in one or a few areas (core areas). Invasion of the core area of one troop by another is rare; but on one occasion when it was observed, the intruding troop fled at the noisy approach of the owner troop. No parts of the troop home range were exclusive or defended against other troops.

Adjacent troops of *Macaca mulatta* in a protected temple area in India showed up to 80 percent overlap in home ranges, but they still usually avoided contact with each other (Southwick et al. 1965). Similar observations were made for the same species in an introduced colony at Cayo Santiago (Altmann 1962) and in another colony off Puerto Rico (Vandenbergh 1967). When a subordinate group failed to retreat fast enough from the dominant group, severe fighting occurred.

Simonds (1965) observed five contacts between groups of *Macaca radiata*. The groups sat and watched each other, then moved away. The only threat behavior seen occurred when a young male dropped out of a tree into the wrong group and was chased away. Mutual staring and withdrawal of troops have also been seen in baboons (Rowell 1967*b*).

The forces that keep primate groups apart in the absence of obvious territorial defense might also be related to dominance relations within troops. For an individual baboon or macaque, another troop is composed of many potential rivals, several of which would have to be fought if he entered their troop since dominance relations based on individual recognition may not have been established. Furthermore, a past history of association or cooperation may not exist for an individual with a strange troop. On meeting such a group it is not surprising that individuals choose to remain in their own group rather than risk fights with unknown rivals. Since the feeling is probably mutual between individuals of both troops, a natural result might be mutual avoidance. Only a highly confident and dominant male would be likely to approach closely another troop under usual conditions. Such a state of affairs could not be called territorial defense, and yet the result can be a dispersion pattern not differing greatly from that found in species that actively defend their home ranges.

The possibility that neighboring troops of baboons and monkeys may have been derived by fission of a single troop shortly before a field study began may also help explain the lack of aggressive actions when troops meet. Since many of the individuals in such troops would be expected to be closely related (e.g., as mother and son), a peaceful relationship between members of some troops may be based on social bonds developed through long and intimate associations. Rowell (1967*b*) has described a case that suggested such an interpretation to her. Peaceful overlapping of troops at waterholes might in some cases be so explained. Only genealogical studies like those done on Japanese monkeys will reveal the answer.

COOPERATION, AID-GIVING, AND KINSHIP

A conspicuous feature of the social life of man is cooperation among members of a group. This may be purposive and "conscious," as in the teamwork of a football team, or "unconscious," as in the division of labor among workers of various kinds in a metropolitan community. In making comparisons between species, it would be futile to try to make such a distinction and it is unnecessarv to do so from an evolutionary point of view.

Parental Behavior / The behavior of primate mothers toward their young is well known and requires no elaboration here. More interesting from the standpoint of kin selection is the predisposition of aunts and other females in the troop to behave in a maternal way towards offspring not their own (Mason 1965; on the chimp, Goodall 1967; on the rhesus monkey, Rowell et al. 1964; Kaufman 1966). This phenomenon is reminiscent of helpers in birds (see Chapter 9). For most species of monkeys, and probably apes, the birth of an infant attracts much attention from all the members of the group, some of which may try to handle it. The mother resists this at first, but in some species she eventually permits some care by other females. All female langurs in a troop aid and protect infants whether or not they are their own (Jay 1963).

Males may also show some interest in infants, although the paternity of an infant can almost never be reliably known in nonmonogamous species. Male marmosets typically carry the young, handing them to the female only for nursing (Eisenberg 1966). In tree shrews (subprimates, according to some) the nest is built entirely by the male (Martin 1966). In other primates, males tend to be protective; they may play with the young, and have been observed to retrieve them in emergencies. In some groups of Japanese monkeys adult males adopted yearlings, clutched, carried, and protected them, and behaved much as a mother would (Itani 1959; Alexander 1970).

Kinship and Dominance / Genealogical studies in groups of macaques have revealed that the influence of kinship extends far beyond the immediate relations between parents and offspring (Kawai 1958; Kawamura 1958; Koford 1963; Koyama 1967; Sade 1967; Missakian 1972). Certain dominance relations established during infancy tend to persist into adulthood. Study of Japanese monkeys indicates that several principles may be recognized that predict the dominance hierarchy of a troop over a period of time as a function of genealogy and of the "original" hierarchy; exceptions occur. These principles are illustrated in Figure 12–3.

1) A mother is dominant over her daughters and all their female descendents. The habit of acting as a subordinate of a mother persists all through life. Apparently the daughters of subordinate mothers learn to behave as subordinates to their grandmothers.

2) The rank order of females of the same age is the same as that of their mothers.

12–3 Genealogy and rank among Japanese monkeys. M = mother; A, B, C, = descendents of dominant mother; a, b, c, = descendents of subordinate mother. (After Koyama 1967.)

RANK AMONG FEMALES

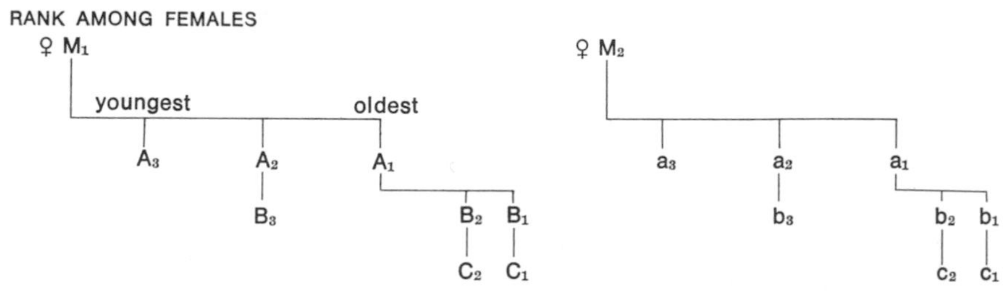

DOMINANCE ORDER:

$M_1 > A_3 > A_2 > B_3 > A_1 > B_2 > C_2 > B_1 > C_1 > M_2 > a_3 > a_2 > b_3 > a_1 > b_2 > c_2 > b_1 > c_1$

RANK AMONG MALES

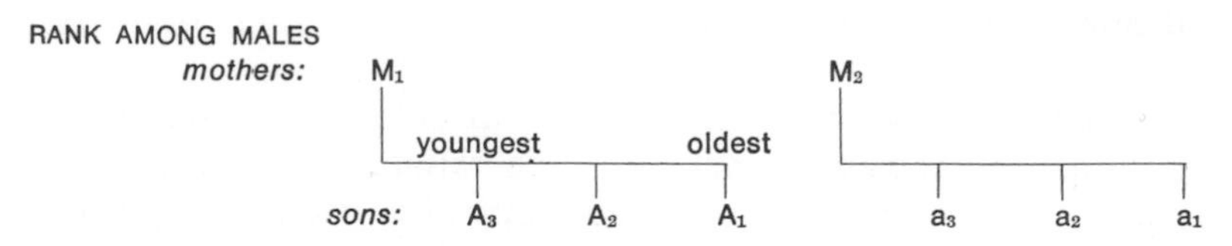

DOMINANCE ORDER:

$A_1 > A_2 > A_3 > a_1 > a_2 > a_3$

3) Females descended from a dominant female tend to be dominant to females descended from a subordinate female, regardless of age; but exceptions tend to occur when the disparity in age is great.

4) Among sisters, younger ones dominate older ones all through life. This habit begins in infancy, apparently because the mother is more protective of her most recent daughter than of older ones, and resolves disputes between them in favor of the younger. This rule is known as the *principle of youngest ascendency.*

5) Males follow the same rules as females for their first two years—until they begin to differ from them in size and behavior.

6) From age three on, older brothers dominate younger brothers. This habit begins when the size differential among brothers is greatest and persists throughout their lives.

7) The rank order of males of the same age is the rank order of their mothers.

8) Males descended from a dominant female tend to be dominant to males descended from subordinate females even when of different age, but exceptions tend to occur when the age disparity is great.

In a stable monkey troop without immigration, the above relationships between genealogy and dominance characterize the dominance structure of the troop nearly completely. If the troop enlarges, the disparity between rank dictated by tradition and genealogy and rank derived from ability leads to the disruption of the troop and the splitting off of small bands from it.

Cooperation / Group living per se reflects some cooperation, since individuals must sacrifice their own inclinations at times to stay with the group. Several activities that benefit all members of a group are probably more efficient when performed in a group than individually. Crook (1966*b*, 1971) has listed some of these. They include several types of defensive behavior against predators, such as throwing sticks (howler monkeys), mobbing leopards or wild dogs (baboons), and making loud noises. Maintenance of exclusive areas or group dominance is probably also aided by a large group size. Cooperation in capturing prey, as when chimpanzees catch a red colobus monkey, has been observed. The spread of culturally acquired traits, such as use of tools by chimps to remove ants and termites from nests, and food-washing by Japanese macaques, is probably also facilitated by large group sizes.

Kin Selection / Before the evolution of cooperative and aid-giving behavior in primates can be understood in terms of Hamilton's (1963, 1964) hypothesis, genealogical and ecological data taken under natural conditions must be analyzed.

It is commonly assumed that the members of a troop of baboons, monkeys, or apes are closely related genetically, but little is actually known of this. Inbreeding in such troops seems very likely. Some evidence of it has been found in baboon troops through biochemical comparisons of individuals in the same and different troops (Buettner Janusch 1963), and in rhesus monkey troops through observations of incestuous copulations (Missakian 1973). Apparently females rarely transfer between troops, but males do so more commonly in many species (Jolly 1972; Crook 1970; Rowell 1967*b*; Hall and DeVore 1965; Struhsaker 1967*a*, *b*; Vandenberg 1967; Lindburg 1969).

Both sons and daughters of dominant females may benefit by rising faster in the hierarchy as they mature (Koyama 1967). Thus altruistic acts by females may frequently benefit their close relatives, but also indirectly aid their grandsons and granddaughters and more distant relatives. The protective and peace-making roles of the dominant male toward females and young are also likely to be favored by kin selection, since he is probably the father in most cases. By contrast, a strange male langur (*Presbytis entellus*) entering a troop may kill all the infants present (Yoshiba 1968).

ENVIRONMENTAL INFLUENCES ON SOCIAL ORGANIZATION

Studies of nonprimate groups have shown that the significance of species differences in social organization often cannot be appreciated except by studies of within-species variations in different environments. One aspect of the environment that has a most important effect on the social system is population density; others are food availability and predation. Often these three factors are inextricably bound together.

Group Size and Carrying Capacity / Some of the first indications of the importance of environment in primate social organizations came from observing baboons. Hall and DeVore noticed that troop sizes tended to be larger in the richer habitats of Amboseli Reserve, Kenya (Figure 12–1C) than in the more barren habitats of South Africa (*P. ursinus*, Table 12–4). That the closely related hamadryas baboon, which lives in a still more arid and barren environment (Figure 12–1A), had even a smaller group size (two to ten individuals) also helps support the hypothesis of a relationship between group size and environmental carrying capacity.

Troop size may also be related to predation pressure. One of the supposed advantages of a large troop is in defense against large predators, such as lions, leopards, hunting dogs, hyenas, and others (see discussion of Figure 12–1).

Density and Territoriality / Vervets provide another example of regional differences in social organization. In Kenya, they were found to be aggressive with strict territoriality and hierarchies (Struhsaker 1967*a*), but in Uganda there was little or no fighting, no rank order, and movements of males between troops were frequent (Crook 1970). Again, population density and carrying capacity may be important for understanding such differences. Struhsaker (1967*b*) reported that in his study area the density of vervets was greater than he saw in any other part of East Africa. It may be that full realization of the capacity of primate species to behave territorially requires a high population density. Similarly, rapidly increasing populations, such as the forest-edge baboons (Rowell 1966, 1967*b*), may have different social organizations than stable populations that have reached their limits.

Another example was found in Indian langurs (*Presbytis entellus;* Yoshiba 1968). In Kaukori, an area of low density, they lived in large, peaceful, mixed troops of both sexes and all ages. In Dharwar, an area of high density, one-male harems and all-male groups were characteristic (see Table 12–7).

Food / The nature of the food supply may also have subtle effects on social organization (Rowell 1967*b*). Where food is abundant, little time is spent foraging and more time in mutual grooming and other social behavior. Where food is scarce and evenly dispersed, foraging is

TABLE 12–7 Variation of group composition as a function of population density in Indian langurs. (From Yoshiba 1968.)

Troop name	Population density (individuals per sq. mile)	Adult		Subadult		Juvenile		Infant	Total
		Male	Female	Male	Female	Male	Female		
Kaukori	7	6	19	2	3	5	5	14	54
West Orcha	16	3	5	—	1	2	2	6	18
North Orcha		6	9	1	2	—	5	5	28
East Orcha		2	4	—	1	—	—	3	10
Dharwar No. 1	43	1	7	2	8	2	3	4	27
Dharwar No. 2		1	7	—	4	—	5	6	23
Dharwar No. 3		1	6	—	1	3	3	—	15
Dharwar No. 4		1	8	—	1	—	—	6	16
Dharwar No. 5		1	8	2	2	5	5	—	23
Dharwar No. 30		1	9	1	—	6	3	5	25

a full-time occupation and social interactions are few. If food is concentrated (as when artificial provisioning is practiced), frequent agonistic interactions at the source are observed and dominance orders become more structured and important to the individuals in the group.

Until we know more about the relative importance of food, predation, behavior, disease, and parasitism in the dynamics of primate populations, the characteristics of social systems in different species of primates will remain difficult to explain.

CONCLUSIONS

There is no "primate type" of social organization; instead, the order has a wide diversity of social systems. Societies of primates range from monogamous pairs to large, promiscuous troops with an infrastructure based on individual recognition. Crook and Gartlan (1966) have sketched a progression from small, paired or solitary, insectivorous, forest-living species, through fruit-eating species living in small groups, to large-sized, open-country species with large troop sizes. They have rightly emphasized the importance of ecological factors for an understanding of these evolutionary changes, but it is not yet clear which ecological factors are critical and how they operate.

Looking back at the three main types of behavior that provide the structure of primate social organizations—agonistic, sexual, and aiding behaviors—we are now in a much better position to describe the role of these factors than was Zuckerman (1932) when he chose to emphasize the role of sexual attraction. The role of agonistic behavior in determining the dispersion patterns within primate species is much better known now. Developmental studies have revealed the great importance of positive social bonds developed through the years of association between parents and progeny in holding together a troop of primates.

The interrelationships among agonistic, sexual, and aid-giving determinants can now be considered briefly. Sexual selection, as Fisher (1958) and O'Donald (1962, 1967) have stated, is a runaway process. That it is important in primates, there is little doubt. The limits to which sexual selection is allowed to proceed are set by other factors, mainly by the need for aid-giving behavior by the male. In primates the male's behavior is still apparently an important determinant of reproductive success (recruitment). In pairs the contribution of the male parent is expected to be of great importance. In species with one-male harems there seems to be an upper limit to the average number of females that the harem-owning male can acquire. This upper limit must in part be due to the necessity for the male to stay with his females throughout the year. (But in the fur seal the males need not stay, and potential harem sizes are much larger.)

It is helpful to speculate why the red deer or fur seal system has not evolved in primates. A system in which males defended mating stations and females lived separately except at the time of copulation is theoretically possible and marks the direction in which uninhibited sexual selection would tend. Teleologically, the hypothesis may be advanced that such a system did not evolve because the presence of one or more males with a group of females raises their reproductive success above what they would have without a permanent male. It has been argued that the primary contribution of males (beyond fertilization) is in defense against predators and, perhaps, against other troops of the same species. Males also are more bold in leading their troops into new areas and in competing with other troops.

Whatever the advantages, a system based on cohabitation of males and females in the same troop requires (1) tolerance of the female for the male, (2) tolerance of the dominant male for subordinate rivals (in multi-male troops) and (3) cohesive social forces in addition to sexual attraction and pair bonding, such as arise through protracted association between mothers and their progeny. Furthermore, in territorial species or in certain nonterritorial, one-male harem species, the males must be capable of channeling their aggressiveness toward outsiders and inhibiting its expression toward insiders. Aggressiveness, then, is important in its presence in certain situations and its controlled absence in others. It plays an obvious role in the spacing patterns of territorial species (with active defense), and a subtle role in the spacing patterns of species with regular dispersions based mainly on mutual (or one-way) avoidance—in addition to its role in intra-troop relations.

In summary, agonistic behavior, when appropriately expressed for a given species, is adaptive both in competitive mating and in setting up the social groups within which the advantages of aid-giving (cooperation, protracted parental care, facilitation of cultural transmission of behavior, etc.) can raise the relative fitness of troop members, one-male harem members, or pair members—as the case may be. The social organization of a primate species is determined in the ontogenetic and phylogenetic perspectives by the interaction of these three primary types of behavior: agonistic, sexual, and aid-giving. Like the various essential organs in the body, it is futile to attempt to rank them in order of importance.

From the comparative point of view, however, these observations are little better as answers than was Zuckerman's emphasis on the paramount importance of sex. The enduring question raised by comparative studies of primates is still this: What are the environmental conditions and selection pressures that have caused some lines of primate evolution to develop large, structured troops, others to maintain a society based on pairs, or even solitary individuals, and still others to evolve toward one-male harems?

The evolution of more complex societies seems to have occurred independently in more than one primate line, as in the Madagascar lemurs, the Old World monkeys, the apes, and the New World monkeys. Since adaptive radiations in social organization have occurred in all of these major phylogenetic groups, it could be argued that certain general features of primate biology preadapt primates for further evolution toward more complex societies when environmental conditions favor such an evolutionary change. It is now up to ecologically oriented primate biologists to identify these environmental conditions through systematic study of primate population dynamics in various species and various environments—so that we may better understand the genesis of "the naked ape."

THE ADAPTEDNESS OF BEHAVIOR IN ANIMAL COMMUNICATION

THE BEHAVIOR OF animals has evolved in so many different ways that questions are often asked about why a particular form of behavior evolved in a particular kind of animal and not in another. The answers depend on two factors: (1) the *genetic material available* to natural selection—which is determined by the phylogenetic past history of the species—and (2) the *selection pressures* that arise in the physical and biological environment and are presently important. The first factor, phylogeny, can be studied with the methods of palaeontology and comparison of living forms. The second factor, which forms the basic theme of Part IV, often requires the study of the animal in its natural environment as well as in a laboratory. The goal in this type of study is to find out how the behavior benefits the individual animal and, consequently, how natural selection might have acted to bring about its evolution—in other words, to find how the behavior is *adaptive*.

The Problem: The Adaptive Nature of Species Differences / Since the selection pressures that have been responsible for the evolution and maintenance of a behavior usually cannot be reproduced faithfully in a laboratory, it is absolutely essential that such studies be done under *natural conditions* as undisturbed as possible from their primeval condition. This is difficult in most parts of the world today, since few unspoiled areas are left, but often a suitable compromise can be made by working in relatively natural areas. Studies of captive animals may often be extremely useful supplements to field studies of unrestrained animals, but they can never completely substitute for them if the adaptive characteristics of the behavior are to be fully understood.

Concepts of Adaptation / The concept of adaptation has various meanings in behavioral biology, and the differences between some of them deserve to be made clear here. Most concepts of adaptation imply a change that benefits the organism (raises fitness). The change may be either of two types. It may be genetic, in which case it is an evolutionary change and occurs in populations over time, but not within a single individual. This is *evolutionary adaptation,* and is the principal theme of Part IV. When the word *adaptive* is used in this book, it refers almost invariably to this kind of adaptation, as is customary in evolutionary biology.

Or the change may be due to nongenetic factors, and occur within a single individual over a period of time. This is *physiological adaptation* or *adaptability*. Examples of this second type include learning and other physiological changes such as faster breathing at higher altitudes, development of calluses on the feet and hands in response to wear, tanning of the skin after exposure to strong solar radiation, and many other responses of individuals to changes in their environments.

In sensory physiology and neurophysiology, adaptation has a third meaning. A preparation becomes adapted when an initial response to a regularly repeated or continuous stimulus has diminished to a relatively low level. For example, a neuron in the optic system may fire in a burst when a light goes on, but then stop even though the light remains on. Similarly, other neurons fire in a burst when the light goes off. This type will be referred to as *sensory* or *neuronal adaptation*. Unlike the previous two types of adaptation, this one does not emphasize benefit to the organism. The *capacity* for physiological, sensory, or neural adaptability can be viewed as an evolutionary adaptation.

Communication Exemplifies Evolutionary Adaptation / Comparative studies of animal communication have been especially fruitful in the study of the adaptiveness of behavior. The functions of many behaviors known to be basically communicative, such as sound production, have been described more precisely. Many behavioral patterns whose function was previously misunderstood have been found to serve communication—for example, the displacement preening of ducks. And some behaviors not previously known at all, such as the release of chemical sex attractants by moths, have been discovered and found to serve communication.

Species differences in communicative behavior are especially conspicuous and have stimulated much interest in how communicative behavior evolved. As a result many species have been studied, many comparisons among species have been made, and a body of theory about the evolution of animal communication has developed. These will be the principal subjects of the chapters in Part IV.

What Is Animal Communication? / Communication is a complicated process involving two or more individuals. Therein lies much of its evolutionary interest. It involves the emission of a signal by one and its reception by another. We know by introspection that communication may occur without an observable response. But in animals the only indication that communication has in fact occurred is in the response of the receiver. "In short, social communication is a process by which the behavior of an individual affects the behavior of others" (Altmann 1967). In the extreme case, the signal may cause a directly observable change in the behavior of the receiver. More commonly, reception of the signal is evidenced through a change in the probabilities of one or more of an array of behaviors.

These social influences on behavior affect the fitnesses of the individuals involved along the lines sketched in Chapter 9 (see Table 9-11). As a working hypothesis we may assume that for a signal to evolve it must benefit the signaler; it may either benefit or harm the recipient, depending on whether the social context is cooperative (e.g., mother and young), altruistic, spiteful, or competitive (e.g., rival males). Similarly, we hypothesize that selection favors responses that benefit the responder; they may benefit or harm the signaler. In short, in an evolutionary context the social influences involved in communication must be considered in relation to their effects on the fitnesses of the participants.

Brain Size and Communication / Because in man communication and large brain size occur together, a tendency exists to extrapolate this association to other species. But large brain size is not necessary for complex communication, nor do species with large brains necessarily have complex communication systems; honeybees possibly have a more complex communication system than some solitary mammals.

Motivation and Communication / In animals signal emission is typically a function of the motivational state of the sender. For example, an observer can predict from the behavior of an animal whether it is likely soon to court, threaten, submit to, or groom another. However, predictive value is not necessarily the adaptive value in all cases of communication.

Similarly, the response of the receiver is typically a function of its motivational state. The success of the signaler depends upon the extent to which the motivational state of the receiver is altered by reception of the signal. In human terms, communication between animals might be said to be more on an emotional or motivational base than on a symbolic one. For this reason and others it is imperative that the methods for studying communication in animals be objective.

Sensory Modes / It is convenient to subdivide the problems of animal communication according to the sensory modes involved. The most common modes are vision, audition, smell, taste, touch, pressure, vibration, and electrical sense; but any type of sense organ can be used for communication. Of these the most attention has been devoted to vision, audition, and the chemical senses; consequently, they are the most rewarding to study from an evolutionary viewpoint.

In an introduction to the study of animal communication it is tempting to try to define animal communication. Definitions tend to be restrictive, however; and in a field that deals with the evolutionary antecedents of communication, restrictions seem dangerous. No formal definition of animal communication will be attempted here, nor will the various definitions in the literature be discussed. As a guideline, however, animal communication will be considered broadly as the processes by which an animal influences others in a social context.

13 Visual Communication: Contexts and Measurement

VISION is useful to animals in a variety of ways that include finding food, detecting predators, choosing a habitat, and becoming oriented rapidly in the environment. In the early part of this century, views of the importance of animal color patterns and visual appearance were largely oriented in these directions. The primary role of animal coloration was thought to be to provide camouflage that made the individual less conspicuous to predators. Many beautiful examples of effective camouflage have been demonstrated (Cott 1957), but the trend went too far when it was suggested that flamingos were pink to make them less conspicuous against tropical sunsets, and that the large and dazzling tail the peacock displays to the peahen was really a type of camouflage in the variegated light patterns of a tropical jungle.

The role of animal coloration in communication was emphasized in studies by Lorenz (1935), Tinbergen (1951, 1953), and Baerends (1953; Baerends and Baerends-van Roon 1950). They and their students demonstrated by observation and experiment the importance of color and pattern in communication in diverse animals, from butterflies to birds. In addition they provided a comprehensive theory that integrated much behavioral information on communication by means of hypotheses about releasers, key stimuli, fixed action patterns, and releasing mechanisms. This theory, which is set forth in Tinbergen's (1951) *The Study of Instinct*, stimulated new research and new hypotheses.

CONTEXTS OF VISUAL COMMUNICATION

The situations in which visual communication between animals occurs will be summarized briefly. Information of this sort has been derived from a long list of comprehensive and detailed studies of communication in single species, particularly in natural settings.

Patterns of coloration thought to facilitate *aggregation*, or the keeping together of a group, are found in some group-living species. The white tail of North American deer (*Odocoileus* spp.) is raised when the deer is alarmed and appears to aid in the coordinated escape of family groups. The white rumps and concealed wing patterns of geese and ducks are exposed when the birds take off and seem to stimulate others to follow. Schooling in the fish *Pristella riddlei* is facilitated by a conspicuous black mark on the dorsal fin (Keenleyside 1955).

Individual recognition is easier for a human observer in dealing with some species than with others, and presumably species differ in the ease with which individuals can recognize each other. Gorillas (*Gorilla gorilla*) may be told apart by their facial features (Schaller 1963), as can various other primates. The facial region is important for individual recognition in hens (*Gallus domesticus*), gulls (*Larus* spp.), flickers (*Colaptes* spp.) and probably other birds, as shown by experimental modifications (Marks et al. 1960; N. G. Smith 1966; Noble 1936).

Alarm is commonly indicated with auditory signals, or simply by the visible escape of the alarmed individuals, but in plains mammals this escape may take a specialized form, as in the peculiar stotting (high leaping) gait of Thompson's gazelle (Walther 1964).

Aggressive threat is indicated in the displays of many species. These often take the form of incomplete motions to attack with the bill, wings, feet, teeth, or whatever weapons are used. Such displays tend to increase the visibility of the weapon to the recipient.

Defensive threat, such as that of a cat to a dog, may also be exhibited to members of the same species (Leyhausen 1956).

Pair formation may be facilitated by elaborate displays in ducks. The long-billed marsh wren (*Telmatodytes palustris*) makes use of a nest as a platform and center of interest for the female (Verner 1964). Male fireflies attract females by emitting specialized light signals (Lloyd 1966).

Maintenance of the pair is thought to be aided by the elaborate greeting ceremonies of herons.

Stimulation of ovulation and synchronization of breeding cycles in male and female are thought to be aided by the displays and vocalizations of ring doves and budgerigars (Lehrman 1959, 1961, 1964).

Precopulatory displays include the presenting of primates and the assumption of a prone or horizontal position in birds.

Long-distance advertisement of the presence of a male is aided by certain displays of the ruff (*Philomachus pugnax*) that lure flying females (Hogan-Warburg 1966).

Appeasement or submission is indicated by certain primate facial expressions, by presenting, and (in dogs) by putting the tail between the legs.

Parent-young relations can be rather simple. Young thrushes will beg in response to very simple stimuli (Tinbergen and Kuenen 1939), but adult estrildid finches may not feed the young in their nest if they lack the proper gape marks (see Figure 11–8; Nicolai 1964).

The ways in which the physical structure of a visual signaler is correlated with the functions of the signal in many of the above contexts have been considered by Marler (1968). For example, in aggressive threat the weapons of the species are commonly displayed, but in contexts requiring close approach or actual contact the weapons and other features used in aggressive threat are hidden or not employed.

The Demonstration of Visual Communication / A variety of experimental methods have been used that illustrate the involvement of visual appearance in communication. One of these is to alter the appearance of a living animal and test responses to it. A simple test of the effect of male red coloration on female jewel fish (*Hemichromis bimaculatus*) was done by Noble and Curtis (1939). A female alone in a large aquarium could see two males, a red one at one end in a separate aquarium, and a duller one at the other end in a third aquarium. The difference in coloration of the males was achieved with drugs or by choosing males as different as possible. The females presented with this choice chose to spawn their eggs at the end of their aquarium nearest to the red males. The red coloration of males appeared to affect the behavior of the females. In the fish *Pristella riddlei*, lone individuals will join a school. This species has a dorsal fin with a dark blotch on it. To test the hypothesis that this blotch facilitates aggregation into schools, Keenleyside (1955) presented lone *Pristella* with a choice between a school of the same species lacking the black spot and a school with the normal coloration. The role of the fin spot in communication was shown by the preference of the lone fish to join the school having the spots.

In birds the facial region seems to be important in individual recognition as well as in recognition of sex and species. Noble (1936) captured a female flicker (*Colaptes auratus*) of a mated pair, painted a black "moustache" on her to make her look like a male, and watched the response of her mate when she was released. The mate rejected her. Similar results were obtained with mourning doves (*Zenaidura macroura;* Goforth and Baskett 1965).

The role of sexual dimorphism in plumage color has also been investigated in the chaffinch (*Fringilla coelebs*). Marler (1955, 1956) found that females in nonbreeding flocks in large cages would approach to within a few cm of other females when in normal plumage, but that when all females were painted with a red breast to resemble males, they tended to stay at least 20 cm apart (see Chapter 5).

In some species it has been possible to gain greater control over the stimulus with the use of models. Noble and Bradley (1933) found that plasticene models could be used to evoke attack in eastern fence lizards (*Sceloporus undulatus*). They showed that a blue throat, present only in males in the population they studied, was essential to evoke attack by a male. In this and other lizards the ventral colors are displayed to other individuals in a push-up-like performance.

Models have been used successfully in studying the role of various aspects of appearance in fishes. Ter Pelkwijk and Tinbergen (1937) presented various models to male three-spined sticklebacks (*Gasterosteus aculeatus*), and studied their success in evoking threat behavior. Male sticklebacks defend a territory around their nest against other males, and a model will be attacked if it has certain characteristics. Size and shape were relatively unimportant, but a bit of red color, like that male sticklebacks normally have on their bellies, was essential. The red had to be on the ventral side (Figure 13–1). A vertical orientation of the model, like that males adopt when threatening, also facilitated a strong response. Other models were used to simulate females; it was found that the principal characteristic of effective female models was a swollen abdomen; males were more interested in crude models with large bellies than more realistic models with normal bellies. Since a female ready to lay her eggs in the male's nest normally has a large abdomen, this preference by the male is natural.

The young of the mouthbreeder, *Tilapia mossambica*, enter their

13–1 Use of models to study visual signals. Male sticklebacks were presented with a series of models of various shapes but all having a red belly (*left*), and a series that were accurate models of a real male stickleback but without the red belly (*right*). The tested males attacked the simple models with red bellies much more vigorously than lifelike ones with neutral bellies (From Tinbergen 1948.)

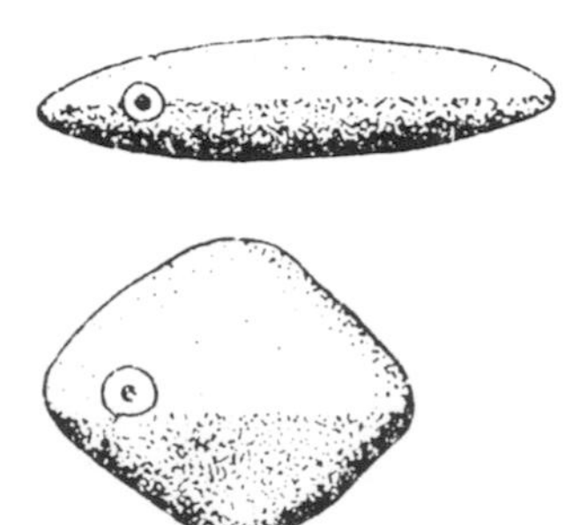

mother's mouth when disturbed by an approaching large object or turbulence. The visual characteristics attractive to the young were investigated with simple models by Baerends and Baerends-van Roon (1950). They observed that the young were attracted to the underside of the model and to dark patches; slowly retreating models were most effective. Further studies of this behavior in *Tilapia* spp. were made by Peters (1963).

The influence of visual signals on animals can be tested crudely with mirrors. Aquarium fishes, such as Siamese fighting fish (*Betta splendens*), perform particularly well and will display, change color, and erect their fins in response to their reflections in mirrors. Primates also respond to mirrors. Some squirrel monkeys (*Saimiri sciureus*), when shown a mirror image of themselves, respond with a display involving phallic erection that typically occurs in agonistic encounters (MacLean 1964).

Chimpanzees, after prolonged experience with mirrors, showed evidence of being able to recognize their own reflections as belonging to themselves, but monkeys (*Macaca* spp.) did not (Gallup 1970).

Colored slides to test visual stimuli of communicative significance have not been widely used, but rhesus monkeys reared in isolation responded differentially to photographs of other monkeys with various facial expressions (Sackett 1966).

An imaginative demonstration of the communicative effects of facial expression on rhesus monkeys was based on a shock-avoidance conditioning technique (Miller et al. 1959, 1962, 1963, 1966; Miller 1967). The first monkey learned to avoid electric shock by pressing a bar when a light went on signaling the coming shock. The second monkey could see the face of the first monkey but not the signal light, and would be shocked at the same interval after the signal unless he pressed a bar of his own. The second monkey learned rapidly to avoid the shock when facial expression alone was available as a means of communication. Similar attempts with food as a reinforcement were less successful.

SOME TERMS AND CONCEPTS

The examples just described show that various acts and color patterns have communicative functions. Some of these, such as the light emissions of fireflies, the push-ups of *Sceloporus*, and the greeting ceremonies of herons, have become highly specialized for communication through evolutionary adaptation. Visual signals that are believed to have become specialized through evolution for communication are known in ethology as *displays.** The question of whether or not a given visual signal qualifies as a display requires a judgment about its evolutionary history; consequently, some uncertainty and disagreement among specialists is to be expected. Nevertheless, there is little controversy over such matters, and most ethologists have little trouble distinguishing displays from unspecialized visual signals. When in doubt, the term *visual signal* can be used for all visual signals whether they are displays or not.

The term *releaser*, from the German *Auslöser* (Lorenz 1935), has been used to designate structures and actions that are particularly effective in eliciting specific behavior patterns. For example, the red belly of male sticklebacks evokes threat from other males; the blue throat of male *Sceloporus* lizards evokes threat in other males; the swollen belly of female sticklebacks acts as a releaser for courtship in territorial males. The *key stimuli* in these cases were determined by experiment.

THE MEASUREMENT OF COMMUNICATION

By observation and by experiments with models and disguised animals it has been demonstrated that animals react to visual signals from others. In this sense, we can say that communication has been demonstrated: the display of the model or animal has influenced the behavior of the reactor. These are highly contrived situations, however, and they reveal little about the degree to which displays and unspecialized visual signals influence the behavior of potential reactors in the context of a natural, unstaged encounter between two individuals. In the latter situation, few variables are controlled, and many are acting simultaneously. It is possible to observe the full *system* of communication in a species *only* in such a situation. Since it is desirable to be able to make statements about the system as it is affected by different environmental conditions or other factors, and to be able to compare the communication systems of various species, it is worthwhile to consider how this has been done.

Although nonhuman animals have relatively simple communication systems compared to man, it should, in principle, be possible to

* It has been proposed that all signals adapted for communication be called displays, *regardless of sensory mode* (Moynihan 1955). However, the word *display* has traditionally been limited to visual effects and is so designated in dictionaries. The term *ritualized signal* is available for all signals adapted for communication through evolution, regardless of sensory mode. The proposed extension of the term *display* to auditory signals would be redundant, since virtually all auditory signals are ritualized, except vomiting, coughing, and secondary sounds like those associated with locomotion. Both usages of *display* are common in the literature; the reader may choose.

apply to animals the modeling techniques that engineers have developed as a part of communication theory. Although there are many difficulties in such an attempt, it seems desirable to examine one such case in some detail, because it suggests to the ethologist ways to analyze communication in rigorous terms.

Few applications of communication theory in ethology have been made. They include a study of visual communication in hermit crabs (Hazlett and Bossert, 1965, 1966; Hazlett 1968), a study of communication by odor trails in the fire ant by Wilson (1962), a study of communication by unspecified sensory modalities in the honeybee by Haldane and Spurway (1954), and a study of communication between mantis shrimps by Dingle (1969). Similar statistical techniques, involving the treatment of behavior sequences as Markov processes, have been employed by Nelson (1964*a*; see also Hauske and Neuburger 1968). Of these, the study on mantis shrimps is most useful as an illustrative example.

The behavior of mantis shrimps (*Gonodactylus bredini*) was first described in detail as observed in finger bowls at a marine station (Dingle and Caldwell 1969). These small shrimps (30–80 mm long in adults) rely heavily on visual communication. They inhabit cavities in rocks, coral, and sponges, and defend them vigorously. All acts by each individual of an interacting pair were recorded in sequence without regard to duration. The acts by each of the two individuals were recorded alternately so that each act by one individual (except the first) was preceded by an act by the other. All pairs of successive acts were then entered in a matrix of inter-individual two-act sequences, like that in Table 13–1.

Each act of such a sequence, except for the first and last, can be considered both as a reaction to the act that preceded it and as a stimulus for the act that followed it. Consequently, each act is entered first as a preceding act and secondly as a following act. The matrix shows how many times each type of act followed each other type of act.

To determine whether or not specific acts were followed by changes in the behavior of the opponent, the distribution of following acts succeeding a specific initial act was compared with the sum of the distributions of the following acts (the column totals), i.e., with the average distribution. The chi-square values and probabilities for the data in Table 13–1 are presented in Table 13–2. The display, *meral spread*, a movement of the appendages used in striking opponents, was effective in altering the following acts of an opponent, as were some of the other acts listed.

Directive effects of the preceding act raised the observed frequency of a following act above its expected value. *Inhibitive* effects lowered the observed frequency below the expected value. Tables 13–1 and 13–2 show that meral spread was directive to the act *avoid* and inhibitive to the act *chase*. (Expected value = row total for that act × column total for that act ÷ number of observations.)

Table 13–1 can also be used to calculate certain quantities used in

TABLE 13–1 Matrix of inter-individual two-act sequences in encounters staged between captive mantis shrimps. In each cell of the matrix the number of observations is followed in parentheses by the expected number. Each row gives the frequency distribution of acts following the initial act stated at the left of the row; each column gives the acts preceding the following act stated above the column. Meral spread is a simultaneous outward spreading of the enlarged raptorial, second thoracic appendages (meri); it functions as a threat. (From Dingle 1969.)

	Following act									
Initial act	Approach	Meral spread	Lunge	Strike	Chase	Grasp	Coil	Avoid	Does nothing	Total
Meet	0(0.16)	3(1.9)	0(0.28)	5(1.8)	0(1.6)	0(0.08)	4(2.7)	2(3.5)	0(2.0)	14
Approach	0(0.46)	15(5.7)	0(0.81)	5(5.2)	1(4.7)	1(0.23)	10(7.9)	8(10)	1(5.9)	41
Meral spread	1(0.66)	10(8.1)	2(1.2)	3(7.5)	0(6.8)	0(0.33)	12(11)	28(15)	3(8.5)	59
Lunge	1(0.36)	2(4.4)	2(0.63)	0(4.1)	0(3.7)	0(0.18)	17(6.1)	9(7.9)	1(4.6)	32
Strike	1(0.73)	4(9.0)	0(1.3)	14(8.2)	1(7.5)	1(0.37)	13(12)	28(16)	3(9.3)	65
Chase	0	0	0	5	0	0	0	0	0	5
Grasp	0	1	0	2	0	0	1	2	0	6
Coil	0(0.32)	6(3.9)	2(0.55)	6(3.5)	0(3.2)	0(0.16)	6(5.4)	8(6.9)	0(4.0)	28
Uncoil	0(0.24)	8(2.9)	1(0.41)	3(2.7)	1(2.4)	0(0.12)	0(4.0)	1(5.2)	7(3.0)	21
Avoid	1(0.95)	0(12)	0(1.7)	2(11)	38(9.7)	0(0.47)	5(16)	2(21)	36(12)	84
Totals	*4*	*49*	*7*	*45*	*41*	*2*	*68*	*88*	*51*	*355*
Expected	(1.1%)	(13.8%)	(2.0%)	(12.7%)	(11.5%)	(0.6%)	(19.2%)	(24.8%)	(14.4%)	

TABLE 13–2 Analysis of the inter-individual two-act sequences in Table 13–1. Less reliable effects are in parentheses. (From Dingle 1969.)

Category	Chi-square	*P*	Directive	Inhibitive
Approach	12.23	<0.01	meral spread	(does nothing)
Meral spread	22.16	<0.001	avoid	(strike), chase, does nothing
Lunge	24.14	<0.001	coil	(strike)
Strike	18.63	<0.001	(strike), avoid	(meral spread), chase, does nothing
Coil	4.35	>0.20		(does nothing)
Uncoil	11.96	<0.01	(meral spread), (does nothing)	(coil), (avoid)
Avoid	54.11	<0.001	chase, does nothing	meral spread, strike, coil, avoid

$H_B = 3.00$ bits/act; $H_t = 1.03$ bits/act; $H_t/IN = 2.09$ bits/interaction.

communication theory. H_B is a measure of the "information" in a frequency distribution, and can be calculated for the average distribution of all following acts using the formula

$$H_B = \sum_j p_j \log_2 \frac{1}{p_j} = -\sum_j p_j \log_2 p_j$$

where p_j is the probability of occurrence of act j (a following act). The estimates of H_B in successive time intervals are shown in Figure 13–2. H_B can also be considered as the uncertainty in the average distribution of displays (Quastler 1958).

The conditional information present in the distribution of acts when the previous act is known is given by

$$H_{B/A} = -\sum_{ij} p_{ij} \log_2 p_{j/i}$$

where p_{ij} is the joint probability of animal A performing act i followed by animal B performing act j (i.e., the overall probability of occurrence of a given two-act sequence), and $p_{j/i}$ is the conditional probability of animal B performing act j given that A has just performed act i (e.g., given that a *lunge* has just occurred, the probability that the next act will be an *avoid*). This calculation yields lower estimates than H_B, as shown in Figure 13–2.

The difference between H_B and $H_{B/A}$ is called *transmission*, H_t, in

13–2 Various values of *H* for agonistic encounters between two mantis shrimps (see text). (From Dingle 1969.)

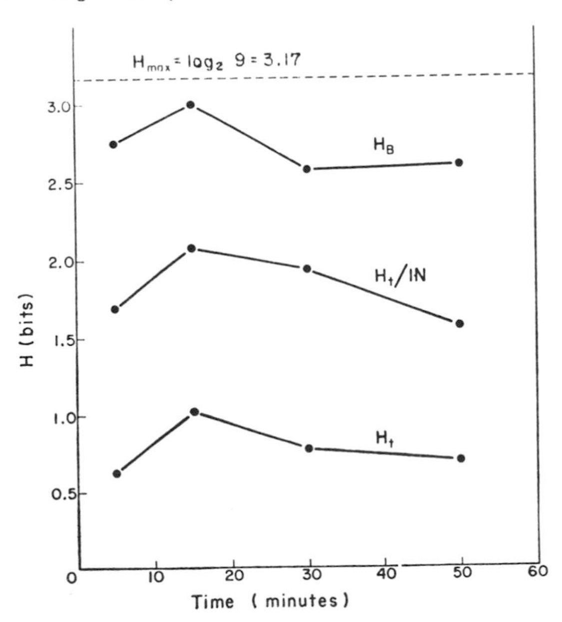

information theory, and is defined as the mutual reduction of uncertainty. It represents the degree to which A's behavior restricts that of B, and can be considered as the amount of information that B indicates by his behavior that he has received from A. It is a minimum estimate that can be used for comparisons among species.

Figure 13–2 shows that H_t rose at first and then declined. This pattern is correlated with the establishment of dominance between the opponents at about the peak of the curve. After a dominance relationship has been established, the uncertainty of the relationship between the two individuals is reduced, and the effect of a given act in altering the relationship is similarly reduced.

Corresponding estimates have been made for a variety of species, as shown in Table 13–3. Not unexpectedly, the values for human speech are highest. While the data for invertebrates are all of the same order of magnitude, it is surprising that the dance of the honeybee rates relatively low by these calculations. However, the factors responsible for species differences in H_t are poorly understood.

In the study of Hazlett and Bossert one hermit crab species, *Pagurus marshi*, had an unusually high H_t (see Table 13–3). This species was unusual among those studied in that it was extremely well camouflaged by pieces of detritus attached to its shell and appendages. For another hermit crab nearby, the contrast in conspicuousness between the stationary cryptic condition and the dynamic displaying condition may be greater for the camouflaged species than for an uncamouflaged one. The lowest H_t among hermit crabs was found in a species, *Clibanarius tricolor*, that often aggregated in dense groups during the day, apparently not responding to each other with agonistic behavior at those times. Still another variable affecting the H_t values obtained was the relative size of the participants; H_t tended to be higher in equally matched opponents than in unevenly matched ones (Haz-

TABLE 13–3 Estimates of information transmission in the communication systems of diverse animals.

Species	Behavior	C (bits/sec)	H_t (bits/act)	Source
Mantis shrimp	agonistic encounter	0.013–5.46	0.78	Dingle 1969
Hermit crab	agonistic encounter	0.4–4.4	0.35–0.44	Hazlett and Bossert 1965
Pagurus marshi	agonistic encounter	1.6	0.516	Hazlett and Bossert 1965
Rhesus monkey	all	5.46	1.96	Altmann 1965
Honeybee	dances	0.1–2.28	—	Haldane and Spurway 1954
Fire ant	trail following	0.4–1.39	—	Wilson 1962
Man	speech	6–12	—	Quastler 1958

lett and Bossert 1966). This stems from the greater uncertainty of the outcome of an encounter between evenly matched opponents.

As the differences in approach of the authors discussed above show, studies of animal communication systems are still in their infancy. The above-mentioned studies are not really comparable, except those done by the same workers under similar conditions. Some workers have carefully distinguished between inter- and intra-individual sequences, while others have chosen not to. Some have restricted their attention to one type of behavior (e.g., agonistic) or to a severely limited number of sensory modes. Still, these studies are valuable in showing that certain aspects of communication in animals can be measured and analyzed quantitatively.

CONCLUSIONS

Visual communication has evolved in animals that depend on vision for orientation in their environment and for hunting or avoiding predators. Visual communication occurs in virtually every social context in advanced phylogenetic groups.

The communicative effects of releasers, which elicit specific behavior patterns, can be studied using models and living animals whose appearance has been modified. Such studies have shown that animals often react to rather few key stimuli; an animal may react strongly to an unrealistic model emitting the key stimuli but ignore a much more realistic model lacking these stimuli. Communication in more natural situations can be studied and measured using some of the techniques of information theory. These have not yet been fully exploited, but they allow preliminary interspecific comparisons of communication systems and identification of some of the factors relevant to species differences in communication.

14 The Evolutionary Origins of Displays

ANY SURVEY OF a large animal group that has a well-developed visual system reveals considerable diversity in the displays that characterize each species. Not only are displays different among species, but some species possess a large repertoire of conspicuous displays while others may be relatively unexpressive. These nearly universal observations from the comparative study of behavior raise the question of how these species differences came about. This is a complex evolutionary problem whose solution requires information of various types. Some progress can be made with knowledge of the form of the displays in related species, their context, their effects on viewers, and the selection pressures to which they are subjected. The problem also requires some knowledge of the physiology, ontogeny, and genetics of the displays.

THE CONTEXTS OF DISPLAYS

Considerable insight into the evolutionary origins of displays has been gained through consideration of the contexts in which they occur. Context can be considered *spatially*, in terms of the location in which the display is given, or *temporally*, in terms of the antecedent and subsequent events.

It is convenient in the following discussions to treat each display as a unit. Although a given display may vary in intensity and frequency, the pattern of coordination is predictable and can be easily recognized by an experienced observer. Variability in the components of displays has rarely been studied quantitatively, since the legitimacy of their status as units of behavior has never been seriously in doubt. One of the few quantitative studies on displays is that of Dane et al. (1959) on the goldeneye duck (*Bucephala clangula*, Figure 15–12), in which the stereotypy of the coordination patterns was emphasized. By contrast, studies of Steller's jays have revealed considerable variability in some display patterns (Figures 5–7, 15–2; Brown 1964). Variability in displays is further discussed in the treatment of graded and discrete displays in the following chapter.

The Spatial Context of Threat Displays / The spatial context of threat displays is simplest in the case of defense of a territory or dominion. In such cases the following idealized schema can often be recognized:

	Behavior of territory owner		
	In own territory	At border	Out of own territory
Relative frequency of attack	high	medium	low
Relative frequency of threat	medium	high	medium
Relative frequency of escape	low	medium	high

This schema reflects the dominance of an individual in a certain area and his subordinate status outside it. The data on dominance hierarchies of Steller's jays in and outside the individuals' dominions (see Figure 5–7) illustrate the schema; a supplanting encounter can be interpreted as an attack by the dominant and an escape by the subordinate, even though a fight does not occur. The high frequency of threat displays at territorial borders can be observed in most species having all-purpose territories. In Steller's jays high intensity threat displays occurred most frequently at the borders of areas of dominance (Brown 1964); they also occurred when evenly matched jays encountered each other outside both their areas of dominance.

The Temporal Context of Threat Displays / Threat displays tend to occur close to attack and withdrawal behavior in time as well as in space. If a series of an individual's behaviors is entered as a series of two-act sequences in a matrix, it tends to follow the idealized scheme shown below, although much variation occurs (e.g., Figure 14–2).

Wiepkema (1961) analyzed the temporal contexts of various acts of the bitterling, *Rhodeus amarus*. The male bitterling defends the

	Frequency of following act by owner in his territory			
Preceding act	Attack	Threat	Escape	Other
Attack	high	moderate	low	low
Threat	moderate	high	moderate	low
Escape	low	moderate	high	low
Other	low	moderate	low	high

mussel in which he induces his female to lay her eggs. The female deposits her eggs in the gills of the mussel by means of a long ovipositor (Figure 14–1), thus guaranteeing that they will be kept clean and well aerated. The activities of the male in the vicinity of the mussel include aggressive behavior toward rival males, courtship toward females, and self-maintenance when alone or when not stimulated by other males or females.

Wiepkema observed male bitterlings and analyzed their behavior as a series of two-act sequences. A factor analysis based on these sequences showed that three factors accounted for 90 percent of the variance. (For details of the statistical methods, consult the original paper.) In factor analysis, the possibility is tested whether a large number of variables (in this case, the different types of two-act sequences) can be described in terms of fewer variables (here termed *vectors*).

The analysis yielded three principal vectors, as shown three-dimensionally in the model in Figure 14–2. The model shows that threat behavior tended to occur in temporal contexts close to and inter-

14–1 The female bitterling spawning in the gills of a mussel. **A.** Just before inserting ovipositor. **B.** Spawning, with ovipositor inserted in gills. **C.** Ovipositor partially withdrawn. (From Wiepkema 1961.)

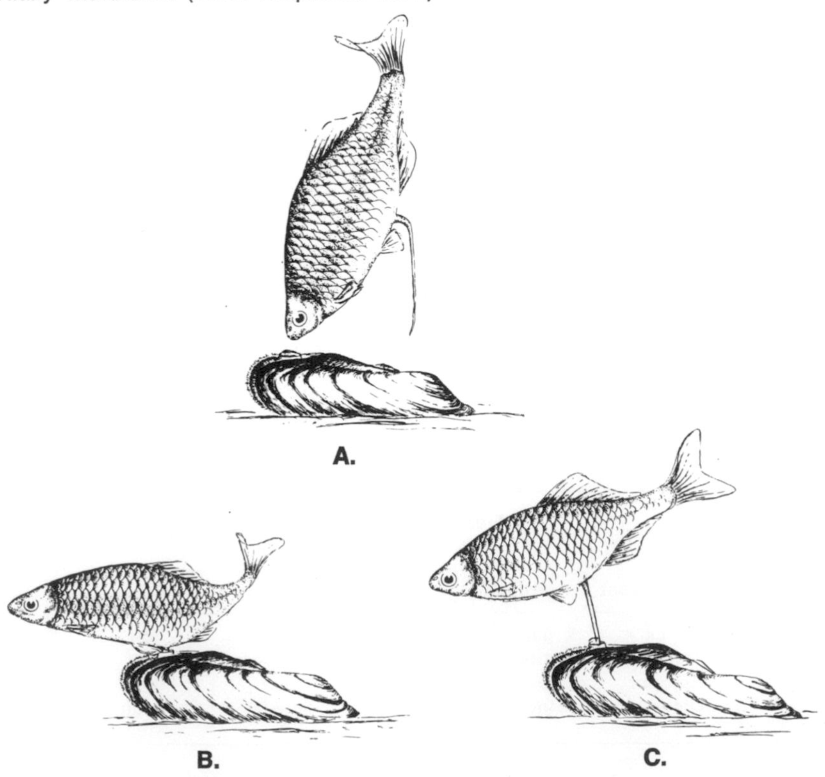

mediate between attack and escape. In the model, attack is represented by *chasing* and *head-butting,* and escape, by *fleeing.* Threat patterns are *turning beat* and *jerking,* which are roughly intermediate between attack and escape. *Fin-flickering* and *chafing* are movements concerned with care of the body surface (*comfort behavior*). Fin-flickering followed fleeing more often than expected by chance. Chafing followed

14–2 Vector model of the correlations of preceding and following acts in male bitterlings in three different situations. The factor analysis was performed on the combined data for the following three situations: (**1**) a male defending his mussel against a rival male, (**2**) a male with his mussel in the presence of an unripe, sexually unresponsive female, and (**3**) a male with his mussel in the presence of a ripe, sexually responsive female. In this model, the cosine of the angle between the vectors representing two behaviors equals the correlation coefficient of the corresponding behaviors in terms of the preceding-following relationship. Therefore, the smaller the angle between two behaviors, the higher the correlation between them and the greater the likelihood of their occurring in succession. The three main factors (1, 2, 3) in this model may be considered hypothetical statistical creations that account for 90 percent of the variability in the observed sequences. Their existence, but not their identity, is revealed objectively by the method of factor analysis. The length of each vector is an indication of how far it can be explained using the three factors in the model. The numerals 1, 2, 3 designate the positive end of the scale for each factor. (After Wiepkema 1961.)

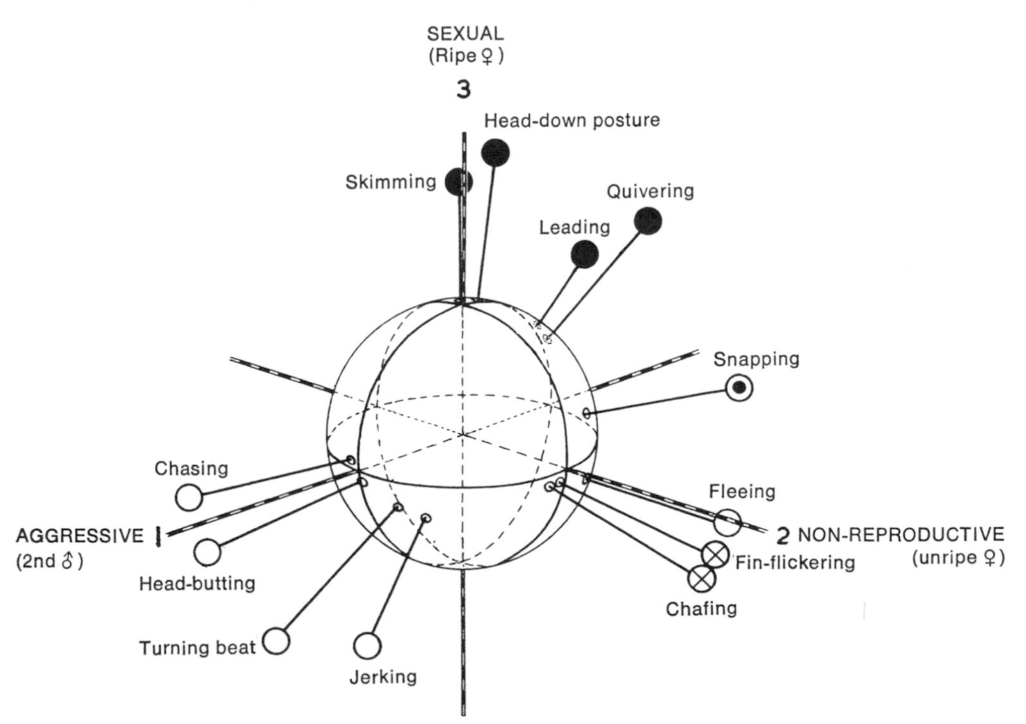

jerking more often than expected by chance. Wiepkema identified the three factors as (1) aggressive, (2) nonreproductive, and (3) sexual; however, an alternative hypothesis is possible. It may be that the three factors correspond to the three stimulus situations used, namely (1) rival male, (2) unripe female, and (3) ripe female.

The contextual intermediacy between attack and flight of a type of threat signal not shown in the vector model is illustrated in Figure 14–3. *Fin-spreading* is a common behavior in agonistic contexts. The temporal context of fin-spreading in relation to the indicated acts is indicated by its frequency. The diagram shows that fin-spreading occurs most commonly in temporal association with acts that are intermediate, in terms of vector position and temporal association, between chasing and fleeing.

The method of factor analysis was also used by Baerends and van der Cingel (1962) on displays of the common heron (*Ardea cinerea*), with similar results.

Many species have more than one threat display, and these differ in their contextual relations to attack and escape. This is shown quantitatively for the bitterling as discussed above. Among the first studies to point this out were those on the black-headed gull (Moynihan 1955). The displays of this gull are illustrated in Figure 15–4.

Seasonal changes in the frequency and intensity of displays are also well known. As the breeding season of an animal approaches, dis-

14–3 Temporal context of a threat pattern, fin-spreading, in the male bitterling. The frequency of occurrence of fin-spreading, before and after the indicated acts is higher in association with other threat patterns than with attack (chasing) or escape (fleeing). Fin-spreading occurs in attack and escape situations but is less common when either attack or escape predominates. *F* = The presumed fleeing vector. *A* = The presumed attack vector. The thickness of the lines from *A* and *F* to the indicated acts shows the closeness of their temporal association, as deduced from the data illustrated in the vector model (see Figure 14–2). (After Wiepkema 1961.)

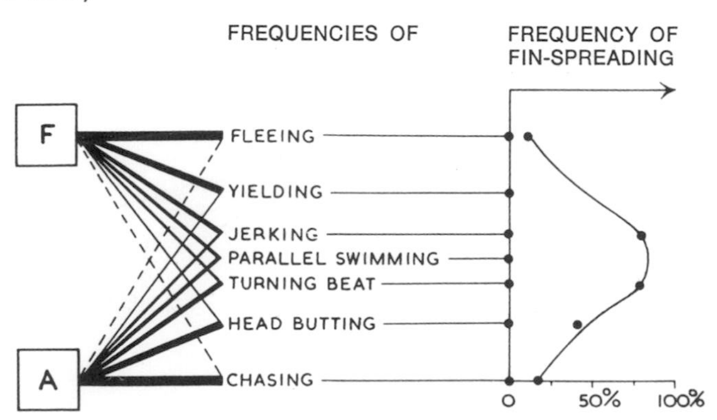

plays related to attracting or stimulating mates and to establishing and holding territories become more common and more intense. Many displays are restricted to the breeding season.

Quality of Opponent / In many animals, as in birds and mammals, adults tend to be of similar size; but in most invertebrates and poikilothermic ("cold-blooded") vertebrates, adults of greatly dissimilar sizes are of common occurrence. In agonistic and sexual contexts relative body size can be a significant factor in determining the kinds of communicative behavior observed. A particularly clear example may be found in Wiepkema's (1961) analysis of the behavior of the bitterling (Figure 14–4). As was found for the temporal context, threat behavior also was intermediate between chasing and fleeing in the context of opponent quality. Figure 14–4 shows that three threat behaviors—jerking, turning beat, and fin-spreading—reached their peak frequencies when opponents were equally matched for size but decreased when the opponent was larger or smaller. Chasing and head-butting, which are representative of attack behaviors, tended to decline in frequency as the relative size of the opponent increased. As might be expected, fleeing showed the inverse relationship.

In another analysis of the importance of the opponents' relative

14–4 Frequencies of various acts of threat (jerking, turning beat, fin-spreading), attack (head-butting, chasing), and escape (fleeing) in relation to size of opponent in territorial male bitterlings confronted with a rival of the same length (x), a greater length ($x + 2$ cm), or a lesser length ($x - 1$ cm). (From Wiepkema 1961.)

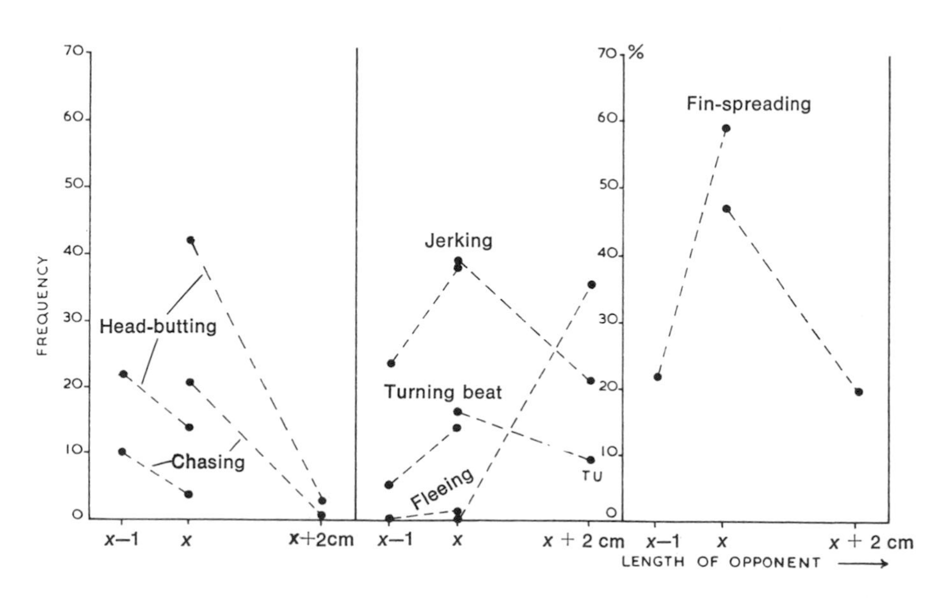

size in an agonistic context, Hazlett and Bossert (1966) showed that the rate of information transmission in encounters between two hermit crabs was higher for crabs equally matched in size than for those unequally matched.

Studies on various fishes have shown that relative size of the partner is an important determinant of body coloration in courtship (e.g., *Lebistes reticulatus*, Baerends et al. 1955; *Etroplus maculatus*, Barlow 1968; *Tilapia melanotheron*, Barlow and Green 1969).

THE POSTURAL COMPONENTS AND ORIENTATION OF THREAT DISPLAYS

The analysis of the postural components of displays was considerably advanced by the studies of Moynihan (1955) on the black-headed gull and by those of subsequent workers on a variety of species. An important conclusion that may be drawn from these studies is that the contextual relationships of the various displays are reflected in their postural components. A display used in spatiotemporal contexts that are near to attack would be expected to expose the species' weapons in a posture and orientation suitable for immediate attack. A display used in spatiotemporal contexts near to escape would be expected to reveal more of a postural readiness to escape.

In general, the postural components of displays may be said to reflect their spatiotemporal contexts. Consequently, it is interesting to discover that some threat displays are in a postural sense intermediate between attack and escape. Many species have threat displays given between two males in which the bodies are oriented parallel to each other; thus body orientation is intermediate between the forward (facing) position used in attack and the opposite position used in escape.

Analysis of the contexts and postural components of displays was greatly advanced through the studies of Tinbergen and his colleagues and students. Their analyses were usually presented in terms of drive levels and the interactions of drives (later, "tendencies"). For the present purposes it seems unnecessary to enter into speculations and controversies concerning the internal states of displaying animals. All that can be objectively observed in these studies are the spatiotemporal and social contexts of the displays and the patterns of muscular coordination. These objective data are adequate for the present evolutionary discussion.

Although the discussion above concerns threat displays, a similar methodology has been used for courtship displays (Morris 1956*b*) and could be used for any type of display.

The courtship display of the male three-spined stickleback can be analyzed using this methodology. The spatial situation is one of leading an approaching female to the nest by the male. As the female nears the border of the male's territory, the male performs courtship display. Then the male shows the nest to the female, who enters and deposits her eggs, which are then fertilized by the male. The courtship display is a mixture of some components that lead the female to the nest and some that threaten the female as if she were a male. The result is a zigzag swimming pattern that eventually lures the female to the nest. The spatial context, the temporal context, and the components of the display and of the antecedent and following acts indicate a transition in the male from threat and attack behavior at the first appearance of the female to, finally, fertilization of her eggs.

DECISION-MAKING

Although conscious awareness of the consequences need not be involved, an animal is at all times engaged in decision-making processes. The program of his life derives from a stream of decisions regarding what activities he will engage in and when. In a context in which threat display is typical, the animal must decide among several alternative courses of action, including attack, escape, threat of various kinds, or some other behavior. The decisions are neural processes that may, in effect, weigh the perceived benefits against the risks for various courses of action depending upon the needs of the moment (Young 1964). Much of the brain is involved in these processes in one way or another, but in vertebrates the role of the limbic-brain-stem system, particularly the hypothalamus, is critical. Particular parts of this system are known to play special roles in agonistic behavior, depending on the kind of behavior that is activated (Brown and Hunsperger 1963; Brown 1970). The neural processes mediating the weighing of risks and benefits for various potential courses of action are complex, and it is doubtful whether external analyses of context and postural components are trustworthy for establishing the main guidelines for analysis. This is a field in which neither ethological nor neurobiological analysis alone suffices.

The Adaptive Value of Displays: Tipping the Balance / Because decision-making processes are involved, the factors that cause an animal to behave one way or another are constantly being balanced against each other. At the territorial border the balance of factors promoting attack and escape tends to be about equal. When the animal is in this condition, any small change in the balance of forces could cause a decision to attack or retreat, depending on the change. Consequently, small changes in the appearance of an opponent might sometimes be significant. *The selective advantage of threat displays and various other displays seems to be that they tend to tip the balance in the viewer in favor of the performer.* Consequently, we should expect them to be most effective when the balance of opposing factors is nearly equal in the viewer, and least effective in circumstances when the viewer is not easily dissuaded from his chosen course of action (as found in hermit crabs by Hazlett and Bossert [1966]). The existence of a relatively even balance between the relevant opposing factors seems to be typical of individuals in most display contexts. Because relatively minor changes in appearance can be significant in such situations, natural selection could conceivably act to exaggerate those changes in appearance that caused a change in the viewer beneficial to the performer. At the same time, in the case of threat displays we should expect natural selection to favor in the viewer an inability to be discouraged by the display of another when such displays are not followed by attack. The ability to respond appropriately should be selected for. Although this conceptualization of the adaptive value of displays has been applied mainly to threat contexts, it should also be useful for other displays and other contexts.

ORIGINS OF DISPLAYS

Because even small changes in appearance may significantly affect the behavior of another individual, a great variety of minor secondary consequences of a social encounter have apparently become modified through natural selection in the direction of increased usefulness in communication. When there is some evidence that a structure or action previously unspecialized for social influence has become specialized for that effect, the unspecialized state is called the *origin* of the specialized state. The specialized state may then be called the *derived* one (Tinbergen 1952). The principal method used to deduce the origin of a display is the *comparative method* in the strict sense (see the introduction to Part I). A critical examination of the method was made by Tinbergen (1962).

Many classifications of display origins have been proposed. Darwin (1872) proposed three principles concerning the evolution of displays. Some of his conclusions are of little use, since they rely on such concepts as "nerve force" and the Lamarckian inheritance of acquired characters. Nevertheless, some of his examples concerning the effects of the autonomic nervous system on visual signals appear valid today. Morris (1956*a*) proposed a distinction between origins concerned with adjustments to a changed *internal environment* and those concerned with adjustments to the *external environment*. Among the former may be included aspects of thermoregulation and respiration; among the latter, pre-flight and protective movements, redirected attack, and out-of-context (displacement) behavior. These categories will be considered briefly.

Thermoregulation / The first relatively modern treatment of thermoregulation as a source of displays was provided by Morris (1956*a*), who reviewed the feather postures of birds and their relations to displays. Feathers can be raised or lowered over the body as a whole or in separate feather tracts or body regions. The control of feather posture depends on the autonomic nervous system (Langley 1903). A role of feather posture in thermoregulation is strongly suggested by the experiments of McFarland and Baher (1968). They subjected Barbary doves (*Streptopelia risoria*) to a range of temperatures and observed that the feathers were raised from a sleeked to a raised position as the temperature decreased from warm to cool (Figure 14–5). The relationship between temperature and feather erection was modified by the activity of the bird; active birds tended to be sleeker, and preening birds more fluffed. In adjustment to temperature changes there were no significant differences among body regions, but in response to an alarming stimulus (at a constant temperature) the pattern of feather erection on various parts of the body was a function of the distance of the stimulus object from the bird.

Morris argued that since changes in body temperature were likely concomitants of various types of social encounters, the strictly thermoregulatory changes in feather posture might serve as visual signals upon which natural selection acted to produce the much more conspicuous feather movements now designated as displays. Since many avian displays consist mainly of the erection of the feathers in particular body regions, it seems likely that these originated from thermoregulatory responses. Some displays featuring feather erection are shown in Figures 8–6 (birds of paradise), 15–2 (Steller's jay crest), and 15–9 (redwing song-spread).

In mammals erection of the fur, or *piloerection*, is also used in thermoregulation and in display—e.g., in the reaction of a cat to a dog and in the displays of the aardwolf (*Proteles cristatus*; Ewer 1968) and various species of ungulates (e.g., Guthrie 1971). In man thermoregulatory perspiration may serve as a visual signal, and blushing may have been derived from vascular changes in thermoregulation.

14–5 Feather erection as a function of ambient temperature in the Barbary (ring) dove. Each part of the body was scored separately, then combined to provide a mean feather index expressing overall degree of feather erection. The index is influenced by the activity of the bird as well as by temperature. Body regions: c = crown, n = neck, d = dorsal, v = ventral, w = wing. The tail was not scored. (After McFarland and Baher 1968.)

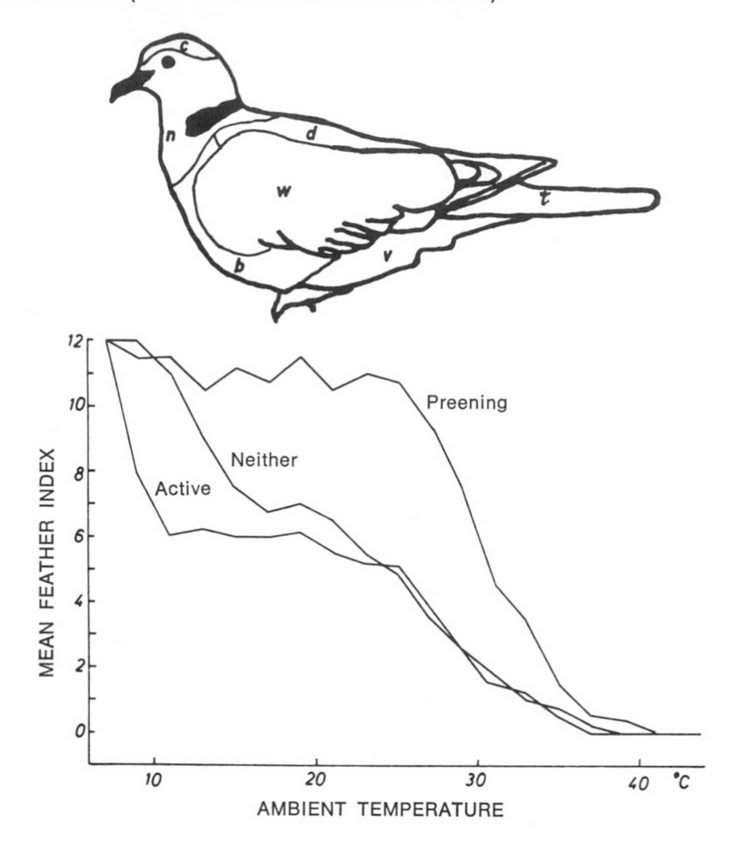

Respiration / Respiratory systems have also given rise to structures used in visual communication. The air sacs of the male prairie chicken (Figure 8–8) and frigatebird are brightly colored and used in courtship. The proboscis of bull elephant seals (Figure 8–4) may be important visually as well as aurally. Among mammals, movable noses may be quite expressive; the elephants (Figure 15–15) are an extreme. In primates, muscles controlling nostril size are important in facial expression (van Hooff 1962, 1967; Andrew 1963*a*, *b*). In fishes the gill covers (opercula) carry dark spots or fleshy "ears" in some species that hold out the opercula in frontal display (Baerends and Baerends-van Roon 1950).

Intention Movements of Locomotion / Incomplete locomotor movements were recognized as an important source of display in birds by Daanje (1951), who followed a tradition dating from Heinroth (1911) of calling them *intention movements.* These are essentially "false starts." They indicate to an observer an increased likelihood of locomotion, especially flight. Daanje pointed out that elements of a variety of displays in many species of birds resembled the pattern of a bird preparing for flight. Such display components as raising and spreading the tail, withdrawing the head and neck, wing-spreading, and leg-flexing can be found in various incomplete locomotor movements that commonly occur when a bird is alarmed but has not taken flight. Some of the displays believed to incorporate elements of locomotor acts are shown in Figure 14–6.

14–6 Displays incorporating elements derived from preparatory locomotor acts. **A.** Gray heron, first and second phases of stretch display (*from left to right*). **B.** European cormorant wing-waving. For further details on the displays see Baerends and van der Cingel 1962 and Van Tets 1965, respectively. (Illustrations from Daanje 1951.)

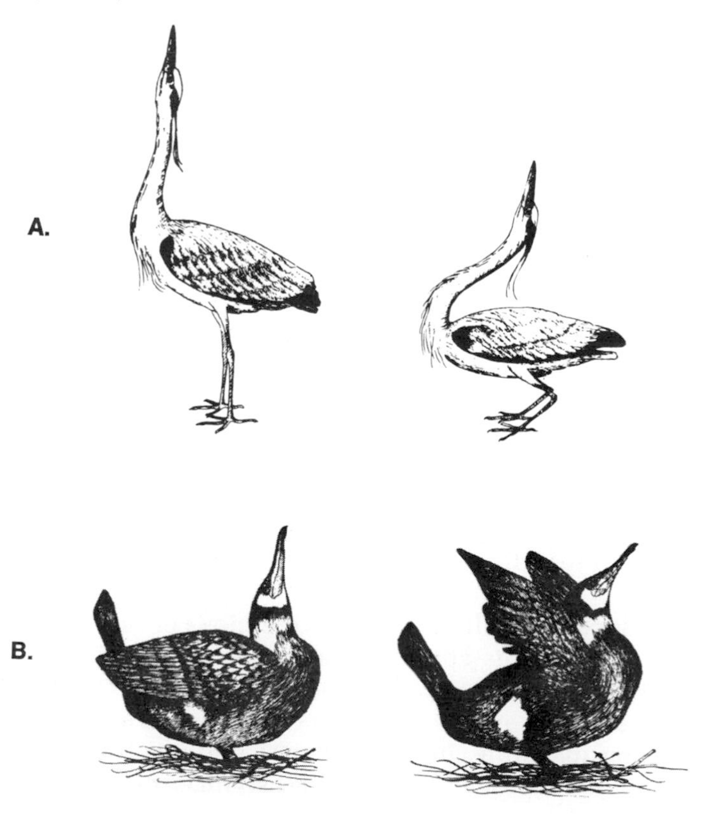

In saturniid and sphingid moths the protective displays given toward predators appear to have been derived from flight movements of the wings. Blest (1957*b*) has compared the protective displays of a large number of these moth species. He classified the displays of each species in response to a disturbing stimulus in the daytime into one of the following principal types: (1) rhythmic, (2) static, (3) mixed, and (4) cryptic. The cryptic type is characterized by a remarkable absence of response to prodding and handling; these species are cryptically colored. In the rhythmic type, the moth typically exposes the hind wings and then performs a series of up to 200 repetitions of a pattern of wing movements. These species are conspicuously colored yet not distasteful to predators. The mixed type is intermediate between rhythmic and static types. Static displays are the most specialized and consist of more than one type; in these, the rhythmic phase is replaced by a posture that reveals certain aspects of coloration to advantage. Blest's analysis of his comparative data suggested that the rhythmic displays are the closest to the original flight movement and that the static displays were the farthest removed.

Two specialized lines leading from the rhythmic type are of interest. In the first, eyespots have evolved on the hind wings that resemble the eyes of a vertebrate predator (perhaps coincidentally); they are revealed when the moth moves its fore wings forward at the start of the display. The resting and display positions of moths having an eyespot display are shown in Figure 14–7.

That artificial eyespot patterns do startle birds that might prey on the moths was shown experimentally by Blest (1957*a*). Whether the eyespot display is effective because it mimics the eyes of a small bird (e.g., an owl) or for some other reason is not known. Many species of butterflies and moths have round spots that resemble eyes to some extent, but they tend to be numerous and small. Selection probably acted to enlarge these and reduce their number.

In the second line derived from the rhythmic condition, a posture is struck that is highly suggestive of a wasp; the moth falls to the ground, and there curls its abdomen under the thorax, raises its wings to the midline, and lies on its side, thus showing the colored abdomen, which tends to resemble that of a wasp.

Protective Movements / In the facial expressions of cats, dogs, and primates, certain elements appear to have been derived from *protective movements* (Andrew 1963*a*, *b*). These include flattening the ears, partially closing the eyes, and withdrawing the corners of the mouth in preparation for biting. In *Galago crassicaudatus* (Lorisoidea), such responses may be evoked by startling, strange, or noxious stimuli.

14–7 Eyespot display in moths. *Above: Nudarelia dione; left,* about to fly; *right,* displaying. *Below: Automeris memusae; left,* at rest; *right,* displaying. (From Blest 1957*b*.)

Confident individuals assume such expressions only when sniffing or grooming a fellow; the less confident the animal, the greater the distance at which these responses are shown. A subordinate individual will show such responses at a relatively great distance and retain them for a longer period.

In other species ear-flattening is less conspicuous than scalp retraction, which also tends to occur in Lorisoidea as a by-product of ear-flattening. Scalp retraction, which in its specialized state in the Ceboidea and Cercopithecoidea may be considered a ritualized condition, may expose conspicuously colored skin normally partly concealed by the eyebrow ridge (Andrew 1963*a*, *b*).

Eyebrow-lowering might also be a protective response in primates. It has become ritualized as a part of the fixed stare used by dominants as a threat against subordinates (Andrew 1963*a*, *b*).

Another type of origin from protective responses might have occurred in certain aquatic groups that can change color to match their body color and pattern with the background. This ability is present in various bottom-living fishes and cephalopods, as well as in some frogs. In cephalopods like the cuttlefish (*Sepia officinalis*), this ability to control body coloration is conspicuous. Many cephalopods change color during escape, while also releasing a temporarily concealing cloud of inky substance. The predator is then unable to match the visual image of the animal he saw before the ink release with that of animals present afterward. In this case a color change acts to confuse the predator, but it is only a short evolutionary step from this capacity to that of being able to make color changes that benefit the performer in intraspecific social encounters. In the squid *Loligo pealii*, color changes are an important aspect of courtship (Arnold 1962). It seems likely that *Loligo's* use of color changes in courtship is derived from the ability to change color to avoid predators.

Redirected Attacks / If a monkey or a bird is unable to act out his aggressiveness on the individual who stimulated him, he may take it out on another, quite innocent individual nearby. This is known as *redirected attack*. In birds an attack may involve a conspicuous aerial dive and is readily visible. It has been argued that certain displays of the black-headed gull (the swoop-and-soar display) and of some terns, which utilize aerial display in courtship, have been derived from redirected attack (Manley 1960; Cullen 1960).

Out-of-Context Behavior / A puzzling aspect of courtship and agonistic behavior in many species is the presence of behavior patterns that appear functionally unrelated to the goals of mating or dominance. For example, a male mallard duck or pigeon in the midst of courting a female may stop to preen his feathers; a stickleback defending his territory may stop to perform movements resembling digging; or a Steller's jay disputing a dominance relationship with an opponent may begin turning aside leaves and other objects on the ground in the manner used in foraging for food. Such activities have long been studied and have become known as *displacement behavior* (Tinbergen 1940; Kortlandt 1940; Armstrong 1947:106).

Displacement behaviors may be further characterized. They tend to be incomplete, and consequently nonfunctional in the sense of their original context. No food is obtained in displacement foraging, no feathers are preened in displacement preening, and no nests are built in displacement nest-building. The vigor, intensity, and rate of performance of displacement activities typically exceed the norms for their original contexts—presumably because of higher arousal levels in threat and courtship situations than in the original contexts. Under natural conditions, displacement activities tend to occur when the probabilities of each of two incompatible acts, such as advancing or retreating, are equally high. These are "choice points" in the animal's social behavior, and the behavior shown in such situations is often somewhat anthropomorphically called *conflict behavior.*

The nature of the neural processes underlying conflict behavior is obscure, although some preliminary attempts to analyze it have been made (e.g., Brown et al. 1969*a*, *b*). There is some support for the idea that when the animal is in the course of decision-making and has not committed himself to approach or avoidance, he may be more responsive to the kinds of omnipresent peripheral stimulation that evoke such common responses as preening, foraging, nest-building, and other displacement behaviors. The role of context in the displacement preening of terns has been emphasized and extensively analyzed by van Iersel and Bol (1958). The role of peripheral stimulation in the displacement preening of the chaffinch was emphasized and studied by Rowell (1961). The "motivation" of displacement behavior has long intrigued ethologists (see, for example, McFarland 1969). Unfortunately the analyses of spatiotemporal contexts and of postural components of displays, to which ethologists have mainly confined themselves, have not proven very useful for examining the neural processes involved. However, an attempt to apply the concepts of neurobiology to the problem of the "motivation" of displacement behavior was made by Delius (1967, 1970).

The role of displacement behavior in the origin of displays was pointed out by Tinbergen (1952) in an influential review. A comparative study of ducks by Lorenz (1941) suggests how displacement (mock) preening has become ritualized in some species. Preening and other movements used in self-maintenance (or comfort) behavior occur commonly in the courtship sequences of ducks (McKinney 1965; Johnsgard 1960). In some species (e.g., Bahama pintail duck, *Anas bahamensis*), preening during courtship closely resembles natural preening. In others it is considerably different. A male mallard after drinking often reaches with his bill behind his partly opened wing as if to preen, but instead he moves the bill quickly against the wing feathers, making a rather loud "rrr" sound. In the gadwall (*Anas strepera*), the pattern is similar except that the drink follows the mock preening. In the garganey (*Anas querquedula*), unlike the other species, the male touches the front side of the wing, which is strikingly colored in a way that enhances the visual effect of the mock preening.

The most extreme specialization of mock preening is found in the mandarin duck (*Aix galericulata*), whose mating system is more of a lek type than in the other species; consequently, the pressures of sexual selection (female-choice type) are greater, and it is perhaps not surprising that specialized feathers are present on wing and head that enhance the display. In this species and in the wood duck (*Aix sponsa*), mock preening is virtually always preceded by drinking; this sequence makes the most of the colorful head and body coloration in these species. The regularity of the sequence, together with the absence of functional preening and the presence of specialized structures that emphasize the movements, indicate that preening the wing in these ducks has been ritualized. The original movement has been modified by natural selection so that it is now a courtship display.

Feeding movements seem also to have been the origin of some

14–8 Courtship displays derived from feeding in the Phasianidae. **A.** Bobwhite quail tidbitting with mealworm to a female. **B.** Domestic cock. **C.** Ring-necked pheasant. **D.** Impeyan pheasant. **E.** Peacock pheasant. **F.** Peacock with peahen. (Birds are drawn to different scales.) (A from Williams et al. 1968, courtesy of American Ornithologists' Union; B–F from Schenkel 1956.)

courtship displays in the avian family Phasianidae, order Galliformes (Schenkel 1956, 1958). The male of some species, such as the bobwhite quail (*Colinus virginianus*), feeds his mate. He commonly precedes the feeding with a food call and display, together termed *tidbitting*, given on discovery of the food (Figure 14–8*A*; Williams et al. 1968). Copulation was seen to follow tidbitting within minutes several times. In other species, mock pecking and manipulation of an object alone are typical, with little or no manipulation or offering of actual food; examples are the ring-necked pheasant (*Phasianus colchicus*) and domestic chicken (Figure 14–8*B*, *C*). The impeyan pheasant (*Lophophorus impajanus*), which normally pecks at the ground for its food, bows low before the female and pecks vigorously at the ground while courting, but in addition he spreads his wing and tail feathers and then holds his head still in a sexual display (Figure 14–8*D*). The peacock pheasant (*Polyplectron bicalcaratum*) displays similarly, but scratches first and will sometimes offer food to the female (Figure 14–8*E*). An extreme specialization is reached in the courting display of the peacock (*Pavo cristatus*), the origin of which could probably not be recognized were the displays of related species not known. The peacock display is similar in form to the displays mentioned above; but the bill is merely pointed at the ground, and no food need be involved (Figure 14–8*F*). While these examples do not constitute an evolutionary sequence, they suggest that similar sequences of ritualization may have occurred. The situation is reminiscent of the ritualized offerings of balloon flies during courtship (see Chapter 1).

CONCLUSIONS

Although the evolutionary origins of displays must remain relatively obscure, it is clear that at some point transitions must have occurred between behaviors and appearances that served communication only coincidentally and those that served communication primarily. Analyses of the spatiotemporal contexts of displays, the relationships between display and relative body size, and the postural components of displays have demonstrated that displays tend to be most frequent in situations where they can *tip the balance* of causative factors in the viewer toward behavior that benefits the displayer. In this situation, slight differences in appearance may be selected for if they benefit the displayer. This concept suggests a way in which displays could evolve from actions that were at first only coincidentally involved in communication. This is the evolutionary process of *emancipation*.

Many different types of origins of displays have been postulated.

These include responses associated with thermoregulation, respiration, protection, attack, escape, foraging, preening, and probably others. The nature of these origins and their subsequent adaptation for use in communication have been revealed by the comparative method, which, despite its shortcomings, is the only one available for such problems.

15 Selection Pressures on Displays

RITUALIZATION, as used by ethologists, refers to an evolutionary process analogous in some ways to the cultural process known by the same name. It is no more nor less than natural selection as it affects the displays and other communicative behavior of animals. The term *ritualized* is perhaps most useful as a short expression for the phrase *evolutionarily specialized for communication.* An ethologist might use it for a behavior that he believed had become specialized through evolution for use in communication. Such beliefs, of course, may not always be correct; and it should be emphasized that using the term *ritualized* does not indicate that the behavior to which it is applied has been proven to have been ritualized. As with many evolutionary concepts, it is necessary to judge wisely before employing the concept of ritualization.

DISCRETE AND GRADED DISPLAYS

In communication systems a *discrete signal* is one that is all-or-none, like a nerve action potential. A *graded signal* may vary in size or intensity. In a system based on discrete signals, information is conveyed by a sequence of symbols, each with its own significance. Intensity of the message is transmitted as some function of the choice of signals or of the temporal pattern of signals, not as a function of signal intensity (which is fixed).* In a system based on graded signals, information may be transmitted by quantitative variations of signal strength. Consequently, graded displays tend to rely more on the relative position of body parts, and often may be continued for indefinite periods. Gradation of motivational intensity is conveyed immediately by rela-

* The terms *analog* and *digital* have been used for *graded* and *discrete* (Barlow and Green 1969), but they have been avoided here because in computer language digital signals are not frequency-graded, while in animals they are (Elias 1961).

tive position (as in graded signaling), but requires a longer time when it must be conveyed by frequency of performance (as in discrete signaling). Animals tend not to display either of these in pure form. Most have what might be called hybrid systems, in which some signals tend to have a discrete character, and others a graded character. Displays may tend toward either type.

Discrete displays are conspicuous in the fiddler crabs (*Uca* spp.), a large group of mainly New World species that inhabit tidal mudflats in marshes and mangroves in temperate and tropical zones. They are especially abundant in the tropics, and several excellent comparative studies have been made of their displays and those of their relatives (Crane 1943, 1957, 1958; Salmon 1965, 1967; Salmon and Atsaides 1968; Schöne 1968; Wright 1968). Fiddler crabs are air breathers. At low tide they feed on exposed mudflats, and as high tide approaches they retreat into their burrows, trap a supply of air by capping a burrow, and wait. Especially in males, one of the claws is tremendously enlarged and used in signal production. Males defend their burrows and attract females with displays of their large claw, which they wave in species-specific patterns, sometimes producing sound (Figure 15–1). These displays tend to be species-specific and constant in form (and hence discrete), but to vary in frequency. Males wave faster (the wave duration is shorter) in the presence of females than in their absence (Salmon 1967).

Other displays that approach the discrete condition are the following: the ruffled courtship display of cutthroat finches (*Amadina fasciata;* Morris 1957); the cartwheel display of two male blue-backed manakins (*Chiroxiphia pareola;* Gilliard 1959*c*; Snow 1963); the light-flashing patterns of fireflies (Figure 18–7; Lloyd 1966); and the push-up displays of *Sceloporus* lizards (Figure 18–6; Hunsaker 1962). These examples suggest that a visual signal is likely to evolve toward a discrete condition, not just to reduce ambiguity, as Morris (1957) suggested, but in cases where the viewer must be able to recognize a *pattern* in the display and discriminate it from other similar patterns, particularly the patterns of closely-related species. The extremes of discreteness in signals, which are found in auditory signals such as the calls of crickets, grasshoppers, frogs, and birds, also suggest the importance of pattern preservation. In those groups in which *species differences* in response to these signals depend mainly on genetic differences, rather than on learned preferences, it would seem especially important that displays be as discrete as possible.

Graded displays are characterized in the extreme cases by the absence of a typical intensity and the presence of much variability in signal intensity. Crest erection is a good example found in many

15–1 Discrete signals in four fiddler crab species. The pattern of claw-waving is species-specific. The display can be frequency-graded to indicate levels of motivation.

Above: The pattern of the display in males is shown in three consecutive stages (*top to bottom*). Arrows below indicate spatial position of tip of chela through one complete movement. Cross marks indicate a "jerk." *Uca speciosa:* (*a*) movement of the chela shown to females; (*b*) movement shown to intruding males.

Below: Graphic level recordings of waves with deflections indicating the beginning and end of each wave. *Uca rapax:* waves produced after arrow are responses to an approaching female. *Uca speciosa:* waves produced after first arrow are responses to an approaching female; after second arrow, those produced when she had moved away. Solid deflections indicate production of a rapping sound. (From Salmon 1967.)

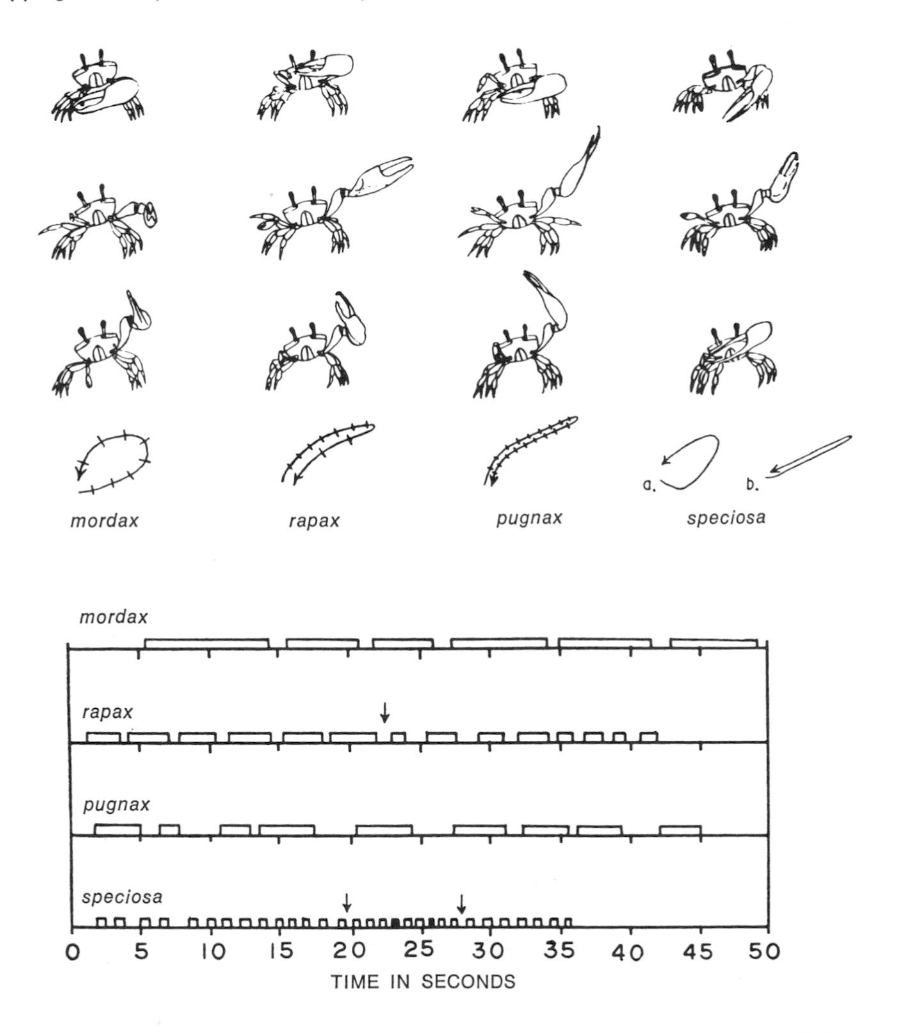

species. It has been studied in detail in Steller's jay, *Cyanocitta stelleri* (Brown 1964). The various possible positions of the crest in this species are shown in Figure 15–2. The angle made by the crest and a line through the long axis of the bill can be used as a measure of crest erection and estimated in the field to within 10 degrees (with practice). When jays were observed in a variety of activities, it was found that the crest assumed characteristic ("typical") positions during different activities (Figure 15–3), ranging from fully depressed during escape-provoking situations and in courtship to fully erect in combat. In general, the degree of erection, or the intensity of the display, tended to correspond to the degree of resistance by an opponent in agonistic encounters, as shown in Table 15–1. The more the opponent resisted, the greater the intensity of display.

Other graded displays are to be found in the facial expressions and limb gestures of primates, the color changes in many fishes and some cephalopods, the tail positions of dogs and wolves, the ear positions of cats, canids, and squirrels, the trunk positions of elephants (see Figure 15–15), and the bill angle of black-headed gulls during upright threat. Graded displays seem to be especially common in threat behavior, although they are also found in courtship and other contexts. In agonistic situations, *immediate information on motivational intensity* would seem to be critical in determining the outcome of closely contested encounters, while information on quality or form would be of secondary importance. For example, in man as in other species, an estimate of how intensely an opponent is likely to fight if provoked can be obtained from facial expressions, gestures, and other threat dis-

15–2 Range of variation in crest elevation in Steller's jay, with method for estimating angle of crest elevation. (From Brown 1964. Originally published by the University of California Press; reprinted by permission of the Regents of the University of California.)

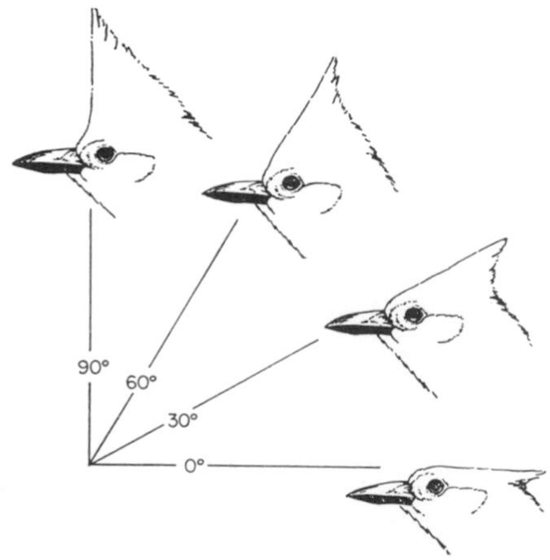

15–3 Gradations in crest-erection display of Steller's jays in various contexts. (From Brown 1964. Originally published by the University of California Press; reprinted by permission of the Regents of the University of California.)

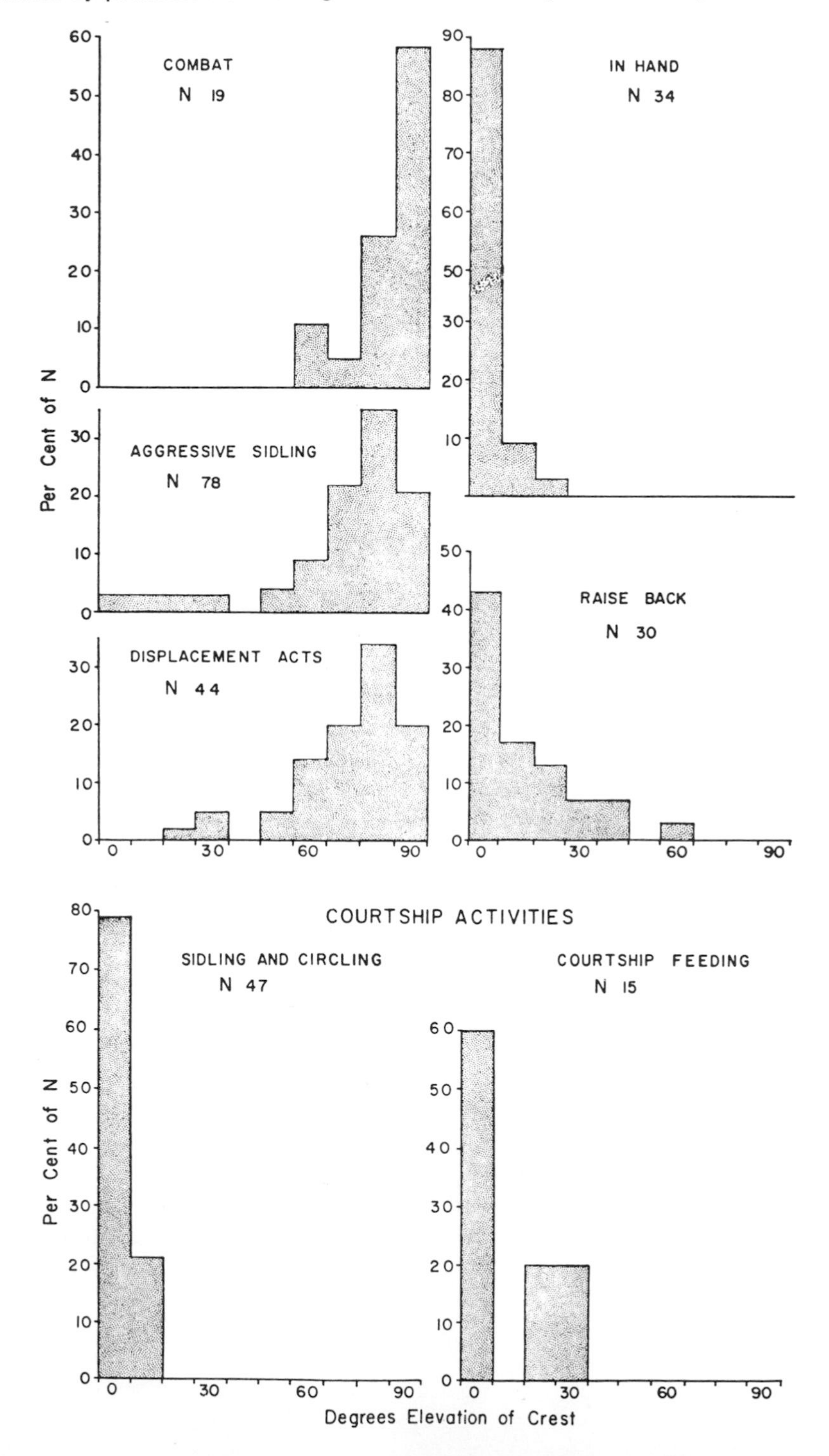

TABLE 15–1 Gradation of crest-erection display by Steller's jays in relation to resistance of opponent. (From Brown 1964. Originally published by the University of California Press; reprinted by permission of the Regents of the University of California.)

Activity	Resistance of opponent	Modal crest angle
Foraging	none	0°
Supplanting		
No increase in crest angle	slight if any	30°
Increase in crest angle	moderate	40°
Aggressive sidling	high	80°
Combat	extreme	90°

NOTE: Crest angle in the dominant bird increases with increasing resistance of the opponent.

plays, depending on the species. The adaptiveness of various types of graded displays has not received much attention and is likely to repay closer investigation.

INTRASPECIFIC DISTINCTIVENESS OF DISPLAYS

In a species that has several displays serving different functions, it is logical to assume that confusion would be avoided by their being as different as possible. This is especially likely to be true in the case of species having both *distance-reducing* and *distance-increasing* displays. A courtship display is likely to be more effective if it does not resemble a threat display and does not contribute to scaring away the potential mate. This was recognized by Darwin (1872) in his *Expression of the Emotions in Man and Animals,* where he referred to it as the *principle of antithesis.* As examples, Darwin chose the aggressive and friendly postures and facial expressions of dogs and cats.

A less familiar but more thoroughly studied example is the mutual display in black-headed gulls known as *facing away* or *head-flagging,* which is shown in Figure 15–4 (Tinbergen and Moynihan 1952). In this species the front of the head is brown and is directed at the opponent during the upright threat display. During pair formation, aggression between the members of the pair must be reduced, and it is important for each bird to avoid the face-forward upright threat posture. In the early phases of pair formation the newly acquainted birds

15–4 Displays of the black-headed gull. The principle of antithesis can be appreciated by comparing *facing away* with *aggressive upright,* in which the bill and face are directed toward the rival. (From Tinbergen 1959.)

go through sequences of displays (the meeting ceremony of the pair in Figure 15–4) in which the face-forward upright position is avoided, and an upright posture with bills pointed away from each other in opposite directions (*facing away* in figure) is common. Facing away thus effectively hides the brown face, which is associated with upright threat (Figure 15–4), from the prospective mate at a critical moment, at least for the duration of the display (a few seconds). At other times in the pair-formation sequence a head-on confrontation is avoided by adopting parallel body positions (as in *forward*).

Another example of the principle of antithesis is in the depressed and elevated crest positions of Steller's jay during courtship and threat, respectively (see Figure 15–3). In displays of the village weaver (*Textor cucullatus*), black parts of the plumage are emphasized during threat displays and yellow parts during courtship displays (Collias and Collias 1970).

There are some notable exceptions. Most fiddler crabs give the same display as a threat to males and as courtship to females, although differences in intensity occur in some species. In the early stages of pair formation, European robins (*Erithacus rubecula*) and other birds in which the sexes resemble each other in external appearance give the same display to neighboring males as to intruding females looking for a mate (see Figure 4–2). In these and other exceptions it seems likely that the subsequent behavior of the female or some other concomitant behavior of the male is important in reducing aggression between the sexes.

Intraspecific distinctiveness is also found in other displays that are not opposites and thus fail to come under the principle of antithesis. The use of several graded components of facial expression in combination, such as the muscles around the eyes, ears, mouth, and nostrils, allows a wide variety of basic patterns and an almost infinite range of intensities in display. Hopeless confusion is avoided by the use of particular combinations of these components in typical "faces" (see Figures 15–5, 15–15).

15–5 Intraspecific distinctiveness in the facial display of monkeys (*Macaca* spp.). Different facial patterns are achieved by combinations of graded components. (Modified from Van Hooff 1973.)

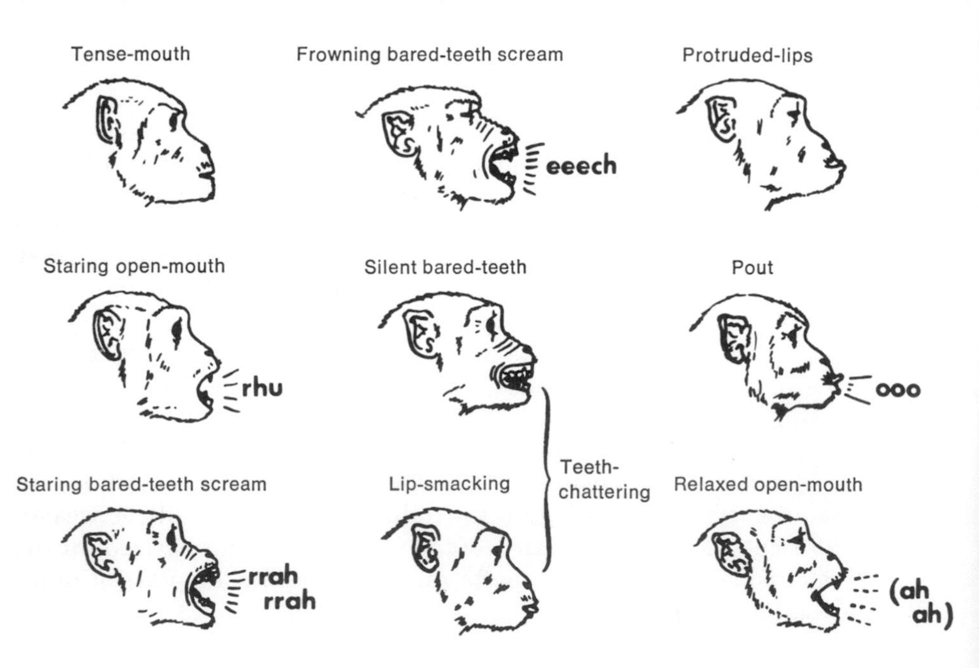

INTERSPECIFIC DISTINCTIVENESS IN VISUAL SIGNALS

Distinctiveness of display is commonly thought to be characteristic among different species. Actually specific distinctiveness is only characteristic of certain types of displays and not of others (Marler 1957). The principal situation in which species distinctiveness is selected for is pair formation and mating. Here it is often necessary that each species have some unique behavior or markings by which individuals can recognize other members of their own species and avoid mating with other species. Displays serving this function may be found in territorial advertisement as well as in courtship. The evolution of this type of display will be considered in more detail in relation to behavioral isolating mechanisms (see Chapter 18).

Species specificity may also be selected for in the visual appearance of group-living species like herding mammals and flocks of shorebirds. These species may have special patches of color or pattern that are revealed when an individual runs from danger (e.g., the "flag" of white-tailed deer) or takes flight (e.g., the rump patch of shorebirds and geese), stimulating other group members to follow (Figure 15–6). Since such groups may often be composed of close relatives, the evolution of these signals might be aided by kin selection.

Species specificity, or species distinctiveness, is also sometimes found in other displays, but these cases seem to have evolved more by chance or by extinction of related species than by selection for distinctiveness.

More commonly there is little or no particular advantage to uniqueness among species in a display. The display elements used by small birds in mobbing owls and other predators often show considerable resemblance among species. Similarly, the tail-flicking patterns of passerines may be nearly identical among several species of a phylogenetic group, or they may show evolutionary convergence in certain situations, as does the up-down pattern of some ground feeders (pipits, *Anthus* spp.; palm warbler, *Dendroica palmarum*) (Andrew 1956).

Close resemblance between the displays or visual appearance of two species may come about in several ways. Many unrelated species of ground-nesting birds have a distraction display in which the bird resembles in some ways an injured bird or running rodent (Figure 15–7). Such displays are employed to lure predators away from a nest. The resemblance here is due primarily to evolutionary *convergence* on a common adaptive pattern. Similarly, several groups of moths seem to have evolved intimidation displays that appear to scare away poten-

15–6 Group-cohesive alarm signals. The tail of the white-tailed deer (*Odocoileus virginianus*) is inconspicuous when the deer is at rest or feeding, but it is raised conspicuously when danger threatens or the animals escape. Similarly in many shorebirds, such as the black-tailed godwit (*Limosa limosa*) and black-bellied (gray) plover (*Charadrius squatarola*), white patches on tail, rump, or wings are revealed when birds take flight.

tial predators (small birds) by revealing a pair of eyespots on the hind wings when the moths are molested (see Figure 14–7).

Resemblance in the displays and visual appearance of unrelated species may also arise through *mimicry*. Two common types of mimicry in butterflies may be mentioned. In Batesian mimicry, an edible species mimics a distasteful one. In Müllerian mimicry, distasteful species may mimic each other. Wasps, being also distasteful and having warning coloration, also serve as models for mimic species of

15–7 Distraction display. Displays that lure predators away from the nest or young are often conspicuous in ground-nesting birds. To a human observer some of these displays resemble injury-feigning, as in the broken-wing appearance of the ringed plover (*Charadrius hiaticula*) at right. The distraction displays of the dunlin (*Calidris alpina*), at left, and other small sandpipers resemble the running of a small rodent in their speed, erratic turns, and rapid, mechanical-toy-like leg movements. (Drawings by Peter Scott; courtesy of Severn Wildfowl Trust.)

moths (Blest 1957), grasshoppers, and flies; and they also show Müllerian mimicry (Wickler 1968).

Other kinds of mimicry are found in fishes. The cleaning fish *Labroides dimidiatus* is mimicked by *Aspidontus taeniatus,* which gains access to large fishes by its resemblance to the cleaning species, but instead of cleaning, takes a bite out of the deceived victim (Figure 15–8; Eibl-Eibesfeldt 1959, Wickler 1963). Many examples of mimicry have been reviewed in a book by Wickler (1968), which is written for the nonspecialist and richly illustrated in color.

15–8 Mimicry of a cleaner fish. The cleaner wrasse, *Labroides dimidiatus* (*above*), which removes ectoparasites from host fishes, is mimicked by the mimic blenny, *Aspidontus taeniatus* (*below*), which feeds on the mucus and fins of other fishes. Both species are widespread in the Indo-West Pacific Ocean. The model and mimic at left are from Aldabra Atoll and Comoro Is. Those at right are from Great Barrier Reef. *Labroides* may be seen cleaning the mouth of a large fish in Figure 9–5. (From Springer and Smith-Vanig 1972. Photos by courtesy of Dr. Victor G. Springer, Smithsonian Institution.)

Cleaner

Mimic

Moynihan (1968) has suggested that interspecific resemblance in the assemblages of various tropical-forest bird species is selected for because it facilitates interspecific flocking. Cody (1969) has suggested that interspecific resemblance in the appearance of various other bird species has evolved for reasons associated with interspecific territoriality.

TYPES OF EVOLUTIONARY CHANGE IN DISPLAYS

The types of change in displays that natural selection may bring about are best appreciated by considering closely related species, so that descent from a common ancestral form provides a background against which relatively recent modifications stand out conspicuously. The types of evolutionary change found in displays and other ritualized behavior have been reviewed by Daanje (1951) for birds, by Ewing and Manning (1967) for insects, by Manning (1965) for *Drosophila*, and briefly by Bastock (1967) for courtship display in various species. Further essays on various aspects of ritualization are found in the symposium volume by Huxley (1966).

Changes of the Whole Display / A given motor pattern, such as a display, tends to be produced in certain situations and not in others. Species differ in which *situations* elicit a particular display. In fiddler crabs the ancestral context of claw-waving (Figure 15–1) seems to have been between males, and the display is still so used; but it has acquired secondarily a role in courtship. Consequently, the *stimulus complex evoking the display has shifted* from primarily males to primarily females, or to both, depending on the species. Some species display differently in response to males than in response to females (Salmon 1967). Similar changes in the releasing stimuli for a given act are found in all examples of emancipation (see Chapter 14 on the origins of displays).

Another type of change affecting the whole display concerns the *frequency and intensity* with which it is performed. *Drosophila simulans* is much like *D. melanogaster* in its courtship, but is more sluggish and seems to require more stimulation to evoke the same behavior (Manning 1959*a*). Genetic differences affecting mating speed in *D. melanogaster* might be interpreted as involving thresholds for male display (Manning 1961). Here, however, the primary effect might have been on competing behaviors; and the "display" does not apparently function as a visual signal. Strain or line differences in sexual responsiveness are known also in rats, mice, and guinea pigs (Bastock 1967). The physiological bases of these differences are unknown, but they suggest differences in the *relationship between thresholds and excitatory state*, and in the case of *Drosophila*, a model based on these relationships has been proposed (Bastock 1956).

Repetition of displays is one of the obvious ways in which they can become more effective. Repetition may be persistent throughout the day at frequent intervals, as in the strutting display of turkeys (Schleidt 1964), or may, in the case of simpler performances, take the form of bursts or bouts of the same act repeated in rapid succession (Figure 15–1). In the latter case there can be variations in burst length and burst frequency.

Species differences in *thresholds* for a display may sometimes be understood in terms of the effect of the display. In saturniid and sphingid moths, species with cryptic coloration generally require a high level of stimulation to provoke protective display, while species with effective intimidation displays (e.g., those revealing eyespots) require less stimulation to evoke display (Blest 1957*b*). A similar relationship between thresholds for protective display in cryptically and noncryptically colored mantids was found by Crane (1952).

Exaggeration of Coloration / One of the better-known ways in which the effect of a display movement on a viewer can become exaggerated through evolution is by the incorporation of a bright patch of color on the moving part or on a revealed part. The male red-winged blackbird (*Agelaius phoeniceus*) has a bright red patch on its wing, most of which is normally concealed. But the patch is extremely conspicuous during the spreading of the wings that is part of the male's display to females and males (Figure 15–9). Male peacocks, by the same principle, have magnificently colored tails that are erected and spread when courting females. In certain species of fiddler crab males may have brightly colored and greatly enlarged claws used in display. Many species of fishes (e.g., in Cichlidae, Centrarchidae) have developed dark-colored "ears" that exaggerate opercular displays (see Figure 18–3). Eyespots in moths very likely were perfected by natural selection because they enhance protective displays. The list of examples is nearly endless. The study of animal coloration has greatly helped to understand the evolution of behavior, and will undoubtedly continue to do so. Further details on the behavioral significance of animal coloration may be found in the general works of Cott (1957), Hamilton (1973), and Wickler (1965).

15–9 Exaggeration of coloration. The song-spread display of male red-winged blackbirds (*Agelaius phoeniceus*) is made more conspicuous by the red patches on the wings. The red areas are inconspicuous normally or when the feathers are sleeked, **C,** but they are maximally exposed during the display, **A, B.** The female is shown in **D.** (From Nero 1956.)

Exaggeration of Structure / Displays and appearances can also be enhanced by exaggeration of the size and shape of structures. The displays of birds of paradise feature highly specialized feathers that have become greatly elongated or notched or have changed in texture (see Figure 8–6). An unusual specialization of the tail feathers is found in the racket-tailed drongos (Figure 15–10) and in the motmots, a tropical family. In the motmots the rackettail enhances the visual effect of the characteristic side-to-side movements of the tail.

Among mammals the evolution of horns and antlers (Figure 15–10)

15–10 Exaggeration of structures thought to be used in communication. **A–C.** Feathers. **D–I.** Antlers and tusks. The fossil series from very small-bodied, tusked Eocene traguloids, **D,** through larger Miocene deer, **E,** to very large, present-day, antlered deer, **F,** parallels the morphological series among living forms from the very small chevrotains, **G,** through larger muntjacs, **H,** to very large, antlered deer, **I.** The modern tusked species fight all year with their tusks. Modern antlered species restrict the use of antlers for fighting to the rutting season. From the parallel between the two series one may infer that (1) in the evolution of deer corresponding changes in behavior occurred in the sequence shown, and (2) fighting behavior in deer evolved in association with the corresponding structural changes from tusks to antlers.

A. Pennant-winged nightjar, *Semiophorus vexillarius.* **B.** Paradise whydah, *Steganura paradisea.* **C.** Greater racket-tailed drongo, *Dicrurus paradiseus.* **D.** *Blastomeryx.* **E.** *Dicrocerus.* **F.** *Cervus.* **G.** *Tragulus.* **H.** *Muntiacus.* **I.** *Cervus.* (A–C: Drawings courtesy of George Miksch Sutton; from VanTyne and Berger 1959. D–I: From Colbert 1958.)

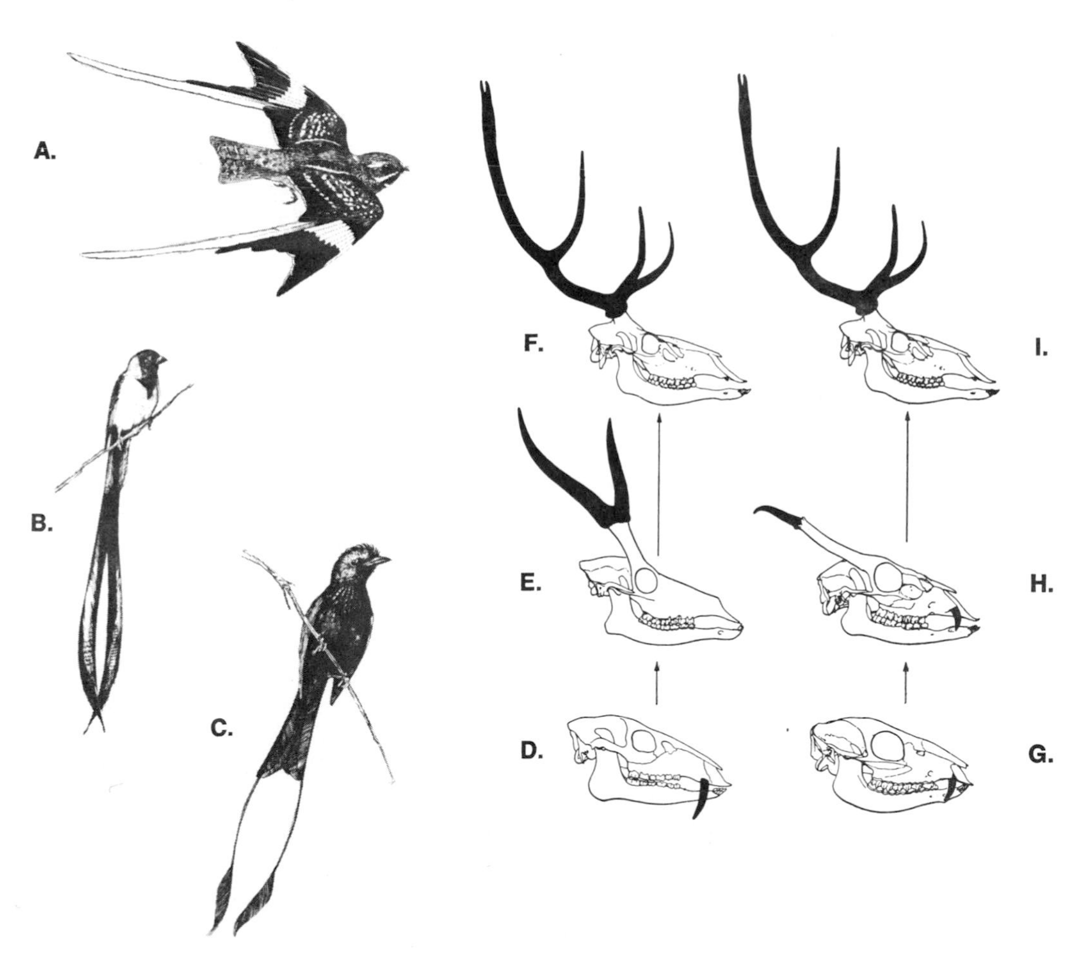

is thought to have been facilitated by their usefulness as visual signals indicating the maturity of the male, as well as their usefulness as weapons (Geist 1966*a*, *b*). Similarly, the manes of lions and baboons probably help to indicate the maturity of their bearer.

Exaggeration of Movement / The effectiveness of a movement of a fin, limb, tail, or facial structure in display has apparently been enhanced in some groups of species by exaggeration of the rapidity, frequency, or extent of the movement. Such changes may simultaneously involve the reduction of movement in other parts so as to bring about an overall change in coordination that, by focusing attention on a particular component of the display, simplifies the visual stimulus to the viewer. The ritualized courtship preening movements in dabbling ducks afford several examples of this sort (Chapter 14). In the grouse family (Tetraonidae) at least one species, the ruffed grouse of the northern United States, has exaggerated the wing and tail movements of display to the point where they make a loud drumming sound. Other examples of changed coordination may be found by comparing the claw-waving patterns of crabs (see Figure 15–1; Schöne 1968; Schöne and Schöne 1963) and the facial expressions of certain primates, carnivores, and ungulates. The evolutionary shifts in coordination of display components in weaver finches have been discussed by Crook (1964).

Transfer of Function / The bowerbirds of New Guinea provide an example of how exaggeration of one component can lead to a drastic reorganization in the display. In these species, which are thought to be polygamous (Gilliard 1969) and thus more subject to sexual selection, individual males maintain isolated display grounds that are typically decorated with a structure made of grass or twigs or with a collection of brightly colored objects such as flowers, fruits, or overturned leaves (Marshall 1953; Gilliard 1969). One species is even known to paint the bower with juice of a violet berry (tool-using). Females visit the display grounds, or bowers, of males for copulation. In general, species with elaborate bowers have inconspicuous plumages, and vice versa (Gilliard 1956). In the genus *Chlamydera*, the bower has a tunnel-like structure through which the female looks to see the displaying male manipulating his "treasures" in his bill (Gilliard 1956, 1959*a*). In *C. nuchalis* an iridescent, erectile, lilac, silver-tipped patch on the back of the head of the male automatically becomes exposed to the female through these actions, and the male does not pay much attention to the objects in the bower. *C. cerviniventris* lacks the bright patch, and with this species the role of the displayed objects—

green berries—is greater. The display patterns of these two closely related species suggest that in their common ancestor a transfer in the exaggerated component of the display took place from display of the back of the head to display of the colored objects, or vice versa. It is not hard to imagine that the selection pressure of predation on conspicuously colored, ground-displaying individuals, combined with the sexual preference of females for object-manipulating males, might have brought about a shift in courtship from plumage display to object display. A similar correlation may be found in two other genera of bowerbirds, *Archboldia* (Gilliard 1959*b*) and *Amblyornis* (Figure 15–11; Frith 1970).

The type of transfer of effective display components that seems to

15–11 Transfer of display function. An inverse relationship exists between crest and bower in the bowerbirds of New Guinea. In some species the crest, which is flashed at the female during courtship, is spectacularly long and colorful; and the bower, where courtship occurs, is relatively simple. In others the crest is shorter and the bower more complex. The crest of *Amblyornis subalaris* extends only to the neck, but that of *A. macgregoriae* reaches the back. The bower of *A. subalaris* is an "extremely ornate structure of twigs forming a hut around the central sapling or maypole and is decorated with blossoms, berries and other objects." "The bower of *A. macgregoriae* is comparatively simple, consisting of a column of sticks placed against and round a central sapling and surrounded by a clear circular display area which is encircled by a rim of moss some millimetres high." (From Frith 1970. Courtesy of Royal Australian Ornithologists' Union.)

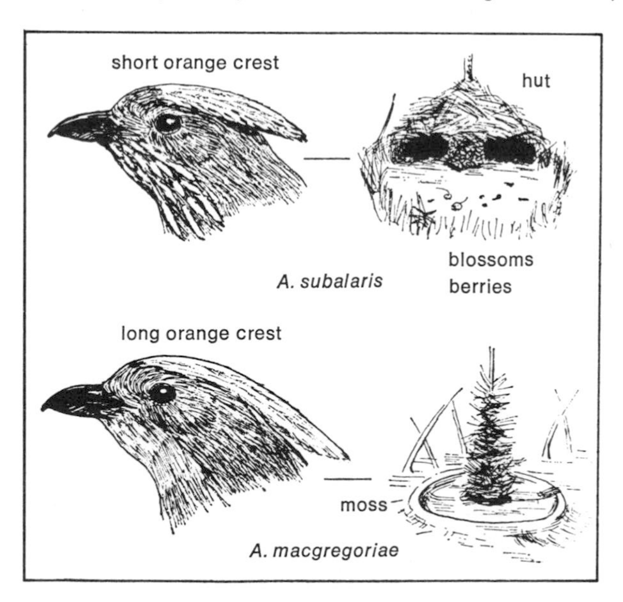

have occurred in the bowerbirds suggests that conditioning plays an important role in the evolution of displays. The success of investigators in classically conditioning the aggressive display of the fighting fish (*Betta splendens*) (Thompson and Sturm 1965) and the courtship behavior of the Japanese quail (*Coturnix coturnix*) (Farris 1967) to a neutral light or tone indicates that animals may come to respond to neutral aspects of the environment or of the behavior of another individual that frequently precede the performance of a reinforcing behavior. In this way an intention movement, displacement activity, autonomic response, or other activity in one animal might come to serve as a conditioned stimulus for the actions of another individual. At this point, natural selection might favor genetic variations that enhanced the likelihood of a given behavior in one individual becoming a conditioned stimulus for a behavior in another individual.

If classical conditioning is important in the evolution of displays, some interesting postulates can be made. Such a process should be most noticeable in animals with a demonstrated capacity for the conditioning of displays, such as fishes, birds, and other vertebrates; it might be less apparent in insects, crustacea, or other invertebrates. Not only is the ability for sensory conditioning necessary for this combination of learning and natural selection to proceed, but also the ability to make fine discriminations in one or more sensory modalities. If differences in this postulated process among phylogenetic groups are important, they may depend more on discriminative and perceptual abilities than on the ability for sensory conditioning.

Another consequence of such an evolutionary process is that the stimulus situation preceding a ritualized behavior changes as fast as the conditioned stimulus becomes ritualized. That is, the hypothetical "borderline" between ritualized and not-yet-ritualized behaviors "advances" as new elements are added on "in front of" the previously ritualized acts. Components at the "hind end" of ritualized sequences might be expected to "drop off" or continue as vestigial traits. The reader is cautioned that attempts to support or deny these hypotheses empirically have yet to be made.

The adding on "in front" of display components would tend to continue indefinitely until blocked by circumstances that somehow prohibit it. What these might be will be left to the reader's imagination, but if the new complex of conditioned stimulus plus original act is not more effective than the unadorned original act, this type of ritualization should come to a halt. In other words, this type of selection can be expected to proceed until a stage is reached that cannot be improved upon with the available genetic variability.

THE PHYSICAL ENVIRONMENT AND THE DIRECTION OF SELECTION

Early evolutionary studies of displays emphasized the concept of gradual improvement of visual signals as elements in a communication system. Consequently, properties inherent in the signal and the system became the focus of interest. With the increase in numbers of field studies of social organization and behavior after World War II, more attention was directed to the environmental features that make certain displays more effective or more likely to be employed than others.

Nest Sites / The nest site characteristic of each species can sometimes be correlated with species differences in displays. Hole-nesting birds in various families have evolved special nest-demonstration displays (Haartmann 1957), as have many colonial nesters (see Figure 15–14). Distraction displays (e.g., Figure 15–7) are widespread and conspicuous among ground-nesting species, but uncommon among species nesting in trees or other sites protected from ground predators (Armstrong 1964). In gulls pair formation on cliff-ledge nest sites in the kittiwake (*Rissa tridactyla*), rather than on nearly level ground as in most species, has apparently resulted in the use of a choking display in typical pair-formation sequences, rather than the oblique-cum-long-call and facing-away displays employed by many ground-nesting species (Figure 15–4; Cullen 1957; Tinbergen 1959). Choking is present in ground nesters as a display given near the nest (see Figure 15–4), but it is not used in pair formation, which usually occurs away from the nest.

Display Perches / Birds often choose conspicuous perches from which to sing, display, and advertise territorial ownership. Where convenient perches are unavailable, as in grassland, marsh, and tundra habitats, species characteristically employing aerial displays are common. The African whydahs (*Steganura*) and bishops (*Euplectes*), for example, are small ploceid finches that inhabit grasslands. The males (see Figure 15–10*B*) have a jumping display or an aerial display that makes them much more conspicuous. Many shorebirds, such as the lapwing (*Vanellus*), have elaborate aerial acrobatics, involving swoops and rolls, used in territorial defense and courtship. Many songbirds, such as larks (*Alauda*), have elaborate song-flight displays. In the red-winged blackbird, even subspecies differ in this respect (Hardy and Dickerman 1965): in *A. p. gubernator* of Mexico display flights are common, while in the neighboring and now overlapping *A. p. fortis*, they are rare. Although both types occur in the same geographical location, it is not known whether the difference is attributable to genetic or to environmental factors.

Aquatic Displays / Splashing displays, which obviously depend on an aquatic environment, are sometimes found in aquatic birds. Some species of grebes have display ceremonies involving pairs skittering over the surface of the water (Storer 1963). In the *head-throw-kick* display of goldeneye ducks (*Bucephala*), a plume of water is sent up by the feet of the male (Dane et al. 1959).

Ground-feeding Species / In species that feed in flocks on the ground the danger of predators is usually present and individuals must be watchful for signs of danger, both from the predators themselves and from the behavior of others in the flock. Intention movements of flight assume special significance in this context. Perhaps this can partially account for the exaggerated tail movements found in the group, as in wagtails, pipits, palm warblers, and others (Andrew 1956).

Among galliform birds the display known as tidbitting (see Figure 14–8) seems to have evolved from the behavior of drawing attention to food on the ground (Williams et al. 1968; and earlier authors).

Visibility / Environments differ in the distance at which one individual can see another. Where vision is blocked, as in thick vegetation, dis-

15–12 Aquatic displays. Male Barrow's goldeneyes (*Bucephala islandica*) kick up jets of water in the head-throw-kick display to a female (*foreground*). (Photo by Philipa Scott. From Severn Wildfowl Trust, 4th Ann. Rept. 1950–51.)

plays become less important and audition and olfaction are used more in communication. In fishes living in muddy water, communication may depend on sound or production of electric pulses, rather than visual signals, and in the depths of the ocean where sunlight fails to penetrate, luminescent displays are used (Crane 1965). In general, birds and mammals living in open environments such as plains, marshes, or lakes tend to have specialized visual signals.

Intraspecific variations in the conspicuousness of visual signals might also be due to geographical variation in visibility. In the Steller's jay, populations living in densely forested areas such as the Pacific coast of Canada tend to have smaller, unornamented crests; populations inhabiting open areas with fewer trees and more clearings, as in Arizona and the mountains of western Mexico, have longer, ornamented crests (Brown 1963*a*). Figure 15–13 shows the relationship between crest length and geography in two forms of Steller's jay. The relationship of crest to vegetation is known in the black-crested form, which is found mainly in the United States; but it has not been studied in the blue-crested form, which is found only in Central America.

SOCIAL ENVIRONMENT AND DISPLAYS

Although it is obvious that a general relationship must exist between animal communication and social organization, the nature of the relationship received relatively little attention until the pioneering field studies by Crook (1962, 1964) on African and Asian birds of the family Ploceidae. In any species, the situations that arise in which communication might be useful to individuals are dependent upon the typical social organization of the species. Of course, the social organization is ultimately adapted to the physical and biotic environment of the species, but an important link in the mesh of evolutionary causal relationships is the social context.

Mating Systems / As we have seen in the comparative study of social organizations, nonmonogamous (polygamous and promiscuous) systems are characterized by more intense sexual selection than monogamous ones. The result of this difference is generally that courtship displays and the plumage specializations associated with them are much more conspicuous in nonmonogamous than in monogamous species. The role of the male in caring for the young is also involved here. Where the male does not associate with the nest or young, as in most nonmonogamous species, selection pressures for cryptic coloration and habits in males are probably less intense, allowing sexual

15–13 Geographic variation of relative crest length (crest:wing ratio) in Steller's jays. The crest is longer in habitats that are more open, with greater visibility, as in Arizona. The range of the species is nearly continuous in coniferous forest from Alaska to Central America; the areas from which measured specimens for each sample were taken are shown in black. The mean, range, and twice the standard error on each side of the mean are shown. Black bars represent black-crested samples; hollow bars represent blue-crested samples with the exception of Michoacan and Mexico-D.F. samples, in which black, blue and intermediate colors occur. Sample size is given with each bar. (From Brown 1963*a*. Originally published by the University of California Press; reprinted by permission of the Regents of the University of California.)

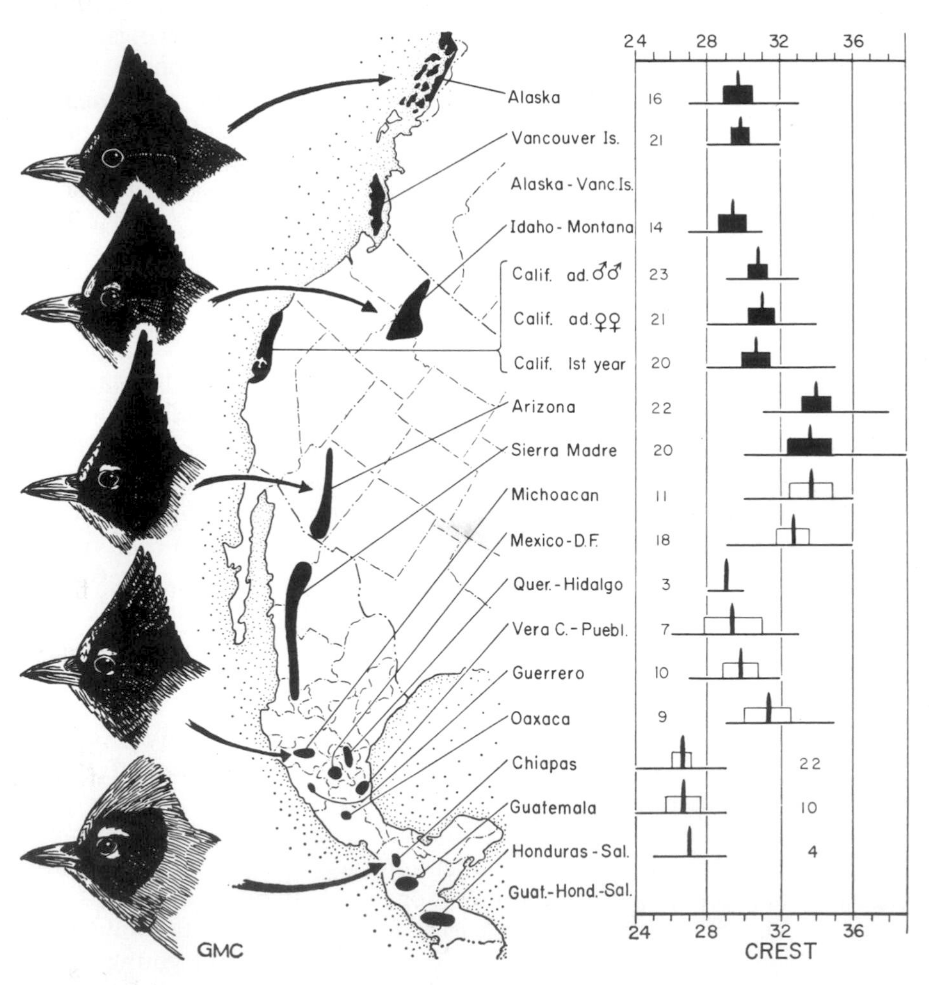

selection to favor more conspicuous male appearances than would be adaptive in a monogamous species having male parental care.

Another influence of social organization on display has been documented by Crook (1962, 1964) in the weaverbird family, Ploceidae. Species living in forest habitats are regularly dispersed, but those inhabiting savannahs usually nest colonially. Because of the competition for nest sites and the robbing of nest materials from each other, males nesting in colonies must stay at their nests to protect them. Consequently, it is more efficient for males to lure females to their nest with nest advertisement displays than to chase and court them away from their nests, as do forest-living species. A nest advertisement display sequence of the colonial village weaver (*Textor cucullatus*) is shown in Figure 15–14. As a result of the colonial social system, the displays used by males in courtship are typically given at the nest rather than away from it. Several species display while hanging from the nest. The sequence of displays also differs from that in the forest species. Colonial species tend to combine courtship with nest demonstration; forest species first court the female away from the nest and then lead her to it. The contrasting location, form, and sequence of displays are correlated with the social systems, which are in turn the result of adaptations to the environment.

Agonistic Systems / Those species that defend large all-purpose territories have different communication systems than those that live colonially and defend only the nest or those that defend only a mating territory. In the latter two cases, the display site is constant, while in the former the male must be prepared to move immediately to any part of his large territory where defense is required. Species differences in display along these lines were found by Crook in the pair formation types just described for weavers. In general, bird species with large territories seem to emphasize vocalization, and species defending only nests or mating sites seem to emphasize display. Perhaps this is because displays are more effective at short range and vocalization at long range, where visibility is impeded by vegetation.

Another type of social system requiring frequent communication at short range is found in group-living primates, ungulates, elephants, canids, and felids. In these species *facial expressions* are conspicuously well developed and rich in variations (Figure 15–5 and Figure 15–15). Facial displays are more diverse in social species like wolves (*Canis lupus*) than in solitary ones like foxes (e.g., *Vulpes fulva;* Fox 1970)

Aid-giving Systems / Species differences in parental care and development of the young are also correlated with display repertoires. The

15–14 Influence of social organization on display. In weaverbirds colonial species emphasize combined nest demonstration and courtship displays at the nest; in regularly dispersed species these functions are separated spatially and temporally. The form and orientation of the displays also differ; colonial species often display while hanging from the nest and orient toward the nest entrance. Here a male village weaver (*Textor cucullatus*) displays to a female, who enters and inspects the nest. In frame **1** the male calls while wing-flapping. In **2–5** the male continues to rotate but stops flapping. In **6** the male sings to the female. (From Collias and Collias 1970. Courtesy of British Ornithologists' Union.)

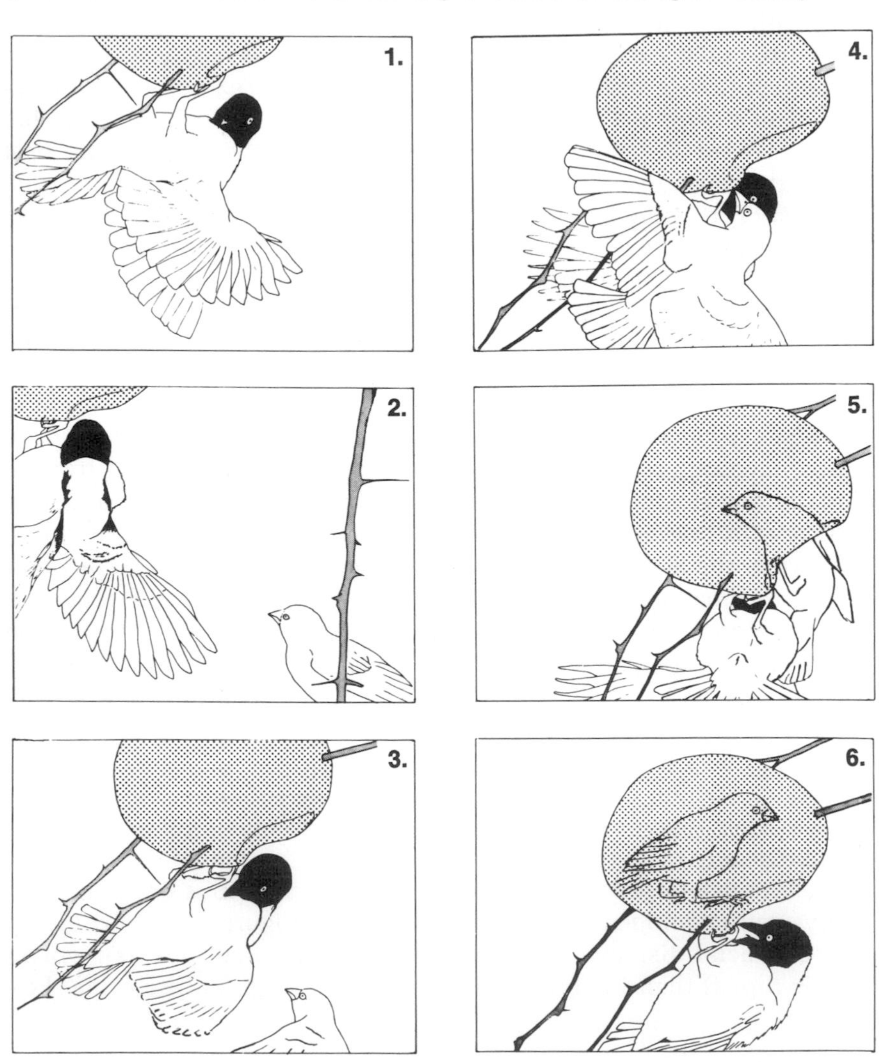

15–15 Communication through facial expression. In group-living species, facial expressions are particularly useful because of frequent close-range social encounters.

African elephants (*Loxodonta africana*) use their trunk and ears expressively. A neutral, resting elephant is shown above for comparison. With ears forward, as in **a,** the elephant is slightly aroused, with no particular mood predominating. In **b,** with head raised, it is a little more nervous. An intermediate mood is shown in **c.** The offensive mood is shown by the forward trunk positions in **f, d, e,** and those below **f.** The defensive mood is indicated by curling up and back of the trunk, as in **k** and below **k.** A cow in heat (**g**) shows a combination of offensive and defensive elements. A male in frontal battle adopts position **i.** (From Kühme 1963.)

Some facial expressions in zebras and horses (*Equus* spp.) are shown below. A series of increasing intensities of threat expressions, with ears down, are shown in **t–v.** Increasing intensities in the greeting face, with ears up, are shown in **w–y.** (From Trumler 1959.)

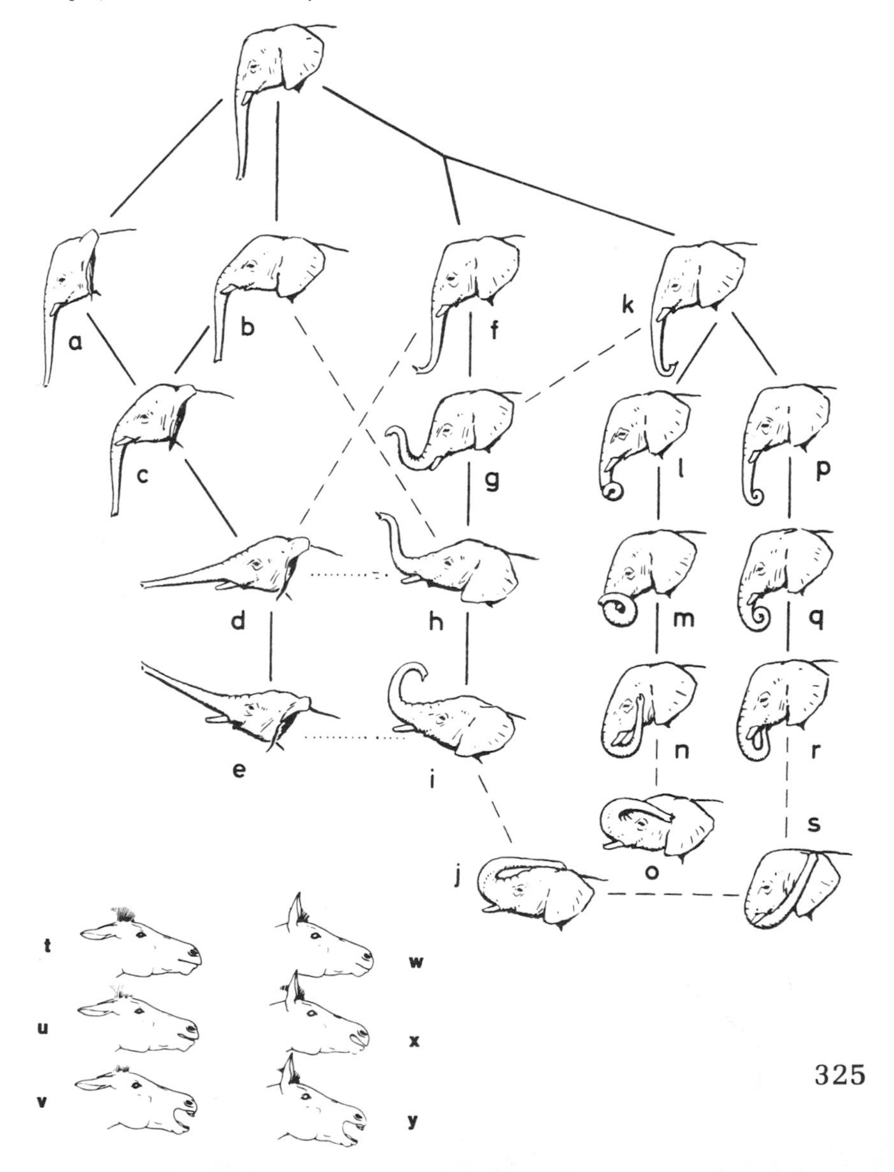

young of altricial species of birds, which must be fed by the parents for a long period, typically have *begging displays* that may in some cases resemble displays of the adult—for example, the precopulatory displays of female songbirds and the wing-spreading display of Steller's jays. Precocial species of birds commonly lack such begging displays.

Invitation displays are characteristic of group-living species that cooperate in mutual grooming or mutual preening, but they tend to be absent in species that characteristically live as solitary individuals or pairs, except for precopulatory displays (Figure 15–16).

Signs of *alarm* tend to be more conspicuous in group-living species. Shorebirds, ungulates, and certain sparrows have conspicuous white areas on the rump or tail that are suddenly revealed when an individual escapes by flight or running, thus exaggerating the alerting effect on other group members and helping to keep the group together (see Figure 15–6). Preparatory movements for escape also have become specialized in some of these species, as in the tail movements of wagtails. Other permanent aspects of appearance may help to keep a group

15–16 Invitation displays. Postures that are appropriate for being groomed are used to solicit grooming in mammals and birds. *Above:* Red avadavat (*Amandava amandava*). (From Sparks 1965; Courtesy of the Royal Society.) *Below:* Lesser kudu (*Tragelaphus imberbis*). (From Walther 1964.)

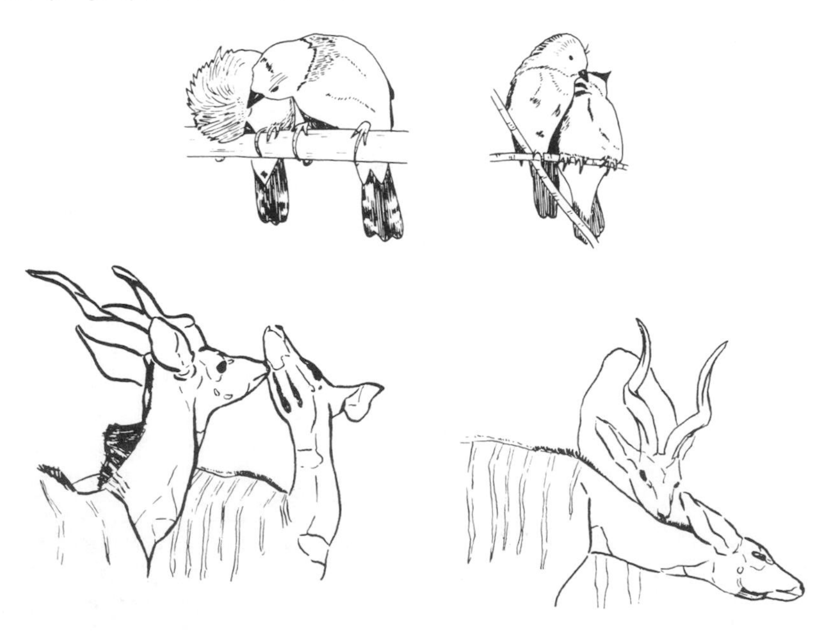

together even when danger does not threaten. Thus some species of fish will choose to school with fishes having a particular appearance.

INTERSPECIFIC DISPLAYS

Displays in response to predators are well known. They include the death-feigning of opossums and hognosed snakes, the snake mimicry of hole-nesting birds, the movements characteristic of mobbing birds, distraction displays (see Figure 15–7), and defensive or protective displays such as the eyespot revelation of moths (see Figure 14–7) and the menacing postures of defensive dogs and cats. Many more could be cited. Another category of interspecific displays includes those performed between symbiotic species, such as displays between cleaner and cleaned fishes.

CONCLUSIONS

In the process of ritualization some visual signals tend to evolve toward the graded type, in which postural gradations are suited for indicating gradations in motivation. Others tend toward the discrete type, in which specificity is retained by preserving a distinctive pattern of movement or light production.

Improvement of visual signals during evolution need not necessarily produce greater distinctiveness in those signals. *Two types of distinctiveness* can be recognized: (1) distinctiveness among the various displays used by an individual or species, and (2) distinctiveness among corresponding displays used by similar species. Darwin recognized that within a species the displays serving opposite functions tended to become opposite in appearance; this is the *principle of antithesis.* Among similar species, the principal selection pressure for species distinctiveness derives from the need to mate exclusively with one's own species. Another involves the advantages of group cohesion. Among different species, selection may also favor visual resemblance. This may occur by convergence on a useful type, as in the broken-wing distraction displays of birds. It also occurs in the nestling mimicry of brood parasites and the mimicry of various kinds of insects.

Evolutionary changes in display occur in many and varied ways. These may be divided into changes of the whole display (not affecting the pattern of coordination), and changes of selected components of the display. The first category includes changes in the quality of the effective stimulus and changes in the display's threshold, frequency, and intensity. The second includes the processes of exaggeration or simplification of the display through alterations in the body's color pattern and form or alterations in the pattern of muscular coordination. During these changes, the component of the display that is the center of visual interest may become transferred from one part of the body to another, or even to inanimate objects, as has occurred in bowerbirds. During the early stages of the transfer process the role of classical conditioning might be significant.

The direction that selection may impose on the evolution of displays is determined ultimately by the environment. The physical environment may bring about species differences in displays related to nest sites (cliff vs. flat ground), display perches (tree, ground, aerial, aquatic), and visibility. The social environment is reflected in the species diversity of displays through its influence on displays involved in sexual selection, territorial systems, and group living. A description of the social organization of a species also specifies the types of social interaction and their relative frequency and importance. These in turn determine which types of displays will be advantageous and to what extent.

16 The Evolution of Auditory Communication

Appreciation of the beauty and variety of animal sounds has stimulated interest in them, but it has also tended to frustrate our understanding of their roles in the lives of animals. Contrary to the inspired thoughts of some romantic poets, animal songs and other sounds cannot reasonably be interpreted as merely the externalizations of joy and happiness. A more realistic view of animal sounds has been revealed by painstaking studies of a great variety of species, especially since about 1950.

The problems that the student of auditory communication attempts to solve center on how the sounds an animal produces may influence the behavior of other animals. Information is needed on the methods and contexts of sound production, the physical characteristics of the sounds, the processes of sound reception and perception, and the ways in which these factors are influenced by natural selection. One of the first questions to be asked was simply, Do animals actually influence each other's behavior by means of sound production? The affirmative has been proven many times, but with each new call and song that is discovered, and for most already known, the answer has not yet been determined. A related question is: Do the different sounds made by an animal influence other animals in different ways, and if so, how?

Our conceptual approach to such problems is greatly confused by our habit of thinking of auditory communication in terms of human language. This unconscious habit tends to make us ask questions of doubtful heuristic value, such as What does that sound mean? as opposed to How does that sound influence the behavior of other animals? Although we can state precisely the message in many human sentences, we find it more difficult to define the message carried by a laugh, snicker, snarl, scream, gasp, or the various gradations of these. These express the motivational state, and their effect on hearers varies complexly. In general, the sounds made by an animal are to an observer also a mirror of the motivational state of the animal rather than a statement of a relationship in symbolic terms. Correspondingly, the effects that animal sounds have on other individuals can be interpreted as influences on their motivational states rather than as "impartial" messages.

From the evolutionary perspective we can hypothesize that any given sound in the repertoire of a species has been favored by natural selection because its influence on the motivational states of other animals is beneficial to (raises the fitness of) the producer or his close relatives. We can also hypothesize that the physical characteristics of the sound produced and the situation in which it is used will be subject to natural selection and will reflect the ecological relationships of the species. In this chapter comparative and experimental studies bearing on these questions will be examined.

Auditory signals and the behavior patterns associated with them have been a rich source of material for comparative studies of animal behavior, probably because the study of sound signals has several advantages for comparative studies over other types of communication. With the aid of the tape recorder, the sounds of animals can be more easily recorded and reproduced than can olfactory, visual, or tactile signals. Analysis and description of sounds has been greatly facilitated by electronic devices, such as the oscilloscope, sound spectrograph, and others, which translate sounds into pictures objectively (Marler and Isaac 1960; Davis 1964; Greenewalt 1968; Marler 1969*c*).

PHYLOGENETIC OCCURRENCE

The use of sounds for communication is commonplace for the human species, but among animals generally it is far from universal. Many of the simpler phyla, such as the various groups of protozoa, coelenterates, flatworms, roundworms, and annelid worms, appear to lack auditory communication. In the largest phylum, Arthropoda, only a few crabs, shrimps, spiders and several families of insects communicate by sound, and most species are essentially silent. Among vertebrates, primates and birds are conspicuous for their use of sound; but other groups, such as salamanders, snakes, and many fishes, make little if any use of sound in communication. The phylogenetic distribution of auditory communication is indicated in Table 16–1.

Sound as a medium for communication has certain advantages over other types of signals. Sound passes around objects and through vegetation that would be opaque to visual stimuli; it can provide

TABLE 16–1 Some animal groups with auditory communication or sound-producing structures. (References to arthropods in Dumortier 1963 and Alexander 1966.)

ARACHNIDS
A few spiders (Rovner 1967) and scorpions (Rosin & Sholov 1961)
CRUSTACEANS
Palinuridae, spiny lobsters
Oxypodidae, fiddler crabs (Salmon & Stout 1962)
Altheidae, snapping shrimps (not used in communication)
INSECTS
Orthoptera, many grasshoppers, crickets, katydids
Homoptera, cicadas
Lepidoptera, a few moths
Coleoptera, a few beetles
Hymenoptera, bees, leaf-cutting ants (Markl 1968)
Diptera, mosquitos
FISHES
Many species (Moulton 1963; Fish and Mowbray 1970)
AMPHIBIANS
Many frogs and toads
Salamanders (Bogert 1960; Blair 1968)
REPTILES
Insignificant in most species (Bogert 1960; Blair 1968)
Tortoises (Campbell and Evans 1967), some geckos
BIRDS
Universal
MAMMALS
Probably nearly universal, but most important in primates (Tembrock 1963)

greater specificity and complexity and can be more precisely timed than can chemical stimuli. Consequently, it is useful for small, inconspicuous, or cryptically colored species such as grasshoppers, crickets, and some birds; for species of all sizes that are nocturnal, or live in murky water, or in thick vegetation; and for species requiring greater precision of stimulus localization and control than is provided by chemical signals.

EVOLUTIONARY ORIGINS OF SOUND-PRODUCING MECHANISMS

Solving the problem of the evolutionary origins of sound-producing behavior is somewhat easier than discovering the origins of display behavior because of the obvious relationships between sound-producing structures and their original, or autochthonous, uses. Sound-producing mechanisms in arthropods, vertebrates, and birds have been reviewed by Dumortier (1963), Kelemen (1963), and Greenewalt (1968), respectively.

From Respiratory Structures / In those animals that produce sound by causing a flow of air through an orifice, the respiratory system is usually involved. The ways in which normal respiratory patterns for particular contexts, such as protective defense, could have been exaggerated by selection are probably numerous. The ability to make sounds with respiratory structures probably evolved (and was lost?) independently in various groups. Given such an ability in one context, as a preadaptation, selection could modify it for use in various other contexts; and the original use might even be discontinued.

Numerous specialized structures have evolved in association with respiratory systems. These include the mammalian larnyx and its counterpart (but not homolog) in birds, the syrinx. Air sacs or chambers that serve as resonators are found in many frogs and toads, howler monkeys (*Alouatta*), frigatebirds (*Fregata*), orangutans (*Pongo pygmaeus*), swans (*Cygnus*), and various grouse. An unusual nonrespiratory mechanism for sound production has been described in the glandulocaudine characid fishes (Nelson 1964*b*); in *Glandulocauda inequalis* a pattern of sound production associated with periodic air-gulping is associated with courtship. The air has no demonstrable respiratory function, and it was thought by Nelson to have been obtained originally while nipping at prey on the water surface. Among the Arthropoda only one species, the death's head moth (*Acherontia atropos*), is known to produce sound by passage of air through the mouth; other species use ejection of air, foam, or phosphorescent vapor from the spiracles (Dumortier 1963). These latter forms are probably of little or no use in intraspecific communication, but they are of interest as potential origins of such use.

From Beating a Substrate / Some animals make considerable noise when moving through vegetation. This helps to indicate locations of animals in a group to each other. Cessation of such sounds may have considerable significance as an indication of danger. Beating of the substrate with a limb is known in fiddler crabs (*Uca speciosa*, Salmon 1967), in various insects (including one which uses its head for this purpose), in many mammals, such as rabbits, ungulates, and in birds (e.g., the drumming of woodpeckers and foot-stamping of kiwis [*Apteryx*]). Beavers (*Castor*) slap the water with their tails.

From Rubbing of Appendages / The most common mode of sound production in arthropods consists of rubbing together various parts of the exoskeleton, a method known as *stridulation*, especially when the structures are specialized for sound production. A stridulatory apparatus is typically composed of two parts; the first, the *pars stridens*, or scraper, by rubbing against the second, or *plectrum*, sets the structure into vibration (see Figure 16–1). The parts of the exoskeleton employed in these roles vary greatly. The following methods exist: antenno-antennary, maxillo-mandibular, rostro-tarsal, cranio-prothoracic, presterno-rostral, prosterno-mesasternal, pronoto-femoral, mesonoto-pronotal, abdomino-pronotal, abdomino-tibial, abdomino-femoral, abdomino-elytral, abdomino-alary, coxo-prosternal, coxo-metasternal, coxo-femoral, tibio-buccal, elytro-abdominal, elytro-femoral, elytro-tibial, elytro-elytral, and alary-elytral. For details see Dumortier (1963).

For multiplicity of methods of sound production the insects are unsurpassed. Although not all the above methods arose independently, sound-producing behavior in insects is clearly polyphylectic, having evolved by many independent routes in various groups. However, in the Orthoptera the behavior is probably very old; the stridulatory apparatus of fossil Tettigoniodea from the Upper Jurassic resembles that in modern forms, and a crude stridulatory apparatus is found in a Palaeozoic fossil (Dumortier 1963).

One such group that has been particularly well studied, the Cerambycidae, reveals a possible origin of auditory communication (Michelson 1964). In nonstridulating species sexual recognition is achieved by contact and/or chemical stimuli. This is followed by licking movements that seem to calm the female, who at first tries to repel the male. Coincident with licking, tapping by the head of the male on the female abdomen develops; and in one case the rhythmic body movements have been specialized to produce sounds by rubbing together meso- and metathoracic plates. This group is interesting because intermediate conditions occur showing that stridulation was probably derived from a courtship behavior based on another sensory modality, in this case olfaction and/or touch. It has been suggested that a similar transfer of courtship function from another sensory modality to the auditory has occurred in crickets (Alexander 1962), namely, that sound production was derived from wing movements used formerly for display or pheromone release, or both.

A process of *evolutionary substitution* has apparently occurred. The essentials of this are shown by the examples of the cerambycid and cricket stridulators. In these groups, stridulation was added to an already effective courtship procedure and subsequently became more specialized, while the antecedent communicative behavior became less important and either dropped out or was modified. This is equivalent to the transfer of function described for bowerbird displays.

The secondary specialization of wing and tail feathers of birds for sound production in aerial displays illustrates the process of *evolutionary addition*. Many species of birds make conspicuous wing noises in flight that have no sure communicative function (e.g., mourning dove, *Zenaidura macroura;* goldeneye duck, *Bucephala clangula*). Others have feathers that are structurally modified in a way which enhances flight sounds during displays, usually in an aerial dive, which exaggerates the effect. These are found in some snipe (*Capella*), woodcock (*Philohela*), nighthawks (*Chordeiles*), little bustards (*Otis tetrax*), and hummingbirds (Van Tyne and Berger 1959). Various members of the manakin family make rattling or cracking noises with their wing feathers. In all of these cases, sound production is not an act that formerly preceded or alternated with display, as in displacement or intention movements. It becomes superinposed on existing methods of communication through a change in structure or coordination or both. In this sense many sound-production behaviors are not derived from nonsignal acts, as are displays; they are *transformations of previously existing signal acts.*

BASIC CHARACTERISTICS OF SOUND SIGNALS IN DIFFERENT ANIMAL GROUPS

Crickets / Insect auditory receptors are of various types, and are more diverse than those of vertebrates. Descriptions of them have been provided by Dethier (1963), Horridge (1968), and others. For the present purposes, their most salient characteristic is their lack of frequency discrimination. Each type of insect auditory organ has a best frequency or best range of frequencies, but information on frequency and intensity are confounded in the message sent along the nerve from receptor to CNS. As a result, insect hearing organs may be characterized as being tuned but "tone deaf." (For an analysis of hearing and a diagram of an insect hearing organ see Chapter 22, Figure 22–2.) There is, in theory, the possibility that by combining information from different hearing organs the CNS could perform a crude frequency discrimination, and records from interneurons suggest that this happens (Horridge 1968); but evidence from behavioral responses to artificial sounds suggests that insects do not use such information even if they possess it. This characteristic is reflected in the sounds used by crickets in communication, which seem not to vary much in frequency within a song.

The stridulation mechanism of crickets is ill-adapted for producing a variety of sounds with rapidly changing or different frequencies, and it does not do so except within rather narrow limits. A brief consideration of the structures and movements involved reveals why. All sounds in cricket repertoires are apparently produced by frictional movements of the partly raised tegmina (modified wings). Normally the toothed pars stridens on the under surface of the right tegmen (above) and the smooth, rigid plectrum, or scraper, on the anteromedial edge of the left tegmen (below) are rubbed together by mesial movements of the partially raised and overlapping tegmina (Figure 16–1). During the calling song of most species, sound is produced mainly on the closing stroke, as shown in Figure 16–2, not during opening (Walker 1962; Davis 1968); but in courtship calling, a weak sound may be produced during opening and closing. In some katydids, however, several different patterns of wing movement are combined in sequence to produce a much more complex calling sound than is found in crickets (Walker and Dew 1972).

16–1 Sound-producing structures in Orthoptera. **A.** Right elytron (modified wing) of the cricket *Gryllus bimaculatus,* showing pars stridens or file (*PS*) and drum (*D*). (From Stärk 1958.) **B.** Stridulatory file of Uhler's katydid (*Amblycorypha uhleri* superspecies). (From Walker and Dew 1972.)

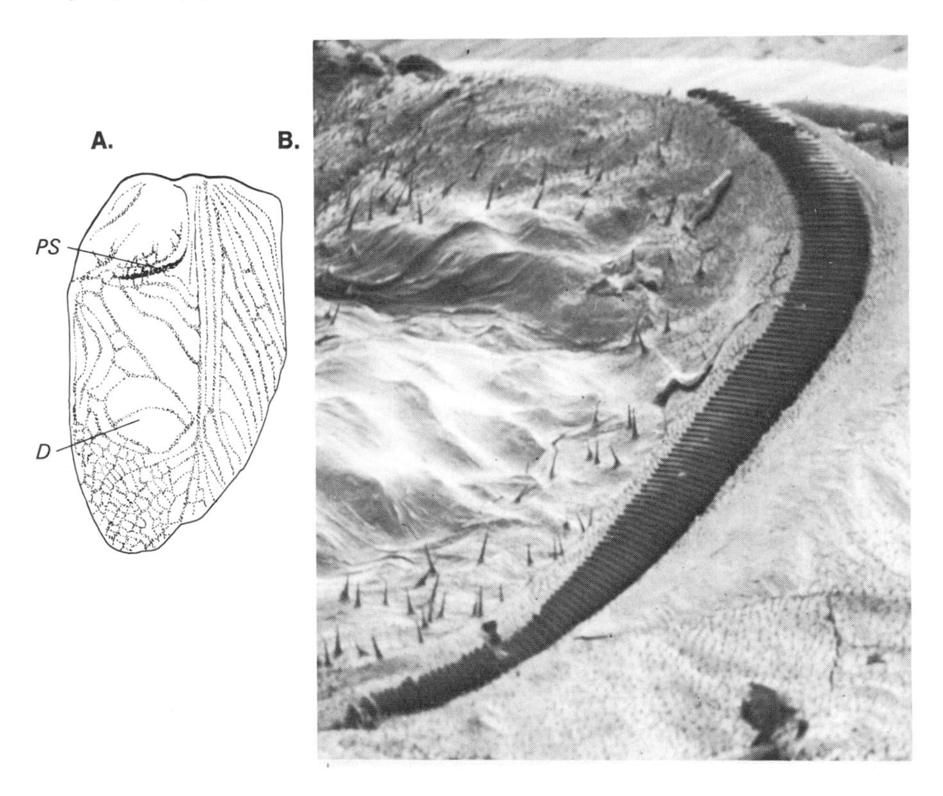

16–2 Wing movement during stridulation in a cricket, *Acheta domesticus*. High-speed filming of wings and oscilloscope simultaneously showed that sound is produced mainly during the closing stroke of the wings (homologous to the downstroke of flight). (From Davis 1968.)

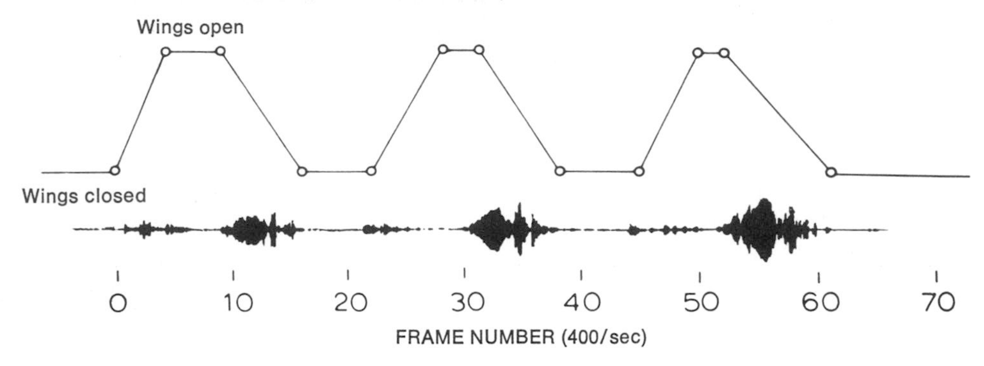

The duration of a sound pulse in the calling song is determined by the length of time it takes for the plectrum to pass from one end of the pars stridens to the other. A second pulse is usually not produced until the tegmina are again opened. The teeth of the pars stridens in many species are slanted so as to be most effective when rubbed in one direction (Figure 16–1). It is partly for this reason that continuous tones are not produced. All cricket stridulation sounds are composed of series of pulses. These pulses are produced in various temporal patterns, depending on social context, species, temperature, and other factors. The frequency or pitch of the sounds is relatively unimportant; it must only be within the cricket's audible range and should include the area of greatest sensitivity. Two temporal pulse patterns common in various species may be defined. A *chirp* is a short group of pulses that is separated from other such groups by pauses. To the human ear a chirp sometimes sounds like a single note, but it is actually a short series of distinct pulses. A long series of pulses produced without pauses is termed a *trill*. Various patterns of chirping and trilling are illustrated in Figures 16–17 and 16–18.

The frequency of the sound in a pulse is roughly the quotient of the number of teeth in the pars stridens divided by the time the plectrum takes to strike them all (Walker 1962). Since the number of teeth is fixed, frequency can vary only with the speed of wing movement. Most species have one fixed rate of wing movement at a given temperature; but some may have two fixed rates (Walker ibid.). The result is that most circkets have an amplitude-modulated (AM) system of auditory signaling: differences among cricket sounds are coded principally (perhaps exclusively) in terms of modulations of amplitude over time. Modulation of frequency (FM) is relatively slight and seems to be of little or no significance in crickets and other insects.

Pulse rate, pulse frequency, and female response all vary as a function of temperature (Figure 16–3). In cases where species differ in these parameters, the species differences are maintained under normal variations of temperature because of the similarity of the regression coefficient among different species. Two species can sometimes be made to call with the same frequency and pulse rate by artificially altering the temperature of one and not the other.

Several studies on the genetics of cricket and grasshopper behavior, based mainly on the behavior of hybrids, F_2s, and backcrosses between species, have revealed various patterns of genetic determination (reviewed by Alexander 1968; Bentley 1971; Bentley and Hoy 1972). In every case, species and subspecies differences in stridulation have been attributable to genetic differences. These have been shown to be multifactorial in some species and monofactorial in others. Various patterns of inheritance were found. Songs of hybrid crickets and grasshoppers are shown in Figures 21–18 and 18–8, respectively.

In many species of crickets all the parents die before the young appear. Consequently, there is no opportunity to learn sound production patterns from the parents. When individuals have been reared in isolation, they have shown normal stridulation behavior, both *producing* sounds appropriate to the behavioral context and *reacting* appropriately to sounds produced by others (Bentley and Hoy 1970; reviewed in Alexander 1968).

Despite the apparently rigid genetic and developmental determination of stridulating behavior, some susceptibility to alteration by experience can be demonstrated in certain species. Alexander (1960) has described and Shaw (1968) has analyzed a case in which katydids (*Pterophylla camellifolia*) will lengthen or shorten their chirps in response to variations in artificial or natural chirps.

Frogs and Toads / Although the calls of frogs and toads are frequently mistaken for those of insects, their methods of producing, receiving, and analyzing auditory signals are very different. The ears of anurans (frogs and toads) can discriminate a wide range of frequencies, and the coding of species identity is thought to be largely a function of frequency (Capranica 1965; Loftus-Hills and Johnstone 1970). The ears of many anurans, although sensitive to a wide range of frequencies, are in part "tuned" to the frequencies represented most strongly in their calls (see Figure 22–8). The importance of frequency changes (FM transients) during a single pulse or song has received little study.

16–3 Pulse repetition rate in cricket song as a function of temperature. Each regression line summarizes the relationship in a separate species. The regression lines for the 19 species shown have similar slopes; but some crossing, which might lead to interspecific confusion at certain temperatures, is evident. Dashed lines represent species in which pulses are produced in short sequences; solid lines, those producing continuous trills. For identification of species, see Walker (1962) Fig. 9.

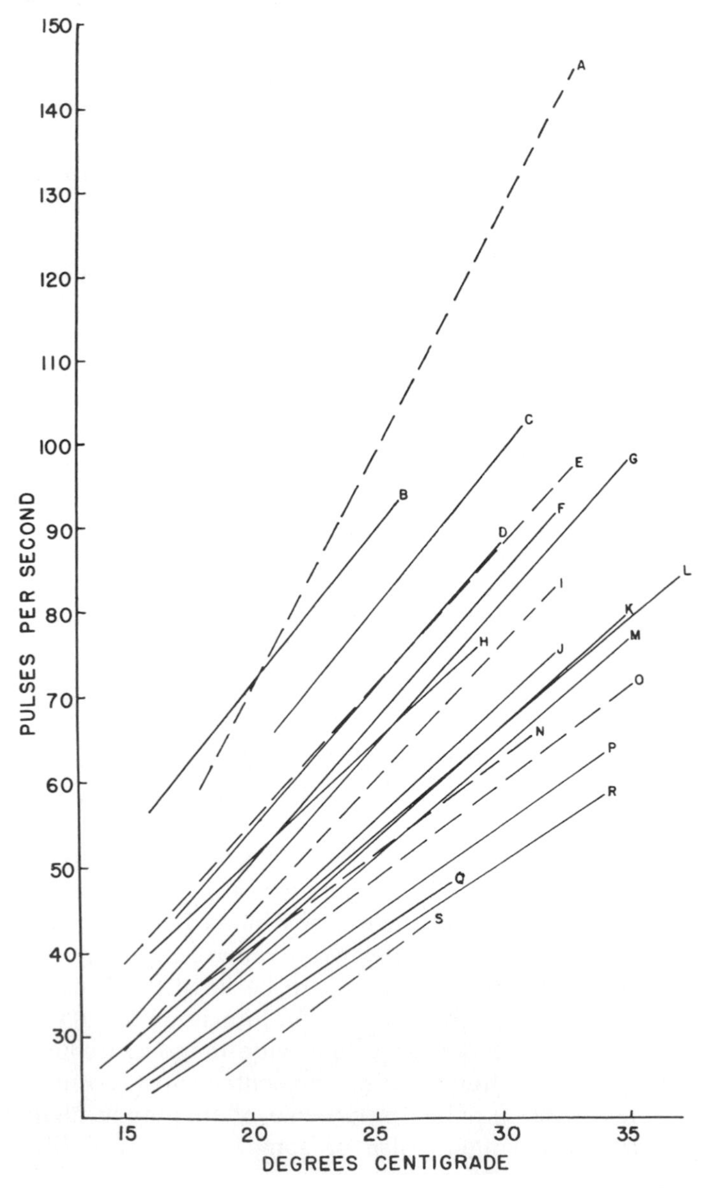

Amplitude modulation is also important, especially in the form of pulse repetition rate (Littlejohn 1971; Loftus-Hills and Littlejohn 1971*a*).

A feature that anurans and insects have in common is their dependence on temperature. This affects both frequency and temporal features of insect sound (Figure 16–3); but in anurans, frequency tends to be more independent of temperature, since the qualities of the vocal and resonating organs are relatively unaffected, while muscular activity is greatly speeded at warmer temperatures. A rise in temperature of 10°C is correlated with a doubling of pulse rate in the toad, *Bufo americanus* (Zweifel 1968). Because of this kind of relationship, comparisons between calls, for instance in studies of geographic variation, must first be corrected for temperature differences.

Many anurans can produce an amazingly loud and long-lasting call for their body size (e.g., 103 dB for *Limnodynastes tasmaniensis*, Loftus-Hills and Littlejohn 1971*b;* 33 sec, *Bufo cognatus*, Kellogg and Allen). The loudness is often aided by resonating sacs which are diverticuli of the buccal cavity (Figure 16–4). These sacs might also aid in producing long calls; since the nostrils and mouth are closed during calling, air is passed under pressure over the vocal cords between lungs and air sacs (McAlister 1961).

In anurans there is a strong correlation between body size and call frequency, larger males having larger vocal cords and deeper voices. Such a relationship in the genus *Bufo* is illustrated in Figure 16–5. As a result, primary evolutionary adaptation of adult body size to ecological conditions results in secondary changes in call frequency.

16–4 Vocal sac of a frog. During calling the vocal sac is inflated with air. Vocal sacs are thought to facilitate long and loud calling. (Photo: J. L. Brown.)

16–5 Relationships between the fundamental frequency of vibration of vocal cords, vocal cord mass, and body size in the genus of toads, *Bufo*. *Above:* Each point shows the mean and range for one species. *Below:* Species with mating call lost or weak are marked with a dot; note the proportionately small sizes of their larynges. Larynx size as a percentage of body size is shown by the guide lines. (After Martin 1972.)

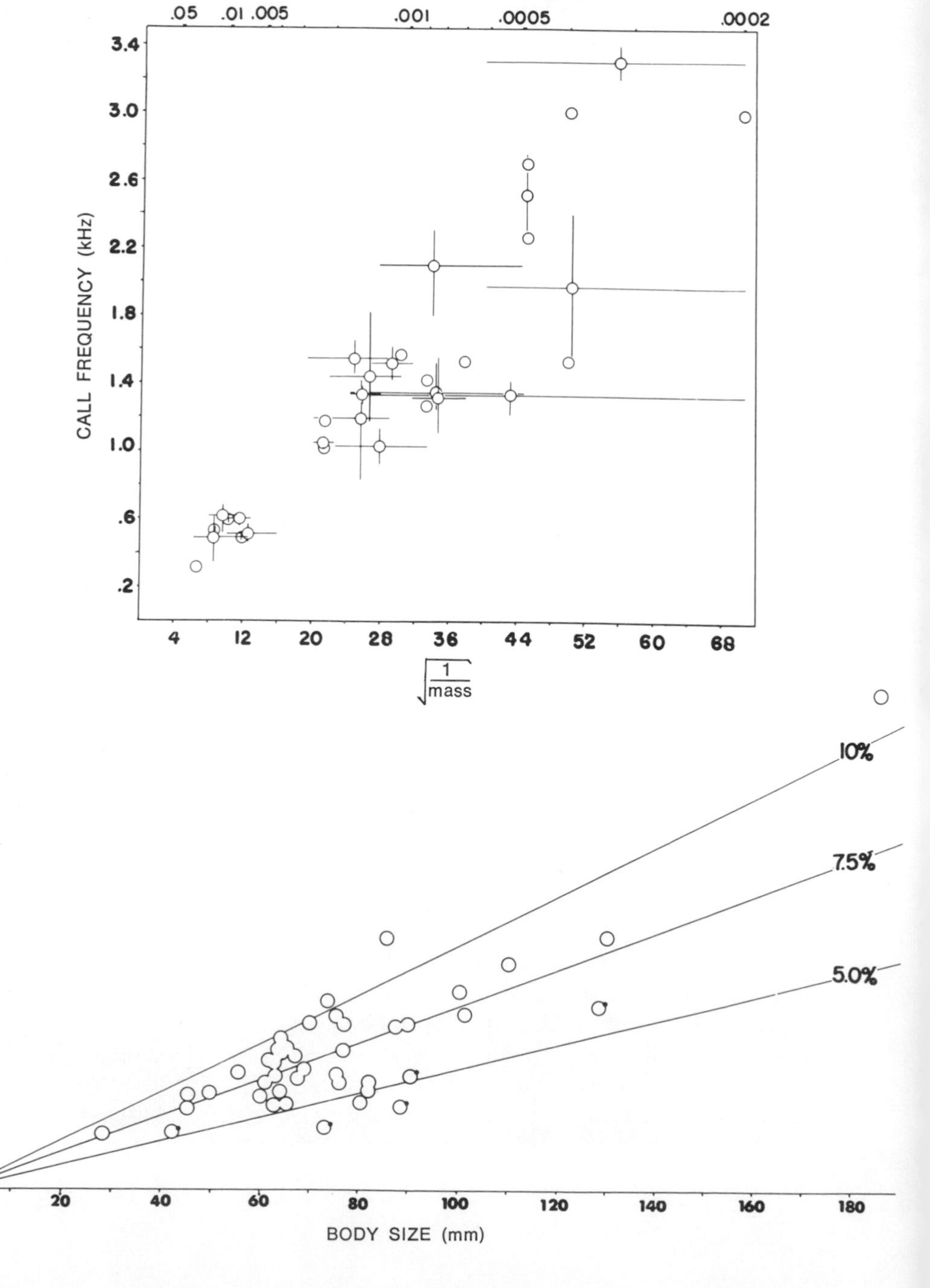

Birds / As a class birds are strikingly similar to primates in several aspects of their communication systems. Both groups are largely diurnal and rely heavily on visual and auditory modes of communication. And many birds share with man the ability to use sounds they have learned during early development. Both groups use complex sounds that are often both frequency- and amplitude-modulated. A counterpart of bird song is even present in some primates, such as the gibbons (genus *Hylobates*) (Ellefson 1968; Marshall et al. 1972); both birds and gibbons use song in territorial advertisement.

Birds differ markedly from most animals by their ability to produce vocally at least two separate frequencies simultaneously (Thorpe 1961:112; Greenewalt 1968). Some songbirds (Passeres) are apparently able to do this by virtue of having two oscillating membranes in the syrinx, one on each side (Warner 1972 and earlier authors). In the chaffinch (*Fringilla coelebs*), cutting the nerve to either side causes the loss of certain parts of the song; apparently one side dominates, but each side participates by producing its own sounds during a song (Nottebohm 1971, 1972).

Mammals / To most persons the vocalizations of mammals, with the exception of pets and domestic animals, are not as well known as those of birds. This is partly because of the greater reliance of temperate-zone mammals on olfactory means of communication, and partly because a fair number of mammals vocalize at frequencies in the ultrasonic range, to which man is insensitive. In tropical forests, however, the cries of arboreal monkeys are a familiar sound. Many bats and rodents vocalize in the ultrasonic range; and these sounds are used in communication in various ways (Gould 1970; Sales 1972*a*, *b*; Okon 1972). The sounds of mammals, which have been reviewed by Tembrock (1963), are generally less patterned and complex than those of birds, although exceptions occur.

BEHAVIORAL EFFECTS OF SOUND PRODUCTION: CONTEXTS AS CLUES

The first step in analyzing the role of any type of social behavior, such as sound production, in the lives of individuals is to determine the social situation in which it occurs. This is best done by observing animals in natural, nonexperimental environments. During such observations one may record (1) the antecedent behavior of both callers and hearers, and (2) their subsequent behavior. These data make possible the framing of hypotheses concerning the types of physiological, behavioral, and environmental factors that "cause" or "motivate" the signal and response. And they enable hypotheses about the effects of the behavior on others. In some cases such observations are sufficient for simple conclusions. In this section, this type of observation will be used to provide an overview of the contexts in which sounds are produced and their effects on hearers in selected animal groups.

Crickets/Three principal contexts have been identified for the stridulation sounds of most cricket species that have been studied (Alexander 1962). The sounds produced in these contexts by *Gryllus veletis* are shown in Figure 16–6, together with some others. Female crickets do not stridulate, and these sounds are produced only by males.

In the first context (top of figure), a male is typically *alone* and the sound produced is termed *calling*. Males tend to avoid other calling males when establishing their calling sites. Females ready for copulation tend to approach calling males of their own species. The second context is that of *two males together*, and the sound produced is termed an *aggressive sound*. In *G. veletis* the chirps in this context are longer than in the solo context. The third context is characterized by the *presence of a female*. The sound produced in this context is softer and less patterned; it is called a *courtship sound*. In some species distinctive sounds are associated with other more specialized contexts, such as interruption of courtship, after copulation, and encounters in burrows that are neither overtly aggressive or sexual.

Sometimes comparative study of the contexts of a behavior results in some plausible ideas about its evolutionary history. Alexander (1962) has argued that it is most likely that calling in crickets originated in the context of close-range courtship, primarily because lifting and fluttering of the forewings, with or without sounds, is a widespread response to females by male crickets, related tettigoniids, and cockroaches. The behavior is frequently known to be associated with the exposure of dorsal glands and the presumed release of chemical stimuli.

From this context it is a logical hypothesis that the addition of sound to the male's courtship allowed him to attract females from a greater distance, which in turn provided the principal selection pressure for specialization of a loud, distinct *calling song*. The physiological factors that influence the production of the courtship sounds are similar to those for calling sounds, and the sounds made in the two contexts differ only in degree in some species. However, the stimulus complexes leading to the two kinds of sounds are different. In courtship the female is present; in calling she is not. Building upon these differences in the sensory and motor neural pathways and mechanisms,

16–6 Sonic repertoire of the cricket *Gryllus veletis.* The principal sounds and their transitions are shown as sound spectrographs (frequency × time), together with silhouettes showing wing position. Antennal stimulation, which occurs during confrontation with rival males, leads to the sounds associated with fighting or transition 1. Stimulation of the anal cerci of the male, as happens during courtship, is associated with the sounds of courting and transition 2. The central nervous system is shown, with arrows showing postulated sites of origin of the pulse and chirp rhythms. The solid rectangle delimits the ganglia needed for calling and fighting sounds. The dotted rectangle surrounds the ganglia necessary for courtship sounds. (From Alexander 1968.)

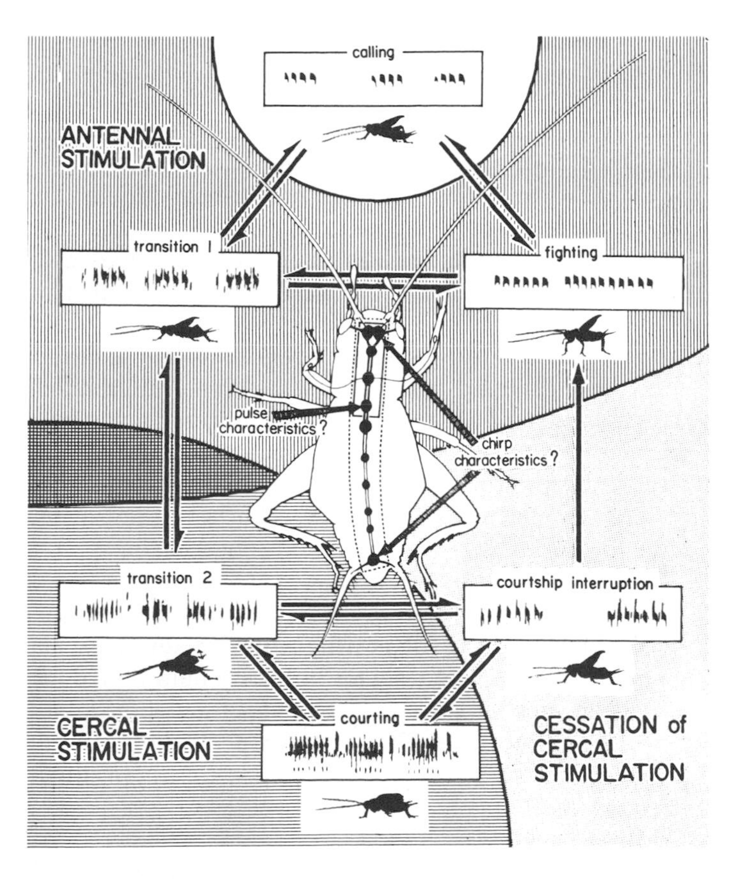

selection could have favored a *branching process* in which a cricket population with separate sounds for solitary calling and courtship could have evolved from a population lacking this distinction. To complete this hypothesis, it would also be necessary to explain why the physical structures of courtship sounds are more effective at close range than are those of calling sounds, and to demonstrate the differences in effectiveness by experiment. Since some species have calling sounds but lack courtship sounds, a process of *substitution*, rather than branching, may also have occurred in different phyletic lines.

By similar reasoning, additional hypotheses for the enlargement of sound repertoire size through evolution can be constructed. Alexander has argued that the following additional branching processes or transformations have occurred: aggressive sounds from calling sounds, post-copulatory sounds from calling or courtship sounds, and "recognition" sounds from courtship sounds. The evidence on which these hypotheses are based may be found in Alexander's (1962) paper.

The basic elements of *evolutionary* increase in repertoire size through branching processes probably apply to birds, primates, and all other animals. They may be summarized as follows:

1) The original stimulus situations must differ, and the pattern of sensory activation of the CNS must differ accordingly.

2) The original signaling behavior must show at least slight differences in different stimulus situations, such as in intensity, completeness, regularity, or other subtle characteristics; and the pattern of motor activation of effector structures by the CNS must differ accordingly.

3) Different selection pressures must be identified for each branch that would favor its divergence from the other. The difference between diverging signals must be appropriate to the different contexts.

The above process bears a superficial resemblance to the process of disruptive selection (Mather 1955), since the extremes in a continuous distribution are selected simultaneously. The important difference is that selection for branching is for both extremes in the same individual rather than in different individuals. A necessary but completely hypothetical part of the process is selection for a *switching mechanism* in the CNS that can associate the appropriate response with a certain stimulus complex. There seems but scant neural evidence about the nature of such mechanisms, but logically they should be present. Localization of function in the CNS with respect to the production of different signals seems to be involved in the switching in invertebrates and vertebrates.

Fishes / Sound production is now known to be widespread in fishes, although most species are silent. The contexts in which sound production occurs in fishes have been summarized by Moulton (1963). These include (1) attraction and/or stimulation between sexes, (2) defense against enemies, (3) intimidation or threat, and (4) schooling. The sounds produced by the noisier species tend to be of a low-pitched grunt-like or knocking quality, and are often repetitive. This is consistent with the hearing abilities of fishes, which tend to be best below 500 *Hz* (Tavolga and Wodinsky 1963). For reviews and further details on various aspects of sonic communication in fishes, see Tavolga (1960, 1964, 1967, 1968 and Winn and Olla (1972).

Amphibians / Vocalization by frogs is well known and has been often studied (see reviews by Bogert 1960 and Blair 1968). Many frogs and toads have highly specialized mechanisms for producing loud songs or calls at their mating stations. Mating often occurs in water, and the cumulative effect of many males calling at the peak of the mating season can be a chorus that can be heard at a great distance. These calls attract females of the same species (see the material in Chapter 18 on isolating mechanisms in frogs) and probably also males.

Frogs and toads also have other calls. Table 16–2 illustrates the repertoire of the northern bullfrog (*Rana catesbeiana*). Besides the mating call (whose function in this species is not entirely clear), bullfrogs have calls associated with territorial spacing (warning and territorial calls 1–3), release from clasping in copulation attempts, and predator evasion (distress calls, presumably). The various calls and their social contexts have been studied by Wiewandt (1969) and Emlen (1968).

Among salamanders only *Dicamptodon ensatus* has vocal cords, and only a few species in this and other genera can make even faint squeaks (Bogert 1960). These sounds are apparently made only in the context of predator evasion.

Reptiles / Sounds of potential use in communication that are made by reptiles have been reviewed by Bogert (1960) and Blair (1968). Alligators and crocodiles produce loud roars and bellows in the context of aggressive spacing among reproductive males; the sounds may also aid in attracting females. Various tortoises produce sounds in courtship. A few geckos regularly make sounds, sometimes during an antipredator display and sometimes in other contexts not yet well known. Rattlesnakes (*Crotalus* spp.) tend to shake the rattle in their tails when threatened by people and other sources of danger.

Birds / Of all animals that use auditory communication, birds are by far the most varied, best known, and most studied. In the study of bird vocalizations a distinction has traditionally been made between *songs* and *calls*. Physically, songs tend to be longer and more complex than calls, especially within one species. A call of one species can, however, be longer and more complex than the song of another. Songs are typically produced by males, but also by females in some species. Most calls, in contrast, are not sex-limited. Songs, unlike most calls, are characteristic of the breeding season and are under the control of testosterone (in males, at least). Some species, such as members of the

TABLE 16–2 Characteristics of all known calls in the vocal repertoire of the bullfrog. (From Frishkopf et al. 1968. Reprinted with the permission of the IEEE from the Proc. IEEE.)

	Mating	Territorial 1	Territorial 2	Territorial 3	Release	Warning	Distress
Made by male	yes	yes	no	yes	yes	yes	yes
Made by female	no	no	yes	yes	yes	yes	yes
No. of croaks in call	3–15	1	1	1	~10	1	1
Croak duration (sec)	0.6–1.5	0.4–0.6	1.4–1.8	3–6	0.5–1.0	0.1	1–5
Croak spacing (sec)	0.5–1.0	—	—	—	1.0	—	—
Croak rise time (sec)	0.2–0.3	0.1	0.4–0.6	0.2	0.2–0.3	0.005	?
Croak fall time (sec)	0.2–0.3	0.1	0.04–0.06	1.0	0.2–0.3	0.04	?
Pulse rate within croak (pulses per sec)	90–110	100	25–30, rising to 90	8–20	60–85	not repetitive	?
Location of spectral peaks (*Hz*)	(1) 200–700 (2) 1500	(1) 200–300 (2) 500–800 (3) 1000–1500	(1) 300–500 (2) 1000	300–500	decreasing from 200–1500	broad peak below 1500	above 700–1000

NOTE: Except for the mating call, the parameter values shown are based on a rather small sample of calls of each type and may therefore not be representative. Considerable data are available for the mating call, and the values given are typical. For more detailed information see Capranica (1968).

Estrildidae and Corvidae, have a relatively soft song that is used primarily in courtship. In species that defend all-purpose territories, the song is typically loud and conspicuous and is given often by unpaired males when alone, and in response to rival males and to females. Thus, song has a dual function of repelling males and attracting females in some species. A phylogenetic distinction involving song has also been made. The suborder Passeres (or Oscines) of the order Passeriformes are commonly referred to as songbirds because of their highly specialized syrinx and their conspicuous vocal abilities; however, birds of other orders have vocalizations that, by the usual physical and functional criteria, also merit the designation of *song*.

The term *song* carries similar connotations when applied to certain vocalizations of frogs and toads, and has been applied also to whales (Payne and McVay 1971) and insects. Although difficulties in classifying individual cases arise, as is proper and to be expected, the distinction between songs and calls is extremely useful in practise for most passerines (Passeres) and many nonpasserines.

The social contexts and functions of avian sounds have been surveyed by Thielcke (1970), from whose account the following treatment is largely drawn. Songs are used in territorial advertisement and defense and in courtship of females, but they are also thought to aid in stimulating and synchronizing reproduction (Lehrman and Friedman 1969). Duet songs, in particular, help to keep the members of a pair in contact and to synchronize their hourly behavior patterns (some duet patterns are shown in Figure 28–2). A species may use one song type most frequently toward males as a threat and another type most frequently toward females as courtship (Morse 1970).

The number of types of calls given by a species varies from 5 or 10 up to 23 (Bremond 1963), depending in part on how finely one subdivides them. One of the commonest categories is *contact* calls. These are given between the members of a pair, family, brood or flock. Besides helping to keep the social unit together, they may have such functions as synchronizing the hatching of eggs in a clutch (Vince 1969), speeding development of the sensory processes of embryos and hatchlings (Gottlieb 1971), and enabling recognition of parents by young and of young by parents (Beer 1970*a*, *b*). The contact calls of freshly hatched turkey chicks prevent aggressive behavior by adults toward the chicks, which may result in their death (Schleidt et al. 1960). Certain calls appear to promote pre-resting and pre-sleeping behavior, especially in species in which individuals roost closely or in contact. Other calls have come to be known as food calls because of their effect (perhaps secondary?) in attracting other individuals to food (e.g., Frings et al. 1955; Stokes and Williams 1972).

Calls given in aggressive contexts are particularly common in species that defend all-purpose territories, but such calls are found in nearly all species. Different intensities and qualities of aggressiveness may be associated with different calls.

Some calls are given between members of a pair in contexts suggesting that their function is mainly to synchronize and coordinate the behavior of the pair, as well as to keep it together. In the context of parental feeding of young, calls are often given by both parties, presumably to alert the young to the presence of the parent and to stimulate feeding of the young by the parent.

Various calls are stimulated by the presence of danger in the form of a predator. In hole-nesting species, which cannot retreat if a predator appears in the nest hole while the owners are in the nest, the calls are thought to intimidate by their resemblance to the hissing of snakes. Ground-nesting birds attempt to lead their young away from danger with soft calls if the predator is relatively distant. But if he surprises them, the chicks scatter and freeze in reaction to the mother's calls. In species in which the young remain for a week or more in the nest, parental alarm calls cause the nestlings to shrink into the nest. Physical attacks on predators are rare in most species, but can be stimulated in some cases by the calls of a nestling or adult that has been caught by a predator. These distress calls can be used to rid a roost of starlings or to clear an airport runway of oystercatchers (Murton and Wright 1968; Pearson et al. 1967).

Although surveys of the type just discussed are useful, they tend to oversimplify the role of sonic signals in avian communication. Because man tends to compare animal communication with human communication, he is prone to pick a particular meaning or message to associate with each discrete sound in the repertoire of a given species. In actuality, however, there is usually no justification for such conclusions. Since birds have no dictionaries, the observer has only a few sets of objective observations with which to evaluate the communication. These include the category of the caller (e.g., age, sex, group member, outsider), the context in which the call (or other signal) is given (e.g., in or out of territory, to dominant or subordinate individual, to male or female, in fighting or courtship), and the response of the hearer. The response is usually not a simple function of the sound because the hearer's response depends on its own context (e.g., whether the hearer is in or out of its territory, whether it is paired or unpaired, whether it is during the breeding season or not). Some idea of the variability in the relationship between call type and context in a field situation can be obtained from Table 16–3. It can be seen that the same call occurs in a variety of contexts and that in any one context a variety of call types may be given.

TABLE 16–3 Contexts in which Steller's jays were observed giving various calls. Aggressive sidling is a high-intensity threat display. (From Brown 1964*b*. Originally published by the University of California Press; reprinted by permission of the Regents of the University of California.)

	Number of observations						
Context	**Rattle**	**Musical**	**Growl**	***Wah***	***Too-leet***	***Shook***	**Total**
By supplanter	25	29	4	8	5	16	87
By supplantee	2	—	—	—	—	—	2
At individual but no supplanting	12	19	5	5	1	22	64
Aggressive sidling	—	2	6	2	7	32	49
In fight	—	—	—	1	—	2	3
After fight	1	—	—	1	—	—	2
Mobbing	1	—	—	31	—	—	32
Just alighted	—	—	—	26	—	2	28
On picnic table	—	—	—	28	—	—	28
Appeasement	—	—	—	1	—	—	1
Answering at a distance	—	—	—	—	—	2	2
Courtship	8	2	—	—	—	—	10
Totals	*49*	*52*	*15*	*103*	*13*	*76*	*308*
Context unspecified	78	88	2	131	17	77	393

Mammals / The contexts in which the sounds of mammals are made have been divided by Tembrock (1963) into four categories: (1) pairing, (2) rearing young, (3) group-related (e.g., contact, alarm), and (4) protection. Reviews of the contexts of sounds made by a variety of mammalian species are available in Sebeok (1968). The contexts of sounds made by a single species that has been intensively studied, the vervet monkey (*Cercopithecus aethiops*), are summarized in Table 16–4, and illustrate some of the relationships between sounds and their social context (Struhsaker 1967). These contexts may be summarized as follows:

Agonistic
- threat
- solicitation of aid
- prevention of copulation
- appeasement

Alarm
- snake near
- major predator near
- minor mammalian predator near
- observer near
- eagle perceived
- sudden movement by predator

Infant Sounds
- weaning
- lost
- reunion with mother
- strange adult male near

Neutral Proximity

Progression

Play

Respiration

Indigestion

TABLE 16–4 Contexts of vocalizations and other sounds in vervet monkeys, *Cercopithecus aethiops*. Corresponding call, "message," and reactions are given for adult male only. Females respond differently in the same contexts, especially to infants. (After Struhsaker 1967.)

Context or stimulus situation	Call of adult male	"Message" and reaction
AGONISTIC		
Intragroup		
1. Intragroup agonism: threats, chases, combat	bark	aggressive threat; disrupts fight
2. Red-white-blue display by dominant to subordinate caller	woof-woof; waa; woof-waa; lip-smack; teeth chatter	expresses subordination; inhibits attack
3. Close proximity of subordinate and dominant	teeth chatter	expresses nonaggression; permits proximity
4. Anti-copulatory	not applicable	
Intergroup		
5. Proximity of foreign group	intergroup grunt	evokes same
6. Intergroup agonism	intergroup chutter B; bark	intergroup fight or chase; may evoke aid
PREDATOR ALARM		
7. Proximity of minor mammalian predator	*uh!*	warn; look
8. Proximity of human observer	chutter to observer	threat
9. Proximity of major predator	threat-alarm-bark	warn; take cover
10. Proximity of snake predator	none	
11. Sudden movement of minor predator	*nyow!*	warn; take cover
12. Initial perception of major avian predator	none	
13. Group progression, initial phase	progression grunt	none described
14. Play	not applicable	none described
PARENTAL AND INFANT		
15. Approach of strange male to infant 1	none	none described
16. Weaning of infant	none	none described
17. Separation of infant and mother	none	none described
18. Reunion of infant and mother	none	none described
19. Ambivalent situation	none	none described
20. Respiration interference	cough; sneeze	obvious
21. Indigestion	vomit	obvious

This listing reflects the relative importance of agonistic and alarm calls in adult vervets, as well as the unimportance of auditory sexual signals. Vocalization appears to be quite important for infants too.

A role of vocalization rather rare among mammalian species studied is one that corresponds to the songs of birds. Some mammals do employ vocalizations in this way. Humpback whales (*Megaptera novaeangliae*) have been shown to produce a series of varied sounds that may last 7 to 30 minutes and be repeated with considerable precision (Payne and McVay 1971). These song sessions may last for hours. Experimentation with whale songs seems impractical at present, and their functions are matters for speculation. Their formal similarities with bird song suggest a role in mating.

Another instance of similarity with bird song has been described in the gibbons of Thailand (Marshall et al. 1972). The calls shown in Figure 16–7 are given in the jungle daily by all pairs, especially at dawn. Judging by their context and the territorial behavior of gibbons, these calls probably help in territorial advertisement. Another important point of similarity with birds is the rather striking extent of the differences between species in the pattern of the calls.

16–7 Sound spectrograms of gibbon sounds used in territorial proclamation. The vocal patterns are similar within a species and differ among different species. Notice the rapid changes in pitch and the relatively pure tones (absence of energy above and below the principal band, except natural harmonics). Gibbon songs resemble bird songs in these respects. These loud, conspicuous sounds are characteristic of the tropical forest of southeast Asia and of zoos throughout the world. Both the male and female of a pair contribute alternately to the performance. In the spectrogram of *H. lar* (female) the vertical note (*fourth from left*) is the last whoop of the male. (From Marshall et al. 1972. Reproduced by permission of American Society of Mammalogists.)

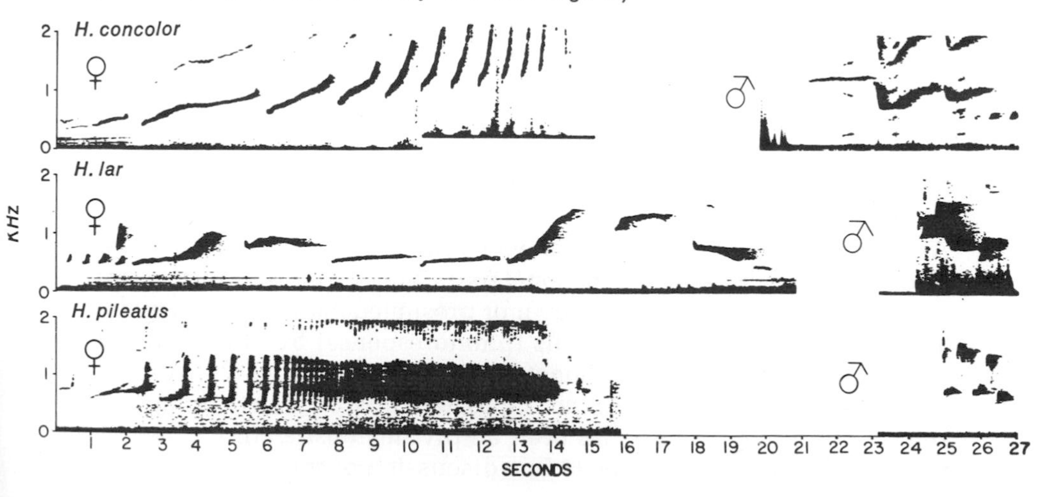

EXPERIMENTAL AND ANALYTICAL STUDIES OF AUDITORY COMMUNICATION

Insects / Many species of insects respond to an imitation of their call by answering (*phonoresponse*) and/or by approaching (*phonotaxis*). Examples of such responses have been reviewed by Dumortier (1963). Since insects discriminate between different audible frequencies poorly if at all, some surprising examples of effective artificial stimuli have been found. In Kansas, females of the cicada *Tibicen dorsata* were observed to sit in large numbers on a tractor whose motor produced sound of the appropriate temporal pattern. Hand clapping, various squeaky noises made by the human mouth, and rubbing the teeth of a comb are all effective means of evoking a phonoresponse or phonotaxis in various species if the timing is correct. The frequency of artificial insect calls must be within the range audible to the insect, but this range can be quite narrow and specific, as in mosquitoes and cicadas (Simmons et al. 1971). It may also be rather broad and include the ultrasonic range, as in certain grasshoppers (Howse et al. 1971). Most authors have emphasized the importance of temporal patterns of amplitude in insects' response to artificial or natural calls (but see Howse et al. 1971). These are thought to be important in species recognition, as discussed in Chapter 18.

The extent of perturbation of the natural call that females of the grasshopper *Chorthippus biguttulus* will tolerate was studied experimentally using pulses of white noise (3–40 *kHz*) by Helversen (1972). The natural call, shown in Figures 16–8 and 16–9, consists of about 3 notes, each of which contains 20 to 60 compound pulses. A compound pulse (called a *main pulse* by Helversen) is a group of 6 or 8 simple pulses. Response of virgin females to male song is a function of the duration of compound pulses and the intervals between them (Figure 16–8). The females responded best to artificial songs resembling the natural ones. More important for species recognition were the durations of notes (Figure 16–9), since the songs of three closely related species differ conspicuously in this parameter (see also Figure 18–8).

Amphibians / Frogs and toads are well suited for experiments to determine the behavior-effective parameters of vocal stimuli. Most experiments so far have dealt with known or presumed mating calls of males. These may cause answering calls (phonoresponse) by other males or orientation and approach (phonotaxis) by females. The ability of females in a variety of species to approach calls of their own species selectively in choice tests has been amply demonstrated. Further details are provided in Chapter 18 in a discussion of species recognition.

16–8 Song of the grasshopper, *Chorthippus biguttulus,* and responses by females to experimental reproductions of it. A typical song consists of three notes (as in **A**) divided into compound pulses and simple pulses. The song is normally produced with both legs (as in **B**), but simple pulses can be seen more easily in a male with only one sound-producing leg (**C**). The songs are depicted as graphs of amplitude vs. time. Experimental variation of the duration of compound pulses and the intervals between them revealed that the optimal durations were interdependent, as seen in the three-dimensional graph (**D**). The graph shows the strength of response of one virgin female at various combinations of stimulus parameters. (From Helversen 1972.)

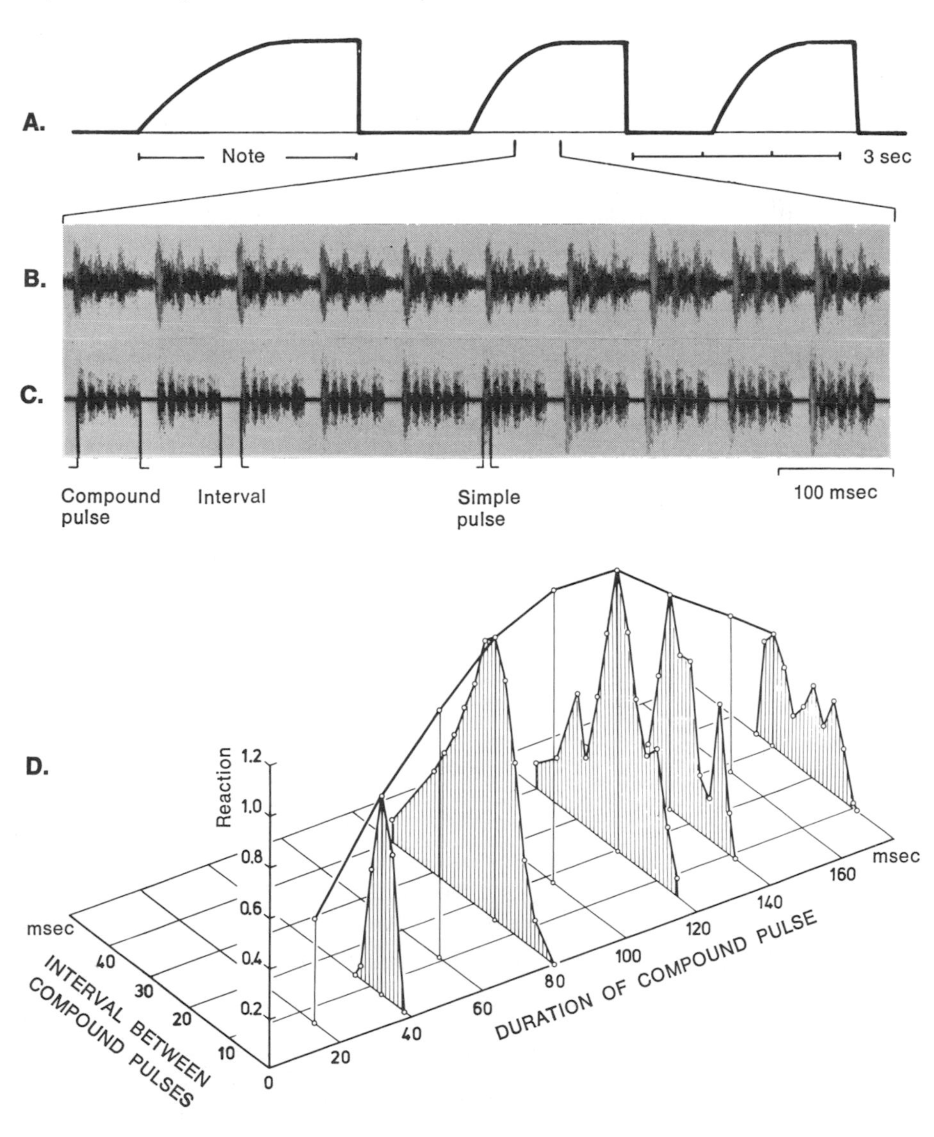

16–9 Oscillograms of songs of the grasshopper *Chorthippus biguttulus* and two closely related species living in the same regions (**A, B, C**), and experimental determination of one critical parameter for species recognition in *biguttulus* (**D**). **D.** Calling reactions of three (▲ ○ ●) virgin *biguttulus* females to artificial songs with various note lengths. The range of variation in normal songs is demarcated by the large black triangles at the bottom. (From Helversen 1972.)

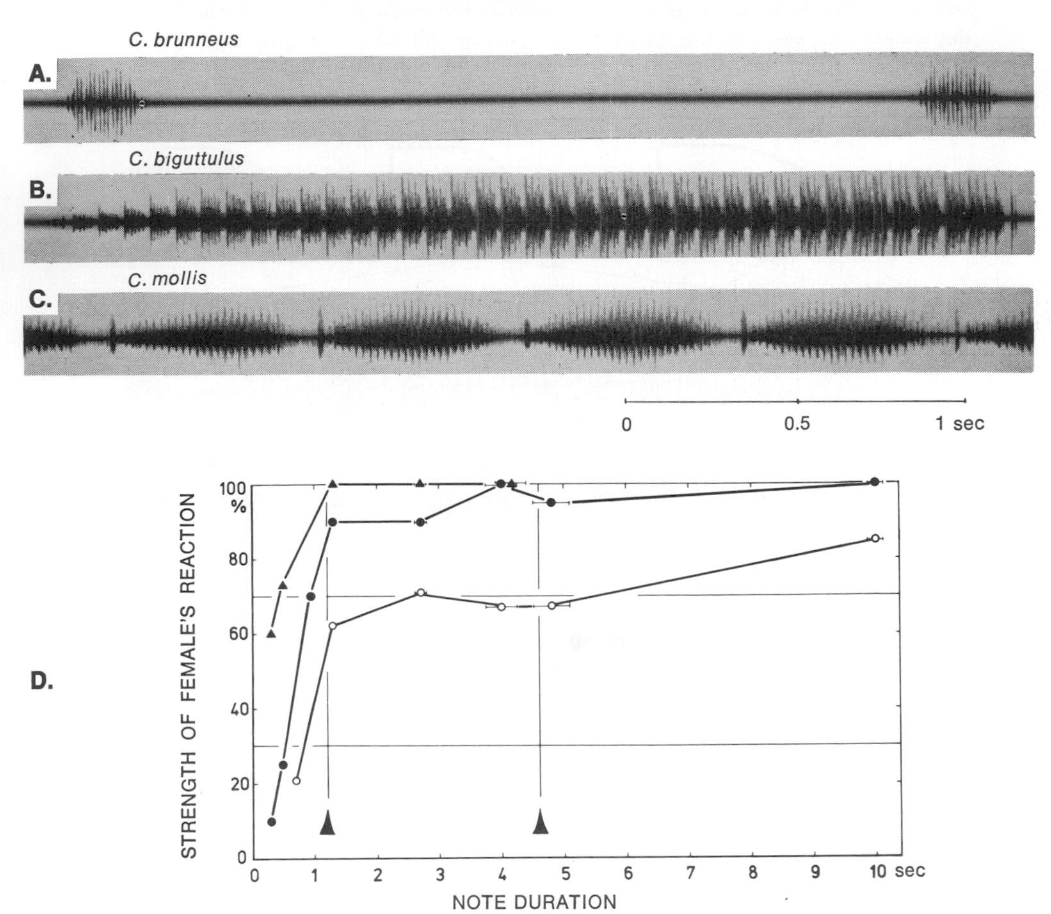

An excellent study of the features of the bullfrog's (*Rana catesbeiana*) call that are needed to evoke phonoresponse has been made by Capranica (1965), and is discussed in Chapter 22 on recognition mechanisms. Capranica's experiments demonstrated the importance of the spectral composition of the calls.

Bird Songs / The songs of most species of birds in a given region are almost invariably distinctly different. Two species do not share the same song as a rule, although exceptions sometimes occur—for instance, occasional individual eastern and western meadowlarks (*Sturnella magna* and *S. neglecta*) that learn songs of both species (Lanyon 1957). Although it is widely assumed that songbirds recognize their own species by their songs, this was not confirmed until experiments were done using tape-recorded songs played back to birds in natural situations. Many such studies have been made, and in nearly all cases the birds respond more strongly to the songs of their own species than to songs of other species. Some examples are described in Chapter 18 in the discussion of behavioral isolating mechanisms.

Other types of information carried by songs can also be demonstrated experimentally. In their study of the ovenbird (*Seiurus aurocapillus*), Weeden and Falls (1959) discovered that responses to tape-recorded songs played to territorial males were quicker and stronger if the songs had been recorded from males who were not adjacent neighbors. One might interpret this result as meaning that the male recognized the song as not being one from a bordering neighbor. Since territorial boundaries with these well-known males had probably stabilized, these individuals should pose less of a threat than a stranger seeking to insert himself into the space of the owner. By this reasoning, an unfamiliar intruder should provoke a stronger response than a familiar neighbor. In the indigo bunting (*Passerina cyanea*), Emlen (1971) also found stronger responses to songs of distant neighbors than to songs of bordering neighbors. In white-throated sparrows (*Zonotrichia albicollis*), Brooks (in Falls 1969) obtained similar results if the playback was done on the side of the territory bordering the neighbor being tested (Figure 16–10). But if the neighbor's song was played from the opposite side of the territory, simulating a deep intrusion into the territory, the full response was obtained. This experiment eliminates the hypothesis that the weaker response to a neighbor is due to habituation caused by hearing the neighbor's songs more frequently than those of other individuals. The interpretation favored by most authors is that these data are evidence of individual recognition among the males in a local area, and that the strength of the response is a function of the position as well as the identity of the neighbor. Apparently strangeness per se is not the factor that correlates best with response strength, for Lemon (1967) found that in cardinals (*Cardinalis cardinalis*) responses were progressively weaker to songs recorded: (1) locally, (2) 7 miles away, and (3) more distantly.

It is interesting to ask which characteristics of a song are critical for species recognition. This can be studied by presenting songs that have been artificially altered or synthesized so as to vary a particular parameter. For instance, songs can be made longer or shorter; certain

16–10 Number of songs given by a male white-throated sparrow in response to playback of songs of a stranger (*filled circles*), himself (*triangles*), and a neighbor with an adjacent territory (*open circles*). (From Brooks, in Falls 1969. From R. A. Hinde: *Bird Vocalizations* © 1969 Cambridge University Press. Used by permission.)

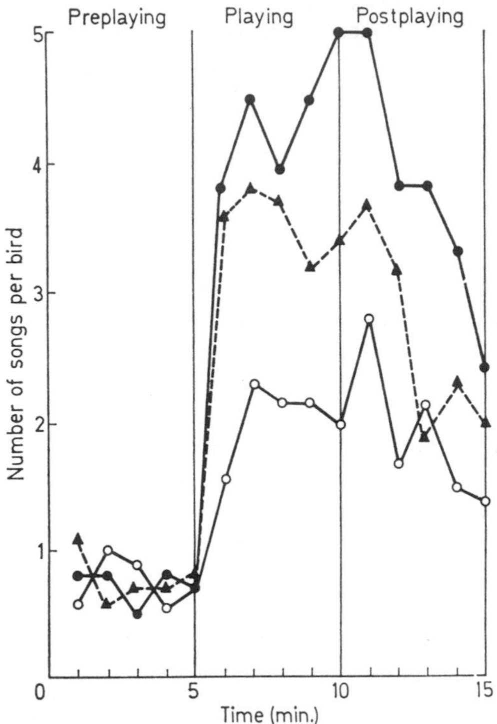

frequency bands can be filtered out; songs can be played backwards; or their components can be assembled in unnatural orders. Studies of this sort have been done with ovenbirds (Falls 1963), white-throated sparrows (Falls 1963, 1969), wood larks (*Lullula arborea;* Tretzel 1965), European robins (Bremond 1967, 1968*a*), European wrens (*Troglodytes troglodytes;* Bremond 1968*b*), *Phylloscopus* warblers (*P. collybita, P. trochilus;* Schubert, G. 1971; Schubert, M. 1971), and indigo buntings (*Passerina cyanea;* Emlen 1972).

The indigo bunting will serve as an example. The operational procedure consists of determining by playback experiments what kinds of alterations of the normal song most reduce the response of the tested male. Responses consist of various combinations of approach, singing, threat display, and calling. To many human hearers, including myself, one of the most conspicuous aspects of indigo bunting song is the fact that most figures or phrases are sung twice in succession before passing to the next, as shown in Figure 16–11. Yet when these phrases are rearranged so as not to be paired, the bird's responses are not reduced. Apparently a feature that is extremely helpful for recognition of indigo bunting song by man is unnecessary for recognition by buntings. This experiment does not eliminate the possibility that the pairing is important; it merely shows that pairing is not the only critical feature of the song. Considerably more effective in reducing response intensity were changes in the intervals between figures, as shown in Figure 16–12. Both shortening and lengthening these intervals were effective

16–11 Indigo bunting songs used to test the effect of pairing of phrases on species recognition by males. The normal song consists of a succession of pairs of figures or phrases. In the experimental (nonpaired) song, the same phrases were used but they were rearranged so as to avoid pairing. Other features of the song remained the same. The males responded equally well to normal (paired) and unpaired song. (From Emlen 1972.)

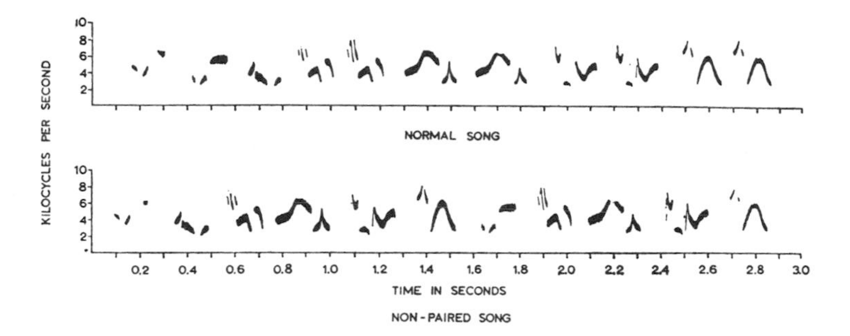

16–12 Indigo bunting songs used to test the effect of varying the inter-figure interval on species recognition by territorial males. (From Emlen 1972.)

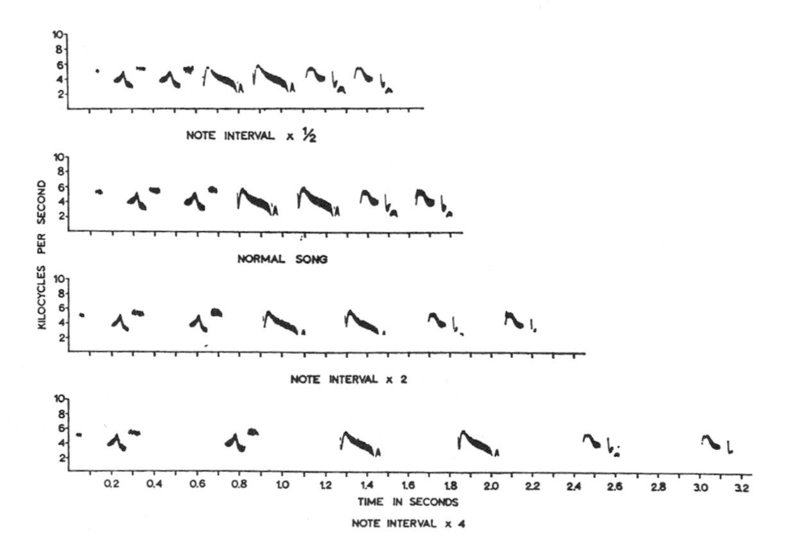

in reducing response score. Response decrements were also obtained using phrases that changed frequency relatively little.

These studies have not dealt with all the possible variables, and they are necessarily incomplete; but they do show that considerable redundancy exists. Some parameters that could in theory be used for species identification (and are used by man) are not crucial for the birds. Some parameters are quite variable in nature and tolerate much variation experimentally, while some are quite invariant and sensitive to manipulation. Most important, when considering this and similar studies of other species, there is no general rule concerning which aspects of a song are critical for species recognition. Tone quality seems important in white-throated sparrows, syntax or ordering of phrases in European robins, and the rhythm and spacing of phrases in indigo buntings. In this respect birds are proving to be much more complicated than frogs or insects. Similar experiments with mammals have not been done.

Recognition within the Family: Birds / Many species seem to have a need for individual recognition within the family. Swans (Scott 1970) and geese may mate for life, and their young may stay in the family group even for a year after hatching, despite migrations of thousands of miles. In contrast, it is commonly assumed that in smaller birds the members of a pair become separated from each other while on migration. For young Canada geese (*Branta canadensis*), staying with the family on the wintering grounds has advantages, since family groups take precedence at food-gathering sites over isolated individuals (Raveling 1970). For adults too, staying together as a pair seems to be advantageous in some circumstances; Coulson (1966) found that kittiwakes (*Rissa tridactyla*) who reunited with their mates of the preceding season (the majority) had better nesting success than the minority who changed mates. For colonial species whose young tend to mix in large flocks or crèches with young of a similar age, parents who can selectively identify their own young from among hundreds or thousands of others should leave more surviving offspring carrying their genes than parents who are unable to make the required discrimination. In such situations, resembling "the species" too closely may hinder parental recognition and hence survival. It is perhaps not surprising in this light that the chicks of colonial species such as the royal tern (*Thalasseus maximus*) tend to be unusually variable in visual appearance (Buckley and Buckley 1970). Such variability is of advantage to the individual; theories based on "the good of the species" are unnecessary. Auditory cues also seem to be important, since field observers have the impression that parents seek their own young first by ear. Although it was formerly guessed that parents were unable to recognize their own young when mixed with large numbers of other young and that feeding of young was indiscriminate, careful studies of individually marked parents and young have shown that faithfulness to one's own young is the rule. The relative roles of visual and auditory cues in such discriminations are poorly known (Beer 1970*a*). The evidence indicates that recognition of family members has selective value in a variety of species.

Experiments to test the role of auditory cues in individual recognition have made use of differential responsiveness to calls played to the subjects with a tape recorder. In species in which a mated pair sings a duet (as in Figures 16–13 and 28–2), the contribution of the second singer is usually precisely timed to follow closely that of the first singer (Thorpe 1963, 1972). Duets are sung only by members of a pair as a rule; and they are sung particularly when the members of a pair are out of visual contact with each other, which occurs often in the dense vegetation that most duetting species inhabit. When the contribution of

16–13 Antiphonal duets of five pairs of eastern whipbirds. The first two notes (the horizontal and vertical lines) were sung by a male and the remaining notes by a female. The second, or vertical, note of the male is said to resemble the cracking of a whip and is very loud. (From Watson 1969.)

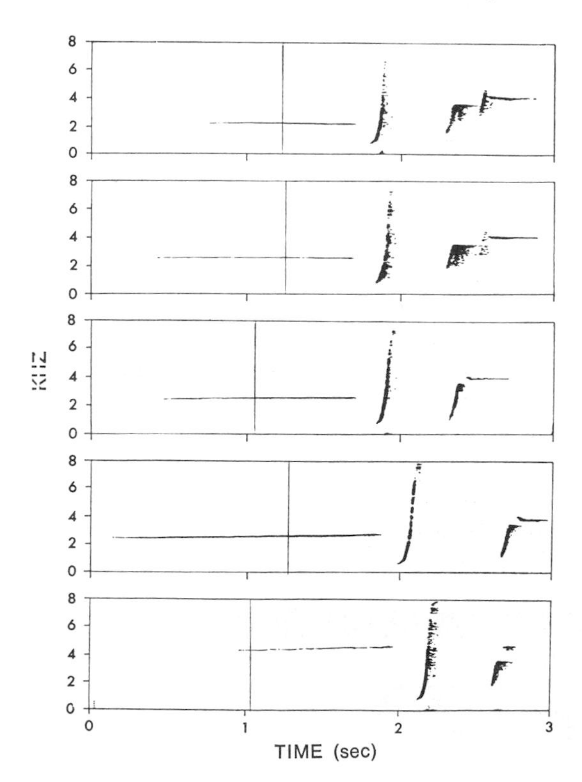

one member of a pair is played, the other member fills in its own part to complete the duet (Hooker and Hooker 1969; Watson 1969). Sample duets of the eastern whipbird (*Psophodes olivaceus*) are shown in Figure 16–13.

The means of recognition have been investigated in some colonial species. The nests of gannets (*Sula bassana*) are closely spaced so that each sitting bird is just beyond pecking distance of its nest-sitting neighbor. There is no neutral ground between nests. A gannet who lands at the wrong nest is viciously attacked by the occupant and must quickly retreat. Mates take turns guarding the nest. By recording calls of known individuals and playing them back later, White (1971) showed that females on their nests responded more strongly to calls of their mates than to calls of adjacent or distant neighbors. The data are shown in Figure 16–14. Responses range from no change in on-going activity such as preening or sleeping, to a strong, sustained reaction consisting of raising the head and looking around. Analysis of the calls used by gannets in this context showed that the amplitude envelope (or profile) (Figure 16–15) had sufficient variability among individuals

16–14 Responses of nesting gannets to played-back calls. Responses to calls of mates were stronger than to calls of neighbors or others ($p = 0.02, 0.03$). Scale of responsiveness: *A*, no response. *B*, perceptible movement, perhaps a response to the playback. *C*, a casual break in activity. *D*, brief head-turning or looking up. *E*, sustained looking up and around, often toward the loudspeaker or the sea. *F*, sustained looking around and increased responsiveness to birds flying overhead. (From White 1971.)

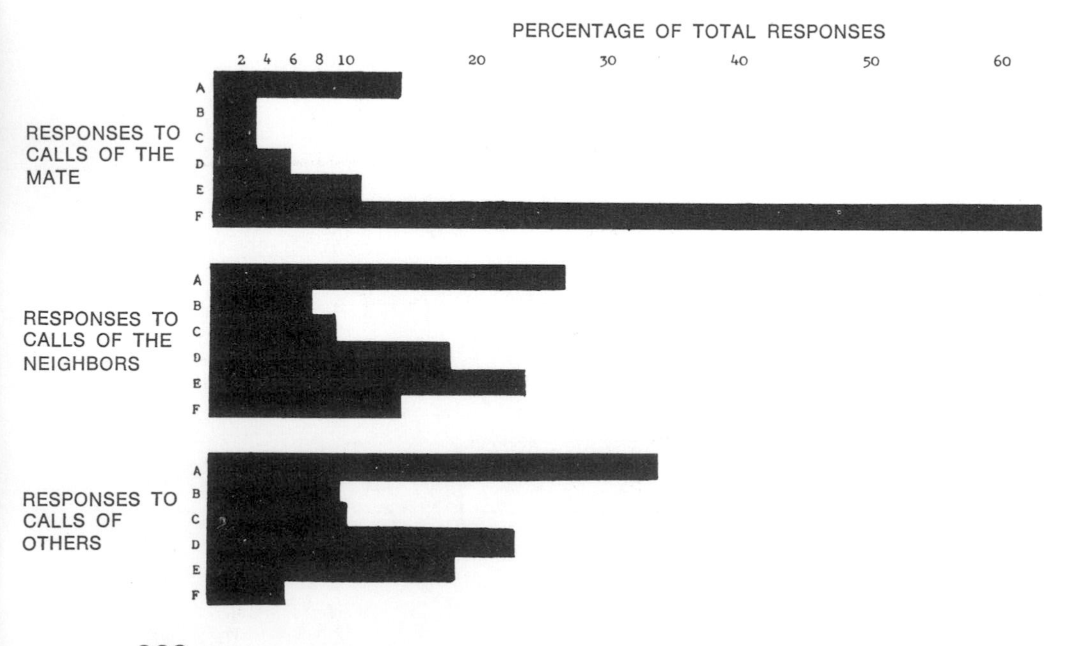

16–15 Sound spectrograms (*above*) and amplitude profiles (*below*) of the calls of two male gannets. (From White and White 1970.)

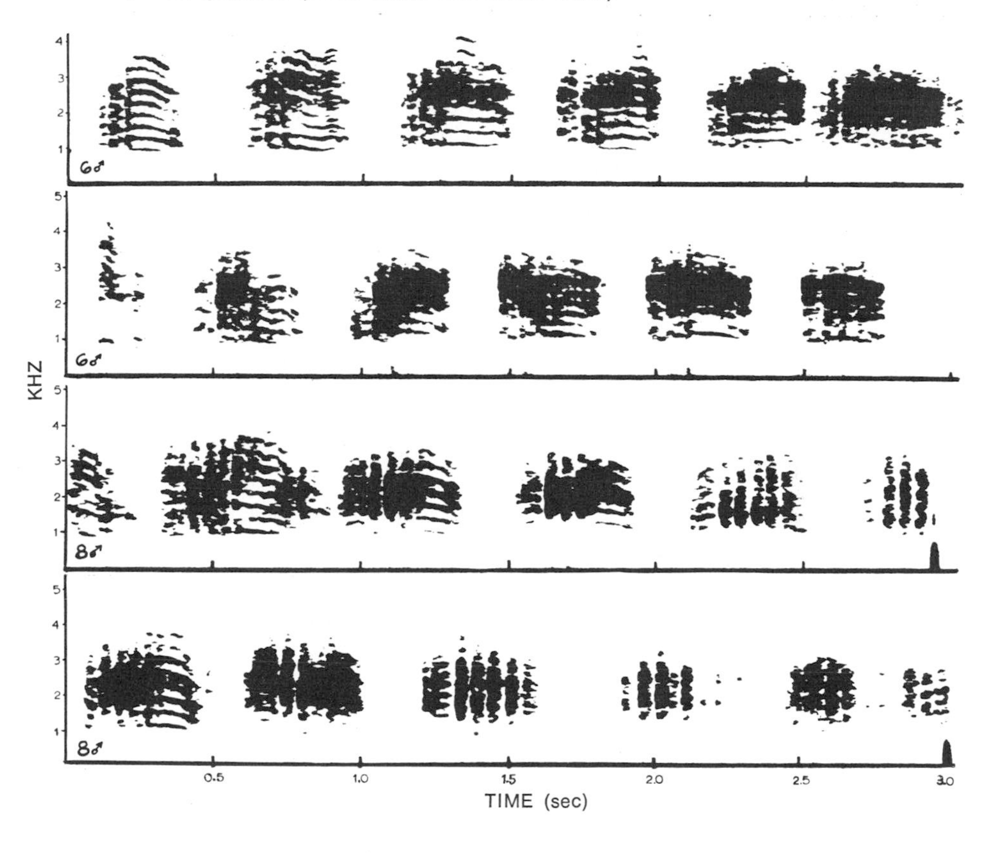

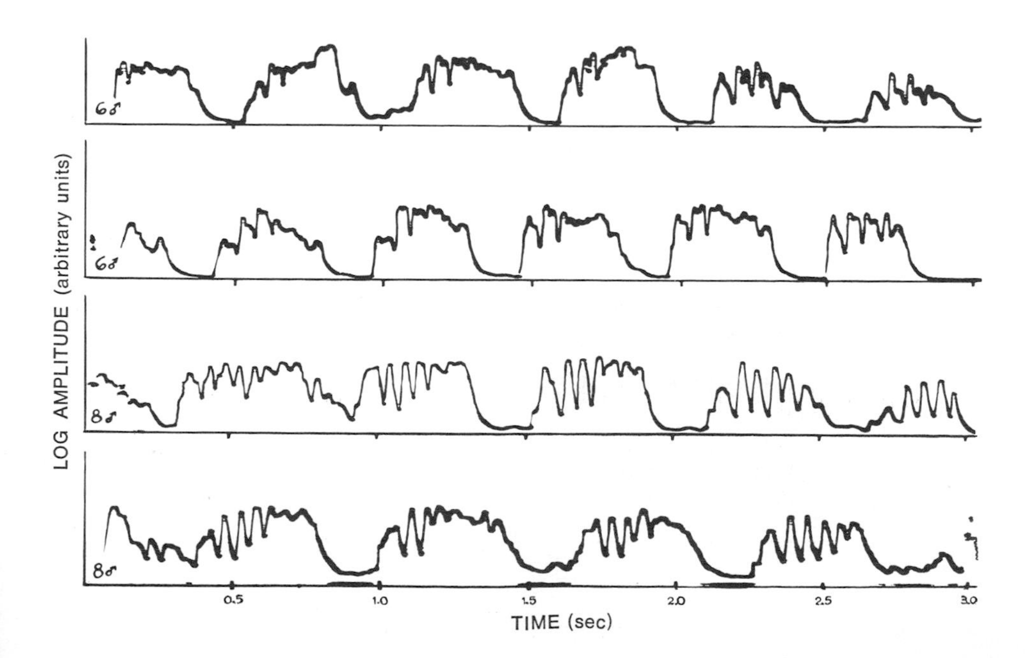

and consistency within individuals to enable discrimination among individuals (White et al. 1970). The feasibility of using the amplitude envelope for individual recognition was confirmed by a computer test. There was sufficient information for computer recognition in the first 0.15 seconds of the 0.50-second call. Further field tests to rule out use of tone quality are needed to confirm this hypothesis.

The young of a colonial species usually must beg from the parents to obtain food. Since a young animal is surrounded by adults who are not its parents, who are unlikely to feed it, and who may even attack it, individuals who beg only their own parents should conserve energy and have a selective advantage over indiscriminating young. That the young of colonial species do in fact discriminate between parents and nonparents has been shown by playback experiments in a variety of species, including the following: common murres (*Uria aalge*, called guillemots in Europe; Tschanz 1965, 1968), black-billed gulls (*Larus bulleri;* Evans 1970*a*), ring-billed gulls (*Larus delawarensis;* Evans 1970*b*), laughing gulls (*Larus atricilla;* Beer 1970*a*, *b*), common terns (*Sterna hirundo;* Stevenson et al. 1970), and adelie penguins (*Pygoscelis adeliae;* Thompson and Emlen 1969).

The use of artificially altered songs and calls has made it possible to inquire more deeply into the features of a sound that are most critical for a given kind of response. There may be more than one type of response to a call or song, and it seems likely that different features of a signal may be critical for different types of response. When variability in songs among individuals is studied, some parameters of the song are found to be more variable than others. For example, in the great-tailed grackle (*Quiscalus mexicanus*) Kok (1971) measured nine parameters of male song and computed means and standard deviations: dividing the standard deviation by the mean, he obtained a number that when multiplied by 100 yields the coefficient of variation. The different parameters had coefficients of variation ranging from 17 to 29. It seems reasonable that the least variable parameters should be most useful in species recognition and that the most variable ones should be most useful for individual recognition, but a clear demonstration of both these hypotheses in the song of a single species has yet to be made.

PROPERTIES OF SOUNDS IN RELATION TO USES

Alarm Calls / Many species of animals have alarm calls, but some that use sound in their social behavior make no use of alarm sounds. Crickets and grasshoppers, for example, seem to lack sounds expressing alarm; on the other hand, some of the social insects do have sonic behaviors that arouse colony members to defense. This contrast is consistent with the idea that alarm signals have evolved by benefiting close relatives, not by "benefiting the species." In a social insect colony it is exclusively colony members or colony genes that are benefited; but in crickets, among the chief beneficiaries would be rival male crickets. By this reasoning we might expect alarm-calling to be confined mainly to occasions when offspring or close relatives would be aided. This is a dificult hypothesis to test for vertebrate species. Alarm-calling resulting in the warning of offspring or mates is nearly universal among birds and mammals. But some calling with a warning effect occurs in situations where the principal beneficiaries are unrelated and even belong to different species (Bergman 1964).

One of the difficulties is specifying what is and is not an alarm call. Some calls that have an alarming effect may be given in various contexts other than the simple proximity of danger. Another complication is that the calling might conceivably benefit the caller directly (Perrins 1968; Trivers 1971) without necessarily aiding relatives.

The problem of the evolution of alarm-calling is one that has attracted considerable attention because of the common assumption that the caller is acting altruistically (J. M. Smith 1965). Yet it remains to be shown for any species by observation or experiment that alarm-calling subjects the caller to greater risk than if he were to remain silent. Field observers have not reported that hawks single out birds giving alarm calls for special attention, and it is doubtful that they normally do, since, with one possible exception—the hawk *Micrastur* (N. Smith 1969)—that is not their method of hunting. Surprise is their chief tactic and it is indeed often prevented by alarm signals.

That the kinds of sounds used in alarm-calling might bear a logical relationship to their function was first suggested by Marler (1955, 1959). He reasoned that sounds which provided relatively few cues about the caller's location would be ideally suited for alarm functions, because the caller would be less likely to be endangered but would convey his message just as efficiently. Surveying the cues that can be used to determine the direction of a source of sound, Marler could predict which type of sounds would be best suited to serve as alarm signals. Such sounds should have the following qualities:

1) They should begin and end gradually rather than abruptly, so as to hinder precise binaural comparisons of the time of start and stop on each side of the head of a predator.

2) They should not be high-pitched, because such sounds cast sharp sound shadows and facilitate binaural comparisons of sound intensity by predators.

3) They should be above the frequency at which the phase of the

sound waves can be usefully compared at the two ears of a predator.

To these can be added two more:

4) They should be narrowly tuned to one frequency band and have no energy in other bands. This property limits the range of frequencies at which the predator can make binaural comparisons, and it hinders analysis of binaural differences in tonal quality.

5) They should not vary in pitch, in the manner of the calls of most echolocating bats (e.g., Figure 23–4), because this would enable more precise comparison of time and loudness cues at the two ears of the predator.

In a sample of British birds, mostly from woodland or forest-edge habitats, Marler (1959) illustrated the differences between alarm calls given in response to a flying hawk and mobbing calls given in response to perched owls. As shown in Figure 16–16, the hawk alarm calls fulfill our five predictions, while the mobbing calls do not. Attempts to repeat this test of the predictions using an unbiased sample of species from another continent or habitat would be worthwhile.

Thorough physiological testing of Marler's hypotheses about the localization of sound sources by bird hawks has not yet been attempted. Konishi (1973*a*, *b*) studied trained, captive barn owls (*Tyto alba*). When pure tones of several seconds in duration were used as cues, the easiest frequencies for the owls to locate were between 6 and 9 *kHz*, just in the range of the passerine alarm calls given in response to hawks (Figure 16–16). Apparently the frequency of the calls is not as critical as their other characteristics. Further testing of the other characteristics is needed.

Advertisement Songs / A singing male bird elicits responses of different kinds from males and females, but both male and female responses benefit the singer more if they are properly oriented. Songs should be easily locatable in addition to indicating the species, sex, and motivation of the singer. As pointed out by Marler (1957, 1959), the physical structure of songs enables both species recognition and location; songs have several to many notes, often modulated over a wide frequency range.

The average frequency of a bird song may also be a function of the vegetation in which it lives. The Fickens (1962, 1963), in a study of warblers in Maine spruce woods and elsewhere, found that species which sing from high in the trees tend to have higher-pitched songs than those which sing close to the ground. Chappuis (1971), in a survey of about 100 African species, found that lower frequencies (<1,500 *Hz*) dominated in the songs of species inhabiting denser habitats. He felt that this could be explained by the greater ability of low frequencies to carry around objects.

16–16 Sound spectrograms of two kinds of alarm calls in various species of British birds. The horizontal type is given in situations of intense and abrupt alarm, as when a hawk flies over. It is characterized by its narrow frequency range (actually narrower than shown because these spectrograms were made with a wide-band filter). The vertical type is given in situations of greater relative security, as when "mobbing" or "scolding" a perched owl or hawk. The horizontal calls resemble a brief high thin whistle; the vertical calls resemble a click, chip, or ratchet sound. (From Marler 1959.)

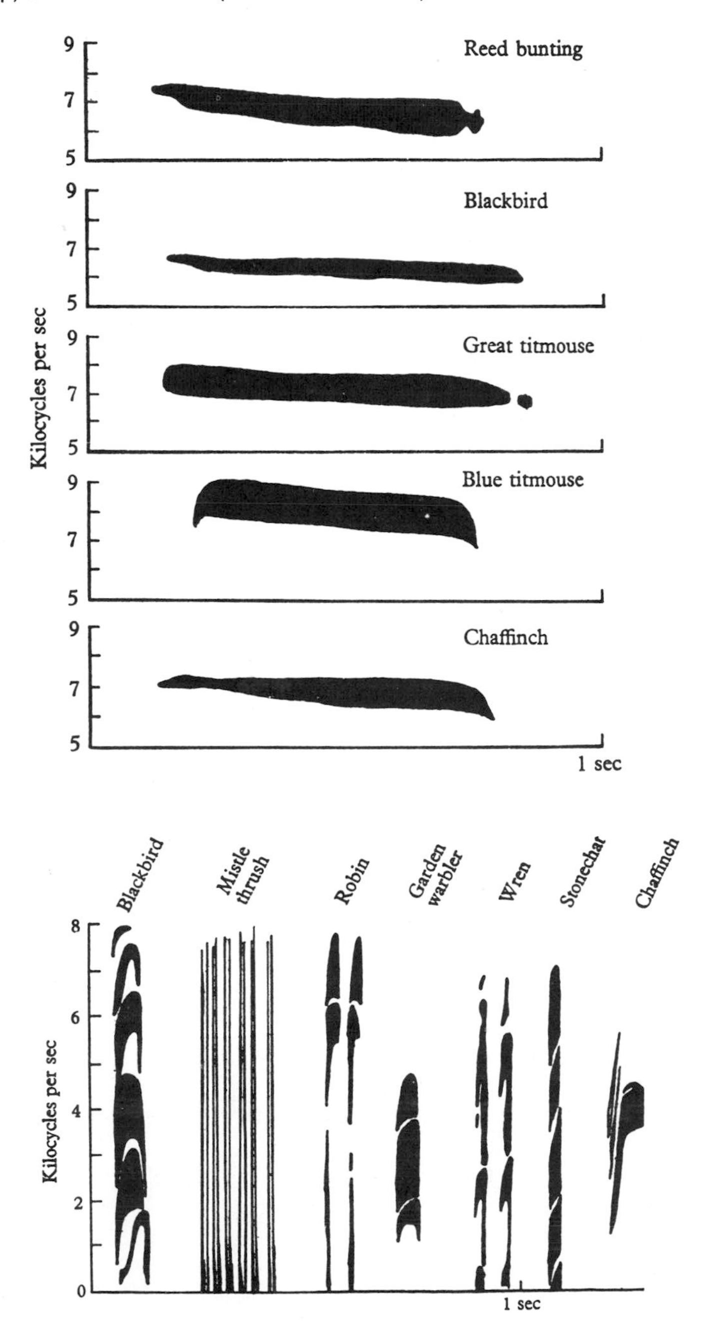

Species Diversity / The evolution of diversity in sounds that are critical for obtaining a mate of the same species is influenced by the means of sound production and the properties of auditory systems. Differences among species that easily produce and discriminate a wide range of frequencies, like birds and mammals, are often to be found in the patterns of ups and downs in frequency during a song. Some species differences in this respect are illustrated in the gibbon calls in Figure 16–7 and the thrush songs in Figure 18–10. That the pattern of highs and lows in frequency can be necessary for species recognition was shown experimentally by Bremond (1967, 1968*a*) in studies of the European robin. Another way in which the songs of species with good frequency discrimination often differ is tonal quality. An excellent discussion of the difficulties of analyzing tonal quality in animal sounds has been written by Marler (1969). Although tonal quality seems to be an important distinguishing characteristic in bird and mammal sounds, little experimental work has yet dealt with it.

In contrast to the flamboyant extremes of bird and monkey songs and calls, songs of insect species differ principally in the *temporal* rather than frequency dimension. Species differences in frequency are conspicuous among insects, but mainly to the extent that the ears of different species are "tuned" to different frequencies; patterns of frequency change within a song among insects are rare. There are no performances like those of gibbons or thrushes.

16–17 Diagrams of trilling song patterns in crickets. Vertical axis indicates intensity. Time is on the horizontal axis. (From Alexander 1962.)

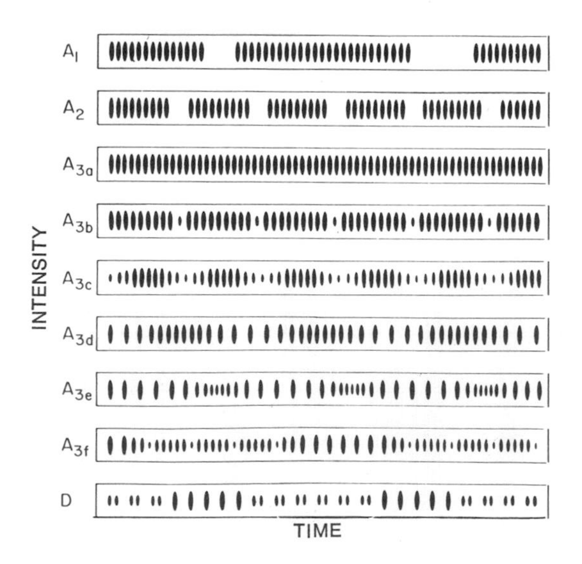

The most common type of temporal difference among insects is in the rate of repetition of a simple pulse. Several species may share the same basic trill, but avoid reacting to each other's songs because of differences in pulse rate. A trill may be broken by pauses at characteristic intervals, as in the cricket songs labeled A_1 and A_2 in Figure 16–17, and in this respect they approach the chirping songs. A trill may also be varied by reducing the amplitude of certain pulses in a regular way (A_{3b}, A_{3c}), or by varying the interval between pulses regularly (A_{3d}), or both (A_{3e-f}, D).

Another dimension of complexity in cricket songs is added by the use of chirps as units of repetition. The number and spacing of pulses within a chirp may vary (Figure 16–18) as well as the amplitude pattern (compare the songs labeled B_{1-3} and C). The patterns shown in Figures 16–17 and 16–18 represent the main types in a sample of about 100 cricket species (Alexander 1962). The songs of most other insects can be matched to one of these patterns, but those of certain katydids are

16–18 Diagrams of chirping song types in crickets. Vertical axis indicates intensity; horizontal axis, time. (From Alexander 1962.)

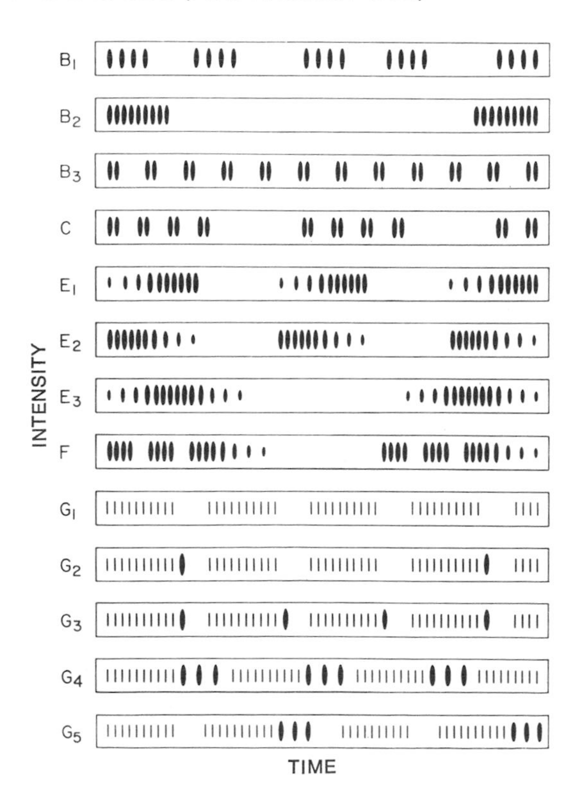

more complex (Walker and Dew 1972). The insects have achieved adequate diversity among songs of different species without relying on frequency modulation.

CONCLUSIONS

The sounds of animals are produced in association with characteristic motivational states, and achieve their principal effects by influencing the motivational states of their hearers. Animal sounds cannot be translated into human language with anything approaching the precision with which one human language can be translated to another. Sound is used socially by mammals, birds, and some other vertebrates, as well as by many arthropod groups; but most animal species are probably mute. The sound-producing structures are especially varied in the arthropods, where they seem to have evolved independently in many groups. Stridulation, the production of sound by rubbing parts of the exoskeleton together, is the prime method in arthropods.

Animal groups differ in their means of sound production and auditory reception in ways that affect the types of sounds they use socially. Insects in general are "tone deaf," and many species produce sounds that vary little or not at all in frequency. Many species, especially in the order Orthoptera, produce sound in pulses by stridulation and cannot produce tones longer than a single pulse without an interruption. An influence of learning in the development of insect songs has yet to be shown. The songs of frogs and toads often sound like those of insects and are also temperature-dependent; but they are produced vocally, often with the aid of resonating buccal pouches. Larger frogs and toads tend to have deeper voices. The calls and songs of birds and mammals often utilize rapid changes in frequency, unlike those of insects and anurans. Many birds have two sound-generating oscillators, one on each side, which are to some extent controlled independently. Consequently, they can produce different sounds simultaneously. Many species of mammals utilize ultrasound (above the upper limit of human ears).

Observations of the social and environmental contexts of calls and songs can be used to form hypotheses about the causation or motivation of the behavior and the effect it has on hearers. Observing the social context and behavior of both caller and hearer before and after a call is especially useful.

The calling, courting, and aggression sounds of crickets are named after the contexts in which they occur. Fishes employ sounds in attraction or stimulation of the opposite sex, in defense against other species, in threat, and in schooling. Use of sound by reptiles is rare but is found in defense against other species and in courtship and threat. Frogs and toads, in addition to having conspicuous mating calls, use sounds in defense against predators, as threats, and when an unreceptive frog is clasped by a male seeking a receptive female. Among birds, the distinction between songs and calls is important. Calls tend to be short and simple sounds, and are given in a wide variety of social contexts. Songs tend to be longer, more complex, and associated with territorial advertisement and/or courtship. One type of call may be used in a variety of contexts, and a variety of calls may be used in a single context. A simple, one-to-one relationship between call and function or context probably rarely exists. Mammals use sounds in pairing, in rearing young, in groups, and in protection. In primates, use of sound in situations of alarm and in agonistic behavior is conspicuous.

Experiments have revealed which characteristics of a sound are important in evoking a response. Using artificial calls or songs played to animals with a tape recorder, features of the call or song can be varied systematically while the response of the tested animal is observed. In insect mating calls, timing and the temporal patterns of amplitude are important. Frequency is not critical so long as it is within the range of hearing, which can be quite narrow in some species. The frequency and spectral composition of frog mating calls can be important in a more complex way than is the case with insects. That a species responds best to calls or songs of its own species and poorly or not at all to those of even closely related species has been demonstrated experimentally for insects, frogs, and birds of many species. Individual recognition is important between neighboring territory owners, between members of a mated pair, and within a family. Experiments have demonstrated that birds use auditory cues for recognition in these contexts.

Sounds used socially might be expected to be optimized by natural selection according to the selection pressures acting on the caller at the time a given type of sound is produced. For example, songs and calls serving for advertisement should be easily locatable by conspecifics. In contrast, calls stimulated by the detection of a hawk should not make the caller easily locatable by the hawk. To some extent this seems to be true.

Other factors that may influence the type of sound evolved for a particular purpose are the methods of sound production and reception in the taxonomic group. For example, the songs of insects, which have simple means of sound production and reception, vary mainly in the temporal pattern of loudness. In contrast, birds and primates employ rapid changes in frequency and are able to produce and analyze such sounds easily.

17 Chemical Communication

THE CHEMICAL SENSES, including smell and taste, constitute one of the most important means of communication in many animal groups, including such numerous and important groups as insects and mammals. Investigation of the ways in which animals are adapted to use chemical information in communication was at first slow because of man's limited chemosensory abilities, and because of technical problems in identifying the chemical stimuli. The limitations of our senses have not been overcome, but many technical improvements have been made that allow chemists to work with micro-quantities. These methods, which have made possible rapid progress in the isolation, identification, and synthesis of behaviorally important compounds, include combination gas chromatography–mass spectrometry, high-resolution nuclear magnetic resonance spectrometry, high-sensitivity infrared spectroscopy, and reaction gas chromatography. At least as important as the technical advances are the teamwork and cooperation among chemists, ethologists, and entomologists that have been so conspicuous.

Chemical communication involves a system that includes a *signal*, in the form of the release of a chemical substance, a *medium*, air or water, and a chemosensory *receptor*—not to mention the brain. The process is comparable in some ways to communication among parts of the body by hormones. Relatively small amounts of chemicals are used in both cases. Most animals rely on these chemical signals to a far greater extent than man can fully appreciate. Many of these species must perceive the world mainly as a chemical one. Specific compounds are employed as signals, and specialized receptors may respond selectively to them.

Chemicals that carry information between organisms have been called *semiochemicals* (Regnier 1971). Such substances when used in intraspecific communication are termed *pheromones* (Karlson and Lüscher 1959*a*, *b*; Karlson and Butenandt 1959). Interactions between different species "involving chemicals by which organisms of one species affect the growth, health, behavior, or population biology of organisms of another species (excluding substances used only as foods by the second species)" are termed *allelochemic* (Whittaker and Feeny 1971:757). Among such chemicals, or *allelochemics*, a principal division has been drawn between *allomones* (W. L. Brown 1968), which benefit the organism producing them, and *kairomones*, which benefit the organism receiving them (W. L. Brown et al. 1970). This classification is summarized in Figure 17–1. How these chemical substances are used in communication is the subject of this chapter. The field has been reviewed by Wilson (1968, 1970), and by various authors in the volume edited by Johnston et al. (1970).

Two modes of action of pheromones on their recipients have been distinguished (Wilson and Bossert 1963). The *releaser effects* of a pheromone are those behavioral responses that occur almost immediately. The *priming effects* are slower, longer lasting responses, often of an endocrine, morphogenetic, or metabolic nature. Priming effects may be of an activating or inhibiting nature. Releaser pheromones might also be predicted to have inhibitory effects; effects suggestive of such action have been found in the form of a masking of the effects of aggregation pheromone in a bark beetle (*Dendroctonus pseudotsugae;* Rudinsky 1969). The distinction between releaser and priming effects is not strictly dichotomous, but it remains useful as a simplification.

The functions of pheromones in chemical communication parallel those of auditory and visual signals. They include aggregation, sexual union, territory marking, nonterritorial spacing, group, individual, or colonial social bonds ("recognition"), recruitment to a food source, the giving of alarm, and others. Examples of most of these functions are discussed below; others are provided in the review of Wilson (1968). Some identified pheromones are listed in Table 17–1.

17–1 Classification of chemicals used in communication between and within species, according to function. For definitions and references see text.

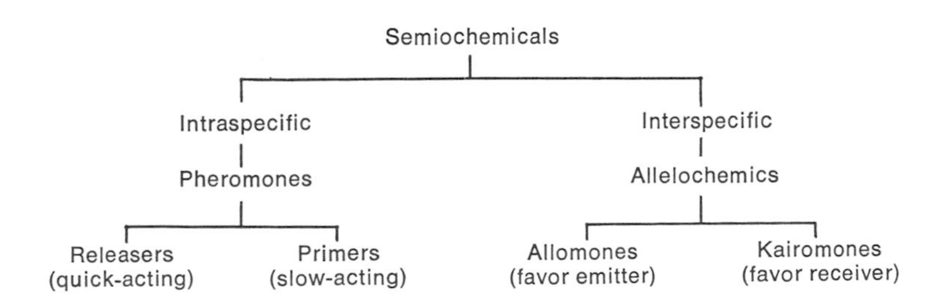

TABLE 17–1 Some chemicals used in animal communication. References in Regnier (1971).

Alarm pheromones	Insect family
Undecane	Formicinae
Tridecane	Formicinae
Pentadecane	Formicinae
Heptan-2-one	Hymenoptera, Dolichoderinae
Nonan-2-one	Hymenoptera
Tridecan-2-one	Formicinae
2-*trans*-Hexen-1-al	Myrmicinae
Isoamyl acetate	Hymenoptera
4-Methylhexan-2-one	Dolichoderinae
2-Methyl-2-hepten-6-one	Dolichoderinae
2-Methylheptan-4-one	Dolichoderinae
4-Methylheptan-3-one	Myrmicinae
Octan-3-one	Myrmicinae
2,6-Dimethyl-5-hepten-1-al	Formicinae
2,6-Dimethyl-5-hepten-1-ol	Formicinae
Citronellal	Formicinae
Citral	Formicinae, Hymenoptera
α-Pinene	Isoptera
Terpinolene	Isoptera
Limonene	Isoptera, Myrmicinae

Sex pheromones	Organism
cis-7-Dodecenyl acetate	cabbage looper (*Trichoplusia ni*)
trans-7-Dodecenyl acetate	false codling moth (*Argyroplace leucotreta*)
cis-8-Dodecenyl acetate	oriental fruit moth (*Grapholitha moleata*)
cis-9-Tetradecen-1-ol	fall army worm (*Laphygma frugiperda*)
cis-11-Tetradecenyl acetate	European corn borer (*Oatrinia nubilalis*)
trans-3-*cis*-5-Tetradecadienoic acid	black carpet beetle (*Ahagenus megatoma*)
trans-10-*cis*-12-Hexadecadien-1-ol	silkworm moth (*Bombyx mori*)
(—)-14-Methyl-*cis*-8-hexadecen-1-ol	grain beetle (*Trogoderma inclusum*)
(—)-14-Methyl-*cis*-8-hexadecenoic acid	grain beetle (*Trogoderma inclusum*)

Aggregating pheromones	Organism
Methyl oleate	beetle (*Trogoderma granarium*)
Ethyl oleate	beetle (*Trogoderma granarium*)
Ethyl palmitate	beetle (*Trogoderma granarium*)
Ethyl stearate	beetle (*Trogoderma granarium*)
Ethyl linoleate	beetle (*Trogoderma granarium*)
2-Methyl-6-methylene-7-octen-4-ol	bark beetle (*Ips confusus*)
2-Methyl-6-methylene-2,7-Octadien-4-ol	bark beetle (*Ips confusus*)
Verbenol	bark beetle (*Ips confusus*)
Verbenone	bark beetle (*Dendroctonus*)
Brevicomin	bark beetle (*Dendroctonus brevicomis*)
Frontalin	bark beetle (*Dendroctonus frontalis*)
Cyclic AMP	slime mold (*Dictyostelium discoideum*)

Common defense chemicals	Insect
Formic acid	ant (Formicinae)
Undecane	ant (*Acanthomyops claviger*)
Citronellal	ant (*Acanthomyops claviger*)
Water vapor	bombardier beetle (*Brachinus*)
Hydrocyanic acid	millepede (*Apheloria corrugata*)
	centipede
Benzaldehyde	millepede (*Apheloria corrugata*)
	centipede
	harvester ant (*Beromesser pergandei*)
p-Benzoquinone	bombardier beetle (*Brachinus*)
Toluquinone	bombardier beetle (*Brachinus*)
Phenol	tenebroid beetle (*Zophobas rubipes*)
m-Cresol	tenebroid beetle (*Zophobas rubipes*)

Recruiting pheromones	Insect
Geraniol	honeybee (*Apis mellifera*)
Neral	honeybee (*Apis mellifera*)
Geranoic acid	honeybee (*Apis mellifera*)
Nerolic acid	honeybee (*Apis mellifera*)
Hexanoic acid	termite (*Zootermopsis nevadenis*)

INTERSPECIFIC CHEMICAL COMMUNICATION

The best-known use of allomones is as repellents, the skunks offering the most familiar examples. Actually, chemical agents are used in many different kinds of interactions between species. Table 17–2 gives some idea of the great variety of chemical effects involved. Not all of these would be classified as communication by all workers, and interactions between plants are also included; but the broad scope of the list provides a good perspective.

Repellents are dramatically specialized in the insects, and considerable effort has been expended in their isolation, identification, and study (e.g., Eisner and Meinwald 1966; Eisner 1970). A variety of species and their repellent secretions are illustrated in Figure 17–2. The southern walkingstick (*Anisomorpha bupestroides*) is highly specialized for the use of repellents. It ejects a defensive spray when disturbed (Eisner 1965). The active ingredient of the spray, anisomorphal (named after the genus), is painfully irritating if inhaled. Unlike many species employing chemical deterrents, this one can spray accurately in the direction of the disturbance, as shown in Figure 17–3. The bombardier beetles (*Brachinus*) are also highly specialized in the production and use of a chemical deterrent (Aneshansley et al. 1969). When disturbed, these beetles eject a fine mist, including water and quinones, at a tem-

TABLE 17–2 Classes of interorganismic chemical effects. Adaptive advantage is indicated by +, detriment by −, and adaptive indifference by 0, for the releasing organism first and the receiving organism second. The virgule (/) indicates that adaptive advantage or detriment is not specified for one side of the relationship. (From Whittaker and Feeny 1971.)

I. Allelochemic effects
   A. Allomones (+ /), which give adaptive advantage to the producing organism
      1. Repellents (+ /), which provide defense against attack or infection (many secondary plant substances, chemical defenses among animals, probably some toxins of other organisms)
      2. Escape substances (+ /) that are not repellents in the usual sense (inks of cephalopods, tension-swimming substances)
      3. Suppressants (+ −), which inhibit competitors (antibiotics, possibly some allelopathics and plankton ectocrines)
      4. Venoms (+ −), which poison prey organisms (venoms of predatory animals and myxobacteria, aggressins of parasites and pathogens)
      5. Inductants (+ /), which modify growth of the second organism (gall-, nodule-, and mycorrhiza-producing agents)
      6. Counteractants (+ /), which neutralize as a defense the effect of a venom or other agent (antibodies, substances inactivating stinging cells, substances protecting parasites against digestive enzymes)
      7. Attractants (+ /)
         a. Chemical lures (+ −), which attract prey to a predator (attractants of carnivorous plants and fungi)
         b. Pollination attractants, which are without (+ 0) or with (+ +) advantage to the organism attracted (flower scents)
   B. Kairomones (/ +), which give adaptive advantage to the receiving organism
      1. Attractants as food location signals (/ +), which attract the organism to its food source, including (− +) those attracting to a food organism (use of secondary substances as signals by plant consumers, of prey scents by predators, or chemical cues by parasites), (+ +) pollination attractants when the attracted organism obtains food, and (0 +) those attracting to nonliving food (response to scent by carrion feeder, chemotactic response by motile bacteria and by fungal hyphae)
      2. Inductants (/ +), which stimulate adaptive development in the receiving organism (hyphal loop factor in nematode-trapping fungi, spine development factor in rotifers)
      3. Signals (/ +) that warn of danger or toxicity to receiver (repellent signals [A, 1] that have adaptive advantage to the receiver; scents and flavors that indicate unpalatability of nonliving food, predator scents)
      4. Stimulants (/ +), such as hormones, that benefit the second organism by inducing growth
   C. Depressants (0 −), wastes, and so forth, that inhibit or poison the receiver without adaptive advantage to releaser from this effect (some bacterial and parasite toxins, allelopathics that give no competitive advantage, some plankton ectocrines)

II. Intraspecific chemical effects
   A. Autotoxins (− /), repellents, wastes, and so forth, that are toxic or inhibitory to individuals of the releasing populations, with or without selective advantage from detriment to some other species (some bacterial toxins, antibiotics, ectocrines, and accumulated wastes of animals in dense culture)
   B. Adaptive autoinhibitors (+ /) that limit the population to numbers that do not destroy the host or produce excessive crowding (staling substance of fungi)
   C. Pheromones (+ /), chemical messages between members of a species, that are signals for:
      1. Reproductive behavior
      2. Social regulation and recognition
      3. Control of caste differentiation
      4. Alarm and defense
      5. Territory and trail marking
      6. Food location

17–2 Some defensive allomones in action. **A.** A millipede, *Narceus gordanus,* being tapped with a metal mallet, has begun to discharge its quinonoid secretion from two of its segmentally arranged glands. **B.** The same millipede discharging profusely from several glands, after persistent tapping. **C.** The whip scorpion, *Mastigoproctus giganteus,* discharging its aimed spray in response to the pinching of one of its appendages with forceps (the arrow points to gland openings); the animal has been affixed to a rod, and the acid spray is rendered visible on a background of filter paper impregnated with alkaline phenolphthalein. **D.** The same scorpion discharging a second time, in response to pinching of the right rear leg. **E.** Caterpillar of *Papilio machaon,* being pinched with forceps, everting its two-pronged postcephalic gland. **F.** An arctiid moth, *Utetheisa bella,* emitting froth from its right cervical gland. **G.** The grasshopper *Romalea microptera* discharging a froth (*arrow*) from its anterior thoracic spiracle; the respiratory tracheal tubes leading inward from the spiracle are beset with glandular tissue and filled with secretion; the froth is a mixture of secretion and respiratory air. **H.** Tenebrionid beetle, *Eleodes longicollis,* ejecting quinonoid secretion in response to the pinching of one of its legs with forceps; the secretion is rendered visible on a substrate of filter paper impregnated with an acidulated solution of potassium iodide and starch. **I.** An oniscomorph millipede (*Glomeris marginata*), coiled into a tight sphere in response to disturbance, has discharged three droplets of secretion (*arrows*) from glands opening on the dorsal midline; the sticky secretion can be drawn into fine threads, as is here being done with the tip of a needle. **J.** Unidentified onychophoran from Panama, discharging its sticky secretion in response to handling. (From Eisner and Meinwald 1966. Copyright 1966 by the American Association for the Advancement of Science.)

17–3 A stick insect, *Anisomorpha,* spraying in response to stimuli applied in various ways. **A–C.** Bilateral discharges, elicited by tapping the dorsal thorax (**A**), touching both antennae with a heated probe (**B**), or pinching rear of abdomen (**C**). **D–F.** Unilateral discharges, induced by pinching respectively the right forleg (**D**), the left middle leg (**E**), and the right hind leg (**F**). (Figure and legend from Eisner 1965. Copyright 1965 by the American Association for the Advancement of Science.)

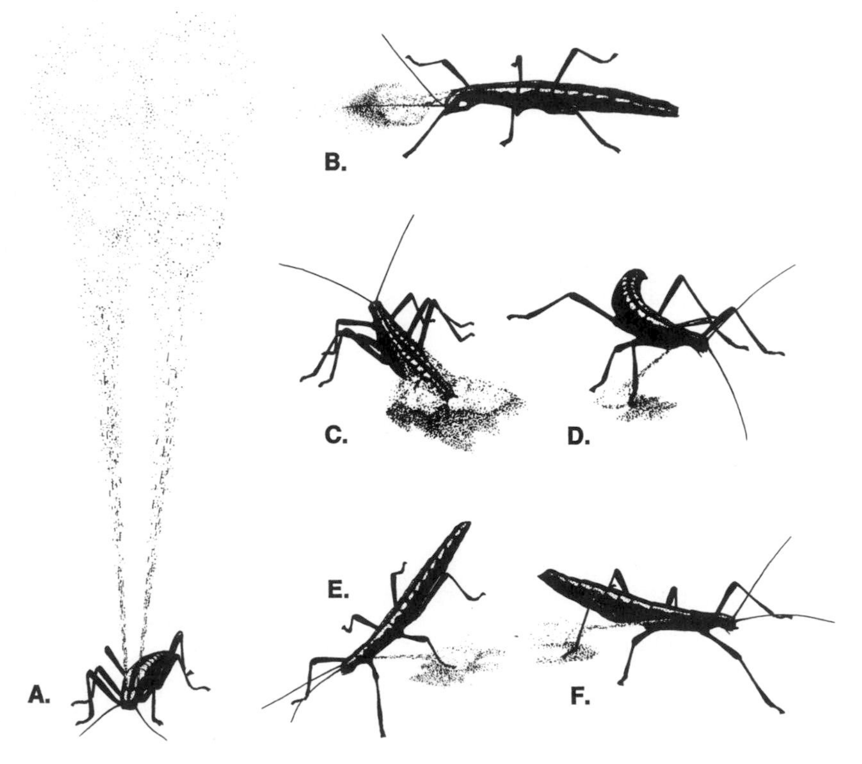

perature of 100°C with an audible detonation. The heat and quinones are produced by mixing together, in the presence of air, the contents of separate compartments containing hydroquinones and catalases.

Kairomones that evoke escape responses from their natural predators have attracted attention. The anemone *Stomphia coccinea* actually detaches and swims away in response to chemical stimuli from its natural predator, a species of starfish (*Dermasterias imbricata*). Scallops also escape by swimming in response to chemical stimuli from other species of starfish that are their commonest natural predators. Escape or defensive responses to chemical stimuli from starfish predators are also found in sea urchins and a variety of other molluscs and echinoderms.

SEX ATTRACTANTS

With about a million described species of animals in the world, virtually all of them capable of some chemosensory discriminations, we can expect the number of different chemical signals to be large. As discussed earlier in regard to other types of signals, species distinctiveness is useful mainly in certain contexts and not necessarily in others. Prominent among these is the attraction or influencing of the opposite sex for the purpose of reproduction. Chemical signals are used as sex attractants by crustaceans, fishes, salamanders, mammals, and many insects (Wilson 1968; Butler 1970; Michael et al. 1972).

Insects / Because mature ovaries tend to weigh more than testes in moths, it is energetically more efficient for the female moths to be the stationary sex and to release the attractant while the males search for the females, rather than for the females to fly to the males. In some groups of insects, including some moths, the distance attractants are released by the females. In others, however, it is the male that "calls" by releasing a pheromone. And in a few species, both sexes call (Wilson 1968).

Among the most studied species and, therefore, the best for comparative purposes, are the moths. When a virgin female is ready for copulation, she releases a sex pheromone from glands at the tip of her abdomen. At a distance from the female, the concentration of the pheromone is so low that the males cannot make use of the density gradient to locate the females (osmotropotaxis). In still air, a male may be unable to locate a female only a few meters away except by chance (Schwink 1954). But if a light breeze is blowing, a male tends to drift downwind until he encounters the scent of a female. He then flies upwind in a zigzag pattern (anemotaxis) until he either finds the female or loses the scent. In the latter case he circles, flies crosswind, or goes back downwind and repeats the sequence. A light breeze (100 cm/sec) offers the best conditions for the attraction of males; stronger winds tend to carry away the scent too fast and to dilute it too rapidly. The distance and area over which a given pheromone is above the threshold of detection for males can be computed by referring to gas diffusion laws and analysis of turbulence (Figure 17–4).

The prevailing belief has been that male moths could locate females only by anemotaxis. But some doubt has been cast on this belief by experiments on the pink bollworm moth (*Pectinophora gossypiella*) done by Farkas and Shorey (1972). They created a "trail" of sex attractant in still air, and demonstrated that males were capable of following it in a zigzag pattern much as an ant follows a recruitment trail on the ground. Having taken flight in moving air, the moths were able to locate the pheromone source as accurately in still air as in moving air from a distance of 183 cm in the laboratory. Some of the moth flight tracks and a characteristic odor plume are shown in Figure 17–5. These

17–4 Active space of the sex attractant of the female gypsy moth as a function of wind speed. Note differences in body size and antennae between male and female. (From Wilson and Bossert 1963, and Marler and Hamilton 1966.)

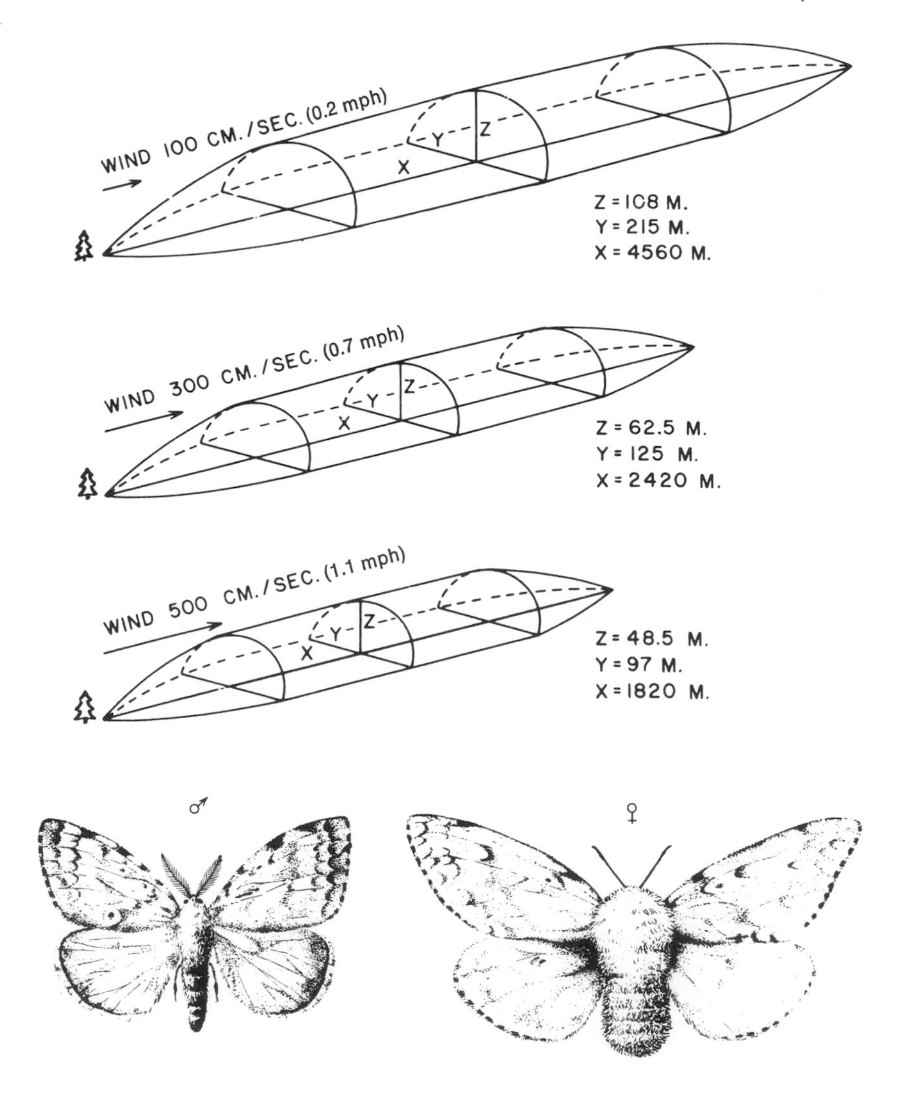

17–5 Moths can follow odor trails in still air. *Above:* Flight tunnel, showing pheromone source (*A*), wire hoop (*B*), and release cage for male moths (*C*). An artist's representation of an odor plume, based on an actual photograph of a visible smoke plume, is drawn in the tunnel. Screens on the ends of the tunnel are represented by cross hatching. The arrow indicates the direction of the wind. *Below:* Tracings of photographic records of moth flight tracks superimposed on the outline of a time exposure of a smoke plume. Photographs were taken from a top view. **A.** Track of a moth flying through an odor plume in air moving at 25 cm/sec. **B.** Track of a moth flying through an odor plume in still air; the plume was produced earlier in air moving at 7 cm/sec. Dots indicate the outside dimensions of a wire hoop within the tunnel. The arrow indicates the direction of the wind. (From Farkas and Shorey 1972. Copyright 1972 by the American Association for the Advancement of Science.)

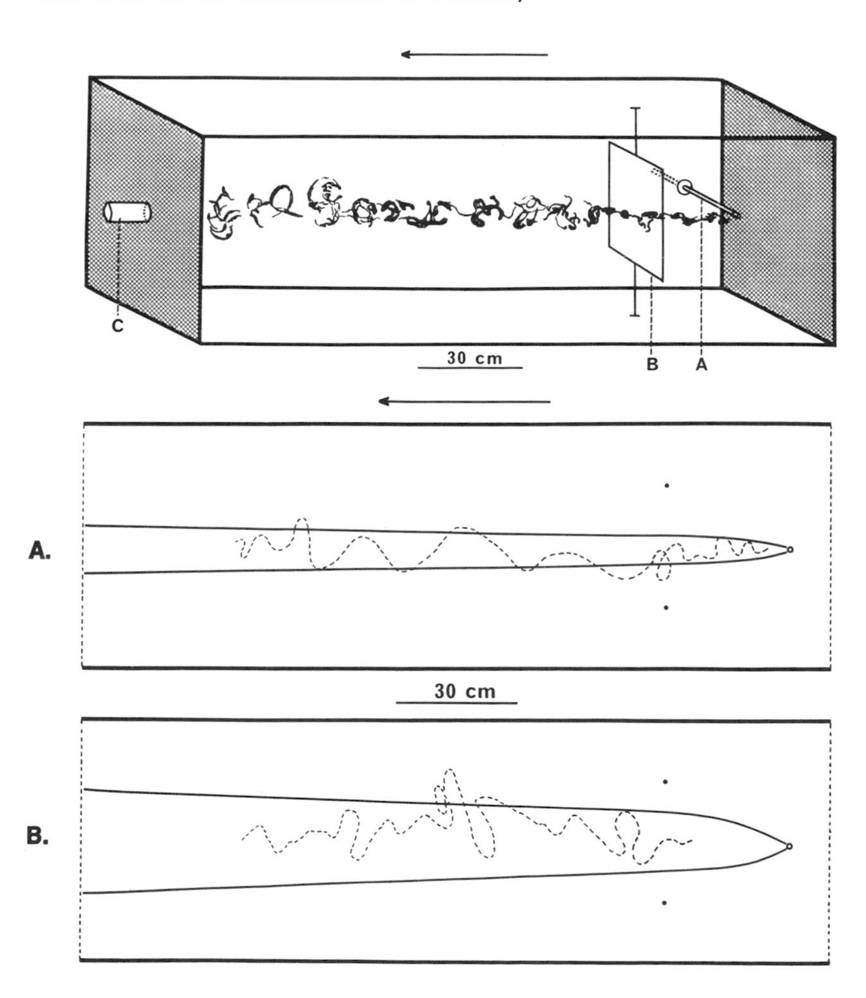

experiments are not conclusive, and the issue seems to be still unresolved.

The sex attractants of moths are impressively efficient. This is mainly due to the astonishingly low thresholds for behavioral response. Boeckh et al. (1965) calculated that one of the receptor cells involved (sensilla trichodea) can be fired as a result of a collision with a single molecule of the pheromone. Wilson (1968) noted that the 1.5 μg of bombykol, the normal quantity of the sex attractant of the silkworm moth (*Bombyx mori*) carried at one time, should theoretically be enough to activate at close range probably the entire world population of males. More realistically, it has been shown that male insects may be attracted to a female as far as 30 to 60 m away; earlier reports of attraction from miles away have probably been misinterpreted (Butler 1970; Sower et al 1971).

In studying *Tricoplusia ni*, Sower et al. (1971) observed that only about half of the females called on a given night and that each called for one or two periods of only about 20 minutes. This tended to reduce the distance from which a male could be attracted by up to one-half, because at slower wind speeds the pheromone would not fill the potential active space in the time available and still leave time for the male to fly upwind to the female. According to their calculations of the high emission rate of females and the minimal threshold for detection by males, the maximum potential distance from which a male could be attracted (disregarding limitations on calling period and other complications) was about 500 m. By similar calculations, the average female should be able to attract the average male at no greater than 20 m with a breeze of 50 cm/sec. (0.1 mph).

The effectiveness of sex pheromones as lures for males has been used in effective pest eradication campaigns. The Mediterranean fruit fly was eradicated in Florida by means of traps baited with a sex attractant. In the northeastern United States an artificial lure, disparlure, is used to bait traps for the gypsy moth (*Porthetria dispar*), a forest pest whose larvae can defoliate an entire woodlot during outbreaks (Beroza and Knipling 1972). The use of such methods is preferable by far to the use of persistent pesticides.

Since sex pheromones are important in mating, a logical hypothesis would be that each species would possess its own and would not react to those of other species. The experiments of Barth (1937) with moths of the family Pyralidae showed that males did respond to the attractants of a few other species. The lack of complete species specificity in the responses of males was also shown in the family Saturniidae by means of electrical recordings from isolated male antennae (Figure 17–6; Schneider 1966). Members of the same genus tended to

17–6 Saturniid moth reactions to sexual lure glands. Graded electroantennogram responses in male moth antennae when stimulated with female sexual attractant glands. Fully black discs are maximal responses observed with the gland of the same species. ½ and ¼ black discs stand for reduced responses, empty discs for a lack of any response. (Inside the genera or even the group of the first three genera, responses were always 100 percent.) The first five genera belong to one subfamily of the Saturniidae, *Automeris* and *Aglia* to different subfamilies of the same family, and *Brahmaea* belongs to a different but closely related moth family. (From Schneider 1966.)

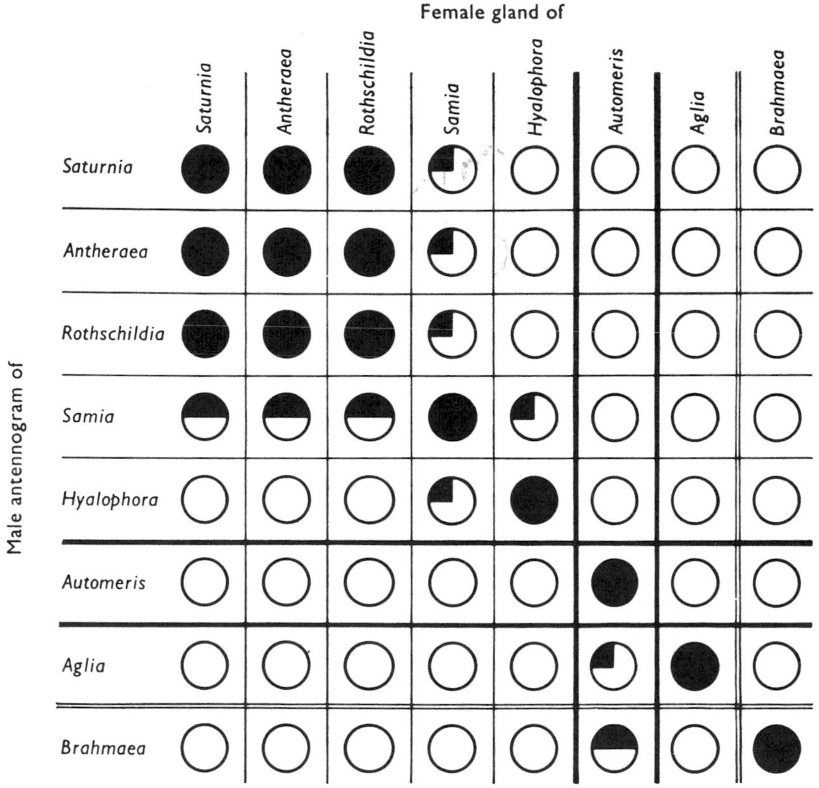

Female gland of

Saturnia *Antheraea* *Rothschildia* *Samia* *Hyalophora* *Automeris* *Aglia* *Brahmaea*

Male antennogram of

Saturnia
Antheraea
Rothschildia
Samia
Hyalophora
Automeris
Aglia
Brahmaea

respond similarly to each other's sex attractants. In the Arctiidae a hydrocarbon, 2-methylheptadecane, is an attractant for males of at least nine species of two genera (Roelofs and Cardé 1971). Other examples of a lack of species specificity are listed by Butler (1970). In other insects (e.g., the beetle genus *Rhopaea* and the mecopteran genus *Harpobittacus*), species specificity in response to sex attractants has been shown (reviewed by Wilson 1968).

The selectivity of insect antennae for certain pheromones is illustrated in Table 17–3 by the effectiveness of the different isomers of bombykol, the sex attractant of female silkworm moths (*Bombyx mori*). The various isomers differ in their potency by a factor of as much as 10^{12}. Clearly, chemical composition is not the only determinant of pheromonal effectiveness. In the pink bollworm moth (*Pectinophora gossypiella*), the *cis-* isomer acts as an inhibitor or masking agent for the *trans-* isomer, propylure; a similar situation was found in *Trichoplusia ni* (Jacobson 1969). In contrast, two isomers of the sex attractant in female summer fruit tortrix moths (*Adoxophyes orana*) were found by Meijer et al. (1972) to act synergistically. Mixtures of the two isomers resulted in high behavioral activity, while the pure compounds were ineffective. This case is an important example of the achievement of species specificity by the use of multiple pheromones rather than the simple alteration of a single compound's chemical structure.

Because pheromones may have dramatically different behavioral effects as a result of a simple change in molecular structure, the possibility of sympatric, saltational speciation has been considered by Roelofs and Comeau (1969). They described two pairs of species that are almost identical in morphology and appearance and are kept from interbreeding only by the females' sex pheromones. In one pair of species (*Bryotopha similis*, B_1, B_2) the sex attractants of the two species are geometrical isomers and are reciprocally inhibitory. As Roelofs and Comeau pointed out, such a simple difference in the structure of one chemical compound could presumably come about by a single mutation in a female, but it could result in isolation of a new species only if a corresponding mutation affected the olfactory structures of a nearby male. The occurrence of both events at about the same time and place would be a rare event at best. But the gradual acquisition of a new pheromone that is a geometrical isomer of an older one is also not easily envisaged. An alternate hypothesis might be that an ancestral

TABLE 17–3 Comparative attractancy of bombykol isomers. Bombykol is the sex pheromone of the silkworm moth, *Bombyx mori.* (From Regnier 1971.)

Compound	Attractancy[a] (μg/ml)
10-*trans*-12-*trans*-Hexadecadien-1-ol	1
10-*trans*-12-*cis*-Hexadecadien-1-ol	10^{-12}
10-*cis*-12-*trans*-Hexadecadien-1-ol	10^{-3}
10-*cis*-12-*cis*-Hexadecadien-1-ol	10
Bombykol (natural)	10^{-10}

[a] Attractancy of bombykol is expressed as the concentration of pheromone in pentane on a glass rod that will attract 50 percent of the *B. mori* males tested.

species formerly existed which produced both isomers, and that each isomer became fixed in a separate isolated population.

If male moths respond to females of other species, do they then interbreed in nature? Apparently not. Several factors are responsible. The geographic ranges of some species responding to the same pheromone do not overlap. In other cases, the habitats or mating seasons are different among related species. In moths, different species tend to fly and mate at different times of the night. Wilson and Bossert (1963) showed that this factor could account for isolation among some of the species and genera reported by Schneider (1966 and earlier) to respond to the same pheromones. In the family Saturniidae, three species of *Callosamia* in the southeastern United States are kept from interbreeding by the time of day of their mating flights. *C. carolina* mates from 10 A.M. to 3 P.M. *C. angulifera* is totally nocturnal, and *C. promethea* mates in late afternoon (L. N. Brown 1972). Interspecies matings within the genus can be accomplished by manipulating the photoperiod.

Primates / Although sex attractant pheromones have received extensive study in insects, they are poorly known in mammals. Among mammals, anthropoid primates (see Table 12–1) have a relatively small proportion of the brain devoted to olfaction and are generally regarded as having poor olfactory powers compared to other mammals. Certainly the conspicuous glands and scent-marking behavior found in many other mammals are inconspicuous or absent in primates (except prosimians). Consequently, it was surprising to find that female rhesus macaques (*Macaca mulatta*) signal sexual receptivity (estrus) chemically to males (Michael and Keverne 1968; Curtis et al. 1971; Michael et al. 1971, 1972). Intact male monkeys can discriminate without tactile cues between ovariectomized females with different estrogen doses applied intravaginally; they choose the one with the higher estrogen level for copulation. Males deprived of their sense of smell do not make such discriminations. Vaginal secretions of an attractive female painted on an unattractive one make the latter more attractive to males. The recipient females were, however, themselves unreceptive, leading to masturbation in one male monkey. The active fractions of the vaginal secretions have been isolated, synthesized, and tested successfully; they include acetic, propionic, isobutyric, butyric, isovaleric, and isocaproic acids.

ALARM SUBSTANCES

If a pike injures a minnow, the chemicals released from the broken epidermis may frighten away the school of minnows for hours or days. Normally this fright-inducing substance, long known as *Schreckstoff*, is released only when the integument is broken. The material is stored in specialized, club-shaped, epidermal cells. The alarm reaction to injury of the skin of the same species is restricted in fishes mainly to members of the Ostariophysi (Siluroidea and Cyprinoidea) and Gonorhynchiformes (Pfeiffer 1967), although it has been tested for in 87 species of 32 families of other orders. It is also found in tadpoles of the toad genus *Bufo*. The fright reaction is not completely species-specific, since related species may respond weakly (Figure 17–7), but a full response is given only to the pheromone of the species tested. The degree of species specificity thus appears to be greater for fright sub-

17–7 Fish alarm substance reactions. Graded fright reaction responses of the teleost fish group Ostariophysi. Single lines separate the genera; double lines separate the family Cyprinidae from the family Cobitidae, which is only represented by a single species. The effect of skin extracts of a species on a swarm of the same species (fully black discs) is compared with the effect of skin extracts of other species (as indicated by graded blackening of the discs). (Simplified and rearranged from Schutz 1956; see also Pfeiffer 1962*b*. Figure and legend from Schneider 1966.)

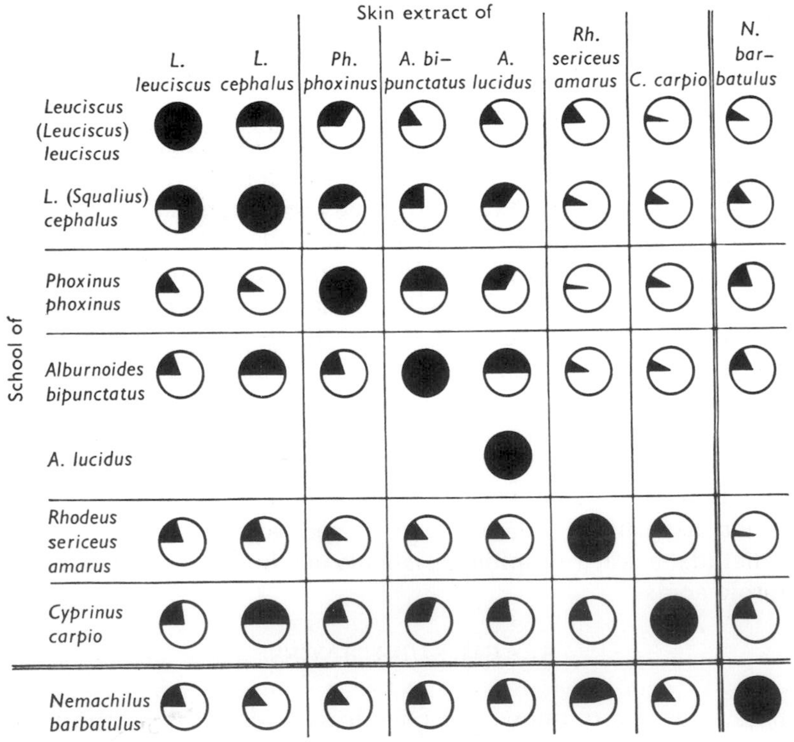

stances than for sex attractants, contrary to what many biologists would have predicted. Schneider (1966) has summarized the original work of von Frisch on this subject, as well as the later works of von Frisch's students, Shutz and Pfeiffer, and of Eibl-Eibesfeldt.

Alarm reactions to crushed tissue or to secretions have also been found in 19 of 30 species of aquatic gastropods tested (Snyder 1967), in earthworms (Ressler et al. 1968), and in a sea urchin, *Diadema antillarum* (Snyder and Snyder 1970). In the experiment on snails, all four species of *Helisoma* tested were reciprocally reactive, but snails from six different families were mostly unreactive to each other.

Social insects release alarm pheromones as glandular secretions that do not depend on tissue destruction. An alarmed honeybee releases isoamyl acetate, along with other compounds, when it stings an enemy (Boch et al. 1962; Ghent and Gary 1962). This may be partly responsible for the tendency of honeybees to cluster their stinging in one area. The alarm substances of ants are not species-specific, or even restricted to a subfamily (Wilson 1968; Dumpert 1972). They show evidence of the influence of phylogenetic relationship (Blum and Brand 1972). For example, attine ants (Attini) share a common chemical heritage based on 3-ketones and 3-alcohols (Crewe and Blum 1972).

The alarm-defense type of chemical communication has been studied comparatively in formicine ants (Regnier and Wilson 1968, 1969, 1971; Wilson and Regnier 1971). In the more typical formicine ants, a minor threat to the nest is repulsed with biting and stinging in which formic acid is released. The latter substance functions as a deterrant to enemies, but not significantly as a pheromone, being insufficiently volatile. The communication of alarm to other workers is accomplished in *L. alienus* mainly by undecane, which is released from the Dufour's gland and aids the penetration of formic acid into the enemy. The size and position of this gland in a related species, *F. subsericea*, is illustrated in Figure 17–8.

In species with relatively small nests close to the surface of the ground, such as *Lasius alienus* and many other Formicinae, alarm pheromones stimulate a rather disoriented but excited running around ("panic reaction") that causes many workers to flee the source of danger. In a small colony, defense by fighting to the death would eliminate a large fraction of the colony and so is not favored by natural selection.

A departure from this rather common mode is found in *Acanthomyops claviger*. This species lives in large subterranean colonies and keeps a "herd" of symbiotic root homopterans. The nests are substantial and cannot be rapidly replaced, nor can their symbiotic homopterans. Therefore, an adaptive response for this species is to stand and

17–8 Major adaptive changes in the formicine alarm-defense system. An asterisk indicates a new character state. (From Wilson and Regnier 1971. Copyright 1971 by the University of Chicago. All rights reserved.)

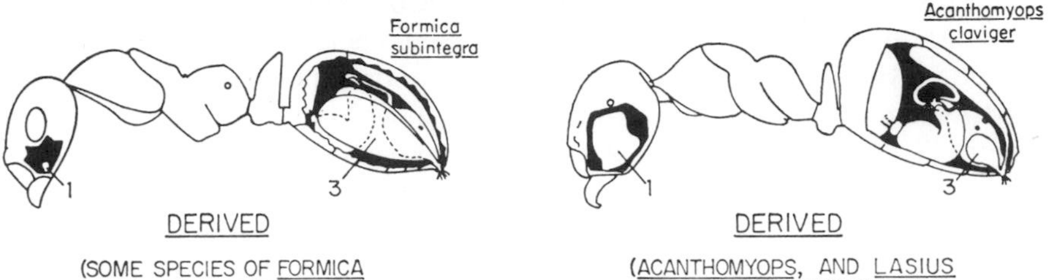

(SOME SPECIES OF FORMICA SANGUINEA GROUP)

1. Small mandibular glands
2. Mandibular gland terpenoids & furans scarce or absent
*3. Enormous Dufour's gland
*4. Dufour's gland contains alkanes, ketones & large quantities of acetates
*5. Aggressive alarm communication with secondary "propaganda" function

(ACANTHOMYOPS, AND LASIUS SUBGENERA CHTHONOLASIUS AND DENDROLASIUS)

*1. Small to large mandibular glands
*2. Mandibular gland terpenoids or furans abundant
3. Small Dufour's gland
4. Dufour's gland contains alkanes & related ketones
*5. Aggressive alarm communication

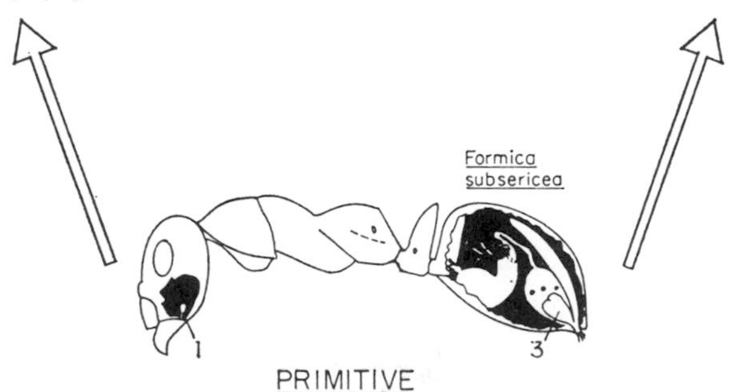

(MOST FORMICINAE)

1. Small mandibular glands
2. Mandibular gland terpenoids & furans scarce or absent
3. Small Dufour's gland
4. Dufour's gland contains alkanes & related ketones
5. Panic alarm communication

fight. An alarmed *Acanthomyops* releases pheromones from both mandibular and Dufour's glands. The response to alarm pheromones in *Acanthomyops* is more aggressive than in *Lasius alienus* and most Formicinae. The ants rush to the source of release of the alarm pheromones with mandibles ready to attack. The mandibular gland is enlarged in such species and contains an abundant supply of terpenoids or furans, including citral and citronellal, which provide the characteristic odor of a freshly opened nest perceptible to man. The large size of colonies in these species means that proportionately few workers are sacrificed in defense. The aggressive defense of *Acanthomyops*, which is stimulated and aided by the abundant mandibular gland products, is consequently an adaptive response. The hypertrophy of the mandibular gland in this species seems to be due to their greater dependence on biting attacks for colony defense than other species. The glands and their contents are shown in Figure 17–9.

A second departure from the more common type of alarm-defense system is found in facultative slave-makers such as *Formica subintegra* (of the *Formica sqnguinea* group). This aggressive species employs its

17–9 Locations of the principal exocrine glands and their alarm-defense substances in a formicine ant, *Acanthomyops claviger.* (From Regnier and Wilson 1968.)

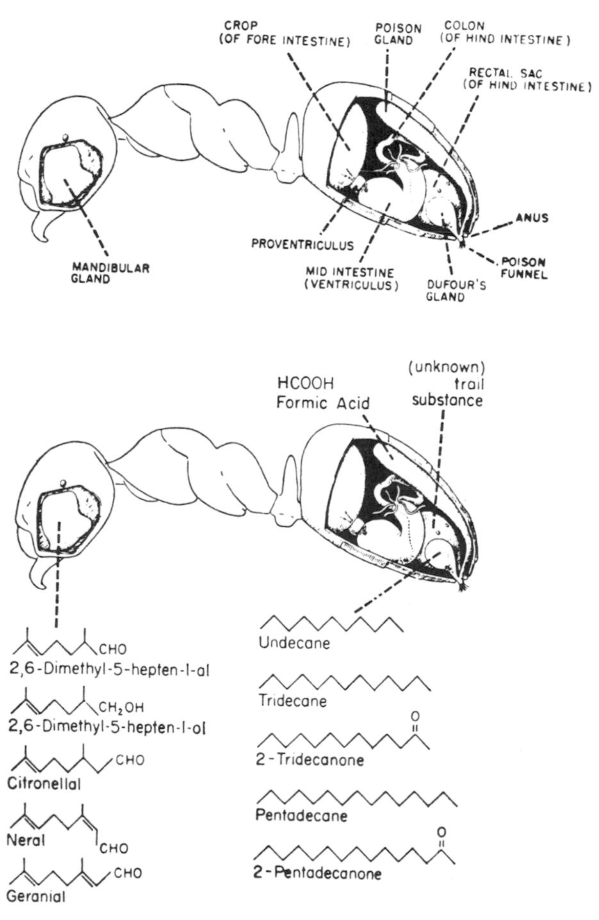

alarm pheromones not only in defense of its own colony but also as an attractant when attacking colonies of other species. The tremendously enlarged Dufour's gland of these species releases large amounts of acetates, alkanes, and ketones. These not only attract the slave-maker workers, but also cause the victimized species to give the panic reaction and to escape. As a result, entry into the victim's nest can be accomplished more easily and with less actual fighting. This super-discharge of alarm pheromones by the slave-maker has been likened to propaganda. The enlarged Dufour's gland is shown in Figure 17–8.

Comparative study of alarm-defense pheromones has revealed that the chemical communication system in use in a particular species is adapted to the specialized needs of that species. The size of the glands and the nature of their chemical products show differences among species that are correlated in meaningful ways with the varying needs for colony defense (and offense) among species. Figure 17–8 summarizes these findings.

TRAIL MARKING

The use of scents in recruitment of species members to a food source is particularly common in social insects. The laying of scent trails is common in ants and some social bees. These trails serve to excite workers and attract them to temporary food sources. Trail following has been investigated in the fire ant (*Solenopsis saevissima*) by Wilson (1962). The trail pheromones are relatively volatile; the natural trail of a single worker fire ant evaporates in a minute or two to below threshold density (Figure 17–10). Consequently, the trails tend to persist only while they are in use by numbers of workers and to fall quickly into disuse when the food source has been consumed. The attractive nature of the trail substance can be demonstrated experimentally by blowing it as a gas toward a group of ants or by drawing it out in artificial liquid trails from an applicator. The ants will approach or follow the scent.

The trail substances of ants tend to be species-specific among species whose home ranges overlap; but at least one species, *Camponotus beebei*, systematically exploits the trails of another species, *Astera chastifex* (Wilson 1968). Moreover, different species of army ants which eat the brood of other species of ants, will follow trails of a variety of ant species.* Many species of insects and even of snakes are known to follow ant trails, but it has only rarely been demonstrated that scent is the critical means of orientation. However, Moser (1964) showed that a small wingless roach (*Attophila fungicola*) that inhabits

* H. Topoff 1973: personal communication.

17–10 Active space of the trail of a fire ant at various times, *t*, after the trail layer reaches the nest. (From Wilson and Bossert 1963.)

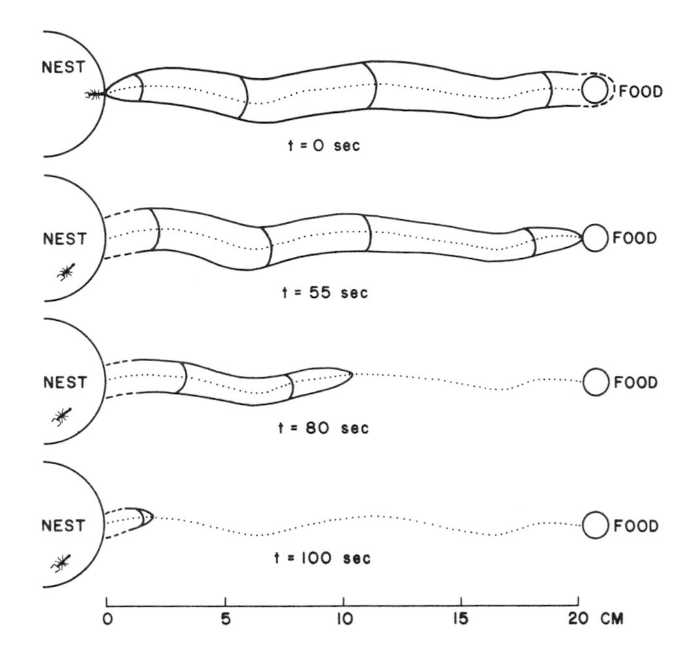

the fungus gardens of the Texas leaf-cutting ant (*Atta texana*) as an inquiline will follow artificial trails made from the trail substance. An interesting sidelight is that these roaches have often been found riding on the backs of ant queens during mating flights.

REPERTOIRE DIVERSITY

Social Insects / Chemical communication systems may evolve through the use of additional compounds for different behavioral effects by the same species. The chemical repertoire of a species may be conceived as the number of functionally different chemical signals that it uses. Probably in no species with a large and complex chemical repertoire is the number of such signals completely known. A few of the better-known species will be considered. The most extensively studied cases are in the social insects. A preliminary idea of the diversity of chemical signals employed in a species of social insect may be obtained through consideration of the various exocrine glands. Some of the exocrine glands of a honeybee and an ant are shown in Figure 17–11.

The behavioral functions of some of the glands shown in Figure

17–11 will be briefly considered by way of illustration. Further details can be found in reviews by Wilson (1965*b*, 1968, 1970). The mandibular glands of the honeybee and several ants are a source of an alarm pheromone. In the honeybee, Nasanoff's gland secretes geraniol and other compounds that function as attractants at food sources, but it

17–11 Exocrine gland systems of (*top*) the honeybee worker (*Apis mellifera*) and (*bottom*) a worker of the ant species *Iridomyrmex humilis*. Glands of the two species are labeled with the same number if they are considered homologous; where different names have been used in the literature, the name used for ants is given in parentheses in the following key; several minor glands of unknown function are omitted. (1) Mandibular gland; (2) hypopharyngeal (= maxillary) gland; (3) head labial gland; (4) thorax labial gland; (5) postgenal gland; (6) wax glands; (7) poison gland; (8) vesicle of poison gland; (9) Dufour's gland; (10) Koschevnikov's gland; (11) Nassanoff's gland; (12) postpharyngeal gland; (13) metapleural gland; (14) hindgut (glandular nature uncertain); (15) anal gland; (16) reservoir of anal gland; (17) Pavan's gland. (Figure and legend from Wilson 1965*b*. Copyright 1965 by the American Association for the Advancement of Science.)

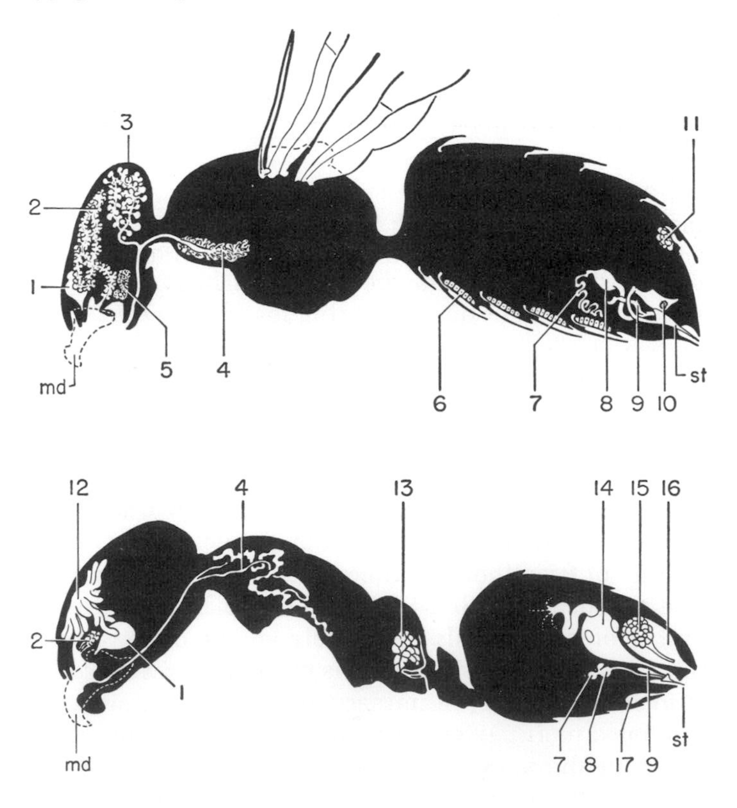

apparently does not serve this purpose in ants. In some ants, however, Dufour's gland is the source of an attractant used in scent trails; the behavioral significance of this gland in bees is unknown. In various other ant species the trail pheromones are produced in the hindgut, the poison gland, or Pavan's gland. Alarm substances are produced in various ant species by the anal gland, Dufour's gland, poison gland, and mandibular gland; in bees, the glands of the sting chamber as well as the mandibular glands are known to have an alarm function.

Other functions and sources of pheromones in social insects are known (see Blum and Brand 1972). In many species the exchange of oral and anal liquids between individuals (trophallaxis) is important. Caste determination, colony recognition, and swarming are also dependent on chemical communication. The variety of glands and compounds produced by them makes possible a high degree of complexity in the communication systems of social insects. It has become recognized that communication in the social insects is to a large degree of a chemical nature, although the importance of tactile and auditory stimuli should not be minimized (see Chapter 19).

Deer / Most species of mammals have a highly developed olfactory sense and rely on chemical communication extensively. The black-tailed deer (*Odocoileus columbianus*), which has been studied by Müller-Schwarze (1971), may serve as an example in which several glandular sources of pheromones are utilized. The glandular sources of pheromones used by these deer in communication are illustrated in Figure 17–12, and the associated behaviors, social contexts, and functions are listed in Table 17–4.

The tarsal gland, located on the inside of the lower hind legs, was often sniffed by other deer living in a group. The tarsal gland of a newcomer to a group was extensively investigated by the group, but the newcomer was reluctant to reciprocate. Possibly the tarsal scent is used in individual recognition.

The smell of urine appears to have social significance in deer, as in many other mammals. Deer normally urinate with hind legs apart, but in certain contexts they urinate on their hind legs. This is done by fawns separated from their mother and by adults during aggression. In adults it seems to intimidate rivals and attract females to males. The scent from the metatarsal glands is given off in fear-inducing situations and evokes alarm in others.

The forehead is richly supplied with glands and in addition picks up scent from the hind legs. Deer, like many mammals, sniff the scents deposited by other deer wherever they go and add their own by rubbing

17–12 Pathways of social odors in black-tailed deer. Scents of tarsal organ (1), metatarsal gland (2*a*), tail (4), and urine (5) are transmitted through air. While reclining the metatarsal gland touches ground (2*b*). Deer rub their hindleg over forehead (3*a*), forehead is rubbed over dry twigs (3*b*). Marked twigs are sniffed and licked (3*c*). Interdigital glands leave scent on ground (*6*). (Figure and legend from Müller-Schwarze 1971.)

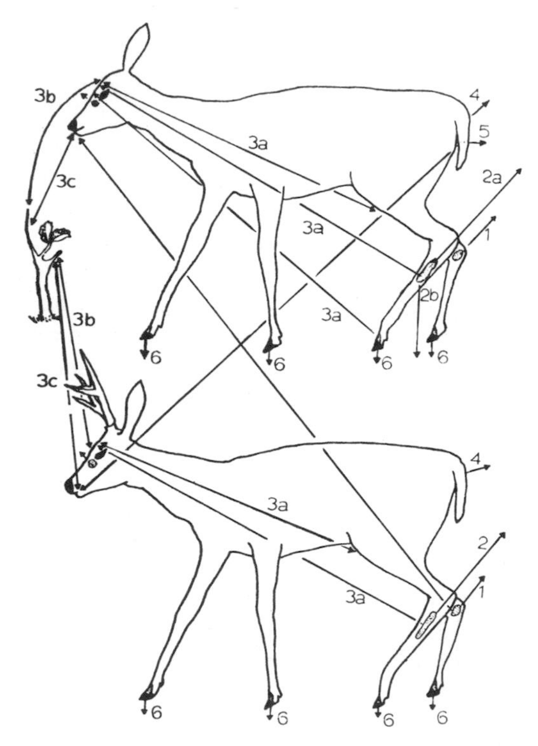

their foreheads on twigs and branches, as shown in Figure 17–13. The scent of an individual becomes most prominent around his sleeping place, where the deer marks most actively; but it also comes to delineate the individual's home range. In an aggressive mammal the scent mark helps demarcate the area of dominance or territory. The scent marking of some mammals is functionally equivalent in terms of spacing to the singing of birds; both are behaviors that may indicate the presence of an aggressive territory owner.

The tail is richly supplied with glands in some other species of deer but only sparsely in the blacktail. The tail glands apparently discharge during excitement, since deer are observed tail-sniffing at such times.

During alarm the interdigital glands on the feet discharge and mark the footprints.

TABLE 17–4 Chemical communication in black-tailed deer. An additional source, the antorbital gland, is poorly developed in this species though prominent in certain other cervids. (Compiled from Müller-Schwarze 1971.)

Gland or source	Marking behavior	Context	Response	Functions
Tarsal	none	unspecialized within group	sniff	recognition
Urine	on legs	aggression in adults	attract male	attraction
			*Flehmen**	intimidation
		separation of fawn from mother		
Metatarsal	none	fear	alarm	alarming
Forehead	rub twig	sleeping site, home range	retreat	marking home range
Tail	none	excited	sniff	?
Interdigital	running	escape	investigation	?

* *Flehmen* is a characteristic facial expression common in ungulates upon sniffing the scent of an animal: the lips are retracted and nostrils dilated.

17–13 Sniffing of a twig (*left*) followed by forehead rubbing (*right*) by a seven-month-old female black-tailed deer. (From Müller-Schwarze 1971.)

Mice / Although the identification of the pheromones used by mammals and of their behavioral effects is just beginning, it is already evident that a sizable repertoire of chemical signals is available to some, and perhaps most, species. The various behavioral and physiological effects of pheromones in mammals can be conveniently investigated in laboratory mice (house mice, *Mus musculus*). A useful review of studies on pheromones in rodents has been provided by Bronson (1971). The various postulated pheromones or pheromonal complexes are listed in Table 17–5.

The variety of behavioral and physiological effects listed in Table 17–5 is considerable. Perhaps the evident richness of chemical stimuli

TABLE 17–5 Summary of chemical communication in the laboratory (house) mouse. (Compiled from Bronson 1971.)

Condition of emitting animal	Body source	Behavioral and physiological effects	Tentative name of substance
Stressed	urine	avoidance	fear substance
Female	urine?	attract males	sex attractant
	urine	reduce frequency of attack	aggression inhibitor
	urine?	increase adrenal cortical function	adrenocortical activator
Grouped females	urine?	suppress or lengthen estrus cycle; spontaneous pseudo-pregnancy; gonadotropin release	estrus inhibitor
Androgen-primed male	urine (preputial glands)	increase female activity; awaken females from sound sleep; preferred in choice test by females	estrus inducer
	urine (preputial glands?)	repels other males; elicits attack in other males	aggression inducer
	bladder urine	shorten estrus cycle	estrus inducer
	male-soiled bedding	accelerate maturation in young females	
	urine	block implantation	
	urine (and preputial gland)	shorten estrus cycle	?
Grouped males	urine (coagulating gland)	increase motor activity	?
	feet (plantar gland)	increase motor activity	?
Strange male	urine (coagulating gland)	increase adrenocortical function and influence prolactin, FSH, LH, ACTH	adrenocortical activator

for mice is to be expected in nocturnal animals with large olfactory lobes. Some of the effects, such as avoidance, attraction, influences on fighting, and increases in motor activity, are of the short-term releaser type. Other effects, which seem to have attracted more attention, are of the long-term primer type. The latter, in most cases, are known to depend on prior action of the chemical stimuli on the endocrine system, which in turn produces the detectable primer effects. Most of the known primer effects concern the estrous cycle.

Although the source of most pheromonal effects in mice is urine, as late as 1972 none of the active substances had been isolated, much less identified, synthesized, and tested. These tasks will probably occupy researchers during the 1970s. In our present state of uncertainty it is difficult to know how many different substances and glandular sources are involved. There is good evidence for the existence of a low-molecular-weight, airborne substance in the bladder of male mice (from examinations of intact adults or androgenized castrates), and for a separate (?) substance in the preputial gland lipids. Other sources have been tentatively identified (Table 17–5). Despite considerable uncertainty at present it seems likely that a fair number of different pheromones with different biochemical and anatomical origins will be found eventually in laboratory mice.

Other Mammals / Marking of frequently used parts of the home range with scents either from feces and urine or from specialized glands is important in the social organization of a wide variety of mammals (Ewer 1968; Eisenberg and Kleiman 1972). Scent marking is highly developed in the Australian phalanger (*Petaurus breviceps;* Schultze-Westrum 1965). This species marks its territory with the sternal gland, plantar glands, glands of the anal region, and with secondary odor mixtures from the saliva and fur. In addition to marking their territories, males also smear secretions of their frontal glands over their partners with specialized motions. Phalangers are able to discriminate among species, groups, and individuals by odor. In Australian rabbits (*Oryctolagus cuniculus*) the activity of specialized chin glands used in territorial marking is correlated with dominance status (Mykytowycz 1965, 1970). In many mammalian species individuals scent mark frequently in situations in which they are both intolerant and dominant to other conspecifics (Ralls 1971). The well-known scent marking of dogs appears to have been derived from the scent marking wolves use to mark territories (Mech 1970, 1973). Scent marking is often a conspicuous feature of the territory in territorial mammals and is also used by species with other aggressive spacing systems.

OPTIMIZATION OF PHEROMONAL PROPERTIES

The properties of chemical signals are subject to natural selection acting on the individuals who produce them. There should be, therefore, a relationship between the physical properties of a pheromone and its use in communication. These relationships have been ably discussed by Bossert and Wilson (1963) and Wilson and Bossert (1963) in a pair of papers that have greatly increased interest in evolutionary aspects of chemical communication. The main points of the following discussion stem from these two papers.

The *active space* of a chemical signal is the space in which its concentration is above threshold for its normal recipient. A pheromone that attracts individuals over long distances, such as a sex attractant, should have a large active space (e.g., Figure 17–4). One that marks a precisely defined spot or line, such as a trail substance, should have a small active space (e.g., Figure 17–10).

The active space of an airborne female sex attractant can be increased by increasing the emission rate (Q), by lowering the threshold of the males (K), and by using a compound of appropriate volatility and persistence. The ratio of Q to K should be, and in fact is, relatively high in pheromones used similarly for attraction over long distances (Table 17–6).

The active space of an ant's trail substance can be reduced most efficiently by releasing only a small amount at a time. It is also advantageous to use small amounts so as not to exhaust the limited supply when laying trails. It would therefore be inefficient to reduce active space by raising the threshold of response. The Q/K ratio for trail substances should be, and is in known cases, relatively low (see Table 17–6).

Alarm substances should occupy an intermediate position. In

TABLE 17–6 The Q/K ratios of three communication systems with different requirements. (After Wilson 1970.)

Chemical signal	$\frac{Q}{K} = \frac{\text{Natural emission rate in molecules/sec}}{\text{Behavioral threshold conc. in molecules/cm}^3}$
Imported fire ant (*S. saevissima*) odor trail	1
Acanthomyops sp. alarm substances	10^3–10^5
Silkworm (*Bombyx mori*) sex attractant	10^{10}–10^{12}

an ant colony the alarm substance released by one alarmed ant should act at a distance great enough to attract nearby ants but not the entire colony: the signal should be graded. The alarm compound should have the property that if many ants are initially alarmed, more ants will be attracted. In the harvester ant, *Pogonomyrmex badius*, release of the mandibular gland alarm substance by one alarmed individual creates an active space described as a sphere with radius 6 cm within 13 seconds; the active space fades completely after 35 seconds. If the alarm stimulus persists, more ants will be attracted and will release more alarm substances as they arrive. If the stimulus ceases, the state of alarm quickly disappears. As expected, in known cases the Q/K ratio is intermediate between those for sex attractants and trail substances.

Another important property of pheromones is their degree of species specificity; that is, the extent to which a species avoids using pheromones that are employed also by other species. For a sex attractant that is broadcast in the air and to which all the insects in the community are exposed, it is necessary to avoid attracting the wrong species. Since there is a limited number of organic compounds of low molecular weight, there is a larger chance that a given pheromone will not be used by other species if the pheromone has a higher molecular weight. The upper limits on molecular weight are imposed by the lowered volatility and greater energetic cost of synthesizing such compounds. Wilson and Bossert predicted, on the basis of these and other considerations, that the molecular weights of airborne insect sex attractants would lie between 80 and 300 and would have between 5 and 20 carbon atoms. The prediction has been remarkably accurate. Specificity can also be achieved by using mixtures or "medleys" of compounds. The extent to which this solution is relied on is poorly known at present.

CONCLUSIONS

With modern chemical methods suitable for dealing with very small quantities, progress in the investigation of chemical communication has been greatly speeded. Chemicals used in communication between organisms are called semiochemicals. Those used between members of one species are known as pheromones. Of those used between members of different species, allomones benefit the signaler and kairomones benefit the receiver. An important class of allomones includes chemical sprays and secretions that repel predators. Among kairomones are substances by which prey can detect predators in time to escape.

Sex attractant pheromones are known in a wide variety of animals but have been extensively studied mainly in insects, where they are important in pest control. They are usually released by females for a brief period during the night and are effective in concentrations as low as 10^{-10} μg/ml. Males locate the "calling" females by flying upwind when they encounter the female's scent (anemotaxis) and using the scent gradient (osmotropotaxis) when they are within about a meter. The sex attractant of a species will attract only males of the same species, with a fair number of exceptions that tend to be closely related members of the same genus. Confusion between species is avoided in such cases by mating at different times of day and in different habitats, as well as by simple geographical isolation. Change in the molecular structure of a sex attractant has been considered as a conceivable mode of saltational, sympatric speciation.

Alarm substances are found in certain fishes, snails, mammals, and social insects. There is little species specificity in the alarm pheromones of ants, and the alarm pheromones of members of the same taxon tend to share certain chemical properties. Nevertheless, evidence of adaptive specialization is present. A species unusually aggressive in its nest defense has enlarged mandibular glands, which release an alarm pheromone. Some slave-making species have enlarged Dufour's glands in their abdomens, which release extraordinary amounts of alarm substances during slave raids on neighboring colonies of other species.

Trail marking is not restricted to ants and occurs commonly in certain groups of bees. Trail substances tend to be species-specific, but the trails are sometimes followed by symbionts and raiders of other species. The trail pheromones must decline rapidly to subthreshold values on trails not in use, so that the animals will not follow old trails.

Chemical communication systems within a species may evolve through the use of additional compounds and glandular sources for different behavioral effects. The social insects use various glandular products in attracting males, recruitment to food, alarm, defense against enemies of other species, controlling the appropriate numbers of each caste in the colony, and colony recognition. Deer use various glands in recognition, intimidation, sex attraction, alarm, and marking the home range. In mice, several effects are attributable to substances found in the urine of one or both sexes but originating from at least three sources. Among these are excellent examples of the priming action of the pheromones; these involve an intermediate metabolic or neuroendocrine change. The estrous cycle is accelerated by the urine of males and slowed by that of females.

The properties of chemical signals are subject to natural selection acting on the individuals that produce them. In this way the active space, the fading time, and the species specificity of a pheromone or pheromonal complex can be adjusted for the particular role it has in the life of the animal. For example, sex attractants should have a large active space to increase the chances of mating, while trail substances, in order to be precise, must have an active space that is limited to the trail. Sex attractants and trail substances should be specific to the species that uses them, but for alarm pheromones specificity is of little consequence.

18 The Origin of Species and of Species-specific Behavior: Behavioral Isolating Mechanisms

DESPITE THE TITLE *The Origin of Species*, and probably to the surprise of many readers, Darwin had little to say about the *origin* of species. He wrote mainly about *natural selection* and its influence in bringing about evolutionary change. The two are not synonymous. The particular branching process by which one species splits into two, or *speciation*, did not receive much attention from Darwin (Mayr 1942), and it was not until the mid-1900s that this process became reasonably well understood. The study of speciation has been a prime concern of twentieth century evolutionary biology, and progress has been marked by a series of summarizing synthetic works and symposia, such as Dobzhansky (1937), Mayr (1942, 1957), Stebbins (1950), White (1954), Blair (1961), and Grant (1963). Much of this literature has been summarized by Mayr (1963) and Dobzhansky (1970).

For understanding behavioral diversity among species, a knowledge of the speciation process is absolutely essential. Although the principal modern theory for the origin of species is not completely universal (since it does not apply to asexually reproducing forms), it is nevertheless one of the most fundamental and generally applicable of all biological theories. This chapter will explore the relationships between evolutionary theories of speciation and the existence of species-specific behavior.

SPECIES-SPECIFIC BEHAVIOR

Behavior that is *unique* to one species of animal is commonly termed *species-specific*. The vast majority of behaviors are not species-specific but are shared by two or more members of a group of species (see Chapter 1 on patterns of species diversity). Behavior that is unique may be quite conspicuous. Species-specific behavior may evolve in at least four ways: (1) by extinction of closely related species sharing the same behavior, (2) by a unique set of selection pressures operating in a unique environment, (3) by selection pressures that favor species differences per se, and (4) by chance.

Extinction / Many species are unique in structure and behavior because other species that shared the same traits have become extinct, leaving only one species remaining from a formerly more numerous group. The duck-billed platypus (*Ornithorhynchus anatinus*) is the only mammal that lays eggs in a nest in a burrow, but it is likely that at an earlier geological time other species having this trait coexisted with each other. For similar reasons, the spiny echidna (*Tachyglossus* spp.) is the only mammal that carries its eggs in a pouch. These are extreme examples, but many less spectacular ones could be cited from the lists of monotypic (containing one species only) genera and families in all groups of animals with a fossil record.

Unique Selection Pressures in Unique Environments / When a species becomes established on an island of suitable habitat that is remote from its main range, the new environment is likely to differ from the old in many ways. Not only will the physical environment be different, but the composition of the plant and animal community will differ. If the ecological island is an oceanic island, it is likely to be deficient in large predators and competing species. In such relatively rare situations unusual forms of life are frequently found. The Galapagos Archipelago has its marine iguanas, woodpecker finches, and nocturnal swallow-tailed gulls. Many oceanic islands have flightless species of birds, which presumably would not have evolved had the normal predators been present. Caves also offer unusual environments, and specialized but unrelated blind cavefishes and salamanders adapted to underground environments are found in various parts of the world. Many other cases could be cited of species that are specialized in some respect through the conventional evolutionary processes of adaptation to unusual or unique environments.

Species Differences Per Se / There is one situation in which behavioral differences between species may be selected for their own sake. When two species hybridize, behavioral and other traits may evolve that tend to prevent such interbreeding. In this way behavioral traits can evolve that separate a species from its most closely related species and are usually unique. Such traits need not enhance the ability of the species to survive or reproduce in any way except by the reduction of interbreeding with other populations. Since the evolution of such traits is closely tied in with the origin of new species, and since it may be the principal evolutionary cause for species-specific behavior, it will be necessary to consider the processes of mating behavior and speciation in relation to each other. To understand this relationship we must first briefly review the biological species concept, the nature of isolating mechanisms, and theories of the origin of species.

THE BIOLOGICAL SPECIES

Among the most common biological misconceptions of nonbiologists and even in some branches of biology is that a species is defined on the basis of appearance and resemblance to a particular type. This fallacy (the typological species concept) is sometimes reinforced through the use of simplified keys and identification manuals, which tend to give the impression that species are invariant. Historically, the typological species concept was a product of pragmatic taxonomy in the post-Linnaean period. Later, as biologists began to study geographic variation *within* species, instead of just identifying and classifying animals, they found that many species differ in their morphology and behavior in different parts of their ranges. The Steller's jay, for example, varies geographically in crest length, facial ornamentation, body coloration, head color, and other characters of potential usefulness in visual communication (see Figure 15–12). The populations having these variations cannot (on the basis of present evidence) be called different species, because the pattern of gradual intergradation between them suggests that no barriers to gene exchange other than geography are operative between neighboring populations.

The unique and biologically most significant property of a sexually reproducing species derives from the fact that it does not exchange genes normally with other species; or if it does occasionally, the hybrids are typically sterile or less fit in some way and so contribute relatively little to subsequent generations. For this and other reasons, species typically remain *genetically distinct* from each other. Genes within the gene pool of one species remain there and do not ordinarily enter the gene pool of another species, and vice versa (exceptions occur). This has consequences of tremendous importance. Each species is different from every other, and most are adapted to particular environmental conditions and life histories. The barrier to gene flow between species *protects these adaptations* by preventing genetic dilution and contamination from other species with different adaptations. This protection allows different species to live together in the same location, each specialized for its own way of life, without interfering genetically with each other.

Now that we have recognized the biologically most significant attributes of a species, we are able to provide a meaningful definition of a species as it exists at any one moment in geological time. (This definition is not applicable to fossil series.) Among sexually reproducing animals *a species is a population or a series of populations characterized by free actual or potential interchange of genes and by biological* (i.e., not just geographic) *barriers to such interchange with other populations under natural conditions* (after Mayr 1963). Notice that with this definition it is not necessary that all individuals of the same species resemble each other in appearance, nor must any two individuals from divergent populations necessarily share any genes in common—although they do for other reasons. The continuous morphological intergradation of Steller's jays from Alaska to Nicaragua suggests (but does not prove) that no significant barrier to gene interchange occurs within this whole series of populations. Consequently, they may be all assigned to the same species, even though the extremes of geographic variation in Alaska and Nicaragua would certainly have been consigned to different species on the basis of the Linnaean or typological species concept.

Notice also that lack of interbreeding is *not* the principal criterion. Interbreeding may occur freely between horses and donkeys, but mules are almost invariably sterile. There is essentially no genetic interchange between horses and donkeys despite their willingness to interbreed in captivity. Even similar-looking species may interbreed freely but fail to exchange genes because of hybrid inviability or sterility, as in certain grasshoppers (White 1968).

In the history of the species concept there has been a long-standing controversy between those who adhere to the species concept as described above and those whose concept is essentially typological. The latter school is represented philosophically today by numerical taxonomy and the study of character variation (phenetics). Historically, the biological species concept grew out of a phenetic approach when the latter failed to accommodate new knowledge of evolutionary processes derived from studies of geographic variation. Today it is clear that phenetics and "computer taxonomy" will make important contributions to evolutionary biology at the levels of local populations and of higher categories (above the species). It is unfortunate that some numerical taxonomists have been slow to accommodate modern concepts of isolating mechanisms in their computer programs.

The intricate relationships between phenetics and the biological species concept are not revealed by our definition. In practice, when patterns of variation in species are analyzed and decisions must be made as to whether or not a group of populations constitutes a species, phenetics is of crucial significance. For the vast majority of species most of our information on presence or absence of gene exchange is derived from phenetic study of specimens. Modern evolutionary biology cannot afford to be without either phenetics or the biological species concept. In the last analysis the two are mutually compatible and indispensable.

For dissenting views on the significance of the biological species concept, see Sokal and Crovello (1970) and Ehrlich and Raven (1969).

ISOLATING MECHANISMS

Types / The properties of populations or species that prevent or hinder genetic exchange between them are known as *isolating mechanisms* (Dobzhansky 1937). A species might be described as a set of populations bounded by effective isolating mechanisms. These can be variously classified; the present classification closely resembles those of Dobzhansky (1970), Mayr (1963), and Littlejohn (1969) (see Table 18–1), with minor departures.

Isolating mechanisms can be classified into those that tend to prevent individuals of two populations from encountering each other in breeding condition (types 1–4 in Table 18–1), those that prevent copulation from taking place even though the individuals of two popu-

TABLE 18–1 Types of isolating mechanisms.

I. Pre-mating mechanisms—prevention of interbreeding
- A. Pre-encounter mechanisms—prevention of contact
    1. Geographic (or locomotor) isolation—inability to cross a geographic barrier
    2. Ecological isolation—preference for different habitats, which may be contiguous
    3. Seasonal isolation—breeding in different seasons
    4. Temporal isolation—mating at different times of day
- B. Post-encounter mechanisms—opportunities for interbreeding present
    5. Sexual ("ethological," "behavioral") isolation—behavioral mating preference for own species

II. Post-mating mechanisms—interbreeding attempted but progeny absent or inferior
   6. Mechanical isolation—copulatory parts of different species do not match, preventing fertilization of eggs or separation of copulating partners
   7. Gametic isolation—gametes die after mating without fertilization occurring
   8. Inviability of hybrids or subsequent generations—developmental abnormalities
   9. Sterility of hybrids or subsequent generations—normal gametes not produced

lations may encounter each other while in breeding condition (type 5), and those that prevent genetic interchange even after interbreeding (e.g., copulation) has occurred or been attempted (types 6–9). Behavior plays an important part in the first five only.

1) *Geographic* (or locomotor) isolating mechanisms are based on the inability or reluctance of a species to cross a geographic or ecological formation, such as the Atlantic Ocean, the Great Plains, the Sahara Desert, the Rocky Mountains, or even a branch of the Amazon River. Clearly a mountain range is not a species property, but the ability of a species to cross a mountain range is certainly a property of the species.

2) *Ecological* isolation tends to keep apart species which live in different habitats. Such species would come into contact only where the habitats meet. Man's alteration of natural habitats has resulted in some cases in massive interbreeding between species that probably were previously kept apart significantly by habitat separation (Cory and Manion 1955; Anderson 1949).

3) *Seasonal* isolation tends to keep apart species that breed at different times of year. Among crickets, some species breed only in the fall and others only in the spring, with little or no overlap of breeders in the summer, as shown in Figure 18–1. Alexander and Bigelow (1960) have termed such species *allochronic* (separate in time). Those which breed at the same time have been termed *synchronic*.

4) *Temporal* isolation tends to keep apart species that breed at the same season and place but at different times of day. Among saturniid moths, temporal isolation is important between some species that have the same pheromone as a sex attractant (as we have seen in Chapter 17). For example, *Hyalophora promethea* mates early in the evening and *H. cecropia* mates mainly toward dawn (Wilson and Bossert 1963). *Drosophila pseudoobscura* and *D. persimilis* also mate at different times of day.

5) *Sexual* isolation has sometimes been called behavioral or ethological isolation, but since behavior is involved in four other types of isolating mechanisms the latter terms are not adequately precise. Although the term "sexual isolation" is inadequately descriptive, it is widely used. In this type, at least one sex can discriminate between its own and other species and mate accordingly even when confronted with individuals of similar species at the same time and place. This is certainly the most important type of isolating mechanism in animals in terms of behavior, and possibly also in terms of evolution.

Coaction of Isolating Mechanisms / Commonly several isolating mechanisms may function simultaneously. For example, the Fowler's

18–1 Seasonal isolation in the crickets *Gryllus* (*Acheta*) *veletis* (*left*) and *G. pennsylvanicus* (*right*) in a brushy pasture in Michigan. The temperature was generally below 50°F. Counts were higher on warm sunny days than cold, cloudy, or rainy days. (From Alexander 1968. Reprinted with permission of the Quarterly Review of Biology.)

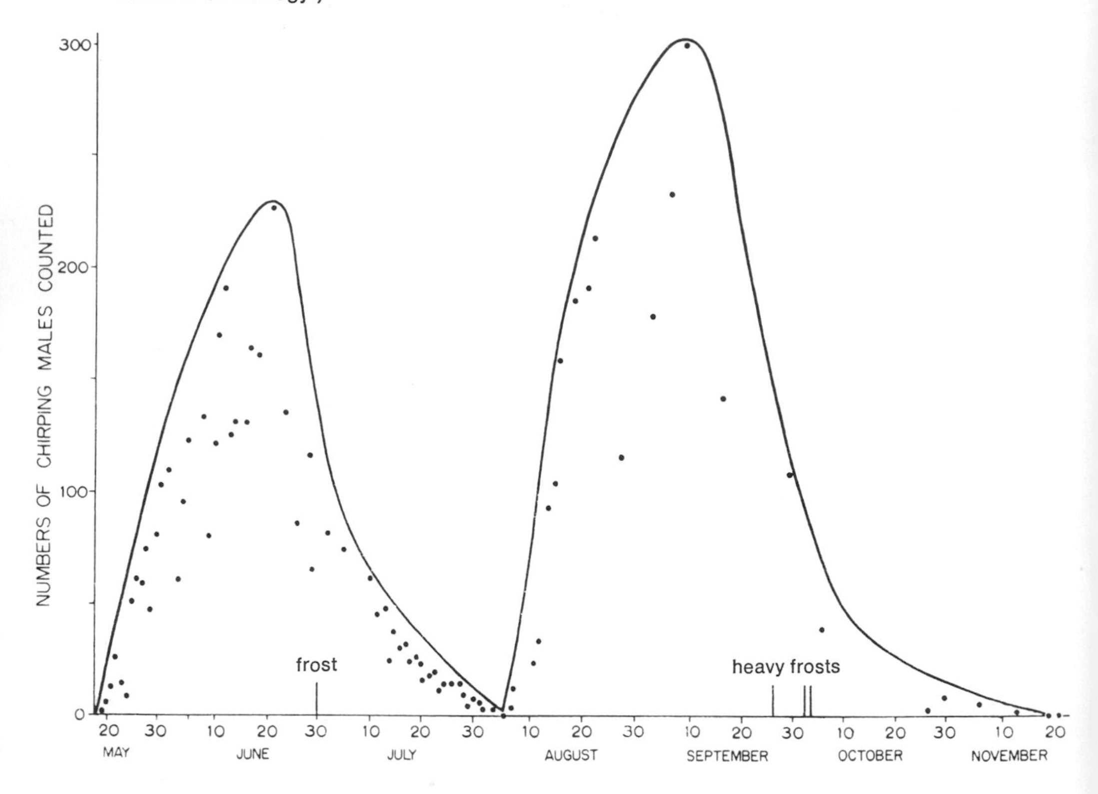

and American toads in Indiana have partly different breeding seasons, live in different habitats, and have different mating calls (Cory and Manion 1955). In saturniid moths there may be seasonal, temporal, ecological, and sexual isolating mechanisms, as well as possible post-mating mechanisms, all operating together. However, in animals ethological isolation is commonly the primary factor, without which reproductive isolation would break down (Perdeck 1958; Davis and Twitty 1964; Littlejohn 1965, 1969; Lloyd 1966; N. G. Smith 1966).

ALLOPATRIC SPECIATION

Darwin's observations in the Galapagos Archipelago that slightly different forms of the same species inhabited each island helped to lead him to the concept of natural selection. Similarly, it was partly through studies of geographic and genetic variation in animals that biologists arrived at the present theory of speciation (Mayr 1942). In the study of geographic variation of behavior, or of any biological trait, two populations are termed *sympatric* in the region where both are found together, and *allopatric* in the regions where only one is found but not the other. Allopatric speciation refers to the origin of a new species while it inhabits a separate geographic range from its parent species. Sympatric speciation is the postulated origin of a new species by division of one population into two without the geographic isolation required by allopatric speciation.

The essentials of the speciation process are outlined in Table 18–2 and Figure 18–2. Two principal phases can be recognized. The first of these concerns genetic events that take place during allopatry; the second concerns genetic events during subsequent sympatry.

Before the process begins, at t_0, one species exists that may or may not show geographic variation. If it is a large, continental population, the forms inhabiting geographically extreme parts of the range may be quite different from each other and may rarely if ever have the opportunity to interbreed (isolation by distance). The essential feature of the allopatric model (Phase I) is the role of geographic isolation. When a population becomes separated into separate parts that no longer have the opportunity to interbreed (t_1) due to geographic and ecological barriers (termed *ecogeographic,* since both are necessarily present at the same time), genetic divergence between the now two populations may occur through adaptation or genetic drift (t_2). Many species became broken into separate populations during the Pleistocene. At that time separate unglaciated areas, or refugia, were cut off from each other by the ice cap; for example, in Europe refugia were located in Spain, Italy, and the Balkans (where the subspecies of the honeybee, *Apis mellifera mellifera, A. m. ligustica,* and *A. m. carnica* probably differentiated). The Great Plains have served to isolate populations of woodland species in eastern and western North America from each other. The freshwater fishes of one river system are effectively isolated from those of another system, except when stream capture occurs. Oceanic islands also provide adequate isolation for speciation to occur.

The environmental conditions of isolation are almost always different for two populations. Consequently, natural selection tends to bring about adaptive differences between them. Genetic differences between populations may also come about through chance, especially if the populations have been very small at some time in their history, such as when the population was founded (here termed *genetic drift*). If these genetic differences include some that could function efficiently as isolating mechanisms when the two populations again meet, the

TABLE 18–2 A two-phase model for the origin of species and of behavioral isolating mechanisms. See also Figure 18–2.

Phase	Stage	Ecogeo-graphic barrier	Genetic divergence among populations	Actual gene exchange	Potential gene exchange*	Sympatry	Character displacement	Fitness of potential hybrids*	Potential frequency of hybrids*
	t_0: one population	—	none or little	free, much	free, much	not applicable	—	normal	high
PHASE I	t_1: geographic separation between two populations	+	little, at first	none	much	no	—	normal	high
PHASE I	t_2: isolating mechanisms may be evolving	+	more, through adaptation and genetic drift	none	may be reduced	no	—	may be reduced	high or a little lower
PHASE I	t_3: isolating mechanisms now adequate	+	present	none	little	no	—	may be reduced	low
PHASE II	t_4: start of secondary contact	—	present	little	little	yes, partial	—	may be reduced	low
PHASE II	t_5: selection against hybrids	—	present	even less	even less	yes, partial	+	must be reduced	even lower

ECOGEOGRAPHIC BARRIER—Ocean, river, mountain range, something that normally can't be crossed by the species concerned.

GENETIC DIVERGENCE AMONG POPULATIONS—Significant divergence before geographic isolation is prevented by free interbreeding between populations. After geographic isolation each population becomes adapted to its local conditions or may diverge through chance.

ACTUAL GENE EXCHANGE—Because of free interbreeding before geographical isolation, gene exchange is then high. After geographical isolation occurs, interbreeding and gene exchange are assumed to be essentially nonexistent. When the two populations come together again after the geographical barrier is gone or overcome, gene exchange or interbreeding is prevented by isolating mechanisms (other than geography).

SYMPATRY—Before isolation the populations intergrade ("primary contact"), and hence are not properly spoken of as sympatric or allopatric. During isolation they are completely allopatric. After the geographic barrier is overcome and the populations again come into contact ("secondary contact") they are partially sympatric and partially allopatric.

CHARACTER DISPLACEMENT—Divergence in region of sympatry due to selection pressures unique to the sympatric condition, e.g., in feeding competition, in behavioral isolating mechanisms.

FITNESS OF POTENTIAL HYBRIDS—Must be lower than that of nonhybrids for the second phase in the evolution of isolation mechanisms to proceed.

* Potential gene exchange—The degree of interbreeding, hence gene exchange, which would be expected if the barrier were removed at that stage.

populations may not interbreed or produce viable or fertile offspring (t_3). There is no selection pressure for the evolution of isolating mechanisms per se in allopatric populations, and those that arise do so for other reasons – as by-products. Frequently, when two formerly isolated populations reestablish contact (termed a *secondary contact*), interbreeding does occur on a wide scale (t_2), and genes from each popula-

18–2 History of one species becoming two as a result of isolation for a long period by a geographic barrier followed by removal of the barrier. Compare with Table 18–2.

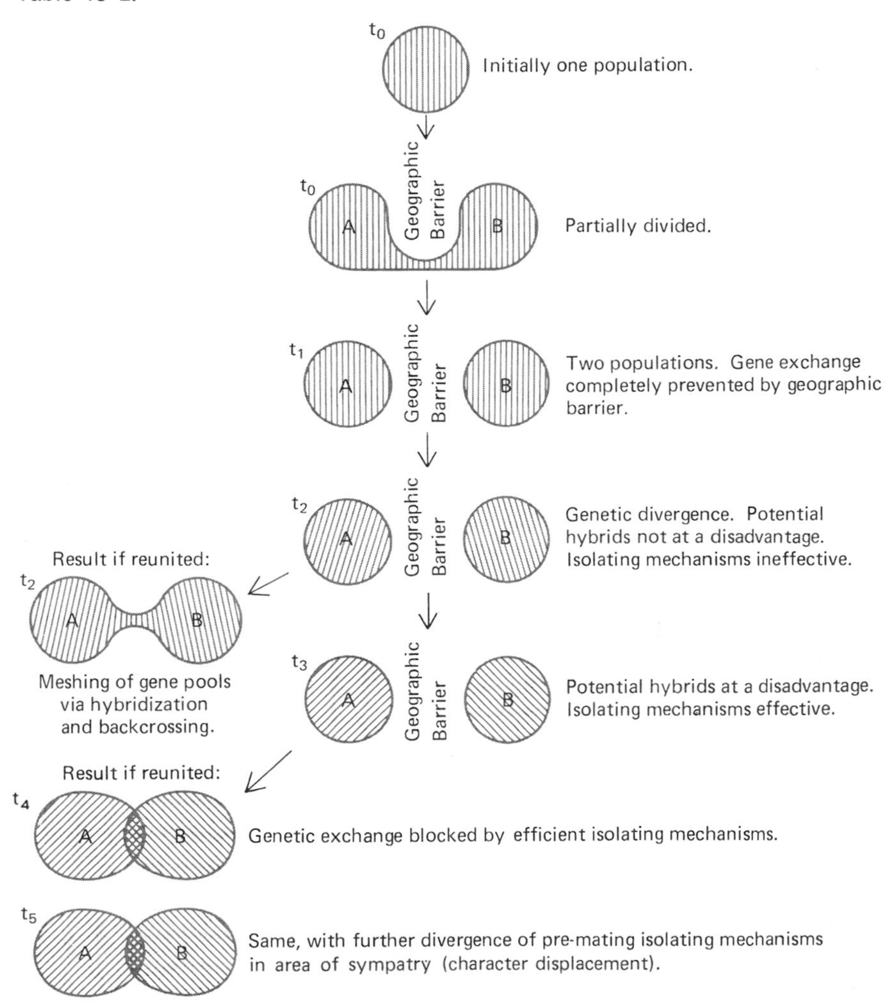

tion may spread into the other (*introgressive hybridization;* Anderson 1949). In some cases narrow "hybrid zones" are found where hybrids are numerous and outside of which traces of introgression are relatively rare (Bigelow 1965).

If the divergence in allopatry has reached the stage in which introgression would be insignificant if a secondary contact occurred, the two populations may belong to separate species. Borderline cases do occur. If the two populations are very different, but still allopatric, then one can only speculate whether introgression would occur if the two became sympatric. In many cases there is insufficient evidence available about the amount of introgression actually occurring, especially with large animals, since the decision is often made on the basis of detailed analysis of large collections of specimens.

There has been considerable uncertainty about how completely two populations must be isolated to allow speciation. If the isolation is insufficient, even relatively infrequent gene exchange may prevent isolating mechanisms between two populations from evolving. Ernst Mayr (1942, 1963) has strongly and consistently maintained that only the isolation provided by the inability of animals to cross ecogeographic barriers is complete and enduring enough to allow speciation. There seem to be no known cases that cannot be interpreted according to this model.

SYMPATRIC SPECIATION

Despite the general acceptability of the allopatric speciation model, belief has persisted that in some cases new species may have arisen in other ways, in the absence of geographic isolation. Seasonal isolation between the sympatric spring and fall breeding propulations (allochronic populations) of certain crickets is apparently complete and might have been the basis for sympatric speciation in one case (Alexander and Bigelow 1960), although this too is disputed. Chromosomal rearrangements occurring within a population resulting in sterility of hybrids have also been thought to be a basis for sympatric speciation in Australian grasshoppers (White 1968; but see Key 1968 for a contrary view).

New plant species may sometimes arise through hybridization or allopolyploidy. This seems not to happen in animals, although a laboratory case has been reported in wasps (Rao and DeBach 1969). Other suggested hypothetical possibilities are speciation from assortative mating (like mates with like) of genetic morphs within a population or from differentiated populations living in different but adjacent habitats (ecological isolation).

Mathematical and experimental arguments for the plausibility of certain postulated varieties of sympatric speciation have been given by J. M. Smith (1966), Thoday (1972), and others (Pimentel et al. 1967; Levins 1964). It has been proposed that the process of imprinting, in the original sense of Lorenz (1935), would facilitate sympatric speciation if imprinting were absolute (Seiger 1967 and earlier authors; see Chapter 27). Persuasive evidence from nature for the actual occurrence of most of these types of sympatric speciation is lacking, although imprinting appears to affect the choice of mate in the blue-snow-goose complex (Cooke and Cooch 1968). We lack this evidence partly because of the difficulties of determining what events occurred in past evolutionary history. With the allopatric model this difficulty is somewhat alleviated, because geological evidence can be used to document the past histories of intermittent land bridges and palaeo-habitat connections. It is difficult at present to assess the relative importance of sympatric versus allopatric speciation in individual cases, and the question should remain open. However, since most writers on the evolution of behavioral isolating mechanisms have employed the allopatric model, it will be convenient to restrict further discussion to it.

MATING SIGNALS AS ISOLATING MECHANISMS

What types of signals a species uses in its mating or precopulatory behavior, and which types might function as isolating mechanisms, seem to depend largely on which of the available sensory modalities is the most efficient. In many species, attraction of a mate is facilitated by the use of signals that can be received and responded to at a long distance from the signaler. In others, short-range signals may be the most efficient, especially when mobility is of a low order (compare crabs, fireflies, and birds) or when conspicuous long-range signals may attract predators.

Visual Signals / The visual appearance of a species is apparently often an important factor in choice of a mate and has been studied in a variety of species. The various species of North American sunfishes (*Lepomis*) differ considerably in breeding coloration, especially of their opercular flaps (Figure 18–3). The importance of these flaps in preventing hybridization was shown in experiments by Childers (1967). He found that in ponds stocked with female bluegills (*L. macrochirus*) and male red-ear sunfish (*L. microlophus*), interbreeding occurred when the scarlet portions of the opercular tabs of the males were removed but did not occur in ponds where the males were intact. These results suggest that visual appearance is critical for reproductive isolation between these species. Sounds are also likely to be important in these two and other species of *Lepomis* (Gerald 1971). The importance of vision in the isolating mechanism of *Lepomis* is also supported by the success of Noble (1934) and Breder (1936) in inducing typical spawning circling in male pumpkinseed (*L. gibbosus*) by manipulating fish-shaped models in the nest to resemble a spawning female.

The ability of breeding males of three sympatric species of sunfish from Ontario waters to discriminate among females of the three species was tested in aquaria by Keenleyside (1967). A male was presented with two females simultaneously, one of his own species and one of another. Each female was housed in a screw-top jar with a plastic screen cover; a female was placed at each end of the aquarium. The data shown in Table 18–3 reveal significant preference by the males for females of their own species in terms of the frequency of frontal displays, bites, and courtship circles given in response to females of each species.

These data establish that males can discriminate among females of the different species. Females also discriminate among males

18–3 Visual signals as isolating mechanisms in four species of sunfish. Note the tabs on the operculum (gill cover) used in species recognition. See also Table 18–3. (From Miller 1963. Artist, Rudolf J. Miller.)

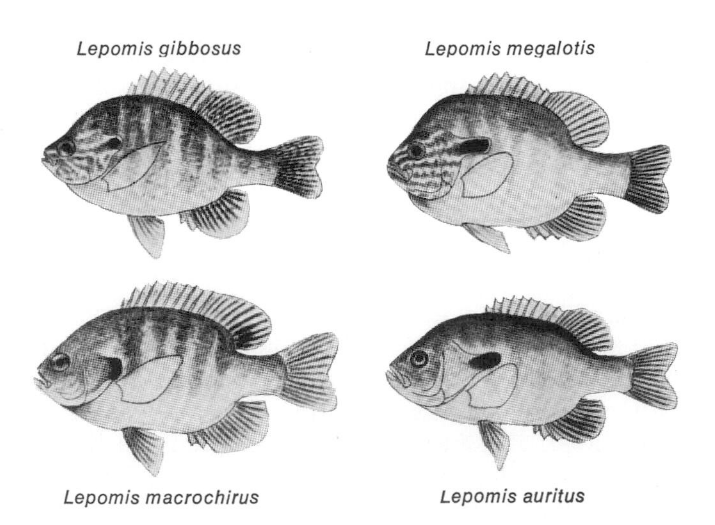

TABLE 18–3 Preference of male sunfish (*Lepomis*) for females of their own species. A male in an aquarium was presented simultaneously with females of his own and one other species in separate glass jars. The mean numbers of courtship and aggressive (biting) acts toward each species are shown. See Figure 18–3. (From Keenleyside 1967.)

Females presented in pairs	*n*	Frontal display	Bite	Courtship circle
MALE *gibbosus*				
gibbosus		81.8	68.9	104.9
and	12	**	*	**
megalotis		12.8	19.4	16.3
gibbosus		60.0	79.5	68.2
and	12	**	**	**
macrochirus		6.8	11.3	18.9
MALE *megalotis*				
megalotis		13.5	57.3	31.8
and	12			**
gibbosus		7.3	37.1	5.2
megalotis		8.8	68.8	45.8
and	12		*	**
macrochirus		3.6	23.9	13.7

* $0.01 < p < 0.05$. ** $p < 0.01$.

(Steele and Keenleyside 1971). In Ontario the breeding seasons, spawning times, and habitats of the different species overlap significantly; and hybridization does occur occasionally under certain conditions, such as extreme and unnatural crowding on the spawning grounds. Judging by survival and further reproduction in the resulting "hybrid swarms," post-mating isolating mechanisms are apparently relatively unimportant. Consequently, the principal mechanisms by which these species are kept distinct appear to lie in the differential behavioral responses of males (and possibly also females) to the visual appearance of the opposite sex in the various sympatric species. Auditory cues are probably also important in some species of *Lepomis* (Gerald 1971).

Among birds the features of the facial region are often important. In experiments under natural conditions in the Canadian Arctic, N. G. Smith (1966) carried out an experimental study of isolating mechanisms in gulls. Normally, glaucous gulls (*Larus hyperboreus*) and Thayer's gulls (*L. thayeri*) do not interbreed, but Smith was able to induce 55 female *hyperboreus* to pair with males of *thayeri* by painting the reddish-purple eye-rings of *thayeri* males to resemble the yellowish eye-rings of *hyperboreus*. These males, however, did not copulate with their mates unless the eye-ring colors of the females were altered to resemble the species of the male. Thus, sexual preference is expressed by both sexes but at different stages in courtship. Pairs captured after mating had occurred broke up when changes in the females' eye-ring color were made. Control pairs remained intact when left alone or captured and released. The method of altering the color of the eye-ring is shown in Figure 18–4, and the results of one experiment on already-mated females of Kumlien's gull (*L. glaucoides kumlieni*) are shown in Figure 18–5. Similar experiments on the other species showed that the appearance of the region of the eye is of crucial importance as an isolating mechanism among all four species studied. Alterations of the appearance of the wing tips, which also differ among species of *Larus*, had no detectable effects on paired female gulls and only a synergistic effect with eye-ring color on unpaired male gulls.

Young gulls are fed in the nest by their parents and so have ample opportunity to learn the facial pattern and other characteristics of their own species. It seemed possible that mating preference could be altered by providing foster parents of a different species. When eggs of the herring gull (*Larus argentatus*) and the lesser black-backed gull (*L. fuscus*) were exchanged, the foster young on reaching sexual maturity paired with adults of the foster species rather than with their own species (Harris 1970). Isolating mechanisms between these two species thus depend in part on learning and in part on differences in coloration.

In some species the isolating mechanisms include not only the ability to recognize a species-specific color pattern, as in the preceding

18–4 Experimental study of eye-ring color as an isolating mechanism in gulls. *Left:* Changing the eye-ring color of a Kumlien's gull. *Right:* A herring gull whose normally orange eye-ring has been darkened and widened to resemble that of a Thayer's or Kumlien's gull. (From N. G. Smith 1966.)

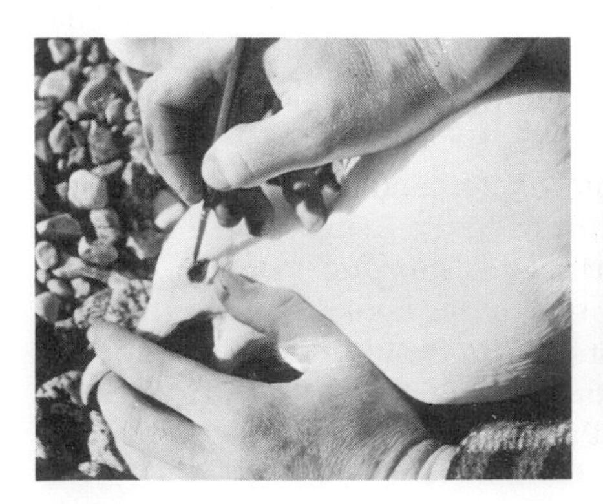

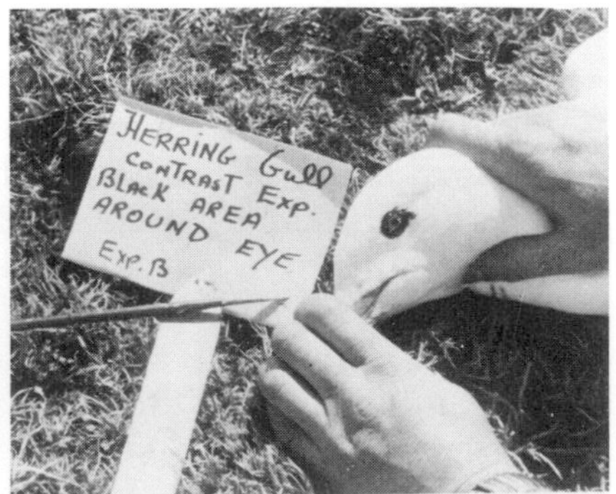

18–5 Isolating mechanisms in gulls. *Left:* Principal visual features thought to function in species recognition: eye-ring, iris, mantle (top of wings and back), and wing tips. *Right:* Results of altering the eye-ring color in mated female Kumlien's gulls before copulation. A change from purple to white resulted in the break-up of 21 of 34 pairs; other alterations and controls were relatively ineffective. (From N. G. Smith 1966.)

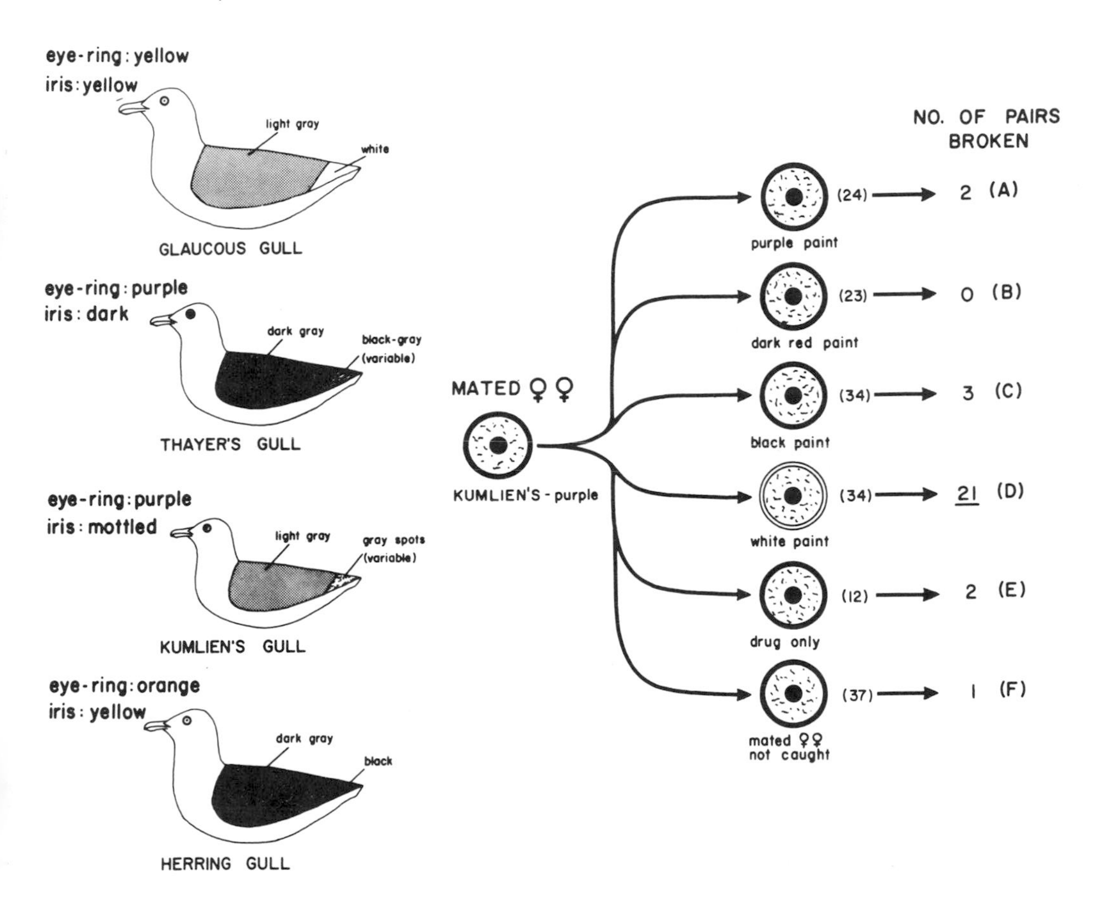

example, but also the ability to perform a species-specific motor pattern. Displays involving complicated pattern movements are found in a variety of species, including fireflies, fiddler crabs (*Uca;* see Figure 15–1), and the large genus of lizards, *Sceloporus.* The push-up and head-bobbing displays of males of several species of *Sceloporus* have been described by Hunsaker (1962). The species-specific head-bobbing patterns are illustrated in Figure 18–6.

In fireflies (Figure 18–7) male flashing patterns tend to be species-specific (McDermott 1917; Buck 1937; Seliger et al. 1964; Lloyd 1966).

18–6 Display action patterns that probably function as isolating mechanisms in the lizard genus *Sceloporus*. Males perform a species-specific head-bobbing movement. The height of the bob is shown on the ordinate and its time course on the abscissa. Preliminary tests suggest that females prefer to associate with models that bob in their own species-specific pattern. (From Hunsaker 1962.)

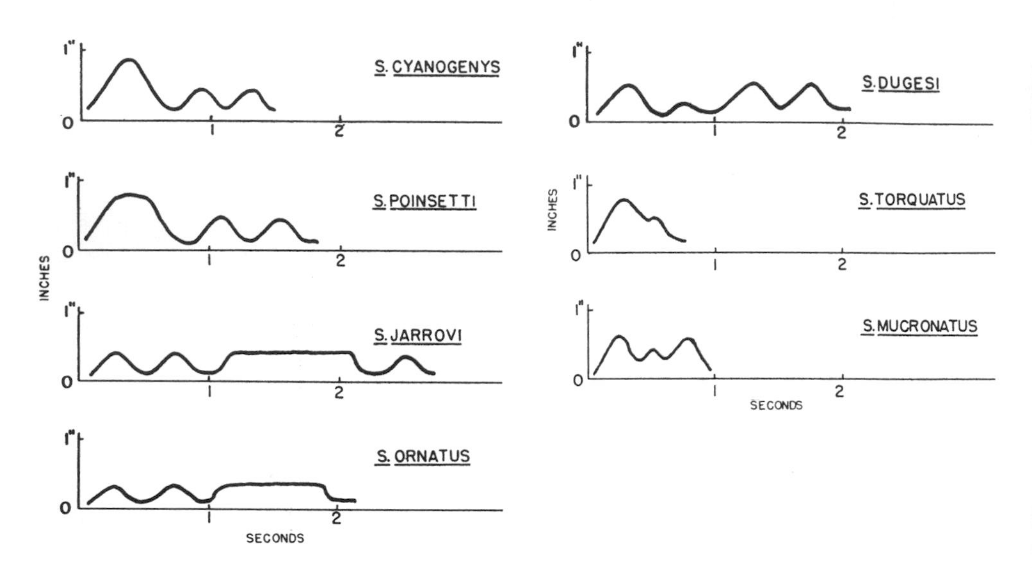

In this group females may flash in response to males; the isolating mechanism may involve successive recognition of two different, relatively complex patterns by two individuals.

Auditory Signals / Interest in auditory isolating mechanisms has probably exceeded that in all other types of species recognition mechanisms combined. Through the study of mating calls in a variety of groups of animals, from crickets to birds, many new species have been discovered that had gone undetected when only morphological characteristics were considered. Species that resemble each other to the extent that they become difficult to distinguish morphologically have been called *sibling species* (Mayr 1942). More recently, the term *cryptic species* has been preferred by some entomologists, since it does not carry the connotation that close morphological resemblance necessarily implies recency of common ancestry (Walker 1964). However, there is little doubt that in many cases sibling species are genetically similar (e.g., Hubby and Throckmorton 1968) and do have a relatively recent common ancestry. Moreover, the term *cryptic* is pre-empted by its use to designate camouflaging colors.

Sibling species occur in most of the major animal groups. Their

18–7 Flashing patterns that probably function as isolating mechanisms in the firefly genus *Photuris*. Females must answer with a delay that is also characteristic of the species for mating to occur. Species 2 and 14 are probably allopatric and are quite different morphologically. (From Alexander 1968*b*, after Barber.)

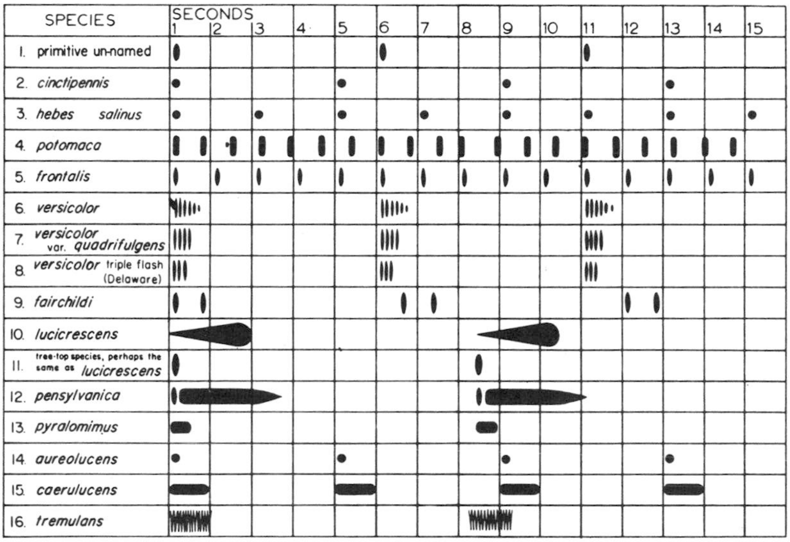

SPECIES	SECONDS 1	2	3	4	5	6	7	8	9	10	11	12	13	14	15
1. primitive un-named															
2. *cinctipennis*															
3. *hebes salinus*															
4. *potomaca*															
5. *frontalis*															
6. *versicolor*															
7. *versicolor* var. *quadrifulgens*															
8. *versicolor* triple flash (Delaware)															
9. *fairchildi*															
10. *lucicrescens*															
11. tree-top species, perhaps the same as *lucicrescens*															
12. *pensylvanica*															
13. *pyralomimus*															
14. *aureolucens*															
15. *caerulucens*															
16. *tremulans*															

phylogenetic occurrence has been reviewed by Mayr (1963), Blair (1962), and W. L. Brown (1959). In some groups they may be especially numerous; for example, among 167 species of sound-producing Gryllidae and Tettigoniidae in the eastern United States, approximately 40 were unrecognized or doubtful until their calling songs and life histories were known (Walker 1964)!

Among sibling species of insects, the European grasshoppers *Chorthippus brunneus* and *C. biguttulus* have been thoroughly studied with respect to isolating mechanisms by Perdeck (1958). Normally these species overlap broadly in geographic range, in habitat occupancy, and in time and season of breeding; yet Perdeck failed to discover any natural hybrids. The calling songs of these two species differ principally in pulse duration and rate (Figure 18–8). Normally the male calls and moves about randomly until a female of his species answers. He then alters his song and approaches her. Successful mating occurs only after male and female have exchanged chirps for a while. In laboratory tests Perdeck found that in the vast majority of instances both males and females would answer only to the opposite sex of their own species. Similarly, among males locomotion, answering song, and approach to an unseen female were much more frequently evoked by

18–8 Auditory isolating mechanisms in grasshoppers. Sound-level recordings of songs of (**A**) *Chorthippus brunneus,* (**B**) *C. biguttulus,* and (**C**) F_1 hybrid between these species. See Figures 16–8 and 16–9. (From Perdeck 1958.)

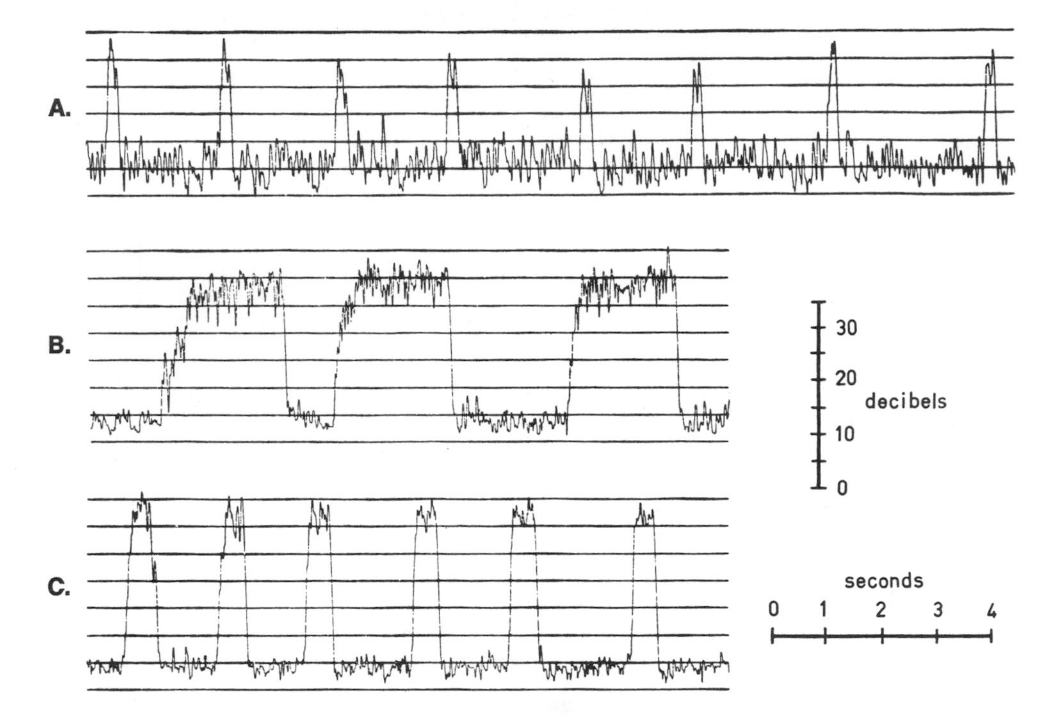

the female of the same species than of the wrong species. Interspecific matings could be achieved only by having conspecifics of each partner sing in the background, and even then they were difficult to obtain. Perdeck's behavioral experiments proved that auditory communication in these two sibling species functions as an important isolating mechanism and normally prevents interbreeding.

In order to establish the relative importance of species recognition as opposed to post-mating isolating mechanisms, further experiments were necessary on the hybrids and their progeny. Perdeck found that the hybrids and backcrosses were as fully viable and fertile as were the parent species. Consequently, it was concluded that behavioral mechanisms of species recognition are the *only* important isolating mechanisms between these two species, and of these, auditory communication was established as the principal one.

Studying the songs of hybrids and backcrosses revealed that the differences between the species were genetically determined. The songs of F_1 males (Figure 18–8), which were intermediate between those of the parent species, were not answered by females of either parent species. F_1 females answered to hybrid males and to males of one or both parental species.

In *Drosophila*, vibration, scissoring, and flicking of the wings are important parts of male courtship. The distinctness of the courtship sounds among various species of *Drosophila* has been demonstrated by Waldron (1964) and Ewing and Bennet-Clark (1968), as shown in Figure 18–9. These sounds are produced by the wings and sensed by females with their antennae. Experimental procedures and mutations that reduce the size or function of the antennae or wings reduce the effectiveness of male courtship (Burnet et al. 1971). The antennae are also necessary for sexual isolation in *D. pseudoobscura* and *D. persimilis* (Mayr 1950; Ehrman 1959; Ewing and Manning 1967). The interval between pulses of sound seems to be critical for species discrimination by females. In *D. melanogaster*, doubling or halving the pulse rate of artificial courtship sounds played to females reduced the mating success of wingless males; doubling or halving the pulse length had little effect. The song of the closely related *D. simulans* failed to induce sexual receptivity in female *D. melanogaster* under similar circumstances (Bennet-Clark and Ewing 1969).

Other studies of acoustical communication in arthropods in relation to isolating mechanisms have been reviewed by Alexander (1967).

Among birds, many pairs or groups of sibling species are known in which the species are distinguished most easily by their songs. Several of these have been investigated experimentally in respect to the value of song as an isolating mechanism. Presumably in most of these cases the important discrimination between individuals of separate species is made by the female, since it is usually only the male that sings. Unfortunately, females do not respond well to playbacks of test songs, but the response of males is conspicuous and easily quantified. Consequently, all studies to date on species recognition of bird song have been done on the responses of males to male song.

One of the earliest and most extensive of these studies was done by Dilger (1956) on six species of thrushes that breed in the northeastern United States. Although these species closely resemble each other morphologically, their songs are strikingly different (Figure 18–10). Since all six of them may breed within a few miles of each other in mountainous areas, and since the more northerly species are exposed to the songs of all the other species either on migration or on their breeding grounds, numerous opportunities for potential confusion exist. The response of the male of each species was tested to models and playbacks of each species in all possible combinations. These experiments established the importance of song as a mecha-

18–9 Courtship sounds of males of eight *Drosophila* species. In the simple songs, the species vary in pulse length and pulse interval. In the complex songs the frequency varies, alternating as shown in *D. micromelanica* and *D. paramelanica*. *D. athabasca* shows amplitude variations on a steady frequency of 440 c.p.s. (*Hz*). (From Bennet-Clark and Ewing 1970.)

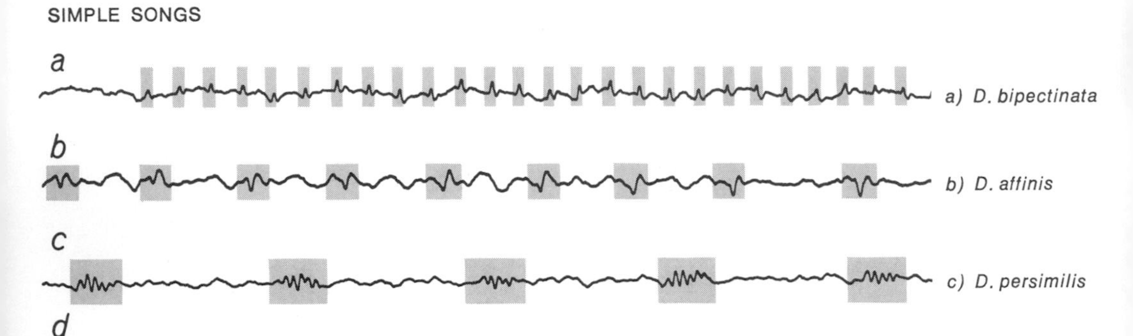

d

d) *D. pseudoobscur*

e

e) *D. ambigua*

200 C.P.S. reference

MORE COMPLEX SONGS

a 440 C.P.S. 385 C.P.S.

a) *D. micromelanic*

b 440 C.P.S. 265 C.P.S.

b) *D. paramelanica*

c 440 C.P.S.

c) *D. athabasca*

440 C.P.S. reference

nism of species recognition among males. When natural models were presented without any songs they were typically attacked by all species, regardless of which species the model resembled. The males did not appear to discriminate on the basis of coloration, even though a trained human observer could do so. When songs were coupled with models, only the species whose song was being played attacked the model, and all models were attacked regardless of their species. Thus each species responded to its own song, even if the model had the wrong appearance; and it ignored songs of other species, even if the model was of the same species (Table 18–4).

In Europe two species of tree creepers (*Certhia familiaris* and *C. brachydactyla*) are found that are even more similar morphologically than the above-mentioned thrushes, but whose songs are easily dis-

TABLE 18–4 Results of playbacks of male songs to males of five thrush species. When a song was being played, males attacked models of any species only if the song was of their own species. Silent models were attacked regardless of species. (From Dilger 1956. Reproduced by courtesy of the American Ornithologists' Union.)

MODELS (columns); TERRITORIAL MALES (rows)

silent

	m	f	g	u	mi
m	X	X	X	X	X
f	X	X	X	X	X
g	X	X	X	X	X
u	X	X	X	X	X
mi	X	X	X	X	X

mustelina song

	m	f	g	u	mi
m	X	X	X	X	X
f					
g	S	S	S	S	S
u					
mi					

fuscescens song

	m	f	g	u	mi
m					
f	X	X	X	X	X
g					
u					
mi					

guttatus song

	m	f	g	u	mi
m	S	S	S	S	S
f					
g	X	X	X	X	X
u					
mi					

ustulatus song

	m	f	g	u	mi
m					
f					
g					
u	X	X	X	X	X
mi					

minimus song

	m	f	g	u	mi
m					
f					
g					
u					
mi	X	X	X	X	X

m = mustelina
f = fuscescens
g = guttatus
u = ustulatus
mi = minimus

The symbols are
X—attack by all territorial males tested
S—some attack by a few territorial males
blank—no overt attack

18–10 Auditory and visual stimuli used in species discrimination by five species of thrushes. (Illustration of thrushes from Dilger 1956*b*. Songs from Stein 1956.)

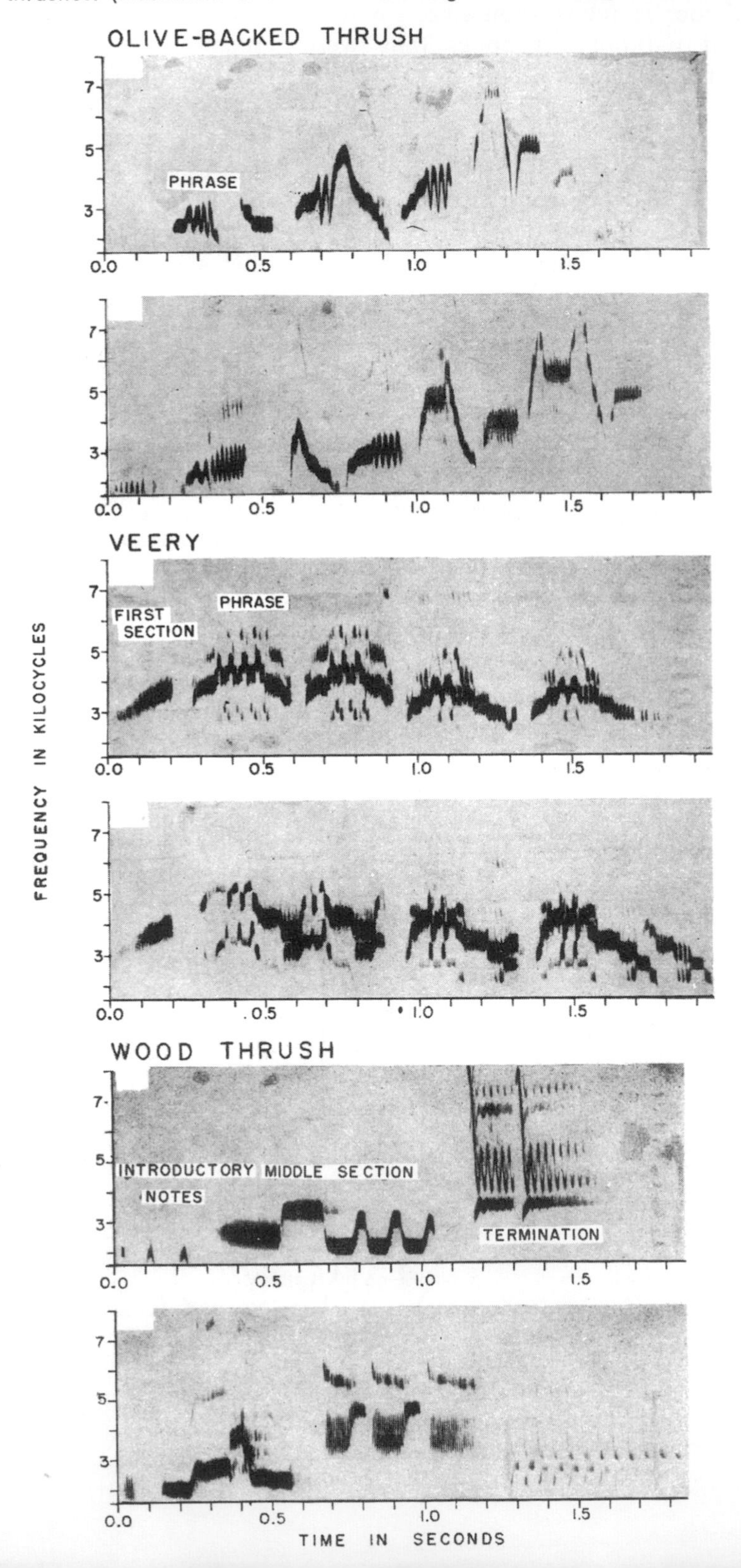

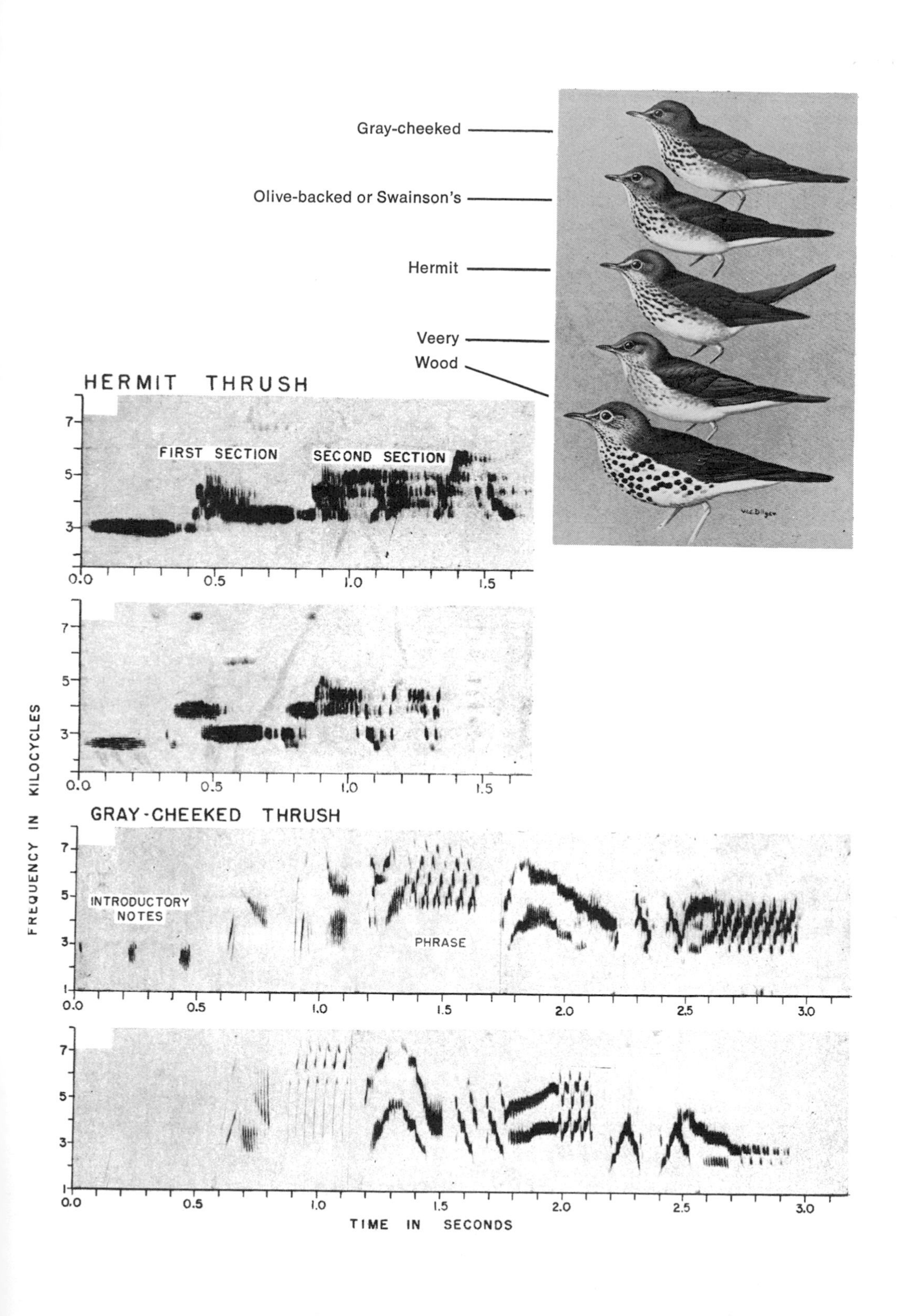

Gray-cheeked
Olive-backed or Swainson's
Hermit
Veery
Wood
HERMIT THRUSH
FIRST SECTION
SECOND SECTION
FREQUENCY IN KILOCYCLES
GRAY-CHEEKED THRUSH
INTRODUCTORY NOTES
PHRASE
TIME IN SECONDS

tinguished (Figure 18–11). When songs are played to them in the field, each species typically approaches, sings, and calls to its own song but not to that of the other, or to the similar-appearing American species, *C. americanus* (Thielcke 1962).

Other groups of birds in which the role of song as an isolating mechanism has been investigated by experiments on species recognition include the following: Traill's flycatcher complex, *Empidonax* spp., Stein (1963); nightingale thrushes, *Catharus* spp., Raitt and Hardy (1970); *Myiarchus* flycatchers, Lanyon (1963); the blue-winged warbler, *Vermivora pinus,* and the golden-winged warbler, *V. chrysoptera,* Gill and Lanyon (1964) and Ficken and Ficken (1969); the white-throated sparrow, *Zonotrichia albicollis,* Falls (1963); *Phylloscopus* warblers, Schubert (1971*a, b*) and Bremond (1968). In all these cases, conspecific males responded by singing and calling and sometimes by attacking a stuffed specimen, if present.

In the study of auditory isolating mechanisms the frogs and toads have been especially suitable subjects. Not only are their calls and songs less subject to the ontogenetic effects of learning and experience than those of birds and mammals, but the animals may be easily captured from nature, and tested and hybridized under controlled laboratory conditions. Moreover, the responses of female anurans can be tested easily, which is not true for birds.

The specificity of response in male bullfrogs was tested by Capranica (1965) by playing the mating calls of 34 species, both sympatric and allopatric. Males responded only to calls of their own species. (Capranica's experiments using artificial frog calls are described in Chapter 22.)

A striking example of a sibling species that was undetected with morphological criteria but was discovered by analyses of mating calls is found in the tree frogs of the *Hyla versicolor* complex. The two species in this complex are completely indistinguishable morphologically. Blair (1958) first pointed out the two song types. The songs of the two species differ in trill rate: *H. versicolor* sings at 17 to 35 notes per second, and *H. chrysoscelis* at 34 to 69 (data uncorrected for temperature) (Johnson 1966). The ranges of the two species overlap broadly. Females were shown to be able to discriminate between the fast and slow songs in laboratory tests (Littlejohn et al. 1960). That the two call types really represent distinct species and not just variants of a single species was shown by the failure of fertilized eggs from hybridizing pairs to develop; there was genetic incompatibility between the song types, but not in crosses within either type (Johnson 1959, 1963; Blair 1958). A similar history of discovery that one "species" was actually

18–11 Songs serving as probable isolating mechanisms in creepers. The two sympatric species of European creepers (the long-toed tree creeper, *Certhia familiaris,* and short-toed tree creeper, *C. brachydactyla)* react to their own song but not to songs of the other sympatric species or to the song of the allopatric American species *(C. americana).* (From Thielcke 1962.)

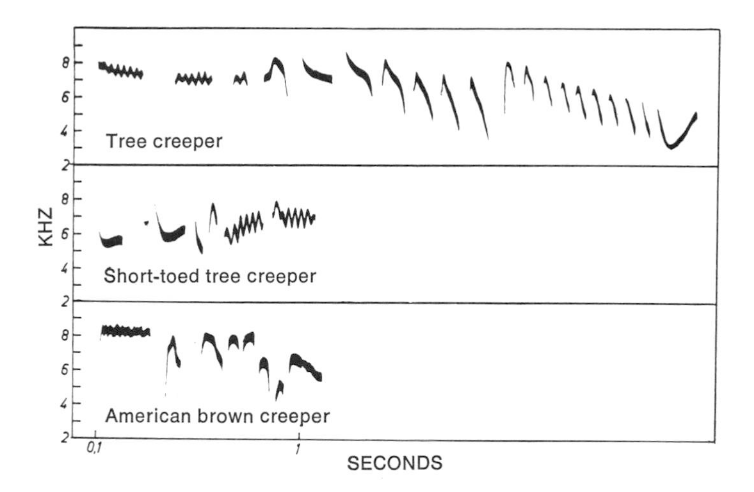

composed of several species with different mating calls occurred in the so-called *Rana pipiens* (Mecham 1968, 1971; Littlejohn and Oldham 1968; Brown, L. E. and Brown 1972).

Other experimental studies of mating calls as isolating mechanisms in frogs and toads include the following: *Pseudacris streckeri* (Littlejohn and Michaud 1959), *P. clarki* and *P. nigrita* (Michaud 1962), *P. triseriata* and *Hyla crucifer* (Martof and Thompson 1964), *H. regilla* and *H. californiae* (Ball and Jameson 1966), *Scaphiopus hurteri* and *S. couchii* (Awbrey 1968), and *P. streckeri* versus *P. ornata* (allopatric populations; Blair and Littlejohn 1960). These and related studies on mating calls and isolating mechanisms in frogs and toads have been reviewed by Bogert (1960), Mecham (1961), Martof (1961), Blair (1964), and Littlejohn (1969).

Although mating calls are potentially important as isolating mechanisms in all vocal species of frogs and toads, other types of behavioral isolating mechanisms may also be important in some cases. For example, the calling site may be important. Johnson (1966) observed that, in the same areas, the fast-trilling *H. chrysoscelis* tended to call from higher up than the slow-trilling *H. versicolor*. Duellman (1967) observed that in one pond where ten species of hylid frogs bred, not only did the songs of all species differ, but also the calling sites of most species. Some species called from the upper sides of leaves of emergent vegetation, some suspended from leaves and branches of bushes over the water, and some from the surface film of the pond. In the spadefoot toad species *Scaphiopus couchii* and *S. hurterii*, Awbrey's (1968) experiments suggested the hypothesis that sexually responsive females of both species orient to calling males of both species but are unable to discriminate between them on the basis of the quite-similar mating calls. However, these two species are separated ecologically by their preference for different soil types, and their hybrids and backcrosses display low levels of viability (Wasserman 1957).

Fishes also use auditory communication, and it may prove to be significant as an isolating mechanism in some cases, as in the minnow genus *Notropis* (Delco 1960; but see Stout 1963) and in sunfishes, genus *Lepomis* (Gerald 1971).

Chemical Signals / Among the most difficult behavioral isolating mechanisms to detect and study are those we cannot perceive directly with our own senses. In the earlier chapter on chemical communication we have seen that in moths (e.g., Saturniidae, Sphingidae, Pyralidae) sex attractants may function as partial isolating mechanisms, especially between genera, but that they are not necessarily by themselves complete isolating mechanisms. Other differences among species of moths—in habitat or in season and time of mating—also function as pre-mating isolating mechanisms.

Gustatory stimuli, in the form of secretions of the femoral pores on the hind legs and tasting with the tongue, are suspected to be important as isolating mechanisms in *Sceloporus* lizards (Hunsaker 1962). Gustatory stimuli have also been suspected to be important as isolating mechanisms in snakes (Neill 1964).

Contact chemical isolating mechanisms are probably important in a great variety of insects, as in *Drosophila melanogaster* and *D. simulans* (Manning 1959*a*, *b*). Ewing and Manning (1967) have expressed the belief that species discrimination throughout the genus *Drosophila* is probably based on chemical differences, often in preliminary contacts with the female by the male's fore-tarsi by means of chemosensory hairs present on both individuals. This view is supported by the experimental work of Narda (1968 and earlier papers cited there) and Shorey and Bartell (1970). Ehrman's (1969) study of the rarity effect in choice of a mate in *Drosophila* also suggests that olfaction is important. Vibratory stimuli may also be involved, as we saw earlier in this chapter.

Olfactory stimuli may prove to be important as isolating mechanisms in mammals; but although much evidence of the importance of pheromones in mice is available (e.g., Whitten 1966; Whitten and Bronson 1970; Bronson 1971), detailed experimental studies of their role in isolating mechanisms is still relatively scarce (Rauschert 1963; Moore 1965).

EVOLUTIONARY DIVERGENCE OF BEHAVIOR IN ALLOPATRY

According to Phase I of the allopatric model of speciation, populations may diverge with respect to certain properties that *later*, after secondary contact, may become isolating mechanisms. This view was favored by Darwin (1859), Poulton (1908), Muller (1940), and Patterson and Stone (1952). The selection pressures, if any, responsible for this divergence are unrelated to the need for discrimination from the stem species. Divergence of isolated populations may occur because of different environmental conditions, and hence different selection pressures, or by chance effects on gene frequencies, especially among founders and in small populations. Divergence, if it occurs at all, may occur in potential pre-mating or potential post-mating isolating mechanisms; and there would seem to be no inherent reason why one should consistently precede the other. The behavioral differences among allopatric populations of a species have been studied in respect to the mating signals of males, the responses of females to the signals, and the relative effectiveness of divergences (often of unknown kind) as isolating mechanisms. Of these, divergence in the signals is most easily studied.

Divergence of Mating Signals in Allopatric Populations: Visual Signals / An appreciation of the amount of evolutionary divergence in potential isolating mechanisms that can evolve among allopatric populations may be gained by examining geographic variation in biological traits which are known to function as isolating mechanisms. Those traits that are most interesting behaviorally in this respect are the signals which have been demonstrated by experiment to be involved in species isolation. For these purposes, visual and auditory mating signals are more easily analyzed that those in other sensory modes.

For species that differ conspicuously in appearance and have adequate visual systems, it is commonly but tentatively assumed that visual appearance is important in species identification. The common occurrence of geographic variation in coloration suggests that this is a possible source of isolating mechanisms. But only for a few species have the aspects of coloration most relevant to species recognition, such as the facial region in gulls, been identified experimentally. And it is known that in some groups differences in appearance are essentially ignored in species recognition—for instance, the several species of thrushes (see Figure 18–10) studied by Dilger (1956).

The facial region may be particularly important. Geographic variation in facial appearance has been documented for Steller's jays (see Figure 15–12), for Brazilian species of marmoset (Figure 18–12), and for many other species (e.g., titmice, drongos). Geographic variation in facial appearance is striking in African monkeys (Dorst and Dandelot 1970). It is known that changing the coloration of the facial region in birds can cause the disruption of established pairs or drastic changes in behavior (flicker, Noble 1936; mourning dove, Goforth and Baskett 1965; gulls, Smith 1966). In the estrildid finches *Pytilia phoenicoptera* and *P. lineata*, Nicolai (1968) observed that interspecific pairing failed to occur even when unpaired and opposite-sexed individuals of each species were placed together. Since these species differ only in bill color, this seems likely to be the basis of the isolating mechanism.

18–12 Geographic variation in faces of marmosets in South America. These races, or subspecies, are allopatric populations of the one species *Saguinus fuscicollis.* The differences among populations are so great that they might conceivably act as isolating mechanisms if the populations were to make secondary contact—although this is purely speculative. (From Hershkovitz 1968.)

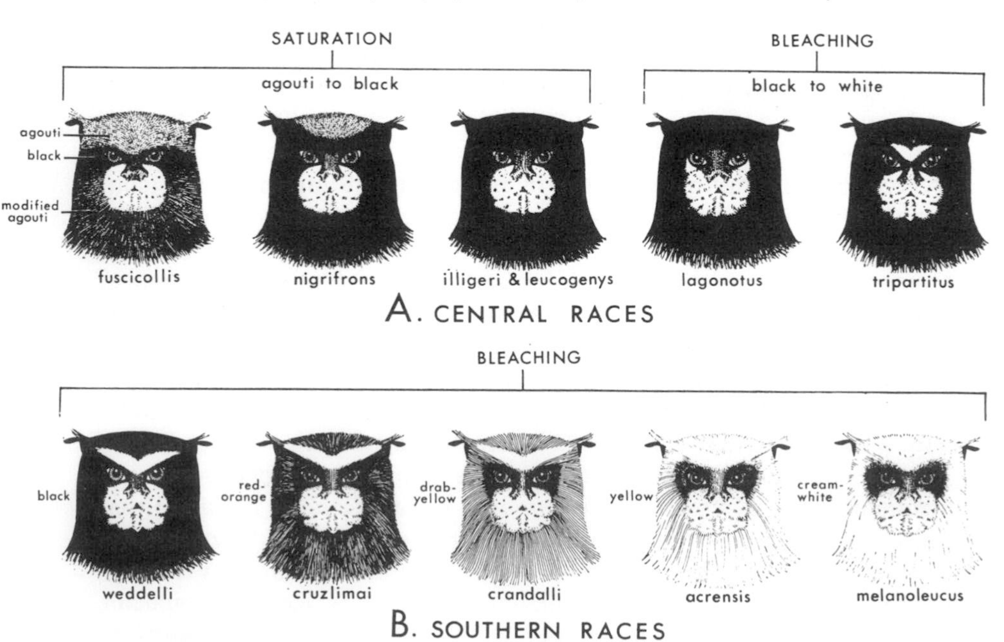

Facial pattern is probably more important than song in species recognition among males in the blue-winged and golden-winged warblers (Ficken and Ficken 1968, 1970; Gill and Lanyon 1964; see also Short 1969).

Examples of intraspecific variation in the frequency of color phases in head color and pattern have also been described (blue and snow geese, *Anser coerulescens*, Cooch and Beardmore 1959; bush-shrikes, *Chlorophoneus*, Hall et al. 1966; murres, *Uria*, Southern 1966).

In some populations assortative or homogamous mating (like mates with like) between color phases occurs. Among gibbons (*Hylobates lar*), Fooden (1969) found that homogamous pairs of the dark and pale color phases were more common than expected by chance in some populations but not in others. The blue and snow geese of the Canadian Arctic were formerly considered separate species, but are now regarded by most as color phases of one species. They interbreed commonly, but homogamous matings are predominant.

Geographic variation in visual signals important in mating is also known in display action patterns. Male lizards of the species *Uta stansburiana* have a push-up display that might be used by females as an aid in species recognition. In other lizards push-up displays are known to influence females (Hunsaker 1962; Jenssen 1970). The push-up pattern varies within and among populations, as shown in Figure 18–13.

Divergence of Mating Signals in Allopatric Populations: Auditory Signals / For studies of geographic variation in potential isolating mechanisms, the mating songs of frogs, toads, and insects are especially useful, because they are easy to record and analyze and relatively free of complications due to learning from adults. Two types of studies can be recognized. In the first, two sets of populations are compared that are completely separated from each other by a geographic barrier; these are referred to as *disjunct populations*. In the second, series of *intergrading populations* that are more or less continuously distributed over a wide geographic range are examined.

An example of the first approach, in which insular populations were compared with those on the mainland, is a study by Littlejohn (1964). He found that mating calls of the frog *Crinia signifera* from Tasmania and nearby Flinders Island were significantly different from those recorded on the adjacent mainland of Australia. The mating call of this species is a short, pulsed, cricketlike chirp, rapidly repeated. Calls recorded on the islands were longer, slower, and contained fewer pulses. There was considerable overlap in the ranges of variation of the island and mainland samples; so it is unlikely that the observed

18–13 Geographic variation in a display action pattern of the lizard *Uta stansburiana. Above:* Push-up displays in various populations. The relative height of the push-up is indicated by the ordinate; the abscissa represents elapsed time in tenths of a second. From one to five display types may occur in a population; the percentage of occurrence of each type in the population is given (observed number in parentheses). *Below:* Modal number of peaks in the push-up display in various parts of the United States. Dash line indicates the limits of distribution of the species; stippling indicates areas of intergradation between subspecies named. (From Ferguson 1971.)

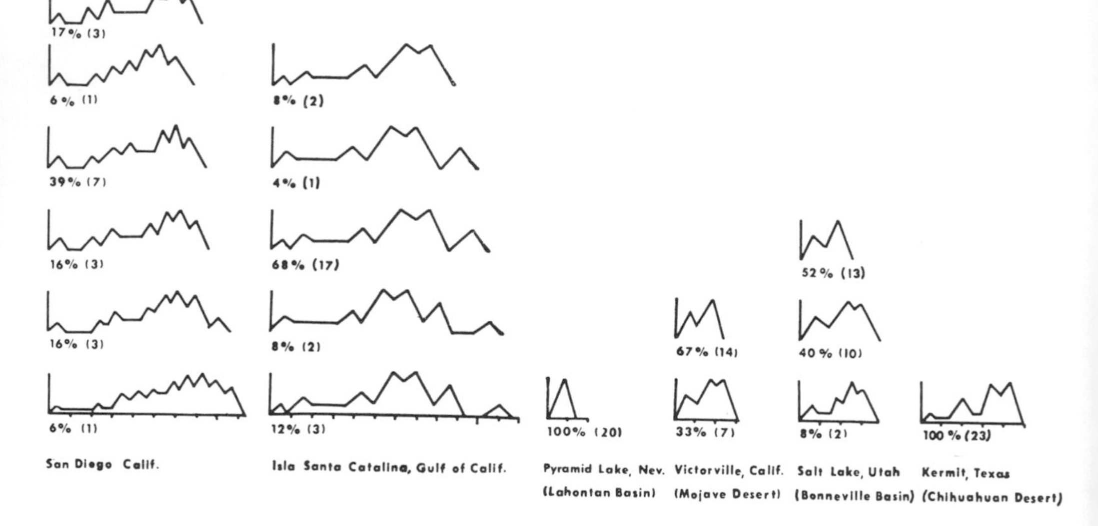

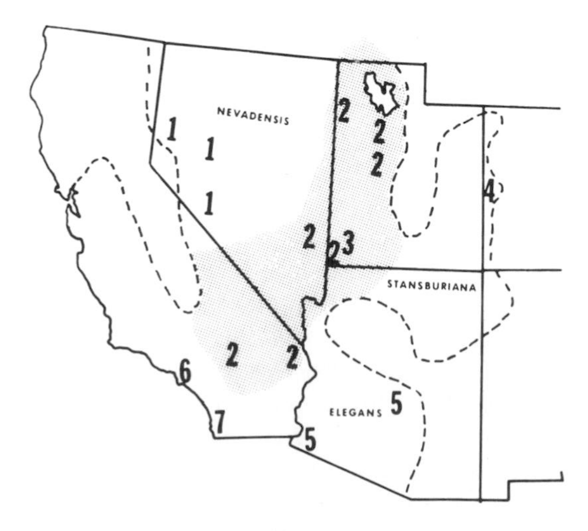

differences would completely prevent interbreeding if the populations were to regain contact now. It is tempting to conclude that these differences evolved in the 12,000 years since the islands became discontinuous with the mainland; but it cannot be assumed that the differences arose in that time, since they might have been present, at least partially, before the isolation of the islands. Differences in mating calls among island populations of *Hyla ewingi* were also reported by Littlejohn (1965).

Evolutionary change in mating call is not a necessary consequence of long isolation. The calls of insular populations of *Crinia insignifera* isolated for 8,000 years show no divergence from those of mainland Australia; another species, *Heleioporus eyrei,* also shows no difference between island and mainland populations (Littlejohn 1959). The spring-breeding and fall-breeding species of cricket *Gryllus veletis* and *G. pennsylvanicus* have identical songs, and the European *G. campestris* is only slightly different despite long isolation. Also, *G. firmus* of the mainland United States and *G. bermudiensis* of Bermuda have similar songs despite presumed separation since Oligocene time (Alexander 1962). Further examples of a lack of difference in mating call between sibling species are given by Alexander (1960). These examples suggest that stabilizing selection may be acting strongly on mating calls and that they may remain unchanged for thousands of generations.

The disjunct populations of frogs known as *Pseudacris streckeri,* of Texas, and *P. ornata,* of the southeastern United States, were compared by Blair and Littlejohn (1960). These species are thought to have diverged during the Pleistocene glaciation, or perhaps earlier. The calls of the two species are quite similar, differing mainly in dominant frequency (*streckeri,* 2,225–2,430 *Hz; ornata,* 2,750–2,850 *Hz*); but tests with *gravid* female *streckeri* revealed a high degree of preference for the *streckeri* call in choice experiments. The differences in frequency in this case appear to be by-products of the size differences between the species (larger body and lower frequency in *streckeri*). In general, among frogs and toads, larger species have lower-pitched calls—although exceptions occur (see Figure 16–5).

The mating calls of a disjunct population of the toad species *Bufo hemiophrys* were found to be significantly longer and higher pitched than those of the nearest populations at least 500 miles away (Porter 1968). A disjunct population of the wood frog *Rana sylvatica* in the same region also shows divergence in call structure from its nearest neighboring population (Porter 1969).

A detailed study of geographic variation in mating calls in an extensive continental area was done by Snyder and Jameson (1965) on

37 populations of *Hyla regilla*, a species of frog widespread in western North America. Using sonagrams of mating calls from 295 individuals, they determined the dominant frequency, the duration of each note in the call, the interval between notes, and the total duration of the call. They also recorded call repetition rate and water temperature. As in other species of anurans, temporal qualities of the call (the measurements of rate, duration, and interval) were correlated with water temperature, and frequency was correlated with body size. Comparisons among geographic areas were corrected for temperature. Using a computer to perform an analysis of discriminant functions, they found significant differences among all seven groups of samples, as follows: (1) Pacific Northwest lowlands, (2) Oregon mountains, (3) California coastal and valley lowlands, (4) central California mountains, (5) southern California lowlands, (6) southern California mountains, and (7) Baja California. Differences among populations were found in all the call measurements. For example, the mountain populations of Oregon and central California have lower dominant frequencies than the lowland populations of California. The largest differences in call characters among populations were found in the southern parts of the range, especially among disjunct populations exposed to different environmental extremes. Taken together, the results indicate a rather high degree of geographic variation in mating-call parameters, of an order of magnitude that in *Pseudacris streckeri* and *P. ornata* was associated with a strong conspecific mating preference by females.

In certain other species of frogs and toads a striking absence of geographic variability in mating calls has been found over large continental areas. Littlejohn (1959) found very little geographic variability over the entire range of *Crinia insignifera* and over 500 miles of the range of *C. signifera* in Australia, and he remarked that this seemed to be a general feature of the calls of all species so far studied.

Geographic variation in the songs of birds has long been known and has been reviewed by Thorpe (1961), Borror (1961), E. A. Armstrong (1963), and Thielcke (1969). Comprehensive studies of geographic variation based on analyzing sonagrams of songs have been done on the European tree creeper, *Certhia* spp. (Thielcke 1961, 1965); the white-throated sparrow, *Zonotrichia albicollis* (Borror and Gunn 1965); the song sparrow, *Melospiza melodia* (Borror 1965); the white-crowned sparrow, *Z. leucophrys* (Marler and Tamura 1962); the cardinal, *Cardinalis cardinalis* (Lemon 1966); and several other species (e.g., Nottebohm 1969; Shiovitz and Thompson 1970; Borror 1967).

In all these species of birds significant geographic variation in song pattern has been found. Interpretating the evolutionary significance of these "dialects" is difficult because of the complex relationships between heredity and environment in the ontogeny of bird song. The observed geographic variations cannot be ascribed entirely to genetic differences, nor can they be dismissed simply as "learned" without verification through experiment. The ontogeny of bird song is discussed in more detail in Chapter 28. The studies cited there suggest that even when learning plays an important role in the perpetuation of song "dialects," as in the white-crowned sparrow, genetic factors still are involved in the development of song pattern. In most cases their importance in song "dialects" cannot be ruled out, nor have they been proven.

Homogamic Mating in Allopatric Populations / The evolution of behavioral isolating mechanisms in allopatry can also be studied experimentally. Two approaches have been employed. In the first, separate laboratory strains derived from a common stock are examined for evidence of isolating mechanisms. In the second, natural populations of a species are sampled at various geographic locations and tested against each other in the laboratory for evidence of isolating mechanisms. According to the theory outlined earlier, varying degrees of reproductive isolation are expected among these samples.

Strains from the same population may be isolated for several to many generations, or they may be selected for some trait that is not of direct importance to sexual behavior. After a period of isolation, with or without selection, the strains may then be tested against each other to determine whether any significant homogamy (like types tend to mate) or *heterogamy* (unlike types tend to mate) has evolved. Various methods for determining experimentally the degree of sexual isolation (due to mating behavior) between populations and expressing it numerically have been clearly summarized by Parsons (1967:79–86). The *joint isolation index*, used by Ehrman, may vary from +1.0 in the case of complete homogamy to zero (random mating) to −1.0 in the case of complete heterogamy. Ehrman (1964) determined joint isolation indices for six populations of *Drosophila pseudoobscura* that had been derived from the same initial population but had been maintained in genetic isolation for over four years. Male-choice experiments of the form

$$10\ \text{A}♂♂ \times 10\ \text{A}♀♀ + 10\ \text{B}♀♀$$
$$10\ \text{B}♂♂ \times 10\ \text{A}♀♀ + 10\ \text{B}♀♀$$

were carried out, where A and B represent flies from different lines. The isolation indices among these lines varied from +0.11 to +0.21; they were relatively low but significantly and consistently greater than zero. Isolation was found among populations kept at the same temperature as well as among those kept at different temperatures. Such experiments have established that at least a small degree of homogamic mating preference can evolve among genetically isolated stocks even after a relatively small number of generations. The behavioral mechanism for homogamy in these lines is unknown.

Similar techniques can be employed to test for *geographic variation* in the effectiveness of behavioral isolating mechanisms (under laboratory conditions). In these experiments a common procedure is to place ten males together with ten females of their own geographical strain and ten females from a strain collected at a different geographic location. The joint isolation index can then be calculated from the observed matings. A summary of such experiments in 21 species of *Drosophila* showed that some departure from zero (either homo- or heterogamy) was found among populations in 19 species and in 123 of the 257 pairs of strains tested (Anderson and Ehrman 1969); however, the authors failed to find indications of homogamy or heterogamy in *D. pseudoobscura* in their own experiments. Here again, the behavioral bases for such performances are unknown. Other experiments purporting to demonstrate homogamy among experimentally isolated laboratory populations have been reviewed critically by Manning (1965).

Allopatric populations of the same or closely related species of rodents and lizards have also been tested. Blair and Howard (1944) demonstrated ethological isolation between *Peromyscus maniculatus blandus* and *P. polionotus leucocephalus*. Godfrey (1958) showed that preference for the animal's own race existed in four disjunct populations of bank voles (*Clethrionomys glareolus*). In the side-blotched lizard *Uta stansburiana*, males on their own home range in a natural population, when given a choice of tethered females, preferred those from their own population (McKinney 1971).

EVOLUTIONARY DIVERGENCE OF BEHAVIOR IN SYMPATRY

Are species differences as we see them today due mainly to genetic changes that accumulated during the allopatric phase of closely related species, or are they due mainly to selection for differences after secondary contact? This remains one of the major unresolved questions in the study of the origin of species and of species-specific behavior. In theory, secondary contact between two formerly isolated populations could result in selection pressures causing them to diverge much more than before contact. This theory has been traced back to the writings of A. R. Wallace in 1889 by Grant (1963, 1966), but has more recently been associated with Fisher (1958) and Dobzhansky (1940). The frequency, extent, and importance of evolutionary divergence in areas of secondary contact are difficult to estimate, and further studies will be necessary before a consensus can be reached.

We will now examine the events that are predicted to occur in Phase II of the speciation model shown in Table 18–2. When two populations that have previously been isolated from each other for a long period come into contact again, the amount of interbreeding that will occur cannot be safely predicted. It should depend on the genetic divergence that has occurred during the period of isolation—not on genetic divergence per se, but only in features effective as isolating mechanisms. The degree of reproductive isolation between such populations in areas of secondary contact should vary from little or none to complete isolation.

The intensity of selection for isolating mechanisms between two populations in the area of sympatry at a secondary contact should depend upon the fitness of the resulting hybrids, if any, and on the frequency of interbreeding that occurs. If the fitness of hybrids is equal to that of the parent population, or better, there should be no selection against hybrids or against the tendency to hybridize; indeed, if hybrids were more fit than the parental species, they would be selected for. Hybrids might theoretically possess superior fitness because of heterosis, or because the environment in which they were produced was in some critical way intermediate between those of the parent populations so that intermediate phenotypes would be best suited to it.

If the hybrids were less fit than the parent species (in the environment of the secondary contact), then selection should work against hybrids and their parents and in favor of individuals that do not hybridize. However, if sufficient numbers of hybrids, backcrosses, and recombinants were produced before selection eliminated most of the individuals prone to hybridize, the population should come to consist of such a variety of intergrading phenotypes and genotypes that sexual isolation would not have a chance to operate frequently or effectively. Distinctive traits of the two populations would be blurred, as would preferences for them by individuals. Consequently, the action of selection for stronger sexual isolation might in theory be much reduced or nullified by the existence of a "hybrid swarm." Quantitative aspects of secondary contacts have been modeled by Bossert (in Wilson 1965*a*), as shown in Figure 18–14.

In an area of secondary contact, pre-mating isolating mechanisms may be strengthened if conditions are favorable, but precisely which

18–14 Prediction of the outcome of secondary contact of two populations in various degrees of reproductive isolation. The outcome depends on the fitness of heterozygotes (abscissa) and the initial error in choosing a mate of the appropriate species (ordinate). If all individuals choose their own species correctly, the two species remain distinct regardless of heterozygote fitness. If most individuals mate incorrectly, even a low heterozygote fitness will not provide a selection pressure strong enough to separate the populations, which then merge to become one species. Many degrees of intermediacy are possible. (Modified from Bossert in Wilson 1965*a*.)

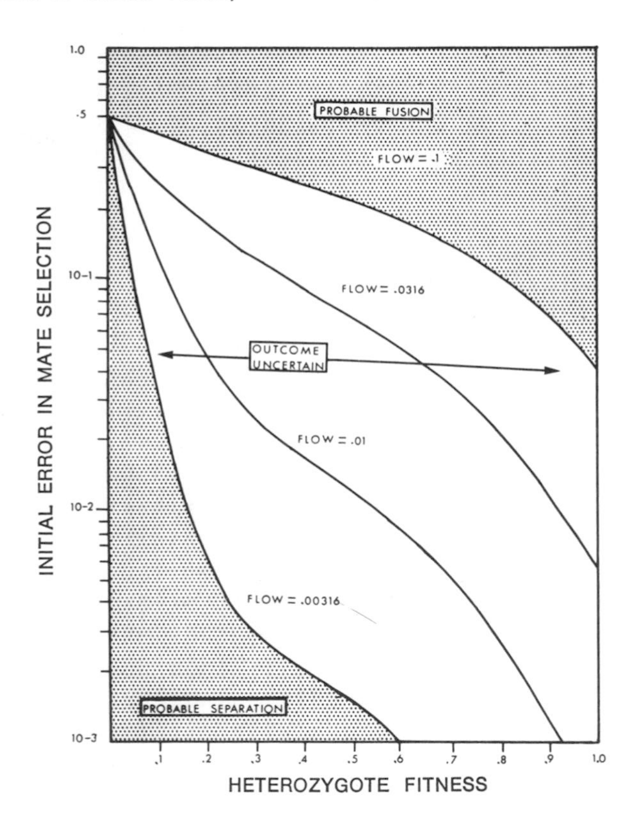

ones are affected cannot be predicted. The theory does not predict strengthening of post-mating isolating mechanisms in the area of secondary contact (Michaud 1962) except when these mechanisms are the consequences of genetic changes stemming from changes in pre-mating mechanisms (Littlejohn 1969). Even if complete reproductive isolation were achieved by post-mating mechanisms, we should still expect selection for improvement of pre-mating mechanisms as long as some individuals attempt hybridization. Of course, if pre-mating mechanisms were completely effective before secondary contact occurred, we should not expect to find further strengthening of them in the zone of sympatry.*

Character Displacement / The importance of character displacement for the study of evolution was emphasized in a stimulating review by W. L. Brown and Wilson (1956). In their words (p. 49), character displacement may be roughly described as follows:

> Two closely related species have overlapping ranges. In the parts of the ranges where one species occurs alone, the populations of that species are similar to the other species and may even be very difficult to distinguish from it. In the area of overlap, where the two species occur together, the populations are more divergent and easily distinguished, i.e., they "displace" one another in one or more characters. The characters involved can be morphological, ecological, behavioral, or physiological; they are assumed to be genetically based.

Character convergence obtains when the characters are closer together in the sympatric area than in allopatric areas. These patterns of geographic variation are more easily studied in morphological traits, but are also known in behavioral traits. Character displacement represents an exaggeration in the sympatric areas of a difference that is relatively slight in the allopatric areas. Commonly, character displacement is attributed either to competitive interaction between the two species after they made secondary contact, or to selection for strengthening of isolating mechanisms. Typically each species becomes more different, but sometimes only one is affected.

Although the classic examples of character displacement concern adaptations to foraging or other contexts in which ecological competition is commonly studied, the phenomenon should also be found in isolating mechanisms. Attention has centered on sexual isolation because of the ease with which geographic variation in mating signals can be studied, but character displacement is also known in mating performance (degree of homogamy in *Drosophila paulistorum*) and even in post-mating mechanisms (Watson and Martin 1968), where it might be fortuitous.

* Greater effectiveness of post-mating isolating mechanisms in the zone of sympatry has been observed by Watson and Martin (1968). Since this is not predicted by current theory, the following hypothesis is offered to account for it. If mutations favoring post-mating isolation occur in the zone of sympatry, they raise the fitness of genes that cause pre-mating isolation. But if they occur in allopatry, they have no such effect. If genes for post-mating isolation are linked with those for pre-mating isolation, they will both be favored in sympatry. Such linkages should occur more frequently in sympatry than allopatry because the linkage raises the fitness of both genes only when in sympatry. Each gene is favored by an increase in frequency of the other. A simpler possibility is that increased post-mating isolation could be simply a by-product of the increased pre-mating isolation in the region of sympatry.

Mating Calls and Songs / Mating calls are favorable material for the examination of geographic variation in isolating mechanisms because they can be easily recorded in the field over a wide geographic area, and because they can be objectively measured by means of the sound spectrograph. Frogs and toads are especially useful because experiments have shown that gravid females of several species do choose between call types in mating; in addition, the young do not need to

18–15 Character displacement in songs of the Australian frogs *Hyla ewingi* and *H. verreauxi*. Map of region of secondary contact. See also Figures 18–15 and 18–16. (From Littlejohn 1965.)

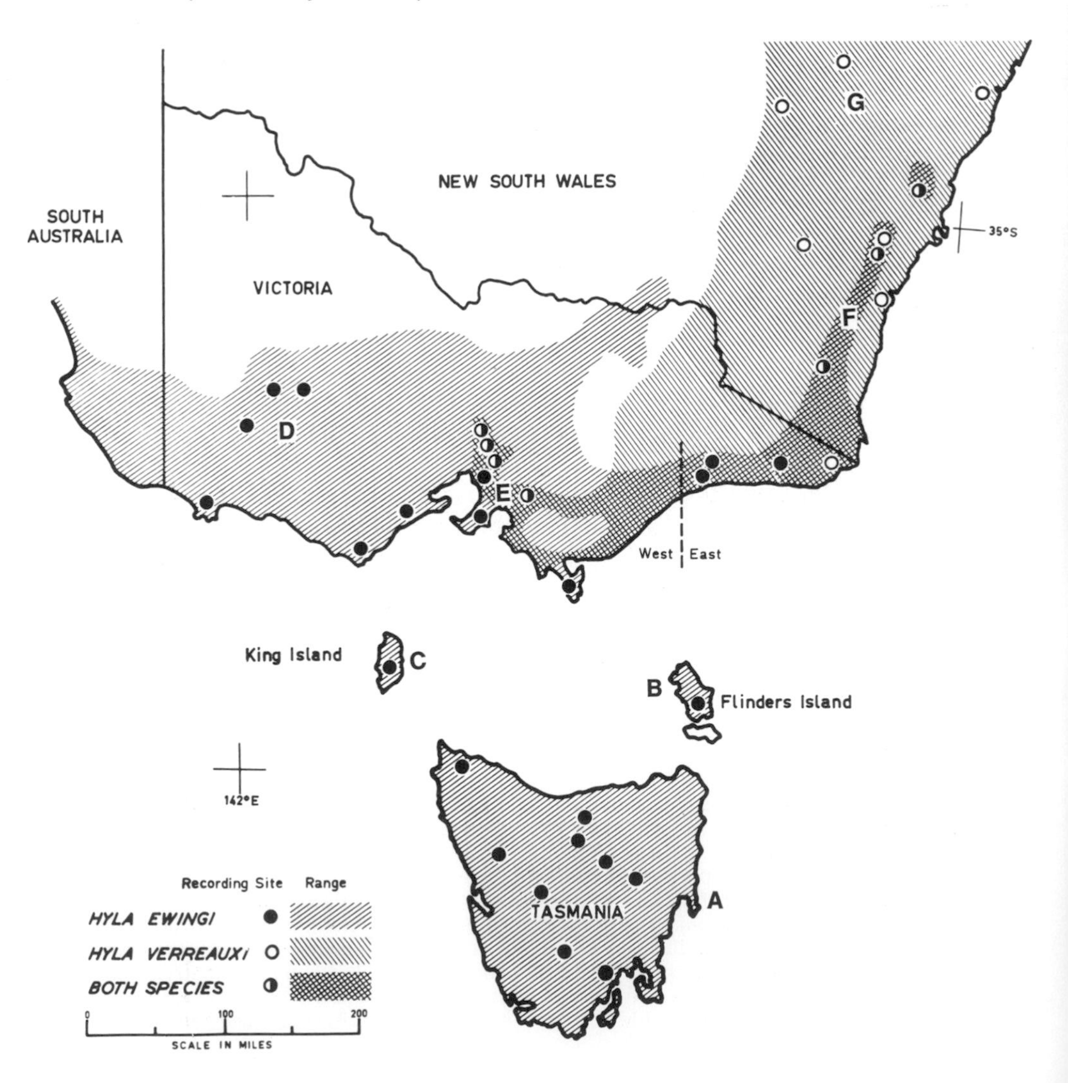

learn the calls from adults. Furthermore, species differences in mating calls of some frogs are known to be genetically determined, as studies of hybrids have revealed.

The most convincing and thoroughly analyzed case of character displacement in isolating mechanisms that has been described concerns the frog species *Hyla ewingi* and *H. verreauxi* (Littlejohn 1965). As shown in Figure 18–15, the ranges of these two species in southeastern Australia are partly allopatric and partly sympatric. The mating calls of both of these species are similar, and consist of a series of rapidly repeated notes, each broken into pulses. *H. ewingi* differs significantly from *H. verreauxi* in having fewer pulses per note, shorter and more rapidly repeated notes, a higher dominant frequency, and a greater degree of amplitude modulation. Some of these differences are shown in Figure 18–16.

Character displacement in pulses per note and pulse repetition frequency is shown in Figures 18–16 and 18–17. The means for *H.*

18–16 Character displacement in songs of Australian frogs *Hyla ewingi, E,* and *H. verreauxi, V.* The oscillograms (amplitude × time) are of single notes from songs. Time is shown by the 50-*Hz* line below each note. (From Littlejohn 1965.)

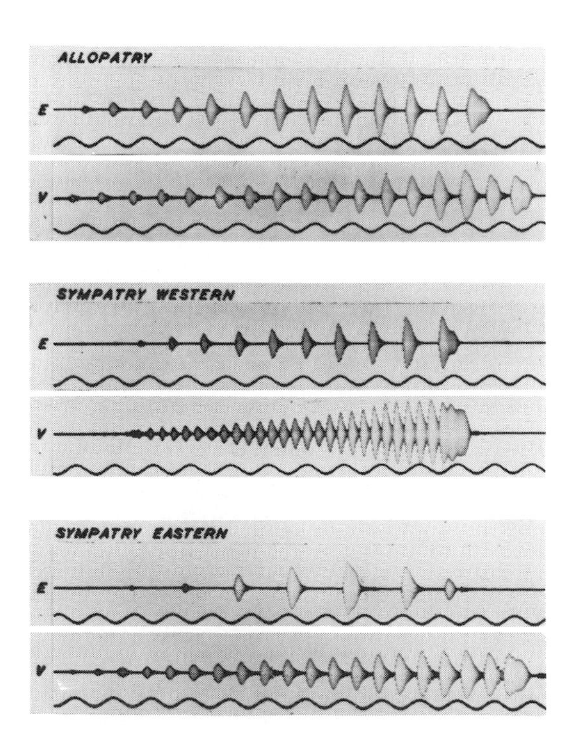

18–17 Character displacement in mating calls of the Australian frogs *Hyla ewingi* and *H. verreauxi*. The letter below each vertical line indicates the location of the sample on the map in Fig. 18–15. E and E′ come from the same place; similarly for F and F′. The long vertical line indicates the range, and the projecting horizontal line, the mean. The inner rectangle includes the 95-percent confidence limits on either side of the mean. Rectangles for *ewingi* are black; those for *verreauxi* are shaded. The outer limits of the rectangle show one standard deviation on either side of the mean. (From Littlejohn 1965.)

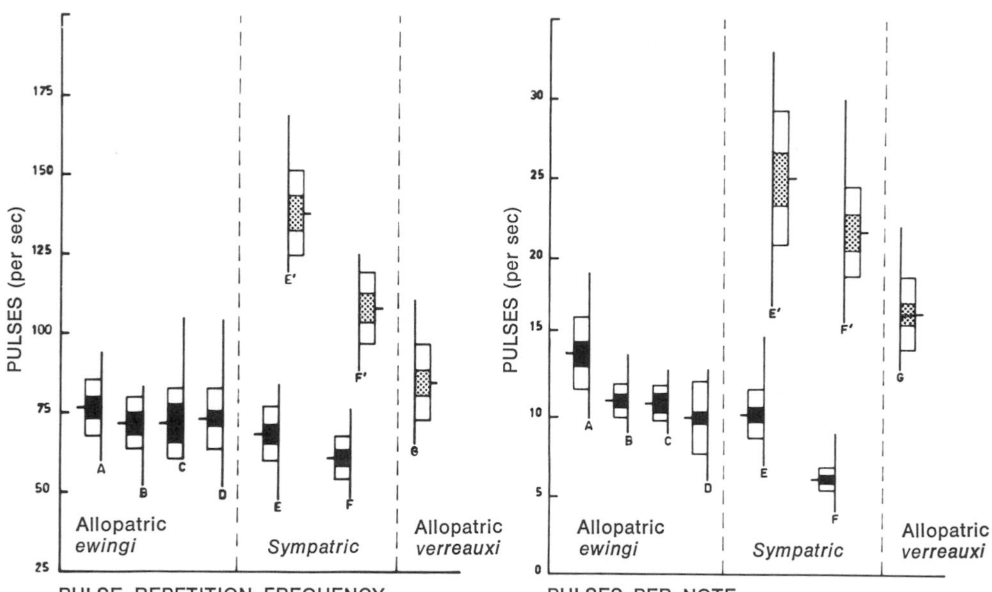

ewingi are lower in the area of deep sympatry (longest contact with *H. verreauxi*) than in allopatric areas, and lower in deep sympatry than in shallow sympatry (shorter, secondary contact). In *H. verreauxi* the opposite trend is found; the means are progressively higher in the series allopatry, shallow sympatry, deep sympatry. Character displacement was also found in two other call parameters that were measured.

Further work (Littlejohn and Loftus-Hills 1968) confirmed a basic assumption of the above analysis, namely, that females of the two species were able to choose their own species purely on the basis of mating calls. This was done by presenting gravid females of each species from the area of sympatry with tape recordings of male song of each species from the area of sympatry played alternately at each end of an arena, and observing which loudspeaker the females approached. In addition, the investigators showed that calls of sympatric *ewingi* and allopatric *verreauxi* were so similar that female *ewingi* did not discriminate between them. This finding supports the interpretation that the observed character displacement was actually important in improving the ability of females to recognize their own species, and not a coincidental by-product of some other cause. The magnitude of the difference in calls in *verreauxi* between allopatry and sympatry was emphasized by the consistent preference of sympatric *verreauxi* females for sympatric *verreauxi* calls in a choice between those and allopatric *verreauxi* calls.

Apparently selection in this area of secondary contact was still operating, since 3 of 66 naturally occurring mating pairs were heterogamic. Although no hybrids have been discovered in nature, preliminary artificial crosses using all combinations of allopatric and sympatric populations of the two species have shown that some viable metamorphosing hybrids can be obtained even from the area of sympatry. Little or no genetic incompatibility in development was detected in crosses involving allopatric populations of the two species (Watson and Martin 1968).

Despite intensive interest in character displacement of isolating mechanisms, surprisingly few cases have been adequately documented. One of the first cases studied was in the frog genus *Microhyla*. Blair (1955*b*) showed that the mating calls of *M. carolinensis* and *M. olivaceus* from allopatric populations in Florida and Arizona, respectively, were similar in duration and dominant frequency, but that the calls differed greatly in a zone of sympatry in Texas. That some natural hybrids were found suggests that selection for strengthening of isolating mechanisms was occurring in the zone of sympatry. This may not be the only explanation, however, since both species show a cline in body size, which increased from east to west, with character displacement of body size in the zone of sympatry (Blair 1955*a*). Since the pitch of the mating call tends to decrease with increasing body size in some anurans (Blair 1958), the character displacement in the mating call could be simply a consequence of the clines in body size, which may have evolved for reasons unconnected with mating calls.

In another pair of species known to mate interspecifically on occasion, *Hyla regilla* and *H. cadaverina* (= *californiae*), Ball and Jameson (1966) failed to find character displacement in the mean dominant frequency of the mating call. They did, however, find that in one of the two species the call was less variable in the zone of sympatry than out of it (a circumstance resembling character displacement). Previous work had shown that gravid female *H. regilla* would discriminate between playbacks of the calls of these species (Snyder and Jameson 1965). Since the mating call of *H. regilla* consists of two notes separated by a brief pause and that of *H. californiae* is only a single note, it would be interesting to determine whether this difference is important as an isolating mechanism. Unfortunately, Ball and Jameson did not tell us whether or not the pause in the double-note call of *H. regilla* is longer in sympatry than in allopatry, nor did they analyze variations in another critical call parameter, pulse repetition rate. As stressed by Littlejohn (1971), this case deserves further analysis of geographic variation and experimentation with synthetic calls.

Other documented cases of character displacement in known isolating mechanisms are rare. Smith (1966) described one involving eye-ring coloration in gulls (*Larus glaucoides* vs. *L. thayeri* and *L. hyperboreus*). No authenticated cases are known involving the vocalizations of birds, despite much interest (Thielcke 1969).

Character displacement in mating *preference* (as opposed to mating signals) has been described in a few cases. In the *Drosophila paulistorum* superspecies (a group of closely related species), Ehrman (1965) found that heterogamic matings were less frequent between flies from sympatric populations than between flies of the same taxa from allopatric populations. In male-choice experiments with fishes of the genus *Gambusia*, naive males chose their own species more often when sympatric populations were tested than when allopatric populations were tested (Hubbs and Delco 1960, 1962).

Experimental Selection for and against Pre-Mating Isolation / It is possible to study selection for or against ethological isolation between species in laboratory experiments. Such experiments have proven that under certain conditions selection for strengthened ethological isolation is effective. The first systematic experiments were done by Koopman (1950). Hybrids between the sibling species *Drosophila pseudoobscura* and *D. persimilis* are virtually unknown in nature, despite intensive search, yet the two species will hybridize in the laboratory at low temperatures. Koopman showed that in his population cages the frequency of hybrids in the first generation was 22.5 to 60 percent, but fell in all cases to about 5 percent within six generations. The course of three populations is shown in Figure 18–18. The selection was performed by the flies themselves. From 60 to 400 individuals of each species were placed in the population cage for about two weeks, during which time mating occurred; then the food and larvae were removed and the newly emerging adults were collected, classified, and accumulated for the next generation. Only flies that mated with their own species contributed to the next generation; females that mated with the other species did not contribute. Hybrids and the two parent species were recognized by marking each species with a different homozygous recessive; only hybrids then appeared as wild type. Since hybrids were either sterile (males) or produced inviable larvae (females), they

18–18 Selection for strengthening of isolating mechanisms between *Drosophila pseudoobscura* and *D. persimilis* in a population cage. Selection against flies that hybridize resulted in lower frequencies of hybrids. (From Koopman 1950.)

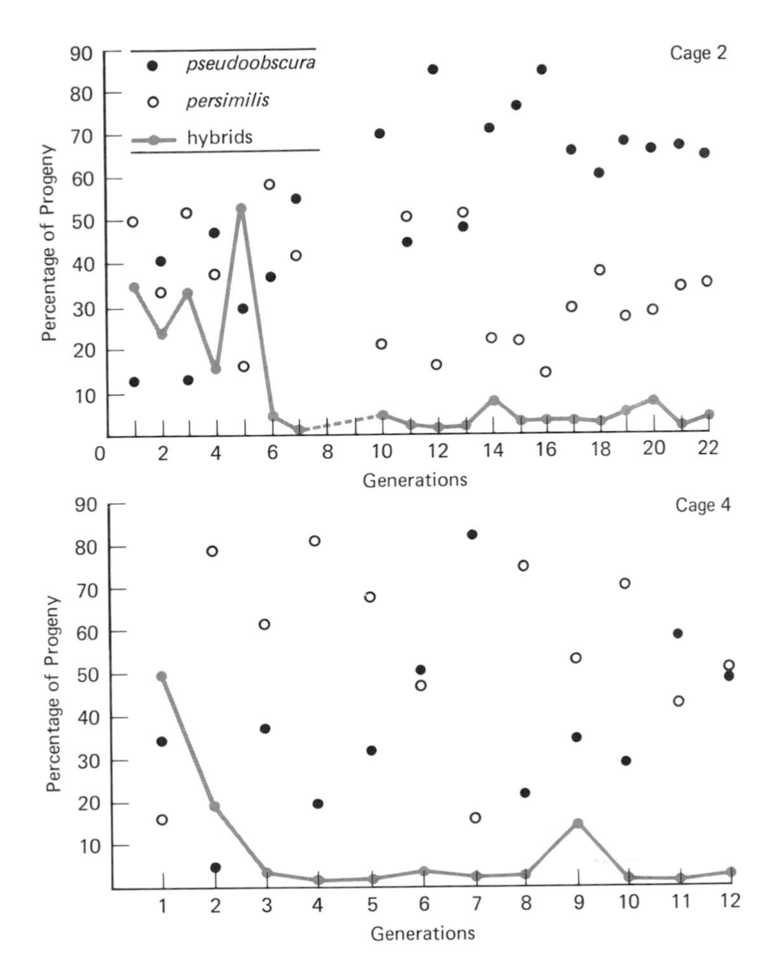

could not contribute to future generations even if left in the population. The flies of the parent species produced by each generation were used as the parents for the next generation; hybrids were discarded. Multiple-choice tests showed significant increases in sexual isolation from the unselected population to the selected populations.

In experiments in which individuals of each sex of one species were given a choice of mating with the opposite sex of the other species, or not at all, Kessler (1966) successfully selected for both increased and decreased ethological isolation between *Drosophila persimilis* and *D. pseudoobscura*. Although his experiments are not comparable to those of Koopman because of numerous procedural differences, his results tend to support Koopman's findings that sexual isolation can be modified by selection in relatively few generations. The major changes occurred in the discriminatory ability of the *pseudoobscura* females, who unlike the males responded well in both the increased and decreased lines.

Experiments somewhat similar to those of Koopman were done by Dobzhansky and Pavlovsky (1971) with two populations (Llanos and Orinoco) of the *Drosophila paulistorum* complex that had developed a partial sterility barrier but lacked pre-mating isolation. The percentage of hybrid females dropped from 41.9 at the first generation to 31.8 after 70 generations, and in direct observation of mating in multiple-choice experiments, a significant increase in the isolation coefficient was found. At the start of these selection experiments the Llanos and Orinoco populations would probably be classified as belonging to the same species. Consequently, these experiments, unlike those of Koopman, may be interpreted as demonstrating a part of the process that could be involved in the evolution of a new species and of new genetic differences responsible for species-specific behavior.

In the case of *D. pseudoobscura* and *D. persimilis*, artificial selection was unnecessary because no gene exchange would have occurred if the hybrids were allowed to remain in the population. But in the case of *paulistorum*, there was no barrier to gene flow before selection; the selection pressures that acted to strengthen pre-mating isolation were necessarily generated by the experimenters.

In both the *D. pseudoobscura* × *D. persimilis* and *D. paulistorum* cases, considerable genetic divergence had occurred before selection experiments were begun. Selection then acted to strengthen existing isolation. It is possible, however, to begin such experiments with minimal initial genetic difference between the selected lines. The selection experiment of Knight et al. (1956) started with stocks of *D. melanogaster* that were "genetically very similar except for the marker genes," *ebony* and *vestigial*. Their selection regime was similar to that of Koopman. "Hybrids" of the *ebony* and *vestigial* lines were discarded, leaving only flies that mated with their own type as parents for each successive generation. Although great fluctuation occurred from generation to generation, there was a more or less steady decline in the percentage of hybrids (from 80–90 percent initially to less than 50 percent). In a repetition of this experiment Pearce (1960) obtained similar results, and found that the increased isolation was due mainly to increased repulsion of heterogamic males by females. Pearce also observed that some isolation between *ebony* and *vestigial* was present at the start due to the nature of these phenotypes. In another experiment of this type, Hoenigsberg et al. (1966) achieved increased homogamy among flies of the *ebony* phenotype but decreased homogamy in the line marked by *dumpy*.

Some failures to achieve reproductive isolation by selection against hybrids have also been recorded in *Drosophila melanogaster* lines marked by mutant genes and started from the same population, although mating preferences were altered "only slightly" (Wallace 1950, 1954, 1968*a*:396). However, the successes must be interpreted as more important than the failures.

Selection against hybrids and their parents has also led to effective reproductive isolation between artificial lines in maize (*Zea mays*), where season of flowering was the main factor affected (Paterniani 1969). Insects whose pre-mating isolation is correlated with flowering time or other aspects of species differences in plants could conceivably be affected indirectly by such genetic changes in plants. It would be interesting to study insect pollination in this respect.

Another selection regime has also yielded reproductive isolation. Disruptive selection for number of sternopleural chaeta (bristles) in *D. melanogaster* (Thoday and Gibson 1962) was successful in the rapid differentiation of high and low chaeta number, even though all matings of both lines took place in the same container in a large population and were not arranged directly by the experimenter. In selecting for high (or low) chaeta number, one in effect selects the progeny of homogamic matings between mates both having high (or low) chaeta number. The result is selection against parents that mate heterogamically. In this work it was not determined whether the isolation was due to pre- or post-mating mechanisms. Other workers have attempted unsuccessfully to confirm these results (Robertson 1970; Scharloo 1971), and the interpretation that disruptive selection has been demonstrated to have resulted in reproductive isolation is controversial. However, Beardmore (reported in Thoday 1972) apparently succeeded in duplicating the original results of Thoday and Gibson. Failure to achieve reproductive isolation in a similar experiment with *cubitus interruptus*, a wing vein mutant of *D. melanogaster*, has been recorded (Wallace 1968*a*:396).

In the various selection experiments and tests on reproductive isolation in *Drosophila*, there is generally no information on the means by which selectivity in mating is modified. Information is needed on the behavioral basis of ethological isolation in *Drosophila* and how it is modified in selection experiments. Although the courtship behavior of many species of *Drosophila* as visible to a human observer has been described (Spieth 1952; Bastock 1956; R. G. B. Brown 1964, 1965), these studies yielded little insight into the behavioral basis of pre-mating isolation in the genus, probably because visual stimuli are not the basis of specificity in mating. Volatile pheromones produced by female *D. melanogaster* stimulate increased activity in males (Shorey and Bartell 1970), but their role in ethological isolation is unknown. Ehrman's (1969) experiments also suggest that airborne olfactory stimuli are involved.

Another possible sexual isolating mechanism in *Drosophila* is provided by diversity in the "songs" of various species. The differences in the sounds produced by *D. persimilis* and *D. pseudoobscura* and other species are shown in Figure 18–9 (Ewing and Bennet-Clark 1968; Bennet-Clark and Ewing 1968, 1969; Ewing 1969).

The genetic bases of pre-mating isolating mechanisms are not well known. Although various aspects of the genetic bases of sexual behavior in *Drosophila* have been studied extensively (Manning 1965; Ewing and Manning 1967; Spieth 1968; Petit and Ehrman 1969), little information on the genetics of *selective* matings is available. Various theories are available that have not been adequately tested (see Chapter 8). In the sibling species *D. pseudoobscura* and *D. persimilis*, the responsible factors were detected to be at least in the X and second chromosome (Tan 1946). This correlates well with the finding of Ewing (1969) that genes controlling differences in the courtship sounds of these species are X-linked. In *D. paulistorum* subspecies (or semispecies), the responsible genes are distributed over all the chromosomes and seem to act additively. In several species of crickets, the X chromosome is important in the control of courtship signals (e.g., Bentley 1971; Bigelow 1960).

CONCLUSIONS

Species-specific behavior is behavior that is unique to one species. Species-specific behavior may evolve by chance, by extinction of closely related species, by selection in unusual environments, or by selection for differences per se. To understand the origin and importance of species-specific behavior, it is necessary to have a basic understanding of the modern biological species concept. A species can be characterized as a set of populations that is bounded genetically by effective reproductive isolating mechanisms. An isolating mechanism has the effect of preventing or reducing exchange of genes between populations. Isolating mechanisms may be characterized as pre-mating (behavior-related) and post-mating (sterility and inviability). Sexual (ethological) isolation, in which interbreeding is prevented by behavioral preferences, is one of the most important isolating mechanisms in animals and is the center of attention in this chapter.

Speciation (the origin of species) is the branching process by which one phyletic line divides and becomes two separate phyletic lines that are reproductively isolated where secondary contacts between the formerly separate populations occur. Most, if not all, speciation seems to require a period of geographical isolation between populations, during which time the populations diverge genetically. Two or more populations that do not overlap geographically are said to be allopatric. Two or more populations that do overlap are termed sympatric. The term *sympatric speciation* refers to speciation without the usual requirement of a period of geographical isolation; whether it ever occurs in nature or not is debatable.

Any sensory modality may be used to receive the signals between the sexes that make possible selective mating with one's own species. Visual signals, such as the coloration of fishes and birds, the head-bobbing of lizards, and the flashing patterns of fireflies, have been shown to function as isolating mechanisms. Auditory signals such as the songs of crickets, grasshoppers, *Drosophila*, frogs, and birds have also been demonstrated experimentally to function as isolating mechanisms. Chemical stimuli may play a similar role, as in certain moths, bees, and mammals, in which females employ sex attractants to lure males.

According to the most popular theory for the origin of new species, the allopatric model, we should expect to find allopatric populations that show all degrees of divergence from each other. In order for these differences among populations to function as behavioral isolating mechanisms, they must involve behavior used in pairing, courtship, or mating. Examining geographical variation in frog and insect calls, bird coloration, and mating preferences of *Drosophila* has revealed sufficient divergence among populations to support the model. In some cases the differences are of "species magnitude," while in others there are none at all; more commonly they are intermediate.

When two populations that have been geographically isolated from each other for a long period contact each other again (secondary contact), they may show any degree of interbreeding and gene exchange, depending on the effectiveness of any isolating mechanisms that might have evolved during the period of separation. If individuals who attempt matings with the other population (the other species, in this case) are much less successful at contributing genes to further generations than those who mate only within their own population, selection will favor individuals who can avoid attempts to breed with the other species. The effect of this kind of selection should be to exaggerate the differences between the species that are used in species recognition. Moreover, this exaggeration should be confined to areas of sympatry and be nonexistent or much reduced in areas of allopatry. This pattern of geographic variation in isolating mechanisms, which is known as character displacement, has been demonstrated in a few cases involving mating calls of frogs. It is possible to produce similar effects in artificially selected populations of *Drosophila*. The amount of sexual isolation between species or populations can be increased or decreased by artificial selection. Consequently, the model that postulates exaggeration of isolating mechanisms in areas of secondary contact is supported to some extent both by studies of geographic variation in mating behavior and by selection experiments done in the laboratory. The rough outlines of a model for the evolution of new species and for species-specific behavior are now available, as described above, and are supported by data. But there are many unexplained problems remaining, and there is a need for refinement and further testing of our concepts of speciation.

19 The Dances of Bees

IF ETHOLOGISTS were asked to list the seven greatest wonders of the natural world, I am sure many would name the "language" of the honeybee as one of them. Communication in the honeybee (*Apis mellifera*) is involved in mating, caste differentiation, aggressive behavior (between queens), colony defense, and founding new colonies. All of these are fascinating, but it is the communication of information about sources of food by means of "dances" for which the honeybee is justly famous. This chapter will survey the principal features of this form of communication in the honeybee and will consider the evolutionary history of dancing. Reviews of communication in bees and its evolution have been provided by von Frisch (1967*a*), Lindauer (1971*a*), Wenner (1968), and Kerr (1969). By far the most enjoyable introduction to the subject that I know of is the little book written by von Frisch in 1971. For detailed references to the material of this chapter, the reader is referred to the comprehensive summary by von Frisch (1967*a*).

Methods / If a dish of scented sugar water is put outside on a warm spring day, it may be visited by foraging bees. If we keep a record of the time of arrival of each bee, it can often be noticed that the pattern is not a random one. For a long time no bees will visit the food, but after the first one or two, a number of bees may arrive relatively close in time. Wasps do not exhibit the same pattern; they arrive more randomly. This kind of observation suggests (but does not prove) that the original forager may have somehow recruited others from her colony. To investigate this possibility, two simple techniques have been developed. To watch what goes on at the colony, special observation hives are used. These are often smaller and always narrower than the usual bee-keeper's hive. They have glass walls that allow nearly all of the face of the comb to be viewed. The colony is normally always dark, but the bees soon become accustomed to light. A direct view of the sky from the comb must be eliminated; this can be done with a canvas shade or, better, with an observation hut.

The second important technique is that of marking the bees. Special paints may be touched to the thoraxes of foragers as they feed at a bait. By using different combinations of colors and locations in marking the bees, many individuals can be recognized. It is also possible to use colonies, all of whose members have been tagged and numbered beforehand. With these simple techniques a great variety of experiments can be performed in which sugar water is offered in various ways at different times and places and the effects upon the bees are noted, both in the hive and at observation stations around the hive.

Some Chemical Cues / When bees discover a rich food source, information about it is communicated when they return to their colony. There these individuals, designated as *scouts* in this chapter, perform a dance repeatedly and then typically return to the food source. After the dance, other bees from the same colony, designated here as *recruits*, may also visit the food. If we arrange to have some foraging bees feed on cyclamen flowers sprinkled with sugar water, we can test whether the scouts convey to the recruits information on the kind of flower from which the food was obtained. After the scouts have fed and returned to the hive, the newly arriving recruits are given a choice between flowers of cyclamen or phlox, neither with sugar water. The newly arriving bees will seek out the cyclamen. If the experiment is repeated but starting with phlox flowers, the bees will seek for these and ignore the cyclamen. Finally, we can dispense with the flowers altogether and use artificially applied scents, such as lavender and peppermint. The recruited bees go to cards scented in the same way as the original food source. Somehow the first bees to find the food, the scouts, have conveyed the scent of the food to the recruits.

There are essentially two ways in which this information could be given to the colony. The scent could be carried on the surface of the bee or in her stomach. Since returning bees typically regurgitate and share the contents of their honey stomach with the colony, the recruits might acquire the scent in this way. Von Frisch tested these possibilities by (1) arranging to have the scouts feed on unscented sugar water on scented flowers, and (2) having them feed on scented sugar water from unscented perches. He found that the recruits utilized odor cues from both external scents and those carried in the honey stomach. At longer distances from the hive, the external scent tended to wear off and the influence of the internal scent predominated. But when the bees were forced to choose between the odor of the sugar water and of the external scents close to the hive, they chose seemingly at random.

Honeybees also utilize a pheromone to help recruits locate a rich food source. If a scout locates a new food source, she flies to the hive with some food, dances, and returns to the food. On her return she exposes the Nassanoff's gland on the dorsal side of her abdomen and extrudes a pheromone (geranial and neral along with other compounds) while hovering over the source. This is apparently detectable by bees several meters away. That it does facilitate the finding of a food source by recruits has been shown by sealing the glands with a touch of shellac. In paired comparisons, feeding places at which the glands of the scout were not sealed attracted ten times as many recruits as those at which the glands were sealed. When the pheromone was collected by dabbing the gland with filter paper, it was shown that the isolated secretion was as attractive as were the naturally applied pheromones. It was also found that it was as attractive to other colonies as to the colony from which it was taken. This pheromone acts as an attractant and is also released by bees at the hive on occasion. The use of pheromones to locate food sources is more highly developed in the stingless bees, the tribe Meliponini (Kerr 1969).

The Dances / When a rich food source is provided close to the hive, it is observed that the bees recruited by the dancers search in all directions from the hive. On the other hand, if the food is offered at a considerable distance from the hive, the searches of the recruits are concentrated in the direction of the food. The forms of the dances used in these cases are different. Close to the hive, the *round dance* is used. At a distance, the *waggle dance* is employed. At intermediate distances, intermediate dances are used, at least one of which has been named, the *sickle dance*. These dances are illustrated in Figure 19–1. The honeybee normally performs all dances in darkness on the vertical face of a comb.

The round dance is thought to convey no information concerning the direction or the exact distance of the food source. It simply announces to the colony that food is available close to the hive.

The waggle dance is potentially more informative. Observers of marked dancers found that the angle between the straight (or waggle) run and the upward vertical equals the horizontal angle between the sun and the food, as measured at the hive. A straight run going straight up indicated food in the direction of the sun; a straight run going straight down indicated food directly away from the sun; and so forth. The relationship is shown in Figure 19–2. Any point of the compass can be indicated by the waggle dance. As long as the recruits have some knowledge of the position of the sun in the sky, some of them are able to fly to the food upon leaving the hive. The necessity of knowing the sun's direction reveals the bees' dependence on cues from the dance.

In experiments with feeding bees at different distances from the

19–1 Dances of the honeybee. **A.** The round dance is used close to the hive. The waggle dance is used at a greater distance from the hive. The waggle dance is named for the side-to-side abdominal movements of the dancer during the straight run. These movements and the accompanying sounds do not occur during the return part of the dance. At intermediate distances transitional forms are employed. *A. m. carnica* lacks the sickle type of intermediate that is found in the other subspecies. **B.** A dancer in the center is followed by eight recruits. The other bees are nonparticipants. (A and B reprinted by permission of the publishers from K. von Frisch, *The Dance Language and Orientation of Bees,* Cambridge, Mass.: Harvard University Press, Copyright 1967, by the President and Fellows of Harvard College.)

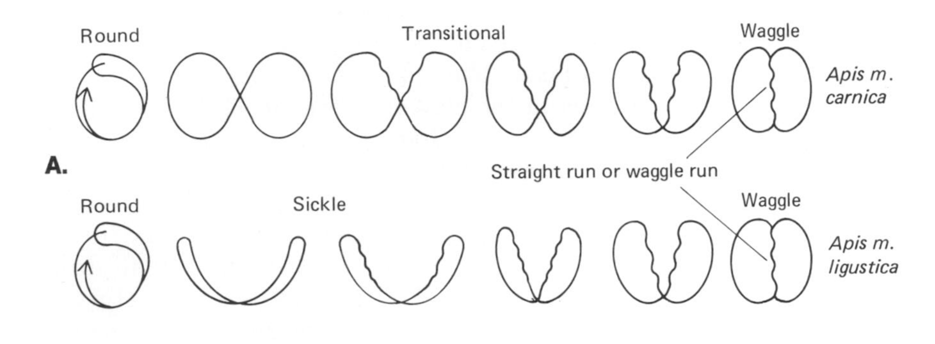

B.

19–2 The direction of the waggle dance of the honeybee corresponds to the direction of the food. At the hive the horizontal angle between the direction of the sun and the food approximates the vertical angle between gravity and the straight-run portion of the dance. F_1, F_2, and F_3 are the locations of feeders. D_1, D_2, and D_3 are the corresponding dances. (Original drawing.)

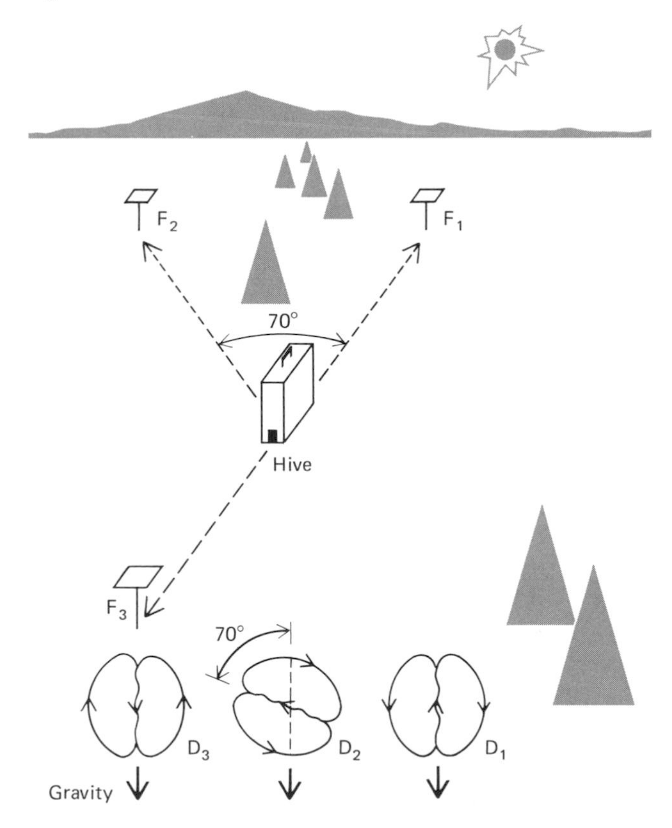

hive, it was noted that the character of the waggle dance varied according to the distance. When the food was relatively distant, the dance was slower, fewer complete dances were performed per unit of time, and more time was taken in each part of the dance (the straight run and the return). The relationship between distance and dance tempo found by von Frisch and his colleagues is shown in Figure 19–3.

With the correlations just noted involving dance form, tempo, direction, and distance, an experienced human observer can predict where a food source newly discovered by the colony is to be found. However, man uses his visual sense to derive this information from the dance whereas the bees must use some other senses, since the dance is

19–3 Honeybees dance rapidly for food close to the hive and slowly for distant food. The relationship between dance tempo and food distance is quite precise. The curve is based on 6,267 measurements of dance tempo. (Reprinted by permission of the publishers from K. von Frisch, *The Dance Language and Orientation of Bees,* Cambridge, Mass.: Harvard University Press, Copyright 1967, by the President and Fellows of Harvard College.)

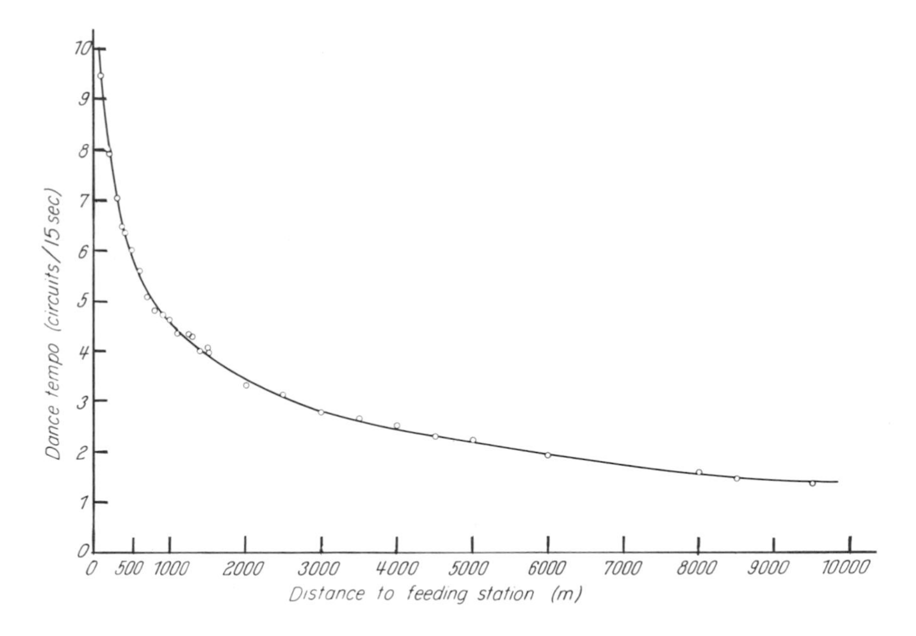

normally performed in darkness. The question of how the recruit bees obtain the information from the dancers remains.

A plausible answer in regard to the determination of distance was obtained through studies of the sounds generated by dancing bees. During the straight run, a dancing bee wags her abdomen from side to side and produces repeated pulses of sound, as shown in Figure 19–4 (Wenner 1962; Esch 1961, 1964). These sounds delimit the duration of the straight run in normal waggle dances. The fundamental frequency of the sounds, about 200–250 *Hz*, corresponds with the frequency at which the antennae are most sensitive to vibration. Bees following a dancer seem to try to keep their antennae in contact with the body of the dancer (see Figure 19–1*B*). Whether these sounds are perceived by the bees from airborne vibrations or only by tactile communication is not certainly known. However, there is no doubt that the distance of the food source correlates nicely with the duration of the straight run as indicated by the duration of pulse emission and number of pulses per run. The duration of the straight run is a more precise

19–4 Sound production by honeybees. **A.** Sound is produced only during the straight run, *SR,* not during the return, *R.* (From Esch 1964.) **B.** Sound pulses recorded during part of a straight run. **C.** Amplified sound pulse. **D.** Pattern of vibration of the thorax during a waggle dance; the trace shows both acoustic vibrations and waggle motions. The recording was made by fastening a tiny magnet to the back of the dancer and monitoring its movement electronically. **E.** Three tail-wagging movements shown by the bracket in D. The vibratory episodes (sound production) occur both during and between waggle motions. Although both occur only during the straight run, they are produced independently of each other. (B–E from Esch 1961.)

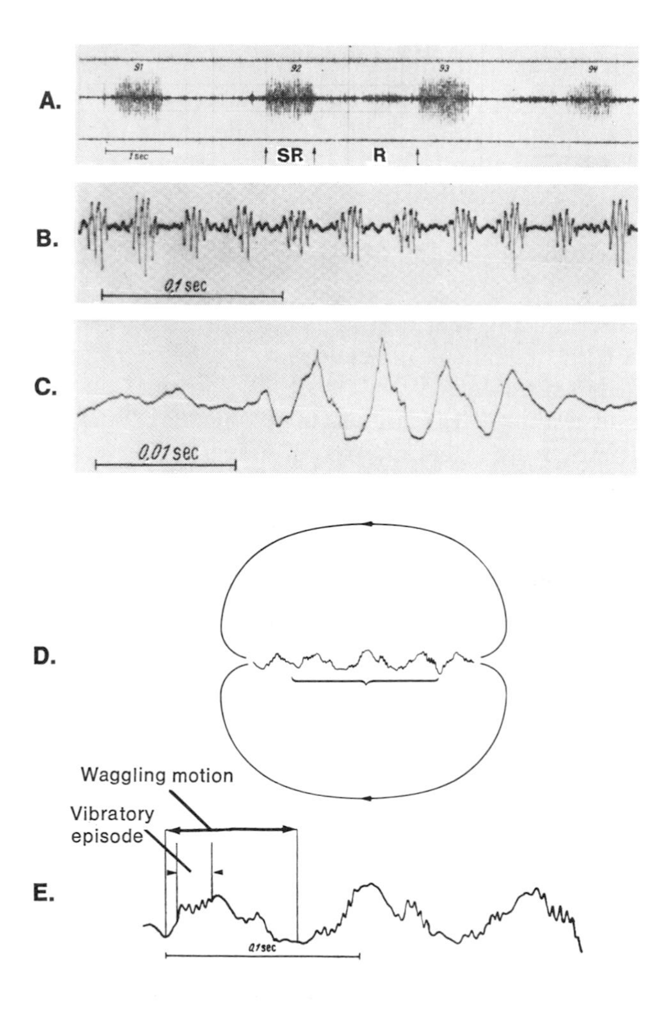

indicator of distance to the food than is the total dance time, the return time, or the rate of dancing or turning (Wenner 1962).

It is important to have some idea of the effectiveness of dancers in recruiting bees to a food source. Esch and Bastian (1970) watched the performance of 70 recruits after a single dancer feeding at a scented feeder 200 m away had danced in the hive. Sometimes the trained scout failed to dance, and then no new bees came to the test feeder. Even when the scout did dance, over half the bees (36) did not follow the dancer and never found the food. These observations suggest that random foraging and olfactory cues alone were unlikely to bring a forager to the test site. Of 34 bees who followed the dancer, only 14 ever found the test feeder, and some of these were unsuccessful at first and had to return to the hive and follow the dancer again. None of the 14 found the test feeder without following at least six dances. A total of 18 bees left the hive after following zero to five dances, but failed to find the food site on the first trip. The failure of bees to find the test feeder using both odor cues to which they had been conditioned and other information from the dance suggests that odor by itself was insufficient.

A puzzling finding was that over half of the 32 recruits who followed from 6 to 36 dances never found the test site ($n = 18$), even though most left the hive several times. No information on the causes of failure is available. It is evident from the flight times of successful recruits that they were not all making a "bee line" from the hive to the test site. Experienced bees made the trip in 20 to 60 seconds. The 14 successful naive recruits took 56 to 660 seconds on their first trips; 7 made the first trip in less than 120 seconds. In two cases the observers could see fresh recruits approaching from a distance (Esch and Bastian 1970:180): "They came straight from the hive in a zig-zag flight which brought them down from a height of approximately 10 meters at the point where they were spotted to 1 or 2 meters near the food site. It was obvious from their behavior that they were *not* searching at random. One had the impression that they knew the location of the food site." In these experiments, the bees used a combination of chemical and other cues obtained from the one dancer. Chemical cues alone were totally inadequate, as judged from the need to follow six or more dances. Even the dance in combination with chemical cues was insufficient in over half the cases.

The causes of failure prompt some speculation. Could the 18 bees who followed dances but did not come to the test site have been distracted by other food? Or were they defective in their sensory or other behavior? In nature, the visual and chemical cues supplied by other foragers for the food would help to attract the new recruits. The observers felt that some bees found the site but failed to land and so could not be counted among the 14 successful ones. In nature, the successful recruits would return to the hive and dance, thus building up the number of dancers. In the experiments, each successful recruit was captured on arrival. Did these bees release an alarm pheromone on capture? Probably much more remains to be learned about the detailed mechanisms by which bees communicate information about food sources to recruits.

How bees handle such disconcerting effects as variations in wind direction and speed, the presence of obstacles, differing routes to and from the food, the carrying of loads, the disruption of their time schedule, and others has been studied extensively, and the reader is referred to the summary of von Frisch (1967*a*) for these details.

The precision with which information on distance and direction of food is communicated by dancing, as well as the ecological significance of dancing, have been questioned (Johnson 1967; Wenner 1967; Wenner, Wells and Johnson 1969; Wells and Wenner 1973), but the original conclusions of von Frisch still appear to be valid (Dawkins 1970; von Frisch 1967*b*, 1968, 1973). Subsequent experiments with careful controls have confirmed the original findings of von Frisch and colleagues and reaffirmed the importance of odors (Gould et al. 1970; Gould 1974; Gonçalves in Kerr 1969; Lindauer 1971*b*).

In focusing on the remarkable dances of the honeybee, the importance of chemical cues has sometimes been underestimated. The early work of von Frisch and the later work of Wenner and his colleagues have shown that chemical cues are important in the conditioning of workers to forage for certain scents at certain places and in the pheromonal marking of locations (Johnson and Wenner 1966; Wenner 1967; Johnson 1967; Wenner et al. 1969).

Population Differences in Dancing / Different geographic populations of honeybees vary in certain aspects of dancing. These variations provide some insight into the evolutionary history of dancing and some of the selection pressures that might be significant. Figure 19–5 shows interpopulation differences in the distances from the hive at which the different types of dances are performed. Three types of dances are involved. Two of these, the round dance and the waggle dance, are performed by all of the populations tested, but the populations differ in the distances from the hive at which they are performed. *Apis mellifera fasciata* begins waggle dancing relatively close to the hive, about 12 m, while *A. m. carnica* begins relatively far away, 85 m. A distinctive

19–5 Variation among subspecies of the honeybee in the distances at which round dances, *R*, sickle dances, *S*, and waggle dances, *W*, are used. (After Boch 1957. Reprinted by permission of the publishers from K. von Frisch, *The Dance Language and Orientation of Bees*, Cambridge, Mass.: Harvard University Press, Copyright 1967, by the President and Fellows of Harvard College.)

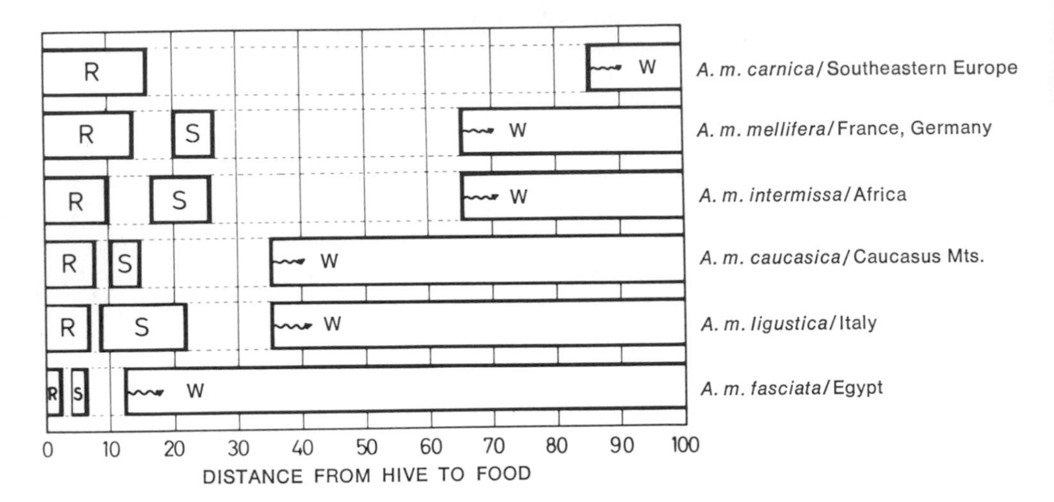

intermediate form, the sickle dance, was used for intermediate distances by all populations except *A. m. carnica*, whose intermediate form resembled a figure 8 (see Figure 19–1).

Another racial difference occurs in the dance repetition rate. Carniolan bees (*A. m. carnica*) danced faster than all other populations tested under comparable conditions, and tropical African bees (*A. m. adansonii*) danced the slowest (Figure 19–6). Hybrids tended to be intermediate. In colonies consisting of artificial mixtures of workers of *carnica* and *ligustica*, the differences in tempo of dancing persisted and caused *ligustica* to search too close to the hive when the dancer was *carnica* and *carnica* to search too far from the hive when the dancer was *ligustica*.

In addition to *A. mellifera* there are four other species of *Apis* known, all of which occur in India. Some of these species differ from *A. mellifera* in dance tempo (Figure 19–7), but they also differ in other ways that are correlated with their social organization and habitat. The dance of the closest relative of *A. mellifera*, *A. indica*, differs no more from that of *A. mellifera* than one geographic race of *mellifera* differs from another. *A. dorsata* apparently dances on the vertical side of its single comb where the sky can be seen; both round and waggle dances have been reported. *A. florea*, the dwarf honeybee and the smallest member of the genus, dances only on a horizontal dance platform on top of the comb. The waggle run of the dance points directly toward the food, and no transition from a gravitational to a solar frame of reference is involved. *A. mellifera* and *A. indica* also dance horizontally if the comb is placed in the horizontal plane; normally their combs are vertical.

19–6 Variation among subspecies of the honeybee in dance tempo (dance repetition rate). Northern subspecies dance at a faster rate; southern ones, slower. For geographic ranges of subspecies, see Figure 19–5. (Reprinted by permission of the publishers from K. von Frisch, *The Dance Language and Orientation of Bees,* Cambridge, Mass.: Harvard University Press, Copyright 1967, by the President and Fellows of Harvard College.)

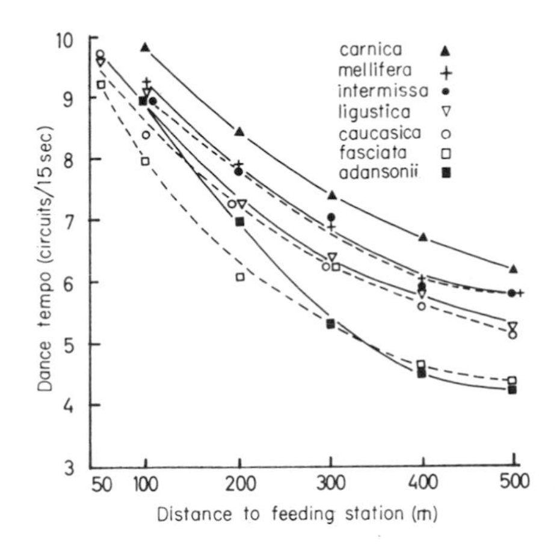

19–7 Variation among species of bees of the genus *Apis* in dance tempo at various distances of the food from the colony. (After Lindauer 1956.)

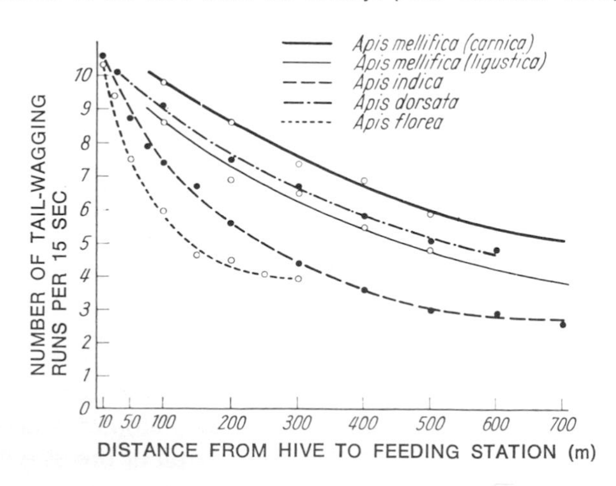

A. indica and *A. mellifera* build multiple-combed nests inside a cavity. *A. florea,* the little honeybee, and *A. dorsata,* the giant honeybee, build exposed, single-combed nests and have colony sizes of 4,000 to 5,000 individuals—considerably smaller than those of *mellifera* and *indica. A. andreniformis* is a distinct species but poorly known.

The communication systems of the different races and species appear to be correlated with their flight ranges in some cases at least. The smallest species, *A. florea,* has the smallest flight range; *A. m. carniola* seems to have the largest one. Von Frisch (1967*a*) suggested that the modest flight range of *florea* and the relatively small *indica* is adaptive for them because of the danger of being caught away from the colony in tropical downpours. The flight range of the large *A. dorsata* is not known. The curves expressing the relation between dance tempo and distance to food are steepest at short distances for the two small species, *A. florea* and *A. indica.* This means that the dances are more precise about indicating distance to food than in the other species when the food is relatively close to the nest (10–100 m). The shapes of the curves in Figures 19–6 and 19–7 can be interpreted tentatively as adaptations for the particular foraging ranges that are most common for each population. A short steep curve would be most efficient for a colony with a small foraging area nearby; a long, gradual curve would allow a colony to communicate distances over a larger foraging range.

Putting together some of the pieces of the pattern of intraspecific and interspecific variation in the genus *Apis,* we can reconstruct a plausible recent evolutionary history of dancing in the genus. In a *florea*-like form, colonies were small and exposed, foraging ranges were small, and oriented dancing could only be done on a horizontal surface in view of the sun or blue sky. Dances on vertical surfaces cannot be induced in this form by changing the orientation of the comb and do not occur naturally. A simple step toward *mellifera*—the switch from horizontal to vertical dancing—is found in *dorsata,* but *dorsata* has not yet reached the *mellifera* stage because it becomes disoriented when it cannot see the sun or blue sky. *Indica* and *mellifera,* however, because they dance in the dark, can utilize the protection of closed cavities rather than exposing their colonies to the elements.

The switch to cavity nesting also probably made possible larger multiple-comb colonies not found in *florea* and *dorsata.* The preadaptations of cavity nesting, large colony size, and larger body size enabled *mellifera* to expand its ecological and geographic range into harsher climates. The unpredictability and harshness of temperate-zone climates, coupled with large colony size, created a need for food that required a larger foraging range. The shape of the curve indicates that the relationship between dance tempo and food distance was then adapted for greater efficiency of communication over a wider range of foraging distances.

Other Bees / Greater differences from the *A. mellifera* communication system can be found by examining their close relatives in the tribe Meliponini. Studies of communication in this group have been reviewed by von Frisch (1967*a*) and Kerr (1969). In this group, scent trails are especially important (Figure 19–8). Pheromones are deposited from

19–8 Scent trail of *Trigona ruficrus,* a member of the Meliponini, stingless bees. The forager made several flights between the nest (*top*) and the food (*bottom*) before depositing a scent trail. Her first two scent trails are shown. Points marked by scent are shown with dots and circles. (From Lindauer and Kerr 1958.)

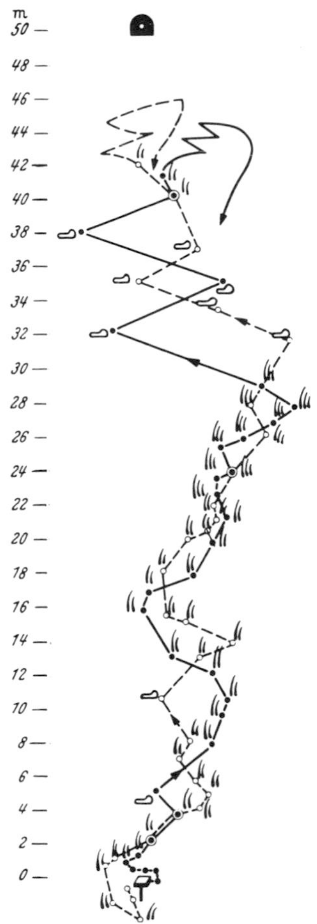

the mandibular glands at short intervals along the route between the colony and the food; these are later followed by colony members.

Also involved in some species are an alerting buzz when the forager enters the nest, zigzag runs in the nest that attract attention, and guide bees that lead recruits. Sound production is correlated with distance to food in some species. The scent of the food is important too; and, unlike *A. mellifera, Trigona silvestris* cannot recruit followers unless a scent has been added to the food.

Bumblebees are not known to communicate about food sources, but males use scent deposits to mark circuits that they patrol regularly. This trait is presumed to be important in mating. Its existence in bees lacking communication about food suggests a possible origin of scent trails used in foraging. However, the diversity of glands used by social insects in communication makes other origins equally plausible.

Figure 19–9 shows the diversity of mechanisms for communicating information about food location, and provides a summary of proposed phylogenetic relationships among the Apinae.

Other Insects / Some of the components of dancing in bees are found in quite unrelated insects. Although they need not be presumed to be homologous, their occurrence is suggestive of some possible origins of dancing.

Tail-wagging, with resultant arousal of colony members, has been reported in a wasp, *Polybia atra*. Chemical communication about food is widespread in ants and termites. Movements suggestive of a crude "dance" were discovered by Dethier (1957) in the blowfly, *Phormia regia* (Figure 19–10). If a drop of sugar water is offered and then withdrawn before the fly has finished with it, the blowfly makes circling movements to left and right more or less randomly on a surface. The "dances" are longer if the food is richer. If the surface is illuminated with unidirectional light, the movements are more parallel with the light rays (but are not oriented toward the food). If the surface is vertical and in the dark, the movements of the fly tend toward the vertical. Thus, after exposure to the food the fly performs movements that in some contexts might be capable of arousing other flies. It is surprising that the movements are oriented—even if they are not oriented toward food.

A curious parallel between bees and saturniid moths was discovered by Blest (1960). These moths, upon settling after a flight, perform side-to-side vibrations of the body. In individuals of the genus *Automeris*, a linear relation was found between the duration of the preceding flight and the duration of vibration. After 2 minutes there were 4 waggles; after 10 minutes, 11; after 30 minutes, 25. The result

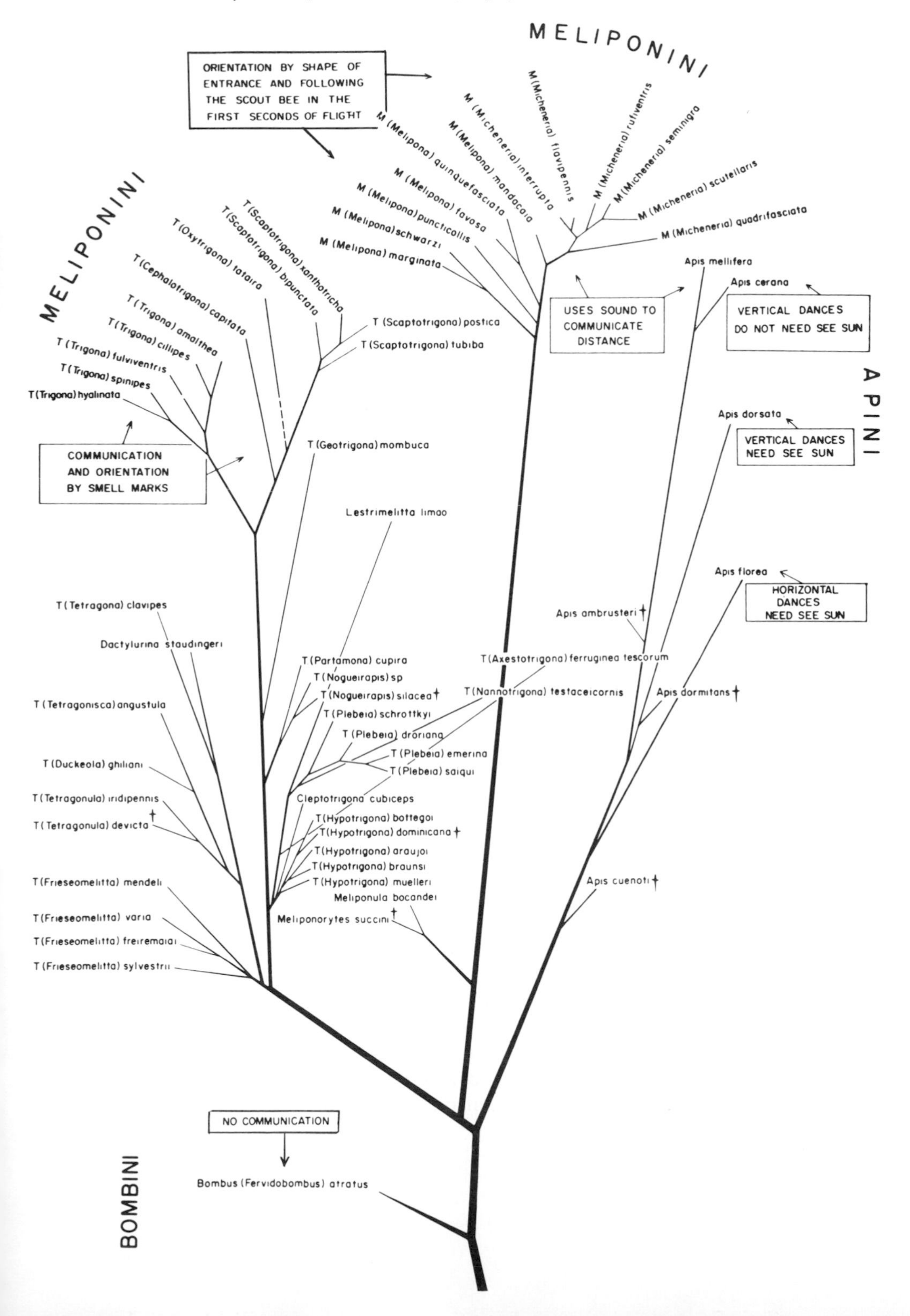

19–9 Communication systems of the Apinae in relation to phylogeny. Daggers, †, indicate fossil species. (From Kerr 1969, by permission of Prentice-Hall, Inc.)

19–10 Oriented flights of the blowfly. If a drop of sugar water is given to a blowfly on a horizontal surface and then removed before the fly has finished feeding, the fly makes randomly oriented searching movements if the lighting is diffuse. **A.** If the light is unidirectional the movements tend to be oriented with the light. **B.** In the dark on a vertical surface the searching movements tend to be oriented with gravity. (From Dethier 1957.)

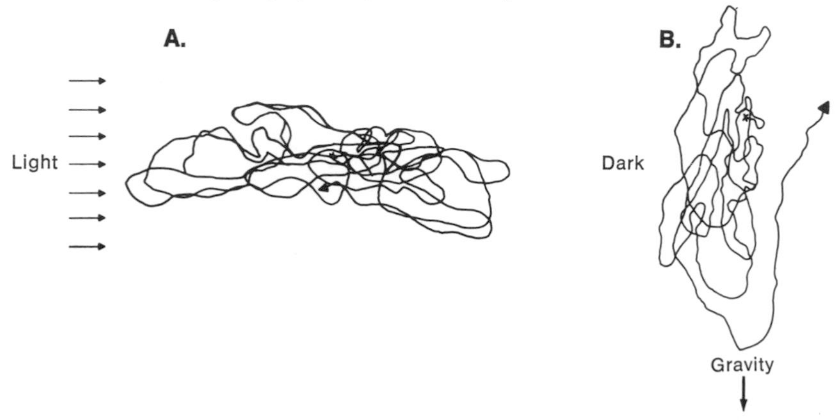

was similar even after an interruption of up to 1.5 hours between the end of flight and the start of vibratory movements. The similarity of these relationships with the waggle dance of *A. mellifera* suggests that the dance may have arisen from similar behavior in bees. It would be interesting to know whether the observations of Dethier and Blest could be repeated in solitary bees.

Another feature of the dance of *A. mellifera* that has a counterpart in other insects is the ability to transfer orientation from the sun to gravity. This ability is shared by some ants, dung beetles, and ladybird beetles. In the dung beetle the direction toward the sun is converted to the gravitational downward direction—the opposite of bees.

CONCLUSIONS

The communication systems of the advanced social bees in regard to food location are among the most complex of any invertebrate species. The honeybee has been intensively studied, and a great many of the factors affecting the system are known—although some problems remain. The complex system of the honeybee can be viewed in phylogenetic perspective as derived from an accumulated series of adaptations found in varying combinations in related species and even in quite unrelated ones.

THE PHYSIOLOGICAL BASIS OF SPECIES CONSTANCY AND SPECIES DIVERSITY IN BEHAVIOR

In Parts I–III, the principles and methods of evolutionary biology have been employed to help us understand the diversity of behavior in the animal world. Comparative and ecological methods were stressed, and little attention was given to the physiology of behavior. Yet implicit in the major conclusions of Parts I–III are certain assumptions about the physiological mechanisms of behavior. Evolutionary interpretations of behavior require genetic determination of behavioral differences. Genetic determination in turn entails corresponding differences in the behavioral physiology or structure of the species involved. Therefore, a postulate of the evolutionary study of behavior is that *species differences in behavior, if they are caused by evolution, must be mediated by corresponding structural and/or physiological differences.* In Part V we shall examine the nature of these physiological and structural differences. Examples will be chosen mainly from the invertebrates, because the relationships between evolution, neurobiology, and behavior are clearest in these animals.

The physiology of behavior is a subject of tremendous inherent interest and rightfully deserves full and separate treatment—not just as a small part of this book. Our purpose in this section is not to provide a general review, but to examine selected physiological concepts and phenomena of interest to evolutionary behaviorists.

To understand the "mechanisms" of behavior and of species differences in behavior, one must use information from all aspects of the biology of the animal, including its skeletal structures, integument, muscles, sense organs, reproductive structures, endocrine glands, and nervous system. Each animal meets the challenges of its environment with a variety of *effectors* (muscles, glands, electric organs, light producers, nematocysts, etc.). The use of these effectors is affected by *stimuli* acting on *receptors* from inside and outside the animal. It is the primary role of the nervous and endocrine systems to process the information from the sensory structures in such a way that the effectors will be activated at appropriate times and places. From this simple summary we can recognize three main problem areas in the control of behavior:

1) *Sensory systems:* How is information received by the sensory structures, and what are the first stages in the processing of sensory data?

2) *Motor systems:* How is behavior produced through activation of effectors and of the neural structures closely associated with them?

3) *Integrative systems:* How is effector output integrated according to receptor input and central programs? This category is a large and heterogeneous one that includes everything not included in the first two categories, i.e., learning, memory, motivation, reflexes, spontaneity, and rhythmicity.

Two principal centers of interest or approaches may be recognized in the physiological study of behavior. On the one hand are workers whose attention is attracted by the ability of animals, especially large-brained species, to adapt to their environment through the mechanisms of learning. These people are interested in the *plasticity* of behavior as manifested, for example, in learning of various sorts. Their prime subject is often that superlative learner, man. On the other hand are workers whose attention is attracted by the opposite phenomenon, the lack of randomness and plasticity in behavior and the presence of *order* and predictability in the behavior of each species. These people are impressed by the *constancy* of behavior within a species and of differences between species. Such behavior is sometimes described as *stereotyped.* The search of these workers is for the mechanisms of species-typical behavior. These two problems can be broadly referred to as the "plasticity or learning problem" and the "instinct problem." Since neither problem can be "solved" completely without reference to the other, it is unrealistic to attempt to limit one's interests to one or the other. Nevertheless, since the learning problem has received so much attention elsewhere, and since we have made no pretense of presenting a general review, we shall confine ourselves here to the instinct problem.

Stereotypy in behavior can arise through learning, but our concern will be with stereotypy derived from underlying features of the nervous system. That stereotypy in behavior is often based on stereotypy of neural structures and processes should probably be recognized as a universal law of behavior. Stereotypy of behavior can stem from properties of sensory, motor, or integrative systems (reviewed in Brown 1969). In the following chapters each of these sources will be considered separately.

20 Reflexes in Simple and Complex Animals

The unit reaction in nervous integration is the reflex, because every reflex is an integrative reaction and no nervous action short of a reflex is a complete act of integration.

—*Charles Sherrington, 1906*

NERVOUS SYSTEMS have evolved because of their usefulness for producing behavior in appropriate relationship to stimuli. Although incredibly complex neural circuitry has evolved to control behavior, many behavioral patterns depend on relatively *simple* neural circuits and are consequently relatively *autonomous* and *rapid*. In this chapter some of the simplest nervous systems will be considered in relation to species differences in behavior.

The simplest types of behavior are associated with the simplest nervous systems and also with simple systems of neurons embedded in more complex nervous systems. We shall examine first some general principles of neurobehavioral systems that evolved in simple nervous systems of the nerve-net type. Then we shall examine the organization of reflexes in more complex animals.

NERVE NETS

Nerve nets are simple not in numbers of neurons but in their lack of organization and lack of a diversity of structural elements. Nerve nets generally occur in a sheet and so can be treated for most purposes as *two-dimensional*—a major simplifying feature in comparison with three-dimensional nervous systems. Nerve nets are typical of the phylum (or subphylum) Coelenterata (jellyfishes, hydras, corals, anemones), but they also occur in the phyla Echinodermata (starfishes, sea urchins, sea cucumbers) and Hemichordata (acorn worms). Nerve nets have been most studied in coelenterates.

The structural characteristics of a nerve net can be appreciated from Figures 20–1 and 20–9. In all coelenterate nerve nets the structural elements, neurons, are *simpler* than those in more complex animals. There are also, typically, *fewer types* of neurons. In Figure 20–1 the neurons are mostly bipolar, but multipolar neurons are common elements of some nerve nets. Typical of the structure of nerve nets is an orientation of the neurons that approaches randomness; nerve nets are not organized into conspicuous nuclei and tracts. It is as if one sprinkled many short wavy sticks over a table top. At the points of contact between two neurons, synapses commonly occur. Unlike most synapses

20–1 Nerve net in a sea anemone. The neurons in a silver-stained whole mount of a thin sheet of tissue (a mesentery) of *Metridium senile* were drawn with the aid of a camera lucida tracing device. The muscle fibers in the mesentery run predominantly vertically. Natural endings of neurons are indicated by forked tips. Dashes and dots are used only to distinguish different neurons. The thickness of the lines is relative to the diameters of the neurons, but is not drawn to scale. This nerve net consists almost entirely of bipolar neurons. Notice that the neurons are not organized into discrete tracts and groups; instead, they cross each other nearly at random, although there is a predominance of neurons running vertically. See Figure 20–8 for orientation in the body of the anemone. (From Batham, Pantin, and Robson 1960.)

1mm

in other phyla, these are often *nonpolarized;* that is, they transmit impulses from one neuron to the other in either direction. Correspondingly, transmitter vesicles are found on both sides of the synapse. These synapses are of the *en passant* type; so there are no specialized synaptic knobs. Since the nerve fibers have both dendritic and axonal functions, they are sometimes called neurites.

The principal functional characteristic of a nerve net is *diffuse spread* of impulses. This property is illustrated in Figure 20–2, which shows that excitation spreads diffusely around cuts in every direction. Because of this property, a nerve-net–mediated behavior is highly resistant to injuries of the nervous system.

Now that we have reviewed the characteristics of nerve nets, we can define them as "anatomically dispersed systems of neurons, so connected that excitation can spread through some considerable number of neurons in any direction and diffusely, bypassing incomplete cuts" (Bullock and Horridge 1965:288). An outstanding and detailed coverage of nerve nets is given by Bullock and Horridge (1965) in their chapter on coelenterates.

The behavior mediated by nerve nets tends to be rather simple, such as the postural changes of anemones associated with feeding, ejection, and shriveling shown in Figure 20–3. In jellyfishes, swimming and mouth movements (see Figure 20–11) may be mediated by nerve nets, although nerve rings may also be involved in some cases. In hydroid polyps, movements of the tentacles mediated by nerve nets bring food to the mouth.

Nerve nets are more efficient than central nervous systems not just for surviving injuries but also for mediating *local responses*. When the receptors and the effectors of a behavior pattern are close together, the distance traveled by nerve impulses may be shorter on a direct line from receptor to effector than via a central nervous system. The neural control of the defensive pincer organs of starfishes known as pedicellariae (Figure 20–4) illustrates this. These tiny pincers are found in large numbers on the backs (aboral sides) of starfishes, particularly around the thin, vulnerable skin-gills. Their actions help to keep very small animals from settling on the backs of the starfishes. Jennings (1907) found that a light touch on the back of *Asterias forreri* caused the pedicellariae within a few millimeters to bend toward the stimulus and to open and close. Touching the outside of the pincers caused them to open, while contact inside the pincer jaws caused immediate closure. Opening excitation can spread to neighboring pedicellariae, but closing does not spread.

Most important in regard to the neural mechanism was the observation that all these responses were shown in small, excised pieces

20–2 Diffuse spread of excitation around cuts in a nerve net was demonstrated beautifully by Romanes (1885). **A.** The medusa, *Aurelia aurita,* is about the size of a soup dish—a few inches in diameter. At the indentations in the margin may be seen the eight marginal bodies, in which neurons are concentrated. In this picture the mouth and neck (manubrium) have been removed. The ovaries (four-leaf-clover-shaped) are in the center, and the dark lines represent nutrient canals, not nerves. The marginal bodies are the sources of rhythmical discharges in the sheet of muscle that comprises the bell of the jellyfish. If all but one are removed, as in **B,** each wave originates at the remaining marginal body and travels in both directions around the bell, meeting at the opposite side. This occurs despite the large cuts. Similarly in **C,** with the inner core, *x,* removed, the contraction wave originating at the remaining marginal body follows the arrows, invades the inner ring only at *z,* and travels in both directions around the inner ring. Cutting the neck of tissue at *z,* which is about ⅛ inch across, prevents the invasion of the inner ring. In **D,** if the end at *a* is gently brushed with a camel's hair brush (too gently to start a contraction wave in the bell) the ganglion at the opposite end, *b,* will initiate a contraction wave in the bell traveling toward *a.* Sometimes the wave passing from *a* to *b* can be seen in the contraction of the tentacles. A slow (25 cm/sec) wave of contraction passes from *a* to *b.* When it reaches the ganglion at *b,* a fast wave of contraction of the muscle of the bell passes back toward *a.* The experiments demonstrate (1) passage of excitation around the cuts in a way that cannot be explained by any simple fiber tract, (2) the existence of a slow and a fast conduction system in the same tissue, and (3) the role of the ganglion (marginal body) as a point of access from the slow system to the fast system. The existence of separate nerve nets in *Aurelia* was confirmed by electrophysiological and histological experiments by Horridge (1954). (From Romanes 1885.)

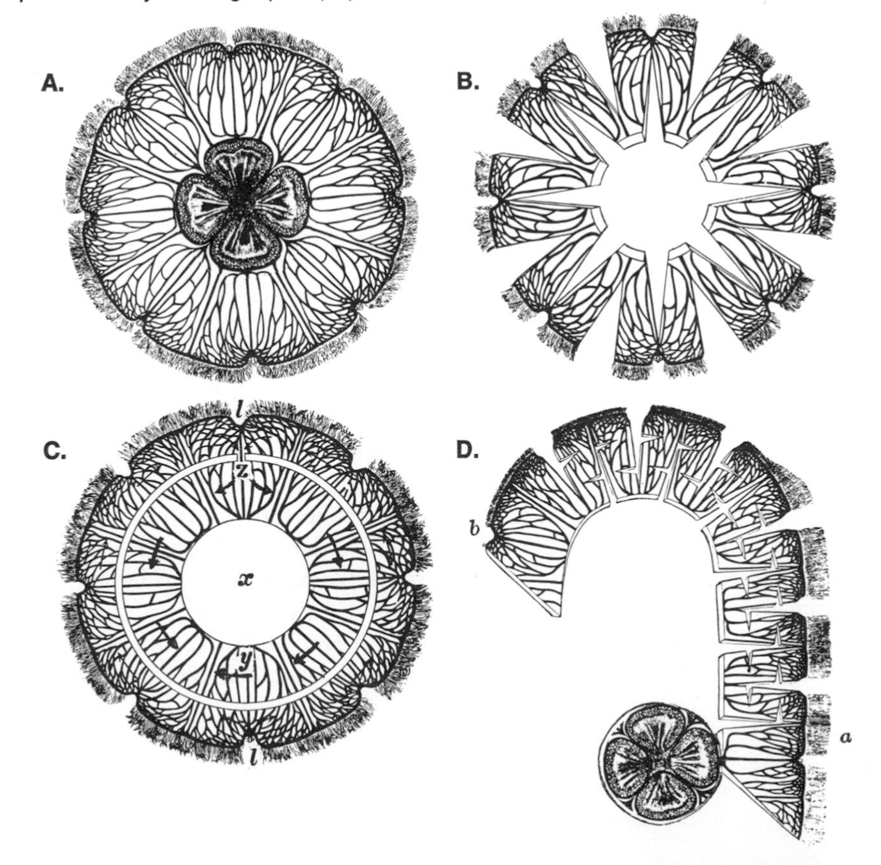

20–3 Behavior of an anemone, *Metridium*. In **A** the stages are: *a*, normal extended posture showing the means of attaching the hook to make the kymograph records shown in **B**; *b*, ingesting food; *c*, ¼ to ½ hour after ingestion; *d*, *e*, swaying movements 6 hours after food-extract stimulus; *f*, *g*, *h*, *i*, successive stages in antiperistaltic constriction, leading to defecation (emptying) and shriveling; *j*, subsequent refilling, showing extreme expansion 4 hours after shriveling; *k*, specimen showing excretory ring of mucus and uric acid (*u*); *l*, state of extreme contraction found in unfed individuals exposed to sunlight. **B.** The light-controlled rhythm is largely abolished by feeding. (From Batham and Pantin 1950.)

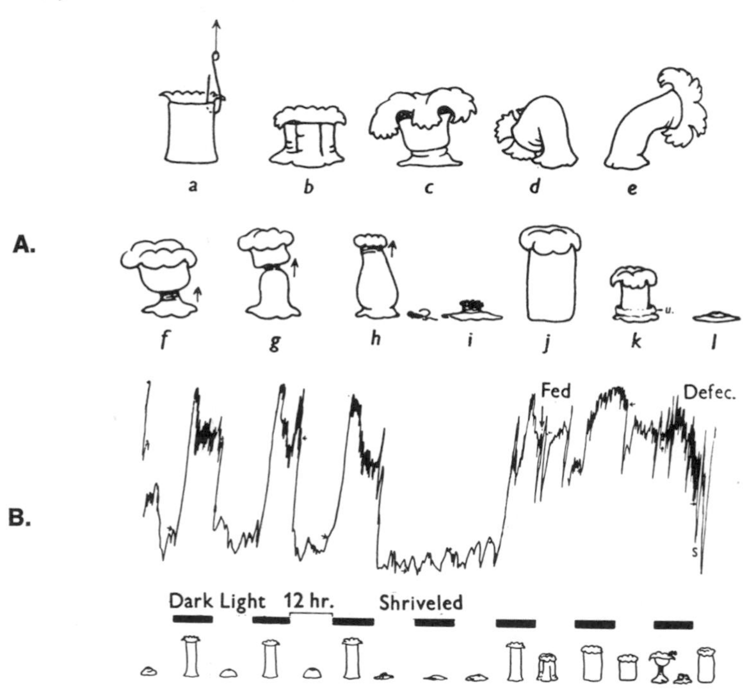

separated from the central nervous cords. This means that within each such isolated piece must exist the complete reflex arc (stimulus-response circuit) for each of the demonstrated stimulus-response correlations. In other words, the receptors, effectors, and their coordinating neurons are all probably present locally at any spot on the back of the starfish; and their reflex action can exist independently of control from the central cords, although it can also be modified through the cords. It is important to notice that for these properties of behavior a random arrangement of neurons without differentiation into tracts is sufficient. This is true only because the appropriate effectors are those nearest to the stimulated receptors. In such cases the appropriate re-

20–4 Pedicellariae offer good examples of local responses that are mediated entirely by peripheral rather than central neurons. Pedicellariae are defensive pincer organs found on the upper surfaces of starfishes and certain other echinoderms. The neural reflex circuit of an isolated pedicellaria can operate independently of nerve cords and the central nervous system. The pincers are normally open when not stimulated. They close on small organisms that settle on the starfish, thus keeping the upper side of the starfish free of barnacles and other hangers-on. The pincer teeth are closed by the adductor muscles when the sensory endings on the inside of the pincers are stimulated. **A.** Typical pedicellaria of a sea-urchin. **B–F.** Other types of pedicellaria found in various sea-urchins. **G–L.** Types of pedicellaria found in starfishes. (Reproduced by permission of the Hutchinson Publishing Group from *Echinoderms* by D. Nichols, Hutchinson University Library, London, 1962.)

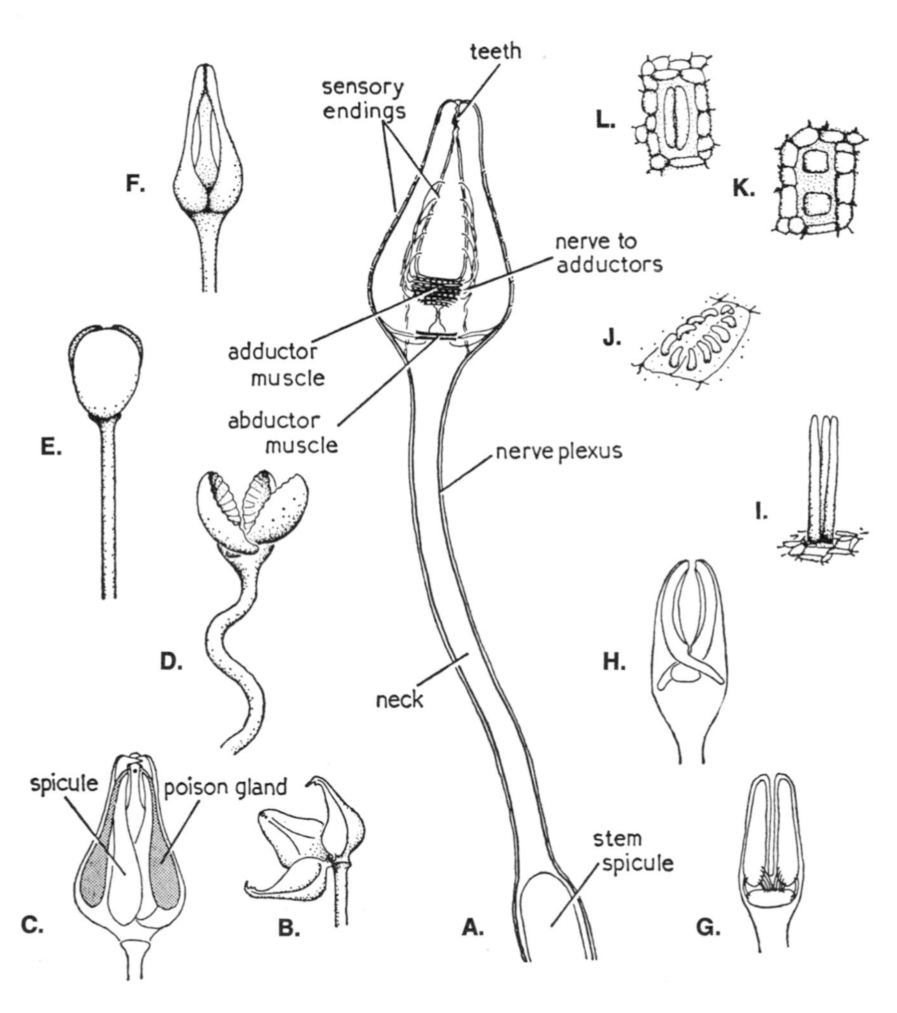

sponse is given to the stimulus because when the excitation caused by the stimulus spreads randomly in all directions from the source, the first effectors to be reached are the most appropriate ones.

This may be an oversimplification where starfishes are concerned, but this type of neurobehavioral control is typical of nerve nets. In the case of pedicellariae, it remains to be shown that all the requirements stated in the above definition of a nerve net are satisfied (Bullock and Horridge 1965:1534–1535; Bullock 1965). The control of pedicellariae is probably more complicated than the above description suggests.

SPECIES DIVERSITY OF BEHAVIOR BASED ON NERVE NETS

If the neurons composing a nerve net are arranged truly at random, how is it possible for the net to mediate specific kinds of behavior? More precisely, two types of general questions arise with random arrangements of neurons. What features of nerve nets allow species diversity in behavior? And what features allow more than one type of behavior by the same individual?

One answer to the species diversity problem is suggested by some systematic observations made by Horridge (1957) on the spread of the withdrawal response from a point stimulus in coral colonies (summarized in Bullock and Horridge 1965). A coral is a colony of individual organisms of the polyp type, called zooids, each of which inhabits a private chamber formed by its own secretions (Figure 20–5). Individuals of a coral colony communicate with each other by means of

20–5 A coral colony, *Porites.* The polyps or zooids are mostly open, but a few in the closed position can be seen in the center of the colony. (From Horridge 1957. Reproduced by permission of the Royal Society.)

lateral basal tissue bridges containing nerve cells. When a zooid is stimulated by electric shock, it retracts into its chamber, and neighboring zooids may also retract, depending on species and circumstance. The contraction of neighboring zooids depends on the number of shocks given, the time intervals between them, and the species. Horridge found that species differ in the number and distribution of neighboring zooids that contracted after successive shocks. The differences are shown in Figure 20–6. In some species, a single shock caused retraction of all polyps of the colony. Neuronal systems that make possible the spread of excitation to all parts of the system if any one element is excited are said to be *through-conducting.* If the net is not through-conducting, the pattern of spread may take one of the forms shown in Figure 20–6.

The failure of impulses to pass a synapse shows that spread in the system is decremental. When two or more impulses are needed to cross a synapse, the synapse is said to require *facilitation.* Synapses differ in their facilitation requirement, and this difference may be an important factor responsible for the species differences in behavior shown in Figure 20–6. Facilitation is often sensitive to the interval between stimuli. This property is shown in Figure 20–7, which also illustrates the basic phenomenon of facilitation. Nerve nets with a facilitation requirement make possible gradation of the strength of the response according to distance from the stimulus, as shown in Figure 20–8. By contrast, in a through-conducting system the response should be all-or-none.

The responses of nerve nets like those in the corals studied by Horridge as functions of several variables have been studied by computer simulations (reviewed in Bullock and Horridge 1965; Fehmi and Bullock 1967). The variables incorporated in the simulation included (1) the proportion of synapses that were through-conducting rather than needing facilitation, (2) the length of the neural elements and hence the number of synaptic contacts per neuron, (3) whether the synapses transmitted in both directions or only one, and (4) the number and density of elements initially excited. All of these variables had some influence on the area of spread in the net resulting from an initial excitation, but (1) was relatively more influential and (3) less influential than the others. Not all of the patterns found in nature and shown in Figure 20–6 could be duplicated by the model. The patterns that involved progressively larger areas with successive stimuli were not obtained, and the authors speculated that these patterns might require the addition to some synapses of the property of after-discharge (more than one post-synaptic spike for each pre-synaptic spike).

The requirement for facilitation at synapses is a property of fundamental significance for understanding neurobehavioral systems. Imagine what would happen in your own brain if all pathways were through-conducting. Any stimulus sufficient to fire one sensory neuron, except for the action of inhibiting neurons, would lead to the

20–6 Spread of excitation (retraction) in coral colonies. **A.** Diagrammatic representation of types of spread of excitation across coral colonies. A succession of electric shocks is given in the center; the circles show the distance of spread of response after each shock. The serial number of the shock is printed beside some of the circles. *Palythoa* shows uniform increments of radius of spread. *Porites* shows decreasing increments, soon reaching a ceiling less than the size of large colonies. *Sarcophyton* appears to show increasing increments of radius, soon involving the whole of a large colony. Apart from through-conduction systems, these are the principal types of spread so far observed. The scale is in units corresponding to the suggested size of the conducting elements.

B.The relations between the number of stimuli (at 1 per sec) and the approximate number of polyps caused to contract for a variety of corals. The different slopes, some increasing, some decreasing, depend upon the different parameters of the nerve net in each case. Although the nerve cells of corals have rarely been seen, and their activity is certainly not known, all the above curves can be simulated by computations based upon the known properties of other coelenterate nerve cells, as investigated in solitary anemones and medusae. Whether coral nerve cells actually operate in the way the explanatory concepts suggest remains to be directly tested. (Figures and legend from Horridge 1957. Reproduced by permission of the Royal Society.)

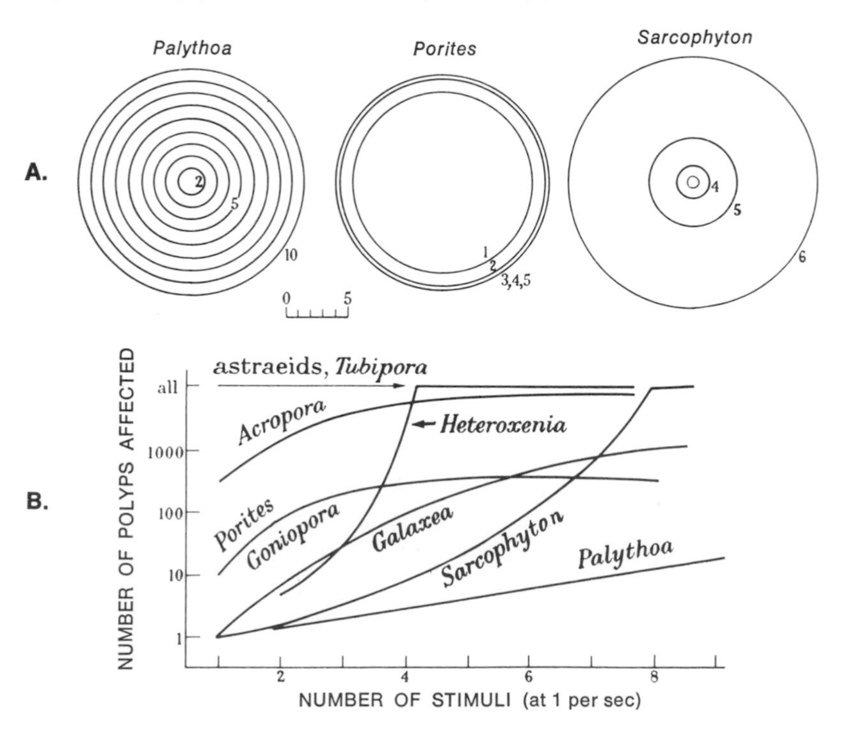

firing of the vast majority of neurons in the nervous system and to the near-simultaneous action of all effectors. You would be in continuous convulsions. The capacity to block the spread of impulses in the nervous system must accompany the capacity for spread of impulses if orderly behavior is to prevail. Excitation must not be allowed to spread without check. To restrict responses to appropriate stimuli, neural excitation must not be allowed to spread beyond certain limits, which vary with each special case. In the entire animal kingdom there is apparently no nervous system that lacks the need for facilitation at some and probably most synapses. Only certain restricted and highly specialized subsystems can afford to be through-conducting.

20–7 Facilitation in the sphincter muscle of a sea anemone, *Calliactis.* The upper line in each pair of records shows the extent of muscle contraction. The lower line shows the time course of the stimuli. From left to right, stimuli were presented every 2 sec, every 1 sec, and every 0.5 sec, for a total of 4 in each case. Closer spacing of the stimuli produces a greater muscular response. (From Ross and Pantin 1940.)

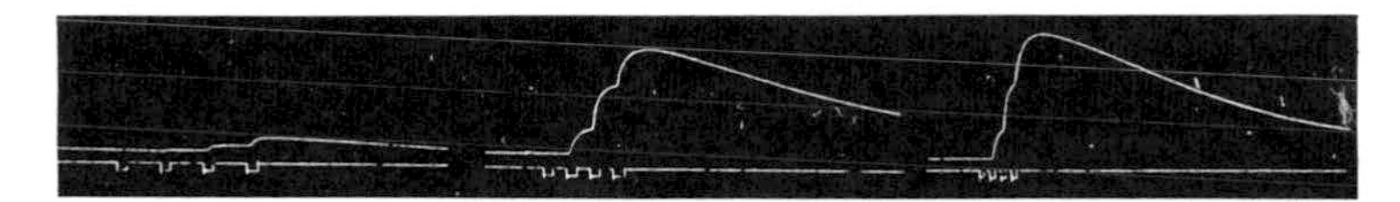

20–8 Hypothetical model of the effect of a simple summation requirement. This model provides a simple explanation of decremental falloff in the strength of a response over distance from the point stimulated (*S*). A stimulus at *S* causes six impulses in neuron *A.* The first impulse opens the path to neuron *B;* the second excites *B,* which opens the path to *C,* and so forth. At least two impulses are required for one to cross the synapse. The strength of the response, shown below, is a function of the number of impulses reaching the muscle, which diminishes with distance from *S.* (From *Interneurons* by G. Adrian Horridge. W. H. Freeman and Company. Copyright © 1968.)

A	B	C	D	E	etc.
6					
5	6				
4	5	6			
3	4	5	6		
2	3	4	5	6	
1	2	3	4	5	6

spread of successive impulses

S

observed movement

Species diversity in behavior, consequently, can arise through species diversity in the facilitation requirements of synapses. Facilitation has been studied in various animals at the level of the synapse, using microelectrodes. Synapses, whether between neurons or between neurons and muscles, differ in the number of impulses required at the pre-synaptic side to cause the generation of one impulse on the post-synaptic side. If the pre-synaptic impulses arrive close enough together in time and space, *spatiotemporal summation* of the post-synaptic potentials may bring the post-synaptic neuron to its threshold for firing. Since transmission across synapses is a function of both relatively stable morphological properties and relatively labile chemical properties, there are many ways in which synaptic function could be controlled genetically and influenced by natural selection. Evolutionary change in synaptic properties could result in evolutionary change in behavior. By such means, species diversity in behavior, like that Horridge showed in corals, could be achieved through general changes in synaptic properties that are independent of specific morphological neural pathways and of the "switchboard concept" of neurobehavioral function.

Up to now, we have emphasized the role of junctions between neurons in restricting the spatial pattern of spread of excitation in nerve nets, but nerve-muscle junctions deserve some special consideration too. Since muscles in the same individual differ in their response to different frequencies of stimulation, differences in behavior can be determined by the frequency of spikes arriving at the muscles from the nerve net (Bullock and Horridge 1965:485). When the various muscles of an anemone, as shown in Figure 20–9, are studied while applying electrical stimulation to the column (body), their "best frequencies" differ. For example, one shock every 10 seconds led to circular muscle contraction, one every 3 seconds to slow parietal shortening, one every 1.5 seconds to a strong mesenteric contraction, and one every 0.6 seconds mainly to sphincter contraction.

In nature the frequency of impulses in the nerve net could be expected to vary with the strength and character of receptor stimulation. A weak stimulus would cause a relatively low frequency of impulses in the net; a strong stimulus would cause an initially high frequency, followed by *adaptation* of the receptors and a slowing of the resultant neural firing. The pattern of behavior that results from a natural stimulus in an animal having a nerve net with random connections can now be seen to be, not the disorganized paroxysm that might at first have been predicted, but a complex function of the frequency response of the receptors, the proximity of various effectors to the site of stimulation, the transmission characteristics of synapses in the nerve net, and finally the facilitation properties of various muscles reached by excitation in the net.

We have now seen several mechanisms through which diversity in behavior can arise both within individuals and between species. Although these mechanisms occur also in species with more complex

20–9 Anatomy of a sea anemone, *Metridium senile.* **A.** Vertical section of a whole animal. A perfect mesentery is a thin sheet that extends from the body wall inward to the pharynx. **B.** Transverse section at the level of the pharynx, showing perfect and imperfect mesenteries. **C.** Enlarged transverse section of a mesentery. **D.** Section of mesentery and body wall showing neurons of the type of net shown in Figure 20–1. (From Batham, Pantin, and Robson 1960.)

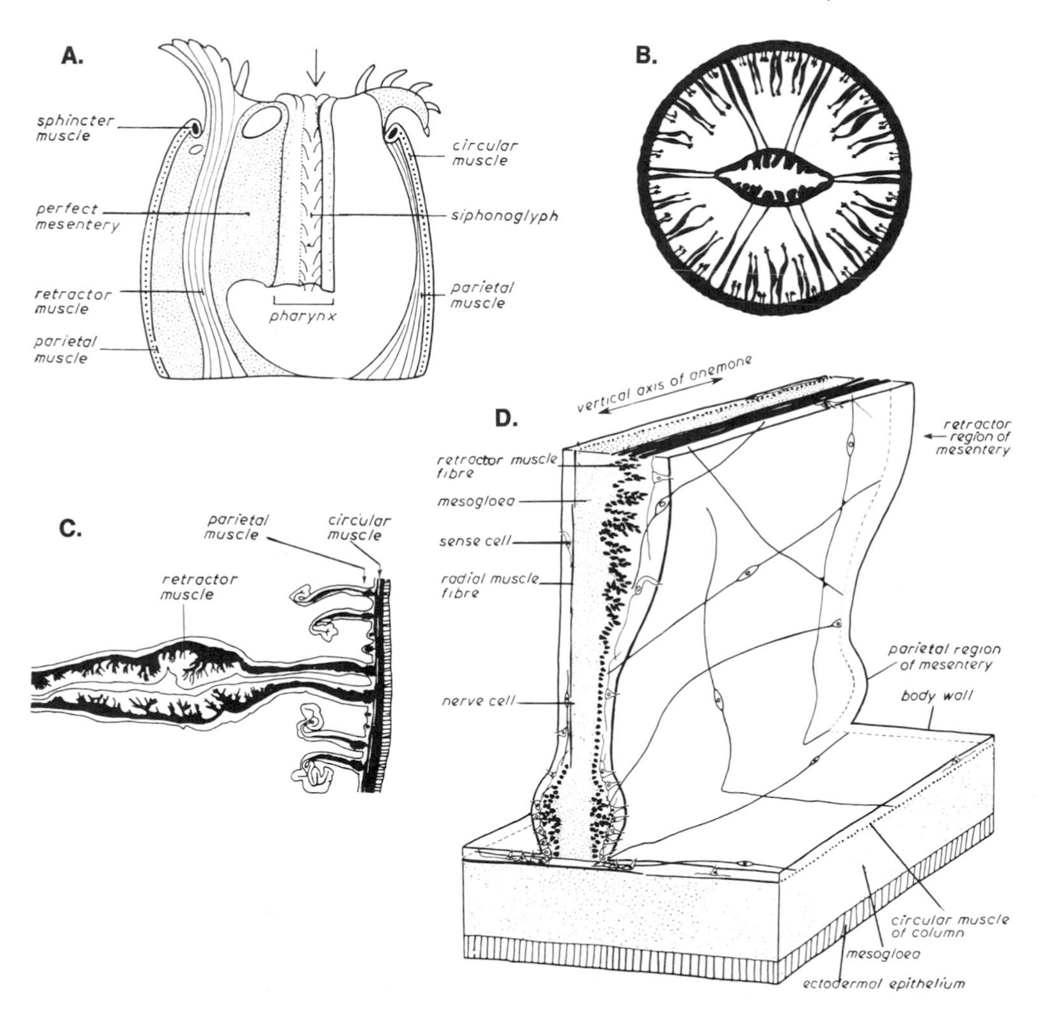

nervous systems, their effects are most clearly displayed in simple species having nerve nets, since specialized, oriented neural pathways are unnecessary.

Having briefly surveyed some integrative properties of random arrangements of neurons, we are now in a position to examine the behavioral effects of some simple and basic types of deviation from randomness. Again, these effects are more easily appreciated in simple animals with simple nervous systems. The following departures (or "advances") from the simplest nerve nets are found in coelenterates: (1) separate nerve nets in the same individual, (2) directionally oriented axons, (3) condensation of neurons into cords, and (4) the presence of both inhibitory and excitatory neurons in the same nervous system. Animals as simple as jellyfishes possess these refinements at the reflex level of neurobehavioral organization.

The existence of two or more *separate nerve nets* in the same individual is known or suspected in several coelenterate species (Bullock and Horridge 1965; Josephson 1965), and may be widespread. An example is illustrated in Figure 20–10. In the medusa, *Geryonia* (Horridge 1955*a*), the separation of nets is associated with the separation of behavioral patterns that might otherwise interfere with each other. In *Geryonia* the contraction of the bell that produces the swimming beat is mediated by the through-conducting nerve net in the bell and velum. When a prey item stimulates the tentacles at one spot, they contract locally (mediated by a net in the tentacles); and if the stimulus is strong enough, excitation extends both to neighboring tentacles (through the

20–10 Two nerve nets in the ectoderm of a jellyfish, *Velella*. The large-fiber and small-fiber nets are anatomically and functionally distinct. Specimen is a silver impregnation. The nuclei belong to ectoderm cells. (From Mackie 1960.)

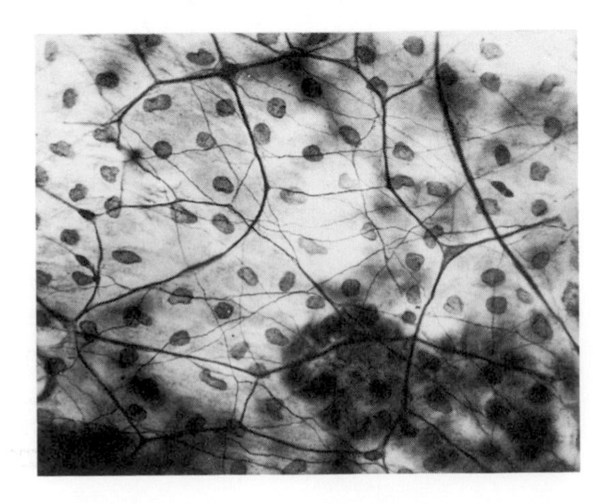

upper nerve ring) and, through a radial centripetal system in the bell, to the manubrium, which terminates in the mouth. If the stimulus is strong enough, the mouth is carried by the manubrium over to the prey (as in Figure 20–11), which is then ingested. The placing of the mouth on the stimulated spot is quite accurate, but if two points are stimulated, the mouth goes to a point equidistant between them. Functional separation of at least two systems in the bell is indicated by the observation that stimulation of the tentacles does not alter the rhythm of bell contraction, nor does the swimming beat pacemaker affect the placing response of the manubrium.

The existence of *directionally oriented* axons is also evident in the tentacle-manubrium system. Short cuts on a line between the stimu-

20–11 Feeding responses in two jellyfishes. *Left:* In *Geryonia,* mechanical stimulation at the points indicated with lower-case letters causes contraction in the area designated by the corresponding capital letter. This brings the mouth to the food. The swimming beat continues. *Right:* In *Aequorea,* mechanical stimulation at any point on the bell causes that part of the rim to bend toward the mouth, thus bringing the food to the mouth. In contrast to *Geryonia,* the bell stops beating during this response. This suggests the presence of an inhibitory effect of the bending system on the swimming system. (From Horridge 1955*a, b.*)

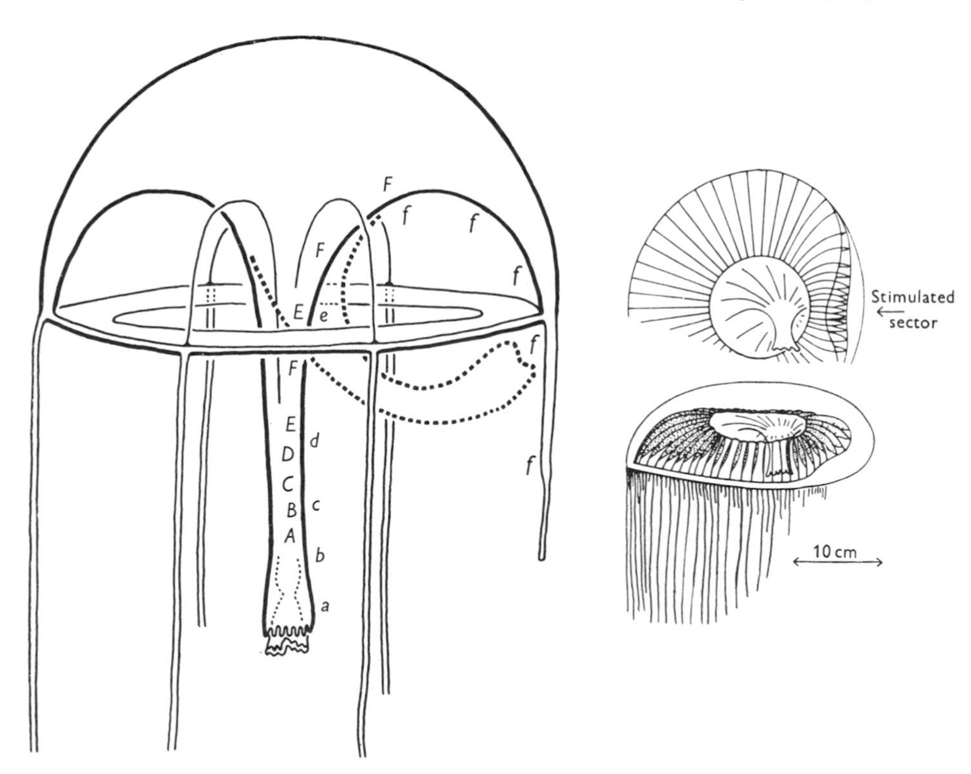

lated spot and the manubrium prevent the placing response of the manubrium. Since the experiments indicate that excitation traveling from the perimeter to the manubrium travels only in straight lines and does not bypass short cuts, the tentacle-manubrium system fails to fulfill all of the requirements of a nerve net as defined above. An additional feature of this system is the one-way connection between the tentacles and the upper nerve ring; excitation can enter the ring from the tentacles, but the ring cannot excite the manubrium. If it could, accuracy of the manubrial placing response would presumably be greatly reduced.

The first instance of *inhibition* in a coelenterate was demonstrated in another jellyfish, *Aequorea*, by Horridge (1956*b*). As shown in Figure 20–11, the manubrium in this species is too short to reach over to the tentacles unaided. The stimulation provided by prey on the tentacles causes that sector of the bell to bend toward the manubrium (unlike *Geryonia*), which moves to meet it. Inhibition of the rhythmic swimming contractions of the bell occurs during this feeding reflex. Since the feeding reflex causes asymmetry in the shape of the bell (unlike *Geryonia*), the jellyfish's direction of travel would be altered if the beat continued. This may be the adaptive significance of inhibition of the beat during feeding.

In brief, we have seen the appearance, in the simplest phylum that has an undoubted nervous system, of several specializations at the reflex level. The adaptive nature of these "advances" is usually quite simple and obvious. The new properties—namely, separation of nets or functional systems, directional orientation of axons, condensation into tracts (rings), one-way transmission, and inhibition—are more highly developed in more complex animals; but their fundamental nature can be appreciated more easily in simple animals with simple behavior.

REFLEXES UNDER CENTRAL CONTROL

Nerve-net systems of neurons are useful for appreciating the simpler stages in the evolution of reflex circuits, but phyla with central nervous systems provide better material for showing certain more specialized features.

In most phyla, sensorimotor circuits involve a central nervous system (CNS), and all sensory excitation must be routed through the CNS before a motor effect can result. This reduces the autonomy of local regions by subjecting elements of their reflex circuit to easy influence from relatively distant parts of the nervous system.

In the coelenterates we saw the simplest departures from a randomly structured nerve net in the direction of nerve tracts with specific connections between certain receptors, neurons, and muscles. In the vertebrates, the specificity of the neuronal circuitry underlying even the simplest reflexes is striking, especially when one considers that nearly all synaptic connections between neurons in the vertebrates are made in the gray matter of the CNS. In other words, the receptors that initiate a certain behavior are connected selectively with effectors that produce the response. They also connect to other relevant structures, such as the longitudinal pathways in the CNS.

The important feature of central control over elements of reflex circuitry will be illustrated with a few examples from among the vertebrates. Because of their relative simplicity (compared to brain-mediated behavior), spinal reflexes have been intensively studied as models of certain properties of vertebrate nervous systems. The studies of Sherrington (1947), summarized in his famous book, *The Integrative Action of the Nervous System* (first published in 1906), served as a foundation and stimulus for much further study.

The stretch reflex, of which the knee-jerk reflex is an example, is a traditional point of departure. Stretch reflexes are primarily concerned with body posture, and they occur in various animals from cats to crayfishes. If a muscle used in maintaining posture is stretched, as would happen if an external force caused a deviation in the posture, the nervous system causes a corrective force to be applied to the same muscle so as to restore the original posture. With a stronger external force, other synergistic (cooperating) muscles may be stimulated to respond, both by their own stretch receptors and by connections with the first reflex.

The stretch reflex is mediated basically by four types of cells: (1) The *receptor* cells are modified muscle cells known as intrafusal fibers, whose central portion is sensitive to being stretched. They are located in specialized muscle-sense organs known as muscle spindles. (2) The *sensory neurons* run from the intrafusal fibers to the spinal cord. Their axons (type Ia) enter the cord by the dorsal root and make synaptic contact with various types of neurons. (3) The *motor neurons* are located in the ventral part of the gray matter of the spinal cord. They receive direct synaptic contact from the Ia axons. Their axons (type α, alpha) leave the spinal cord in the ventral root and run to (4) the regular *muscle fibers*. The arrangement is shown diagrammatically in Figure 20–12.

Since only one synapse inside the spinal cord is involved, this reflex is often termed *monosynaptic*. Almost all other reflexes in the vertebrates involve more than one central synapse and are therefore

20–12 Stretch reflex. The basic circuit involves a stretch receptor (muscle spindle) in the muscle, a sensory neuron, a motor neuron, and the muscle. **A.** Anatomical arrangement. **B.** The circuit at the cellular level. *IF* = intrafusal fiber. *SN* = sensory neuron. *MN* = motor neuron. *C* = contractile. *NC* = noncontractile. *M* = muscle cell. *IN* = interneuron.

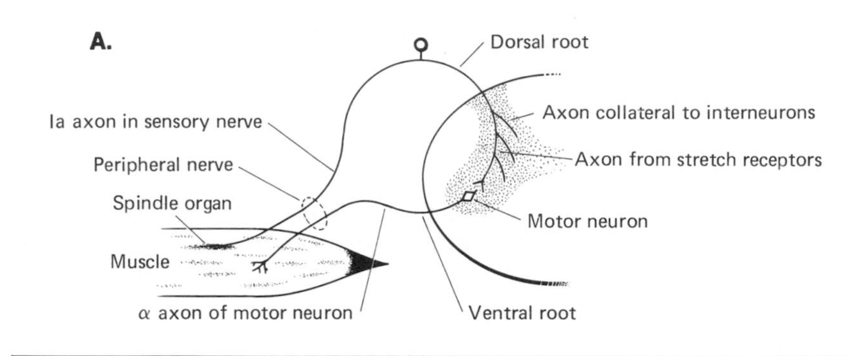

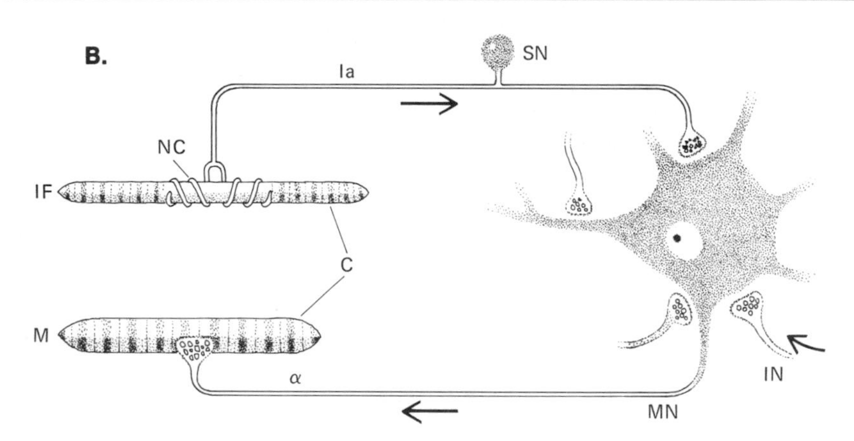

termed *polysynaptic.* The neuronal connections in this reflex were described by Ramón y Cajal (1909) from observations of silver-stained neurons and are illustrated in Figure 20–13*A*.

This four-cell circuit operates as a simple control system (see Chapter 25, especially Figure 25–8) for maintaining a muscle at a constant length. If the muscle is stretched beyond the set length, the sensory and noncontractile portions of the intrafusal fibers in the muscle are stretched. This distortion causes the sensory axons (Ia) to excite the motor neurons, whose axons (α) cause the same muscle to contract. The contraction continues as long as the stretch persists on the intrafusal fibers. By this means the muscle is returned to the originally set length (at which the intrafusal fibers are no longer stretched enough to fire the

20–13 Neurons involved in vertebrate stretch reflex. **A.** Monosynaptic connection between incoming sensory neurons and outgoing motor neurons. **B.** Polysynaptic circuit involving an interneuron interposed between sensory and motor neurons. The interneuron might be excitatory or inhibitory, depending on the particular circuit involved. (From Ramón y Cajal 1909.)

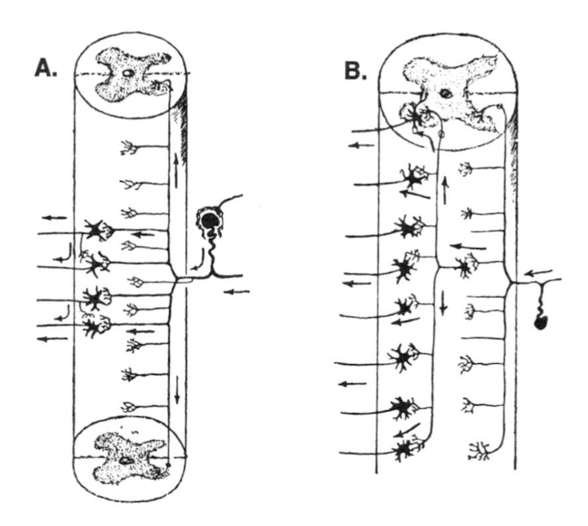

sensory neurons). The system operates as a simple servo-mechanism involving negative feedback (reduction of stretch) to the receptor.

Several methods for central modulation of this stretch reflex exist. One will be described. The intrafusal muscle fibers are composed of two main parts, a sensory, noncontractile part near the middle of the fiber, and contractile parts near the ends. The contractile parts are activated by small diameter axons known as gamma efferents.

The stretch reflex tends to maintain the muscle at a constant length. To adjust to a new length, it is necessary to readjust the contractile part of the intrafusal fiber. The intrafusal fibers are few in number, and their contraction has only a negligible effect on the length of the muscle. Consequently, the principal direct effect of contraction in the intrafusal fibers is not to contract the muscle but to stretch the stretch receptors. This, of course, initiates contraction of the whole muscle by the circuit already described. The result is different; in this case the muscle length becomes shorter rather than remaining constant. The new shorter length is determined by the amount of activation of the intrafusal fibers by the gamma efferents. The muscle shortens until the stretch on the stretch-sensitive part of the intrafusal fibers is eliminated. By this means the length of the muscle comes under the influence of the gamma efferents, which are under central control. Additional modulation of this stretch reflex is possible by means of both excitatory and inhibitory inputs to the motor neurons from the brain and via interneurons (Figure 20–13*B*) from other receptors. These brief statements just begin to describe what is really a complex system for the control of posture and muscular activity.

For reflexes to function effectively in central nervous systems, their neural circuits must be precise and specific. The sensory neurons that initiate a given response must bypass inappropriate neurons during their growth and end selectively on the "correct" neurons only. When neurons terminate on two or more types, as is usually the case, the proportion ending on each type must be appropriate. The precision of these connections does not occur by chance. It has been shown experimentally that nerve cells "insist" on developing and maintaining connections to their proper "goals" even when such connections produce maladaptive behavior because of surgical rearrangements of the muscles.

In review, it is worth considering some properties shared by both coelenterate and vertebrate reflexes. In both groups the threshold and extent of the spread (or "irradiation") of excitation to various muscles are dependent on the strength of the stimulus. In both groups a particular configuration of muscular response (or "reflex figure") is associated with a particular kind and location of stimulus. In both groups muscular movement in certain reflexes may be directed toward the source of stimulation (showing "local sign"). And in both groups inhibitory relationships between certain neuromuscular systems governing antagonistic or competing behavior can be demonstrated. The list of similarities does not end here, and the reader can probably think of others.

THE DEFINITION OF A REFLEX

The concept of a reflex has a tortuous history (Efron 1966) that cannot be reviewed here. It originated with Descartes, who imagined fluids from the periphery being reflected off the pineal organ in the brain to the muscles. Under Sherrington, after the turn of the century, the idea of *reflex* prospered. Pavlov (1927) broadened the concept beyond usefulness to include all conditioned responses and, by the implications of his school, all behavior. For example, he once wrote, "It is obvious that the different kinds of habits based on training, education and discipline of any sort are nothing but a chain of conditioned reflexes" (p. 395).

In true anthropomorphic spirit, reflexes were at one time defined in part by their involuntary and unconscious nature; and under Pavlov these features became the only ones of importance. But scientists found the concepts of voluntariness and consciousness difficult to work with and began to seek ways to study behavior without using them. It then became exceedingly difficult to define *reflex*, but the obvious value of the concept led authors to continue using it without defining it. Pavlov and many of the early workers suffered from a lack of knowledge of the neural circuitry underlying reflexes. Today we can see very clearly the dependence of the properties of various reflexes on their neural circuitry, so that the idea of a reflex being defined without reference to neural circuitry seems preposterous. In an attempt to supply an operational but heuristic definition that allows realistically for fuzzy borderline cases, I offer tentatively the following definitions:

A reflex is

1) a type of *behavior* that is characterized by

2) a highly consistent correlation between a particular stimulus and a particular response (*stimulus-response specificity*), and

3) that is mediated by *neural mechanisms that are relatively simple*, and are

4) alike in all members of the species (*species-constant*).

For distinctions between reflex and fixed action patterns, see the following chapter.

REFLEXES AS INTEGRATIVE MECHANISMS

In our treatment of the physiological mechanisms relevant to the problem of instinct, reflexes have been treated first because of their relative simplicity and basic importance. Because of this simplicity, reflexes provide some of the clearest examples of the dependence of stereotypy in behavior on properties of nervous systems. As emphasized by Sherrington in the quotation at the beginning of this chapter, a reflex is the simplest integrative mechanism. It is neither a purely sensory nor purely motor phenomenon, for it combines both sensory and motor elements into a stereotyped pattern of behavior. All animals with nervous systems need such simple integrative mechanisms, and man is no exception. Primates are richly endowed with a variety of reflexes, especially at the spinal level; but they may be difficult to demonstrate in conscious, intact animals because of the conspicuousness of more complex neurobehavioral mechanisms. In many invertebrates, on the other hand, reflexes are prominent. The behavior of sea urchins (Echinoidea, Echinodermata) has even been described as "a republic of reflexes" (von Uexküll 1909). This, of course, is an oversimplification, because even in the simplest of animals integrative phenomena higher than reflexes can be demonstrated.

Reflexes can be considered as units for purposes of experimental analysis, as Sherrington did; but reflexes are not to be thought of as completely isolated integrative units. Probably in all animals the neural circuits that mediate any given reflex also connect with those of other reflexes and with interneurons involved in other sensory, motor, and integrative processes. In this respect the concept of a reflex as an isolated neurobehavioral mechanism is a fiction, as Sherrington himself has admitted. The experimental analysis of the central modulation and control of reflex behavior is a subject of considerable importance, but it is beyond the scope of this book.

CONCLUSIONS

Fundamental properties of simple neurobehavioral systems are well represented in the nerve-net systems of coelenterates. Nerve nets are essentially two-dimensional nets of randomly ordered, rather similar neural elements. The diffuse spread of effector (behavior) excitation around cuts in any direction characterizes a nerve-net system. In a nerve-net system, local behavioral responses can be produced simply by excitation traveling over the most direct route between the stimulus and the nearest effectors; detour through a central nervous system is unnecessary. Consequently, a high degree of local behavioral autonomy is present in animals with nerve nets, as can be demonstrated by the survival of behavior in isolated pieces of an individual. The pedicellariae of starfishes, for example, are excited mainly by nearby stimuli on the skin.

Diversity of behavior between species or within an individual is achieved in nerve-net systems through variation in synaptic physiology and in neuronal orientation. Synaptic function has an important effect on behavior in that it requires spatiotemporal summation of incoming neuronal spikes, which manifests itself behaviorally as the phenomenon of facilitation. Inhibition of certain behavior patterns is also a function of synaptic physiology. Some species differences in behavior that might be interpreted as due to such differences in synaptic function have been discussed for corals and anemones.

The second way in which the nerve-net systems of coelenterates reflect species diversity in behavior is in the addition of order to the randomness inherent in the simplest nerve nets. Separate nets in the same individual, directionally oriented fibers, and the existence of nerve cords have behavioral correlates in coelenterates.

Reflexes have been most often studied in animals with central nervous systems (one or more nerve cords, ganglia, and a brain). These systems make possible a greater degree of central control over the reflex than can occur in a nerve net. The control of stretch reflexes by gamma motor innervation exemplifies the addition of central control to a simple reflex. Centralization of neural function also allows a greater complexity of neural circuitry and behavior.

With the benefit of modern knowledge of the neural circuitry underlying reflex behavior, a reflex can be defined as (1) a class of behavior that is characterized by (2) a strong correlation between a particular stimulus and a certain response, which is (3) mediated by simple neural (and/or chemical) circuits that are (4) basically alike in pattern and structure in all members of a species. Behavior with these characteristics is involuntary and may occur with or without consciousness, but the concepts of volition and consciousness are peripheral to the fundamental properties of reflexes. Similarly, reflexes are felt to develop without the aid of learning, although they may require practice, but this also need not enter into the definition. Although the definition is necessarily arbitrary in the requirement for "simple" neural circuits, it is at least operational and realistic. It does not include conditioned responses, since these need not have simple neural circuits nor are they necessarily the same in all members of a species.

21 Stereotypy in the Mechanisms of Coordination of Motor Patterns

FROM THE VIEWPOINT OF the evolutionary behaviorist, one of the most conspicuous features of animal behavior is that certain motor patterns of behavior are confined to particular species or larger phylogenetic groups. Often these behavior patterns are performed in standard ways with relatively fixed patterns of coordination, as with many displays, vocalizations, comfort movements, or locomotor actions. The *stereotypy*, or low degree of variability, in such cases is conspicuous. In this chapter some neural mechanisms of stereotypy in motor patterns will be examined.

The motor patterns to be treated in this chapter differ from those in the preceding chapter in their lower degree of *stimulus-response specificity*. It is a feature of reflexes that each is activated by its own characteristic stimulus, often by a localized subset of one type of receptor. In contrast, the motor patterns to be treated in this chapter have a low degree of stimulus-response specificity. Such behaviors as the flight of insects, the emergency escape of crayfishes and squid, and the songs of insects may occur as responses to a variety of different stimuli arising from a variety of different receptors; or in some cases they may even occur spontaneously.

The motor patterns discussed in the preceding chapter are parts of integrative units; reflexes are integrative in the sense of bringing about a certain relationship between a given stimulus and a given response. The motor patterns treated in this chapter are not viewed as integrative units in this sense; they are units of motor coordination. Their integration with stimuli and programming in relation to other possible behaviors are controlled by other neural mechanisms, some of which will be discussed in Chapters 22 and 25.

The goals of this chapter are to provide the reader with a general feeling for the nature of the neuronal circuits that may underlie species-typical, stereotyped motor patterns by examining a few well-studied cases. In these few favorable cases the relations between motor patterns and neural organization are unusually clear. These investigations exemplify modern studies of one facet of the instinct problem.

The scheme adopted here of separating species-typical, stereotyped motor patterns into reflex and nonreflex patterns and of dividing them on the basis of sensory input is, of course, an oversimplification. I do not wish to imply that a dichotomy is involved, except in the focus of our attention. The specificity of the stimulus-response relationship likely varies on a continuum and is certainly not easily quantifiable. Many cases are intermediate and cannot be easily categorized. Moreover, a given act, such as crayfish emergency escape, could be classified differently depending on one's point of view. This behavior has some reflexlike properties, especially its anatomically traceable reflex arc from sensory to motor structures; but the same motor neuronal configuration can be triggered by a variety of other neuronal inputs. This kind of duality is characteristic of many and perhaps all reflexes on closer examination. Nevertheless, there is a real difference of degree. The cases to be discussed in this chapter tend to show greater complexity in their underlying neuronal circuits and greater availability to a variety of inputs.

SOME CHARACTERISTICS OF SPECIES-TYPICAL, STEREOTYPED MOTOR PATTERNS

Like reflexes, the action patterns discussed in this chapter tend to have the following characteristics:

1) The patterns of motor coordination are *stereotyped*, although they may show variability in orientation.

2) The distribution of action patterns among species is correlated with taxonomic and *phylogenetic* groups (see Chapter 1).

3) The pattern of motor coordination is the same in all members of a species or larger taxonomic group (see Chapter 1). The *behavior* is *species-typical.*

4) The motor patterns are subject to *genetic* manipulation by hybridization and artificial selection (see Chapter 2).

5) The motor patterns develop without the aid of learning from other individuals, although other kinds of experiences may be important (see Chapter 28). In this sense they are *self-differentiating.*

6) In favorable cases, the fixed pattern of behavior can be attributed to the organization of the underlying *fixed neuronal circuits.*

7) In these cases the neuronal circuit is typically the *same in all* members of the species.

Some action patterns that share the above characterization *differ* from reflexes in one or more of the following respects:

8) Some motor patterns can be evoked by a wide variety of stimuli, including two or more sensory modes; they have a *low* degree of *stimulus-response specificity.*

9) The animals show *motivation* to perform the act, and this fluctuates.

10) The acts are performed *spontaneously* in some cases (see Chapter 25).

Obviously not all of these statements apply to every nonreflex pattern. Similarly, not all of them apply to the nonreflex patterns treated in this chapter.

The existence of nonreflex motor patterns with the characteristics listed above was pointed out by Lorenz (1932, 1937*a*). These papers helped to break the stranglehold of reflexology that prevailed at the time, since they pointed out important aspects of behavior that were inconsistent with the contemporary reflex theory. At the same time they became the nucleus of a still influential and controversial theory of instinctive behavior. Motor patterns that eventually became known as *fixed action patterns* (FAPs) were described by Lorenz and Tinbergen in a manner similar to, but not identical with, the above characterization. Their treatment was necessarily less explicit about the role of the nervous system, and they devoted considerable attention to the problem of identifying and separating genetic and environmental influences on the development of FAPs. It would be beyond the scope of this chapter to enter into the controversy that still surrounds this area. Interested readers are referred to Eibl-Eibesfeldt (1970) for a recent Lorenzian treatment of FAPs, and to Moltz (1965) for a critique of the concept of the FAP. Readers may wish to decide for themselves whether or not the more recent studies described in this chapter overcome the principal objections of Moltz to the concept of FAP.

A classical example of an FAP is the egg-retrieving pattern of the greylag goose (*Anser anser;* Lorenz and Tinbergen 1938). In this example two components can be recognized, namely, a *fixed pattern* that is relatively inflexible and an *orienting component* that is relatively flexible. When an egg has been displaced a short distance from the nest, the incubating bird could theoretically roll it back with its foot or wing, but invariably the rolling is done with the underside of the bill as the bird stands on or near the nest and stretches out its head to roll the egg, as shown in Figures 21–1 and 7–6C. The movement of pulling the head in toward the body seems always to be completed even though the bill may slip off the egg; the pattern is performed as a complete entity even though logically the bird should not continue it after losing contact with the egg. If the egg rolls slightly to one side,

21–1 A classical example of a fixed action pattern: a greylag goose rolling an egg back into its nest. The goose invariably attempts to roll the egg back using its bill in the manner shown, rather than using its foot, wing or bill in some other way. The neural mechanisms of this behavior are unknown. (From Lorenz and Tinbergen 1937.)

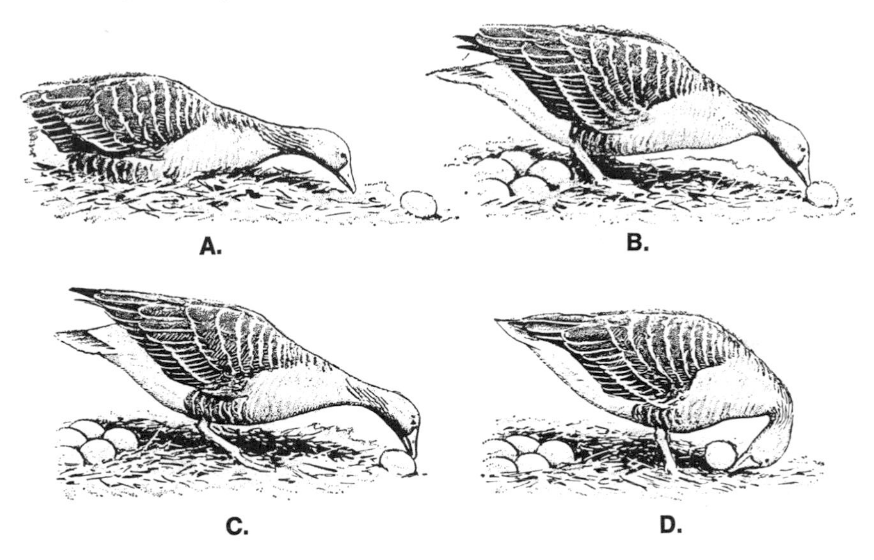

the bill is moved sidewards to correct for the displacement. Correction of the course of the head and egg is possible as long as the bill is in contact with the egg; the behavior is not fixed to the extent that no corrections of this sort are possible. It is not uncommon for a fixed action pattern to be accompanied by such orienting components.

This is not a chapter about fixed action patterns, since the examples have been chosen without respect to the criteria used by Lorenz to separate FAPs from reflexes. Treatment of the phenomena of spontaneity and motivation, so important in the recognition of FAPs, has been deferred to later chapters. Nevertheless, at least one of the examples, that of cricket song, is a perfect example of an FAP, and others are not far off. In this chapter our primary concern is with the neuronal bases of species-typical, stereotyped motor patterns other than reflexes. We shall focus on the pattern-forming processes.

GIANT NEURONS AND FIXED BEHAVIOR PATTERNS

The Importance of Giant Neurons / For the purpose of experimenting on the neural basis of stereotyped motor patterns, invertebrate animals are excellent subjects because of their relatively simple nervous systems and small number of neurons. Of critical importance is the presence of individually recognizable giant neurons and their giant axons. These are also found in vertebrates (the Mauthner and Müller neurons of fishes, amphibian larvae, and lampreys; see Figure 26–6), but they reach their greatest diversity and evolutionary specialization among invertebrates. A "giant" neuron or axon is merely one that is conspicuously larger than most other neurons in the animal.

The principal importance of giant neurons for the analysis of behavior, aside from their obvious facilitation of experiments, is the ease with which they can be individually recognized anatomically and electrically and their connections traced. These features allow several to many individual neurons to be mapped. When this is done, it is typically found that each giant neuron has a *characteristic size, shape, position, and projection* and that these are *virtually constant for all individuals* in a species. Figure 21–2 illustrates the appearance of some giant neurons as they are seen *in situ*. The diversity of shapes and projections of giant axons is probably greatest among annelid worms, where they are especially useful in the rapid coordination of escape contraction over long distances from head to tail (Figure 21–3).

Giant Neurons in Cephalopods / Certainly the most famous giant axon is "the giant fiber" of "the squid," which has been the subject of Nobel Prize–winning experiments on electrical potentials and conduction in axons. Actually, there is not one giant axon but an integrated system of giant neurons in squids; and there is impressive diversity among genera of cephalopods in systems of giant neurons and the escape behavior patterns they mediate. Figure 21–4 shows a common squid, *Loligo*, in various postures.

The giant fiber system of the genus *Loligo* is illustrated in Figures 21–5 and 21–6. The system allows a small number of neurons to coordinate a large number of muscle cells into an effective escape response. The escape is produced by forcing a large volume of water out of the mantle cavity through the funnel to produce a jet that pushes the animal backwards. When the mantle is contracted, the body contour is simultaneously streamlined by retracting the head, placing the arms together, and folding the fin. In addition, a cloud of a mucus-ink mixture may be ejected, and a neurally controlled color change may occur. Many of these components are simultaneously coordinated by the giant fiber system.

At the highest level in the system are the two first-order giants, one on each side. These are located in the magnocellular lobe of the brain in a position where they receive afferent input from the major

21–2 Cell bodies (somata) of giant axons in crayfish. **A.** Third abdominal ganglion as it looks in the dissecting microscope. **B.** Whole mount of a cleared ganglion photographed with fluorescence microscopy after injection of the LG (lateral giant) with Procion Yellow. The cell body is on the contralateral side; the dendritic branches, in the ipsilateral neuropil. **C.** Electrical junction between fast flexor motoneuron and LG axon. The fluorescent dye crossed this junction, but on the other side remained confined to the motor giant cell. **D.** Double injection of LG axons to illustrate the commissural synapse. In all figures the width of the ganglion (1.0 mm) provides a size reference. (From Remler, Selverston, and Kennedy 1968. Copyright 1968 by the American Association for the Advancement of Science.)

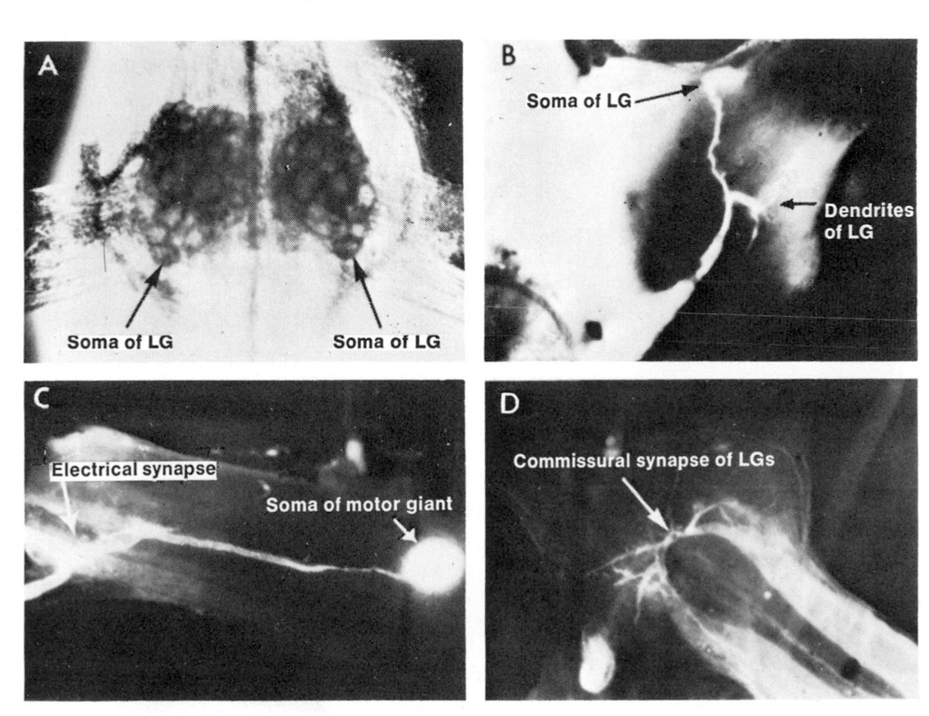

sense organs (in eye, skin, statocyst) and from higher parts of the brain. They are connected by a protoplasmic bridge which ensures that impulses will pass symmetrically down both sides, since there is no synapse between the two sides to delay conduction in one side relative to the other.

The second-order giants number about seven in *Loligo* and more in the cuttlefish, *Sepia*. They are located in the palliovisceral lobe of the brain and receive their primary input from the first-order giants, but they also receive other inputs. Thus the system can be triggered at two levels in the brain. Five second-order giants project to the muscles

21–3 Diversity of giant axons in annelid worms. Most species have elongate bodies and a need for rapid longitudinal contraction as an escape behavior. Giant axons with a variety of shapes and sizes have in common the ability to mediate such behavior. The earthworm is shown in 5 and 6 (in 5, the lateral giants and in 6, the medial giant). The genera represented are: 1 and 2, *Euthalenessa;* 3, *Sigalion;* 4, *Lepidasthenia* and *Euthalenessa;* 5 and 6, *Lumbricus;* 7, *Euthalenessa;* 8, *Eunice;* 9 and 10, *Nereis* and *Neanthes;* 11 and 12, *Arenicola;* 13, *Nereis* and *Neanthes;* 14 and 15, *Halla* and *Aglaurides;* 16, *Mastobranchus;* 17, *Sabella* and *Spirographis;* 18, *Myxicola.* (After Nicol 1948; from *Interneurons* by G. Adrian Horridge. W. H. Freeman and Company. Copyright © 1968.)

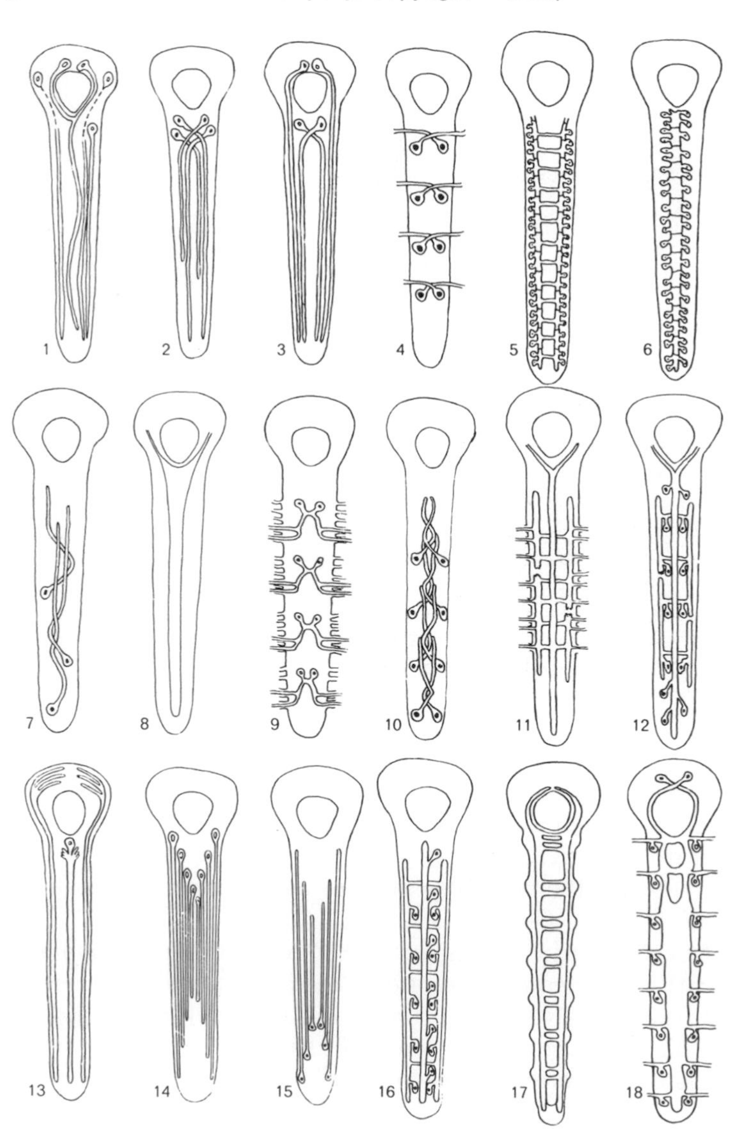

that retract and position the head and funnel during escape. The others project to the stellate ganglion outside of the brain.

In the stellate ganglion are located the cell bodies of the third-order giants. Their principal input is from the second-order giants, and the main synapse transmits impulses without facilitation on a 1:1 basis (one input spike gives one output spike). There may be 12 third-

21–4 The squid, *Loligo vulgaris.* The lower two pictures show animals seen from the side, swimming towards the right of the figure; the arms are used as vanes to keep the head up. The animal in the bottom right-hand corner shows the lateral white flecks and the white ventral line characteristic of a male in breeding condition. (Figure and legend from Wells 1962.)

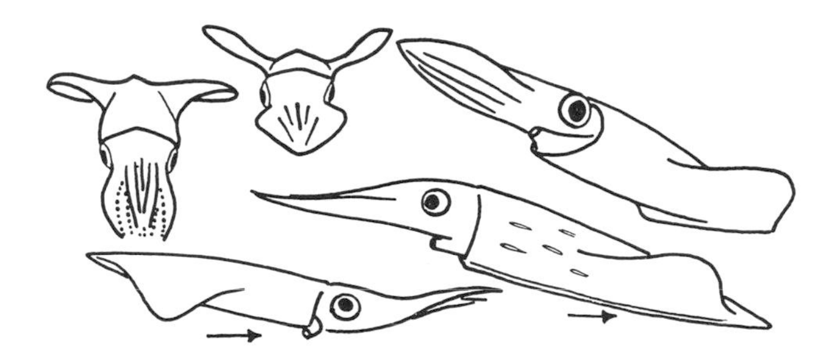

21–5 The giant nerve fiber system of *Loligo.* 1, 2, and 3 show the first-, second-, and third-order giant nerve fibers, running respectively from the magnocellular to the palliovisceral lobe of the subesophageal part of the brain, from the palliovisceral lobe to the stellate ganglia, and from the stellate ganglia to the mantle. There are only two first-order giant fibers, fused together at one point, so that both fire simultaneously. The other giants are more numerous, and are not all shown. The largest, fastest-conducting fibers run to the most distant parts of the mantle, ensuring simultaneous contraction of all parts of the mantle wall. Second-order giants also run to the muscles holding head to abdomen, and to the funnel retractor muscles. (After Young; figure and legend from Wells 1962.)

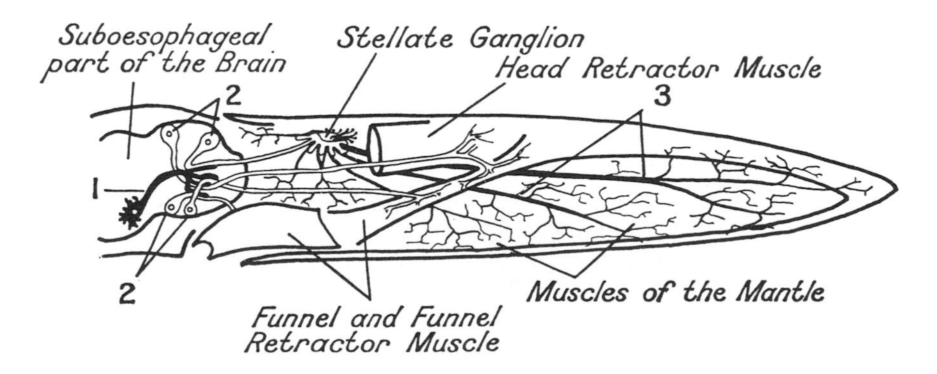

21–6 Schema of the giant axon system in the squid, to show the cascade arrangement.

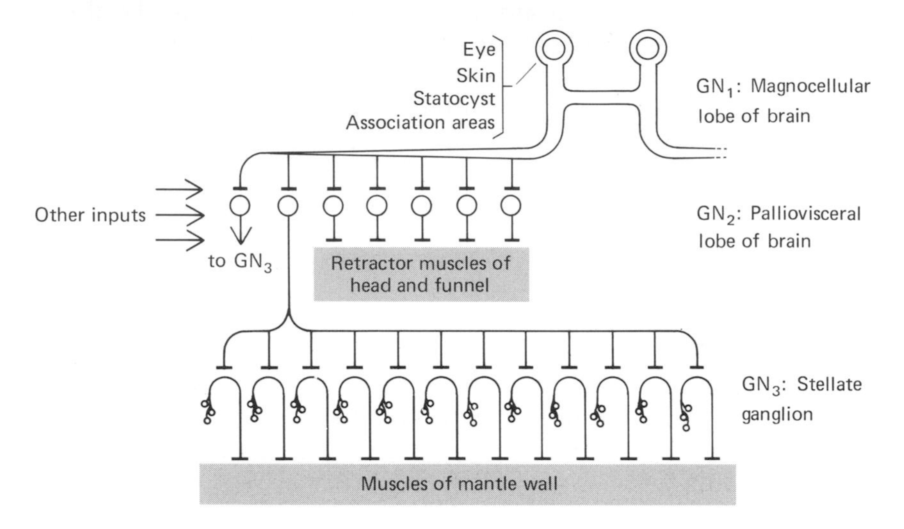

order giants on each side, and they branch profusely to reach all parts of the mantle musculature.

The various giant axons are graded in size so that the largest and fastest ones project to the most distant parts. This helps to ensure virtually simultaneous contraction of the mantle muscles. The neuronal assembly that produces the motor pattern of escape in *Loligo* can be seen from the above description to be *organized hierarchically as a cascade.* The motor pattern can probably be initiated by excitation of a single neuron, the first giant. By activation of greater numbers of neurons at two, successive, more peripheral levels, this neuronal assembly produces a relatively complex, precisely coordinated, stereotyped motor pattern. This general pattern of organization is found in many giant-axon-mediated motor patterns.

The presumed adaptive value of this system derives from its short latency and efficient coordination. Young (in Bullock and Horridge 1965, II:1481–1485) has estimated that with the enlarged axon diameters the escape latency should be half that with normal axons. The system is not developed to the same extent in all cephalopods. It is most specialized in the pelagic genera, such as *Loligo,* in which there is typically no danger of collision during "blind" backward escapes. These squids have a separate giant fiber lobe in the stellate ganglion and possess the largest giant axons (1.5 mm in the giant squid). Bottom-living genera such as *Sepia* have a giant fiber system, but the cells of origin in the stellate ganglion do not occur together in a specialized lobe. In *Octopus*, which characteristically occupies a "house" and lives in close relationship to spatial features of its rocky environment, the giant fiber system is lacking, and the escape behavior less swift.

Giant Axons in Crayfish / One of the systems of large axons most thoroughly investigated in terms of behavior is found in the American crayfish *Procambarus clarkii*. These axons have been mapped morphologically, and their sensory inputs and motor outputs have been determined by experiment. More than 75 axons are now individually known through these painstaking investigations. Much of this work has been done by groups in the laboratories of Wiersma at the California Institute of Technology and Kennedy at Stanford University. The nervous system of arthropods generally consists of a series of *ganglia*, in which the nerve cell bodies are located, joined by *connectives* composed of axons and typically lacking neuron cell bodies. A cross section of a crayfish abdominal connective is shown in Figure 21–7.

The role of single axons in the integration of relatively complex motor patterns has been investigated by stimulating and recording from individual axons in the abdomen of the crayfish (e.g., Kennedy, Evoy, and Hanawalt 1966). These experiments revealed that in many cases

21–7 Cross section of an abdominal connective in a crayfish, showing the relative sizes of the giant axons in the schema of Figure 21–9. *LG*, lateral giant; *MG*, medial giant; *A, B, C*, interneurons. (From Zucker et al. 1971. Copyright 1971 by the American Association for the Advancement of Science.)

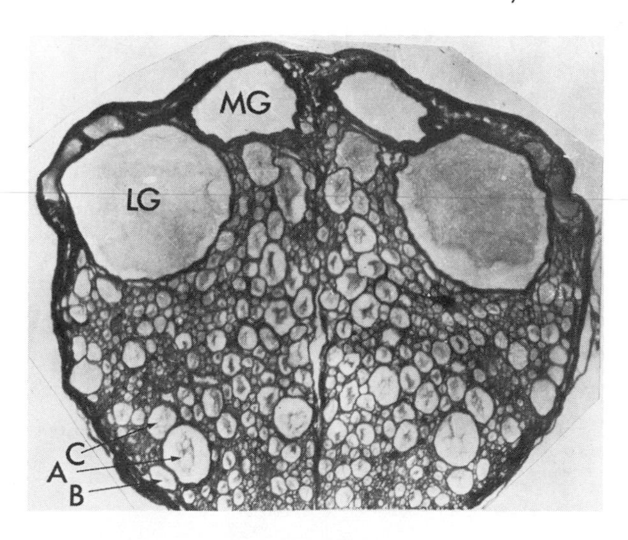

each single axon produces a characteristically different pattern of response, involving combinations of various degrees of flexion or extension in the different segments. It was possible to dissect out strands from the ventral nerve cord in which conclusive controls could be run to ensure that only one axon was being effectively activated. A flexion in one segment of the crayfish's abdomen is produced by excitation of the five excitatory neurons on each side going to the flexor muscles and of the one inhibitory neuron going to the extensor muscles, as well as inhibition of the excitatory motor neurons to the extensors and of the inhibitor of the flexors. When this pattern of neuronal action is replicated in the several abdominal segments, at least

21–8 The motor patterns of LG- and MG-mediated tail flips in an unrestrained crayfish; action runs from top to bottom. A sharp tap on the abdomen fires the LG and causes an upward escape. A sharp tap on the anterior part of the body fires the MG and causes a rapid, rearward retreat. These are tracings of frames, about 10 msec apart, from a high-speed motion picture. The prod used as a stimulus and the electrode leads used to verify the firing of the MG and LG are shown. (From Wine and Krasne 1972.)

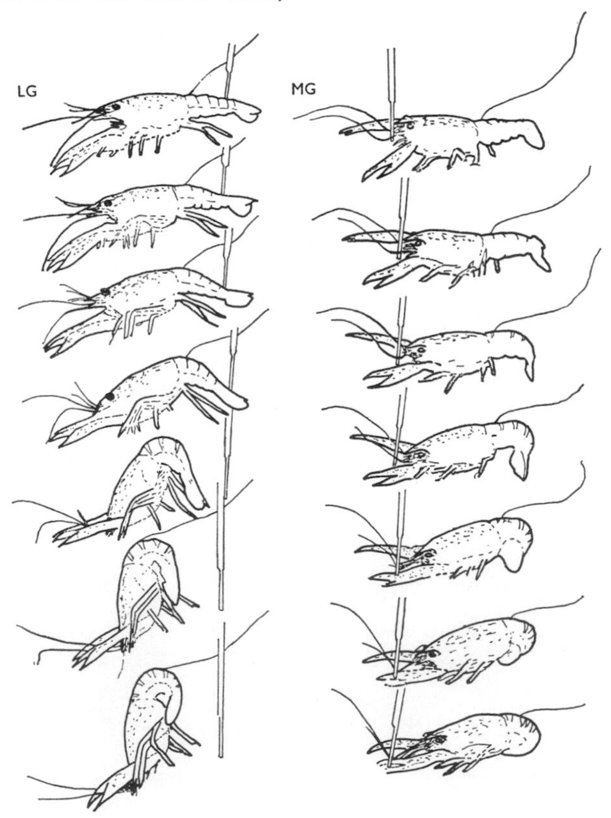

100 separate motor neurons are affected by one command fiber. The coordination brought about by a single interneuron affecting abdominal position is thus considerable.

The neuronal circuits mediating the tail-flip escape response (a rapid abdominal flexion) to phasic, tactile stimuli have been studied in detail in the crayfish (Zucker et al. 1971; Wine and Krasne 1972; Zucker 1972). Abdominal flexion provides the main motive force for the backward locomotion that is often used by crayfishes and lobsters. The behavior patterns generated by stimulation of command fibers in a free-swimming crayfish are shown in Figure 21–8. The reflex circuit underlying escape behavior elicited by the lateral giant is shown in Figure 21–9. The circuit comprises a set of tactile sensory neurons, two levels of large interneurons, and two kinds of motor neurons organized as a cascade from one of the interneurons (LG). The movement is generated by nine large motor neurons in each abdominal half-segment. One of these, the motor giant, innervates all the fast flexor muscles in its half-segment, and is driven by the lateral giant axon through an electrical synapse. The other eight motor neurons are smaller and receive impulses from giant axons and the neuropile. In contrast to the motor giant, each of the eight innervates only a restricted fraction of the flexor musculature of its half-segment. Thus the flexor muscles are doubly innervated. Some idea of the structure of an abdominal motor neuron in the crustacea can be gained from the reconstruction in Figure 21–10.

The lateral giant has the largest axon in the abdominal cord (see

21–9 Diagram of identified neurons involved in the LG escape response circuit in the crayfish abdomen. Squares refer to populations of elements, whereas circles represent single neurons. *TR,* tactile receptors; *A,* an unisegmental tactile interneuron; *B* and *C,* multisegmental tactile interneurons; *LG,* lateral giant neuron; *MoG,* motor giant motor neuron; *FFMN,* fast flexor motor neurons exciting the phasic flexor musculature. Only excitatory connections are shown. Bars indicate electrical junctions, filled circles represent facilitating chemical synapses, and open circles are antifacilitating chemical synapses. Minor pathways are shown by thin lines. (Figure and legend from Zucker 1972.)

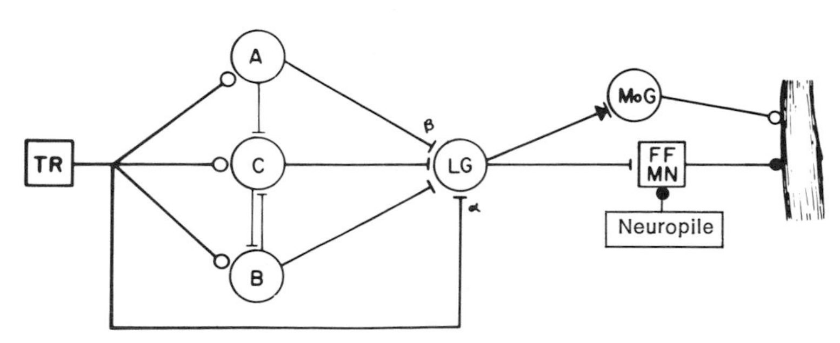

21–10 A giant motor neuron from the third abdominal ganglion of a lobster. The ghost drawing was made from a reconstruction based on a dye-injected neuron. **A.** Three-dimensional model. **B.** As seen from the front. **C.** From the top. **D.** From the side. The axon can be seen passing out of the ganglion to the right in A and B; to the left in C. The size calibration is for B–D. (From Davis 1970. Copyright 1970 by the American Association for the Advancement of Science.)

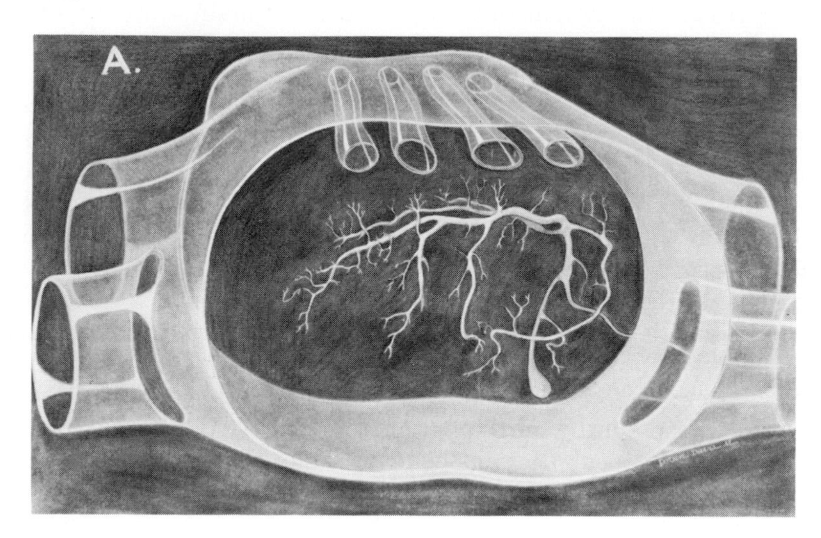

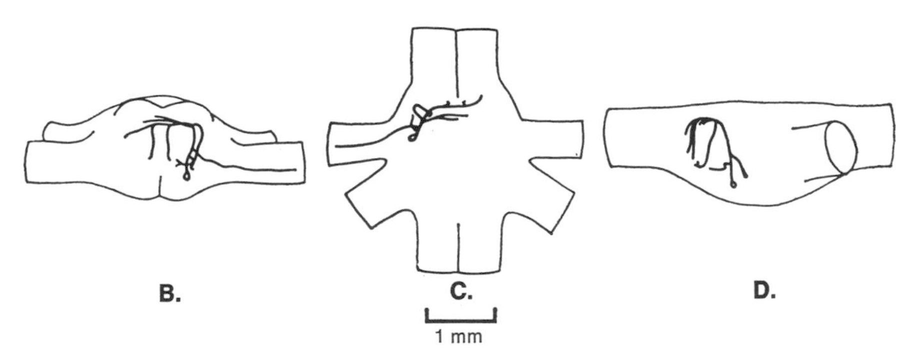

Figure 21–7). It has dendrites in each abdominal segment; its axonal segments are connected anteriorly, posteriorly, and across the midline with each other by fast, electrical synapses, thus ensuring rapid and bilaterally synchronous, anterior-posterior conduction (see Figure 21–2).

The input to the lateral giant, the *decision unit* in the system, comes from tactile sensory receptors (exoskeletal hairs) on the dorsum of several abdominal segments and the tail fan. Sensory nerves from these receptors reach the lateral giant either directly or via three interneurons designated *A*, *B*, and *C* in Figures 21–7 and 21–9. Interneuron *A* receives afferents only in the sixth ganglion and only from the tail fan. Interneurons *B* and *C* receive afferents from sensory neurons of several abdominal segments and from other sensory interneurons.

The two motor neuron paths to the flexor muscles differ in their temporal stability. The escape response habituates rapidly, and the motor giant fatigues rapidly. The response can still be performed after the motor giant fatigues but at lesser strength because of fatigue of the motor giant and because of certain properties of the chemical synapses on sensory interneurons *A*, *B*, and *C*. Thus, the circuit has built into it the possibility of behavioral habituation dependent upon synaptic physiology. It provides a fixed pattern of muscular coordination combined with opportunities for plasticity of response.

CENTRAL PATTERN GENERATORS

Complex rhythmic patterns of coordination are characteristic of many types of behavior such as insect songs, flight, walking, and swimming, as well as the simpler acts of respiration and heartbeat. Two basic types of theories have been proposed for the physiological generation of rhythmic behavioral patterns: (1) the cyclic-reflex hypothesis and (2) the central-pattern-generator hypothesis. Combinations of these two types of mechanisms also occur.

Under the cyclic-reflex hypothesis, as shown in Figure 21–11, a stimulus excites a motor mechanism in the CNS that produces a behavioral act; this act then provides stimuli that reflexly excite the same

21–11 Schematic cyclic reflex.

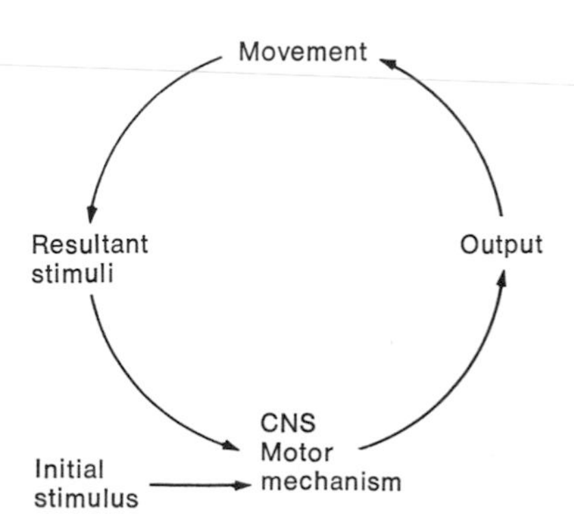

mechanism again. This circular process continues repetitively until inhibited by other reflex paths or by the brain. Until the 1960s nearly all rhythmic behavior, particularly locomotion, was thought to depend primarily on cyclic-reflex mechanisms. Since then it has been proven in several cases and strongly indicated in others that the cyclic-reflex hypothesis is untenable and that the pattern is generated by the inherent network organization of the CNS. These cases include cicada song, crustacean heart beat, crayfish swimmeret beat, insect respiration, insect copulation, swimming and siphon withdrawal in molluscs, cricket song, and cricket flight (references in Wilson 1966 and in Bentley and Hoy 1970); and cockroach walking (Pearson and Iles 1970), rhythmic mussel opening and closing (Salanki and Varanka 1972), lobster scaphognathite movement (Mendelson 1971), and others. It is now felt that both reflex and central-pattern mechanisms are commonly involved in many types of behavior and that neither theory alone can adequately account for all cases of rhythmic or repetitive behavior patterns. The once heated controversy can now be regarded as "clearly a sterile one, for it is already clear that there is every shade of intermingling of the extremes as well as shifts from one to the other during a specific activity. . . . The emphasis must be switched away from the examination of either actions which can be reflexly evoked or those which can be observed to occur spontaneously, to the study of the functioning of nerve cells as producers of motor neural patterns" (Hoyle 1970:426, 428).

One of the first cases of central pattern generation to be examined was that of flight control in locusts and grasshoppers, as studied by Wilson, Weis-Fogh, Gettrup, and Wyman (summarized in Wilson 1964, 1968). To investigate the possible influences of reflex control on insect flight, a desert locust (*Schistocerca gregaria*) was suspended from a pendulum in front of a wind tunnel in such a way that the position of the pendulum determined the wind speed. If the locust flew faster, the forward movement of the pendulum would activate the controls of the wind tunnel to blow harder; and if the locust lagged, the pendulum would slow the wind. In this way the locust remained nearly stationary although in flight, and electrical recordings could be made of nerve and muscle potentials. The locust (and most insects) can be induced to fly by removing its feet from contact with a substrate (the tarsal reflex inhibits flight) and by blowing air on the sensory hairs of its forehead. The muscle potentials (which reflect their motor neuron firing pattern on a 1:1 basis) for the wings, together with the temporal pattern of wing movement, are shown in Figure 21–12. Not surprisingly, activation of wing depressors alternates with activation of wing

21–12 Pattern of muscle contractions during locust flight. The records show the angular positions of the two wing pairs during 1½ wingbeat cycles, and a composite of action potential records from the flight muscles. The number of motor units recorded in each horizontal line is indicated at left. Many of the motor units fire synchronously. Note that the hindwings lead the forewings and that depressors alternate with elevators. (Reprinted from "The origin of the flight-motor command in grasshoppers," by Donald Melvin Wilson, from *Neural Theory and Modelling,* edited by Richard F. Reiss, with the permission of the publishers, Stanford University Press. © 1964 by the Board of Trustees of the Leland Stanford Junior University.)

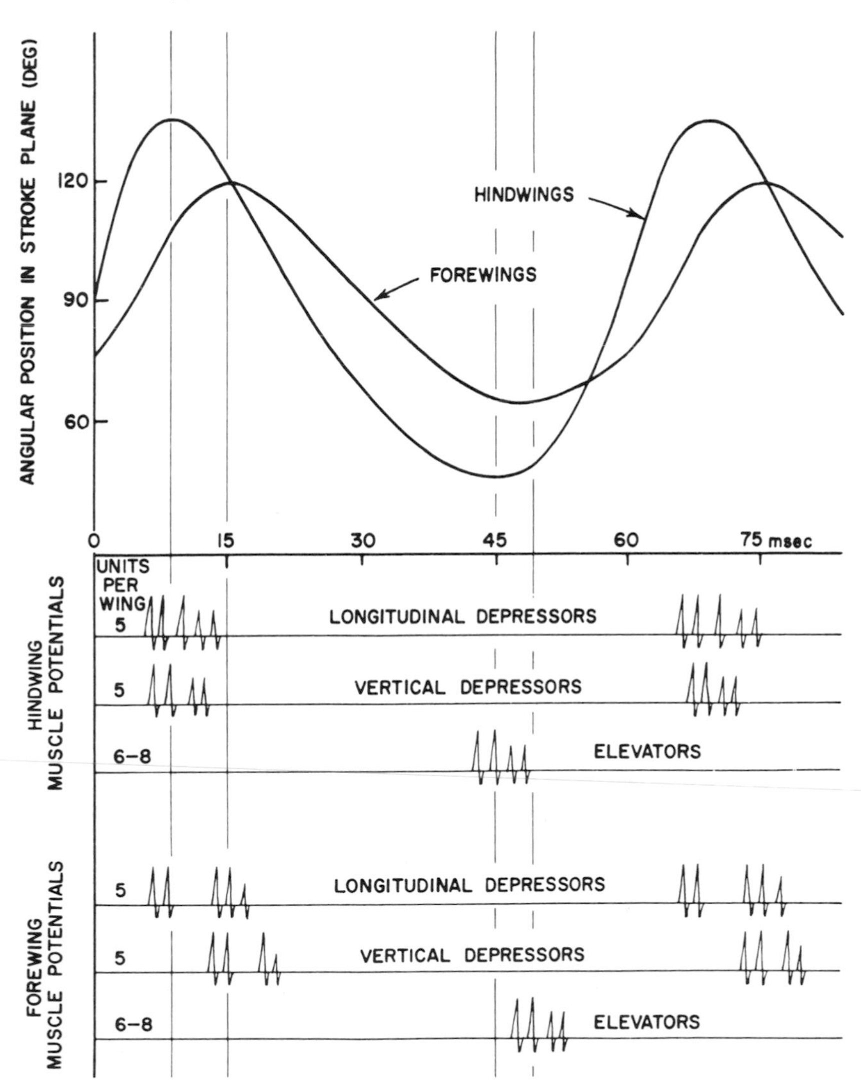

elevators; synergists are nearly synchronous. The hindwing leads the forewing.

The sensory consequences of wing motion for the rhythmic pattern are relatively simple. Lift receptors on the wing veins usually discharge during the downstroke, and wing position receptors at the wing hinge discharge one to several impulses near the end of the upstroke. The preparation and electrical records are shown in Figure 21–13.

According to the cyclic-reflex hypothesis, the sensory feedback from these receptors, which is activated with each cycle of wing beats, should activate the wing movements of flight. This was tested as follows: Fixing the wings of a grasshopper in a permanent elevated or

21–13 Locust mounted for recording from stretch receptor and downstroke muscle. (Reprinted from "The Origin of the Flight-Motor Command in Grasshoppers," by Donald Melvin Wilson, from *Neural Theory and Modelling,* edited by Richard F. Reiss, with the permission of the publishers, Stanford University Press. © 1964 by the Board of Trustees of the Leland Stanford Junior University.)

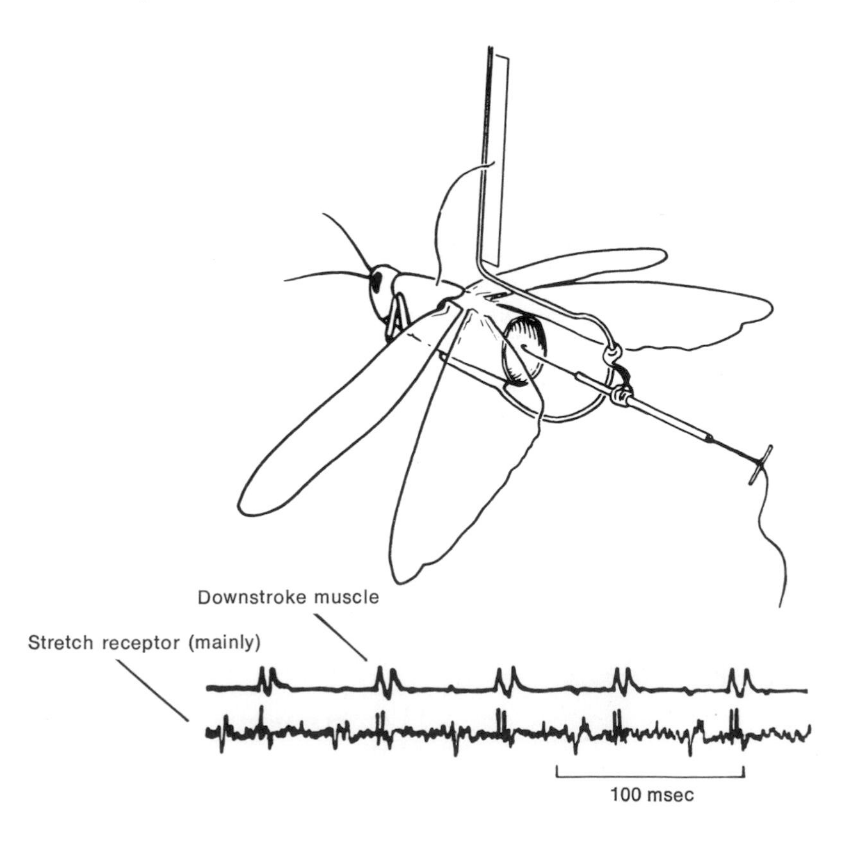

depressed position so as to prevent their activation by wing movement during "flight" in the wind tunnel did not prevent rhythmic contraction of the flight muscles (Figure 21–14). Removal of the wings and wing-hinge position receptors similarly failed to eliminate the rhythm. Cutting the sensory nerves from the lift receptors affected certain corrective maneuvers but not the basic rhythmic pattern. Destruction of the wing position receptors drastically reduced the frequency of wing beats but did not disrupt the pattern. Complete blocking of all phasic sensory feedback (that was in phase with the wings) failed to stop the characteristic rhythmic pattern of wing movements in flight and the characteristic phase relationships between synergists and antagonists. These experiments established the central-patterning

21–14 Muscle firing patterns for locust flight during various kinds of sensory restriction. **A.** Normal performance of metathoracic subalar muscle of *Melanoplus.* **B.** Same, but with wings waxed in the upward position. **C.** Record from a tergocoxal elevator muscle (*upper trace*) and the subalar muscle (*lower trace*) of metathorax after cutting away the wings and wing-hinge organs. (Reprinted from "The Origin of the Flight-Motor Command in Grasshoppers," by Donald Melvin Wilson, from *Neural Theory and Modelling,* edited by Richard F. Reiss, with the permission of the publishers, Stanford University Press. © 1964 by the Board of Trustees of the Leland Stanford Junior University.)

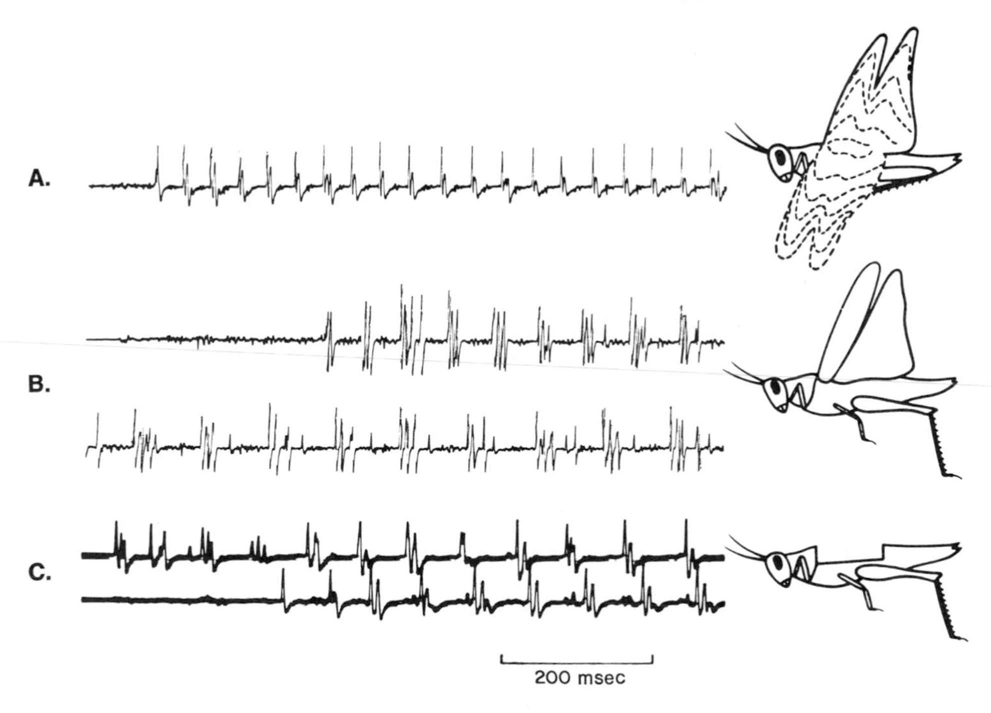

hypothesis as the accepted mechanism for control of wing coordination and temporal pattern in locust flight.

If the sensory inputs during flight do not provide the rhythmic pattern, what are their functions? Four sensory inputs to the "flight motor" are prominent. Normally flight is begun with a jump. While the insect is on the ground, proprioceptors from the legs inhibit flight, but after the jump the flight motor system is disinhibited. In addition, there are three excitatory inputs; two are from the wings, mentioned above, and one from the hairs on the forehead. When the locust takes flight, the breeze against its head provides an excitatory input to the flight motor mechanism that tends to maintain flight. Removal of any of the three excitatory inputs reduces the level of excitation of the flight motor mechanism. In addition to general excitation, some of the receptors are used in corrective flight maneuvers to maintain a straight course.

Further evidence of the independence of the rhythmic motor output pattern from its rhythmic input pattern was provided by Wilson and Wyman (1965). After decapitation and removal of the wing receptors, stimulation of the main nerve cords leading posteriorly to the thoracic ganglia with random pulse patterns caused the normal motor output pattern to be produced in the motor nerves.

The neural circuitry that produces and coordinates the flight pattern has not been fully clarified. Oscillators composed of pairs of neurons with reciprocal connections have been popular as models or neuromimes of the locust flight system (Harmon 1964; Wilson 1966). One of the various possible models is shown in Figure 21–15. At this writing the task of locating, mapping, and recording from the single neurons involved in flight control is under way, but little is yet known

21–15 One of several neuromimes (neuronal models) that are capable of yielding the motor output required for insect flight. Arrows are excitatory; dots, inhibitory. Thus excitation of Neuron I excites III and inhibits II. (From Wilson 1966.)

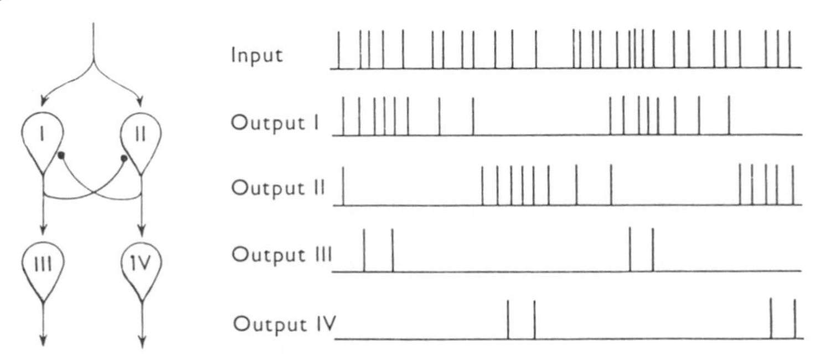

about the circuitry of the mechanism except that some of the motor neurons are consistent among individuals in relative size and location, that both excitatory and inhibitory potentials can be recorded from them, and that positive excitatory coupling exists between some synergistic motor neurons (Hoyle 1970). Generator neurons that produce rhythmic oscillations in neural activity have been observed in crickets, but they have not been investigated with regard to behavior (Vereschschagin et al. 1971).

In other arthropods (hermit crabs and lobsters) it has been shown by Mendelson (1971) that a behavioral rhythm can be driven by a single oscillatory neuron; consequently, this kind of model must be considered a serious alternative to the two-neuron model in Figure 21–15 favored by Wilson's group. Mendelson found, in the deafferented subesophageal ganglion, a single neuron whose rhythmic fluctuations in membrane potential drove the motor neurons which control the muscles moving the structure (scaphognathite) that propels a stream of water through the gill chambers. This oscillator neuron could be driven at the normal speeds of 80 to 120 bursts per minute by a command neuron that turned the oscillator on but did not provide its rhythm. In the ganglion controlling cicada song, Hagiwara and Watanabe (1956) found an interneuron that fired at precisely double the firing rate of the single motor neuron which excites the sound-producing tymbal muscle. Since the tymbal motor neurons on each side fire alternately, the interneuron actually fired at the pulse rate of the resultant sound. It was established that the firing rhythm of the motor neurons was determined by a pre-motor interneuron and that it was independent of the rhythm and firing pattern of the sensory nerve used to excite the song circuit. Although in the case of cicada song it has not been conclusively proved that the song rhythm originates in a single interneuron, there is at least considerable evidence for that conclusion.

A behavior of particular interest to evolutionary ethology is cricket song. Analysis of the role of the CNS in cricket song was begun in Huber's (1962 and earlier) studies of the behavioral effects of brain stimulation in *Gryllus campestris*. He found that electrical stimulation of points in the mushroom bodies, central body, and elsewhere in the cricket brain could evoke various sounds (Figure 21–16). In further work by Otto (1971), the calling song, aggressive song, and courtship song were evoked by electrical stimulation in the same areas (Figure 21–17). However, the songs could also be evoked by stimulation of the cut connectives leading into the thoracic ganglia (isolated from the brain). Consequently, the pattern of the song must be generated in the thoracic ganglia rather than in the brain, and the brain probably acts

21–16 Technique and anatomical results of stimulating points in the brains of crickets electrically.

A. Male cricket ready for brain stimulation, showing holder (*H*), cork ball, and electrodes (E_1, E_2).

B. Diagram of the brains of *G. campestris* and *A. domesticus* with localized areas of stimulation shown sagittally: triangles for normal calling songs; half black circles for aggressive sounds and associated behavior; squares for inhibition of stridulation; black circles for previously undescribed (atypical or preflight?) sounds. Normal symbols indicate *G. campestris*, symbols with superscript points indicate *A. domesticus*. *PC*, protocerebrum; *DC*, deutocerebrum; *TC*, tritocerebrum; $gl_{1,2}$, globuli neurons of mushroom body; *pe*, pedunculus of mushroom body; *ca*, α-lobe of mushroom body; *b*, β-lobe of mushroom body; *cc*, central body; *pc*, bridge; *nt*, nervus tegumentarius (afferent nerve coming from the dorsal region of the head capsule); 7, tractus olfactorio globularis; 11, ascending and descending fibers from and to the ventral nerve cord; 14, fibers of the tegumentary nerve.

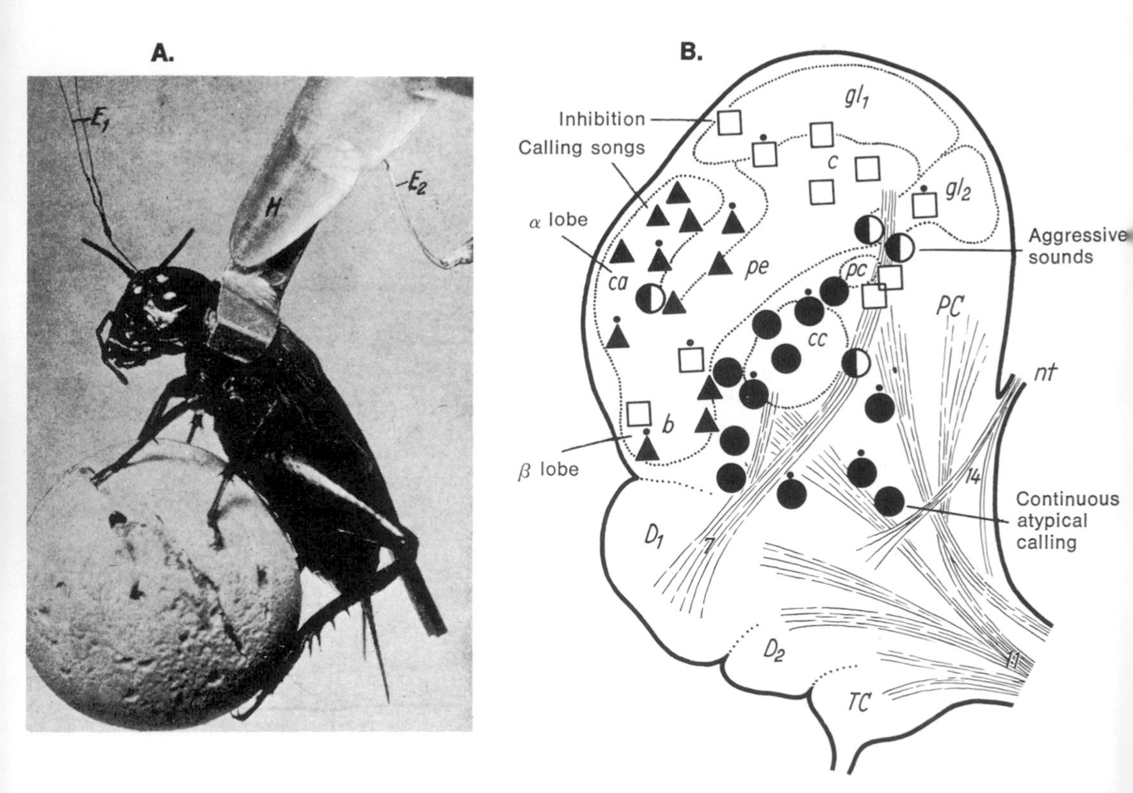

C. Diagram of the brain of *G. campestris* and *A. domesticus*, seen frontally, showing protocerebrum (*PC*), deutocerebrum (*DC*), and tritocerebrum (*TC*); *mb*, mushroom body; *cb*, central body; *b*, bridge; *mn*, motor neurons for antennal muscles; *nol*, ocellar nerve; *nop*, optic nerve; *na*, antennal nerve; *nl*, labral nerve; *fco*, frontal connective; co_1, first connective; 1–3, fiber bundles of the optic nerve; 4, optic commissure; 5, fiber bundle of the ocellar nerve; 6, antennal commissure;

7, tractus olfactorio globularis coming from the first sensory area of the deutocerebrum and ending in the mushroom bodies of both sides; 8, afferent antennal nerve; 9, efferent antennal nerve; 10, fibers connecting lateral protocerebral neuropile with deutocerebrum; 11, ascending and descending fiber bundles from and to the ventral nerve cord; 12, fiber decussation between bridge and central body. In these two species, the brains show only slight differences in structure of some fiber bundles. (Figures A–C and legends B–C from Huber 1962.)

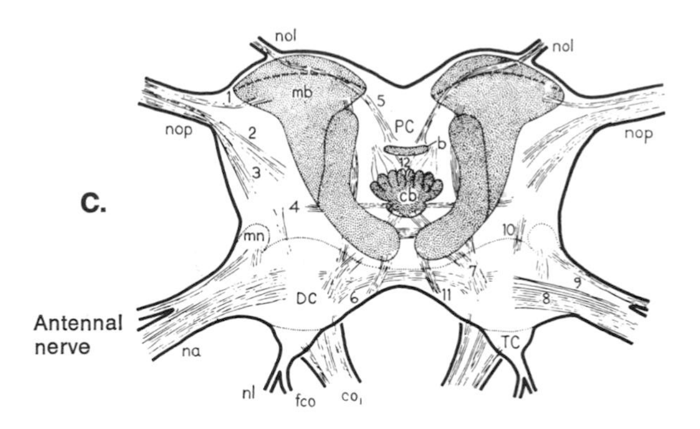

21–17 Comparison of naturally elicited sounds and those evoked by electrical stimulation of the brain (ESB) in the cricket, *Gryllus campestris*. Time marker: 200 msec. (From Otto 1971.)

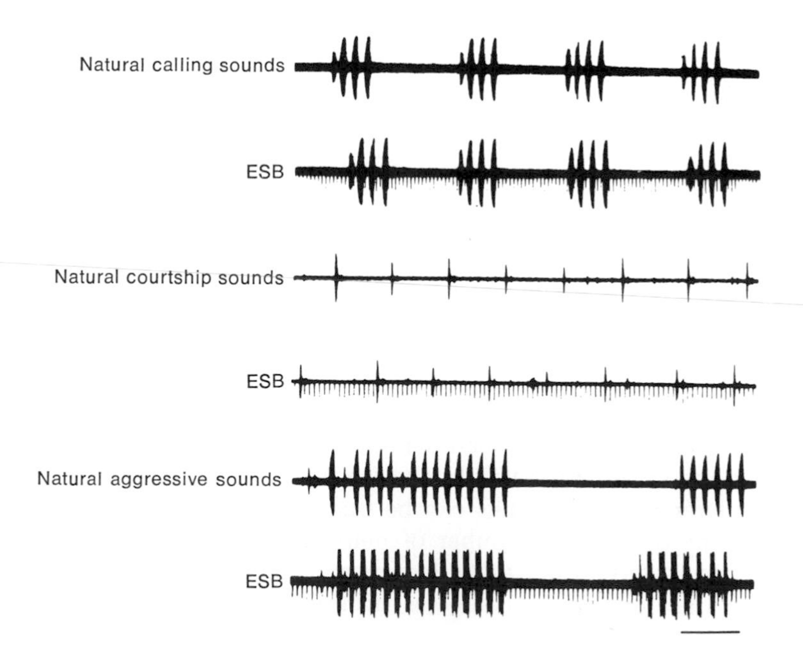

to turn it on and off and to influence transitions from calling song to either aggressive song or courtship song.

Unlike flight, a model of the song of *Gryllus campestris* requires two oscillators, one for the slow chirp rhythm and one for the faster pulse rhythm. The slow rhythm is the same as that for respiratory ventilation and, at times, for walking, which suggests that the slow oscillator could be coupled to a variety of behaviors. In a study of intracellular activity in the ganglion controlling calling, Bentley (1969) found neurons that follow the chirp rhythm and others that followed the pulse rhythm. Appropriately timed inhibition (IPSP) was found in both the opener and closer motor neurons but was probably derived from interneurons rather than direct reciprocal inhibition. In Bentley's model, the chirp oscillator neuron excites the opener motor neurons, which provide the pulse rhythm by inhibitory coupling with the closer motor neurons (which are excited alternately with the opener by rebound from inhibition). Details remain to be confirmed by experiment.

Cricket song offers suitable material for genetic analyses of neurobehavioral mechanisms in species that are well suited for neurobiological analysis (unlike *Drosophila*). The firing patterns of motor neurons controlling song in hybrids and backcrosses of *Teleogryllus oceanicus* and *T. commodus* were studied by Bentley (1971). The firing patterns of the motor neurons used in song (and flight) in *T. commodus* develop normally in an ordered sequence in younger instars before they are actually used (Bentley and Hoy 1970). These younger crickets cannot sing or fly because the wings are not full grown, and they do not attempt the corresponding movements. Song is presumably inhibited by the brain in young crickets, since localized brain lesions allow the nymphs to produce the stridulation movements. The songs develop normally in individuals that have never heard songs of their own species.

The songs and muscle potentials of the genetically different forms are shown in Figures 21–18 and 21–19. Various differences in the song motor patterns can be correlated with genetic differences. The transition interval (indicated by the bar) between the initial chirp part of the song and the terminal trill part decreases in passing from *T. oceanicus* to *T. commodus*. Also, the number of motor unit bursts per trill increases from two in *oceanicus* to 14 in *commodus*. In some cases the firing patterns between genetically different forms differ by as little as one spike in the trill (e.g., *A* vs. *B* or *B* vs. *C*). Therefore, the genetic difference in the songs can be traced in part to effects on the firing patterns of a small number of neurons—very likely the motor neurons or the interneurons that affect them. These findings lend strong support to the concept that species differences in certain relatively

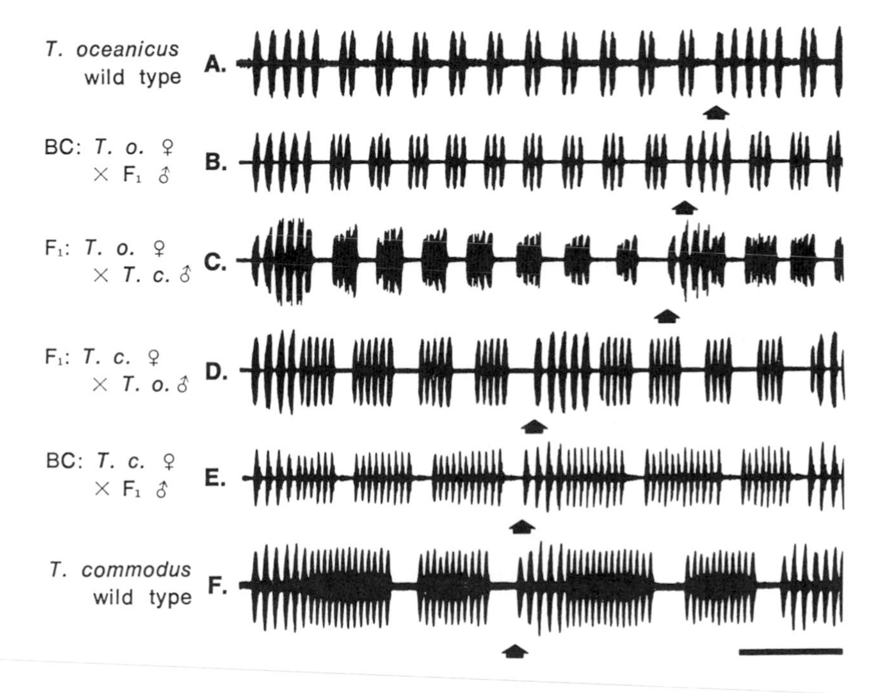

21–18 Sound pulse patterns in the calling song of *Teleogryllus* wild types and hybrids. Each record begins with a single 4- to 6-pulse *chirp,* followed by a series of *trills* containing from 2 to 14 sound pulses, depending on genotype. Chirps and trills are arranged in a repeating *phrase.* Records start with a complete phrase, and arrows mark the onset of the second phrase. Records are as follows: **A,** wild type, *T. oceanicus;* **B,** backcross, *T. oceanicus* ♀ × F_1 ♂ (shown in C); **C,** F_1, *T. oceanicus* ♀ × *T. commodus* ♂; **D,** F_1, *T. commodus* ♀ × *T. oceanicus* ♂; **E,** backcross, *T. commodus* ♀ × F_1 ♂ (shown in D); **F,** wild type, *T. commodus.* Song patterns are strictly determined by genotype. Most hybrid features are intermediate between corresponding parental features—for example, number of sound pulses per trill and number of trills per phrase. The bar at bottom shows 0.5 sec. (Figure and legend from Bentley 1971. Copyright 1971 by American Association for Advancement of Science.)

fixed patterns of behavior are narrowly determined by genetic differences controlling the neural networks that produce them.

CONCLUSIONS

The nervous systems of animals are products of natural selection. They are organized structurally and functionally to provide the most useful combination of plasticity and stereotypy for a particular behavior and

21–19 Motor unit firing patterns responsible for the calling song of *Teleogryllus* wild types and hybrids. Records are as follows: **A**, wild type, *T. oceanicus;* **B**, backcross, *T. oceanicus* ♀ × F_1 ♂ (shown in C); **C**, F_1, *T. oceanicus* ♀ × *T. commodus* ♂; **D**, backcross, *T. commodus* ♀ × F_1 ♂ (shown in C); **E**, wild type, *T. commodus.* Traces show muscle action potentials recorded from single, identified fast units, and elicited by single motor neuron impulses (triangles, D and E). The upper trace shows a wing opener unit (unit 1, subalar muscle), and the lower trace shows a wing closer unit (unit 2, promotor muscle); wing closing produces the sound pulse. The portion of the phrase shown is the transition (bars) from the chirp (left of bars) to the sequence of trills (right of bars). Motor unit activity corresponds to, and determines, the sound pulse patterns (see Figure 21–18). Details of differences in firing patterns of homologous single motor neurons from different genotypes are seen. (i) The transition interval (bar) decreases steadily from *T. oceanicus* (A) to *T. commodus* (E). (ii) Motor unit bursts per trill increase from 2 (A) to 14 (E). Genetic information is precise enough to specify a difference of one burst between the trills of *T. oceanicus* (A) and the backcross (B). For many units, this difference is only a single impulse. In some traces, the smaller promotor unit (arrow) partially obscures the recording. The bar at the bottom shows 100 msec. (Figure and legend from Bentley 1971. Copyright 1971 by American Association for Advancement of Science.)

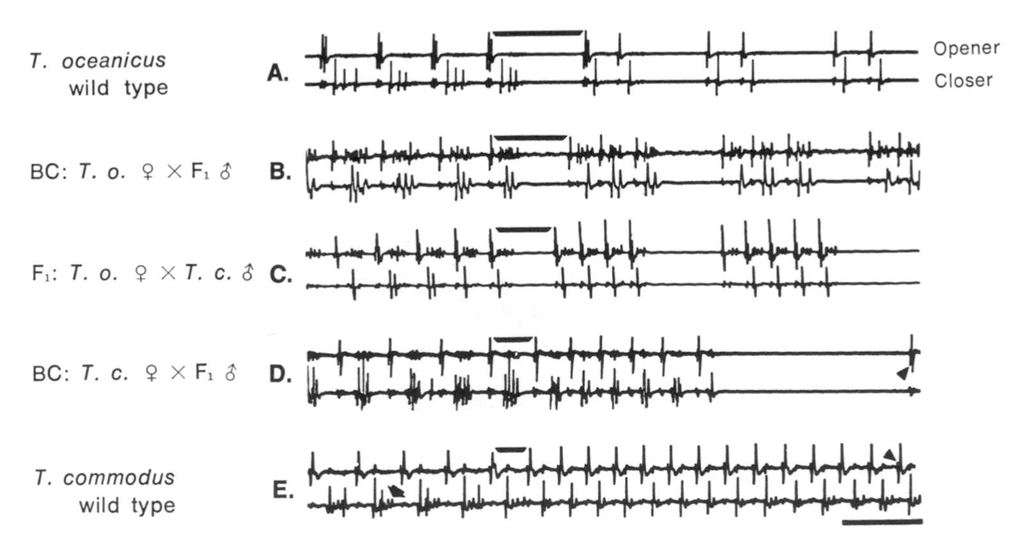

a particular species. Although it is much easier to demonstrate nuances in species diversity of behavior than in neural structure and function, it is, nevertheless, possible in favorable cases to single out principles of neural organization that underlie species diversity in behavior. One of these principles is that certain patterns of behavior may be produced by the output patterns of organized circuits of neurons and that these circuits may differ from species to species, or at least from genus to genus. Little has been said about the organizations of these circuits, or about circuit properties that might be responsible for the temporal patterning of behavior, because little is yet known. Discovering the mechanisms of production of patterned neural output is a basic problem of the future neurobiology of behavior.

The success of recent studies along these lines using the methods of unit stimulation and recording suggests that we are on the threshold of a new level of understanding of the neural control of stereotyped behavior.

22 Sensory Systems: Feature Detection

MUCH OF BEHAVIOR can be interpreted as consisting of responses to stimuli. The manner in which the animal senses and analyzes stimuli is, therefore, of great importance. The constancy of behavior within a species and the diversity of behavior among species are in part attributable to features of their sensory analyzing systems. In this and the following chapters on orientation and navigation, we will examine the diversity of behavior among animals as a function of their sensory analyzing systems.

For the sake of simplification, we can recognize three main types of stimulus-response relationships. In the first case, emphasis is on some *quality* or combination of qualities present in the stimulus that provides a cue to the animal to perform an appropriate behavior. For example, a scallop swims away when it is touched by a species of starfish that feeds on scallops, but it ignores a similar tactile stimulus from a harmless species. Or a female frog swims toward a male of her own species uttering a mating call, and bypasses frogs of other species who may be closer and louder. These are examples of stimulus *recognition*. Often they depend on detection of some key qualitative feature of the stimulus. In the second case, the emphasis is on *quantitative* aspects of the stimulus. The animal judges the strength of the stimulus and *adjusts* its response to an appropriate strength and direction. For example, a lizard will sit in the early morning sun until its internal temperature warms to an appropriate level. A frog released on land away from its home pond will orient its direction of travel in relation to celestial and other stimuli. In the third case, the stimulus has a predisposing or *priming* effect. For example, the display of the male dove does not directly bring about precopulatory behavior in the female; but it helps to bring her into reproductive condition by stimulating ovarian growth.

This chapter deals with the first case, namely *recognition by fea-*

ture detection; the following two chapters deal with selected aspects of the second case, *adjustment*. Together, these chapters provide a rather selective view of the role of sensory analyzers in behavior. Their goal is not to delve into the details of sensory behavioral physiology in a comprehensive or reductionist manner, but rather to try to impart some feeling for selected concepts that are of special interest in comparative behavior studies.

Of special interest is the way in which an animal recognizes stimuli and objects that are of critical importance to it, such as mating calls, threat displays, predators, and food. Some species must recognize and react to these stimuli correctly without benefit of any training or previous experience. What features of sensory systems make possible this kind of stimulus recognition?

The problem is an old one but cannot yet be said to be solved, although answers for some of the simpler cases are available. A kind of lock and key arrangement was suggested by Lorenz (1970) in 1935. He pointed out that because individuals had to be able to respond selectively to signals given by their own species, their sensory analyzers might be expected to be adapted specially for this function. The postulated analyzers for such critically important stimuli were designated *innate releasing mechanisms* or IRMs. The close relationship between the motivational state of the animal and the workings of the recognition mechanism was emphasized by the use of the term *releasing* and by the incorporation of the IRM concept into a hydraulic model (which has proven to be rather unpopular). Nevertheless, the concept of a key-stimulus-recognition mechanism, despite its difficulties, has continued to serve, unconsciously perhaps, as a sort of "holy grail" for feature-detector physiologists.

That sensory analyzers have become specialized through natural selection for certain species-typical stimuli has been a generally attractive concept, but objections have been raised about the term used to designate this concept, *IRM*. In the first place, in physiological terms nothing is actually "released." Neurons and muscles are excited or inhibited but not released, although they may be disinhibited. Secondly, the term *innate* is controversial (see Part VI of this book). Lastly, it has been objected that the term IRM implies a "unitary" mechanism (Hinde 1970:123), which in most cases is likely to be an oversimplification. In the light of modern physiology the term *recognition mechanism* seems preferable, but this too has its faults. There seems to be no unanimity concerning terminology in this field.

The general problem of recognition of biologically significant stimuli, and some common research strategies for tackling it, are outlined as follows.

The first step is the identification by the experimenter of a *biologically significant stimulus-response relationship.* In this context, "biologically significant" is usually interpreted as meaning that the stimulus-response relationship is so important to the animal that we may expect natural selection to have specialized the sensory analyzers in some way so as to be highly or selectively responsive to the normal stimuli. Intraspecific signals may also be specialized to activate the recognition mechanism. The recognition of members of one's own species, recognition of predators, recognition of potential food, and recognition of the signals used in communication are examples of problems of this type.

Secondly, after a particular "recognition problem" is chosen, it is necessary to determine by experiment precisely which aspects of the presumed stimulus are necessary to evoke or release the response. These key or "sign" stimuli are typically produced by a motion, sound, chemical, or structure termed a *releaser.* Demonstration by experiment that the species *responds selectively* is necessary to define the problem. The results of these experiments should then be compared with the natural stimulus. The hypothesis that the characteristics of the most effective stimulus are the same as those provided by the stimulus most frequently effective in nature is often, but not always, confirmed. In some cases, a "supernormal" stimulus has been discovered. In the chemical mode certain odorants may be necessary while others have no effect. In sound signals, a given frequency range or combination of frequencies may be critical. And for visual communication, a particular color may provide the key *releasing stimulus.* Investigations of this sort are purely behavioral, and they often require considerable ingenuity and sophistication. The long history of study of the pecking response by gull chicks toward food held in a parent's bill may be considered representative of this type of study (Hailman 1967, 1970).

The third step involves the application of similar stimulus-testing procedures to *responses of the nervous system,* both peripheral and central. Typically, electrical potentials generated by a sense organ, by a peripheral nerve, or by neurons in the CNS are studied as a function of the stimuli presented to the animal. Ideally this third step should be carried out with the benefit of knowledge of the second step: to interpret the physiological findings, one must compare them with the behavioral findings. Only a few studies have been carried to the point at which meaningful comparisons between the behavioral and the physiological studies can be made.

A question often asked in such studies is whether the recognition mechanism is located peripherally in the sense organs or centrally in the CNS. The *central-versus-peripheral question* may in some cases force the problem into an unnecessarily strict framework, since interactions or combinations of central and peripheral processes may sometimes be important. It must also be recalled that the processes of analysis of sensory input go on at various levels in most complex nervous systems; so the mere identification of a recognition process as being located in the CNS is only an early step in the investigation.

RECOGNITION OF BIOLOGICALLY SIGNIFICANT AUDITORY STIMULI

Moths / Nocturnal moths are avidly eaten by bats. It was observed by Treat (1955) and others that moths change their flight course in response to high-frequency sounds. Roeder (1962) demonstrated that free-flying noctuid moths in the field would take evasive action in response to artificial ultrasonic pulses resembling those of a bat. Two kinds of response were observed: moths turned away from the sound if they were at a moderate distance from it. But if they were close, they immediately adopted an unoriented evasive pattern such as spiraling or dropping downward. Some examples are shown in Figure 22–1. These observations constitute our steps 1 and 2 in the analysis of the moth's bat-recognition mechanism, namely, choice of a biologically

22–1 Tracks of free-flying moths in the field. The reflection of the moth is interrupted every 0.25 sec by a rotating shutter. The isolated white dots in the photos are made by small insects close to the camera. The arrows indicate the point at which an ultrasonic signal from the loudspeaker on the pole was switched on. *Left:* The meandering track of an unstimulated moth. *Center:* Power dive by a moth exposed to an intense ultrasonic signal. *Right:* Turning reaction of a moth receiving a faint ultrasonic signal. (From Roeder 1966.)

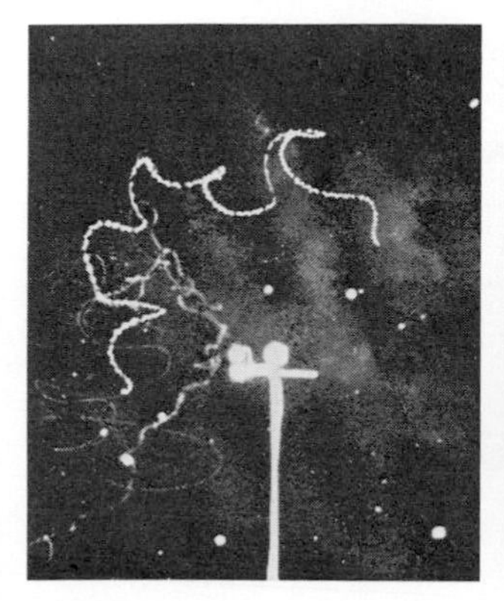
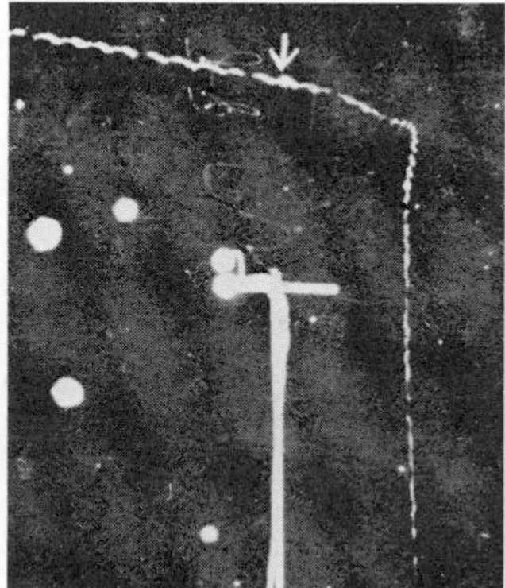
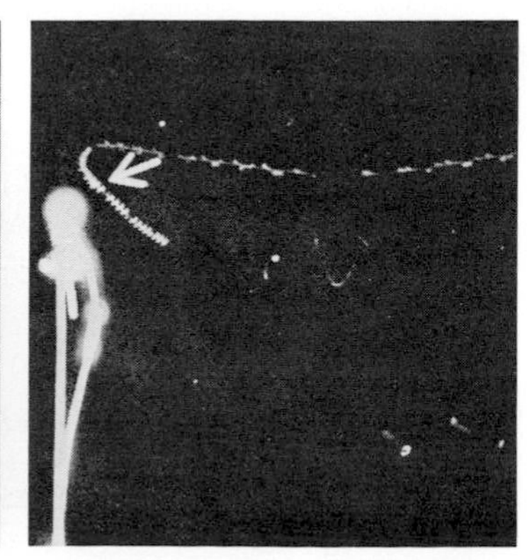

significant stimulus-response relationship to study, and experimental specification of the stimuli necessary to cause it.

Step 3 in our research strategy for "the recognition problem" has been summarized by Roeder (1966, 1971). A key finding was the simple structure of the noctuid moth "ear" or tympanal organ (Figure 22–2). The cells sensitive to sound in one ear are only two in number, each with a different threshold. A third cell was inhibited by sound. The structural simplicity of the moth's hearing organ is at the heart of its behavioral simplicity in response to bats. A second major finding was that the moths were "tone deaf"; their auditory nerve carried no information that would allow them to discriminate between different frequencies independently of loudness. They could hear sounds in the ultrasonic range, 17–100 *kHz*, but could not discriminate among them in regard to frequency. Ultrasonic receptors have also been found in lacewings (Miller and MacLeod 1966), some sphingid moths (Treat and Vandeberg 1968), and hawkmoths (Roeder et al. 1970).

Roeder recorded the activity of single neurons at various places in the moth's auditory system. He used as stimuli artificially generated ultrasonic pulses resembling bat calls, and found that the two sensitive neurons of the hearing organ fired repetitively during each pulse. Neurons with similar firing patterns, the "repeater neurons," were found in the CNS; but other types were also found which suggest that

22–2 Hearing organ (tympanic organ) of a noctuid moth. The auditory nerve (*IIINIb*) contains only three axons, two A fibers from the sensillum (*S*) at the tympanic membrane (*TM*), and a B cell (*BAx*). *TAS,* tympanic air sac. *B* and *SP,* skeletal supports. (From Roeder and Treat 1961.)

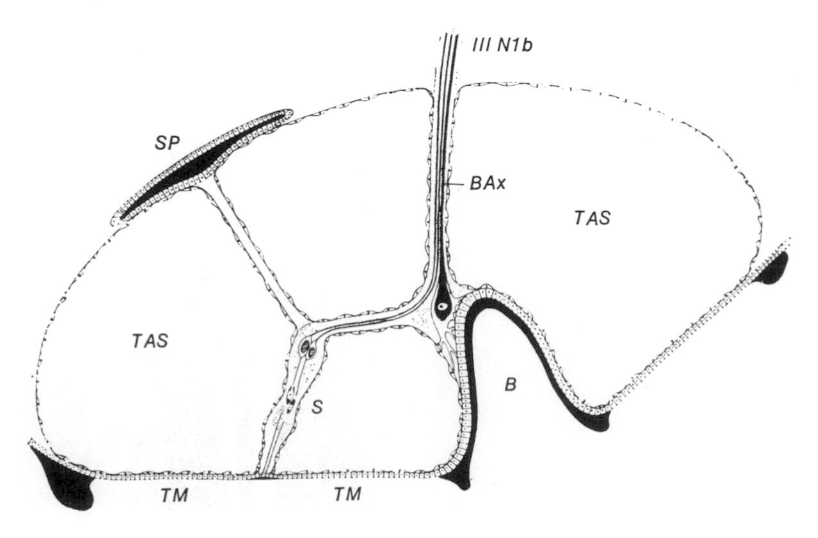

some central processing of the auditory signal occurs. In regard to the problem of bat-call recognition, the two most interesting types of central neurons were the pulse-marker and train-marker units. The pulse marker tended to fire only once for each ultrasonic pulse regardless of pulse length. The train marker fired at its own intrinsic frequency throughout a train of ultrasonic pulses, at the same rate through both pulses and pauses.

Now let us reconsider the main question. Considering the great variety of sounds on a summer evening, how is it possible for moths to react selectively to bats, disregarding other sounds in their environment? The answer would seem to be simple. The ears of moths are sensitive only to sound in the 17–100 *kHz* band. Since the only significant source of ultrasound in the moth's environment is bats, immediate avoidance by a moth in response to any ultrasonic stimulus would seem to be an adaptive response. In this case, the selectivity in the recognition mechanism would seem to be entirely peripheral (in the ear).

22–3 Feature detection by the hearing organ of a noctuid moth. The relative acoustic sensitivity of a moth, *Feltia subgothica,* at various frequencies is shown by the dotted line. The solid line represents the summation of relative sound intensities recorded in the field in the moth's environment at night. The sounds were taped and then integrated with a spectrum analyzer. Sounds below 15 *kHz* came mostly from insects; those from 25 to 50 *kHz* came mainly from passing bats. Note that the ear is tuned to the frequencies used by the bats and is relatively insensitive to other sounds. (From Roeder 1971.)

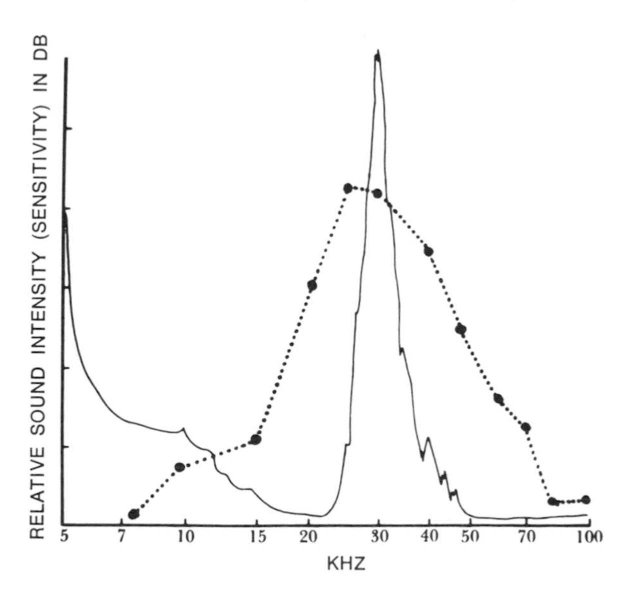

The ears of moths are tuned to the calls of bats, as shown in Figure 22–3. What then are the functions of the pulse-marker and train-marker neurons? Are there other significant sources of ultrasound that moths distinguish from bats? Or are these two sorts of neurons necessary only to convert the phasic input from the ear to a tonic output to the "flight motor"?

It is worth noting that some moths, when shaken or exposed to ultrasonic pulses, produce trains of clicks with ultrasonic components. Dunning and Roeder (1965) found that playing a train of ultrasonic pulses to a bat engaged in pursuing insect prey reduced its effectiveness. Moth sounds were more effective deterrents than bat sounds. Perhaps some moths protect themselves by emitting ultrasonic pulses.

Frogs / A model investigation of the sort described above has been carried out by a team of researchers investigating how frogs recognize and respond selectively to calls used in mating and other contexts (Capranica 1965; Frishkopf et al. 1968).

Step 1 in the sequence described above was the choice of the species-identifying croak of the northern bullfrog (*Rana catesbeiana*). This is a sound commonly heard in marshes, swamps, and ponds during summer. It probably functions in species recognition by females, for mating, and by males, for aggressive spacing. The response of males was chosen for study; that of females remains unstudied. Capranica observed that one bullfrog would croak in response to another's croak. In step 2 of the strategy described above, he investigated the stimulus specificity of the evoked vocal response by playing recordings of the mating calls of 34 species of frogs and toads to male bullfrogs. Only the bullfrog call evoked a vocal response; the calls of 33 other species were ineffective. This clear selectivity by the bullfrog in response to calls of its own species encouraged further investigation. (Bullfrogs also have a few other calls [see Table 16–2], but the recognition mechanisms, if any, for these calls have not been studied.)

Carrying the study of stimulus specificity further, Capranica next attempted to determine which artificial auditory stimuli were most effective in evoking the croak and to compare them with the natural croak. Various sounds were synthesized electronically and tested for their effectiveness in evoking croaks. Spectral structure was the primary aspect investigated. The test calls consisted of three to eight croaks with normal silent periods. To evoke calling, the signal had to have the following spectral characteristics:

1) A pulse repetition rate of 100 *Hz*
2) Energy in the *L* (low) region (centered at 200 *Hz*)
3) Energy in the *H* (high) region (centered at 1,400 *Hz*)

4) Less energy in the *M* (medium) region (centered at 500 *Hz*) than in the *L* region.

Other relevant aspects of the croaks, such as number of croaks per call, silent interval between calls, croak duration, and rise and fall times were not studied. Of particular interest is the complexity of the minimum requirements for an effective artificial call. The results of experiments that demonstrate the necessity for energy in two frequency ranges (*L* and *H*) are illustrated in Figure 22–4. The quite unexpected inhibitory effect of sound in the *M* range is illustrated in Figure 22–5.

When the experimental findings were compared with the spectral structure of the natural call, the agreement was impressive. Figure 22–6 shows the characteristics of a typical call and a croak given in response. The energy peaks in the *L* and *H* regions of the natural call and the trough in the *M* region agree strikingly with the corresponding features of the optimal artificial call. Another important resemblance of the natural and optimal artificial calls was in the fundamental pulse rate of about 100 *Hz*.

In the third step, the response of the neurons of the auditory nerve to tones of different frequency was studied. For each neuron tested, the threshold was determined at a series of frequencies. As is usual in peripheral systems, each unit (neuron) was found to have one "best" frequency, namely, the frequency at which the threshold for firing in response to a pure tone stimulus was lowest. The thresholds of two units at various frequencies are shown in Figure 22–7*A*. The relative abundance of units with various best frequencies is shown in Figure 22–7*B*. The most important point is that the pattern of peaks in the *L* and *H* region with a trough in the *M* region was again found. The *L* units were found to originate from the amphibian papilla of the inner ear, and the *H* units from the separate basilar papilla. It was also found that sound in the *M* region had the effect of inhibiting units in the *L* range. In short, the spectral structure of the sound that excited the auditory pathways to the brain most effectively and efficiently had energy in the *H* region to excite the *H* units, and energy in the *L* range to excite the *L* units, and had considerably less energy in the *M* than *L* range so as to avoid inhibiting the *L* units.

The correspondence seems to be too good to be attributable to chance. The ideal sound for exciting the auditory input to the brain is the same as the ideal sound for evoking a vocal response, and is the same in its spectral composition as the species-identification call of a fully adult male bullfrog.

This case is more complex than that of the moth's bat detector. In the moth, it appears that the critical stimulus-filtering is performed entirely by the ear; since the ear only hears sounds in the ultrasonic

22–4 Evoked calling in bullfrogs as a function of the spectral characteristics of artificial calls. Note that energy must be present in both the high and low frequency bands and that either band alone is not effective. (Reprinted from *Vocal Response of the Bullfrog* by R. R. Capranica by permission of the M.I.T. Press, Cambridge, Massachusetts. Copyright © 1965 by the Massachusetts Institute of Technology.)

A.

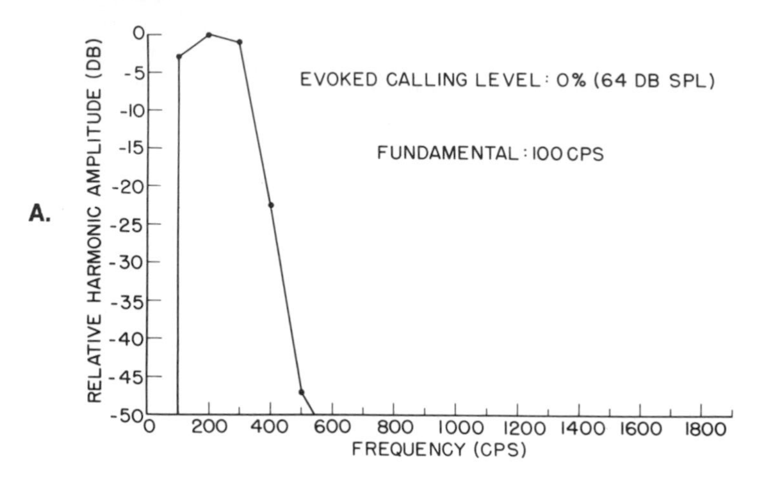

B.

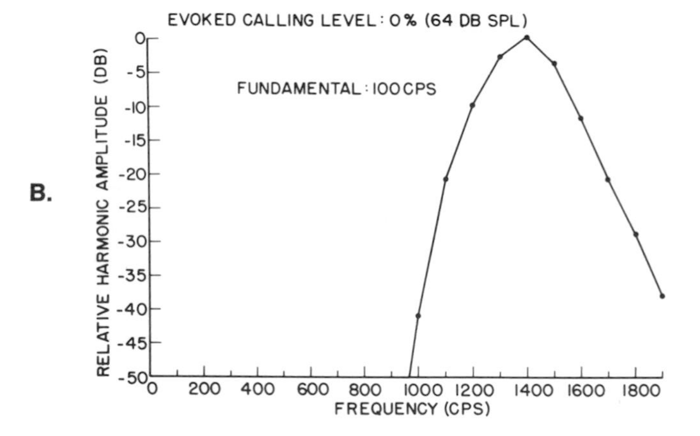

C.

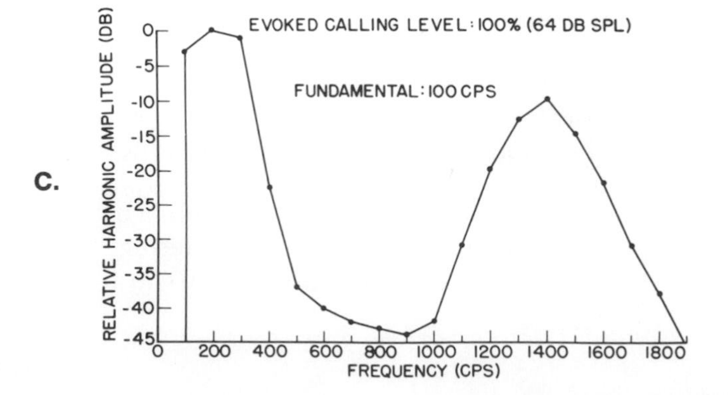

range of bat calls, any sound that excites the moth's auditory nerve is likely to be a bat. Consequently, no further central analysis seems to be necessary—although some occurs anyway. (It would be interesting to know which, if any, ultrasonic sounds in the frequency range of bat calls do *not* cause evasive behavior in the moth and do *not* excite the pulse-coder and train-coder interneurons.) In the bullfrog, it is not

22–5 Inhibitory effect of sound in the M (middle) range of frequencies on evoked calling in bullfrogs. *Above* and *below:* Two types of artificial calls are shown. Both types of experiment show an inhibitory effect of M frequencies. (Reprinted from *Vocal Response of the Bullfrog* by R. R. Capranica by permission of the M.I.T. Press, Cambridge, Massachusetts. Copyright © 1965 by the Massachusetts Institute of Technology.)

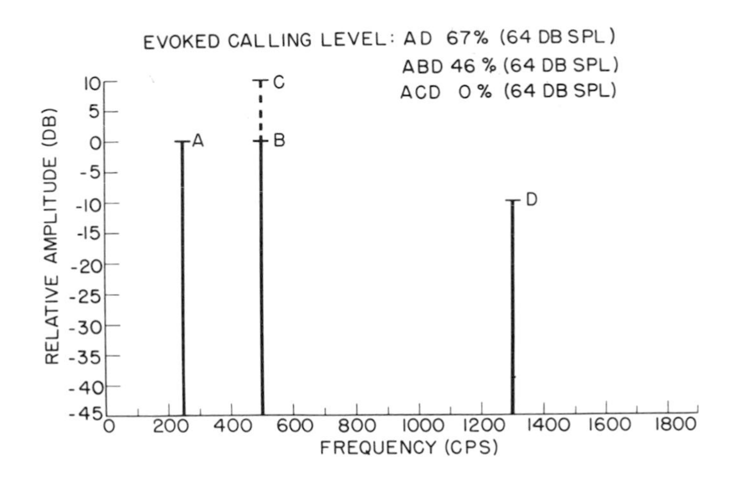

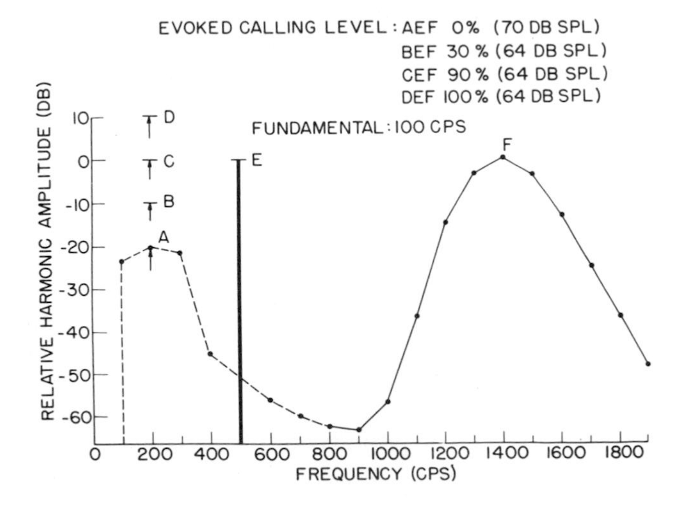

22–6 Spectral structure of a natural bullfrog call. The call is composed of 3 to 15 pulses of sound, here termed croaks. The call in **A** has six croaks. The photo shows an oscilloscope picture of amplitude vs. time. In **B,** a part of one croak is greatly expanded to show the fundamental period of 10 msec. **C.** A histogram showing amplitude at each frequency. Notice the peak in the L and H regions at 200 *Hz* and 1,600 *Hz,* respectively, and the trough in the M region at 500 *Hz.* (From Capranica 1966.)

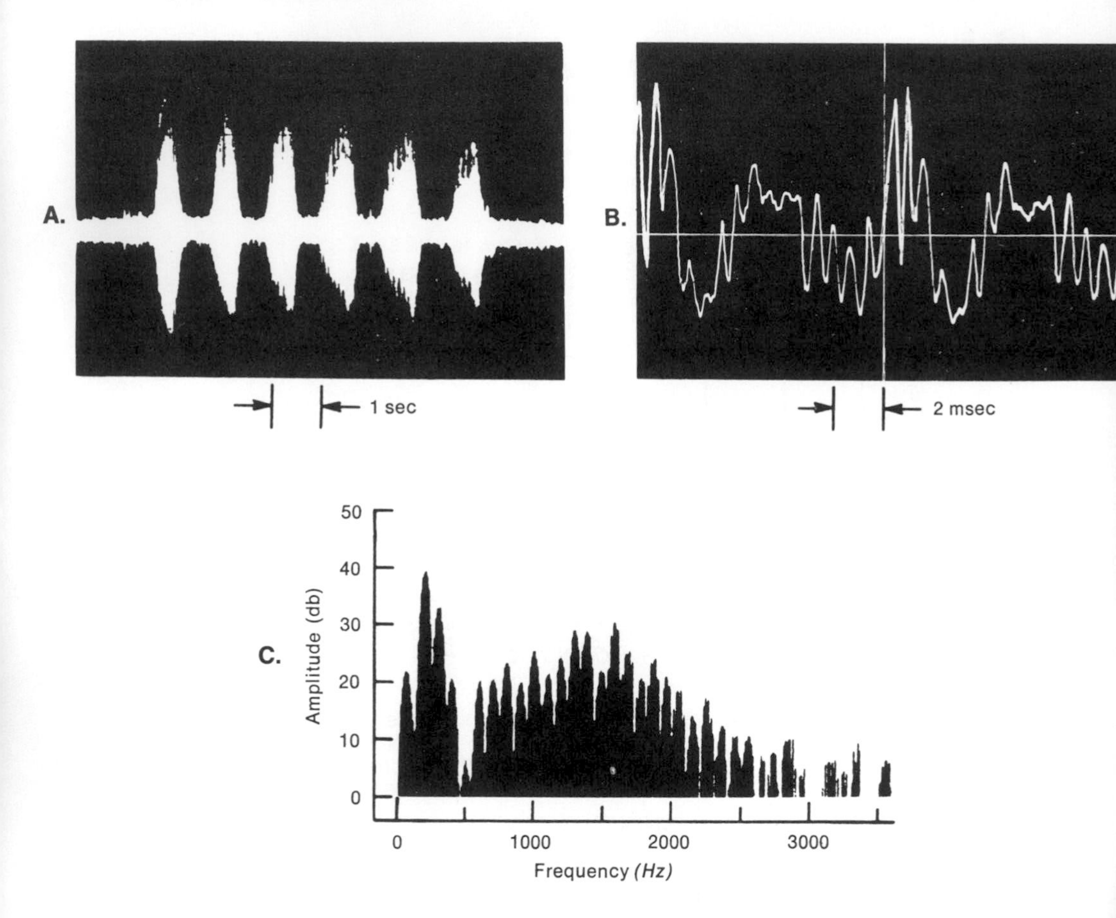

true that any sound which will excite the auditory organ will evoke a behavioral response. Even loud sounds, if they have the incorrect spectral composition, will be behaviorally ineffective. Some central processing of the auditory input appears necessary to distinguish between sounds that will evoke (1) no response, (2) a species-identification call, (3) one of the other calls, or (4) some other response, such as escape. It is true that the ear performs a critical role in the stimulus-filtering that occurs, but without central processing it would be insufficient.

22–7 Responses of neurons in auditory nerves of bullfrogs to sound stimuli. *Above:* Tuning curves of a simple unit (neuron) in the H range and of a complex unit in the L range. The best frequency for each unit is the frequency with the lowest threshold. *Below:* Histogram showing number of units with best frequencies in the indicated intervals. Simple units showed no evidence of being inhibited by any sound stimulus. Complex units were those that could be inhibited by a sound stimulus with frequency in the M range. (From Frishkopf et al. 1968. Reprinted with permission of the IEEE from *Proc. IEEE* 56:969–980.)

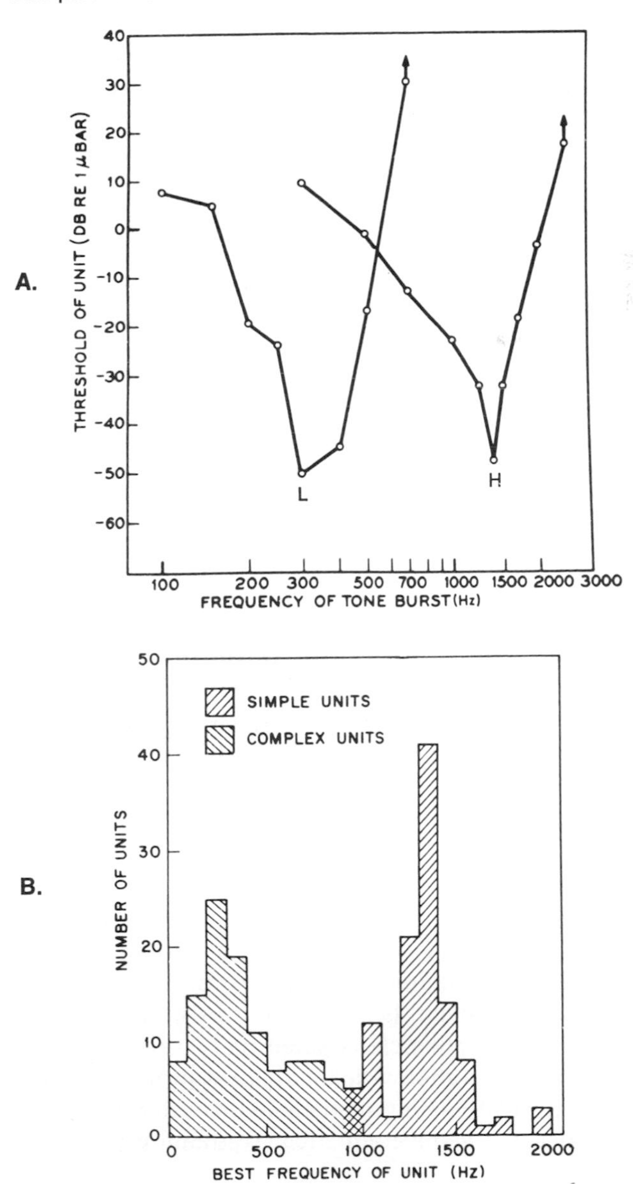

Analysis of CNS processing of auditory information has traced the auditory pathways in the brain (e.g., Potter 1965; Loftus-Hills 1971) and has characterized neurons in the midbrain according to their best frequencies. "OR-units" have been found by Frishkopf et al. (1968) and Loftus-Hills (1971). These midbrain neurons have two best frequencies, one in the *L* and one in the *H* range. Probably they are points of convergence of *L* and *H* neurons. More information about the nature of auditory units in the frog brain is needed to construct a neuronal model of auditory analyzers in frog auditory communication.

The findings of Frishkopf et al. (1968) regarding bullfrogs, and similar observations by Sachs (1964) on the green frog (*Rana clamitans*), suggested that the pattern of having two best frequency ranges separated by an inhibitory range might be general in frogs and toads, and that it might form the basis of species differences in recognition mechanisms. This idea was tested by Loftus-Hills and Johnstone (1970) in a study of midbrain auditory evoked potentials in six species of frogs that breed sympatrically. The earlier finding of a bimodal best-frequency sensitivity (see Figure 22–7) suggested that this feature might be general. But Loftus-Hills and Johnstone stated that many species of frogs have only a single dominant frequency in their mating calls, and so may rely on just one frequency for spectral distinctiveness rather than on two. In support of this, they demonstrated that the dominant frequency in the mating calls of many species corresponds to the upper of the two frequencies of greatest sensitivity in midbrain evoked potentials (Figure 22–8). As far as species identification is concerned, the upper sensitivity peak might turn out to be the more important, but only careful experimentation can confirm this. In any case, the function of the lower sensitivity peak poses an unsolved problem.

Other Animals / The use of auditory receptors that are selectively sensitive to biologically significant sounds has been reported in some other insects. The antennae of male mosquitos (*Aedes aegypti*) are constructed so that they vibrate in response to a sonic stimulus and have a resonant frequency, 300 to 350 *Hz*, close to that of the wings of the female mosquito (355 to 415 *Hz*). The wings of males vibrate at 495 to 565 *Hz* (Tischner and Schief 1954; Roth, Roth and Eisner 1966).

In the termite species *Zootermiopsis angusticollis*, soldiers alarm other individuals in the colony by banging their (own) heads against the roof and floor. The signal is transmitted by vibration and picked up by the subgenual organ, which is selectively sensitive to the signals of the soldiers (Howse 1964). Cricket species produce from one to six sounds of communicative significance (Alexander 1962). Since all of these are at about the same frequency, frequency coding (shown by

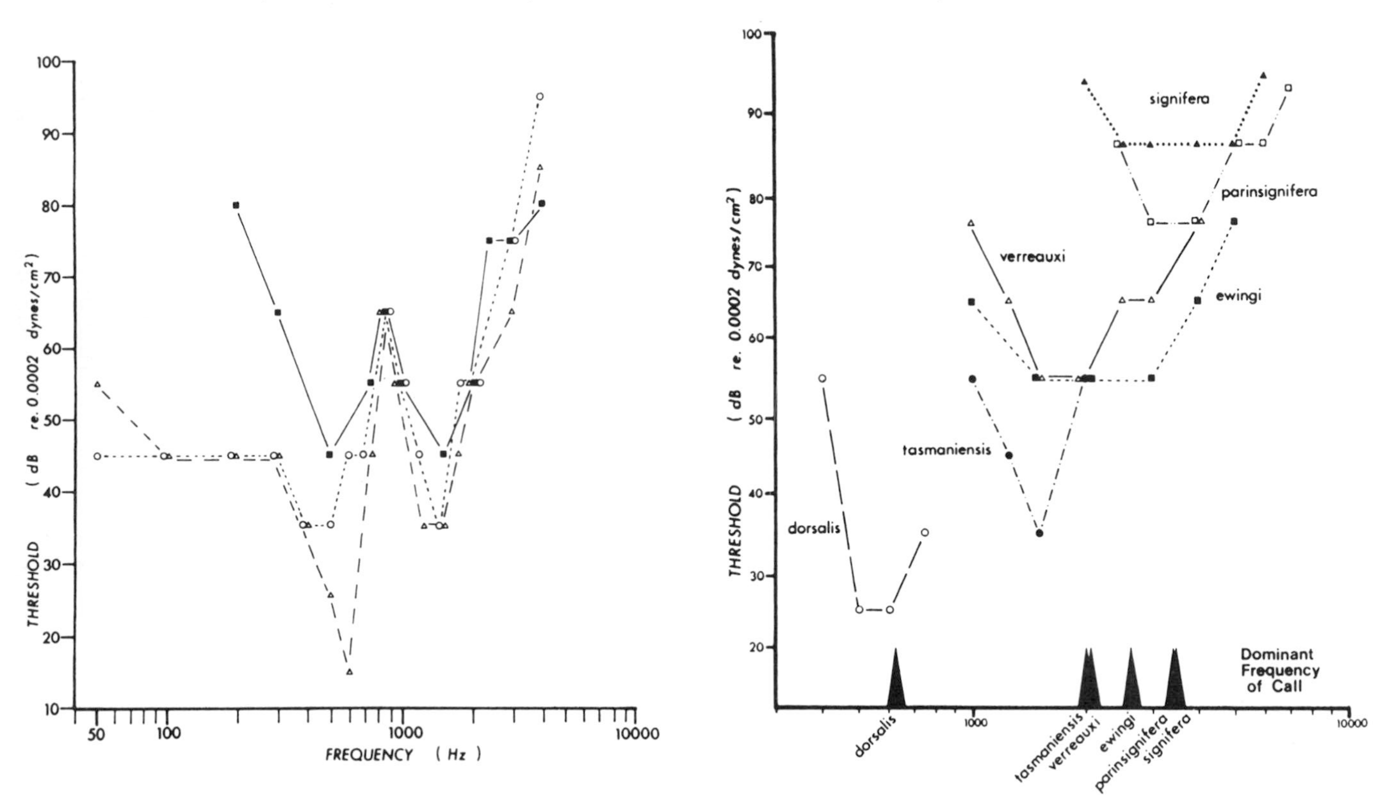

22–8 Frequency ranges of hearing and calling in Australian frogs. *Left:* Threshold curves (audiograms) of three individuals of *Limnodynastes tasmaniensis.* The different symbols are for different individuals. Thresholds are for detection of an electrical response in the midbrain. *Right:* Comparison of upper sensitivity peaks for hearing and dominant frequencies in the calls for six species of frogs. Note that the frequency emphasized in the call is the one to which the frog hearing system is most sensitive. (From Loftus-Hills and Johnstone 1970.)

Loftus-Hills 1971) by having ear and call tuned to the same narrow frequency band must be supplemented by temporal coding. A start in the analysis of cricket song-recognition mechanisms was made by Stout and Huber (1972) with female *Gryllus campestris*. They found that certain units fired continuously during each cricket sound pulse (somewhat like Roeder's repeater neuron), and others fired continuously during all the pulses and pauses of a chirp. Still others were cyclically responsive.

The complexity of auditory communication in mammals has tended to discourage the types of experiments emphasized in this chapter. But some attempts have been made to study neuronal types in response to natural stimuli. O'Keefe and Bouma (1969) found neurons in the amygdala that responded selectively to simulated bird and mouse calls. They were notably unsusceptible to habituation and changes in arousal state of the animal. Wollberg and Newman (1972) played natural calls to squirrel monkeys and recorded the responses of neurons in the superior temporal cortex. Some cells responded to many types of calls, others only to one. Mammals cannot be expected to function like insects in these respects, but this should not deter analysis of how their brains analyze natural auditory stimuli.

RECOGNITION OF BIOLOGICALLY SIGNIFICANT VISUAL STIMULI

The analysis of biologically meaningful visual stimuli received a major impetus with a well-known study of the frog retina titled "What the frog's eye tells the frog's brain" (Lettvin et al. 1959; see also Lettvin et al. 1961; Maturana et al. 1960). The analysis of stimulus recognition in frog vision began with physiological studies designed to identify the visual stimuli necessary to cause firing in single axons in the optic nerve. Surprisingly, the stimuli necessary to excite a frog optic-nerve axon were far more complex than had been suspected. When electrical recordings were made from the axons that carry information from the retina of the eye to the optic tectum of the midbrain, it was found that the frog retina was doing more than just relaying to the midbrain the pattern of visual excitation in the eye. The retina accomplished a significant amount of processing of visual information. Five types of neurons were found to occur among the optic fibers going to the tectum. These were at first identified purely on the basis of the kind of visual stimuli required to cause firing in the axon. Morphological correlates were found later.

In these experiments, conscious, immobilized frogs of the *Rana pipiens* complex were placed with one eye at the center of an aluminum hemisphere with a radius of 14 inches. The stimulus objects were moved on the inner surface of the hemisphere by a magnet controlled by hand on the external surface. The principal stimuli and a diagram of the experimental situation are shown in Figure 22–9. Microelectrodes were used to record from axons originating in the retina. The different cell types are distinguished as follows:

Class 1) Sustained edge detector. These neurons respond to moving or stationary edges in their receptive fields. They respond neither to the *on* nor the *off* of general illumination. The receptive field is 1–3°.

Class 2) Moving dark spot detector. Called *convex edge detectors* by Maturana et al. (1960), these neurons respond to small, dark, moving

22–9 Situation and test objects for experiments on the stimuli that excite axons in the optic nerve of frogs. **A.** Situation: The eye of the frog is at the center of an aluminum hemisphere 14 inches in diameter. **B.** Test objects and backgrounds: scale drawings of some test objects are above; backgrounds, below. Their sizes, in degrees, when in place on the inner surface of the hemisphere are indicated. Some of the backgrounds used: uniform gray; checkerboard patterns (*shown*); dotted patterns (*shown*); and a color photograph of grass and flowers from a frog's-eye-view. The test objects were on the inside of the hemisphere, and were moved by means of a magnet on the outside. (From Maturana et al. 1960. Reproduced from the *Journal of General Physiology, Suppl.* 43:129–175, 1960, by copyright permission of The Rockefeller University Press.)

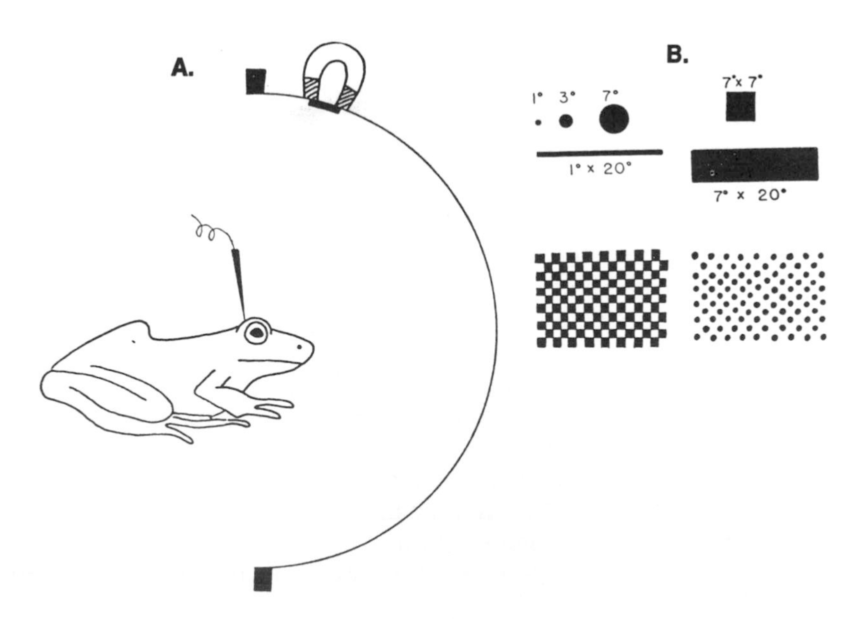

spots and to moving convex edges, but not to stationary edges. They also fail to respond to the *on* or *off* of general illumination. They are hypothesized to function as "bug detectors" (Lettvin et al. 1959). The receptive field is 2–5°.

Class 3) Moving contrast detector. These neurons respond to moving but not stationary objects and to both the *on* and *off* of general illumination. Since they detect all relatively large pale or dark moving objects against any type of background, they were thought by Pigarev et al. (1972) to be concerned with detection of predators. The receptive field is 7–12°.

Class 4) Dimming detector. These neurons respond to the *off* of general illumination but not to the *on*. They respond to moving and stationary objects to the extent that they darken the receptive field. The receptive field is up to 15°.

Class 5) Dark detector. The activity of these neurons is inversely proportional to light intensity. They fail to respond rapidly to movement or to abrupt *on* or *off* stimuli. The receptive field is large and difficult to determine.

In addition to the five neuronal types just described, others have been found that project elsewhere in the brain and that respond to stationary light areas (Muntz 1962*a*, *b*) or to stimuli with directional properties (Norton et al. 1970). We shall concentrate on a few of the first types to be described.

Anatomical correlates of two types are important, one concerning the endings of the optic nerve axons in the tectum and the other concerning the shape of the dendritic trees of the ganglion cells in the retina. The endings of each of these first five fiber types were found to be specific to a given cell layer in the optic tectum. The five types ended in four different layers, except that one layer received cell classes 3 and 5. Thus each tectal layer is specialized in respect to different types of retinal processing.

The origin of the properties of the five cell types from the retina is not completely understood, but it seems likely that the characteristics of each cell type are determined by interactions with neighboring cells and that these interactions differ depending on the structure and branching of the dendritic tree, particularly of the ganglion cells. A cross section of the retina, shown diagrammatically in Figure 22–10, shows that the optic nerve fibers are axons of the ganglion cells. Because excitation originating in the receptors passes through layers containing bipolar, horizontal, and amacrine cells before reaching the ganglion cells, it was felt that these cells are partly responsible for the processing; and this has been confirmed by experiment (e.g., Werblin 1972). From this information, we can conclude that it is extremely

22–10 The frog retina. *Above:* Schematic diagram of idealized pathways in the vertebrate retina. Inhibitory synapses are shown by arrow points and excitatory synapses by inverted Vs. (From Dowling and Boycott 1966.)

Below: Diagram of a frog retina. *RT,* receptor terminal. *H,* horizontal cell. *B,* bipolar cell. *A,* amacrine cell. *G,* ganglion cell. (From Dowling 1968. Reprinted by permission of the Royal Society.)

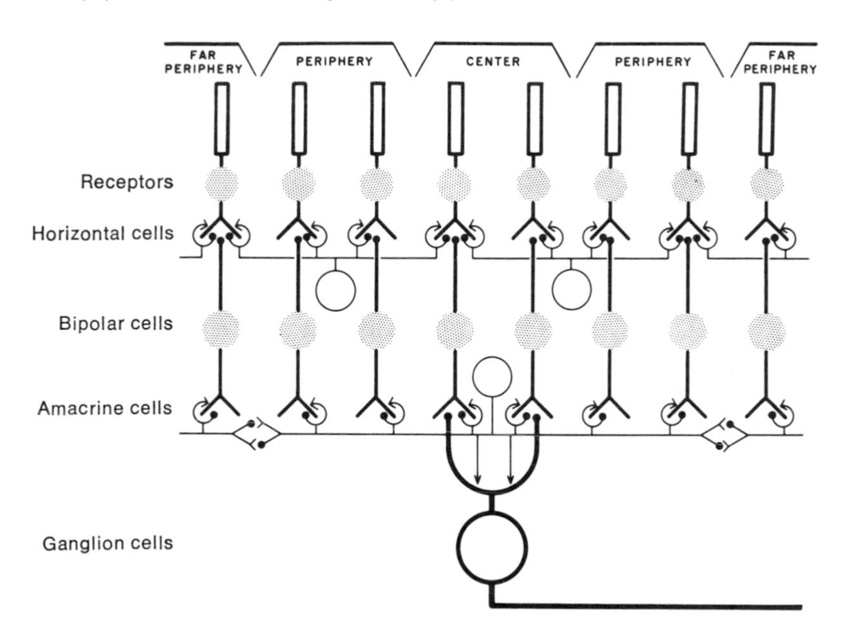

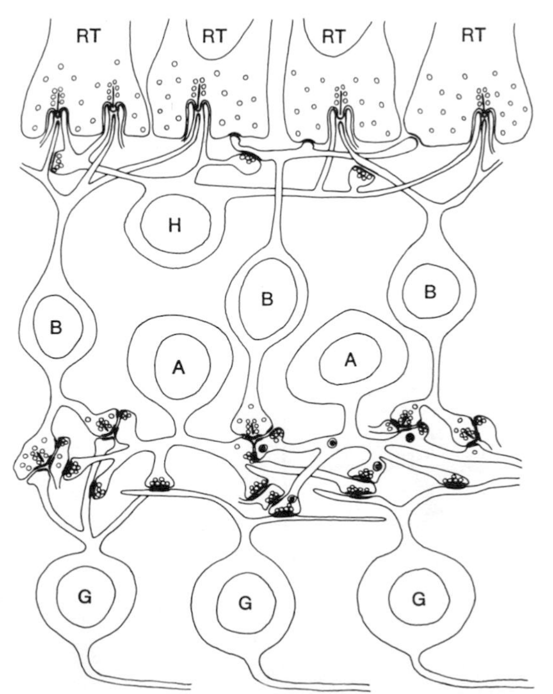

likely that the properties of the axons observed by Maturana et al. (1960) are determined at least in part by the shape of the dendritic tree of the ganglion cell, since this is the part of the ganglion cell that recieves input from the bipolars and amacrines. The correlations that Lettvin et al. (1961) suggested were supported by evidence from similar studies on tadpoles (Pomeranz and Chung 1970); the scheme is shown in Figure 22–11.

No attempt will be made here to explain just how the shape of the dendritic tree of a ganglion cell determines its response to visual stimuli, but some indication of the processes involved can be gained by considering the phenomenon of lateral inhibition. This was discovered by Hartline (1949) in the compound eye of the horseshoe crab (*Limmulus*), but it has since been found to be of basic importance in vertebrate eyes. The process is explained in Figure 22–12.

Up to now we have considered steps 1 and 3 of our strategy for analyzing species-typical mechanisms of stimulus recognition. For several years after Lettvin et al. (1959) had raised the problem, step 2, the behavioral approach, received little attention in studies of frogs, although some excellent work was done with toads (*Bufo;* e.g., Ewert and Hock 1972). The hypothesized behavioral function of the "bug

22–11 Correspondence of ganglion cell morphology with physiological characteristics. Cell classes are numbered as in the text. **A.** Adult frog. Correspondence in adult proposed by Lettvin et al. (1961). **B.** Tadpole (larval form of frog). Correspondence found by Pomeranz to be same as in adult. Class 1 responses and constricted tree cells are both lacking in tadpoles. (From Pomeranz and Chung 1970. Copyright 1970 by the American Association for the Advancement of Science.)

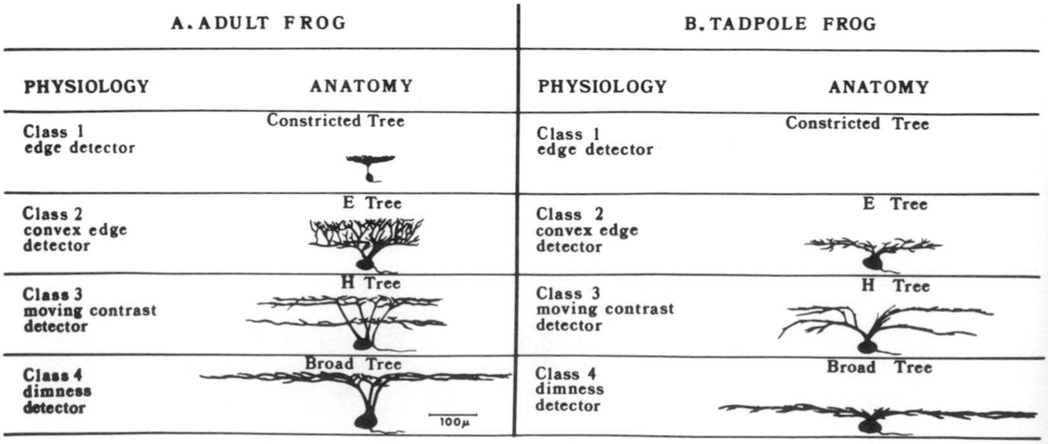

A. ADULT FROG		B. TADPOLE FROG	
PHYSIOLOGY	**ANATOMY**	**PHYSIOLOGY**	**ANATOMY**
Class 1 edge detector	Constricted Tree	Class 1 edge detector	Constricted Tree
Class 2 convex edge detector	E Tree	Class 2 convex edge detector	E Tree
Class 3 moving contrast detector	H Tree	Class 3 moving contrast detector	H Tree
Class 4 dimness detector	Broad Tree (100μ)	Class 4 dimness detector	Broad Tree

22–12 Lateral inhibition. The *receptive field* of a ganglion cell is the area on the retina where light can influence the firing of the ganglion cell. For convenience it can be divided into two zones, *center* and *periphery.* It can be shown that excitation of one ommatidium in *Limmulus* or of one ganglion cell in a vertebrate retina causes inhibition in neighboring cells. The area around a cell from which it can be inhibited is known as its *inhibitory surround.* The effect of this is to *sharpen edges,* as can be appreciated from the diagram. The firing rate of each cell will be determined by how much excitation it receives on its center and how much inhibition in its periphery. For the above cells, the relative strength of excitation will be $B > A > D > C$ (rather than $A > B > C > D$, which would occur if there were no lateral inhibition). Here, *B* fires more than *A* because it is less inhibited; for the same reason *D* fires more than *C.* (After Horn 1962.)

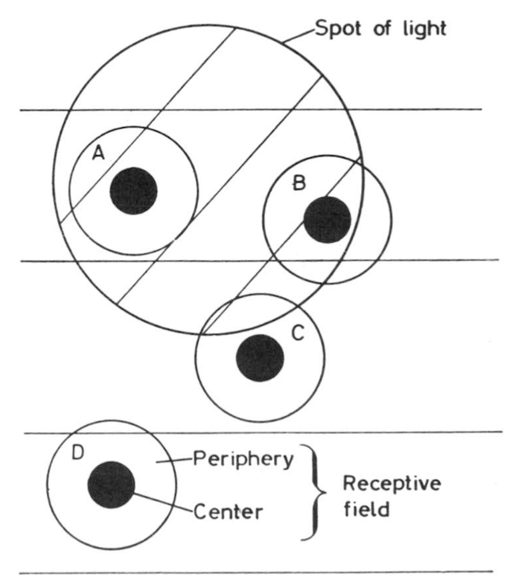

detector" was novel and stimulating, but not really very convincing because of the lack of behavioral studies, despite an attempt to correlate data gathered in field experiments with the physiological and anatomical data (Grüsser and Grüsser-Cornehls 1968).

The first successful approach to this problem involved the use of the same procedures for presenting stimuli in the physiological experiments as were used in the behavioral experiments (Pigarev and Zenkin 1970). By this means it was shown that, in *Rana temporaria,* moving black dots or rods of the same size that fire the spot detector ganglion cells also elicit feeding behavior. A black rod fires the spot detector and evokes feeding behavior only if moved onto the field, not during withdrawal. Real insects that evoke feeding behavior also fire the spot detectors. A comparison of the firing rate of spot detectors and the feeding responses of normal frogs (with implanted electrodes) to the same visual stimuli is made in Figure 22–13.

These experiments were carried further by releasing frogs with microelectrodes in a "natural environment" and recording from the various kinds of ganglion cells (Pigarev et al. 1972). The frogs were connected by a 1.5-m wire to a portable amplifier carried by the observer, who followed and watched the frogs. Spot detectors (class 2), the type most numerous in the frog retina, did not fire when the frog was sitting still, crawling, or jumping. They responded actively to flies, beetles, and a crawling earthworm, but not usually to large, moving objects such as bushes or trees waving in the wind.

Moving contrast detectors (class 3) were likewise unexcited in a sitting frog, but activity occurred during crawling and jumping, and a

22–13 Comparison of the responses of spot detectors and a frog's feeding behavior to the same stimuli. In the behavioral experiments, a jump toward the stimulus represents a feeding response. When three dark spots were presented together, jumps were made to one of the lateral spots four times. The broken circle denotes the relative size of the receptive field. (From Pigarev and Zenkin 1970.)

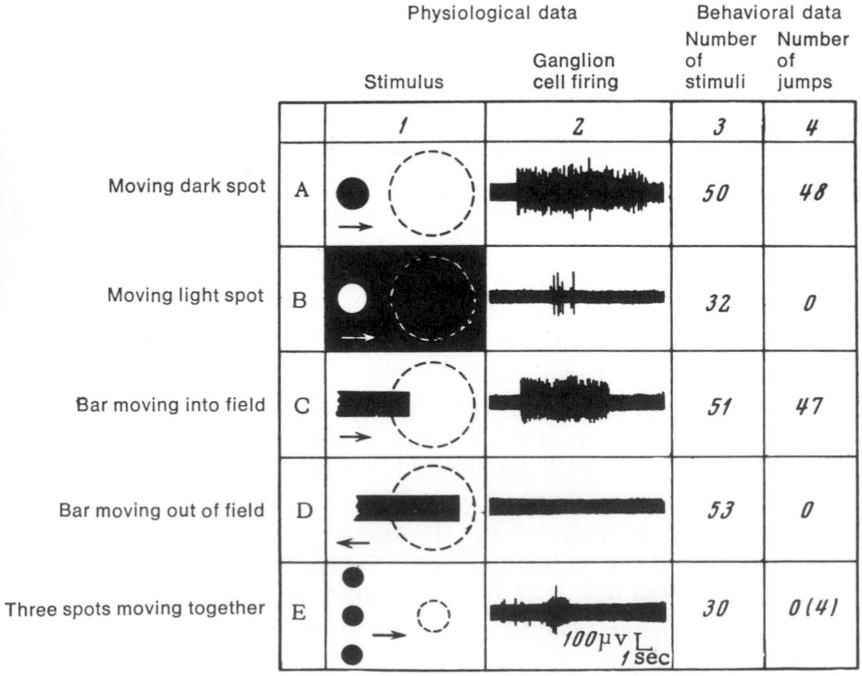

		Physiological data		Behavioral data	
		Stimulus	Ganglion cell firing	Number of stimuli	Number of jumps
		1	2	3	4
Moving dark spot	A			50	48
Moving light spot	B			32	0
Bar moving into field	C			51	47
Bar moving out of field	D			53	0
Three spots moving together	E		100μv 1 sec	30	0 (4)

strong response was obtained to large moving objects lighter or darker than the background. It has been suggested that these neurons could serve to detect an approaching predator.

Dark detectors (class 5) did not fire in a sitting frog unless a large, dark, moving object entered the field. It has been suggested that these cells might indicate good hiding places to an escaping frog (Pigarev et al. 1972).

Evolutionary theory suggests that the nervous systems of animals have been specialized through natural selection for the performance of tasks relevant to the way of life of each species. Are the properties of the frog retinal ganglion cells to be explained on this basis? Or are these properties merely incidental by-products of a more comprehensive machinery for analysis of visual stimuli? A definite answer may not be possible now, but the correspondence between visual stimuli that excite certain cell types and those that evoke certain types of behavior now seems impressively strong.

If animals do have species-typical neural mechanisms for recognizing biologically important stimuli, comparative study of the mechanisms of visual processing in other species should bring this out. Studies on cats, rabbits, pigeons, ground squirrels, catfishes, and mud puppies have revealed a bewildering variety of detector types, not just in the retina but also at various places in the brain (Table 22–1 and Figure 22–14). Each species has been found to have a different array of detector types, and these are located at various places in the nervous system. At present it is difficult to know the evolutionary significance of the many differences that have been found among these species. For most of these detectors a function as simple as "bug detector" or "enemy detector" seems extremely unlikely.

TYPES OF RECOGNITION

The word *recognition*, which has been used freely in this chapter, requires some discussion. "Recognition of a stimulus" can be gauged only by an animal's response. The method employed has been to vary the stimulus and record the strength of a particular type of species-typical response. When this response (and only this response) is given to a stimulus, we say that the animal has "recognized" the stimulus even though it may never have encountered the stimulus previously; it has made a biologically appropriate and species-typical response. Our use of *recognition* means no more than this, and is not intended to include various anthropocentric connotations of the word.

In the examples considered here, the response is given only to a

TABLE 22–1 Trigger features of neurons at different anatomical locations in various vertebrate species. (From Barlow et al. 1972. Copyright 1972 by the American Association for the Advancement of Science.)

Anatomical location	Trigger feature
	GOLDFISH
Retina	local redness or greenness; directed movement
	FROG
Retina	convex edge; sustained edge; changing contrast; dimming; dark
Optic tectum	newness; sameness; binocularity
	PIGEON
Retina	directed movement; oriented edges
	GROUND SQUIRREL
Retina	local brightening or dimming; local blueness or greenness; directed movement
Lateral geniculate body	color coded units
Optic tectum	directional units; oriented slits or bars; complex units
	RABBIT
Retina	local brightening or dimming; directed movement; fast or slow movement; edge detectors; oriented slits or bars; uniformity detectors
Lateral geniculate	greater directional selectivity
Tectum	habituating units
	CAT
Retina, lateral geniculate, cortex	for main types see Figure 22–14
	INFREQUENT TYPES
Retina	directed movement; uniformity detectors
Lateral geniculate	local blueness or greenness; binocular, directional, and orientational units
Optic tectum	directed movement; complex units
	MONKEY
Retina	local brightening or dimming; local redness, greenness, or blueness
Lateral geniculate	various forms of color coding
Cortex	similar to cat; some color coded
Inferotemporal cortex	very complex; possible hand detector

restricted set of stimuli and these stimuli elicit one type of response or a restricted set of responses. *Stimulus-response (S-R) specificity* may be said to obtain in these cases. This S-R specificity is of two types. In Class I, any moderate activation of a particular sensory mode is adequate to evoke the response in question, provided that other permissive restrictions have been met. This is exemplified by the noctuid moth's response to a bat call: any sound picked up by the tympanal nerve probably can evoke an escape response. In this class, *within-mode S-R specificity is lacking.*

In Class II, the response in question cannot be evoked by just any pattern of stimulation in the proper sensory mode; the pattern of stimulation within one sensory mode is critical. This case is exemplified by the selectivity of bullfrogs in answering auditory stimuli and by the

22–14 Receptive fields and trigger features for samples of cells from six places in the visual pathway of the cat. Plus signs indicate that light in that region excites a response; minus signs indicate inhibition of response, but are somewhat ambiguous. They may mean that light in that region prevents responses that would otherwise occur as a result of excitation elsewhere in the same receptive field. This diagram is also simplified in other ways, for movement and disparity are often specific requirements, and there are some rarer types of units (see Table 22–1). The visual cortex is divided anatomically into regions, areas 17 and 18. (Figure and legend from Barlow et al. 1972. Copyright 1972 by the American Association for the Advancement of Science.)

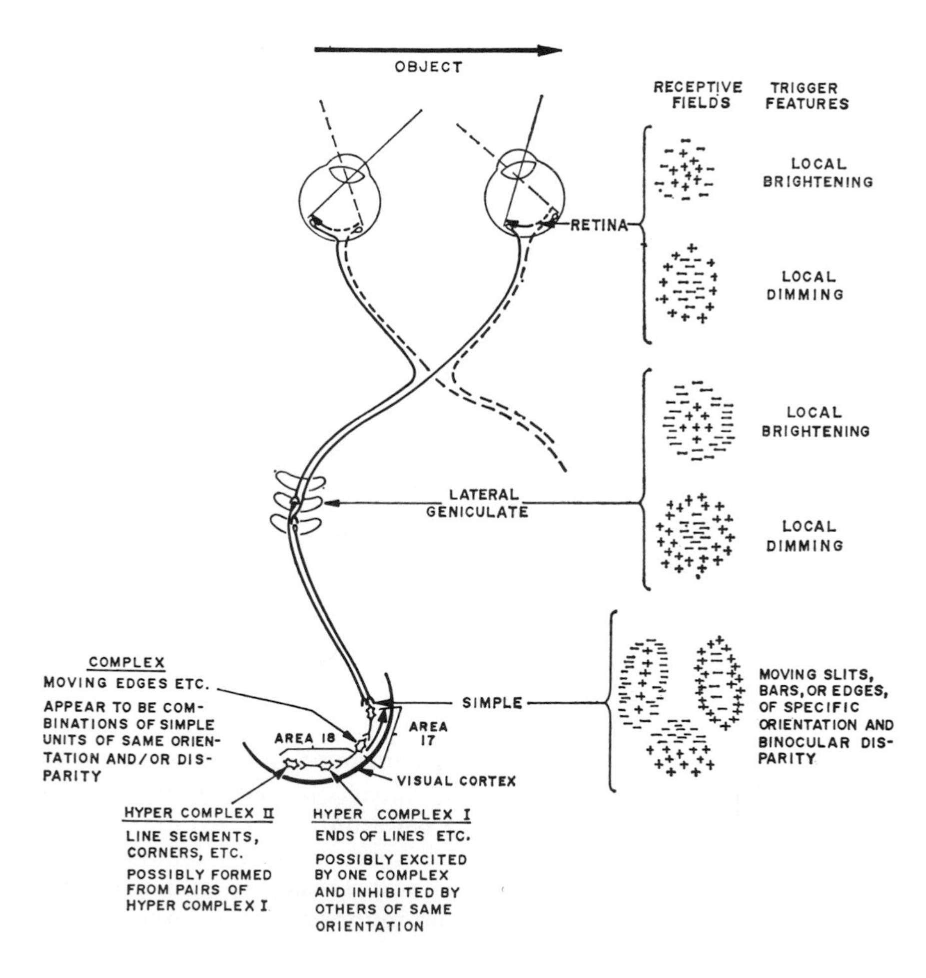

selectivity of *Rana* in making feeding actions in response to visual stimuli. *Within-mode S-R specificity is present.*

The origins of within-mode S-R specificity are of two types. The response may be limited to one stimulus because a neural analyzer has detected a key property, as is suspected in the example of the frog retina. Or the key property may be detected by a subset of receptors in one sensory mode, as in reflex flexion of a limb away from a painful stimulus. In the latter case, the subset consists of the pain-sensitive receptors in the skin of the flexed limb. Stimulation of a different subset, say on the skin of another limb, evokes a different pattern of response.

CONCLUSIONS

Species-typical behavior is often made possible by specialized sensory abilities of particular groups of animals. A stimulus may *predispose* an animal toward a particular behavior, or it may serve to *adjust* a posture or direction of behavior, or it may *evoke* a particular pattern of behavior. This chapter deals with the mechanisms by which a particular response is evoked by a particular stimulus, and with the physiological mechanisms that mediate species-typical *stimulus-response (S-R) specificity*. The focus is on the recognition problem: What features of sensory systems bias an animal toward species-typical S-R specificity?

The following strategy is suggested as an approach to this problem: (1) Choose a biologically significant S-R relationship, between, for example, mating, eating, or escape and the cues that evoke these behaviors. (2) Determine, by experimentally varying the stimuli, precisely which aspects of a stimulus are critical for evoking the response. Then compare the results with the characteristics of the natural stimulus. (3) Determine, by experimentally varying the stimuli, precisely which aspects of a stimulus are critical for evoking a response of the receptors and of neurons in the sensory analyzers. Then compare the results with the characteristics of the natural stimulus and with the results of step 2. Some cases have been discussed in which all three of these steps may be recognized.

Many nocturnal moths are sensitive to sound in the range from 15 to 70 *kHz*, and therefore to the calls of bats. This enables them to perform escape movements in flight if a bat approaches. Although some processing of auditory information in the nervous system of moths can be demonstrated, the critical information necessary to initiate escape behavior by a moth from a bat is present in the tympanic nerve coming from the hearing organ, since the latter is sensitive mainly in the range in which bats call.

When one male bullfrog gives its species-specific mating call, another is likely to answer it. This ability is probably necessary to defend a mating station. An answering call can be elicited in a bullfrog by an artificial call with energy in a high and a low frequency band. These are the same frequencies as those in the natural call. In the bullfrog's auditory nerve, two types of neuron are found, one that is most sensitive in the low and the other in the high frequency band. These bands match the frequencies in the call as well as those in the optimal artificial call. In this instance, not all sounds that excited axons in the auditory nerve are able to evoke calling; only sounds with a certain relationship among high, low, and middle frequencies are adequate. Therefore, the recognition mechanism must include both the ear and the brain; it is both peripheral and central.

In frogs, a visual stimulus is sufficient to evoke food-getting movements such as jumping at the stimulus. A small, dark, moving spot will evoke feeding attempts. The same stimulus will also activate a particular kind of neuron that carries information from the eye to the brain. This "spot detector" could be critical in the recognition of food by frogs and in the activation of a feeding attempt. The properties of the spot-detector neuron appear to be determined in part by the branching of its dendrites. The branching in turn partly determines the neuron's synaptic relationships with other cells in the retina concerned with visual processing.

These examples show how species-typical stimulus recognition tasks are met in part by species-typical sensory analyzers. The main problem is to determine how the sensory analyzer biases the animal to respond to a specific stimulus (bat call, frog croak, insect) in a specific way (escape, answering croak, feeding movement). The bias may be mainly at the *receptor* or peripheral level: it may be caused mainly by the restricted range of sensitivity of the receptor cells, as in the moth-bat relationship. Alternately, the bias may be mainly *neural:* it may be caused mainly by processing in the neurons that are activated by the receptors, as in the frog retina. Since the retinal neurons are outgrowths of the brain, this example might also be classified as *central.* A third alternative is that a bias may operate *both at receptor and neural* levels, as in the mechanisms that cause a bullfrog to answer a mating call of its own species.

The biases imposed on an animal's responses to stimuli by the nature of its sensory analyzers differ markedly from species to species, even within the mammals. Biases due to restricted receptor sensitivity ranges are well known because they are most easily studied. Biases due to the organization of neural analyzers of sensory input are more difficult to study, but they are becoming increasingly better known. When studies at the neural and behavioral levels are coordinated, we may expect further progress on the problem of stimulus recognition or feature detection.

23 Sensory Systems: Orientation, Echolocation

ANIMALS ORIENT THEMSELVES in their living space in a great variety of ways. The concept of orientation adopted here is a broad one and includes such phenomena as choice of surroundings and micro-environment, maneuvering around obstacles and dangers and toward food and shelter, and returning to a favored familiar place. The problems of behavior that are related to orientation are mainly sensory ones. What information from the environment is used, what sensory receptors are involved, and what processing of sensory information by the CNS is involved? How do these systems develop and evolve? How is the diversity in behavior among species due to differences in their mechanisms of orientation? In this chapter we shall first take a brief look at a variety of rather simple types of orientation, and then concentrate on one of the complex types, echolocation. A series of reviews and reports on various topics in the field of orientation is available in Adler (1971).

TYPES OF ORIENTATION IN INVERTEBRATES

Orientation of invertebrates has been studied in a large number of species utilizing a variety of stimuli and sensory analyzing systems. These were conveniently summarized in 1940 by the classification of Fraenkel and Gunn (1961), which is reproduced here as Table 23–1. Some examples from the categories of Table 23–1 show the relative simplicity of many of these orientation mechanisms. For details, the reader is referred to the original book and to more detailed and critical treatment by Carthy (1958), Marler and Hamilton (1966), and Hinde (1970).

Kineses are movements that orient the animals in relation to favorable or unfavorable surroundings without involving a directed response to the stimulus. In *orthokinesis*, as found in woodlice (*Porcellio*), greater *locomotory* activity in dry than humid conditions causes

TABLE 23–1 Classification of orienting reactions. (From Fraenkel and Gunn 1961.)

(1) General description	(2) Form of stimulus required	(3) Minimum form of receptors required	(4) Behavior with two sources of stimulation	(5) Result of unilateral removal of receptors	(6) Formerly called	(7) Examples
KINESES Undirected reactions. No orientation of axis of body in relation to the stimulus. Locomotion random in direction						
ORTHOKINESIS Speed or frequency of *locomotion* dependent on intensity of stimulation	Gradient of intensity	A single intensity receptor	Reaction to whole of the gradient	No effect? Reduced intensity of reaction?	Simply kinesis	*Porcellio*
KLINOKINESIS Frequency or amount of *turning* per unit time dependent on intensity of stimulation. Adaptation, etc., required for aggregation	Gradient of intensity	A single intensity receptor	Reaction to whole of the gradient	No effect? Reduced intensity of reaction?	Phobotaxis; avoiding reactions; *Unterschiedsempfindlichkeit*	*Dendrocoelum, Paramecium?*
TAXES Directed reactions. With a single source of stimulation, long axis of body orientated in line with the source and locomotion towards (positive) or away from (negative) it						
KLINOTAXIS Attainment of orientation indirect, by interruption of regularly alternating lateral deviations of part or whole of body, by *comparison of intensities of stimulation which are successive in time*	Beam or steep gradient	A single intensity receptor	Orientation between the two, curving into one when close to sources	Usually impossible; no effect? Reduced intensity of reaction?	Part of tropotaxis; avoiding reactions; phobic mechanism	Fly larvae, *Euglena*, larvae of *Arenicola, Amaroucium*
TROPOTAXIS Attainment of orientation direct, by turning to less- or to more-stimulated side, by simultaneous *comparison of intensities of stimulation on the two sides.* No deviations required	Beam or steep gradient	Paired intensity receptors	Orientation between the two, curving into one when close to sources	Circus movements in uniform field of stimulus and often also in beam or gradient	Tropotaxis, but klinotaxis excluded, and cases with two-way eyes added from telotaxis	Woodlice, *Ephestia* larvae, *Eristalis, Notonecta*
TELOTAXIS Attainment of orientation is direct, without deviations. Orientation to a source of stimulus, as if it were a goal. Known only as response to light	Beam from a small source of light	A number of elements pointing in different directions	Orientation to one at a time; animal may switch over to the other at intervals, giving a zigzag course	Not known, because often the same animal can behave tropotactically, and the circus movements which occur may be the result of tropotaxis	Telotaxis, but excluding cases which always orientate between two lights before curving into one of them and excluding reactions to specific objects and form reactions	*Apis, Eupagurus*
TRANSVERSE ORIENTATIONS Orientation at a temporarily fixed angle to the direction of the external stimulus or at a fixed angle of 90°. Locomotion need not occur and in any case is seldom directly toward or away from the source of stimulation						
LIGHT COMPASS REACTION Locomotion at a temporarily fixed angle to light rays, which usually come from the side	Beam from a small source	A number of elements pointing in different directions	If each light affects one eye only, orientation is to one alone; if one eye is affected by both lights, orientation to both combined	No effect, except in limiting the possible angles of orientation	Menotaxis	*Elysia*, ants, bees; caterpillars of *Vanessa urticae*
DORSAL (OR VENTRAL) LIGHT REACTION Orientation so that light is kept perpendicular to both long and transverse axes of the body; usually dorsal, but in some animals ventral. Locomotion need not occur	Directed light	Paired intensity receptors	Not known	No effect in some species; produces screw-path in others	Dorsal light reflex (*Lichtrückenreflex*)	*Argulus, Artemia, Apis*
VENTRAL EARTH (TRANSVERSE GRAVITY) REACTION Orientation so that gravitational force acts perpendicularly to long and transverse axes of body. Dorsal surface usually kept uppermost. Locomotion need not occur	Gravity (or centrifugal force simulating gravity)	Statocysts with a number of elements, as is usual with statocysts	A statocyst combines the forces mechanically and orientation should be determined by the resultant	No effect in some species; produces rotation or lateral tilting in others	?	*Leander*, crayfish

individuals to spend more time in damp than dry areas and to congregate in damp areas. In *klinokinesis*, as found in *Paramecium*, the animals change course by a fixed angle when an object or unfavorable environment is encountered. The result is that unfavorable environments are avoided and animals congregate in favorable places. Klinokinesis differs from orthokinesis in that *turning* is increased while rate of locomotion is unchanged.

Taxes are directed movements, usually rather simple. In kineses

23–1 Klinotaxis in a maggot (fly larva). By turning the head alternately from side to side, the maggot can compare light intensities on each side. At position *d*, the light at *m* is put out and the light at *n* is put on. The larva turns to the darker side until the light intensity at each side is balanced. (From Fraenkel and Gunn 1961.)

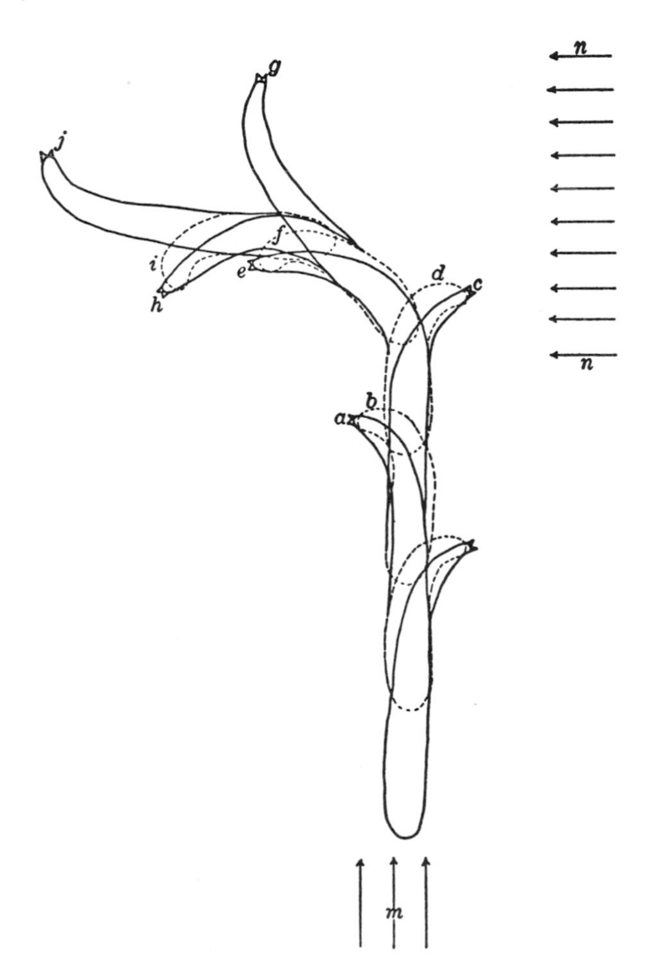

the animals do not react directly to the directional properties of the stimuli, while in taxes they do. *Klinotaxis* is illustrated by the fly maggot in Figure 23–1. This type of orientation is found in animals that are unable to make a simultaneous comparison of the intensity of stimulation on two sides because they lack appropriate bilateral receptors. Instead, they turn from side to side alternately, making successive comparisons with simple median receptors that need provide only a minimum of directional information. This type of orientation is commonly seen in flatworms (e.g., *Planaria*) approaching food. With light-mediated reactions, klinotaxis can be identified by turning on a diffuse light briefly each time the head is turned to a particular side; the animal then circles continuously. *Tropotaxis* is a simple response shown by animals that are able to make a simultaneous comparison between intensities at receptors on opposite sides in order to determine their direction of locomotion. It is illustrated in the pillbug, *Armadillidium*, by the two-light test. When released equidistant from two lights (as in Figure 23–2), individuals behave as if they were trying to maintain an equal intensity of stimulation at each eye; they do this by traveling a path that goes between the lights until finally they fall under the influence of one of them. Unilateral stimulation causes continuous circling, since the animal never succeeds in balancing the stimuli on both sides. Animals orienting by *telotaxis* react differently in the two-light test; they tend to zigzag first toward one light, then the other, as with the hermit crabs shown in Figure 23–2. They do not circle in response to unilateral stimulation.

23–2 The two-light test can be used to separate tropotaxis (**A**) from telotaxis (**B**). The isopod *Armadillidium* in A tends to go between the two lights, behaving as if it were trying to balance the light intensity at each side (tropotaxis). The hermit crabs in B go straight toward one or the other light, sometimes alternating in zigzag fashion (telotaxis). (From Fraenkel and Gunn 1961.)

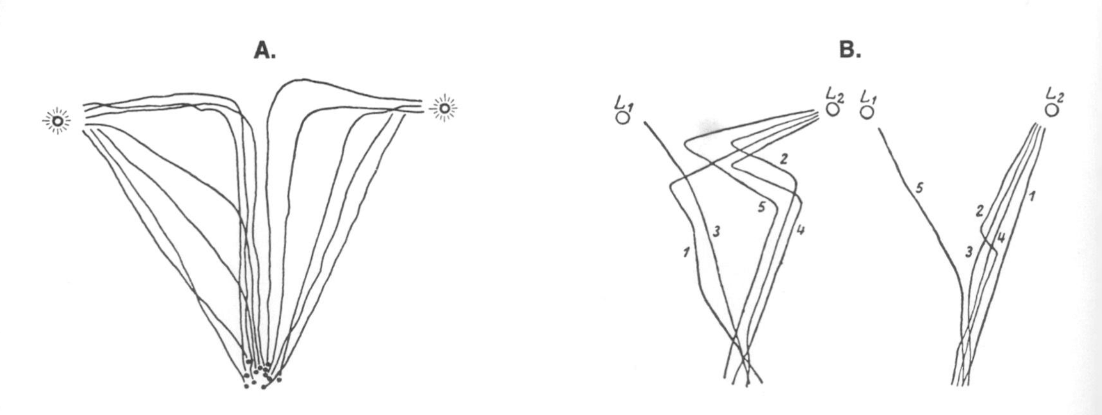

Transverse orientations are briefly described in Table 23–1. They refer mainly to position with respect to the stimulus rather than to movement toward or away from it. Light and gravity are used by some species as references in maintaining posture.

ECHOLOCATION BY ANIMALS

From the many types of complex orientation processes in animals, echolocation (EL) has been selected here for special attention because it is so far removed from man's own sensory world, because of its striking demonstration of species differences in behavior due to orientation abilities, and because the mechanisms of EL remain an unsolved set of problems whose solutions are potentially useful to man. In addition, convergent and parallel evolution in behavior are nicely illustrated by EL behavior. A detailed review covering EL physiology and behavior has been written by Erulkar (1972).

History / Although experiments were done in the eighteenth century by Spallanzani and Juvine showing that bats needed their ears but not their eyes to avoid obstacles and catch food, it was not until the experiments of Griffin and Galambos in 1941 that the scientific world accepted the unorthodox idea that animals could "see with their ears." Much of the history of this concept is covered in Griffin's (1958) valuable book, *Listening in the Dark*. The experiments of Griffin and his colleagues established that the little brown bat (*Myotis lucifugus*), common in the United States, could fly through a "fence" of vertical wires (1.2 mm in diameter) spaced a foot apart without striking wires only if its ears and mouth were unobstructed. It was found at about the same time that the bats were making clicking sounds while negotiating the fence and that these sounds were largely above the upper limit of frequencies that could be heard by man, i.e., in the ultrasonic range. Although to our ears these sounds are faint or inaudible, in absolute terms they are incredibly loud. For example, the calls of a little brown bat at a distance of a few cm from the mouth contain sonic energy comparable in absolute terms to the sound that reaches our ears from a pneumatic drill breaking up pavement on a city street.

Functions and Characteristics of Echolocation Systems / Species vary in their ability to orient using reflected sound. A well-developed EL system serves the following functions:

1) Detection: determining presence or absence of object.
2) Direction finding: estimating direction of target.

3) Ranging: estimating distance of object.

4) Anticipating position: estimating future position of moving object.

5) Resolution: noticing differences in size, shape, and quality of objects.

6) Identification: Should the object be eaten, avoided, or ignored?

Animals that rely heavily on EL can use it for all these purposes. A flying bat can detect objects, estimate their direction and distance precisely, predict the future position of moving objects accurately enough to intercept them, distinguish between objects, and learn to identify certain objects as food while rejecting inedible objects (such as pebbles thrown up by casual naturalists). Bats can do all of these things with amazing precision, speed, and reliability. People can do most of them too, but only in the crudest imaginable ways.

Who Echolocates? / The best-known EL animals besides bats are the porpoises. However, other lesser-known species are known to be able to use EL. Oilbirds (*Steatornis caripensis*) (Griffin 1953) and cave swiftlets (*Collocalia brevirostris*) (Novick 1959) roost and nest in caves that in some cases are too dark for vision. These birds can avoid large obstacles while in flight even in total darkness, but plugging their ears causes them to lose this capacity. They utter clicking sounds while flying in darkness. Certain seals and sea lions are also suspected to use EL. Shrews (*Sorex, Blarina*) can locate a large object in darkness by a simple form of EL (Gould et al. 1964), but being terrestrial they lack the dramatic locomotor and orientation abilities of animals that can orient and travel in three dimensions in darkness. The terrestrial tenrecs (Tenrecidae) of Madagascar are also thought to use EL (Gould 1965).

Surveying the species that use EL, a few general statements can be made. These species tend to have well-developed hearing abilities. The sounds they produce for EL tend to come in brief *pulses* and to be *high-pitched.* Brief pulses provide discrete cues for timing the echos. High frequencies can be beamed more precisely and allow resolution of smaller objects than do lower frequencies. For detecting cave walls and obstacles in a tunnel, the relatively low-frequency sounds of birds and shrews would seem to be adequate.

ECHOLOCATION BY BATS

Species Diversity in Foraging and Flying / The order Chiroptera is divided into two suborders, the Megachiroptera, which include many fruit bats (such as the flying foxes), and the Microchiroptera, which include all the species familiar in America and Europe. The Megachiroptera tend to be rather large and to have well-developed vision. Only one genus, *Rousettus* (Griffin et al. 1958), is known to use EL, and unlike the Microchiroptera it uses its tongue to produce its EL clicks. The Microchiroptera are generally smaller and employ EL extensively. The foods of Microchiroptera vary from flies to horses (vampire bats); some species are nectar feeders, and others catch fish in tropical pools.

The degree of specialization in the Microchiroptera is not fully appreciated by persons familiar only with northern species. The state of San Luis Potosí in Mexico has at least 33 species of bats, most of which differ in their foraging or roosting habits. Dalquest (1953) has provided detailed field notes on most of these species. In this locality, *Micronycteris* plucks fruit from trees, *Glossophaga* imbibes flower nectar, *Choeronycteris* prefers cactus flower nectar, *Diphylla* laps the blood of sleeping chickens, and *Pteronotus* and other genera catch insects. The manner of hunting insects varies considerably among species. *Molossus aztecus* flys "at almost incredible speed" (Dalquest 1953:69) in circles 100 or 200 feet in diameter, 30 or more feet above the ground, in clearings. *Pteronotus rubiginosa* flies 3 feet from the ground among river-edge vegetation. Bats differ also in their foraging periods. *Pteronotus davyi* and *P. rubiginosa* (= *parnellii*) begin their evening flights at sunset, but *P. davyi* fills its stomach with insects after 10 minutes and retires for the night, while *P. parnellii* forages for 30 minutes before quitting. In contrast, *Mormoops* seems to confine its foraging to a period about two hours after dark, and *Carollia* is active at all hours of the night. All of these species are believed to rely on EL, yet their orientation requirements are diverse.

Prey Capture / When methods were developed for keeping bats in captivity, they could be fed mealworms that were shot into the air. The bats were able to zero in on these prey items effectively in total darkness. Under these conditions, the method of capture could be photographed using flashes, and the sounds made during approach and capture could be studied. Bats were found to be quite flexible and dexterous in capturing prey, as shown in Figure 23–3, although species differences in use of the tail pouch for prey capture were observed. Some mealworms were caught in the tail pouch, some were caught in the wing membranes, and some were caught directly in the mouth.

Capture Sequence / Most species that capture insects on the wing were found to share certain general features of the pursuit-and-capture

23–3 Prey capture by bats. **A.** A typical mealworm catch by *Myotis lucifugus.* Positions of the bat (*B*) and mealworm (*M*) at successive intervals during flash photography are shown. Note the use of the tail membranes. Prey is also caught by deflection off the wing and directly in the mouth. **B.** Use of the wing by the greater horseshoe bat *Rhinolophus ferrum-equinum* in catching a moth. The tail membrane is not used by this species. (From Webster and Griffin 1962.)

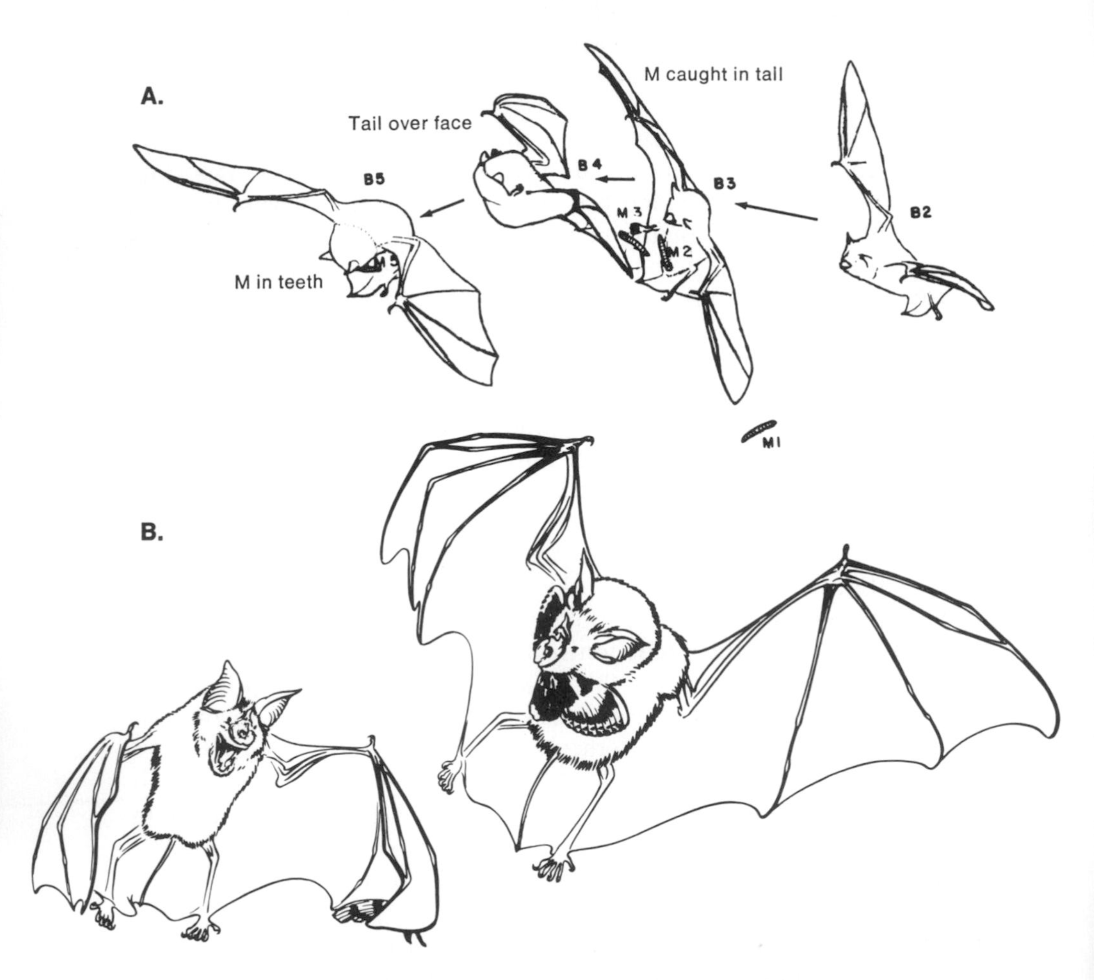

sequence. The first such studies, which were done on *Myotis lucifugus*, revealed that a bat increases its rate of calling as it approaches an object or prey item (Griffin, Webster and Michael 1960). Most other species studied have also been found to speed their rate of calling when approaching an object, whether edible or inedible. An example involving *Pteronotus psilotis* is shown in Figure 23–4.

The sequence of a capture can be divided into three phases: (1)

23–4 Sound spectrograms of the calls of a *Pteronotus psilotis* capturing a fruit fly (*Drosophila*). The top two lines show the end of the search phase and the entire approach phase; the bottom line, the terminal phase. The last pulse of each of the top two strips is repeated in the following strip. Note the changes in pulse duration, interpulse interval, and the slope of the frequency drop. Echoes off the walls of many of the pulses can be seen as faint duplicates immediately following a pulse. During each pulse the frequency drops from about 50 to 40 *kHz* (Novick 1963). The second harmonic drops from about 106 to 76 *kHz.* (From Novick 1965.)

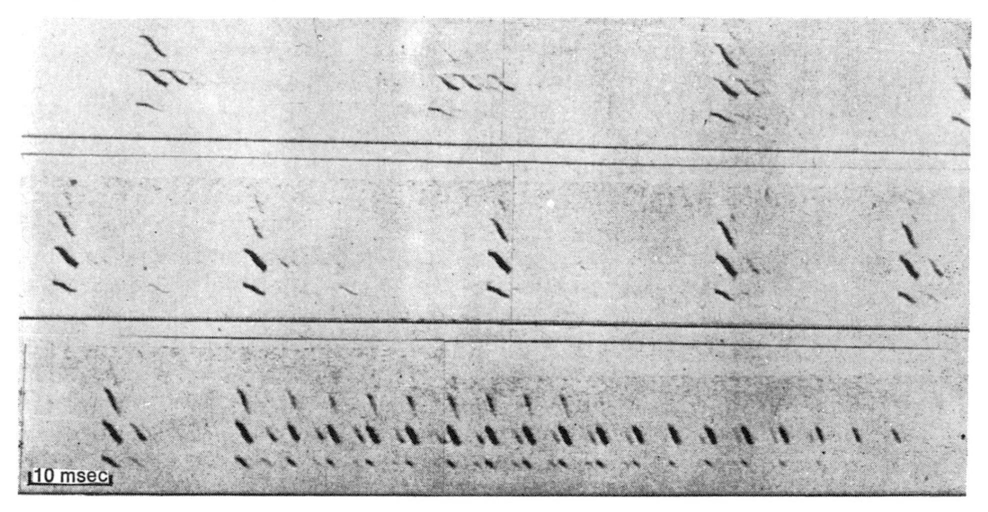

search, (2) approach, and (3) terminal. These are illustrated in Figure 23–5 for *P. psilotis*. The *search* phase consists of normal cruising, presumably while the bat is not attending to any particular object. The *approach* phase, which presumably begins when the bat starts to attend to the object being approached, is signaled by a shortening of both pulse duration and interpulse interval. A change in flight path has often been observed at this time. The *terminal* phase is marked by a sharp increase in pulse rate and a further decrease in pulse duration, and typically ends with capture. It is followed by a long pause before the search phase is resumed. Notice that the entire terminal phase occurs in about the same duration as an interpulse interval of the search phase.

EL Sounds / The sounds used by bats in EL are perplexingly diverse. A common feature of nearly all of them is that during at least a part of the call the frequency descends rapidly over a wide range. This property seems to be of fundamental significance for the echo-processing mechanism. In some species, such as *P. psilotus* (see Figure 23–4), *Myotis*

23–5 Variation in calls of the bat *Pteronotus psilotis* during pursuit of a fruit fly. *Above:* Pulse duration and interpulse interval are plotted against time (msec) and distance (mm) to the end of the pursuit. Note that time reads from right to left. *Below:* Similar plot for pulse-echo overlap. The arrow shows the start of the terminal phase. During the approach phase, the overlap for each pulse is shown; during the terminal phase, there were too many calls for each to be plotted separately, so only representative values are shown. (From Novick 1965.)

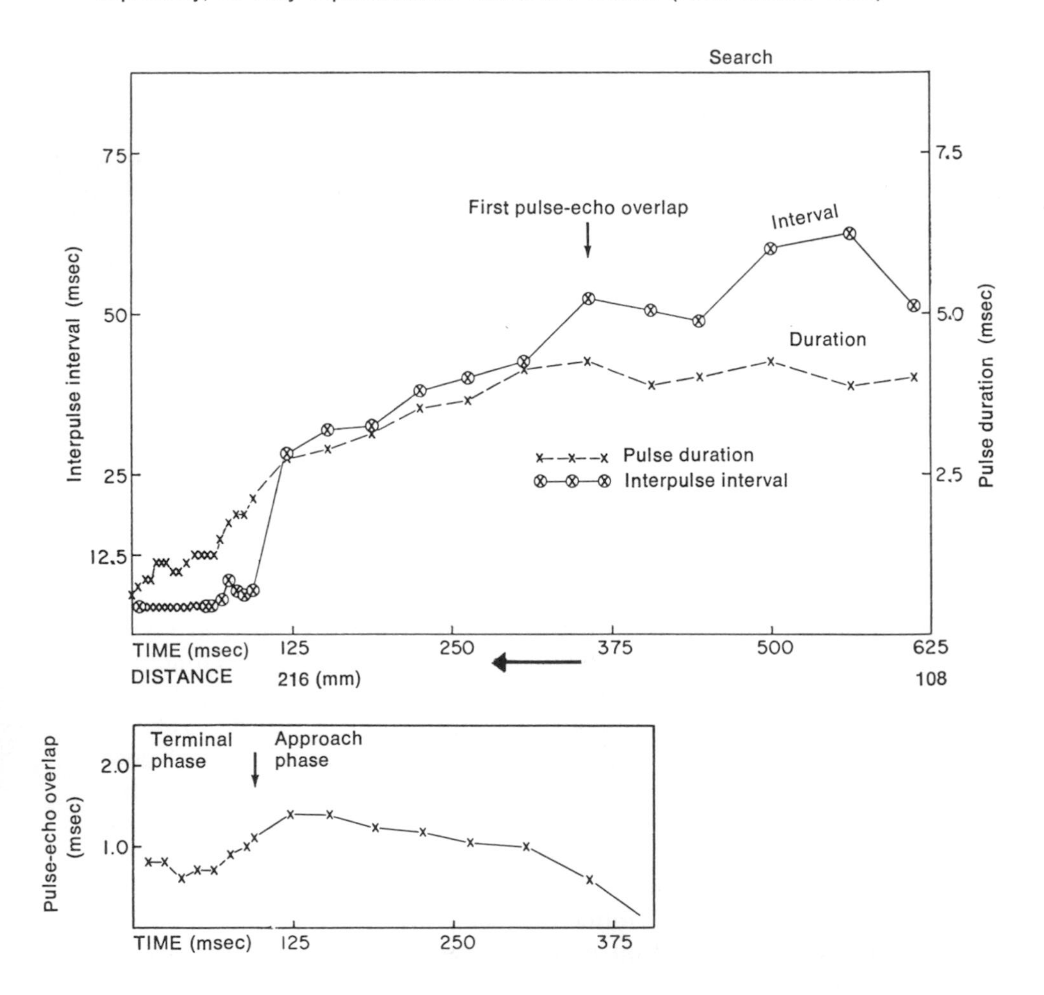

lucifugus, and *Eptesicus fuscus*, the entire call consists of a rapidly downslurred note. These are referred to as *FM calls* in reference to their *frequency modulation*. As a group they are characterized also by being relatively short, less than 15 msec and usually less than 5 msec. A minority of species has calls that are characterized by a long (15–65 msec) tone of constant frequency (CF) followed often by a short FM sweep. These are referred to as *CF/FM calls*. The mechanisms of processing echoes from CF and FM sounds are probably quite different.

Echo Processing / In EL orientation the mechanism by which the echo is compared with the pulse that produced it is the crux of the problem. At least three modes are suggested by behavioral data: (1) pulse-echo overlap exclusion, (2) constant pulse-echo overlap, and (3) variable pulse-echo overlap.

1) In all Vespertillionidae studied (*Myotis, Eptesicus, Lasiurus,* etc.), the pulses are short in the first place, and overlap of the echo with the end of the outgoing pulse is further avoided by shortening pulse duration as the target is approached (Novick 1971).

2) In *Pteronotus psilotis* and *P. davyi* pulses are shortened about when the approach phase begins, but the amount of shortening is adjusted not to avoid pulse-echo overlap but to maintain it at a constant value, about 1.0 msec in *P. psilotis*, as shown in Figure 23–5*B* (Novick 1963, 1965). Could the first pulse-echo overlap serve as a cue for the bat that an object was within a range worth pursuing, as Novick (1971) suggested? Or is the explanation of constant pulse-echo overlap that a greater amount of overlap is detrimental to signal processing (Novick, ibid.)?

3) In a few species, such as *Pteronotus parnellii* (Novick and Vaisnys 1969) and *Rhinolophus* spp. (Schnitzler 1970), long CF pulses are emitted, and pulse-echo overlap is extensive. In these species a Doppler shift in the echo (altered frequency due to target movement) would be especially easy for the bat to notice. That *Rhinolophus ferrumequinum* uses Doppler-shift information is strongly suggested by its pattern of hearing sensitivity at different frequencies (Figure 23–6) and by its behavior. This species adjusts its call so that the frequency of the echo is 83 *kHz* (Neuweiler 1970). Consequently, the echo (for an object being approached) falls in just that narrow part of the hearing range which is the most sensitive, while the call falls in the sharp peak of insensitivity. Another CF/FM bat, *P. parnellii,* has an even more finely tuned region of maximum sensitivity, just above the frequency of the CF pulse where the Doppler-shifted echo would be (Pollack et al. 1972). Doppler-shift information is of no use for determining target range, but it provides immediate information about relative target velocity and should be useful for quick and precise interceptions of rapidly flying insect prey.

The study of EL by bats will probably be strongly influenced by the mathematical theories developed in connection with electronic sonar systems. In such systems there is a trade-off between information on target distance and target velocity. The emitted pulse can be designed

23–6 Auditory sensitivity curves for two species of bats. *Rhinolophus* is sharply tuned to a narrow range of frequencies. By adjusting the pitch of the emitted call, *Rhinolophus* can place the Doppler-shifted echo in the most sensitive part of its hearing range while the emitted call falls in the adjacent high-threshold area. *Phyllostomus,* whose call and echo do not overlap and which therefore cannot use Doppler-shift information as easily, does not adjust the pitch of its call. This species difference in auditory sensitivity curves reflects the adaptation of *Rhinolophus* for analyzing the pulse-echo overlap. (From Simmons 1971, based on data of Neuweiler and Grinnell.)

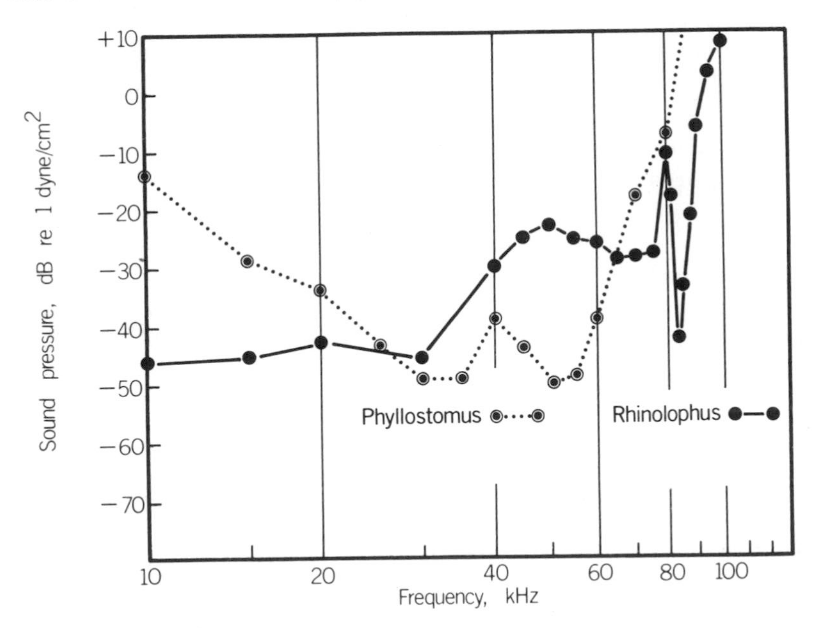

to increase information in the echo about distance at the expense of information about velocity, and vice versa, but both cannot be maximized simultaneously. CF pulses maximize velocity information in the echo, and FM pulses maximize distance information. If an artificial pulse is designed so as to maximize ability to detect a target and its distance regardless of its velocity, the result is similar to the cruising pulse of *Myotis lucifugus* (Altes and Titlebaum 1970). Simmons (1971) has shown that other vespertillionids (*Eptesicus, Phyllostomus*), which use FM pulses much like those of *M. lucifugus*, exceed *Rhinolophus* (CF/FM pulses) in ability to estimate target distance precisely; and he has compared the bats' EL systems with an ideal sonar system.

Adaptations / The pervasive effects of the adaptation of bats to EL are indicated in many aspects of their morphology and physiology. The sounds are produced by a highly specialized larynx. They are emitted through specialized "horns" formed by the nostrils, as in *Rhinolophus* and the leaf-nosed bats (Phyllostomatidae), or by the lips, as in *Pteronotus*. Beams of ultrasound may be produced by these horns. The CF/FM bats, *Rhinolophus* and *P. parnellii*, move their ears in synchrony with pulse emission.

Since the sound of the loud pulse must be compared with its relatively faint echo, mechanisms to reduce the disparity in loudness have evolved. The inner-ear muscles reduce the sensitivity of the ear during each pulse (Henson 1965, 1966), and neural mechanisms reduce the electrical response to an outgoing pulse at higher levels in the brain (Suga and Schlegel 1972; and see Chapter 25 below under "Attention"). To detect echoes, the ears are large in certain species, especially those that feed on terrestrial insects. A fingerlike projection in front of the ear, the tragus, apparently sharpens the ear's directional sensitivity (Grinnell and Grinnell 1965). The cochlea of the bat ear is large, and the auditory portions of the brain stem are relatively larger than in other animals of comparable size. Details of the neurophysiology of hearing and EL in bats have been reviewed by Erulkar (1972), who noted numerous specializations related to EL.

ECHOLOCATION BY PORPOISES

EL has been proven to exist in porpoises, and is suspected in many other cetaceans (whales, dolphins). Kellogg (1961) has provided a readable introduction to EL in porpoises. For the finer points and more recent studies, the reader is referred to the reviews of Norris (1969) and Erulkar (1972). The sounds produced by dolphins during EL are much shorter than those of bats, being on the order of 0.1 msec. They are also higher-pitched than bat sounds, running to over 200 *kHz*. Because the velocity of sound in water is about five times greater than in air, the requirements on the CNS for temporal resolution of echoes from pulses are more stringent. Analysis is aided by the ultrashort pulses used by porpoises. The EL sounds of porpoises are produced not by vocal cords, which are totally lacking, but by structures along the nasal air passage. The sounds apparently are sharply beamed by the structure of the head, especially in species that are blind or that frequent habitats where vision can rarely be used. External ears are lacking, but the auditory system is highly developed, and the auditory portions of the midbrain, neocortex, and other brain areas are greatly enlarged.

CONCLUSIONS

The problem of orienting the body in space in relation to external stimuli is of basic importance for animals. Solutions range from the use of simple receptors for simple types of orientation, as in the taxes and kineses of many invertebrates, to the employment of sophisticated receptors as parts of unbelievably intricate sensory analyzing systems. Of the latter, orientation by echolocation was chosen for special attention.

Echolocation is used by a variety of species, including oilbirds, swiflets, shrews, and tenrecs, but it is most highly specialized in bats of the suborder Microchiroptera, and in porpoises. The sound pulses used by bats for echolocation are perplexingly diverse in frequency, duration, and loudness. It is suspected that information from the Doppler shift in the echo is used to provide information about target velocity in the few species that produce long, constant-frequency (CF) pulses. The short, rapidly descending (FM) calls used by most Microchiroptera seem well suited for providing precise information about target distance.

Porpoises live in a medium in which sound travels five times as fast as in air. Their echolocating clicks (about 0.1 msec) are considerably shorter than those of bats (mostly > 1.0 msec). The frequency of these pulses (to 150 *kHz*) is also well above that of bats (up to 100 *kHz*). The ability of porpoises to negotiate obstacle courses, detect small objects, and make fine discriminations using echolocation (with vision excluded) is as impressive as the abilities of bats. A high degree of morphological and physiological specialization underlies the echolocating abilities of both bats and porpoises. Studies of the physiology and morphology of these animals have shown how species differences in orientation abilities may mediate species differences in behavior.

24 Long-distance Migrations

THROUGHOUT recorded history man must have wondered about the migrations of animals. In earlier times the arrival of migratory animals signaled periods of abundant food, while now migration challenges the ability of modern man to find rational explanations for the natural world—or what is left of it. Although in certain parts of Africa and Asia, such as the steppes of Afghanistan, man himself still exists as a long-distance migrant species, the ability of individuals to travel long-distance routes regularly and accurately is much better developed among other species. Regular or irregular seasonal migrations are a feature of the life histories of many groups of animals, including African desert locusts, butterflies, dragonflies, fishes, amphibians, reptiles, mammals, and birds. Perhaps the most spectacular are the migrations of birds, although eels, salmon, sea turtles, and certain bats (e.g., *Lasiurus* spp.) are nearly as impressive. Since birds are relatively easily studied and have received by far the most attention, this chapter will be concerned almost entirely with them. The following reviews are useful for further information: Dorst (1962), Matthews (1968), Griffin (1969), Orr (1971), Jarman (1972), Emlen (1974), and Keeton (1974).

THE PHENOMENA OF MIGRATION

Before designing sophisticated laboratory experiments on behavior, it is generally wise to survey the general characteristics of the primary phenomena to be explained. This enables the experimenter to ask meaningful questions and provides hints as to how to go about answering them. Consequently, a brief review of the general characteristics of bird migration is in order.

Distance / Estimates of the minimum distances an individual must travel between its summer and winter ranges may be obtained by large-scale marking of individuals in one region, followed by chance recoveries of marked individuals in another region. Probably the champion long-distance migrant is the arctic tern (Figure 24–1). This species breeds in the Arctic and winters among the pack ice in the Antarctic oceans (Salomonsen 1967). Three individuals given metal leg bands on the arctic coast of Russia and in Sweden were recovered in Australia (Gwynn 1968). Considering that these terns probably flew over sea rather than land, the minimum distance the Russian bird could have flown between these two points is about 9,000 miles. This is only an extreme example from a large number of avian species that regularly cross the equator twice yearly on long-distance migrations.

Accuracy / The regularity and precision of migration vary greatly among species. Among small passerine birds, such as the European pied flycatcher, *Ficedula hypoleuca* (Figure 24–2), the precision of individuals in returning to their breeding grounds in Germany after a winter in trans-Saharan Africa was studied by Berndt and Sternberg

24–1 Fall migration of the arctic tern, and eastward drift in the strong west winds found in the winter range. Solid circles represent selected records of the species outside its breeding grounds. A few arctic terns winter near the coasts of southern South America and southern Africa, but most utilize the abundant macroplankton or "krill" found in the oceanic upwellings around Antarctica. Many individuals completely circumnavigate the antarctic continent in their first year or two of life before starting north for the breeding grounds. (After Salomonsen 1967.)

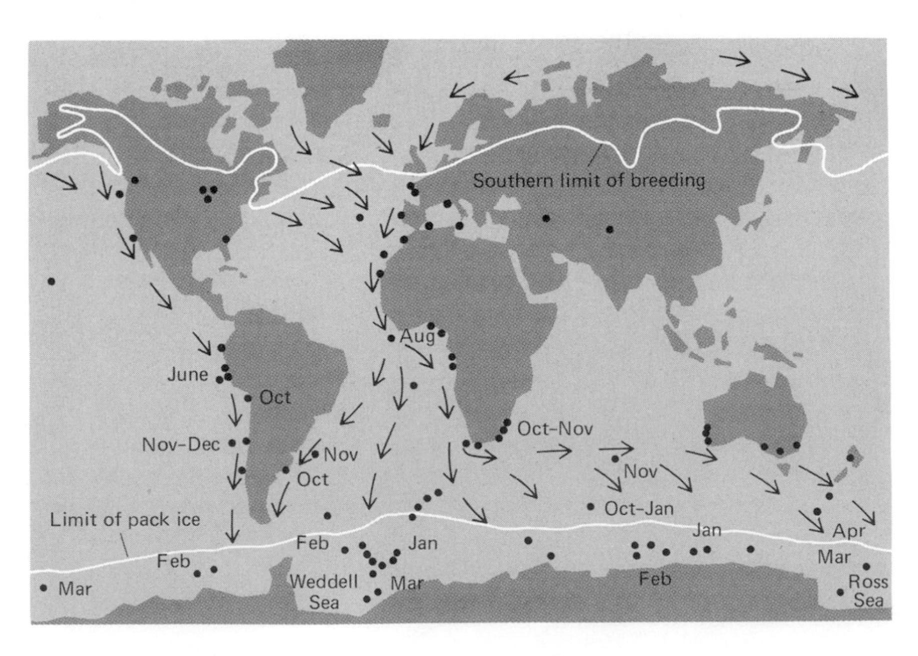

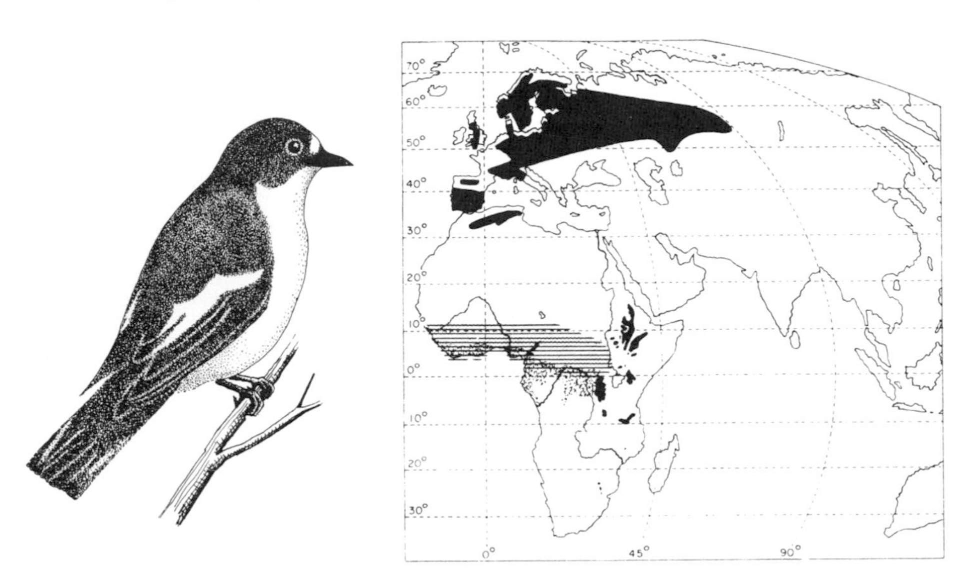

24–2 The pied flycatcher, a common European species used in research on bird migration. (Map from Moreau 1972.)

(1969). Adults and even individuals less than a year old typically returned to the region where they bred or were hatched. Their precision in doing so is indicated in Table 24–1. Of 829 females hatched the preceding year and recovered in 25 study areas in various parts of Germany, about half were found within 1 km of their hatching site, and the average distance was 4.5 km. The mean movement distance of adult males was only 140 m. This phenomenon is referred to as *Ortstreue*, from the German for *faithfulness to a place*.

Length of Nonstop Flights / A general picture of the speed of migration can be obtained by mapping arrival dates for migrants in the spring. Such maps have been prepared for several species (Lincoln 1950; Dorst 1962). Arrival dates of blackpoll warblers (*Dendroica striata*) are shown in Figure 24–3. More precise information about the exact distance covered and the time involved for a population of birds has been obtained for the eastern population of the snow goose (*Chen coerulescens*), which is characterized by a high frequency of the blue morph (see Figure 27–6; Cooch 1955). The main population of this morph breeds in the eastern Canadian Arctic and gathers several hundred miles to the south at James Bay, Canada, at a stopover on its way to the wintering grounds in the coastal marshes of Louisiana (Figure 24–4). By comparing notes on the time of departure from James Bay, arrival

TABLE 24–1 Distance from hatching place to first breeding place for yearling pied flycatchers and distance from breeding place one year to the next for adults. Notice the difference between yearling and adult males and between adult males and females. These data illustrate the phenomenon of *Ortstreue,* or *place faithfulness.* (From Berndt and Sternberg 1969.)

Distance from birthplace or preceding breeding-place	Number of recoveries				Cumulative total (%)			
	One-year-old		Adult		One-year-old		Adult	
	♂♂	♀♀	♂♂	♀♀	♂♂	♀♀	♂♂	♀♀
0 m	2	3	7	55	2.0	0.4	5.4	4.1
1– 100 m	8	50	57	411	9.9	6.4	49.2	34.8
100– 200 m	9	82	35	337	18.8	16.3	76.2	60.0
200– 300 m	11	96	20	132	29.7	27.9	91.5	69.8
300– 400 m	5	53	7	54	34.7	34.3	96.9	73.9
400– 500 m	4	31	1	22	38.6	38.0	97.7	75.5
500– 600 m	3	34	2	18	41.6	42.1	99.2	76.8
600– 700 m	5	22	1	18	46.6	44.8	100.0	78.2
700– 800 m	—	15	—	13	46.6	46.6		79.2
800– 900 m	—	16	—	8	46.6	48.5		79.8
900–1,000 m	—	12	—	3	46.6	49.9		80.0
1– 2 km	25	99	—	69	71.3	61.9		85.1
2– 3 km	12	94	—	59	83.2	73.2		89.5
3– 4 km	4	56	—	26	87.1	80.0		91.5
4– 5 km	—	10	—	10	87.1	81.2		92.2
5– 6 km	4	23	—	13	91.1	84.0		93.2
6– 7 km	1	8	—	3	92.1	84.9		93.4
7– 8 km	—	4	—	1	92.1	85.4		93.5
8– 9 km	1	1	—	5	93.1	85.5		93.9
9– 10 km	—	7	—	4	93.1	86.4		94.2
10– 20 km	—	56	—	49	93.1	93.1		97.8
20– 30 km	5	34	—	12	98.0	97.2		98.7
30– 40 km	—	6	—	5	98.0	98.0		99.1
40– 50 km	2	5	—	2	100.0	98.6		99.3
50– 60 km	—	7	—	4		99.4		99.6
60– 70 km	—	3	—	3		99.8		99.8
70– 80 km	—	2	—	2		100.0		99.9
80– 90 km	—	—	—	—				99.9
90– 100 km	—	—	—	—				99.9
143– 144 km	—	—	—	1				100.0
Totals	*101*	*829*	*130*	*1339*				

in Louisiana, and sightings by commercial airline pilots along the way (who had been alerted because of a collision between a plane and a goose), the conclusion was reached that the entire flight, an air distance of over 1,700 miles, was made in one nonstop trip of about 60 hours at an altitude of 3,000 to 8,000 feet and average speed of about 30 mph.

With the use of radiotransmitters that were attached to birds captured and released in the middle of their migrations Cochrane, Montgomery, and Graber (1967) were able to follow individuals of various thrush species with airplanes or trucks. Such individuals were tracked on northward flights of up to 350 miles in a single night.

24–3 Spring migration of the blackpoll warbler. This species winters in central South America and breeds in stunted trees in subarctic North America. The isochronal lines indicate times of arrival of migrants in spring. The average speed for the vanguard of the migrants increases from about 30 miles per day in early May to 200 miles per day in late May. (From Lincoln 1950.)

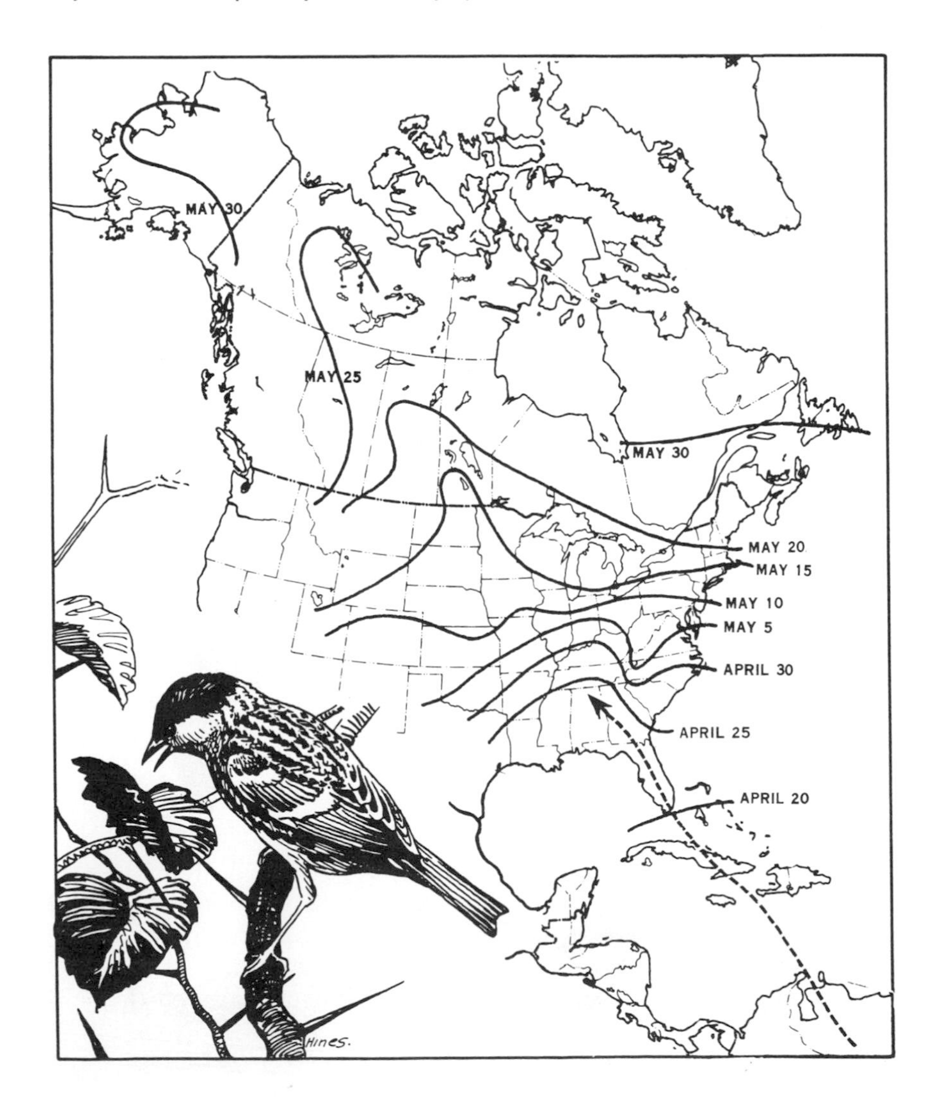

24–4 Fall migration routes of some populations of blue and snow geese. The trip from the stopover at James Bay to the coastal marshes of Louisiana in 1952 was apparently covered in one nonstop flight lasting about 60 hours. See also Figure 27–6. (Map from Johnsgard 1973*b;* sketches by T. M. Shortt from Kortwright 1943.)

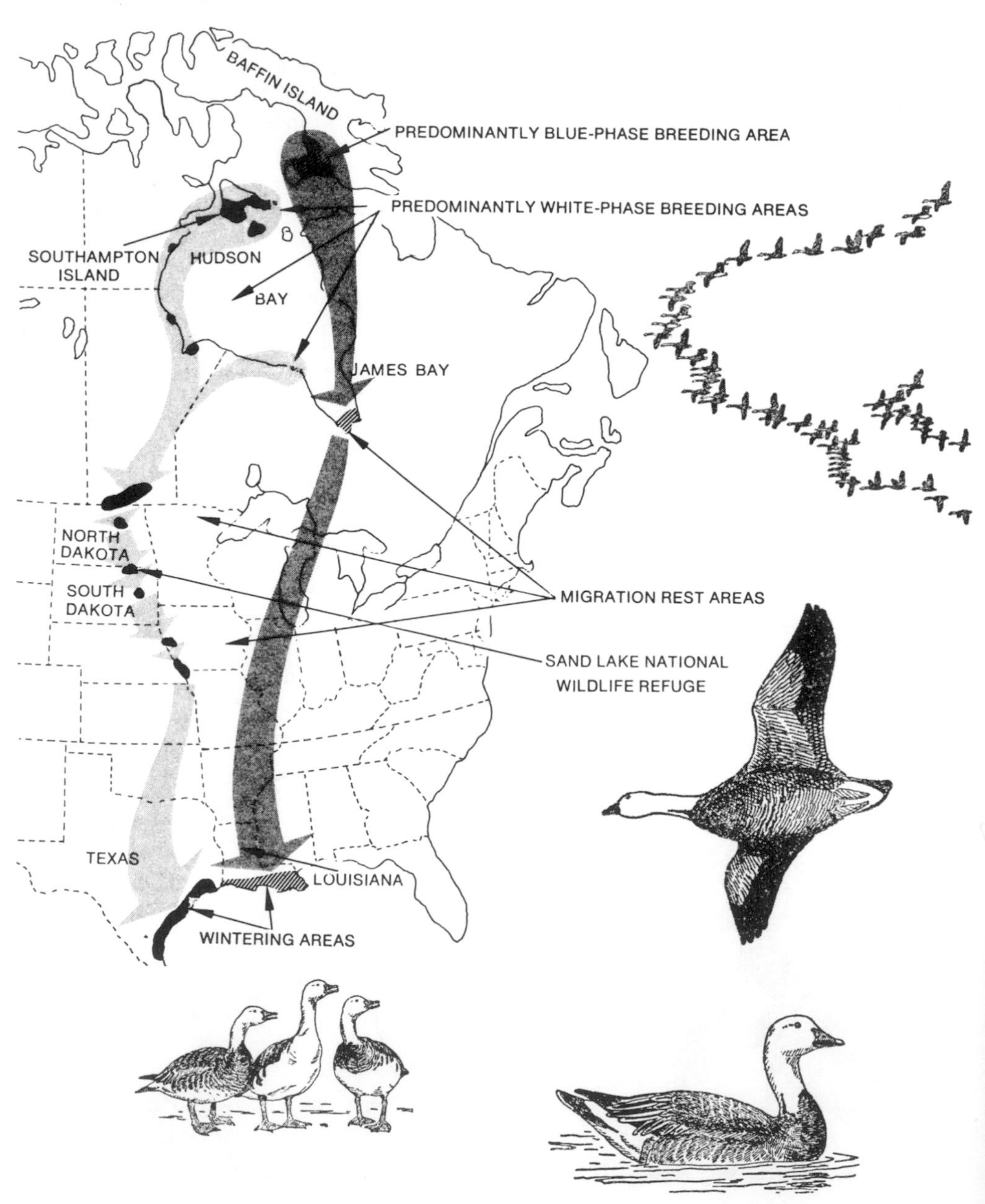

Natural Barriers and Pathways / Many species of small birds regularly cross large expanses of inhospitable habitat on their migrations. For example, many European species winter in Africa and arrive there by nocturnal broad-front migration across the Mediterranean Sea and the Sahara Desert (Moreau 1972). In North America, many species of small land birds regularly cross the Gulf of Mexico and Caribbean Sea on the way to and from wintering grounds in South America (e.g., blackpoll warbler—Figure 24–3). Even greater flights over open ocean are found in the species that migrate from the Asian mainland, Australia, or New Zealand to small islands in the South Pacific Ocean. The Alaskan population of the golden plover is thought to fly nonstop over 2,000 miles of ocean to Hawaii. Many apparent barriers are regularly crossed by species that seem least fit for crossing them.

On the other hand, geographic features may also serve to guide migrations and may concentrate certain species in large numbers at certain places. Small birds crossing the Swiss Alps are concentrated at certain mountain passes. Coastlines are followed by many species of waterbirds. Hawks may be conveniently observed on migration at places where a large body of water is situated so as to concentrate migrants. Lake Ontario, for example, acts as a barrier to hawks coming north in the spring, so that large numbers may be seen along the shore flying east and ultimately around the east end of the lake, northward into Canada (Haugh and Cade 1966). Barriers that concentrate and guide migration are known as *leading lines.*

Time of Day / Although the species that migrate in the daytime, such as hawks and waterfowl, are most conspicuous and easily studied, many more species migrate only at night. Many species probably migrate in long nonstop flights that include both night and day.

Altitude / Reliable information on the height at which migrant birds fly has come from radar observations. Large birds have been observed up to 20,000 feet on overseas flights (Nisbet 1963). In a radar study of nocturnal passerine migration, Able (1970) observed birds up to 11,000 feet above the ground but found 75 percent of them below 3,000 feet. Bellrose (1971) found most nocturnal migrants at an altitude of about 1,000 feet. Altitudes averaged higher at night than in the day, higher in spring than fall, higher before midnight than after, and higher under overcast than clear skies (Eastwood and Rider 1965). At night, birds are easily heard migrating overhead during spring and fall in eastern North America. On arrival from a flight across the Gulf of Mexico in spring, most small birds, in flocks of about 20, were seen to dive straight down to the ground (Gauthreaux 1972).

Weather / In general, small birds wait for a favorable wind before beginning a migration flight and fly with the wind (Gauthreaux and Able 1970). There is disagreement and debate over the extent and occurrence of correction for wind drift. Stopovers between flights allow time to build up fat deposits to fuel long flights. So in the northern hemisphere a north wind brings a wave of migrants in the fall; and a south wind has a similar effect in the spring. Temperature may initiate movement in autumn by killing insect food or by freezing the feeding areas of waterfowl, but temperature and humidity are now agreed to be secondary in importance to wind speed and direction as factors initiating flight on a particular day. The role of barometric pressure changes, to which birds are highly sensitive (Kreithen and Keeton 1974), is uncertain. Reverse or wrong-way migrations, such as southward movements of robins in spring, correlate well with wind direction. The influence of weather on migration has been of great interest to bird watchers (e.g., Nisbet and Drury 1968; Lowery and Newman 1966; Richardson 1971, 1972).

INTERACTION OF LEARNING AND INHERITANCE IN MIGRATION

Although migration is typically species-specific and has evolved in response to ecological pressures, the ability and predisposition to migrate according to a species-specific pattern are inextricably related to the learning experiences of the individual. Experiments by Perdeck (1958, 1964, 1967) on European starlings in nature illustrate what is probably a common pattern.

The design of Perdeck's first experiment is shown in Figure 24–5. Starlings in northern Europe migrate westward in the fall. The starlings passing through Holland in the fall originate to the east, in the Baltic countries and Russia, as established by banding of migrants and recovery of banded birds on their breeding grounds (Figure 24–6). They winter on both sides of the English Channel. By transporting to Switzerland, and releasing there, starlings captured during migration in Holland, it could be tested whether the birds would correct for their displacement by a change of course or would continue migrating along a course parallel to their original one. Results for combined releases of adults and of young on their first migration are shown in Figure 24–7*A*. The adults showed *true goal orientation;* their recoveries after displacement were along a line that would bring them to their normal winter range along the Channel. The young showed *one-direction orientation;* their recoveries after displacement were along a line in the same direction as, and parallel to, their original course. The young consequently ended their first migration in an area where others of their breeding population were not normally found. The directional differences between the headings of adults and young are summarized in Figure 24–7*B*. In subsequent years the birds that had been displaced earlier continued to return to their new winter range in Spain and southern France.

The development of directional preferences in young starlings has not been studied, but the implication is strong that inheritance is influential. An alternative hypothesis is that the young learn the appropriate direction from experienced adults before or early in migration. The physiological basis for such a directional preference is also completely unknown.

In later experiments (Perdeck 1964, 1967), starlings were displaced from Holland to Spain in the middle of their migration. The intention was to determine whether the birds would continue their migration even though they were released in a suitable wintering area. Again the results for adults and young differed. Adults were subsequently recovered on a northerly course toward their normal wintering area. Young were recovered on a southwesterly course. The results indicate that young starlings in mid-migration will continue migration even if transported to a suitable wintering area.

These experiments, together with others, suggest that migration in starlings follows the following scheme: (1) Young on their first fall

24–5 Design of Perdeck's first displacement experiment on the navigational abilities of migrating starlings in autumn. Birds captured in mid-migration were displaced laterally from their normal migratory route. By this means, two hypotheses could be tested. Would the starlings continue migrating in the same direction, thus showing one-direction orientation; or would they alter their course to correct for their displacement, thus showing true goal-orientation? (From Perdeck 1958.)

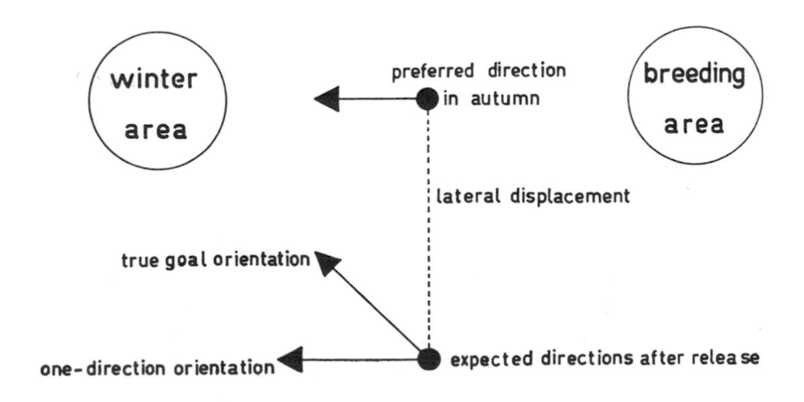

(On facing page)

24–6 Summer and winter ranges of the population of starlings used in Perdeck's (1958) displacement experiments. The distribution of recoveries of starlings banded in Holland in October and November is shown according to date of banding, which reflects the time of passage through Holland. The concentration of band recoveries is shown by the density of hatching. Small dots indicate recoveries outside the normal area; the accompanying numbers show the month of recovery. Note the correlation between date of passage through the Netherlands and the summer and winter ranges. Remember that starlings breed and winter over nearly all the area on the map as well as many other parts of the world; these data are only for Dutch migrants. (From Perdeck 1958.)

migration fly in a fixed direction for a fixed period of time; this normally brings them to the regular wintering area for their population. (2) The young learn the characteristics of the wintering area. (3) On subsequent fall migrations, their previous experiences with the wintering area allow them to navigate to it even after being displaced off course. For further studies of birds displaced while on migration see Emlen (1974).

The results of many experiments on a variety of species from sparrows to albatrosses have demonstrated that birds have the ability to return from unfamiliar parts of the world over impressive distances to learned wintering or breeding areas. The longest distance covered in a homing experiment was 4,120 miles, covered in 32 days by an albatross in the Pacific Ocean (Kenyon and Rice 1958). The critical importance of learning of the breeding area has been shown in experiments in which collared flycatchers (*Ficedula albicollis*) and other species taken from one area and reared in another returned only to the area where they had been reared and not to their original population (Löhrl 1959).

Experimental evidence supporting the hypothesis that inheritance plays a role in setting migratory directional preference has been gathered in experiments on white storks, *Ciconia ciconia* (Schüz 1949, 1950). Storks breeding in western Europe fly to their wintering grounds in Africa by going around the west end of the Mediterranean Sea. Those from eastern Europe skirt the eastern end of the Mediterranean. There is, therefore, a "migration divide" in central Europe, as shown in Figure 24–8 (Schüz 1963, 1971). The developmental basis for these population differences was tested by releasing naive young storks raised in captivity in eastern Europe at a locality in western Europe after all members of the local population had departed. Instead of taking the southwesterly direction appropriate for the population where they were released, they took a southeastern course, appropriate for the population from which the eggs had come. The results support the hypothesis that the population differences in preferred direction of

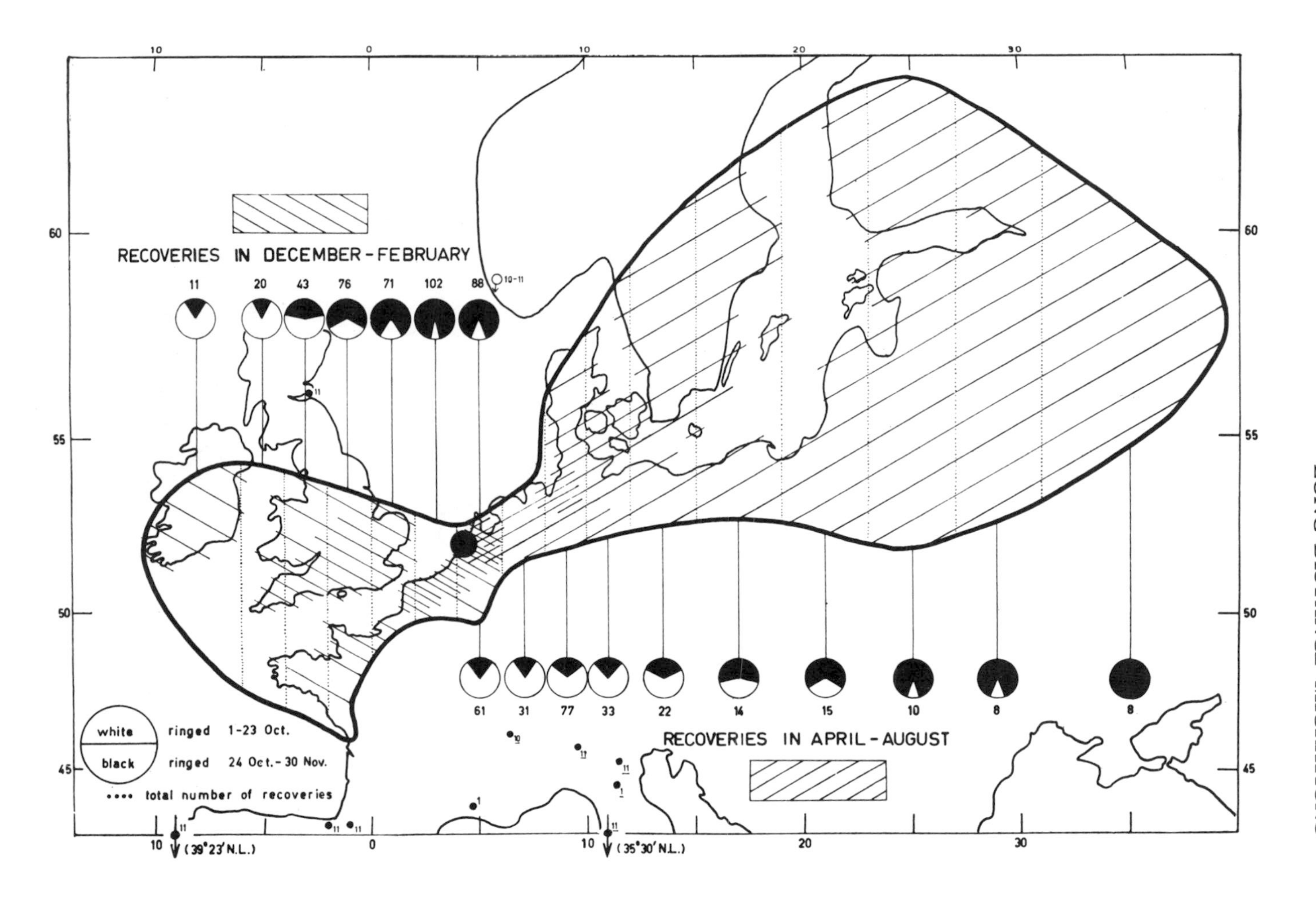
RECOVERIES IN DECEMBER - FEBRUARY
11
20
43
76
71
102
88
10-11
61
31
77
33
22
14
15
10
8
8
RECOVERIES IN APRIL - AUGUST
white ringed 1-23 Oct.
black ringed 24 Oct.- 30 Nov.
.... total number of recoveries
(39°23'N.L.)
(35°30'N.L.)
60
55
50
45
10
0
10
20
30

24–7A Locations of recoveries of displaced starlings in the same autumn as the release. Starlings were captured while on fall migration at The Hague, Netherlands, transported by air to Switzerland, and released at the airports in Zürich (Z), Basel (B), and Geneva (G). Adults and young were released simultaneously, but tended to take separate courses.

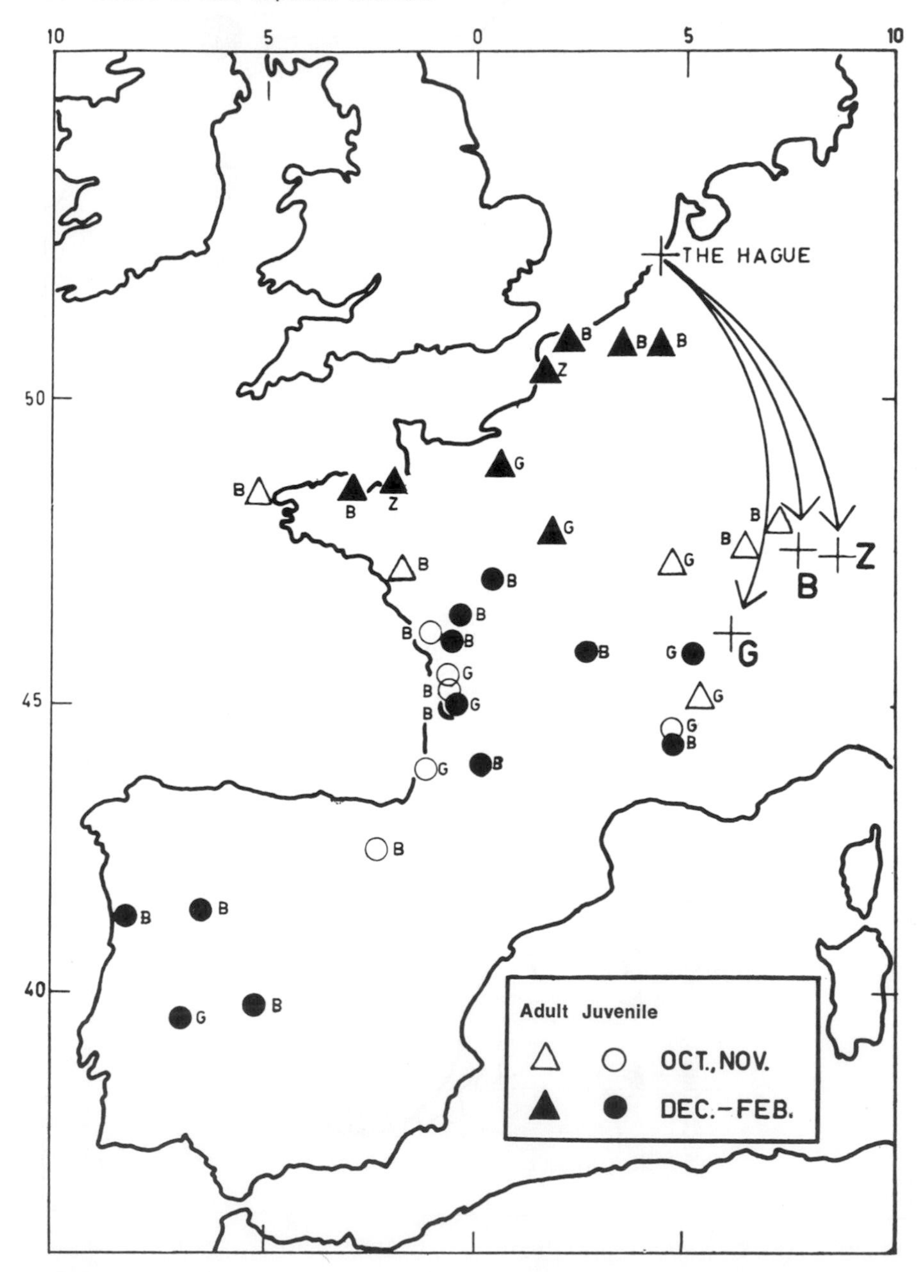

migration of naive young storks are genetically determined. Further experiments would seem desirable, but since the storks of western Europe are declining precariously in numbers it is unlikely that they can be carried out.

24–7B Recoveries in Figure 24–7A plotted as if from one release point, and combined with recoveries of juveniles released separately without adults. Note the clear separation of adults and juveniles. (From Perdeck 1958.)

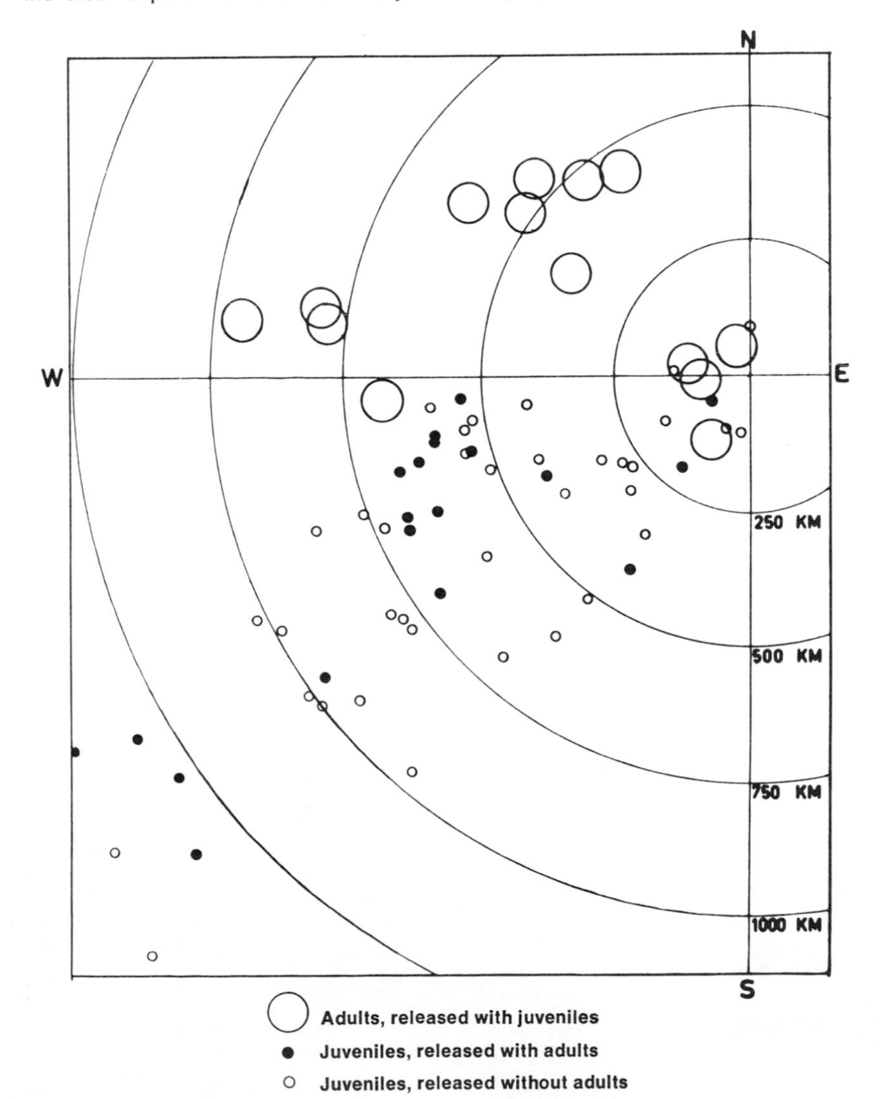

24–8 **A.** Migration divide of European white storks. The four rectangles in Germany show the zone of overlap of eastward- and westward-migrating populations. The few hundred storks that nest in France and Germany west of the line migrate southwestward in fall around the western end of the Mediterranean Sea. Storks of the much larger population nesting east of the line migrate southeastward and around the eastern Mediterranean. The circles designate points of dense concentration of migrating storks. The European population winters mainly in western Africa. Another migration divide in eastern Turkey separates populations wintering in Africa from those wintering in Asia. A migration divide in North Africa separates breeding populations migrating in the indicated directions. (From Schüz 1971.)
B. White stork with young at their nest in Denmark. (Photo by Henning Skov.)

A.

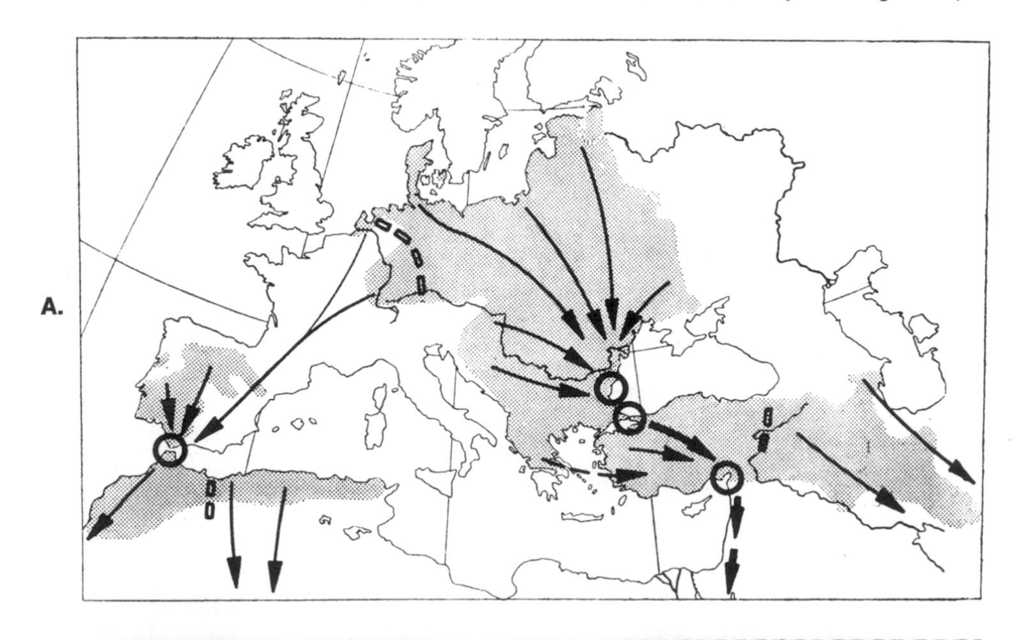

B.

NAVIGATIONAL CUES

Sun / Despite much theorizing, the sensory basis of the navigational abilities of birds did not begin to be put on a firm experimental basis until the pioneering experiments of Kramer just after World War II. Using a single, tame starling, Kramer (1951) first showed that the direc-

24–9 Experimental proof of a bird's use of the sun as a "compass." A hand-reared starling was placed in a circular cage, and the directions of its migratory flutterings, or *Zugunruhe,* were recorded by an observer viewing the bird from beneath. When the starling could see the sunny sky, the direction of its movements in spring was mainly to the northwest (**A,** *left*) as shown by the dots and heavy arrow. Mirrors placed so as to change the bird's view of the sun by 90° caused appropriate shifts in the direction of *Zugunruhe* (**A,** *center and right*). When the sun was obscured by dense clouds, orientation became random (**B,** *left).* When the sky cleared, orientation reappeared (**B,** *right*). The bird could see the sky but no landmarks outside the test apparatus. The cage and surrounding screen were rotated randomly to prevent the bird from using them as landmarks. (From Kramer 1951.)

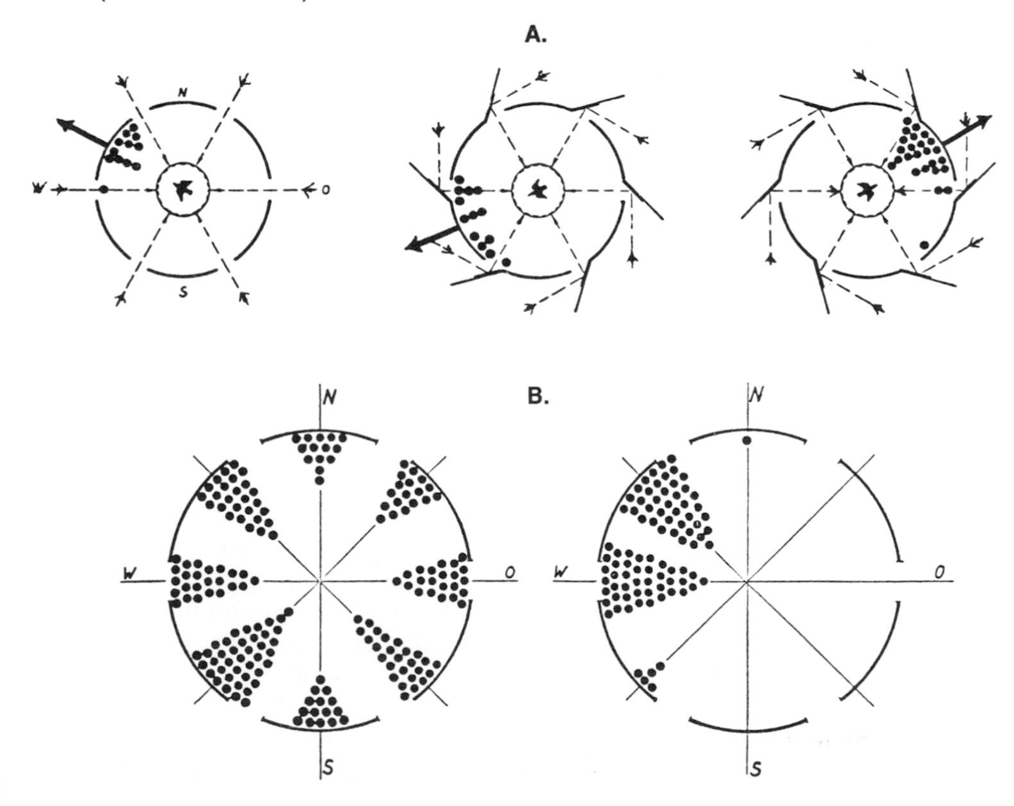

tion of *Zugunruhe* (migratory restlessness) could be changed by changing the apparent direction of the sun with mirrors (Figure 24–9*A*). The bird lost its ability to orient properly when the sun was obscured by clouds (Figure 24–9*B*). Later, Kramer (1952) showed that starlings could be trained to choose a given direction for a food reward in the absence of any cues except from the daytime sky. A view of the sky was essential for success and the chosen direction was thrown off to the predicted extent by the use of mirrors. Training experiments by Hamilton (1962) showed a similar capacity in ducklings. In experiments in which birds were released in unfamiliar terrain in daytime, it was reported for a variety of species that a view of the sun was necessary for homing and for one-direction orientation. These and other experiments have established that many species are capable of using the sun in navigation. Just how the sun is used remains a puzzle and has been the subject of various theories (see Matthews 1968). In view of the recent evidence implicating the earth's magnetic field, these theories will require critical reexamination.

A requirement for using the sun as a reference in navigation is an accurate knowledge of the time of day. It has been shown in numerous experiments (e.g., Hoffman 1960) that setting a bird's light-dark cycle ahead or behind (*clock-shifting;* see Chapter 25) causes a deflection in the bird's directional heading if the bird can see the sun (Figure 24–10). If the sun is not visible, clock-shifting may be ineffective (Matthews 1963; Keeton 1969).

Stars / Although some early observations by Kramer (1949) have been cited in the literature as suggesting that the stars might be used by nocturnal migrants to aid their navigation, it was entirely the experiments of Sauer (1957, 1961; Sauer and Sauer 1955, 1959, 1960) with hand-reared European warblers that really confirmed the existence of this mode of orientation (Matthews 1968; Marler and Hamilton 1966; Sauer 1972). Sauer observed that his warblers (*Sylvia borin, S. atricapilla, S. curruca*) oriented correctly only if the stars were visible. Using star patterns projected in a planetarium, he was able to manipulate the directional preferences of birds showing migratory restlessness. Later experiments on several other species also used the orientation of migratory restlessness of birds in registration cages of various designs to establish that star patterns were used for orientation both outdoors and in planetaria (references in Matthews 1968:41–42).

An ingeniously simple testing method (Emlen and Emlen 1966), involving a central ink pad surrounded by a circular blotter where the test bird leaves its footprints, has greatly facilitated testing of caged migrants (Figure 24–11). Using this method, Emlen (1967*a, b*) clearly

24–10 Effects of "clock shifts" on orientation. Mallard ducks (*Anas platyrhynchos*) were released individually under sunny skies (**A, B**) and starry skies without a moon (**C, D**). Members of this population of mallards typically fly northwest upon release, regardless of season or direction of home (*shown by arrow*). This particular type of one-direction orientation has been called nonsense orientation because its advantage to the birds is unknown. By controlling the photoperiod of the mallards, their "clocks" were shifted 6 hours ahead or behind local time. This caused a deviation of about 90° in the vanishing-point bearings of mallards released under the sun (A, B), but failed to affect the orientation of mallards released under the stars at night (C, D). Each white square represents the vanishing bearing of one control mallard (not clock-shifted); black squares represent clock-shifted mallards. The central fans show the mean deviations of the two groups from their respective medians. Vanishing bearings at night were obtained with the aid of a small light attached to each mallard on release. (From Matthews 1963.)

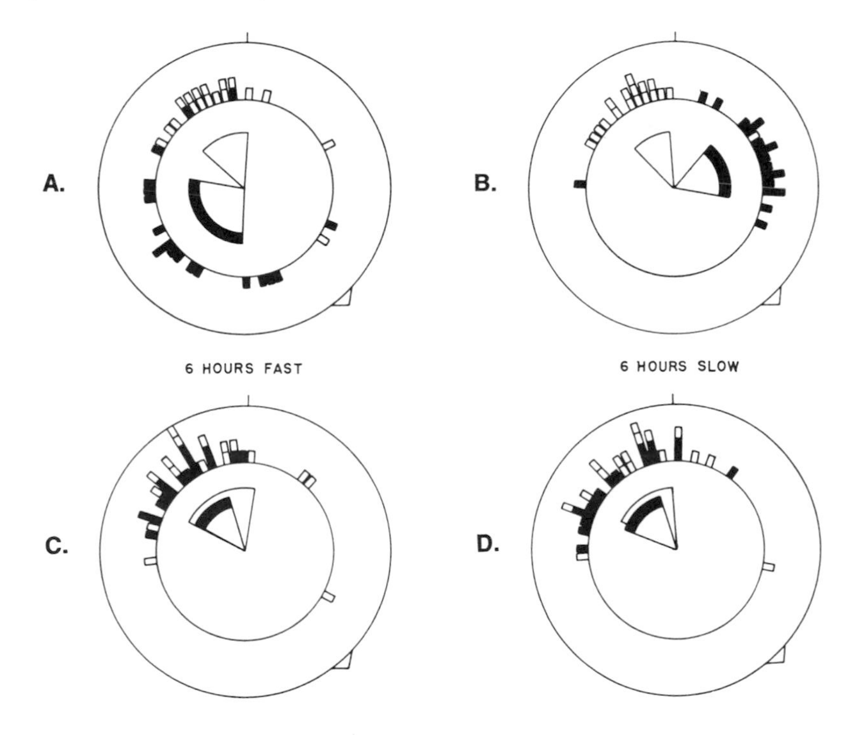

demonstrated an influence of the stars on nocturnal orientation in indigo buntings (*Passerina cyanea*). Figure 24–12 shows some of his results.

A significant finding for nocturnal migrants, such as the indigo bunting, was that clock-shifting them by advancing or retarding their light-dark cycle does not cause errors in orientation (see also Figure 24–10). This contrasts with the orientation of birds using the sun. The

24–11 The indigo bunting is a moderately long-distance, nocturnal migrant. Its summer and winter ranges are shown in **A.** Experiments on nocturnal orientation of *Zugunruhe* in buntings have used the "Emlen funnel," shown below with a bunting in it (**B**). During migratory restlessness, the birds jump or flutter upward and slide down the blotter, thus leaving their footprints in ink on the blotter with each attempt. For sample results using this method, see Figure 24–12. (A, from Emlen 1967*a*. B, from Emlen and Emlen 1966. Courtesy of American Ornithologists' Union.)

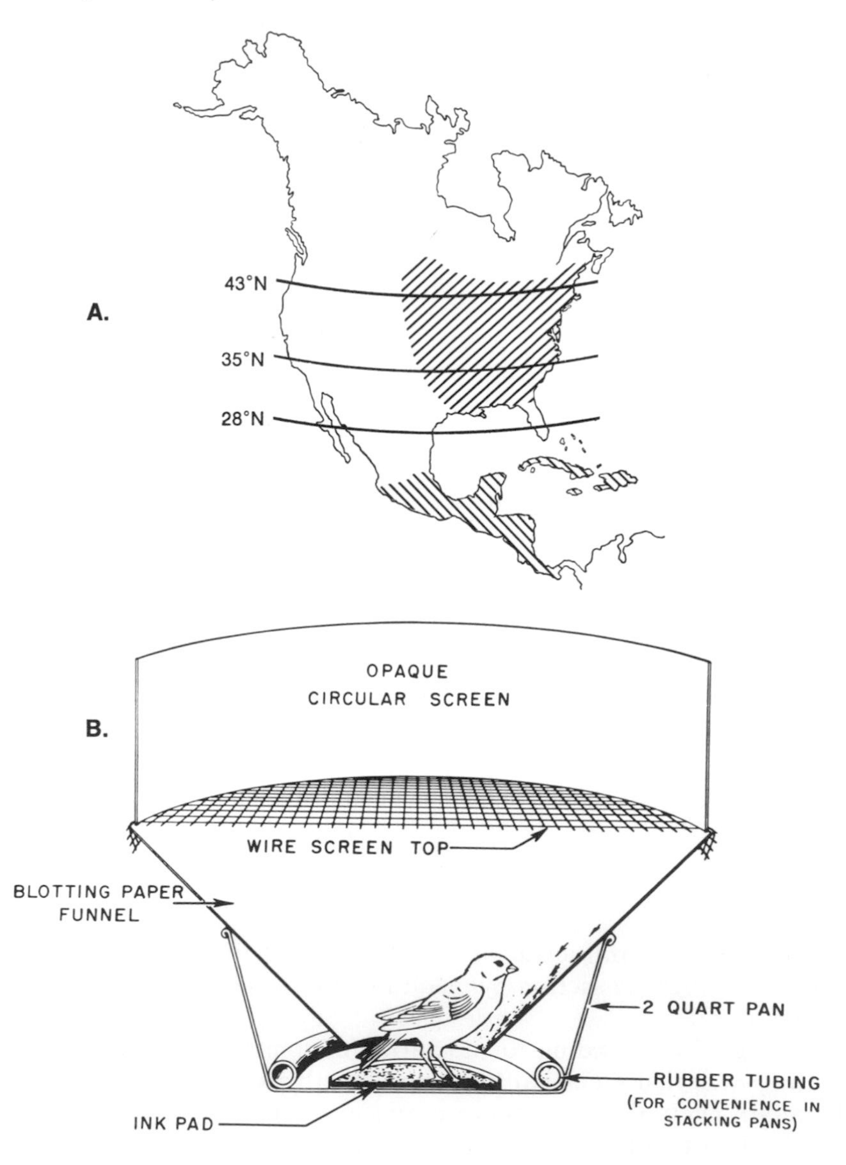

particular part of the sky used by individual indigo buntings seems to vary individually. Some individuals persisted in orienting correctly under each of the following conditions: (1) all southern hemisphere stars off, (2) Cassiopeia off, (3) Milky Way off, (4) Polaris (North Star) off, (5) Ursa Major (Big Dipper) off, (6) all stars within 35° of Polaris off (this eliminated orientation in all but one bird), and (7) all planets off. In buntings, no one star configuration, or constellation, seems to be essential for correct orientation. Rather, the birds seem to know most of the circumpolar (northern) night sky and orient even when relatively small parts of it are visible. Some of the experimental results are shown in Figure 24–13.

With a given star pattern buntings could go north or south, apparently depending on their hormonal state (Emlen 1969*a*). Sauer, however, reported that a spring sky led to northward orientation and a fall sky to the reverse. Other differences between species have also been observed, and the relative importance of a variety of mechanisms in different species has not yet been critically determined.

The moon seems to cause disorientation in many nocturnal experiments. There is no evidence for its use in navigation.

The developmental history of buntings is critical for species-typical behavior (Emlen 1969*b*, 1970*a*). Buntings raised indoors without ever having seen the night sky or a planetarium sky did not orient correctly (Figure 24–14*B*), while those who did orient correctly had been raised under "natural" or altered planetarium skies. Even those raised under an artificial sky of naturally arranged stars rotating around an unnatural axis oriented "correctly" if this star pattern was present (Figure 24–14*D*). A critical requirement was the presence of an axis of rotation of the night sky during development. These results suggest that indigo buntings must learn the appearance of the night sky in order to orient properly on migration. In contrast, Sauer (1957, 1971) reported that his hand-raised European warblers oriented correctly upon first exposure to the night sky even though they had never seen it before. The ontogeny of migratory orientation is still not well known. For example, the role of magnetic fields in the development of star orientation has not been investigated.

Nonvisual Cues: Magnetic Fields / The idea that animals might be able to navigate using information from the earth's magnetic field is quite old. But until recent years the evidence against it has been overwhelming. The recent tentative acceptance of a role of magnetism is due mainly to four types of observation: (1) evidence of persistent, correct orientation by free-flying migrants and homing pigeons without key visual cues, (2) observations on free-flying birds carrying magnets,

24–12 Emlen funnels in operation and sample footprint records. *Below:* S. Emlen in planetarium with funnels in operation. *Right:* Blotter records and the vector diagrams derived from them. The circular record is divided into 24 sectors of 15° each. The amount of activity in each sector is then scored by comparing the density of ink prints with densities on a standard reference graded from 1 to 20. The vector reaching the perimeter represents the sector having the highest score; the other vectors are scaled proportionately. **A.** Swamp sparrow (*Melospiza georgiana*) under night sky in spring. **B.** Indigo bunting in spring under planetarium sky set for local conditions. **C.** White-crowned sparrow (*Zonotrichia leucophrys*) tested in spring under a full moon in the south. The *n* values refer to the highest score for any one sector. North is toward the black triangle. (*Right:* From Emlen and Emlen 1966. Courtesy of American Ornithologists' Union. *Below:* From Cornell Alumni News Feb. 1971.)

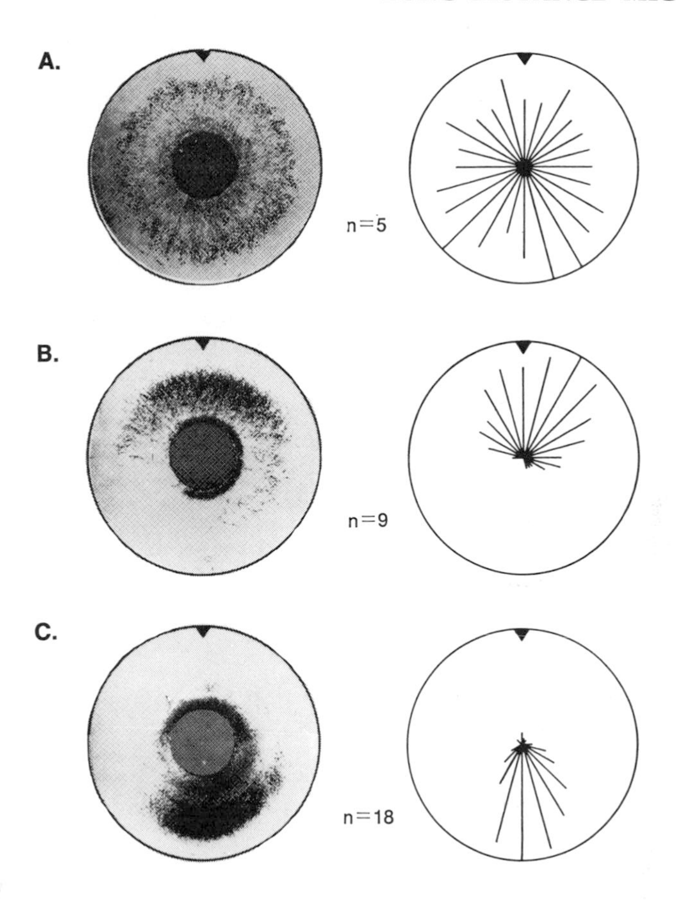

(3) evidence of properly oriented *Zugunruhe* in caged migrants without the critical visual cues, and (4) experiments on *Zugunruhe* in caged birds in artificial magnetic fields.

Radar observations of migrating birds have sometimes revealed persistent correct flight directions under conditions that were almost certainly completely overcast (e.g., Drury and Nisbet 1964). In the most convincing cases, radar revealed large numbers of birds maintaining correct migratory flight at night—despite having the sea or a layer of clouds below, so they could not use visible landmarks, and a layer of clouds above, so they could not use the stars or other features of the night sky. Such observations were not completely convincing, however, because it could never be firmly proven that the birds were not getting occasional glimpses of the stars or landmarks as they flew.

Objections of this type were overcome in a set of experiments by Keeton (1969). Homing pigeons that had been accustomed to flying

24–13 Planetarium experiments on orientation of *Zugunruhe* in indigo buntings in Michigan. Vector diagrams as in Figure 24–12.

A. Fall. 43° N. Outdoor tests were done on moonless nights. The buntings oriented southward under natural skies, southward under the planetarium local sky, and reversed their orientation appropriately when the star field of the planetarium was rotated 180°.

B. Spring. 28° N. Shutting off the southern celestial hemisphere did not eliminate northward orientation in any of the four birds tested (only two are illustrated). Shutting off the northern half caused all birds to cease *Zugunruhe.* Eliminating Polaris, which is the North Star and the only star to indicate true north constantly, did not reduce northward orientation. Elimination of Polaris plus the nearby and conspicuous Ursa Major (Big Dipper) was likewise without effect.

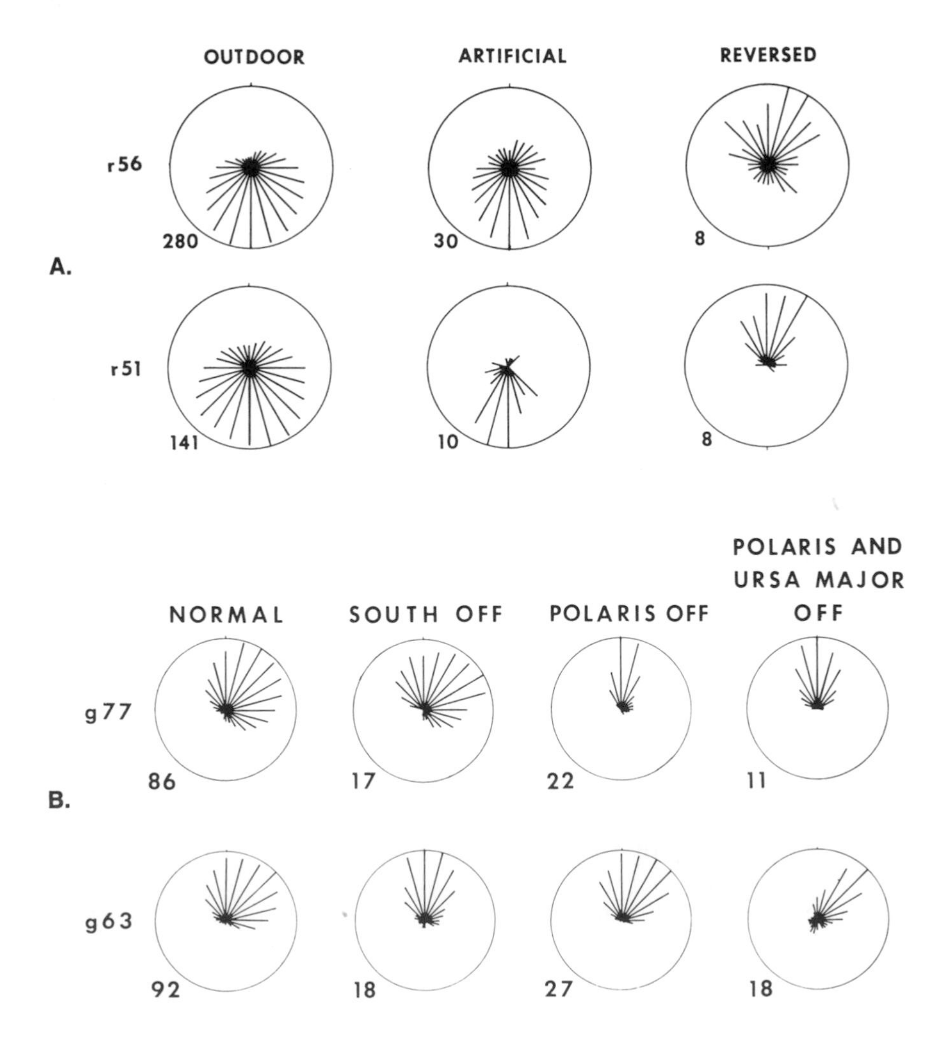

C. Fall. 35° N. Shutting off the southern celestial hemisphere led to random orientation in one (r56) of four birds tested; the others (including r58) maintained their southward orientation. This result suggests individual variation in use of star configurations by the buntings. Loss of Cassiopeia or the Milky Way did not eliminate correct orientation in any of the five birds tested.

D. Fall. 35° N. Elimination of Polaris, Ursa Major, Cassiopeia, and all other stars within 35° of Polaris led to randomly oriented *Zugunruhe* in two buntings (not shown) and to deterioration of orientation in another (r56), but one bird (r58) managed to maintain a southward heading. Next to shutting off the entire northern celestial hemisphere, this procedure caused the greatest disruption of migratory orientation. It is interesting that r56, which was the only individual to persist with oriented *Zugunruhe* with *35° Blocked,* was the only bird to be affected by *South Off;* this suggests that r56 was influenced more by southern stars and less by northern stars than were the other individuals (evidence of individual variation). Since many bright stars are located in a doughnut-like band between 25° and 35° from Polaris (*D*), it is interesting that addition of this band was associated with resumption of correct orientation in two buntings (including r58) of the three that had lost it under *35° Blocked.* Somewhat surprisingly, elimination of the 25°–35° "doughnut" alone allowed roughly southward orientation to persist in four out of five buntings. (Figure 24–13 continued on next page.)

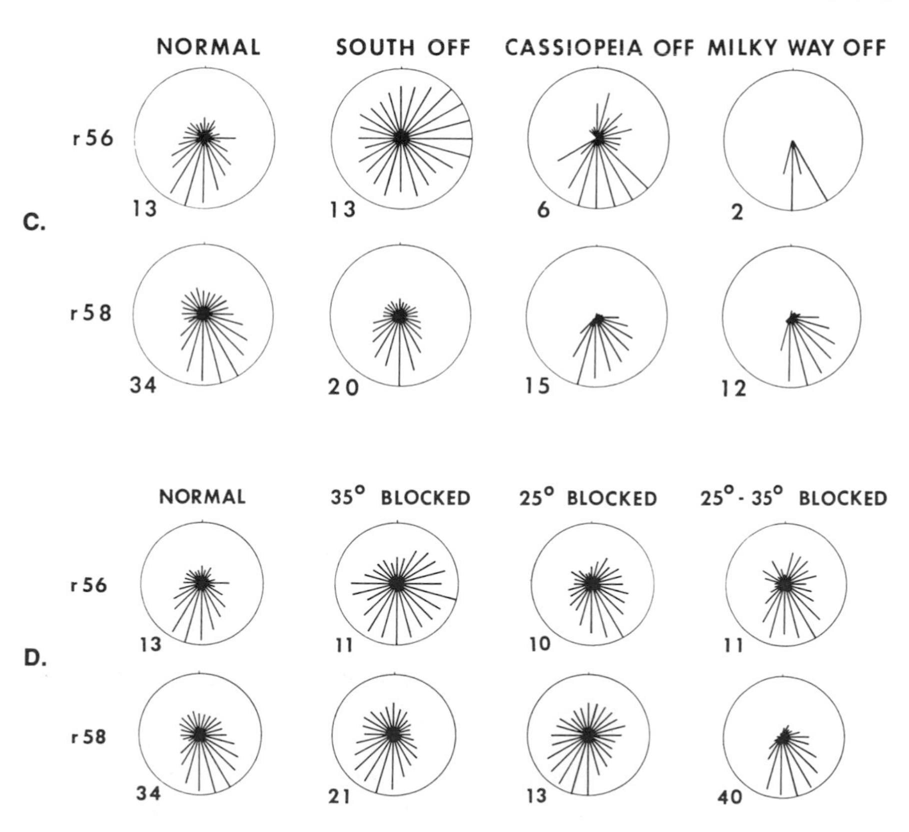

E. Constellations and the 25° and 35° rings.
(All from Emlen 1967*a*, 1967*b*. Courtesy of American Orinthologists' Union.)

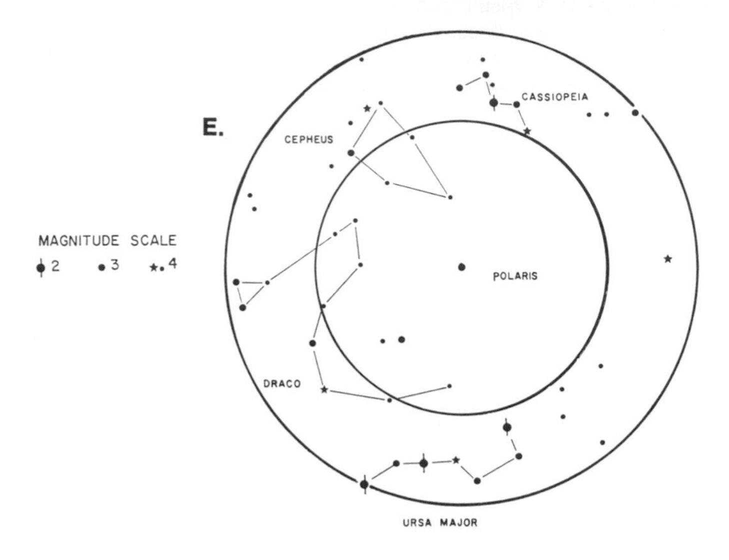

under overcast skies near their loft were clock-shifted by six hours, then released 21 miles from the home loft. Under sunny skies a time shift of six hours causes pigeons to deviate from their correct initial bearing by 90° (Figure 24–15) and may delay their return considerably. So if the pigeons in Keeton's experiment were using occasional glimpses of the sun to orient themselves, they should have showed these deviations. Instead, they took correct initial headings (Figure 24–15) and returned to their home loft in the normal time. In another series of releases, pigeons carrying magnets strapped to their backs returned home significantly more slowly than controls carrying non-magnetic metal bars (Keeton 1971). Further support came from experiments by other workers in which direction-taking of pigeons and gulls was hindered by magnets carried by the birds. In the most dramatic of these, reversing the polarity of an electromagnet carried on the pigeon's head caused a reversal of initial heading under overcast skies but not in sun (Walcott and Green 1974).

A convenient way to experiment on the ability of birds to orient with the aid of magnetic fields should be to employ caged migrants and study the effects of artificial magnetic fields on the orientation of *Zugunruhe*. In indigo buntings (Emlen 1970*b*) and in a variety of other species, removing a view of the night sky or planetarium sky seemed to prevent correct orientation. But in European robins (*Erithacus rubecula*), correct orientation in the absence of key visual cues was

24–14 Ontogeny of star orientation in nocturnal migrants: two views. **A.** Correct southwest orientation of *Zugunruhe* by a young garden warbler upon first exposure to a natural starry sky. (From Sauer 1957.) **B.** Correct southeast migratory orientation of *Zugunruhe* by a lesser whitethroat upon first exposure to a night sky (in her second fall), in this case a stationary planetarium sky. (From Sauer 1957.) **C.** Lack of correct orientation of *Zugunruhe* by young indigo buntings upon first exposure to a night sky, in this case a stationary, autumnal planetarium sky set for 42° N. Vector diagrams as in previous figures; the dot-diagrams show the distribution of mean directions on separate tests. **D.** Correct orientation of *Zugunruhe* by young indigo buntings that had been allowed regular viewings of a normal, rotating, planetarium sky during their early development. Black triangles show mean directions of orientation if the distributions depart significantly from random. (Figure 24–14 continued on next page.)

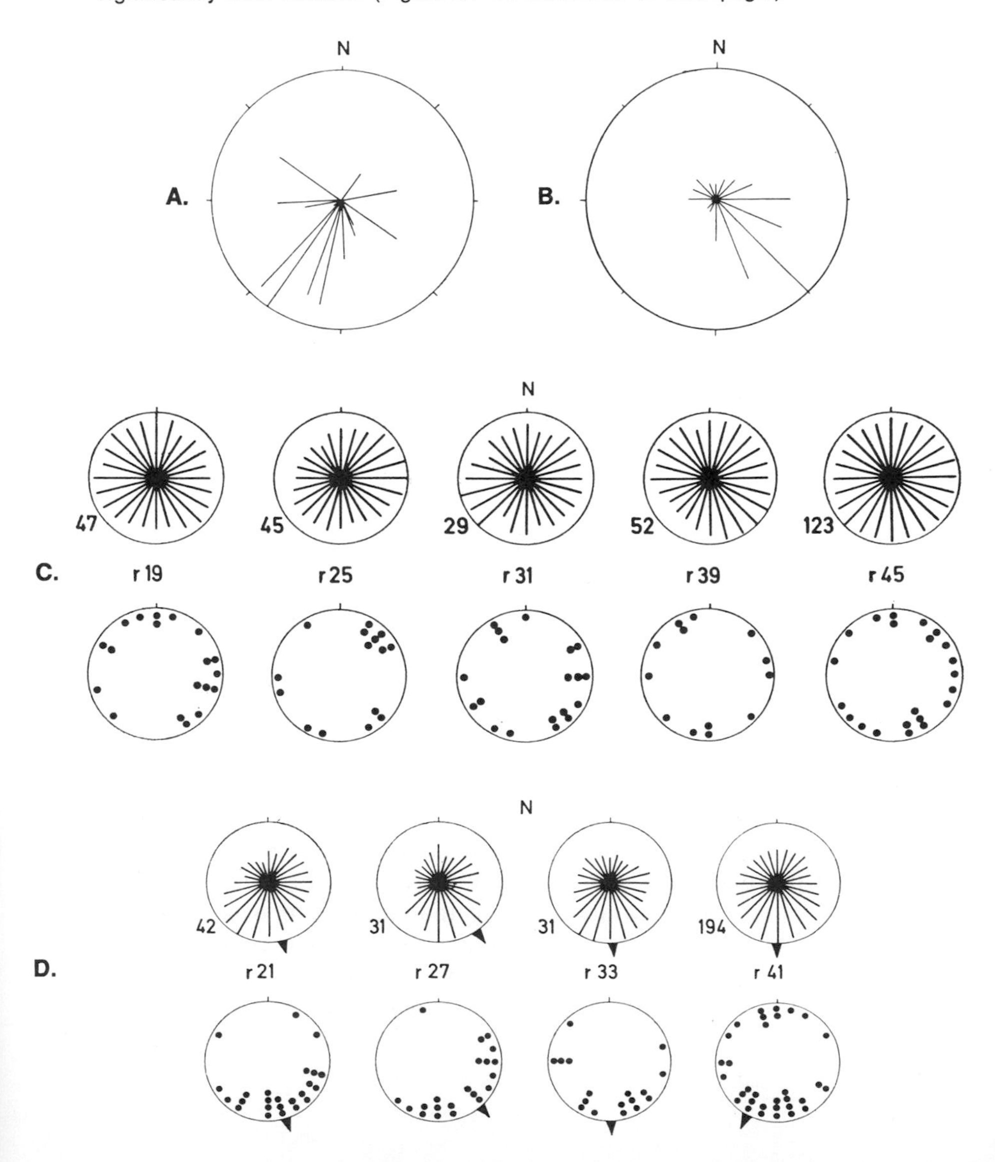

E. "Correct" orientation of *Zugunruhe* by young indigo buntings exposed during early development to a planetarium sky that rotated about Betelgeuse (in the constellation Orion), rather than about Polaris. The difference between Polaris and Betelgeuse in these tests was from 110° to 180°. The buntings were tested on a stationary sky like the one to which they had been exposed. The data are plotted with Betelgeuse (*B*) at "north," and the birds oriented as if they regarded it as north. (C, D, and E from Emlen 1970*a*. C–E: Copyright 1970 by American Association for the Advancement of Science.)

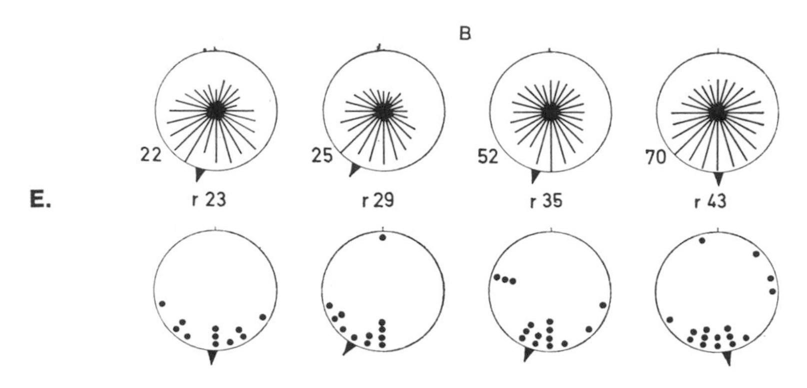

reported by Fromme (1961). The first attempt to duplicate Fromme's findings failed (Perdeck 1963), apparently because of differences in design of the testing cage. But later attempts by others were successful. An effect of the earth's magnetic field on orientation in such "registration cages" is rather subtle and can be easily overlooked unless special statistical methods are employed. European robins can orient their *Zugunruhe* in the appropriate direction in spring and fall without visual cues if the magnetic field surrounding them approximates the strength, direction, and inclination (from the horizontal) of the earth's normal magnetic field (Merkel et al. 1964; Merkel and Wiltschko 1965; Wiltschko and Merkel 1966; Wiltschko 1968; Wiltschko et al. 1971; Wiltschko and Höck 1972; Wiltschko and Wiltschko 1972). The robins were tested in a small octagonal cage around which Helmholtz coils were placed to control the magnetic field. The normal field strength where the robins were tested was about 0.46 gauss. Orientation failed if the field strength was decreased beyond 0.14 gauss or raised above 0.81 gauss; and even at these strengths orientation was only possible after at least three days of acclimation to the new level.

In a surprising series of experiments, Wiltschko and Wiltschko (1972) found that the inclination of the magnetic field was more important than its compass direction (Figure 24–16). When the inclination was changed from the normal 66° below horizontal to 65° above horizontal while retaining the same compass direction (horizontal

24–15 Orientation by homing pigeons without the sun or stars. Open symbols are for controls; solid symbols, for clock-shifted experimentals (6 hours fast). The dashed arrow shows the homeward direction. The solid arrows show the means of control (*MC*) and shifted (*MS*) pigeons. Each point represents the vanishing bearing of one pigeon. **A.** Overcast. Both control and clock-shifted cock pigeons oriented homeward. Clock-shifted birds were *not* deflected 90° as expected if they could see the sun. **B.** Sun. Clock-shifted birds are deflected appropriately while controls are not. **C.** Overcast, Petersberg, N.Y., 102 miles from home. In A and B the birds were released at Marathon, N.Y., 21 miles from home, where most birds had had previous experience. The release at Petersberg was made with relatively inexperienced pigeons that had had no previous experience in the Petersberg area. (From Keeton 1969. Copyright 1969 American Association for the Advancement of Science.)

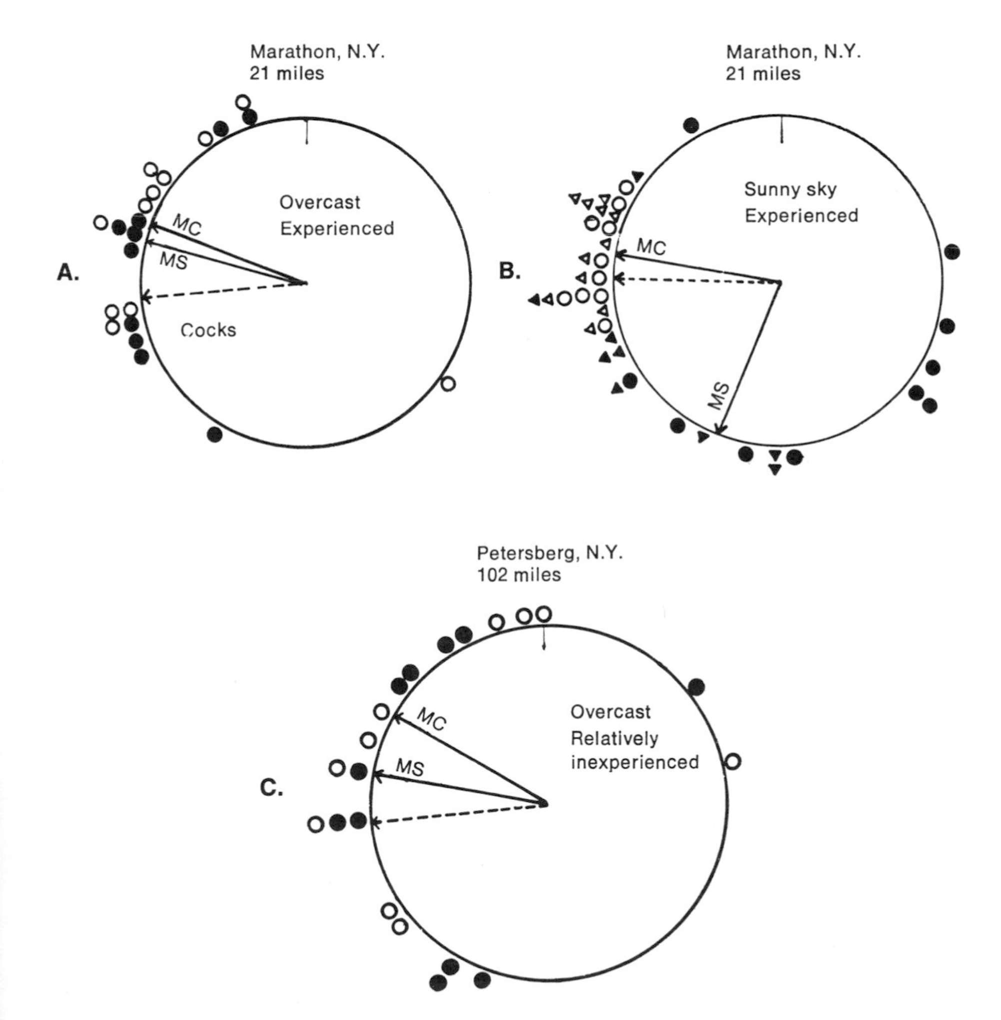

24–16 What feature of the earth's magnetic field is used by European robins to orient their *Zugunruhe?* In these experiments, the horizontal and vertical components of the magnetic field surrounding the bird were varied independently and the orientation of the *Zugunruhe* recorded in spring by automatic counters. Key visual cues were absent. The magnetic field for each test is shown below in the vector diagrams. The results are shown in the circles on the facing page.

The vector diagrams (*below*) are in the vertical north-south plane. $\overleftarrow{H}_h$ = horizontal component. $\overleftarrow{H}_v$ = vertical component. $\overleftarrow{H}_e$ = local earth's magnetic field vector. γ and γ' = angles between magnetic field vector and gravity. $\overleftarrow{g}$ = force of gravity.

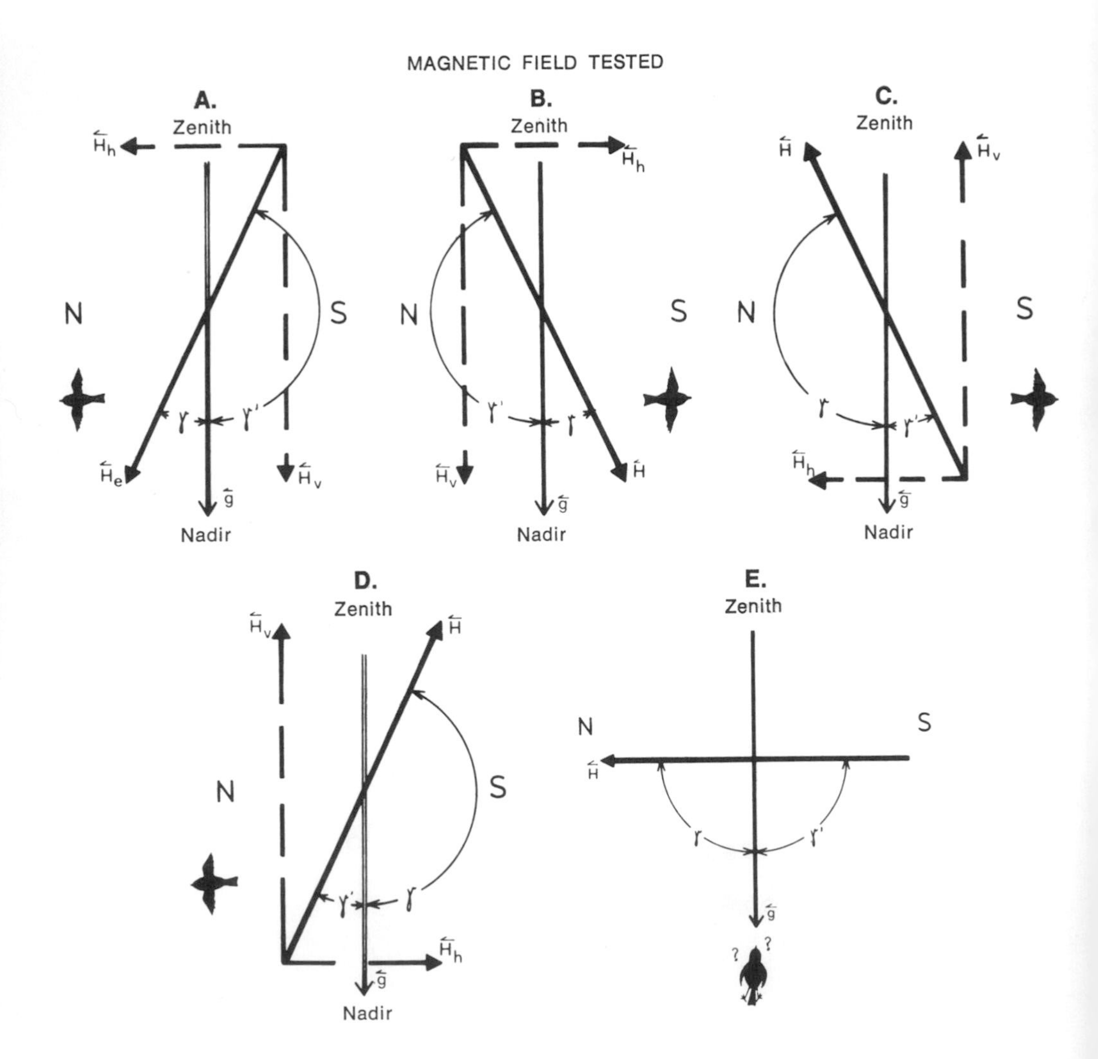

Below: The triangles are the means for single test nights for one bird. The heavy arrow is the mean vector. If the mean vector reaches the dashed inner circle, it is significant at the 5-percent level; if it reaches the inner unbroken circle, it is significant at the 1-percent level. All tests were done at the earth's normal field strength at this locality, 0.46 gauss. Note that **A, C,** and **E** yielded roughly different directional tendencies, although in all three cases the compass direction (horizontal component) was the same. Pairs having the same vertical component (**A** and **B, C** and **D**) also had different tendencies. Only the angle of the magnetic field vector with gravity (regardless of the direction of the field) correlates consistently with the birds' directional tendencies. (From Wiltschko and Wiltschko 1972. Copyright 1972 by the American Association for the Advancement of Science.)

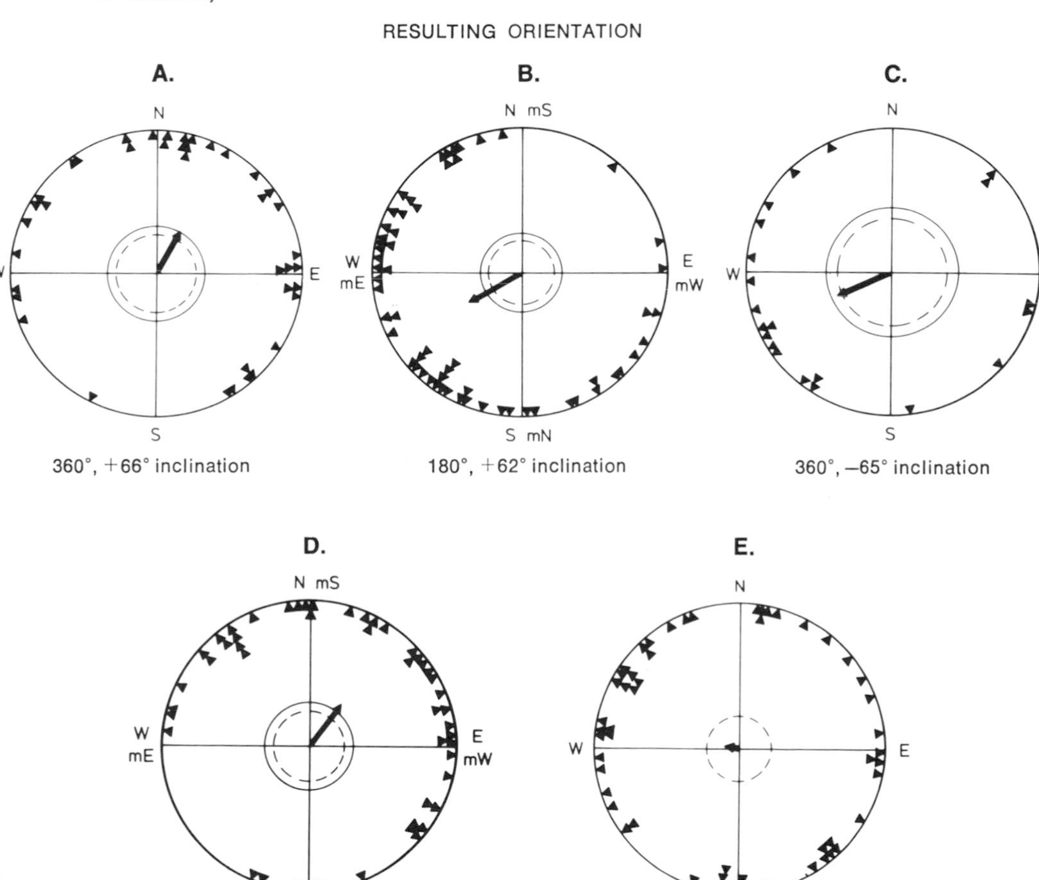

component), the orientation of the robins was roughly reversed (in Figure 24–16, compare *A* to *C*). If the compass direction was reversed and the inclination kept at the normal angle but with polarity reversed, orientation was not reversed (compare *A* to *D*). Robins tested with the normal compass direction but an inclination of zero degrees were unable to orient in any direction (*E*).

Much remains to be learned in this exciting and rapidly developing area of ethology. What species orient by the earth's magnetic field? Where are the magnetic receptors located? How is the information from them processed? Since the robin's system cannot be used by transequatorial migrants, what system do the latter use? What aspects of genetics, morphology, physiology, and developmental experience cause a naive bird to choose the proper direction of migration? Birds can tell where north is by a variety of methods (sun, stars, magnetic fields), but how do they know the direction of home when they have been displaced?

CONCLUSIONS

Of the many hard-to-believe feats of animal navigation and migration, the travels and abilities of birds are preeminent and worthy of special attention. The ecological advantages of breeding in an area that is highly suitable for raising young but not for wintering, and of living the rest of the year in another area that is excellent for survival (but not necessarily for breeding), have provided selection pressures for traits that facilitate traveling between, and finding, these areas. The selection pressures that favor migratory abilities are probably balanced by the rather high mortality arising from the risks of becoming lost over inhospitable habitats. Because the balance of selection pressures acting on one breeding population may differ from that on another, a species may come to have some populations that are highly migratory and some that are sedentary (nonmigratory). Such a situation is found in the white-crowned sparrow and European robin. The ability to migrate is an adaptation that is present in many species but that is shaped into various migratory patterns according to the ecological conditions of the species and its local populations. Many species are nonmigratory, especially in the tropics.

When investigating the physiological bases of migration, it is desirable to take into account the diversity of migratory routes, distances, altitudes, speeds, weather conditions, and habitats used by the various species. Since even one species may have some populations that are migratory and some that are not, the ability to migrate could have evolved, been lost, and re-evolved many times in various phyletic lines. As a result we should be prepared to find diversity among species in the physiological adaptations for migration. The observation that species seem to vary in their dependence on the sun, stars, landmarks, and magnetic fields as navigational aids hints strongly at the existence of underlying physiological diversity.

A central problem in the study of animal navigation has been to characterize the sensory processes involved. The cues must be identified, the sense organs located, and the methods of data-processing discerned. The first major cue to be identified was the sun. It has been demonstrated experimentally that the sun could be used in direction-finding by day-migrants, night-migrants, and nonmigrants. In most cases when the direction of the sun could not be discerned by the test bird, its orientation failed; so a clear effect of the sun was well established.

Since the direction of the sun shifts regularly according to time of day, animals using the sun for direction-finding must compensate for its movement. Experiments showed the existence, use, and surprising precision of such a "clock." Shifting the animal's clock by changing its light-dark schedule caused predictable shifts in the direction of orientation. The means by which information on the sun's position is processed by the brain to yield the correct direction for migration or homing have been debated and require further study.

The fact that many species of birds migrate at night suggested that cues other than the sun might also be used. In experiments under the night sky and in planetaria, it was shown that nocturnal celestial cues can also be used, and that in some species correct nocturnal direction-taking fails without a view of clear sky. Planetarium experiments, in which the planets and stars could be turned off selectively, revealed that no single star, planet, or constellation was necessary. All the planets and most of the stars could be turned off with little effect. The region of the sky close to the North Star seemed to be most important. Clock-shifting does not apparently affect star-navigation, in contrast to sun-navigation. The key features of the star cues used in nocturnal navigation are not well known. Learning of the night sky seems to be important in some species.

More recent experiments have supplied convincing evidence that nonvisual cues must also be used by some birds in some situations. The only nonvisual cue so far identified that might form the basis of a sophisticated navigational ability is the earth's magnetic field. It seems that some species can use the sun, stars, landmarks, magnetic fields, and other cues singly or in combination, depending on circumstances and experience. The nature of the interactions among these different mechanisms in a variety of species is currently being investigated.

25 The Programming of Behavior: Biological Rhythms, Attention, and Motivation

THE PURPOSE OF this chapter is to consider briefly some of the ways in which the various sensory and motor abilities of an animal are programmed in a species-typical and adaptive manner. The mechanisms for producing an integrated overall pattern of behavior are exceedingly diverse; they range in complexity from sensory-motor coordination performed within a single cell to the complicated brain processes involved in learning, attention, and motivation. While simpler integrative mechanisms are reflexlike in character, the more complicated ones are noted for their wide-ranging effects on a variety of behavior patterns. Mention will be made of the simpler stimulus-response-type integrative mechanisms, but emphasis will be on neural processes with a more global influence on behavior, namely, on behavioral rhythms, attention, and motivation.

Probably the simplest mechanism that integrates sensory and motor events in a multicellular animal is that occurring in independent effectors, such as nematocysts. An *independent effector* is a single cell that performs both sensory and motor functions. Nematocysts are produced only by coelenterates and are characteristic of that phylum. Depending on the type of nematocyst, it responds to stimuli such as touch or scent by discharging a threadlike entanglant, or harpoon-like structure (Figure 25–1). These cells are an integral part of the food-catching mechanisms of sea anemones, jellyfishes, corals, hydroids, and other coelenterates. The prey is entangled, harpooned, or poisoned by these structures, which paralyze the prey and enable the coelenterate to move the food into its mouth. These cells have been thought to act independently of neural control. In recent years it has seemed more likely that their thresholds are modulated by neurons, but in any case it is established that they can be fired by direct stimulation.

The type of integrative mechanism next in complexity should be

25–1 Nematocysts as independent effectors. **A.** Cell types in the body wall of *Hydra*. Nematocysts, or "sting capsules," are interspersed among other cells of the outer body wall (ectoderm) but not in the inner body wall (endoderm). **B.** Nematocyst before and after discharge. (From Buchsbaum 1948. Copyright 1938 and 1948 by the University of Chicago. All rights reserved.)

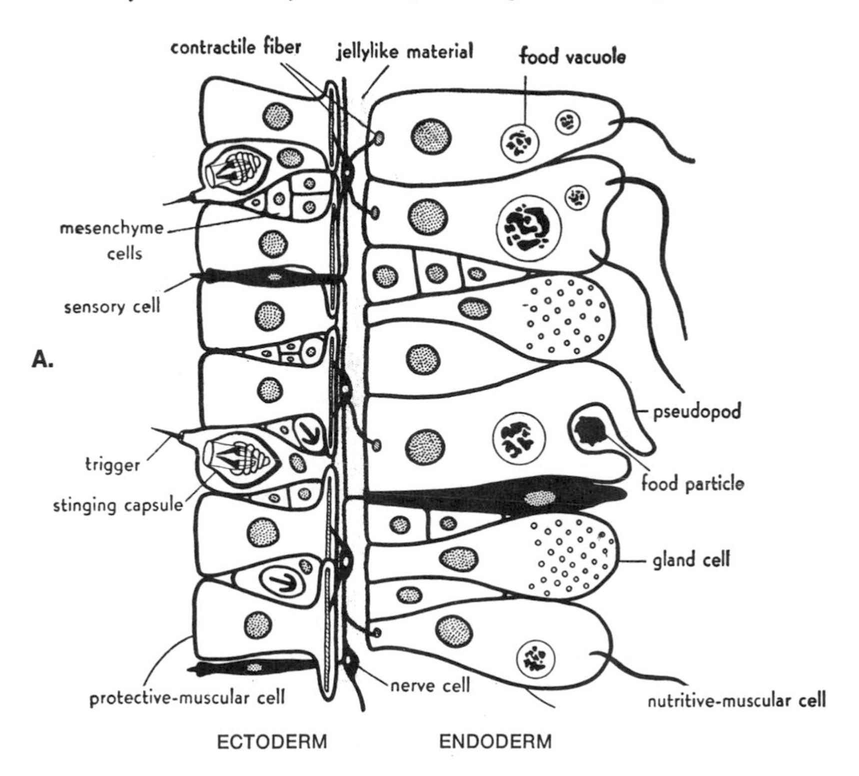

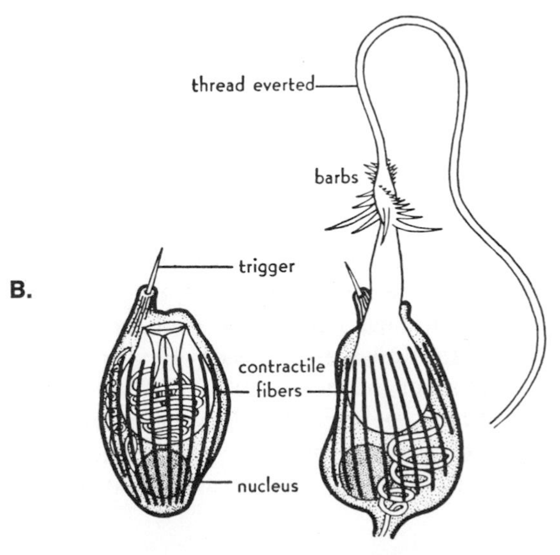

that in which a receptor cell directly activates an effector cell. Rather surprisingly, there seems to be no good evidence that such direct connections exist regularly; they may perhaps occur randomly in nerve nets.

Going up another step in complexity, in the simplest conceivable *neural* integrative mechanism at least one neuron would be interposed between sensory and effector structures. With the exception of the axon reflex, which is of minor importance and will not be discussed here, this type of integration is also unknown as a regular feature in any animal.

In the vertebrates, the simplest integrative hookup involves at least two neurons, one sensory and one motor. This is known as a monosynaptic reflex. A good example is the stretch reflex. (For a discussion of reflexes, see Chapter 20.)

With the introduction of neural integrative mechanisms, it is well to consider that even single neurons have an array of integrative properties (Bullock 1959, 1961). Many neurons do not fire on a one-to-one basis, that is, one output spike for each input spike. Commonly, the output of a neuron is a more complex function of its input. A neuron typically receives inputs from more than one other neuron, and its output often reflects opposing or additive effects of different inputs. Many neurons operate as "decision units"; they fire or remain silent depending on whether the algebraic summation of diverse inputs of different weight has brought the neuron to its threshold for firing. Others have thresholds that are affected by their chemical milieu or that may fluctuate spontaneously. For these reasons, the outputs of even small systems of neurons cannot be predicted from their "wiring diagrams" alone.

BIOLOGICAL RHYTHMS

Many aspects of behavior are rhythmic. The periods of these rhythms may be measured in seconds, minutes, days, or months, and the neurobiological mechanisms that underlie them are diverse. Some are obvious to our perception, such as the swimming beat of an anemone or the pulsating flashing in unison of a tree full of fireflies (Buck and Buck 1968). Others require controlled experimental conditions for their detection, such as the rhythm of gobbling in turkeys (*Meleagris gallopavo;* Schleidt 1964). The best known are probably the near-24-hour cycles of sleep and wakefulness. Even yearly rhythms are known that affect reproductive behavior (Menaker 1971). In this section, a variety of rhythms with different periods will be considered, with special reference to their physiological bases.

Short-term Rhythms / The origin of locomotory rhythms of jellyfishes and other coelenterates has been traced to localizable neural structures (summarized in Bullock and Horridge 1965:503). For example, in *Aurelia* the swimming beat is controlled by eight pacemakers that are located in the marginal bodies on the fringes of this circular animal, as shown in Figure 20–2. Each pacemaker discharges at its own frequency unless caused to discharge earlier by the firing of another pacemaker. Contraction of the muscles that produce a swimming beat is controlled by the fastest pacemaker by means of a through-conducting nerve net. The fastest pacemaker triggers all the other pacemakers, resulting in a pattern of muscle contraction that is nearly synchronous. This results in regularity and symmetry in the swimming pattern. If the fastest pacemaker is surgically removed, the next fastest takes over. The rhythm of one pacemaker can be transferred to another animal by grafting marginal bodies. Pacemaker systems that influence behavior are known also in other coelenterates (Josephson and Uhrich 1969; Lerner et al. 1971; Mackie 1968; Passano 1965).

In many pacemakers, such as the one just described, it is difficult to localize the source of the rhythm to a single cell. In a few cases, however, rhythms can be traced to identifiable nerve cells. One such case is the rhythm of ventilation of the gills in hermit crabs (*Pagurus pollicarus*) and lobsters (*Homarus* sp.; Mendelson 1971). In these crustaceans water is moved through the gill chamber by a pair of external, scoop-shaped appendages called *scaphognathites* at the rate of about 1.5 scoops per second. Each scaphognathite is innervated by 11 motor neurons that are divided among levator (raising) and depressor (lowering) muscles. These neurons originate in the subesophageal ganglia. These are the anteriormost ventral ganglia in the nervous system of an arthropod. Activity in these motor neurons is rhythmic, those firing the depressors alternating with those firing the levators (Figure 25–2). Another neuron can be found within the subesophageal ganglion whose membrane potential moves up and down in exact correspondence with the timing of the motor neuron bursts, yet there is no sign of action potentials (spikes) in the oscillator neuron itself. To test whether the oscillator controlled the motor neurons or merely reflected a rhythm generated elsewhere, current was passed through the oscillator. Depolarization of the oscillator silenced one group of motor neurons while inducing long-lasting, continuous firing in the other group; hyperpolarization of the oscillator had the opposite effect. Therefore, the fluctuating membrane potential of the oscillator neuron can be said to drive the motor neurons. According to Mendelson (1971:1172), "These data show that change in the membrane potential of a single neuron is sufficient to determine the entire rhythm of the

25–2 Source of a behavioral rhythm in a single neuron. Crabs and lobsters move water through their gill chambers with a scoop-like appendage called a scaphognathite. In **A**, the motor neurons that raise (levator) and lower (depressor) the scaphognathite of a hermit crab are shown to fire in alternate bursts. In **B**, the resting potential of an oscillator neuron is shown to fluctuate in synchrony with the alternating bursts of levator and depressor motor neurons in a lobster. **C** and **D**. Experimental depolarization and hyperpolarization of the oscillator neuron have opposite effects on the motor neurons. Calibrations: intracellular records B, C, and D, 34 mv; extracellular voltages are not calibrated. Time: A, 250 msec; B, C, D, 310 msec. (From Mendelson 1971. Copyright 1971 by the American Association for the Advancement of Science.)

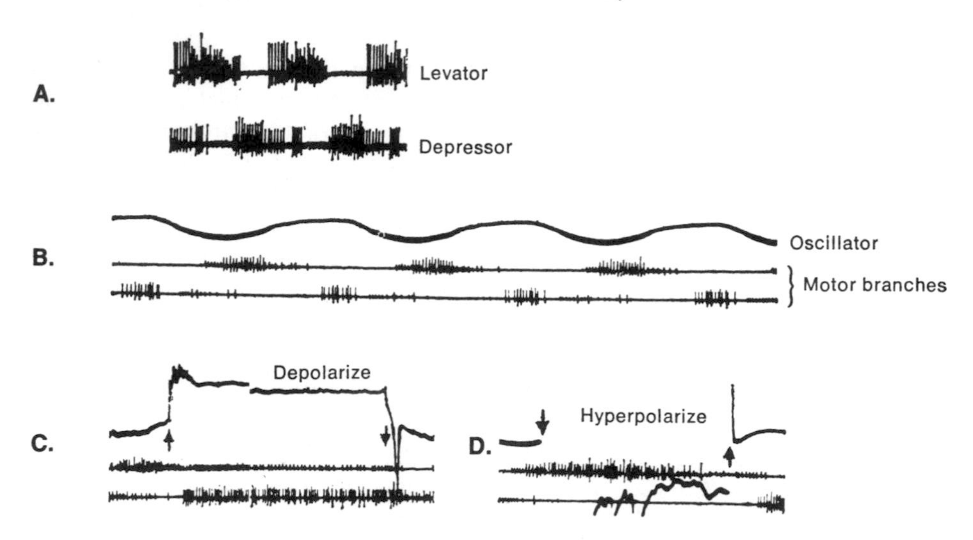

local neural system and that the observed oscillations of membrane potential are the causative agent for the motor neuron activity, not caused by it. They are consistent with the hypothesis that a single oscillator neuron in each hemiganglion controls the scaphognathite beat pattern."

Other examples of short-term rhythms in arthropod behavior, including those of singing and flying, have been analyzed in Chapter 21.

Circadian Rhythms in Some Invertebrates / The term *circadian* (from the Latin, *circa*, about, and *dies*, day) has been applied "to all '24-hour' rhythms, whether or not their periods, individually or on the average are different from 24 hours, longer or shorter, by a few minutes or hours" (Halberg 1959). Circadian rhythms are found in many physiological and behavioral phenomena in insects. A circadian rhythm that has received intensive study is found in the locomotor activity of cockroaches (*Periplaneta americana* and *Leucophaea madeirae*).

Cockroaches are normally active at night and inactive in the day. That the cue, or *Zeitgeber*, which sets the rhythm is the light-dark cycle is shown in Figure 25–3, which also shows that the receptor involved is the compound eye rather than the ocellus (Roberts 1965). Observe that the period of the rhythm is precisely 24 hours until the compound eyes are covered, and that afterward it decreases slightly, causing the onset of activity to begin a little earlier each day. In the former case the rhythm is said to be *entrained* by the light-dark cycle. When the rhythm continues in the absence of an effective *Zeitgeber*, it is said to be *free-running*. Free-running rhythms often are not precisely 24 hours and, consequently, tend to drift.

The persistence of the rhythm in the absence of a known *Zeitgeber* does not prove that the source of the rhythm is *endogenous* (from within), since one is never sure that some other *exogenous* source (from outside the animal) is not involved. In the past, investigators have gone to great lengths (and great depths underground) to exclude known and suspected *Zeitgebers*. A more direct way of establishing the existence of an endogenous source is to locate and remove it. The early work of Harker (summarized in 1964) on cockroaches convinced behaviorists that the approach was feasible, even though her conclusions must now be reinterpreted in light of new and contradictory evidence. The following account is taken from the works of Nishiitsutsuji-Uwo et al. (1967), Nishiitsutsuji-Uwo and Pittendrigh (1968*a*, *b*), Roberts (1960, 1962, 1965, 1966), Roberts et al. (1971), and Brady (1967*a*, *b*, 1969, 1971).

The first and simplest experiments to locate the circadian pacemaker in the cockroach involved cutting off the head and sealing the neck with wax. Such insects lived for several days, but they were both inactive and arrhythmic. They would run when nudged but otherwise remained stationary. The operation removed both the brain (supraesophageal ganglia) and the subesophageal ganglia. When only the brain was removed, the animals were active but arrhythmic. Therefore, as a preliminary hypothesis we may now suggest that the brain is the seat of the pacemaker and that the subesophageal ganglia supply a factor necessary for activity.

A candidate for the pacemaker locus was a group of neurosecretory cells (hormone-secreting neurons) in the front center of the brain known as the pars intercerebralis (PIC) (Figure 25–4). Injury to these cells disrupted or eliminated the rhythm, depending on the completeness of the operation. Since the secretion of these cells was known to influence the neural activity in the ventral nerve cords, an attractive hypothesis was that PIC acted by periodically releasing a hormone that suppresses the activity of the subesophageal ganglia. This has been

25–3 Entrainment of a free-running activity rhythm in a cockroach maintained in a light-dark cycle for 83 days. On day 20, the compound eyes were painted with black lacquer and the rhythm became free-running. On day 50, the paint was peeled off, allowing entrainment to the light-dark cycle again. Removal of the ocelli on day 68 did not disturb the established entrainment. The times of light and dark during each day are indicated by the bars at the top. Periods of high activity are recorded as dense clusters of short, vertical lines on the horizontal lines for each day. (From Roberts 1965. Copyright 1965 by the American Association for the Advancement of Science.)

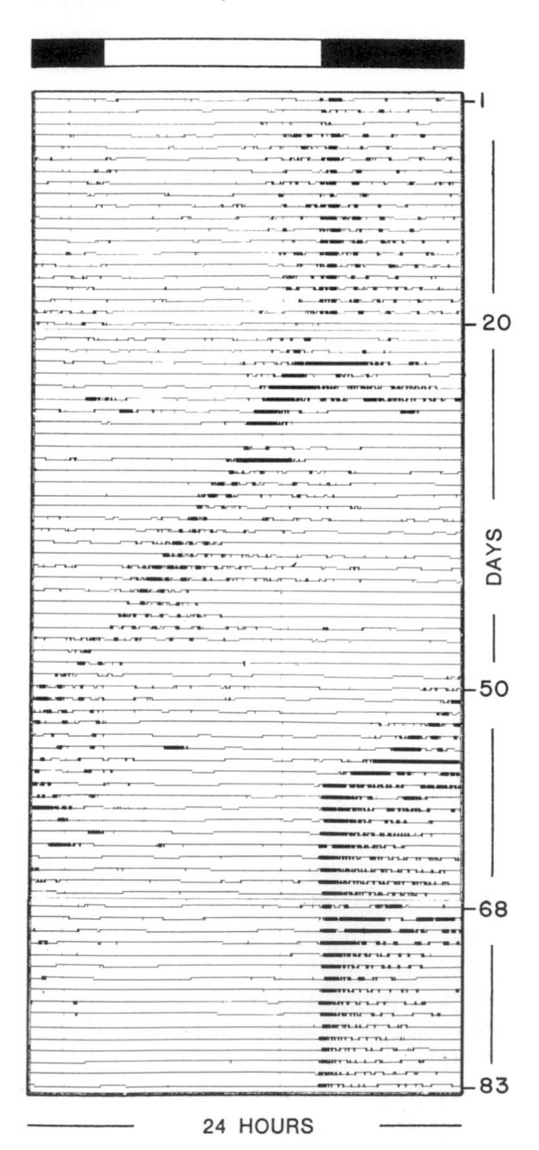

25–4 Brain of a cockroach. **A.** Head, showing location of some cuts made in experiments. The first cut, I, made a 1.5-mm window; the integument was removed before, and replaced after, the operation. The second cut, II, severed the optic nerves. **B.** Brain, showing optic nerves. **C.** Brain, showing pars intercerebralis. *AN,* antennal nerve. *AS,* antennal sclerite. *Br I,* protocerebrum. *Br II,* deutocerebrum. *CE,* compound eye. *OpL,* optic lobe. *OpN,* optic nerve. *OpT,* optic tract. *PIC,* pars intercerebralis. *SC,* scape. (From Nishiitsutsuji-Uwo and Pittendrigh 1968*a,* and Nishiitsutsuji-Uwo et al. 1967.)

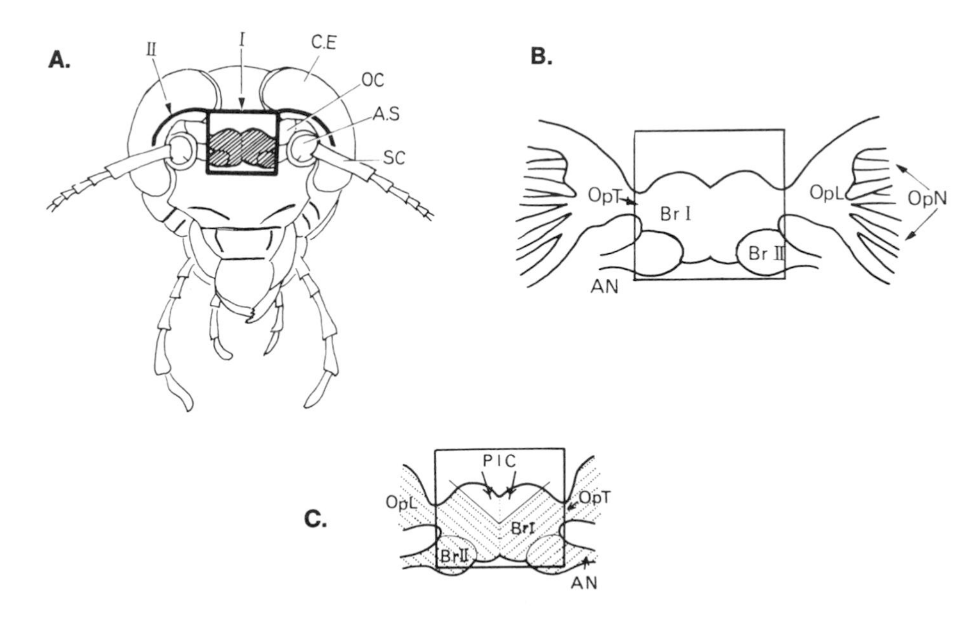

questioned, however; and it is uncertain whether the link from brain to nerve cords is hormonal or neural. But was PIC the source of the rhythm, or did some other brain area activate PIC periodically?

When the entraining pathway was cut between the receptors of the compound eyes and the optic lobes of the brain, the cockroach's activity rhythm became free-running: it was uncoupled from the rhythm of the environment. If the optic lobes were separated from the brain, a different effect was observed. The animal lost its rhythm; it was active but arrhythmic. The experiments showed that the PIC was not the source of the rhythm. The rhythm appears to be generated in the optic lobes or by some interaction of the optic lobes with PIC. These experiments are significant because for the first time they localize, to a particular brain structure, the source of a circadian activity rhythm. The processes that generate the rhythm from within the optic lobes remain to be discovered.

Other evidence on the physiological basis of circadian rhythms comes from *Aplysia* (a sea slug or sea hare). When kept in the dark, an isolated, detached eye shows a circadian rhythm of optic nerve impulses (Jacklet 1969). If the animal has been on a light-dark cycle previously, the first activity of its isolated eye comes at "dawn." If it has not, then the eye shows a free-running rhythm. Three hypotheses may be distinguished: (1) the rhythm of the isolated eye comes from a single pacemaker cell or region with a circadian rhythm, (2) all neurons of the eye are circadian oscillators, and (3) the circadian rhythm arises from a population of noncircadian oscillators. These hypotheses were tested by surgically removing parts of the cell population of the retinas (Jacklet and Geronimo 1971). The first hypothesis is eliminated by the absence of any one part of the eye that is critical for the rhythm. The second hypothesis predicts that even small pieces should possess a circadian rhythm, but they do not. The third hypothesis is supported by the finding that the rhythm fails when fewer than 20 percent of the 950 retinal neurons remain (Figure 25–5). These observations favor the mathematical theory that circadian rhythms may arise from a population of coupled oscillators with noncircadian periodicities (Pavlidis 1969). The relationship of the eye rhythm is interesting with regard to the role of the optic lobe in insects, but its significance for behavior in *Aplysia* is uncertain.

Another line of evidence that points to endogenous control of activity rhythms in invertebrates is their control by inheritance. Konopka and Benzer (1971) were successful in isolating three *Drosophila* mutants with abnormal activity rhythms, as shown in Figure 25–6. The genes affected rhythms of both locomotor activity and eclosion (emergence from pupal case). Arrhythmia was recessive to long, short, and normal periods. Crosses between animals having long, short, and normal periods produced phenotypically intermediate progeny. The genes were mapped on the X chromosome. Artificial selection can also alter circadian rhythmicity in *Drosophila* (Clayton and Paietta 1972).

Circadian Rhythms in Vertebrates / Circadian rhythms in the vertebrates show many of the properties of circadian rhythms in other organisms. They free-run in the absence of *Zeitgebers*; they can be entrained by a variety of stimuli, particularly light, but also sound; their ease of entrainment depends on the phase relations and timing of the stimuli in predictable ways. It is to be expected that the mechanisms of vertebrate circadian rhythms will prove to be complex. Consequently, the success of investigators in localizing key circadian processes in the avian pineal gland was entirely unexpected.

The activity rhythms of the English sparrow (*Passer domesticus*)

25–5 Dependence of a circadian rhythm in electrical activity of an isolated eye of the sea hare, *Aplysia,* on a critical mass of neurons. **A.** Compound action potentials (CAP) from an isolated eye; response to onset of light (indicated by deflection in upper line). **B.** Spontaneous activity in darkness. **C.** Structure of the eye. The retina, *R,* is composed of receptor cells and support cells (in black). Around it is a layer of secondary cells (neurons), fiber tracts, and neuropile (cross-hatched); and outside is connective tissue, *CT.* The optic nerve goes to the cerebral ganglion. The dotted lines indicate the parts of the eye that were cut away. **D.** Free-running rhythm of whole eye (triangles) and one-eighth eye (circles) in constant darkness. **E.** The period (τ) of the free-running rhythm as a function of the fraction remaining of the original population of 3,700 receptor cells and 950 neurons; 0.1 of the population is 380 receptors and 120 neurons; 0.2, is 42 receptors and 73 neurons. Each point represents one complete period in culture. (From Jacklet and Geronimo 1971; Jacklet 1969. Copyrights 1969 and 1971 by the American Association for the Advancement of Science.)

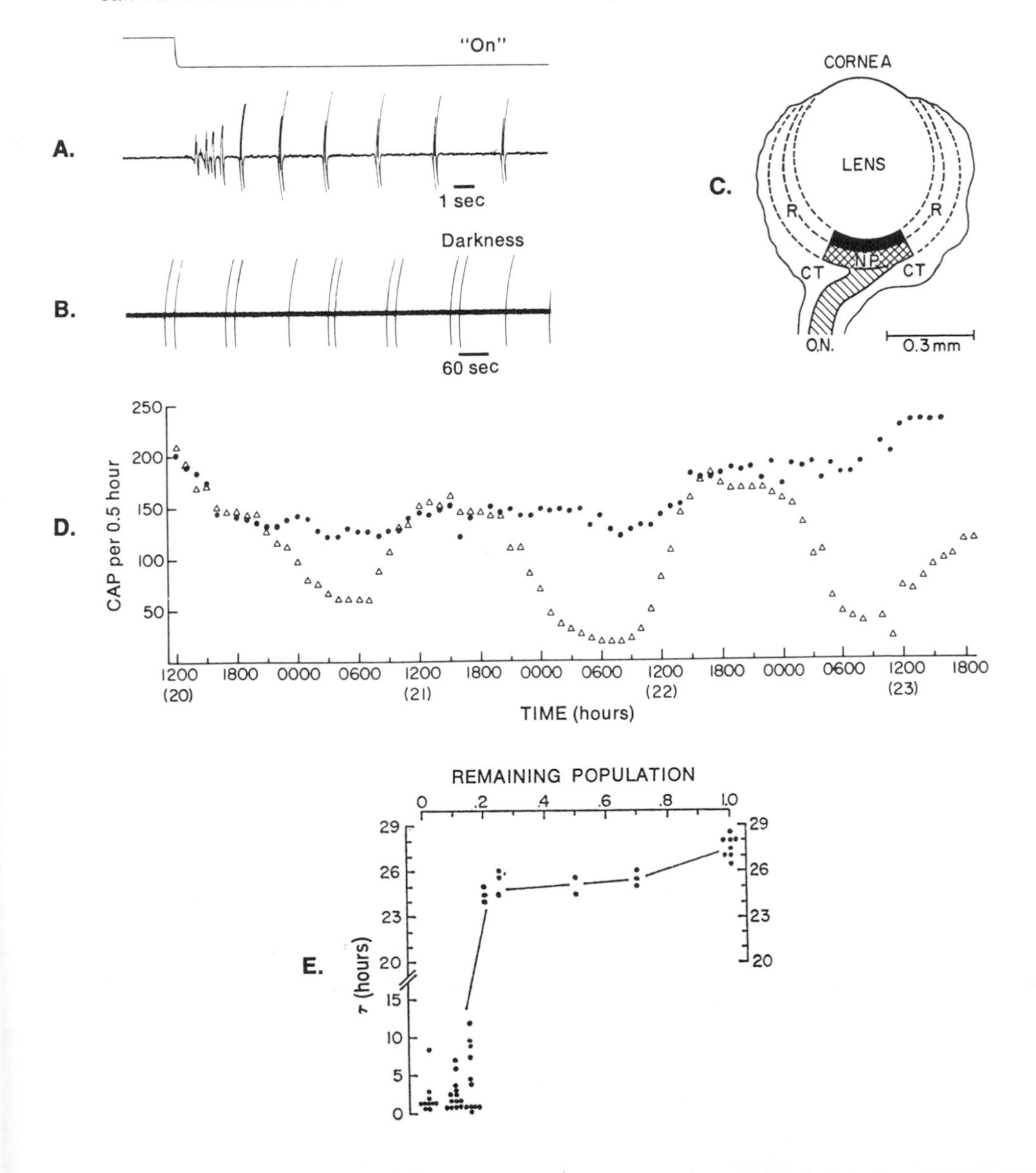

25–6 Genetically controlled differences in circadian locomotor activity rhythms in *Drosophila melanogaster.* Activity was recorded on an event recorder from left to right and top to bottom. Each record is for one individual fly. (From Konopka and Benzer 1971.)

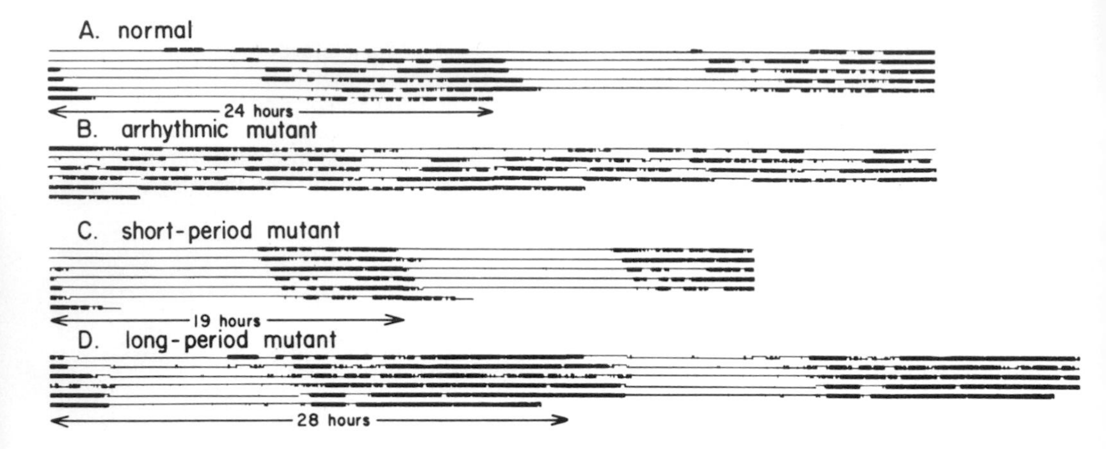

can be recorded in small cages with perches that register a short, vertical line on a moving tape each time a bird lands on them. Normally, activity will be synchronized with the light phase of a light-dark cycle; but in constant darkness without access to a *Zeitgeber*, a free-running rhythm occurs with a period a little greater or less than 24 hours, as in Figure 25–7. When the pineal gland of a sparrow kept in constant darkness was removed, its free-running rhythm disappeared and was replaced by arrhythmia, as shown in Figure 25–7 (Gaston and Menaker 1968; Binkley et al. 1971; Gaston 1971). That the effect of pinealectomy was not simply to eliminate or reduce motor activity was shown by the persistence of activity during arrhythmia and subsequent entrainment to light (Figure 25–7).

The receptor for this entrainment was not in the eyes or in the pineal, since entrainment persisted after their removal (Menaker 1968). Entrainment by *extraretinal receptors* sensitive to light has also been demonstrated in lizards (Underwood and Menaker 1970*b*), salamanders (Adler 1969*a*), and frogs (Adler 1969*b;* Taylor and Ferguson 1970). Extraoptic photoreception is also known in crayfishes, insects, ducks (references in Adler 1969*a*), and perhaps rats (Wetterberg et al. 1970). When tiny luminescent pellets were placed in various parts of the brain in quail (*Turnix turnix*), two general regions of the brain were found whose stimulation by light caused gonadal growth (Homma and Sakakibara 1971). In quail, therefore, the extraretinal receptors for photoperiodic stimulation of gonadal growth, at least, appear to be in the

25–7 Effect of pinealectomy on circadian perching activity in house sparrows. **A.** Constant darkness; removal of pineal on day 19 caused arrhythmic activity. **B.** Constant darkness; sham pinealectomy phase-shifted but did not eliminate the rhythm. **C** and **D** show entrainment in two pinealectomized birds to light-dark cycles; the light periods are bracketed by arrows and extend from days 16 to 29 in C and from 5 to 26 in D. (From Gaston and Menaker 1968. Copyright 1968 by the American Association for the Advancement of Science.)

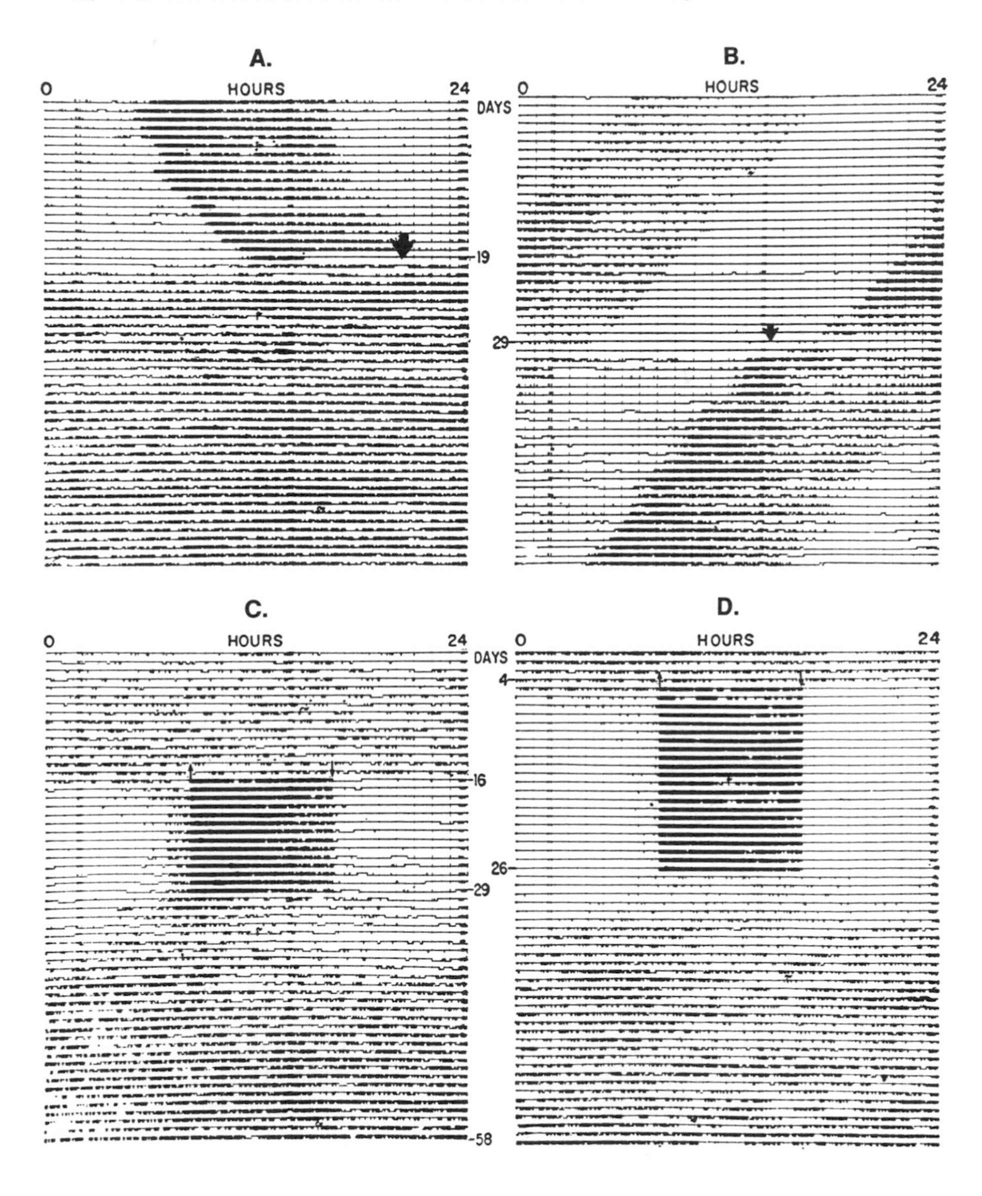

brain. Many questions remain to be studied in respect to the role of the pineal and the brain in the control of circadian rhythms and gonadal growth cycles.

ATTENTION

The concept of "paying attention" implies more than a heightened state of alertness. Typically an animal is attentive *to* something; it has a focus of attention and ignores "irrelevant" stimuli. Specifically, attention is a *selective* process or state in which responsivity to one set of stimuli is raised relative to other sets. For example, we are usually unconscious of the sensations of breathing or of saliva in our mouth. Yet, by focusing our attention on them, we perceive them. In fact, we know by introspection that we ignore most of our sensory input most of the time, yet we have the ability to sense even very weak stimuli if we attend to them. The mechanisms of attention are diverse even within one species. Various peripheral and central attention processes can be identified.

The existence of central mechanisms of attention can be illustrated by the well-known experiment of Sharpless and Jaspar (1956). While recording electrical activity from the cerebral cortex and observing behavior, they presented brief tones of 1 *kHz* to sleeping cats. At first, this stimulus produced a behavioral orienting response (the cat "paid attention" to it) and an electrical pattern of cortical desynchronization (an objectively recognizable pattern). After repeated tones had been presented at irregular intervals, the cat no longer responded behaviorally or electrically as recorded at the cortical level. That the cat's failure to respond or to attend to the 1-*kHz* tone was selective was demonstrated by presenting a similar tone which differed only in that it was at 5 *kHz*. The cat reacted to the 5-*kHz* tone in the same way it had reacted at first to the 1-*kHz* tone. Since the initial responses of naive cats to the two tones were the same for each tone, the difference in response is due to some process in the brain; in this case the process is known as *habituation*, which refers to the gradual loss of responsiveness to repeated stimuli.

In bats using echolocation for orientation, attention mechanisms are important in echo-processing by the auditory system. To use echolocation it is necessary for the auditory analyzers to make precise comparisons between the emitted ultrasonic pulse, which is very loud, and its echo, which is very faint and which arrives at the ear only milliseconds after the emitted pulse. Since auditory systems have difficulty recognizing a second sound immediately following a first, especially when the first is much louder than the second, analysis of the echoes should be aided by mechanisms that reduce the effective intensity of the emitted pulse within the auditory analyzer. Both peripheral and central mechanisms for this have been discovered.

In the middle ear of vertebrates are tiny bones (ossicles) that conduct sound from the tympanic membrane at the outer ear to the oval window at the cochlea. The tension of these bones, and their sound-transmission characteristics, can be varied by contraction of the middle-ear muscles. In *Myotis*, the activity of the middle-ear muscles is timed so as to reduce sound transmission across the ossicles during each pulse and increase it between pulses (Henson 1965, 1966).

Further attenuation of the signal representing the emitted pulse takes place in the bat brain (Suga and Schlegel 1972). Comparison of the firing rates of auditory neurons entering the inferior colliculus (which is greatly enlarged in echolocating species) in response to the bat's own call and a recording of its call showed a stronger response to the playback than to the emitted call. That the difference was not due to the strength of the stimulus or to the middle-ear muscles was shown by monitoring the auditory nerve and adjusting the intensity of the playback so that the response of the auditory nerve was equally strong to the playback and the emitted call. These experiments show that mechanisms of selective attention are probably significant components in the orientation systems of echolocating bats.

Attention implies selective alertness, which is separable only with difficulty from the behavioral and physiological concept of *arousal*. The functions of parts of the brain concerned with specific and relatively nonspecific arousal, particularly the reticular formation, have been extensively studied in mammals. The reticular arousal system runs longitudinally through much of the brain, receiving diverse sensory inputs and ultimately causing cortical activation. The reader is referred to current textbooks of physiological psychology for details, and to reviews (e.g., Weinberger 1971; Horn 1965).

In invertebrates the concept of attention has been neglected, but operational parallels may be present. In flying locusts, wind-angle changes evoke sideways abdominal movements, but on the ground the same wind-angle stimuli fail to evoke the abdominal movements (Camhi and Hinkle 1972). The locust could be said to be attending to these sensory inputs in flight but not on the ground. The fact that a neuronal explanation on the reflex level is available does not completely invalidate the parallel, but it does caution us not to expect fully comparable processes in animals as disparate as locusts and mammals.

Selective attention to certain stimuli should have considerable ecological importance. Prey selection by predators may be influenced by the past experience of the predators and by the relative abundance and palatability of different kinds of prey. An animal having a "search image" (L. Tinbergen 1960) for one type of prey might tend to overlook other types of available prey.

MOTIVATION

The major neuronal subsystems causing animals to actively seek or avoid certain sets of stimuli can be referred to as the motivational systems of the brain. Traditionally they include the nervous mechanisms controlling eating, drinking, pain, fear, sexual behavior, thermoregulation, and aggression, but they also are used for more specific needs, such as gnawing by rodents.

The motivational systems of a brain, regardless of species or phylum, are thought to share certain general properties. Of primary importance is their *organizing* action. A motivational system can be thought of as biasing a set of response systems so as to cause behavior appropriate to that motivational system, and often to cause the animal to disregard other less immediate or urgent motivational goals. For example, during hypothalamically elicited attack motivation the threshold of the mouth-opening reflex in response to tactile stimulus around the mouth is greatly lowered and the sensitive area increased (MacDonnell and Flynn 1966). This reflex participates in the biting used by attacking cats.

Motivational systems have a role in *reinforcement* during learning. The strength of activation of certain neuronal populations within a motivational system may determine the effectiveness of a certain "reward," such as food or heat, in motivating learning.

Motivational circuits also have an *activating* or energizing influence. A female baboon comes into heat through the activating influence of hormones participating in a neuroendocrine motivational system. Considerably more difficult to prove, but perhaps more profound for the reader, is the effect of a motivational system on *conscious desire*. Subjective feelings of desire certainly occur in some animals (man) as a consequence of the activation of motivational systems.

Since motivational systems are complex and interact with each other, control theory can be profitably applied to them. An excellent introduction to the use of control theory in animal behavior is McFarland's (1971) book, *Feedback Mechanisms in Animal Behavior*. A basic distinction in control theory is between open-loop and closed-loop systems. These are diagrammed in Figure 25–8. In an *open-loop system*, the output is a function of the input without correction for disturbance by external factors, and the input to the box (see the figure)

25–8 Open-loop and closed-loop control. (From McFarland 1971.)

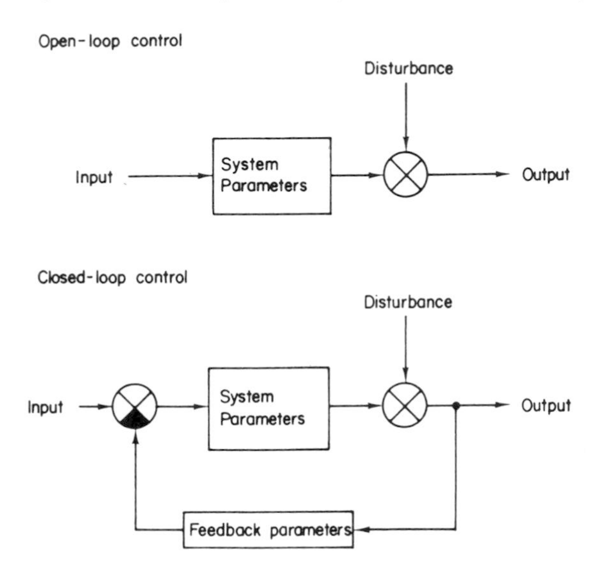

is not influenced by the output. In a *closed-loop system*, a mechanism is added that allows disturbance of the output to influence the input, usually by reducing the effect of the disturbance. In control theory such an addition is known as a *feedback loop*. If it reduces the effect of the disturbance, the feedback is negative; if it amplifies the effect, the feedback is positive.

Open-loop behavioral systems are found where movement is so rapid that effective modification by feedback would be difficult or impossible. The prey-catching strikes of a mantis or squid are examples. The pecking of young chicks at grain has been shown experimentally to be influenced slowly, and only to a small degree, by visual feedback (Hess 1956; Rossi 1968). In the area of motivation, the production of sex hormones is not known to be dependent on success or failure in mating, although mating is a primary output of the sexual motivation system.

Closed-loop systems are characteristic of the motivational systems controlling eating, drinking, and thermoregulation. The general pattern of these control systems is shown in Figure 25–9. In each case there is a "physiological" part that is concerned more with conservation (e.g., of energy, water, heat), and a "behavioral" part that is concerned more with acquisition (e.g., of food, water, heat). Critical for the operation of feedback in the system is a monitor. In the vertebrates, the monitors for these systems are believed to be specialized neurons

25–9 General plan of homeostatic motivational systems. (From McFarland 1971.)

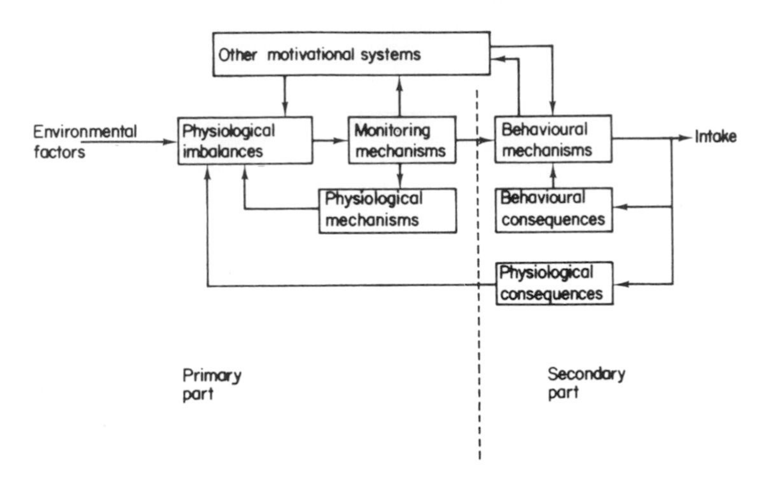

that occur mainly in the hypothalamus and preoptic regions of the brain. There are neurons sensitive to temperature, which are thought to function as thermostats; neurons sensitive to salt concentration, which are thought to function as osmoreceptors in the control of water balance; and neurons sensitive to blood glucose ("glucostats"), which are thought to control food intake. The scheme in Figure 25–9 is only a crude approximation of the actual web of relationships. The "drive" to eat, drink, or thermoregulate can be seen from Figure 25–9 to represent a state of the system in which input mechanisms are activated. Since these phenomena are better conceived in relation to control theory than in relation to "drive theory," the latter will not be considered in this book. For critical discussion of drive concepts, see Valenstein (1968), Hinde (1970), and McFarland and Sibly (1972).

The motivational systems of even the simplest animals share certain operational properties. A jellyfish, a blowfly, and a rat, if recently fed, respond only sluggishly to food. In all three cases there is, as a consequence of having eaten, negative feedback on eating behavior. Of course, the mechanisms are quite different in the three species. In the jellyfish, the lowered response to food may be mainly to physical factors concerning the geometry of the food chamber; there is no brain and probably no central nervous control of responsiveness to food. In the fly, distension of the crop, where the ingested food is stored, causes a signal to be transmitted via the recurrent nerve to the central nervous system, where the feeding responses are inactivated or inhibited. Cutting the recurrent nerve causes loss of feedback regulation of eating (Dethier 1964). The rat brain receives information about recent ingestion of food from the mouth, gut, and glucose-sensitive neurons in the brain; its control mechanism is provided with more inputs and is organized more complexly, both in terms of control theory and neuronal components. Furthermore, in the rat the homeostatic control of food input is closely connected functionally with an impressive capacity for learning and for modification of behavior.

The motivational systems controlling sexual, aggressive, and self-protective (avoidance, escape, defense) behavior cannot be modeled as simply as the homeostatic systems just outlined. Although aggression may at times seem spontaneous, it serves mainly as a means to satisfy needs concerned with food, water, sex, shelter, and other necessities of survival and reproduction. Therefore, only in relation to these latter systems can the role of aggression in motivation be fully understood. Interactions among neural motivational systems are complex and cannot be treated here. A fuller treatment of the neural and hormonal control of aggression and reproductive behavior can be found in textbooks of physiological psychology.

CONCLUSIONS

The "programming of behavior," or the overall organization of behavior in time and space, could, in theory at least, be performed by an array of reflexlike mechanisms in which each stimulus provoked a corresponding response "automatically" and unvaryingly—in short, the "republic of reflexes" of von Uexküll (1909). Apparently no organism has yet been studied whose behavior is quite that simple. In all cases, the relationships between stimulus and response are modulated by mechanisms at "higher" levels than that of the simple reflex. We have considered in this chapter all mechanisms that intervene between sensory and motor processes, but have emphasized those with a more global and "unreflexlike" influence on behavior, such as the mechanisms underlying the phenomena of activity rhythms, attention, and motivation.

The simplest integrative mechanism, in the sense of a mechanism that selects a response appropriate to the stimulus, is the one-celled independent effector, of which nematocysts are the best example. Other integrative assemblages of two or three cells could in theory exist, but the simplest one known to occur regularly involves at least four cells—namely, the monosynaptic reflexes of vertebrates.

Rhythms in behavior have periods ranging from a fraction of a second to a year. The source of short-term rhythms can sometimes be traced to single oscillator neurons or to small groups of neurons. The physiological basis of circadian (≅ 24-hour) and other long-term rhythms is more complex. In cockroaches, the origin of circadian activity rhythms has been traced to the optic lobes; in birds, the pineal gland is essential for circadian activity rhythms. In no animal is the mechanism fully understood.

Attention is a selective process in which responsivity to one set of stimuli is biased relative to other sets. It is mediated by both peripheral and central mechanisms. For example, the middle-ear muscles control transmission of sound from the outer to the inner ear; and the brain controls the effectiveness of auditory stimuli in evoking responses.

The major neuronal systems causing animals to actively seek or avoid certain sets of stimuli are referred to as the motivational systems of the brain. These neural systems organize, reinforce, and activate behavior in accordance with the biological needs of the animal. Since most or all motivational systems are complex and involve negative feedback, the use of control theory is appropriate, particularly in the treatment of eating, drinking, and thermoregulation. Control theory is applicable to animals as diverse as jellyfishes, blowflies, and rats; but the physiological mechanisms involved are, of course, quite different.

THE DEVELOPMENT OF SPECIES-TYPICAL BEHAVIOR

The so-called "nature-nurture" (genotype-environment) problem is not to distinguish which traits are genotypic and which are environmental, for all traits are genotypic and environmental.

Theodosius Dobzhansky, 1950

SPECIES DIVERSITY AND SPECIES CONSTANCY of behavior depend upon predictable patterns of development. In some species, the course of development is relatively inflexible; learning and experience appear to play minor roles in many insects, for example. In other species, the structure and function of the brain make possible a strong reliance on learning and experience; in monkeys, certain kinds of early social experience can render an individual essentially unfit to compete in the natural world. Such differences in the patterns of development of behavior have long fascinated evolutionists and behaviorists. In Part VI some of the issues, problems, and biological phenomena in the development of behavior will be examined in evolutionary perspective.

Undoubtedly, the greatest conceptual problem in the development of behavior is in understanding the interaction of the *two fundamental determinants* of the course of development, *inheritance* and *experience*. The developing animal is constantly exposed to stimuli from the environment, and these may have far-reaching and subtle effects on his subsequent behavior. How the developing individual responds to environmental stimuli may determine whether he lives or dies, or, more subtly, it may determine how successfully he lives. At every developmental stage, certain ways of reacting or not reacting to crucial stimuli will increase the individual's chances of a successful reproductive life. Individuals who are genetically predisposed to react in these ways will survive longer and leave more offspring than others. In short, natural selection acts at every phase in the development of behavior, not just on adults. As a consequence, every phase in development may be regarded as a product of evolution. This does not mean that every response made by a developing animal is predetermined solely by its evolutionary history. It simply means that the animal's abilities to act and react are determined through evolution. In some species, selection has favored individuals with wide-ranging abilities to adapt to local conditions. In others, the capacity of individuals to adapt to varying environmental conditions is greatly restricted, with the result that the species is successful only in a narrowly limited set of environmental conditions. Of course, neither of these types, the generalist or specialist, should be regarded as superior to the other; each is adapted in its own way to its own environment and neither would prosper in the environment of the other.

Of fundamental importance for understanding the development of behavior is the interaction between inheritance and experience. This may be expressed in the following way: Starting from the beginning of development, with the fertilized egg or zygote, P_1, its phenotype at the next stage of development, P_2, will be determined jointly by the genes that are active in guiding its growth and differentiation during the intervening interval, G_1, and by the environment in which the development takes place, E_1:

Zygote + genes + environment →
phenotype at next stage of development

$$P_1 + G_1 + E_1 \rightarrow P_2$$

The phenotype at the next stage of development, P_3, will be determined by the way in which P_2 has been changed by the genes, G_2, and environmental influences, E_2, affecting development between P_2 and P_3.

$$P_2 + G_2 + E_2 \rightarrow P_3$$

In the general case, the phenotype or behavior at any one time, P_t, will be determined by the preceding influences of genetic (G_{t-1}) and environmental (E_{t-1}) factors acting on the preceding phenotype, P_{t-1}.

$$P_{t-1} + G_{t-1} + E_{t-1} \rightarrow P_t.$$

A little reflection on this schema should reveal the truth of the following provocative statements:

1) No phenotype (e.g., a behavior) of an individual can be attributed entirely to genetic factors.

2) Similarly, no phenotype of an individual can be attributed entirely to environmental factors.

More specifically, it is wrong, strictly speaking, to say that a behavior is learned or that it is innate. For example, it is incorrect to say that speaking German is a learned behavior or that the song of a cricket is innate (if by *innate* is meant *genetically determined*). The fallacy in these cases is that the assertions ignore the fact that every behavior must develop in an environment and its development must unavoidably be influenced in one way or another by genetic factors. Anything else is impossible. This leads us to the next statement.

3) It is wrong to ask a question about the behavior of an individual that assumes a *dichotomy* between inheritance and environment in the ontogenetic determination of a behavior. In other words, it is wrong to ask whether a behavior is innate or learned. Such a dichotomy does exist in the *origin* of the influence, i.e., either inheritance or environment, but it does not exist in the *product*, which must be viewed as the result of accumulated interactions between inheritance, environment, and preceding phenotype.

If all these approaches are wrong, how then should we proceed to study the development of behavior? A slightly different approach is needed. Instead of looking at the behavior of an individual, we may *compare* his behavior now with his behavior earlier. Or we may compare his behavior with the behavior of another individual. The following statements apply:

4) A *difference* in behavior between individuals may in theory be wholly attributable to genetic factors (if they are reared in identical environments).

5) A *difference* in behavior between individuals may in theory be entirely due to environmental factors, such as learning (if they are genetically identical twins).

6) Differences in behavior among individuals may be due to both genetic and environmental factors acting together.

It is through analysis of differences among individuals or populations that the effects of inheritance and environment on development of behavior can be most reliably separated. Another useful approach is to analyze differences over time in the same individual. We commonly conclude that learning has occurred when a change in behavior is correlated with a change in the stimuli coming from the environment, usually in the form of a reinforcing experience. An experienced hungry rat quickly depresses a lever and obtains food; a naive rat does not. From these and other observations of the rat's experience, we conclude that the differences in behavior between the naive and experienced condition is due to learning. We assume that genetic factors have not acted to contribute to the difference. It is a pretty safe assumption, but it may not always be true. Some behavioral changes in time are quite obviously due to genetic factors. The differences in behavior between a tadpole and a frog cannot be attributed only to environmental influences. Genetic factors can cause changes in behavior over a period of time through their effects on structure and function. Despite the complications they entail, the following statements also must be correct:

7) A difference in the behavior of an individual at different times in his life can be due entirely to learning or other environmental influences.

8) A difference in the behavior of an individual at different times in his life can be due essentially to genetic factors acting on development (i.e., maturation).

9) Differences in an individual's behavior at different times of life can be due to both genetic and environmental influences acting together.

In practice, when it is stated that a behavior is learned or that it is innate, one finds that the factual basis of the statement is actually an analysis of differences. Most behaviorists are aware of this, but many have become accustomed to routinely making statements of the type prohibited by our statements 1 and 2. This is unfortunate, but the practice is so widespread that it is probably best to simply regard such statements as convenient oversimplification. No behaviorist really believes in preformationism, nor, I trust, does anyone believe that an animal's brain is a *tabula rasa* or "blank slate."

Perhaps much of the confusion comes from unconsciously mixing population-level concepts, such as heritability and genetic determination, with individual-level concepts, such as learning and maturation. Readers should satisfy themselves that heritability has virtually no bearing on development and that developmental studies reveal nothing about heritability.

Probably the most frequent type of experiment used in analyzing the development of behavior involves rearing individuals in controlled environments that differ in regard to the kind of experience being investigated (e.g., isolation experiments). Much controversy has arisen about the interpretation of such experiments (Lorenz 1935, 1965; Lehrman 1953, 1970). They have been used in the past to justify statements about whether or not a particular behavioral trait is "innate," meaning in this case "not learned from another individual" (rather than the other meaning, "genetically determined").

The use of developmental experiments to make statements about the genetic basis of behavior is very risky. In order to firmly establish the *genetic nature of behavior* it is necessary to perform *genetic experiments* or to work with differences that are known to be genetic. Developmental experiments can reveal the role of the experimental variables that are manipulated in the experiments, but *not* of the genetic variables (unless they are also manipulated). For example, if the song of a cowbird develops normally when individuals are raised from the egg in auditory isolation, this allows the conclusion that auditory experience with other individuals is not necessary for the development of a normal song, but it does not allow the conclusion that the song itself is genetically determined. It shows that the *interaction of environmental and genetic factors* under the conditions of this experiment allows development of the normal song. Only one class of experience has been manipulated in the experiment, not all classes, and that is the only class about which conclusions can legitimately be drawn. It is important to emphasize in this context that because the phrase *genetically determined* applies to genetic *differences* (or variances) and not to traits or individuals, it cannot usually be applied in a purely developmental experiment.

The kinds of behavior whose development has been studied are extremely varied, including reflexes, locomotion, vocalizations, and various types of social behavior. It is not possible to cover all these fairly here. I have chosen topics that are of special interest in regard to the evolution of behavior: (1) The behavior of embryos illustrates with particular clarity the important role that maturation of the nervous system plays in behavior. In some animals, development of behavior proceeds even before somatic sensory connections to the CNS have been established, and hence, before learning or sensory stimulation can be significant. (2) Sexual imprinting is of considerable interest in relation to behavioral isolating mechanisms and the choice of a mate. (3) The development of song in birds illustrates the diversity of developmental patterns of a trait that shows conspicuous species constancy and diversity in the adult and also is important as an isolating mechanism.

26 The Beginning of Behavior in Embryos

THE STUDY OF behavior in embryos, although a specialized field, is one with far-reaching implications for ethology and psychology. Some of the major problems and controversies in behavior are illuminated by the study of embryos much more clearly than they could be in adults. For example, the existence of spontaneity in behavior is more easily and convincingly demonstrated and studied in embryos than adults. This facet of embryonic behavior shows the independence of one aspect of behavior from sensory inputs and helps us to evaluate the relative roles of genetic and environmental factors in development. Similarly, the dependence of behavior development on the age of maturation of the underlying neuronal circuits can be more easily studied and appreciated with embryos.

Two main research strategies may be discerned in the study of behavioral development in embryos: (1) the *correlation of maturation* in the nervous system with ontogeny of behavior, and (2) the experimental *manipulation of prenatal experience* suspected to influence later behavior. In embryos, the role of neural maturation is conspicuous and has been a center of attention. Its role in development has been reviewed by Jacobson (1966, 1971), Hughes (1968), Sterman et al. (1971), Gottlieb (1973), and others. The role of prenatal experience was for a long time relatively neglected, but has recently received greater attention (e.g., Gottlieb 1968; Joffe 1969). In this chapter the primary emphasis will be on neural maturation, since its role is more clearly seen in embryos than in later developmental stages.

The First Behaviors / The first discernible actions of vertebrate embryos that are generally acknowledged to be "behavior" consist of lateral bending of the spinal column at the level of the anterior trunk, commonly at the cervical (neck) level. These movements involve only a few myotomes (muscle segments) at first; but as more segments mature, more are included. Movement occurs first in the cervical region, undoubtedly because this region is more advanced in its morphological development than are other regions of the body at this stage.

The stage of morphological development at which these cervical flexions first appear varies from species to species (Figure 26–1). In salamanders and other animals whose larvae are free-swimming, movement begins at a relatively early stage when limbs or even limb buds may be lacking. In contrast, the first cervical movements in placental mammals, whose embryos are completely protected in the uterus and need not swim, occur when limbs are well developed and even when rudiments of fingers and toes may be seen. Perhaps the advanced stage of the digits at first movement in primates is an adaptation to their use by the newborn in grasping the mother. Avian embryos are intermediate in regard to the state of limb development when cervical flexions first occur.

Later additions to this pattern vary with the species. In forms with early swimming stages, S-type undulations sweep backward over the body. These are really alternating left and right flexions that flow posteriorly in waves. In protected embryos, such as those of mammals and birds, movements of the limbs appear relatively early.

Spontaneity / The question of whether or not behavior can be spontaneous—in the sense of not being dependent on a particular sensory stimulus—is a fundamental one in the study of behavior. It relates to the long-standing controversy about the relative importance of endogenous (internal) and exogenous (external) factors. In embryos of many nonmammalian species, the first movements to appear do not seem to be the consequence of any recognizable stimulus. The physiological basis of these movements varies with the species.

In embryo sharks and teleosts (references in Hamburger 1963; Abu

26–1 Comparison of morphological development at start of embryonic movement in salamander, chick, and human embryos. (From Hamburger 1963. Reprinted with permission of the Quarterly Review of Biology.)

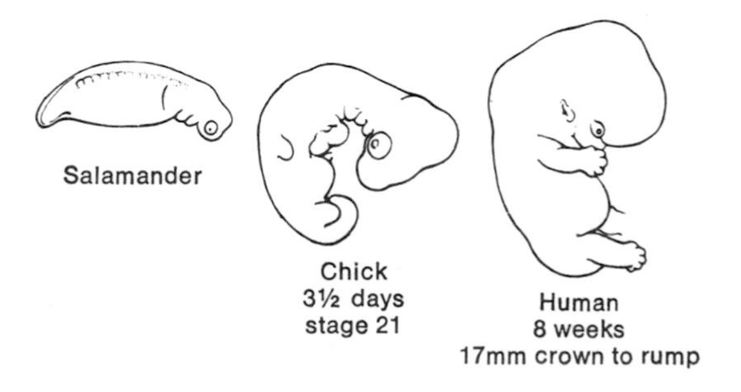

Gideiri 1969), a period of motility has been observed in embryos that lack functional motor neuronal connections to the muscles. The event initiating motility in this case must occur at or in the muscle—hence the term *myogenic*. This type of movement can be distinguished in practice by its persistence after treatment with curare, a drug that blocks transmission from nerve to muscle. It is also recognizable when motility is present before motor neurons have grown to contact the muscles involved or when motility persists after the (motor) innervation of the muscles has been severed. Muscles may be mechanically or electrically stimulated to contract in the absence of their innervation. Myogenic motility is not known in mammals, birds, amphibians, or reptiles.

Nonmyogenic motility is dependent on motor nerves from the spinal cord to the muscles. Consequently, it has been termed *neurogenic*. In adults, muscle contraction can be initiated by appropriate sensory stimuli that make functional contact directly, via sensory neurons that synapse on motor neurons, or indirectly, via interneurons. Such activation has been termed *reflexogenic* (or reflexogenous). In embryo chicks motility is present before the sensory neurons have made functional contact with the motor system (either directly or via interneurons). The motility depends on the functional integrity of neuromuscular synapses. This condition, in which activity is generated in the nervous system independently of sensory input, has been termed *self-generated* or *automatic neurogenic motility*. Nonreflexogenic motility can also be caused by a deficiency of oxygen in the embryo's tissues, or by abnormal concentrations of other substances; but these endogenous stimuli can only be regarded as the "cause" of the spontaneous motility when they deviate from the normal level. In the normal embryo, both reflexogenic and nonreflexogenic motility depend upon normal physicochemical conditions.

Of what importance is nonreflexogenic behavior to the embryo? This is difficult to say because no experiments seem to have been devoted to the question, but the duration of nonreflexogenic motility in the embryo's program of development may give some idea of its importance. In the toadfish (*Opsanus tau*) this period extends for two and one-half weeks. In reptiles it is intermittent during much of the embryonic period (Hughes et al. 1967; Decker 1967).

The evidence for a continued role of nonreflexogenic motility in the chick even after reflex connections have been established may be summarized as follows: (1) The cyclic pattern of activity and inactivity present before reflex connections are complete (days 3 to 7½) continues on to 13 days, when activity becomes mostly continuous (hatching occurs at 18 days). The increase in motility during development is shown in Figure 26–2. (2) It might be thought that some of this motility is stimulated by movements of the amnion (a contractile membrane in the egg), but the motility is not in phase or coordinated with amniotic contractions or with any other evident external stimulus. Self-stimulation by passive contact of body parts with each other might also be regarded as a source of motility, but as Hamburger (1963:347) wrote, "Our observations over long periods have never given any indication that head movements are elicited when the legs in their vigorous movements touch the head, or vice versa." (3) Further evi-

26–2 Chronological development of motility in chick embryos. The lengths of the active and inactive periods in each cycle vary during development, as does the duration of a complete cycle, which is shown by the top line. (From Hamburger et al. 1965.)

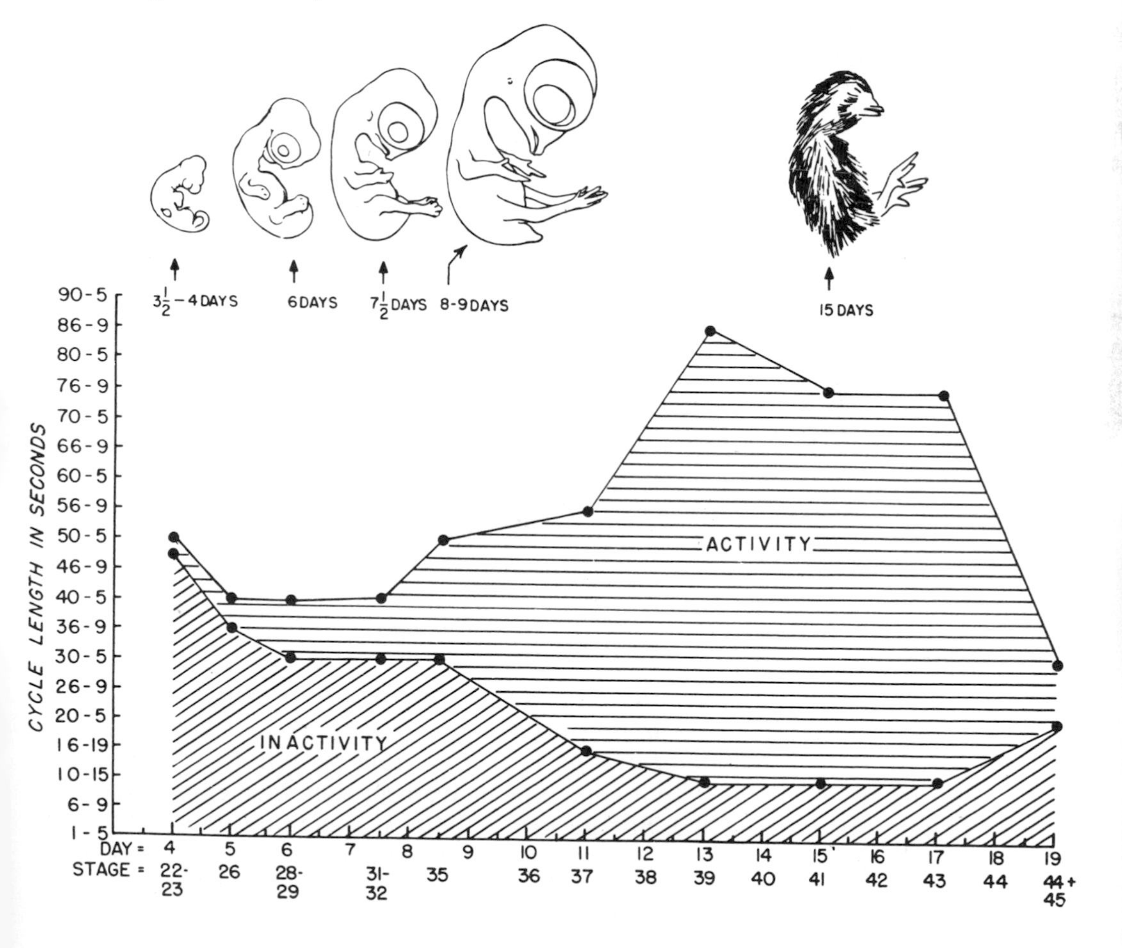

dence involving operations on the nervous system (see below) supports the concept that nonreflexogenic motility continues to be important to the embryo long after reflex circuits have been established.

Despite the lack of obvious causation of motility by self-stimulation, the question of self-stimulation has warranted further experimentation. The question was investigated by stimulating 11-day-old chick embryos with a baby's hair as a tactile stimulus. The evoked behavior was unlike spontaneous motility in that it was localized to the region of stimulation. Also, when applied during an inactive period, it did not trigger a bout of generalized motility—nor did the sensory feedback (mainly proprioception) from the evoked reflex behavior. When applied during active or inactive periods, it did not extend the average duration of motility. The independence of the activity and inactivity phases of embryonic motility from tactile stimulation is illustrated in Figure 26–3.

But perhaps some unknown internal stimulus might be responsible for the motility. That this is probably not the case was shown by experiments in which a segment of the spinal cord was removed in early embryos (Hamburger 1963; Hamburger et al. 1965). In these experiments, isolated parts retained rhythmic motility but lacked synchrony, as shown in Figure 26–4. The lack of synchrony is important because it indicates that each segment can generate its own motility, and because it shows that there is not just one source of rhythmic motility.

More conclusive evidence of independence from peripheral sensory stimulation was provided by the persistence of comparable rhythmic motility of the legs in a completely deafferented lumbar segment (one having no sensory nerves and no connections to the anterior spinal cord; Hamburger et al. 1966). The operations on the nervous system in these experiments are shown in Figure 26–5.

A decrease in activity of 10 to 20 percent in segments isolated from the brain indicates a contribution from this source (Hamburger and Balaban 1963). The contributions of different brain areas to embryonic motility at different stages of development were described by Decker and Hamburger (1967).

Evidence from experiments in which parts of the nervous system were removed strongly implicates the nervous system itself as the source of the spontaneity. But locating the source within the CNS required recording the activity of single neurons (units). In such experiments (Hamburger 1970; Sharma et al. 1970), it was found that, after all the sensory nerves to an isolated lumbar segment of spinal cord were cut, active neuronal firing persisted only in the region where the motor neurons were located. Activity in the sensory area virtually ceased after deafferentation or anesthesia of the muscles on the same side as the electrode, and was reduced by transection of the cord anterior to the recording site. Summarizing, Hamburger (1970:146) wrote, "One cannot escape the conclusion that, although higher cen-

26–3 Is the activity rhythm of embryo chicks caused by sensory stimulation? **A.** The activity cycle of an 11-day-old chick embryo is clearly divided into periods of activity and inactivity. The durations of each phase in seconds have been written on the record. The horizontal base line shows inactivity; the vertical marks indicate activity. **B.** To see if the length of activity and inactivity phases could be altered by sensory stimulation, an 11-day-old chick embryo was stroked 7 times at 4-second intervals with a loop of baby hair during inactivity periods. This procedure consistently caused reflex responses to touch, but did not affect the cycle of motility, as shown in the graph. These results suggested that sensory stimulation was unimportant as a cause of embryonic behavior. (From Hamburger 1963. Reprinted with permission of the Quarterly Review of Biology.)

A.

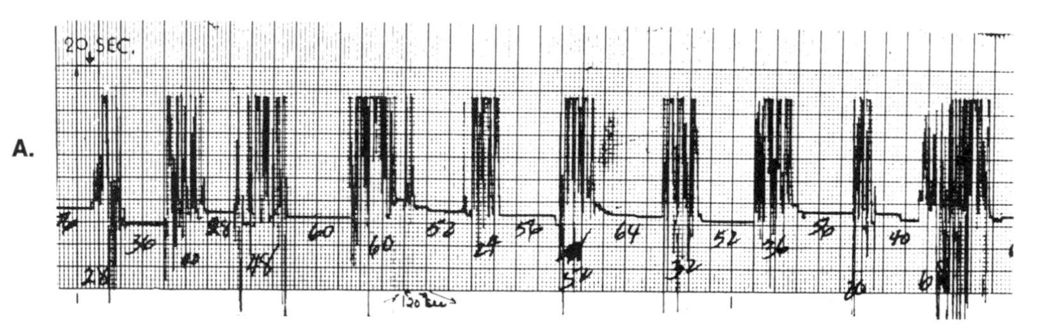

B.

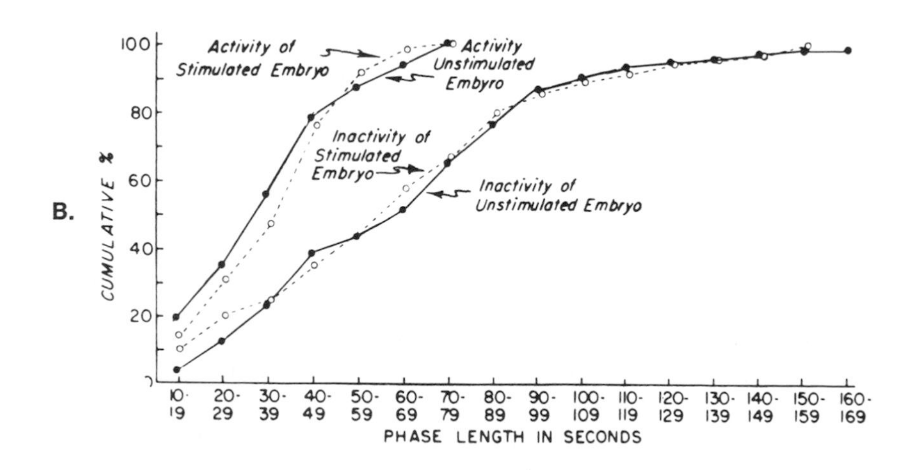

26–4 Is the nervous system the source of activity rhythms in chick embryos? If the source of the rhythm is external to the nervous system, then transection of the spinal cord should not alter the rhythm of the isolated segments. A gap in the spinal cord was made at the thoracic level very early, at stage 11 or 12 (*above left*). The embryo is shown at stage 30, 7 days old (*above right*). The effect of the operation on motility of wing- and tail-segments is shown in an embryo at stage 32 (*below*). Both segments retained a motility rhythm, but the two rhythms were not in phase. These results suggested that the capacity to generate the rhythm spontaneously is located in the nervous system, is found in both segments, and is not restricted to one location. (From Hamburger 1963. Reprinted with permission of the Quarterly Review of Biology.)

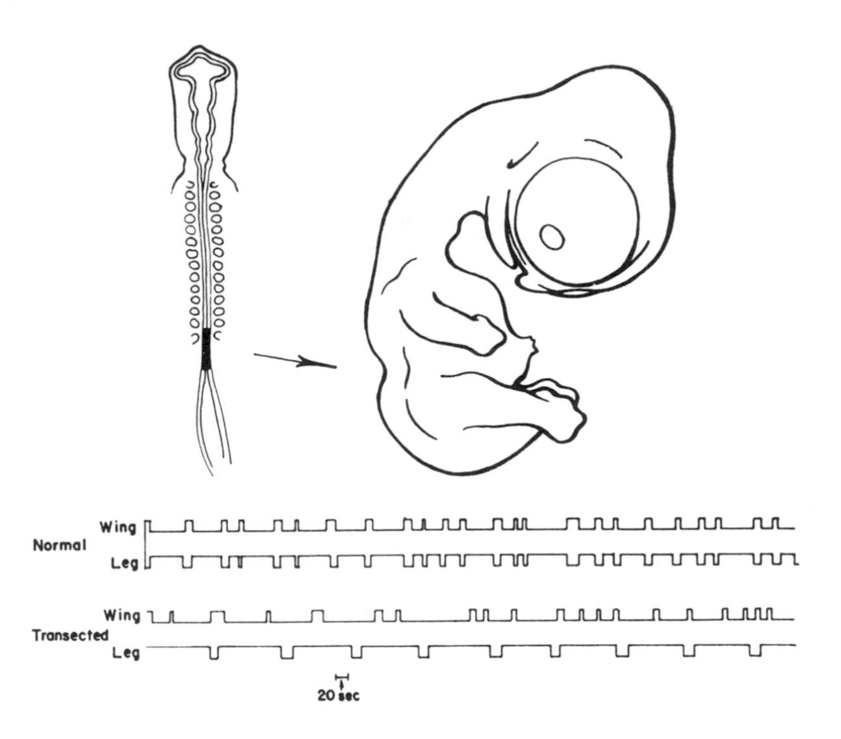

ters and sensory input contribute to the overall firing level of this region, the latter contains elements that continue to initiate discharges of nerve cells in the absence of these sources of input."

The Functions of Motor Neurons / The main thrust of the preceding findings and arguments has been to establish the importance of the motor neurons as a source of spontaneity in the behavior of embryos and as the major activating mechanism underlying most behavior in embryos (Hamburger 1971). This factual evidence for the importance of spontaneity has been obtained for all the vertebrates studied except

26–5 The operation used for deafferentation. This procedure enabled a critical test of the hypothesis that the spontaneous motility of the leg in chick embryos is independent of sensory input. Sensory input from segments anterior to the leg was prevented by the cut shown in the upper left and right, in which segments 23–27 inclusive were removed entirely. Sensory input from the leg segments was prevented by removing the dorsal half of the spinal cord in this region, thus preventing the growth of sensory nerves. The operation was performed in two-day-old embryos. (From Hamburger et al. 1966.)

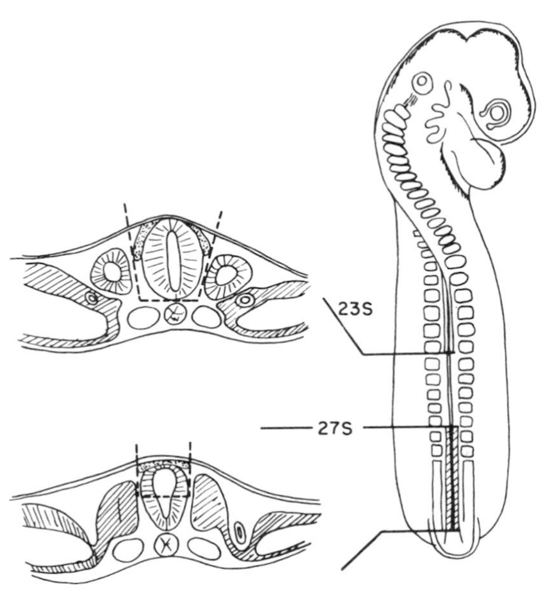

mammals; and even in mammals, despite the theoretical bias of earlier authors against even looking for evidence of spontaneity, some indication of spontaneity has recently been found (Hamburger 1970; Narayanan et al. 1971). It has also been established that, in all the vertebrates, the neural circuits mediating many reflexes mature in the embryo. These reflexes also depend on the motor neurons. Consequently, the motor neurons play a *double role:* on the one hand, they *activate* movements of all the muscles of the embryo; and on the other, they stand ready to mediate *reflex* responses to environmental stimuli.

Evolutionary Adaptations / This dual role seems to have its adaptive value in two contexts. In the relatively constant environment of the embryo, stimuli that might be depended on to activate all muscles are few. There are indications that normal development of the skeletal system, especially the joints, and maintenance of certain neural circuits depend on use of the muscles and neurons involved (Hamburger

1970; Gottlieb 1970; further references in Narayanan et al. 1971). Spontaneous firing of most motor neurons is a reliable way to ensure adequate exercise of all muscles for proper development. Further experiments are needed to establish the adaptive value of embryonic spontaneity of behavior.

Any neurobehavioral system in an embryo must necessarily play a dual role in another sense. It serves the needs of the moment, but it also prepares the embryo for life in a more challenging environment as a larva or neonate. The second of these functions seems to explain many aspects of embryonic development. Quite simply, the embryo develops first those circuits of immediate use to it, and then the circuits that will be needed first at hatching or birth. Thus in most mammals the first area to develop reflex sensitivity is the snout; this will be needed in nursing and as a sense organ to locate the teat. But in primates, grasping with the digits develops equally early. Since most newborn primates must hold on to their mothers or fall (the mother does not always hold them), the adaptive significance of early development of this ability is evident. The diversity of reflexes that can be elicited in embryos is apparently far in excess of what the embryo really needs, especially in view of the hypothesized role of spontaneity. On the other hand, as soon as he lands in the neonatal environment he must be fully equipped to respond appropriately and without comparable prior experience to a much greater range of stimuli than he ever knew. In brief, the motor neurons play a dual role adaptively as well as physiologically.

The Program of Development / Although it is inconceivable that behavior could appear in an embryo before the neural circuitry that mediated the behavior had reached functional maturation (except in the case of myogenic motility), it does not necessarily follow that the first appearance of a particular behavior pattern must be closely correlated with maturation of its underlying neural mechanisms. Conceivably, the neural mechanism could develop long before it was needed or used. Furthermore, if learning were important in the development of a behavior pattern, it is likely that no correlation with neural circuits would be found (except in those influenced morphologically by the learning experience). Consequently, one of the principal generalizations that resulted from the work of Coghill (1914–1936) and other developmental neuroanatomists was that the correlation between first appearance of behavior and neural mechanism in embryos was in fact very good. This type of result is strong testimony for the major importance of factors other than some type of learning in the development of behavior. Since considerable controversy has existed on the importance of learning in vertebrate behavior, this is an important point.

The development of locomotory patterns in the embryo bullhead (*Ictalurus nebulosus*) illustrates this approach (Armstrong and Higgins 1971). Swimming in this species is in part activated, coordinated, and directed by a pair of unusually large neurons called Mauthner cells (Figure 26–6). These neurons, whose cell bodies are located in the

26–6 Mauthner neuron and embryonic stages in the bullhead. **A.** Mauthner neuron of an embryo in the optic-midbrain phase of development. The axon, *a*, arches to the opposite side to descend in the median longitudinal fasciculus to the spinal motor neurons. The ventral dendrites, *v*, extend toward inputs from the optic tectum. The lateral dendrite, *l*, extends toward inputs from the labyrinth. The root of cranial nerve VII (*f*) is not involved. **B.** Embryo in spinal phase. **C.** Embryo in optic-midbrain phase. (From Armstrong and Higgins 1971.)

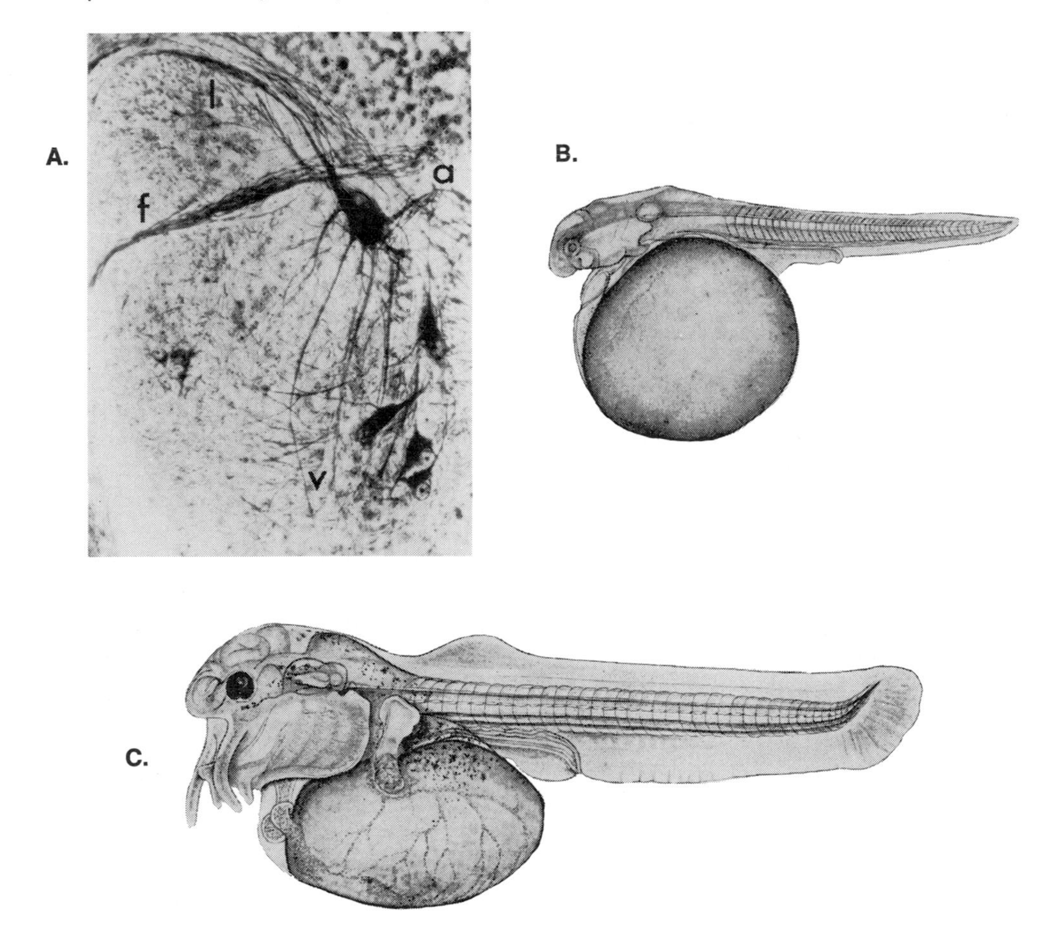

midbrain, send axons down the spinal cord that synapse on the primary motor neurons of the body somites. Their dendrites receive information directly, from the labyrinth (semicircular canals, utriculus, sacculus), and indirectly, from the eye via the optic tectum, as well as from the cerebellum. In the bullhead embryo, three phases of development may be recognized corresponding to the phases of progressively greater control of locomotion by the brain through the Mauthner cells as their synaptic connections form and begin functioning.

In the *spinal phase*, the embryo is essentially a "spinal animal"; that is, its behavior is completely controlled by spinal mechanisms without apparent influence from the brain, which has just begun to differentiate. The behavior at this stage consists of C-shaped lateral flexions—sporadic, nonmyogenic, and spontaneous—that are produced by contractions of the somites of one side. Severing the brain from the spinal cord at this stage does not affect this behavior. In the latter part of the spinal phase, the Mauthner axons reach the motor neurons but do not alter behavior. This is apparently because they lack functional sensory input at this stage, because if the receptive field of the Mauthner dendrites is stimulated electrically, a contraction of the somites of the opposite side is obtained, indicating the presence of functional efferent synapses.

The second or *bulbar phase* is characterized by the appearance of undulatory swimming—which remains, however, sporadic and undirected. Its appearance coincides with the development of sensory connections from the labyrinth to the dendrites of the Mauthner cells. Unlike the spinal phase, severing the brain from the spinal cord in the bulbar phase abolishes this swimming activity, as does cauterizing the labyrinth. The labyrinthine input activates the pair of Mauthner cells which, according to Armstrong and Higgins, plays a major role in coordinating and activating the motor neurons controlling undulatory swimming.

In the third or *optic-midbrain phase*, visual control is introduced via recently matured afferent connections of the Mauthner cells from the optic tectum. The optic input makes possible directed responses to optic stimuli. At first a photokinesis occurs in response to sudden changes in light intensity. Soon after, a negative phototaxis appears. This response is useful to the embryo because it helps to keep it in the shadows of its nest or in depressions in the mud for a long period while it is still dependent on its yolk sac, vulnerable to predation, and relatively helpless. Later, when more precise visual orientation is made possible through the photomechanical orienting properties of the retina and optic tectum, the embryo becomes positively phototaxic. This switch is adaptive because it occurs when the yolk sac has been used up and the fish must look for its own food.

In summary, the program of development of early behavior in embryo bullheads follows closely the program of maturation of functional synaptic connections from sensory inputs to the Mauthner cells that activate, coordinate, and direct locomotion.

Individuation versus Reflex Integration / The earlier years of study of embryonic behavior were marked by a controversy between two points of view. The first, known as *individuation,* is best stated in the words of its author, Coghill (1929:989):

"The behavior pattern from the beginning expands throughout the growing normal animal as a perfectly integrated unit, whereas partial patterns arise within the total patterns, and by a process of individuation acquire secondarily varying degrees of independence . . . always under the supremacy of the individual as a whole."

The opposing view, led by Windle (1940), claimed that behavior was not integrated at first and that the partial patterns emerged first and were later incorporated into coordinated behavior patterns.

As pointed out in an excellent review of the controversy by Hamburger (1963), the disagreement was needless and would not have come about had each worker studied a greater range of species. Coghill had chosen the spotted salamander, *Ambystoma maculatum,* for his studies in the mistaken belief that it could be used to establish principles which would hold for vertebrates generally. Windle was concerned mainly with mammals. Furthermore, he followed the dominant emphasis in his day on reflexes and stimulus-response analyses of behavior.

Coghill's concept was valid for salamanders. Salamander larvae live in ponds and must be able to swim rather early, since their yolk sac is not large and they must soon find their own food. The simple, undulatory swimming movements are integrated nearly from the start by the early maturation of a longitudinal coordinating system (fasciculus longitudinalis medialis) that carries neural activity down the spinal cord to the motor neurons of each segment. Walking, stalking, and catching prey emerge also in a coordinated and functional form as appropriate neural connections mature. Local limb reflexes develop when their connections mature, even if deprived experimentally of connections with the longitudinal coordinating system (Holtzer and Kamrin 1956).

When applied to birds, Coghill's concept of individuation breaks down. After the first C-shaped movements, coordination starts to break down. Chick embryos do not show the degree of coordination in behavior that salamander embryos do. Rather, all the muscles of the body, limbs, and head seem to contract together for a brief period (as in Figure 26–4C), with no semblance of any pattern of coordination.

In mammals, the longitudinal conducting system matures at about the same age as the first local sensorimotor reflex arcs, at least in some species (Hamburger 1963). Spontaneous motility appears to be a normal part of embryonic behavior in rats, following a pattern similar to that in other vertebrates (Narayanan et al. 1971), although this remains to be proven by experiments like those done on chicks. The concept of development of behavior by integration of reflexes fails as a general principle, not only because of the difficulty presented by salamanders, but more importantly because it fails to incorporate the spontaneity we have seen in various vertebrates.

If generalizations are to be made, now that the concepts of individuation and reflex integration have been rejected, two points may be ventured. First, embryonic motility that is known or suspected to be spontaneous is widespread in the vertebrates, although it varies in conspicuousness and pattern in different species. Second, the maturation of reflexes is superimposed on spontaneous motility when sensorimotor reflex circuits mature, in patterns that are adaptive for each species. No one general rule can predict these patterns because of the diversity of embryonic adaptations. Embryonic environments vary from warm wombs to cold mountain streams. We can expect the pattern of behavior development in embryos of a given species to reflect the adaptation of that species to its own embryonic environment.

Inheritance and Environment in the Growth of Nerve Cells / The regularity of the species-typical pattern of development of embryonic behavior and its neat correlation with the timetable of completion of the underlying neuronal circuits have led to a contemporary consensus among biologists that the developmental programs of embryos are dominated by genetic factors, with the environment playing a permissive or weakly stimulative role. The development of vision and retinotectal connections in the toad *Xenopus* illustrates the precision of such a developmental program and its permanent effects on behavior (Jacobson 1969). When the eye of a *Xenopus* embryo is inverted before neural connections have formed between retina and brain, it will establish normal connections with the optic tectum in the midbrain if the operation is performed before larval stage 29. There is a critical period of about ten hours for the formation of these connections, during which mitosis and DNA synthesis in the optic nerve neurons is completed, but about a day before their axons begin growing from eye to tectum. After this period, eye inversion leads to inversion of the retinotectal connections and to permanently inverted vision. If the optic nerve is cut in an adult toad and allowed to regenerate, the same pattern of connections is established. That is, for each point on the surface of the retina there is a corresponding point on the optic tectum that receives the axons from its retinal counterpart. A map of the retina is thereby reproduced on the tectum. If the tectal map is inverted, the animal's visual orientation and coordination are permanently disturbed.

The role of environment in the development of embryonic behavior has been controversial. The extreme environmentalist position regards the role of sensory stimulation of the embryo as being of crucial importance for development. Its most extreme proponent was Kuo (1932, 1967). Despite the fact that some of his extreme claims have been proven wrong, his theories have stimulated interest in the role of sensory stimulation in embryonic behavior. Striking experimental results in this area have been achieved by Gottlieb (1968; 1971*a*, *b*) and others, showing that embryos deprived of their normal sensory stimulation develop more slowly, whereas those receiving more stimulation develop more rapidly (see Chapter 27 for further discussion of these experiments).

Another type of environmental influence on development comes from the chemical environment. The nutrition of the mother, and other chemicals in her blood such as hormones, may affect the structure of the neurons and the behavior of the young. Such prenatal effects are known in mammals, in which they are facilitated by the placenta, and they are suspected in birds. For references, see Denenberg (1972) and Tobach et al. (1971).

CONCLUSIONS

The study of behavior in embryos has been valuable for its role in establishing that neural spontaneity is a basis of spontaneous behavior. It has illustrated, in a way that would be difficult with older animals, the dependence of the program of behavioral development on the program of neural development. And it has shown that even in embryos evolutionary adaptation has been important.

The movements, or motility, of embryos may be myogenic, if they arise in the muscles independently of the nervous system (as in fishes), or neurogenic, if they require the nervous system. Neurogenic motility may arise from spontaneous activity in the central nervous system or from reflex action.

Several lines of evidence have been used to confirm the existence and neural origin of spontaneous behavior in embryos. (1) In some fishes, myogenic motility is present before motor nerves have established neuromuscular junctions. (2) In chick embryos and others, neurogenic motility occurs for a period before sensory connections to the motor neurons have been established. (3) The cycles of activity and inactivity are not correlated with any known rhythmic source of sensory stimulation, such as the myogenic movements of the amnion or heart. (4) Tactile stimulation causing reflex behavior in embryo chicks did not alter the activity rhythm. (5) Parts of the chick innervated by segments of the spinal cord that were surgically isolated from each other maintained their own different rhythms. (6) Deafferentation by cutting off all sensory and central input to a lower spinal area did not prevent rhythmic motility of the deafferented part. (7) Neural activity in similar deafferented preparations was restricted to the ventral horn of the spinal cord, and showed a rhythmic pattern of activity similar to the rhythm of behavior. This evidence has led to the view that motor neurons play a dual role, mediating both reflex and spontaneous excitation of the muscles of the embryo.

The adaptive value of spontaneous motility in embryos remains unclear. The embryonic nervous system is adapted for two functions; it must meet the needs of the embryo, and it must anticipate the needs of the newborn in its growth and differentiation.

Attempts at general theories of embryonic development, such as Coghill's *individuation* and Windle's *reflex integration*, failed because of the diversity of patterns of development of behavior in the vertebrates. The pattern of embryonic behavioral development is better understood as an adaptation to the needs of the embryo and young animal.

The dependence of the program of behavioral development on the schedule of neural development is illustrated by the Mauthner cell of catfishes. The specificity of neuronal connections in the developing embryo is illustrated by the growth of axons from retina to brain in tadpoles.

27 The Naive Young Animal: Imprinting and Species Recognition

It is a fact most surprising to the layman as well as to the zoologist that most birds do not recognize their own species "instinctively," but that by far the greater part of their reactions, whose normal object is represented by a fellow-member of the species, must be conditioned to this object during the individual life of every bird.

Konrad Lorenz, 1937

IN THE ABOVE WORDS, Lorenz began a discussion of imprinting that served to reacquaint the English-speaking world with its astonishing characteristics. Since choice of a mate often functions as a reproductive isolating mechanism, imprinting may prove to play more of a role in this context than had previously been thought. The above quotation can best be regarded as a provocative hypothesis. In the present chapter, we will examine imprinting mainly in the context of species recognition by adults. A more general discussion of imprinting can be found in works by Bateson (1971, 1966), and Hess (1973), and in numerous other reviews by various authors.

Because isolating mechanisms between species are under genetic control and subject to natural and artificial selection, previous discussion has tended to disregard the role of learning in their development. In 1935, Konrad Lorenz, in a paper that was to become a milestone in the study of behavior, emphasized that in some birds the preference of an individual to mate with a member of its own species is subject to learning shortly after hatching. In some species, if a bird is raised by a foster mother—either another species of bird or a human keeper—the bird will become socially attached to the foster mother and on reaching sexual maturity will court individuals of the foster mother's species in preference to those of its own species. As examples, Lorenz cited studies of his own and others on ducks, geese, owls, parrots, herons, and other birds, including both precocial and altricial species.

The process by which these important and long-lasting first social bonds were formed was referred to by Lorenz as *imprinting*. In nature, a young duckling follows its mother and learns about its species characteristics from her. If separated from her and given a choice of following her or another object, the duckling typically chooses its mother. But if the duckling has been imprinted on a foster mother of another species, it may choose to follow the foster species in preference to its real mother. This behavior has come to be known as the *following response*, whether in naive or imprinted ducklings. Because the following response of a duckling to a stimulus object can be tested shortly after initial exposure of the duckling to the object, experimenters have seized on it as a reflection of the success or intensity of the imprinting experience—so much so that, in the minds of many, *imprinting* apparently came to connote only the following response. While convenient, this method of experimenting has a serious limitation from the viewpoint of an evolutionist, namely, that there is no reason to be sure that the object preferred in the following response is necessarily the object which the mature duck will prefer to court. Consequently, in this chapter primary attention will be devoted to experiments dealing with the effects of early experience on choice of a mate, which was the emphasis originally given by Lorenz.

The learning process involved in the acquisition of sexual preference has some unusual characteristics that were pointed out by Lorenz (1937*b*:264–265). "(1) The process is confined to a very definite period of individual life, a period which in many cases is of extremely short duration. . . . (2) The process, once accomplished, is totally irreversible, so that from then on, the reaction behaves exactly like an 'unconditioned' or purely instinctive response. . . . (3) The process of acquiring the object of a reaction is in very many cases completed long before the reaction itself has become well established. . . . (4) . . . the individual from whom the stimuli which influence the conditioning of the reaction are issuing, does not necessarily function as an object of this reaction." Subsequently, each of these points has been debated and qualified (Moltz 1960, 1963; Bateson 1966). Whether imprinting may still be described fairly as "a special kind of learning, different from conventional association learning" (Hess 1964:1138) is debatable, but it does offer unusual opportunities for the study of early learning and perception.

EXPERIMENTAL STUDIES OF SEXUAL IMPRINTING

To distinguish between the early learning of preferences that are tested by the following response to mother-objects and the early learning of preferences that are tested in courtship, the former will be referred to as *filial imprinting*, and the latter, as *sexual imprinting*. Since studies on filial imprinting have far outnumbered those on sexual imprinting, the extent to which these two processes are formally similar is still uncertain. For example, information on critical periods (see below) and on reversibility is moderately abundant for cases of filial imprinting, but sparse to nonexistent for most presumed cases of sexual imprinting. Consequently, use of the term *imprinting* is only a presumption in some of the cases discussed below. Nevertheless, it is clear that these studies deal with phenomena that Lorenz ascribed to imprinting. Following Lorenz, both precocial and altricial species will be discussed, even though a following response in altricial species may be nonexistent or much less evident than in precocial species. Sexual aspects of imprinting have been reviewed by Immelmann (1972).

Although an accumulation of trustworthy anecdotal reports had suggested the nature of sexual imprinting, no systematically controlled experiments were reported until those of Warriner et al. (1963). These studies showed that choice of a mate within a species can be largely determined by early experience. The investigators chose black and white varieties of the common domestic pigeon (*Columba livia*) with which to test for a possible influence of the parents' color on the mating preferences of their young. Sixteen pigeons, eight of each sex, were placed together in a large cage, and their subsequent courtship and pairing behavior were observed. None of them had been previously mated, and each individual had been raised exclusively by either black or white parents without view of any other pigeons. By cross-fostering, half of the white pigeons were raised by black parents and half of the black pigeons by white parents (Table 27–1). Four replicates of this design were used.

TABLE 27–1 Characteristics of pigeons used in each replication of the sexual imprinting experiments of Warriner et al. (1963). Sixteen previously unmated pigeons of the sex, color, and rearing-parent type indicated were placed together in a large cage and allowed to form pairs. The results are shown in Table 27–2.

		Rearing parents	
Color of pigeon	**Sex**	**White**	**Black**
White	male	2	2
	female	2	2
Black	male	2	2
	female	2	2

The pairs that were formed among the 64 pigeons involved are shown in Table 27–2. The males paired preferentially with females of the same color as the parents of the male in 26 of the 32 pairs. In five of the six exceptions, the female paired with a male of the same color as that of her foster parents. Apparently, in most pairs the preference of the male predominated over that of the female. There were no indications of choice according to a bird's own color by either sex. The experiment shows a clear influence of early experience on choice of mate in males. With respect to females, the data are inconclusive; a preference in accordance with parental color might be present in females, but if so it is masked by the dominance of the male in pairing. If the males' preferences were absolute and complete, then exactly half the females in this experiment would have to pair with males of a color differing from that of the females' parents. To detect an influence of parent color on mating preference in female pigeons, a different experimental design would be necessary.

The experiments of Warriner et al. (1963) are admirable because of their systematic control of some relevant variables, and their positive results. But they deal only with color variants within one species, and therefore do not fully test the potency of early learning as an isolating mechanism. The latter problem was studied by Schutz (1965, 1971) in a series of experiments using various species of ducks. When mallard ducklings were raised with a foster mother or a duckling of a different species and then turned free on a lake with over 30 species of ducks and geese, many subsequently tried to pair with the species they had been raised with. For example, of 12 male mallards (*Anas platyrhynchos*) raised with various domesticated forms of the mallard, nine

TABLE 27–2 Choice of mate by imprinted pigeons in experiment by Warriner et al. (From Warriner et al. 1963.)

		White parents		Black parents	
		White ♂	Black ♂	White ♂	Black ♂
WHITE PARENTS	White ♀	1a,b,c,d	4a	3b,c,d	0
	Black ♀	2b	1c,d	1a,b	4a,c,d
BLACK PARENTS	White ♀	5a,c,d	3a,b	0c,d	0b
	Black ♀	0	0b,c,d	4a	4a,b,c,d

a. cells in which male mated with female whose color was same as his parents ($p < 0.001$).
b. cells in which female mated with male whose color was same as her parents (not significant).
c. cells in which male mated with female of own color (not significant).
d. cells in which female mated with male of own color (not significant).

showed pairing with these forms and three did not. Of six male mallards raised with the red-crested pochard (*Netta rufina*), four paired with this species and two did not. Of five raised with Chilean teal (*Anas flavirostris*) two paired with this species and three did not. All three mallard males raised with muscovy ducks (*Cairina moschata*) subsequently courted that species. And both mallards imprinted on geese courted geese. Similarly, all four domestic roosters raised with mallards were sexually imprinted on mallards. Male coots (*Fulica atra*), red-crested pochards, sheldrakes (*Tadorna tadorna*), wood ducks (*Aix sponsa*), and muscovy ducks were successfully imprinted on mallards. These results were expected in view of Lorenz's earlier statements, but the exceptions were of even more interest. The data are shown in Table 27–3.

The results differed importantly between the sexes. Of the 34 male mallards raised with other species or domestic forms, 22 mated with their imprinted species and 12 mated with wild-type mallard females in spite of their upbringing. Among the 18 females raised with another species, only 3 actually mated with that species, and these pair bonds were unusually weak. The species and domestic forms that the females had been raised with were present on the lake at pairing time and were the same as those on which males had been successfully imprinted. Females of red-crested pochards (2), shelducks (4), muscovy ducks (8), and wood ducks (2) likewise were not successfully imprinted on mallards. The main exception to the finding that female ducks are not sexually imprintable was the sexually monomorphic Chilean teal (*Anas flavirostris*); of seven female Chilean teal imprinted on mallards, all subsequently paired with mallard males.

Young ducks are typically cared for only by the females. In nature, one might expect some learning by male ducklings about the appear-

TABLE 27–3 Sexual imprinting in mallard ducks. A summary of findings by Schutz (1965).

	Controls raised with mother and siblings of same species		Experimentals raised with another species	
	Males	Females	Males	Females
Paired with or interested in the other species	0	0	22	3
Paired with or interested in own species	28	39	12	15
Totals	*28*	*39*	*34*	*18*

ance of adult females, but little or no learning by female ducklings about adult males. Where several similar appearing species of ducks live together, the males usually differ conspicuously in coloration while the females of all species look very much alike. Although females can presumably use conspicuous plumage differences among species in choosing a mate, males cannot. Under such circumstances, the accuracy of males in mating with their own species would seem much more susceptible to improvement by early experience with the mother than would the accuracy of females. In monomorphic species, such as the Chilean teal, both sexes have a female-type plumage, and females presumably would not be able to use conspicuous plumage differences in choice of a mate. Where monomorphism occurs in ducks, similar-appearing species are often not present (Sibley 1957). Under these circumstances, reliance on conspicuous plumage differences would be unnecessary and would not be expected to be favored by natural selection (see Chapter 18).

The observations of Schutz illustrate the complicated interactions of imprinting and other factors in choice of a mate. The existence of other factors is indicated by the following evidence. (1) Mallards of both sexes (three of each) raised for nine weeks in visual and auditory isolation from other ducks mated normally with their own species. (2) Most females, and a third of the males, mated with their own species despite being imprinted on another. (3) Mallards are selective in the objects they can be imprinted on; they could not be imprinted on coots and only poorly on domestic chickens. (4) Of 15 male mallards and red-crested pochards reared with their own species and one other (muscovies, red-crested pochards, or mallards), none mated outside its own species; all "chose" to be imprinted on their own species. (5) Some ducks imprinted on another species switched to their own species or vacillated between their own and the imprinted species in different years or in the same year.

As a result of these findings, the view that mating preference in imprintable species is wholly determined by learning to follow the nearest moving object of appropriate size must be revised. Imprinting appears to modify an existing predisposition to mate with one's own species. The effectiveness of an imprinting experience is partly determined by these other predisposing factors (presumably genetic), which remain to be explored.

In an extensive investigation of the effects of early experience on mating preference in doves (Brosset 1971), many of the complications found by Schutz in ducks were missing. In two large aviaries, dozens of pairs of doves belonging to four species (*Streptopelia risoria-roseigrisea*, *S. senegalensis*, *Zenaidura macroura*, and *Nesopelia galapagoensis*) were allowed to nest. Three of these species were then cross-fostered by exchanging eggs from the nests of different species and allowing the foster species to incubate the eggs and raise the young to independence. Among 37 males of three species cross-fostered on four species, 30 oriented their sexual behavior toward the fostering species; 4 courted various species indiscriminately, going from one to the next several times in an hour, and 3 showed no sexual behavior. Of 27 females, 19 confined their soliciting to their fostering species and 8 showed no sexual behavior. None of these males or females paired with members of their own species. Over 100 control subjects of the same species were raised normally by their own parents in the same cages at the same time, but none of these courted any species other than their own. Unlike the situation in ducks, in these experiments none of the cross-fostered young mated later with their own species, nor was a sexual difference in imprintability noticed in these sexually monomorphic species.

Similar experiments were done by Immelmann (1969, 1970) on estrildid finches: the zebra finch (*Taeniopygia castanotis*), the Bengalese finch (*Lonchura striata f. domestica*), and the African silverbill (*Euodice cantans*). A total of 102 male zebra finches, 17 male Bengalese finches, and 2 male silverbills were cross-fostered in the egg stage. These birds directed their courtship exclusively or preferentially to their fostering species. Sexual behavior directed at a member of their own species was found in only nine of these males. Similar results were obtained with the domesticated color forms of the zebra finch; the young preferred to mate with birds of the same color as their fostering parents. Data for females were not reported.

It is not always the male who is most imprintable. In experiments with cross-fostered gulls, Harris (1970) found that the females mated selectively with males of their fostering species while cross-fostered males mated about equally often with females of their own as of the fostering species.

These studies have demonstrated for the first time with extensive, systematic, controlled experiments that early experience can have a dramatic effect on choice of mate, and they have revealed some of the factors involved. Many questions remain. Are there differences between the sexes in imprintability, and are these differences correlated with sexual dimorphism? If so, how does a female recognize her own species if early experience is unimportant? Do the young imprint selectively on their mother or their father? What about vocalizations? Do the young become imprinted on the calls of their parents in a way that influences their choice of a mate?

Some information is available about auditory imprinting and learning of vocalizations (see Chapter 28), but the role of auditory imprinting seems to have been relatively neglected in studies on ducks. Will mallards deafened as hatchlings or earlier mate with their own species in spite of imprinting with a foster species?

Many other variables affecting the early experience of precocial animals might affect sexual preferences and performance. How do they act? A start has been made at analyzing the processes comprising imprinting, as will be seen in the next section of this chapter, but much more remains to be done, particularly on the interaction of early experience with genetic factors in mating behavior.

Further studies on sexual imprinting include the following: Harris 1970; Morejohn 1968; Lill and Wood-Gush 1965; O'Donald 1960; Immelmann 1969, 1970; Klinghammer and Hess 1964; Klinghammer 1967; Mainardi et al. 1965; Schein and Hale 1959; Schein 1963; Guiton 1962.

ANALYSIS OF THE IMPRINTING PROCESS

The vigorous and provocative characterization of imprinting by Lorenz (1935, 1937) stimulated widespread interest among behaviorists. At first, interest centered on the remarkable ability of precocial birds and some other animals to socially attach themselves to an unexpectedly wide range of foster mothers shortly after hatching (or birth). This ability had, of course, been long known in folklore (e.g., Mary's little lamb), but it was probably first carefully described and analyzed by Spalding (1873). Normally, in ducks, geese, quail, grouse, chickens, and other precocial species, the young all hatch on the same day (within 3 to 8 hours in the mallard; Bjärvall 1967), and are called from the nest by the mother within a day or two of hatching, usually never to return. The exodus from the nest is shown for the wood duck in Figure 27–1. The downy birds are capable of locomotion and of feeding themselves at this early age, and require guidance and protection from their mother. The father, who is often brightly colored in a species-specific pattern, is not a part of the family group in many precocial species. Therefore, the young birds in their early life see only the female of their species, who is protectively colored and therefore less likely to attract predators to the young than would be the male. It was not generally appreciated that the young of many (though not all) precocial species could readily learn to follow a foster mother of a different kind until Lorenz published his work on the following response of geese and ducks, and appeared himself in photographs leading a family of goslings.

Critical Period / Probably the most enduring of the characteristics of imprinting pointed out by Lorenz (1935, 1937*b*) was the observation that a social attachment by the young to an object occurs most easily, and perhaps mainly, during a relatively short period just after hatching. This became known as the *critical period*, or sensitive period, for imprinting. Perhaps a partial result of this concern was that the general concept of critical periods for acquiring various types of behavior began to be more widely appreciated in mammals and other animals (e.g., Scott 1962).

Probably the best-known experiments on critical periods in imprinting were performed by Hess (1959) on mallard ducklings (see Figure 27–2). Eggs were taken from nests in a semi-wild population in Maryland. Newly hatched birds were isolated in small cardboard boxes (5 in. × 4 in. × 4 in.). At a certain age, each duckling was placed with a moving model of a male mallard uttering a human rendition of "gock, gock," resembling the call of a duck. The duckling was allowed to follow the model for a while and then was returned to its box. About 24 hours later, the duckling was given a choice of following the model male emitting the "gock, gock" or a model female giving a mallard call (this may not have been the maternal call). Figure 27–2 shows that more of the ducklings exposed at ages 13 to 16 hours gave positive responses to the male dummy than did younger and older ducklings. The most vigorous and complete responses were found only when the ducklings were imprinted at this most sensitive age. Since over 80 percent of the ducklings exposed at 13 to 16 hours of age later chose the male in preference to the female when tested, it is clear that they had attached themselves socially to the unnatural male rather than to the more natural female model. In these tests, ducklings were most sensitive to the model only for a relatively short period. The age and duration of the period of greatest sensitivity were later shown by many workers to vary greatly with the conditions of rearing and testing (Bateson 1966). No subsequent investigators have exactly duplicated Hess's results. More important, however, is the general agreement of all experimenters that, for any given set of conditions, an optimal age for social attachment to a parent-object can usually be found, and that it is early and brief.

This optimum age is defined by an increase in imprintability as the animal nears it, and by a *decrease* afterwards. The identification of the separate processes responsible for the increase and decrease has caused some controversy (reviewed by Bateson 1966). Immediately after hatching, the duckling has trouble walking, cannot keep up with the model, and so may receive less visual, auditory, tactile, and thermal stimulation from it, as well as less proprioceptive stimulation from its own locomotion. It spends more time drowsing and seems to be less easily aroused. The retina may still be undergoing changes that improve visual function. The relative importance of these factors in the early increase in imprintability remains undetermined.

The decrease in imprintability immediately after the critical or

27–1 Exodus of a brood of wood ducks from their nest box. **1.** A female wood duck on her nest before hatching. **2.** Eggs. The young hatch over a period of as much as 18 hours. **3–6.** The mother makes a reconnaissance one to two days after the young have hatched. **7.** The mother drops to the water below the nest and calls to the young. **8–11.** Responding to the calls of the mother, the young climb up the inside of the box with the aid of their unusually sharp claws, then leap to the water below. A whole brood may leave within four minutes. The ducklings and the mother are very vocal at this time. **12.** The mother stays by the nest until no further calls emanate from the hole. She then leads her brood to an area of thick cover. (Photos by Gus Martin and G. Gottlieb. From Gottlieb 1971. © 1971 by the University of Chicago. All rights reserved.)

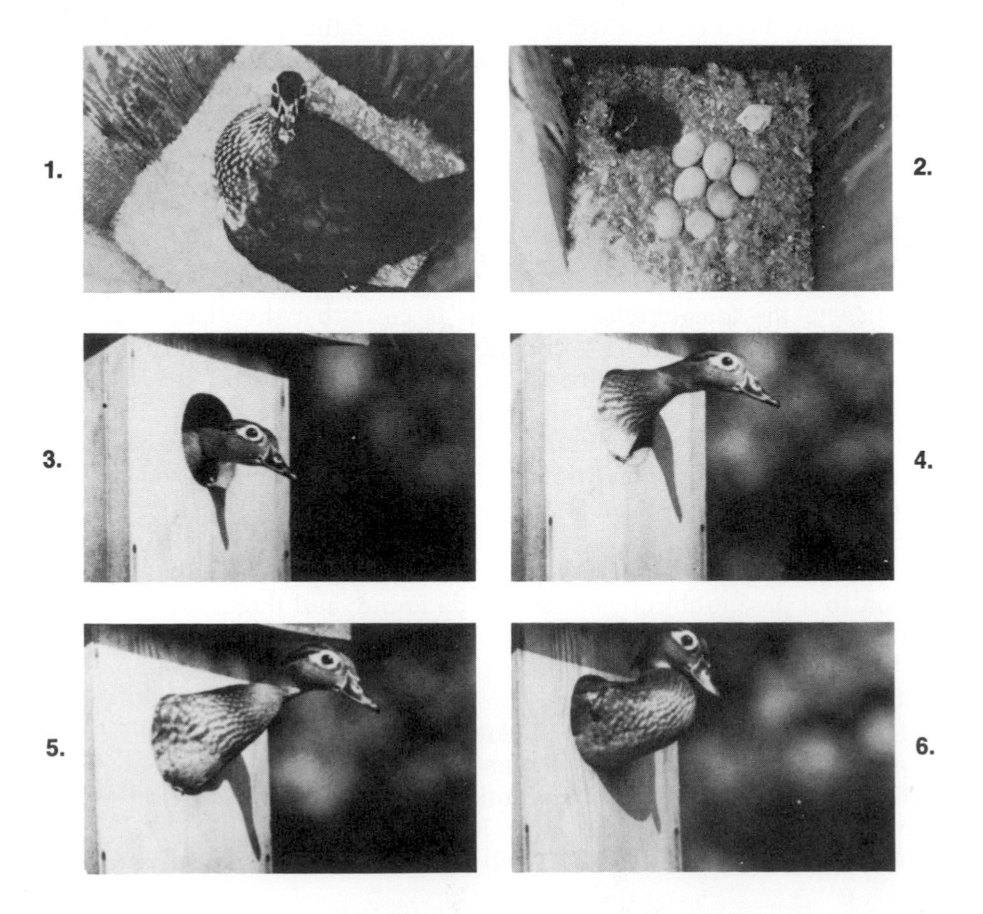

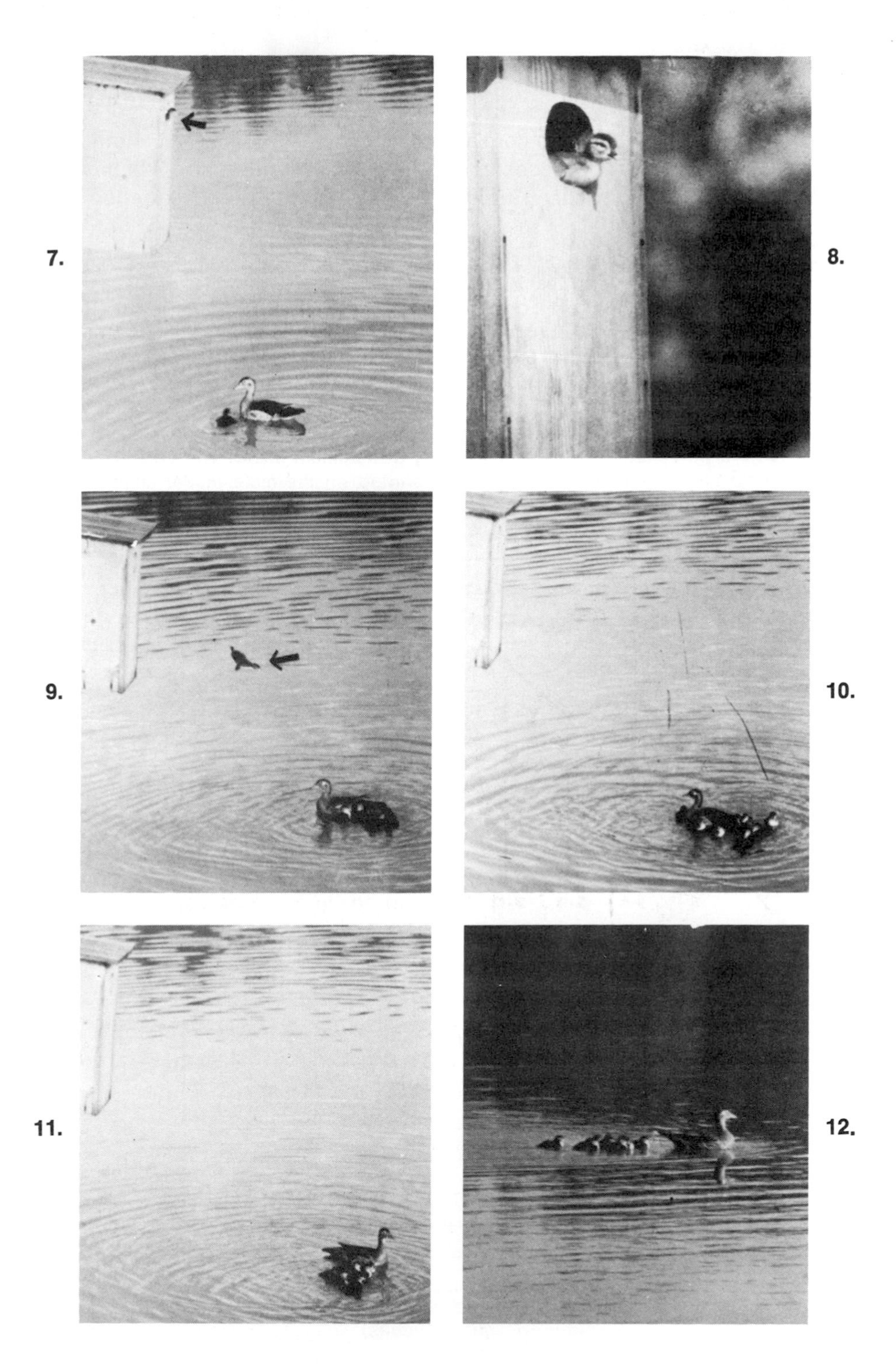

7.
8.
9.
10.
11.
12.

sensitive period has commonly been attributed to a complex of variables subsumed under the concept of fear (Bateson 1966). At about this age in many species of precocial birds, avoidance and distress behavior, such as crouching or cheeping, increase dramatically in frequency and

27–2 Critical period for following preference in mallard ducklings. At the age specified on the abscissa, each duckling was exposed individually to a moving model of a male mallard emitting a human imitation of the call of a female duck. The duckling followed the model 15 to 200 feet in a ten-minute period. Later, each duckling was tested for preference in a choice between the previously used male model and a female model under various conditions of movement and "calling." Scores in terms of the percentage of a full response (to the male) were obtained for each individual. *Above:* The average test scores for each age; *below:* the percentage of ducklings having a maximum score. The critical age at which ducklings were most effectively imprinted on the male model under these specific conditions is shown by the curves. (From Hess 1959.)

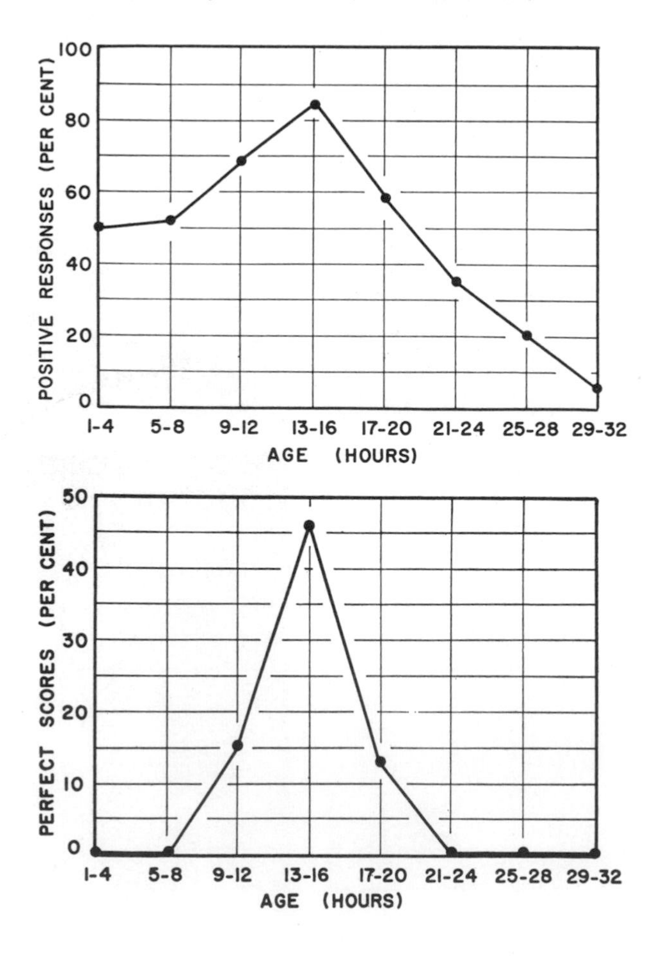

elicitability (Schaller and Emlen 1962). Adrenergic stimulants begin to interfere with imprinting at about the end of the critical period. Aversive stimuli such as electric shocks do not become effective for inducing escape and avoidance until the decline of the critical period (James and Binks 1963). It would appear that fear-related maturational processes, which are not yet precisely identified but which are susceptible to experimental manipulation by control of the experience of the young, are primarily responsible for the ending of the critical period.

Some of the other properties of imprinting are probably dependent on the sensitive period. The difficulty of changing adult preferences established during a critical period in the young may be partly due to the impossibility of re-creating in the adult the conditions of sensitivity and fearlessness that prevailed in the naive young. The establishment of a continuing social bond between mother and young makes possible other social learning processes that might occur at a later age, such as the acquisition of sexual preferences. Hess (1959) found that primacy (the first learning experiences) was more effective than recency (the last learning experiences) in learning to follow a foster mother. Unlike conventional learning in the adult, which depends mainly on the most recent reinforcement schedule, imprinting (whether filial or sexual) depends on the reinforcement schedule that prevailed during the sensitive period (or periods), with later reinforcement schedules having relatively little effect. Similarly, trials massed during a critical period should be more effective in imprinting than trials spaced so that most trials fell outside the critical period.

Is the period of development in which sexual preferences can be modified by social experience different from the critical period for following? Early workers, if they considered the question at all, seemed to doubt that the two were different. Relatively little work has been done on the problem. In the study on color preference in mating of pigeons by Warriner et al. (1963), the young were isolated at 40 days of age, thus fixing the time of learning before that time. In working with ducks, Schutz (1965, 1971) also found that, where mating preferences were modified by learning, the process occurred before the young were about 40 days of age. One-third of the ducklings held with their own species from hatching to 21 days and then placed with a different species mated with the second species. In contrast, Fabricius and Fält (1969) successfully imprinted female mallards between the ages of 8 and 23 weeks. More experiments will be necessary to characterize the sensitive period for sexual imprinting in mallards. In either case, these results would place the critical period for formation of sexual preferences later than the critical period for following. In estrildid finches, Immelmann (1969) found an imprinted sexual preference for the species of the foster parent in birds isolated from their foster parents as early as 33 days of age (when the young are just barely independent).

DEVELOPMENT OF SPECIES RECOGNITION

The development of species recognition in the neonate has been studied mainly in precocial hatchling birds. For convenience, the following discussion will be restricted to mallards and their domesticated derivative, pekin ducks. "The development of species identification in birds" has been extensively investigated by Gottlieb, and much of his work is included in a book by this title (1971) on which the following account is mainly based.

Many workers attempting to imprint ducklings on an object noticed that the addition of a sound resembling the maternal call of a mallard caused a considerable increase in the attractiveness of their imprinting object. Subsequent work showed that the auditory stimulus was more attractive to a pekin duckling than the visual stimulus (Figure 27–3). Less than 10 percent of a batch of naive pekin ducklings would follow a silent, stuffed, pekin mother duck; but about half would follow the source of maternal calls played from a speaker concealed behind a screen. The combination of a stuffed duck with ma-

27–3 Auditory versus visual choice test used by Gottlieb (1971). The mallard duckling is following the mallard maternal call coming from a moving loudspeaker that it cannot see in preference to a visible but silent mallard hen. (From Gottlieb 1971. © by the University of Chicago. All rights reserved.)

ternal call attracted nearly all pekin ducklings tested. The calls of a variety of species were attractive to pekin ducklings in the absence of visual stimuli. These species included the mallard, pintail, wood duck, and chicken, but the first two were a little more effective. This suggests that, in the context in which following normally occurs, selection has favored a relationship between the call of the mother and the duckling's auditory analyzer that results in the strongest attractiveness of the call to the duckling.

Selection pressure for selectivity on the part of the duckling in following maternal calls is probably not strong, since the duckling would only rarely be faced with a choice between mothers of two species in nature. Nevertheless, it is interesting that when offered a choice between an unseen, moving source of mallard maternal calls and one of wood duck maternal calls or pintail maternal calls, naive mallard and pekin ducklings followed calls of their own species in 90 to 100 percent of trials involving simultaneous presentation of the calls of two species. Tests with silent stuffed mallard, pekin, and wood duck females revealed no selective response to the visual stimuli involved.

If naive, newly hatched ducklings can respond preferentially to calls of their own species, perhaps this ability is present before hatching. To test this, Gottlieb set up experiments in which the behavior of the embryo could be observed while it was still in the shell (Figure 27–4). He found increases in heart rate in response to maternal calls of mallards, wood ducks, and chickens, and to the cheeping of ducklings. The response in heart rate was not selective. Most surprising was that significant changes in bill-clapping (opening and closing the bill) were restricted to the mallard maternal call. The ability to react discriminatively to the mallard maternal call was present up to five days before hatching. It preceded development of the ability of the duckling to call, and so did not require self-learning. It occurred in eggs isolated from the mother, and so did not require previous experience with the call. An interesting reversal in the nature of bill-clapping occurred at the time when the embryo first put its bill into the air space of the egg, a few days before hatching, and began to breath air and vocalize. Prior to this time the maternal call evoked a reduction in the rate of bill-clapping; afterward, the rate was increased. The maternal call also had effects on vocalization by the embryo, but these were less significant.

In further study of selective responsiveness of the duck embryo to the mallard maternal call, Gottlieb was able to demonstrate an effect of sensory stimulation on the time of first appearance of this ability in the embryo and on the degree of discriminatory ability. He manipu-

27–4 Recording the behavior of a duck embryo on the day before hatching by means of electrodes placed on the embryo. The top line shows heart beat. The second line records bill movements (bill-clapping). The third shows vocalization. The embryo can vocalize without detectable bill movements. (From Gottlieb 1968. Reproduced with permission of the Quarterly Review of Biology.)

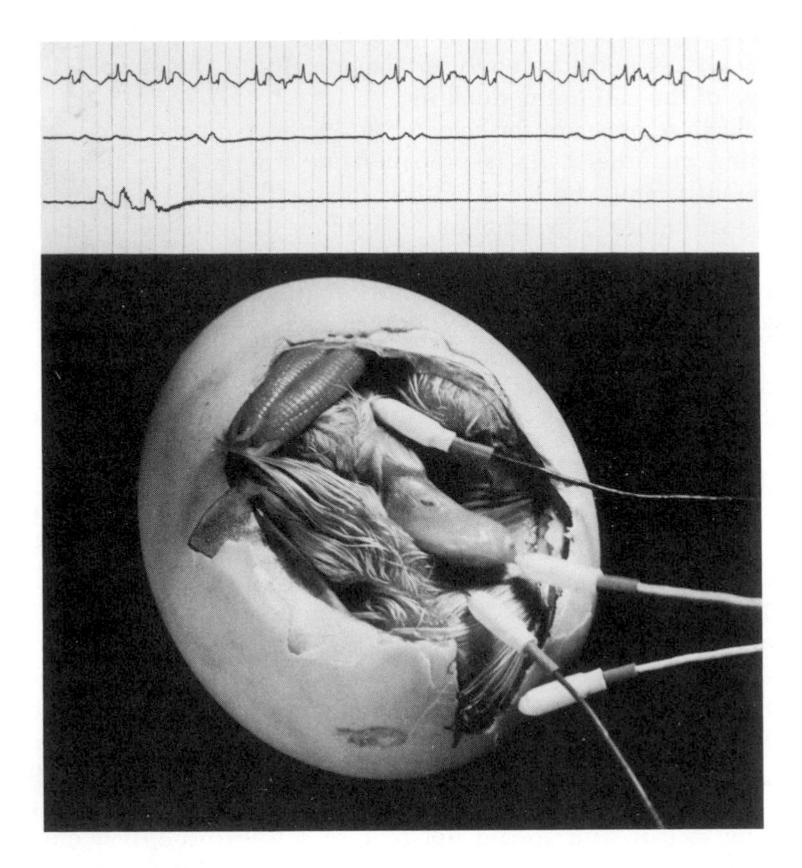

lated the auditory environment by exposing some ducklings to more auditory stimulation than control eggs received, and others to less. Those receiving more stimulation were placed with more advanced eggs so that they would hear calls from other embryos at an earlier stage than they would in a batch of same-age embryos. Those receiving less auditory stimulation were raised in individual sound-shielded incubators to exclude sounds from neighboring eggs, and were devocalized to eliminate sounds made by the embryos themselves.

The results of these experiments were that the age of appearance of auditory responsiveness to the maternal call could be speeded up by adding auditory stimulation from sibs, and that it could be retarded by preventing auditory self-stimulation through devocalization. These results are important because they illustrate that sensory stimulation in the late embryo participates in the timing of the normal program of development, a possibility that has sometimes been ignored in previous studies of prenatal ontogeny. Thus, sensory experience can be shown to affect behavior even in the egg.

SEXUAL IMPRINTING IN EVOLUTION

The experimental studies of sexual imprinting have demonstrated that it can be made to occur in confined conditions. But what are the consequences of sexual imprinting in nature? Two types of effects will be discussed. In the first, some effects of imprinting will be examined by looking at situations where behavior has been changed in wild birds through fostering. In the second, a hypothetical role of sexual imprinting in sympatric speciation will be examined, particularly with reference to polymorphic species.

Disruption of Isolating Mechanisms via Imprinting / Doves of several species have been induced to hybridize by employing cross-fostered birds as parents (Whitman 1899, 1919; Lade and Thorpe 1964; Brosset 1971), but will cross-fostering produce the same effect in wild birds? Harris (1970) showed that it will in gulls. He exchanged eggs in nests between the herring gull (*Larus argentatus*) and the lesser black-backed gull (*Larus fuscus*) on islands off the Welsh coast (Figure 27–5). Two effects on the imprinted birds were conspicuous. The herring gull in Britain is a sedentary species that normally does not leave the British Isles in winter, while the lesser blackback migrates to the Atlantic coast of Spain, Portugal, and Africa. The cross-fostered herring gulls behaved like lesser blackbacks, and were discovered wintering in Spain and Portugal, where herring gulls were never before reliably reported. Some cross-fostered lesser blackbacks, on the other hand, are known to have abandoned their foster species and migrated to the same area. Apparently, it is easier to entice herring gulls to migrate through social pressures than to suppress the migratory restlessness of lesser blackbacks.

Although hybrids of these two species have been reported in colonies where it seemed that the lesser blackbacks might not be able to find a mate of their own species because there were so few of the species present, hybrids in areas well populated by both species were virtually nonexistent. But when the cross-fostered birds matured, at least 79 gulls were involved in mixed pairs in the colonies where the

27–5 Mixed pair of gulls with hybrid young at its nest. *Left:* Lesser black-backed gull raised by herring gulls. *Right:* Herring gull raised by lesser black-backed gulls. Both adults are in the throwback phase of the long-call sequence. Note the darker mantle of the lesser blackback. The head of a hybrid chick may be seen in the nest. (From Harris 1970. Reproduced with permission of the British Ornithologists' Union.)

cross-fostering was done. Analysis of the composition of the mixed pairs suggested that males would settle for a female of either species whether the male had been cross-fostered or not. Females, on the other hand, were more selective; they would mate only with their fostering species. This situation differs from that in mallards or doves in that it is only the female who is effectively imprinted. The result of these many mixed pairs was, of course, widespread hybridization. The consequences of this mass hybridization for the populations involved are not yet known, nor are the mating preferences of the hybrid young.

Effects of Sexual Imprinting on Gene Frequencies / Aside from its role in maintaining isolating mechanisms between species, imprinting can probably also influence the selective advantage of genotypes within a population and has been postulated as a mechanism that could lead to sympatric speciation. In a species in which choice of a mate is influenced by visual characteristics, a complication arises if the population contains two or more visually distinct adult forms of one sex (or both). This condition is widely known as *polymorphism*, or morphism; and the different forms are termed *morphs*, or color phases (see

Chapter 2). Perhaps 3 or 4 percent of the species of birds and mammals show visible polymorphism, and it is also not rare in other vertebrate and invertebrate groups. Figure 27-6 illustrates one polymorphic species.

When choosing a mate, members of a polymorphic species may follow various strategies. (1) They may choose mates who match their own phenotype—positive assortative mating. (2) They may choose mates who differ from their own phenotype—negative assortative or disassortative mating. (3) They may choose a mate of the same phenotype as their mother—imprinted on the mother. (4) They may choose a mate of the same phenotype as their father—imprinted on the father. Or (5) they may mate at random with respect to the polymorphism in question. This list of strategies is incomplete, but it includes the principal ones to be considered here. These alternatives are not mutually exclusive, for imprinting may commonly be a cause of assortative mating.

Positive assortative mating is known to occur in gibbons (*Hylobates* spp.; Fooden 1969), blue and snow geese (*Anser caerulescens;* Cooch

27–6 Polymorphic species. The blue–snow goose complex. See also Figure 24–4. (From Scott, P. 1951. Severn Wildfowl Trust, Slimbridge, Gloucestershire, England.)

and Beardmore 1959; Cooke and Cooch 1968), parasitic jaegers (*Stercorarius parasiticus;* O'Donald 1959), and feral domestic pigeons (*Columba livia;* Goodwin 1958). Negative assortative mating has been recorded in white-throated sparrows (*Zonotrichia albicollis;* Lowther 1961) and in gibbons (Fooden 1969). Further possibilities have been discussed by Owen (1964).

Where two or more color morphs exist in a species, imprinting may affect their frequencies in various ways depending on such factors as (1) the strength of the imprinted mating preference, (2) which sex (if only one) is decisive in pair formation, (3) which sex (if only one) is imprintable, (4) which sex (if only one) carries the color polymorphism, (5) whether or not the heterozygote is visibly distinct (in one-locus polymorphisms), (6) whether selection acts differentially on the morphs, and (7) which sex (if only one) the young are imprinted on. As long as all individuals can mate with individuals of the same average fitness, imprinting should have no effect more drastic than increasing the amount of inbreeding within each morph (O'Donald 1960). When some individuals are deprived of mates because of imprinting, then selection may alter the frequencies of the morphs depending on the way in which individuals are deprived of mates.

The consequences of imprinting for populations have been considered by Mainardi (1964, 1967), Mainardi et al. (1965), Seiger (1967), Scudo (1967), Kemperman (1967), and Kalmus and Smith (1966). These hypotheses will now be considered.

If imprinting is absolute—so effective that no individual mates with a morph on which it has not been imprinted—then certain phenotypes are selected against because more of them are produced than there are mates available for them. A situation for a polymorphism involving a dominant and recessive gene at one locus is shown in Table 27–4. Matings between homozygous recessive individuals produce only more homozygous recessive offspring who are imprinted on the phenotype of their parents and themselves. Similarly, offspring of a mating between homozygous dominants are imprinted on the dominant phenotype. Complications arise in matings involving heterozygotes, or both dominant and recessive homozygotes. In these matings, some of the progeny may become imprinted on a parental phenotype that differs from their own. For example, in Table 27–4 type 4 can mate only with type 3; if more of type 4 than type 3 are produced, some individuals of type 4 will not be able to mate and are thus selected against. This will reduce the frequency of the gene a in the population, but will not eliminate a. As shown in Figure 27–7, the two types of homozygotes in categories 1 and 5 of Table 27–4 tend to receive individuals from the progeny of crosses involving heterozygotes;

TABLE 27–4 Genotypes and mating preferences of all types of progeny from all possible matings under a system of complete and irreversible imprinting of all progeny on the mother (the father's phenotype is irrelevant). The mating preference is indicated as follows: "on *a*" means "imprinted on the *a* phenotype and will mate only with *aa*." Matings that cannot occur because of imprinting are indicated with an *X*. Note that individuals of the type "*AA* on *a*" cannot be produced and so are not included in the table. *A* is dominant to *a*.

MALE PARENT	FEMALE PARENT				
	(1) *AA* on *A*	**(2) *Aa* on *A***	**(3) *Aa* on *a***	**(4) *aa* on *A***	**(5) *aa* on *a***
(1) *AA* on *A*	*AA* on *A*	*AA* on *A* *Aa* on *A*	X	X	X
(2) *Aa* on *A*	*AA* on *A* *Aa* on *A*	*AA* on *A* *Aa* on *A* *aa* on *A*	X	X	X
(3) *Aa* on *a*	X	X	X	*Aa* on *a* *aa* on *a*	X
(4) *aa* on *A*	X	X	*Aa* on *A* *aa* on *A*	X	X
(5) *aa* on *a*	X	X	X	X	*aa* on *a*

27–7 Sources of genotypes with specified mating preferences. (From Table 27–4. Drawing of snow geese by Peter Scott, from Severn Wildfowl Trust Third Annual Report.)

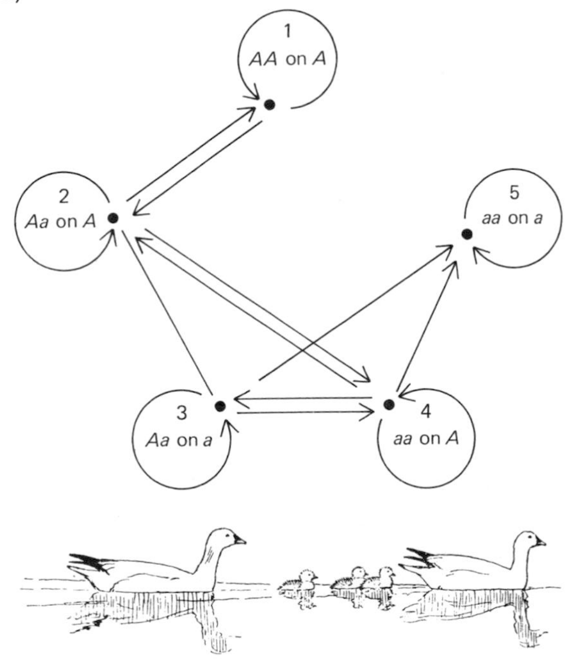

while the only homozygotes that produce heterozygotes are those (type 4) that have a heterozygote as parent (3×4, 2×2). Thus by selection against heterozygotes the population might, in theory at least, evolve to consist of two pools of homozygous individuals that have achieved species rank through sympatric speciation.

It must be remembered that this is an unnatural situation because imprinting is unlikely to be so absolute, so suddenly imposed, or so independent of the father's phenotype. If two species did form in this way, they would at first be virtually identical genetically, except at one locus, and identical in their ecological relationships, except for effects due to the pair of genes separating them. Such genetically and ecologically similar species with visible differences are not known, but they have been suggested (Goodwin 1958), and should be sought so as to assess the plausibility of this mode of speciation, which at present has the status only of a cute model.

An important theoretical objection to this mode of speciation is that when color mutations appear their bearers would have no chance of finding a mate—first, because they would be imprinted on parents lacking the mutation, and second, because the rarity of any mutation would make it unlikely that two individuals with the same mutation would encounter each other in mating condition. Even if they did meet, they would prefer not to mate with each other.

When imprinting is partial, less than absolute, and at least some individuals of each type can mate with any type, the "protected" condition of the homozygous subpopulations is lost and sympatric speciation is not a necessary outcome. Among the possible outcomes, depending on assumptions, are fixation of one allele or a stable or unstable polymorphism. Some of these possibilities were shown by Seiger and Dixon (1970) in a model that started with five colonies of various sizes connected spatially in a linear order, each containing only homozygous recessives, except for the second, which had only homozygous dominants. For the five colonies taken together, the frequency (p) of the dominant gene, B, was 0.54; for the recessive, b, $q = 0.46$. An exchange of 4 percent of the males of each colony to neighboring colonies occurred in each generation. Protection of the pools of homozygotes was further reduced by assuming no imprinting and no mating preference in the females. In matings involving two phenotypes, the young were imprinted on only the mother. To describe the relative effectiveness of imprinting, a coefficient of imprinting, which can be designated I, was defined as the probability that a male will reject a prospective mate of a color unlike that of his imprinted preference. The model was inspired by the authors' study of the data of

Cooch and Beardmore (1959) on the blue–snow goose complex, so *B* will stand for the blue goose phenotype and *b* for the snow goose phenotype (see Figures 24–4 and 27–6). A total of 200 generations were followed by computer.

With no imprinting ($I = 0$), the genes gradually became distributed evenly through the five populations. With moderate or high amounts of imprinting ($I = 0.4$, $I = 0.8$), after 200 generations the original blue colony (number 2) remained over 99 percent blue in phenotype, but maintained some *b* genes. Its neighboring snow goose colonies (1 and 3) also became over 99 percent blue; but more distant colonies (4 and 5) remained over 98 percent snow. Why the latter colonies were not successfully penetrated by the blue gene is unclear, but it might be related to the "protection" of the snow colonies by imprinting. When imprinting was relatively weak ($I = 0.05$), the frequency of the snow phenotype fell to less than 10 percent in all populations and was still falling at the 200th generation. Thus, the dominant gene tended to displace the recessive faster with a low *I* than with a moderate or high *I*.

The advantage enjoyed by the dominant gene is indirect. Because the heterozygote is blue, the proportion of blue phenotypes in the population will be higher than the proportion of the blue gene in the population. Similarly, the frequency of young imprinted on blue will be higher than the frequency of *B*; and the frequency of adults that prefer to mate with the blue phenotype will be higher than the frequency of the blue gene. Consequently, blue phenotypes should find it easier to find a mate that prefers them than should snow phenotypes. The differential favoring a dominant gene should be determined by (1) the number of heterozygotes in the population and (2) the fraction of adults that fail to mate. If all adults mate, the differential is zero, regardless of the number of heterozygotes; and if there are no heterozygotes, the differential is also zero and the two morphs will persist in the population (as two species) in exact proportion to their gene frequency, other things being equal.

There are many differences between the model and the situation in nature. The heterozygote is easily recognizable in nature (see Figure 27–6) but it has not been proven that goslings reared by heterozygotes develop mating preferences for heterozygotes. The genetic situation, although examined in nature by Cooke and Cooch (1968), has not been investigated experimentally; it is more complicated than assumed by the model, since continuous gradation occurs between the blue and snow phenotypes. The distribution of blue and snow morphs along their breeding grounds on arctic coasts also differs from the model. The blue phase reaches 97 percent of some colonies in the eastern

Canadian Arctic (Baffin Is.), and declines in frequency westward in a ratio-cline reaching zero percent in the western Canadian Arctic (Cooch 1963). The frequency of the blue phase is known to have been steadily increasing at about 1 to 2 percent per year in virtually all mixed colonies over about 60 years, but the increase cannot be attributed simply to imprinting because of differences in fitness of the two morphs under various ecological conditions, differences that are independent of mating preferences and imprinting (Cooch 1961).

Nevertheless, indirect evidence that imprinting is affecting gene frequencies in nature has been found. Cooch and Beardmore (1959) demonstrated positive assortative mating, showing that blues tend to mate with blues, and snows with snows; but their data could also be explained on the hypothesis that the birds were choosing mates to match their own color. Further data were necessary to eliminate this hypothesis and implicate imprinting. Cooke and Cooch (1968) analyzed data from 771 nesting pairs and 2,928 goslings from one colony. When the birds were classified into seven phenotypic categories ranging from lightest to darkest, it could be shown that even within most of the seven categories mating was assortative. It appeared that the geese were being quite discriminating in their color preferences. The observation that heterozygous blue geese were mated more frequently with snows than were homozygous blues suggests that imprinting is involved, since many of the heterozygotes must have had one snow parent. This would make heterozygotes more likely to mate with a snow than would be a homozygous blue goose, who could not have had a snow parent. Similarly, the observation that the first blues found in snow colonies are usually heterozygotes may be due to the greater likelihood of a heterozygote than a homozygous blue having a snow parent. Imprinting also explains the finding that among mixed pairs the number of blue males mated with white females significantly exceeds the number of white males mated with blue females. In these geese the male plays the dominant role in pair formation (as in the model); therefore, if imprinting is involved the number of blue males with a white parent should exceed the number of white males with a blue parent. In agreement with this hypothesis, it was found that 8 percent of the white goslings had a blue parent whereas 24 percent of the blue goslings had a white parent.

CONCLUSIONS

The effects of imprinting on evolution may be divided into two areas. First, as the quotation from Lorenz at the beginning of this chapter indicates, sexual imprinting is a necessary participant in the behavioral isolating mechanisms of a surprising number of species. Second, sexual imprinting in a polymorphic species may be partly or wholly responsible for certain evolutionary changes or equilibria.

The role of sexual imprinting in behavioral isolating mechanisms between species of ducks, finches, doves, gulls, and various other birds has been clarified through experiments involving cross-fostering of one species by another. Both in nature and in captives, it has proven possible to disrupt the normal behavioral isolating mechanisms and to bring about courtship, or even pairing, copulation, and rearing of young, by causing one or both species to be reared by the other.

The events in the development of such experience-dependent sexual preferences are still unclear. Although doubt persists, many workers agree that precocial birds are especially liable to develop strong social attachments to foster-mother objects in the first few days after hatching—the sensitive or critical period. The intensity and consequent effectiveness of the learning experience and social attachment that occurs at this time may be difficult to counteract at a later age even with prolonged exposure to such seemingly preferable objects as the natural mother (in early life) or a conspecific member of the opposite sex (at sexual maturity).

The relation of these first social experiences in following the mother to mating preference and other social choices later in life requires further clarification. It used to be thought that the first few days determined the sexual preference once and for all. Later, a different sensitive period, later in "adolescence," was postulated. But whether or not there is a recognizable and discrete sensitive period for sexual imprinting and how it differs from the earlier sensitive period for choice of a mother-figure are problems for the future.

In view of the importance of early learning for species identification it was surprising to learn, mainly through the experimental work of Gottlieb, that ducklings could react discriminatively to the maternal call of their own species just after hatching and even when still in the egg. That this pre-hatching discriminative ability acts to bias the newly hatched duckling toward following its own mother was also shown. Compared to auditory stimuli, visual characteristics of mothers proved relatively ineffective—a finding that suggests that imprinting on and attraction to auditory stimuli in neonatal animals deserves more attention. The overall development of social preferences in neonates can be clearly seen from these studies to be the result of several different kinds of experiences in certain normal sequences, each building on the preceding combination of physiological-anatomical maturation and experience. Studies designed to illuminate only experiential or only genetic factors in development of social experience can never, because of their own one-sided design, give a full understanding of the factors guiding development. The approaches and concepts of both methodologies need to be combined, especially since neither really excludes the other.

The evidence for effects of imprinting on gene frequencies in polymorphic species comes mainly from two types of sources, namely, natural populations and mathematical models. In natural populations the presence of assortative mating has been demonstrated in several species, which is consistent with effects of imprinting; but only in the blue–snow goose complex has evidence been presented against the alternate possibility of individuals choosing a mate because it matches their own phenotype. Since the hypothesis that imprinting is the principal basis for assortative mating has only recently been considered, the paucity of evidence in its favor from natural populations should not yet be held against it.

The mathematical models of imprinting allow various possible outcomes. If absolute imprinting were imposed suddenly on a polymorphic species, heterozygotes would tend to be eliminated because some individuals would become imprinted on phenotypes other than their own. The theoretical result would be sympatric speciation. If imprinting were only partially effective in determining mating, various other outcomes would be possible, including fixation of one morph, and polymorphism.

28 The Development of Songs and Calls

IN THE GENERAL AREA OF the development of behavior there are both theoretical and practical reasons for giving special attention to sound production in a book on the evolution of behavior. On the theoretical side, auditory communication is often a critical interspecific isolating mechanism. Consequently, the mode of development of mating calls and songs is of crucial importance for theories concerning their evolution and operation. For example, evolutionary explanations for patterns of geographic variation in song must take learning into account.

On the practical side, there are two advantages. First, the subject is an unusually rich source of developmental data involving a variety of species with widely divergent ecologies. Thus it is especially useful for comparative and evolutionary studies. Second, because of the availability of electronic devices for recording, analyzing, describing, and manipulating sound, studies on auditory communication can be done more precisely, easily, and reliably than corresponding work involving visual, chemical, or other sensory modes.

Another reason for giving special attention to song-learning is that the process is of unusual neurophysiological interest. It shows evidence of severe restraint on what is "chosen" to be learned by the animal, thus suggesting the operation of genetic factors in a learning process. Song-learning also has imprinting-like qualities: a critical period and a tendency toward irreversibility. It differs from imprinting in that it is the learning of a motor pattern of coordination rather than of a sensory-perceptual preference. Finally, song-learning is not without parallels in human speech-learning (Marler 1970; Nottebohm 1970).

IN WHAT SPECIES IS VOCAL LEARNING IMPORTANT?

The role of learning in the development of songs and calls varies from one group to another. In the invertebrates that use auditory communication, there appears to be no good evidence that normal song development depends on learning. Insect songs are susceptible to temperature changes and other environmental influences but not, apparently, to learning. Crickets raised in isolation not only produce normal sounds but make them in the appropriate contexts to the proper stimuli (Alexander 1968; Bentley and Hoy 1970). Amphibian calls also seem relatively free of the effects of learning. Many frogs, toads, and insects undergo natural isolation experiments; as larvae they may have no opportunity to hear the sounds of their own species because the adults have died or stopped calling. Nevertheless, these species produce normal mating calls at the appropriate season.

The situation in mammals is unclear. There has been little interest in raising mammals in acoustic isolation to study their vocalizations. Even when descriptions of the vocalizations of individuals reared in isolation are available, it may be difficult to compare them with sounds of wild animals. The review of Marler (1963) suggests that aside from man there is little evidence of an important role of learning in the development of normal vocalization in mammals. Even with chimpanzees, where serious efforts have been made to teach individuals human words, little success has been achieved (reviewed in Ploog and Melnechuk 1971). Porpoises can mimic certain sounds quite well (Lilly 1965), but whether most of their normal vocalizations would develop in individuals raised in isolation is problematic.

Only in birds is there abundant evidence of the importance of learning and tradition in the development of normal vocalizations in many species. The parallels between birds and man in this respect are striking (Marler 1970; Nottebohm 1970), including such phenomena as sensitive periods, dialects due to cultural transmission of acquired habits, learning by young from adults, and some evidence of lateralization in the neural structures involved in sound production. Every gradation is found, from species with no apparent influence of learning from others on song development (e.g., cuckoos, cowbirds, doves) to those that are conspicuous for their learning ability (e.g., mynahs, parrots, mockingbirds). It is this diversity in the role of learning in normal song development that makes the birds so attractive from an evolutionary standpoint, for it raises many still-unanswered questions about the adaptive value and physiological mechanisms of the different modes of song development.

GENETIC FACTORS IN SONG DEVELOPMENT

Although the emphasis in developmental studies of vocalization has been on what is learned, genetic factors are also important. They have been reviewed by Marler (1963; Marler and Mundinger 1971). Most studies on genetic factors belong to one of two types, namely, studies of hybrids between species and studies of selected breeds. Songs and calls of hybrids have been studied in grasshoppers, crickets (see Chapter 21), frogs, doves, and *Drosophila* (see Chapter 2). Various degrees of intermediacy have been found in hybrids. A frequent finding is that genes affecting mating calls tend to be localized on the X chromosome. Songs of a hybrid and its parent species are shown for grasshoppers in Figure 18–8, for doves in Figure 2–6, and for crickets in Figure 21–18. Genetic studies of factors involved in song-learning apparently have not been attempted.

Selection for vocal ability has long been carried out by breeders of domestic birds, such as canaries. One of the resulting breeds, known as the roller canary, retains its distinctive features even in the songs of isolates (Metfessel 1945). Breeds of pigeons are also known to differ in voice (Levi 1965). A biometrical analysis of different breeds of roosters showed that variation in duration of crowing was greater among breeds than among individuals (within breeds) or within individuals, even though the breeds were not selected for vocal traits, thereby implicating genetic factors in the crow differences (Siegel et al. 1965). These and other analyses of the sounds of hybrids and selected lines have shown that differences between species and breeds may in some cases be attributed mainly to genetic factors. In most of the genetic studies on sound production, species with little dependence on learning have been used. But the canary is a notable exception, and illustrates that genetic factors operate even in species in which song-learning is important.

TRANSITIONS IN THE DEVELOPMENT OF VOCALIZATION

Since the remainder of this chapter will focus on species differences in modes of song-learning, it will be useful to know the general pattern of development in songbirds. This has been outlined by Lanyon (1960). Many passerine birds and other altricial species with a recognizable song spend their first few weeks of life in a nest. Consequently, they lack the "distress" and "contentment" calls characteristic of precocial chicks, which leave the nest within a day or two of hatching. Unlike precocial species, which vocalize conspicuously even before hatching, altricial species do not apparently begin to call until they begin to beg for food from their parents shortly after hatching; and at first their calls are so faint as to be nearly inaudible. Their vocal activity for the first week is typically restricted to begging for food and distress calls if roughed up; but by three weeks of age, European blackbirds (*Turdus merula*) have developed all 12 of their calls that are not dependent on being in breeding condition (Messmer and Messmer 1956).

By three months of age in most passerines most of the calls have appeared. Many of these later calls develop through simplification and alteration of earlier calls. In many passerines the begging call differentiates into a low-intensity form, which is given at intervals and allows the parent to find the young, and a higher-intensity form, given on sight of the parent, which stimulates the parent to feed the young. The begging call can be elicited in many adult songbirds as part of submissive or precopulatory displays. The location note in the eastern meadowlark (*Sturnella magna*) changes to a social call (Lanyon 1957). In the European blackbird, the juvenile location note changes so that it has two forms, a social call and an alarm call.

In the chaffinch, the nestling begging call develops into the fledgling begging call by partial suppression of the activity of the right half of the syrinx, and then by elaboration becomes a chirp found in subsong, out of which the rattle song element emerges (Nottebohm 1972*a*). Other transitions from the few calls of the young to a larger number of calls in the adult have been traced in a variety of species (see review by Lanyon 1960).

The first attempts at song, which are termed *subsong*, may come in the first spring, the first autumn, or in some tropical estrildid finches even when only about eight weeks old. These first efforts roughly resemble normal adult song (*primary song*) in length, pitch, and tonal quality, but lack standardization and the characteristic elements, or motifs. Subsong differs from primary song in being much more variable, including various call notes characteristic of the species, having a warbling quality (unevenness of pitch and tone), and in lacking those elements or motifs that are acquired by learning. The transition from subsong to primary song includes an intermediate stage, termed *rehearsed song* (Lanyon 1960) or *plastic song* (Nottebohm 1970), which has the characteristics of subsong with the addition of learned primary motifs. The transition from plastic song to primary song involves dropping out of the calls and a refinement and rigidification of pitch and tonal quality so as to lose the warbling quality. These three phases in song development are illustrated in Figure 28–1.

During the transition from subsong to primary song, the singer perfects or "crystallizes" his song. In some species, such as the white-crowned sparrow (*Zonotrichia leucophrys*), each individual has only one song type. After crystallization, every song follows essentially the same pattern for the rest of the bird's life. Attempts to teach such birds to sing a new theme after crystallization has occurred typically fail.

28–1 Stages of development from subsong through primary song in three male chaffinches. Elements of early subsong in males GY/Y (**75**) and RWB/Y (**76**). Elements of later subsong in male RWB/Y (**77–78**) showing first appearance of full song pattern. **79–80.** Subsequent integration and "tightening" of song in male H/P. (From Thorpe 1958. Reproduced from *Ibis* by courtesy of the British Ornithologists' Union.)

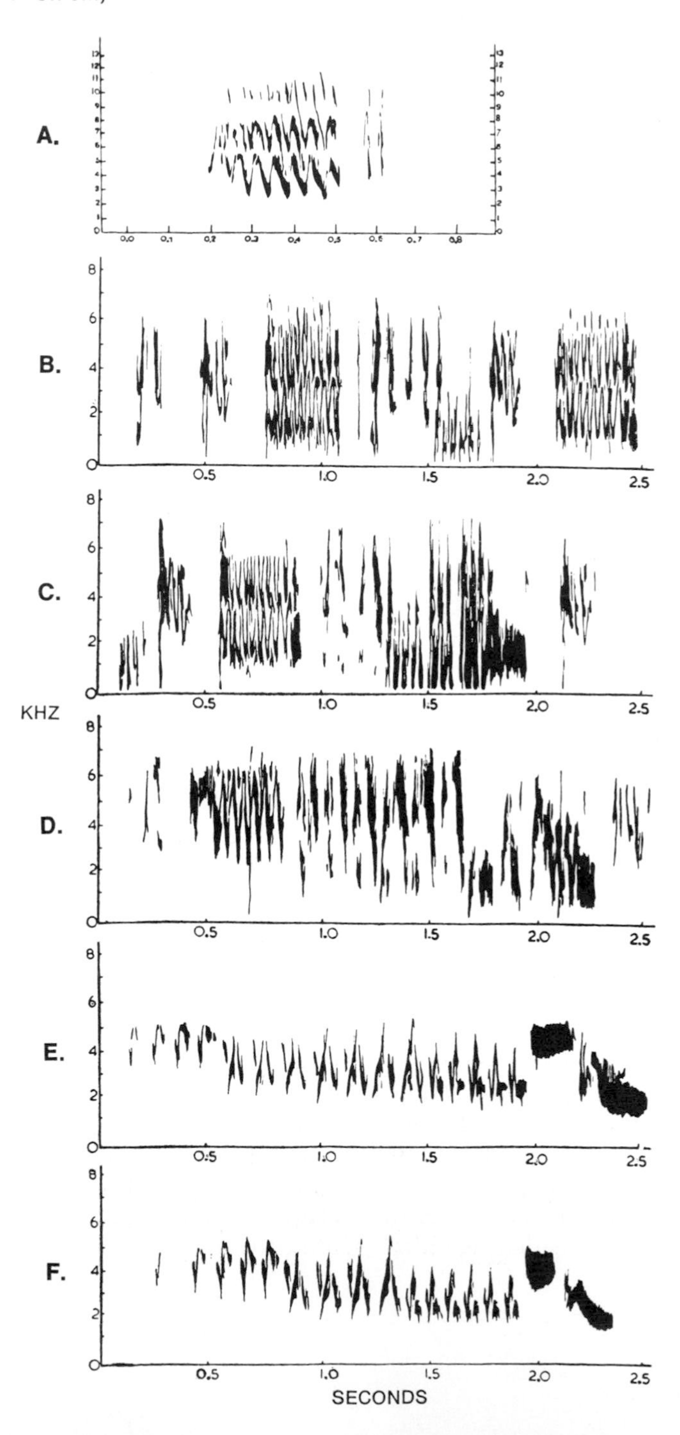

In some other species, each singer typically has more than one song type. For example, each individual male chaffinch (*Fringilla coelebs*) has a fixed number of different songs, typically three to six, which it alternates in various sequences. It is a curious fact that experienced adults of many species pass through a similar sequence from subsong through plastic song to primary song each breeding season when they resume singing after the nonbreeding winter season.

SPECIES DIVERSITY IN SONG LEARNING

The question is often asked by laymen, "Is bird song learned or innate?" The question itself is invalid and cannot be answered, for two reasons. First, the question assumes a dichotomy between learned and innate, but we have seen in the introduction to Part VI that such an assumption is unwarranted on theoretical grounds. Second, the facts are that all degrees of intermediacy exist in birds, from species whose song development shows no trace of learning from others, to species which display a strikingly obvious ability to learn to mimic various sounds, including even human speech. Consequently, the study of song development in birds is especially suitable for illustrating the various ways in which genetic factors constrain what the individual is "allowed" to learn. The interaction between genetic factors and experience in the learning process is probably nowhere more conspicuous and so easily studied.

In many species a clear distinction exists between calls, which are brief and simple, and songs, which are relatively long and complicated (see Chapter 16). As a general rule most calls tend to develop normally in isolated birds, but exceptions occur (Lanyon 1960). Consequently, in developmental studies attention has focused on songs. The most complex songs are found in the order Passeriformes, especially its suborder Passeres, the so-called songbirds. Relatively few nonpasseriforms have songs comparable to those of passeriforms. Vocal learning in birds has been reviewed by Marler and Mundinger (1971) and other earlier writers.

Little or No Evidence of Learning / Species that are brood parasites regularly perform natural experiments in which their young are raised by foster parents. Although the young of such species can usually hear males of their own species singing, they probably have considerably less exposure to songs of their own species than do the young of non-brood parasites. The songs of the European cuckoo (*Cuculus canorus*) and the brown-headed cowbird (*Molothrus ater*) are relatively simple and stereotyped, and are known to develop normally in isolation. They certainly show no trace of learning from the foster parents. This is not true of all brood parasites, however. The African widow birds (*Viduinae*) that parasitize nests of estrildid finches normally learn the songs and calls of their hosts and reproduce them during courtship of their own species (Nicolai 1964). They also have certain call notes, not found in the host species, that are characteristic of the viduine-euplectine phylogenetic group and are given primarily in agonistic contexts. (The social behavior of the Viduinae has been discussed further in Chapter 11, on social parasitism.) It used to be thought that the songs of estrildid finches would develop normally in isolation, but this has been disproved for some species (Immelmann 1969; Konishi and Nottebohm 1969).

Among non-brood parasites the songs of various species of doves show no evidence of learning. Lade and Thorpe (1964) found that doves raised from the egg by a foster species developed normal songs of their own species with no evidence of learning from the foster parent. Hybrids developed songs having characteristics of both their natural parents regardless of their rearing parent, as shown in Figure 2–6.

Learning from Neighboring Conspecifics / In this group, individuals raised in auditory isolation sing a fairly normal song except for the lack of certain motifs or syllables characteristic of their home population. They do not normally learn to produce motifs of other species. The group includes such species as the chaffinch, white-crowned sparrow, indigo bunting (*Passerina cyanea;* Rice and Thompson 1968), and European blackbird. Some of these species have been shown to be resistant to learning songs of other species, even closely related ones (Thorpe 1958; Marler and Tamura 1964). Further discussion of these species will be found in the section on deafening experiments later in this chapter, and in reviews by Marler and Mundinger (1971) and others.

The ability of some species to learn syllables or phrases from their neighbors probably accounts for the observation in nature that neighboring individuals tend to share more elements of their song repertoire than do individuals separated by a long distance. This has been observed in the songs of cardinals (*Cardinalis cardinalis;* Lemmon 1966), indigo buntings (Thompson 1970), black-crested titmice (*Parus atricristatus;* Lemmon 1968a), song sparrows (*Melospiza melodia;* Borror 1965), and other species. Similarly, one variant of a call of Steller's jay (*Cyanocitta stelleri*) tended to be given most often by near neighbors (Brown 1964). In some species, individuals singing in response to a neighbor's song tended to use the phrases or song types used by that particular neighbor, as in the black-crested titmouse (Lemon 1968*a*), cardinal (Lemon 1968*b*), and pyrrhuloxia (*Pyrrhuloxia sinuata;* Lemon and Herzog 1969). Similarly, captive shama thrushes (*Copsychus malabaricus*) answered a sound stimulus poor in harmonics with a song poor in harmonics, and a stimulus rich in harmonics with a song rich in harmonics (Kneutgen 1964). A captive chaffinch having two song types tended to answer a song played to it with another of the same type (Hinde 1958). Bertram (1970) observed that mynahs tended to produce at once the call just given by a neighbor. The response to local song types was greater than to more distant ones in the cardinal (Lemon 1967). White-crowned sparrows responded more to playbacks of their own dialect than to those of other dialects (Milligan and Verner 1971). And mynahs responded more strongly to calls of neighbors than of strangers. Interpreting the relevant experiments on response to neighboring and distant individuals may be complicated by the factor of individual recognition (Falls 1969).

Learning from the Father / Bullfinches and meadowlarks (*Sturnella* spp.; Lanyon 1957, 1960) do not normally mimic songs of other species in nature, but they are capable of it when raised with a tutor of another species (unlike the chaffinch and white-crowned sparrow). Male bullfinches learn only the sounds made by their father (Nicolai 1959). In a study of zebra finches (*Taeniopygia castanotis*) raised by other species, Immelmann (1969) showed that the males learned the song of the foster father even when other zebra finches nearby sang normal songs. Thus in the bullfinch and zebra finch the material to be learned is normally constrained by the social bond between father and son.

Learning from Other Species in Normal Song Development / In this group are the mimics. They include the mockingbirds and their allies (Mimidae), the lyrebirds (*Menurus menura*), some bowerbirds (Ptilonorhynchidae), some starlings (Sturnidae), some corvids (Corvidae), and others (references in Bertram 1970:182). Their impressive imitations of other species pose an evolutionary problem. What could be the adaptive value of such an ability, and why is it found in these so divergent groups of birds and not others? It has been suggested that imitation of the songs of other species helps to keep members of these species out of the mimic's territory, thus reducing interspecific competition. The large size and aggressiveness of many of these species would enable many of them to attack other competing species successfully. Evidence for this theory is so far only anecdotal, and there is no indication that birds of any species are deceived by imitations of mimic species. An alternative might be that such songs are more effective notice to other males of the mimic species that a territory is being defended.

Mimics should have no trouble recognizing their own species by its song. Only a proportion of the song of a mimic is composed of imitations of other species; the rest is composed of species-specific notes and phrases. Even when the imitations are perfect, they are produced in a temporal program that characterizes the mimic rather than the model. For example, mockingbirds (*Mimus polyglottos*) typically sing each phrase three to five times or more; brown thrashers (*Toxostoma rufum*) sing a phrase only twice; catbirds (*Dumatella carolinensis*) tend not to repeat at all.

Learning from the Mate / The "talking" birds, such as mynahs, parrots, corvids, starlings, and others, constitute another evolutionary problem. How could selection have favored the evolution of the ability to mimic the human voice? It is interesting that these species do not normally mimic the sounds of other species in their environment. Their ability seems related to their social organization. These species tend to form strong pair bonds that are retained, even in flocks, both in and out of the breeding season. There is considerable individual variation, and members of a pair appear to develop certain notes in common, in some but not all species (e.g., in ravens but not mynahs). This enables them to recognize their mate by its call at a great distance or in a large group (Thorpe and North 1965). In captive ravens (*Corvus corax*) and shamas, members of a pair, on hearing their "shared" notes, returned at once to their mate, if possible (Gwinner and Kneutgen 1962; Gwinner 1964). Apparently, in most species notable for their ability to mimic human speech when raised and kept by man the ability is used in nature for learning and mimicking the calls of their mate or of individuals of the same sex and dialect (as in mynahs, Bertram 1970).

Another striking vocal behavior of birds involving a special relationship with the mate is known as *duetting* (reviewed by Hooker and Hooker 1969; Thorpe 1972). In these species, members of a pair may give their songs together or in such rapid succession that they sound like one song. The phrases of the male and female may be the same or different. The participants may sing alternately (antiphonal singing) or together (unison singing). Some examples are shown in Figures 28–2 and 16–7. Individual recognition of tape-recorded calls has been shown experimentally in duetting whipbirds (*Psophodes olivaceus*, Watson 1969). The occurrence of dialects in the female of this species but not the male suggests an important role of learning. In a duetting

28–2 Duetting in six species of birds. Antiphonal singing is shown in cases **1–5,** unison singing in case **6.** The different members of the pair, a and b, alternate their notes in **1–5** but produce them together in **6.** The first two species are cuckoos (Cuculidae); the rest, honeyeaters (Meliphagidae). All six are from the southwest Pacific region. (From Diamond 1972. Reproduced with permission from the American Ornithologists' Union.)

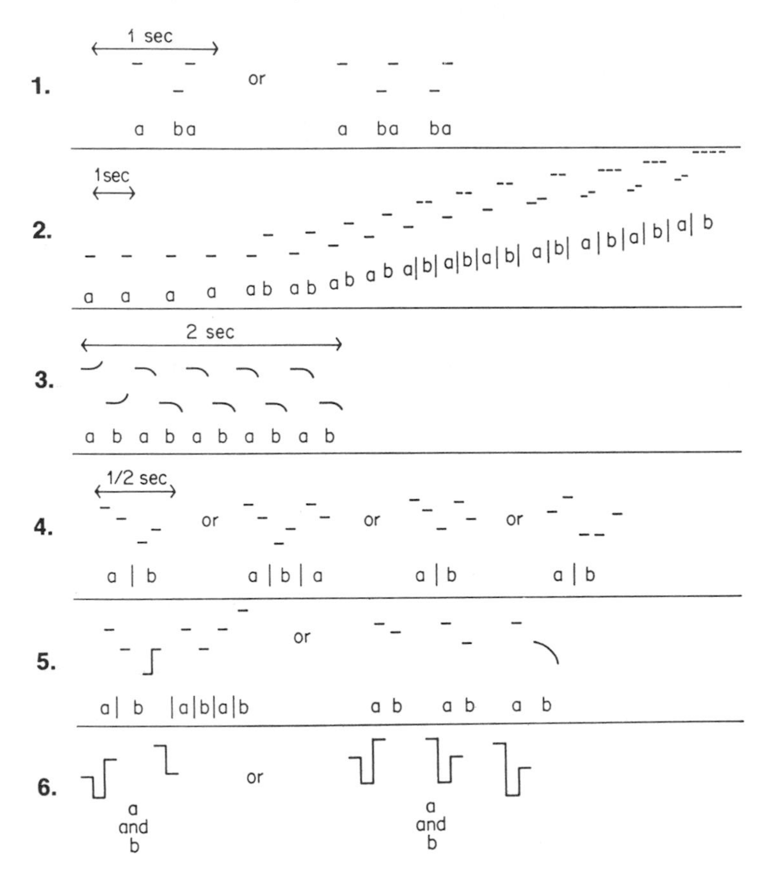

species, each singer learns to incorporate phrases of the other into a joint effort. Here again learning to recognize and produce certain sounds made by the mate functions in maintaining the pair.

Since loss of a mate in many species is likely to cause a lowering of fitness, the role of vocalization in the pair bond is probably important and may exert strong selection pressures on the ability to recognize and produce fine gradations in call and song structure—and, in the case of duetting, to coordinate them with calls of the mate. This may have been the primary selection pressure responsible for the "talking" ability of birds and for duetting (Thorpe and North 1965).

Dialects / The term *dialect* is used by ethologists to designate populations or demes within a species that differ in their vocalizations from other populations or demes. The term refers to differences in phenotype and is not intended to imply that the difference is necessarily due to learning alone. The occurrence of dialects is widespread in birds (Armstrong 1963; Thielcke 1969), but their developmental basis has been analyzed in only a few species. In these, however, learning

28–3 Dialects in the white-crowned sparrow populations inhabiting the San Francisco Bay Area. Time marker is half a second. (From Marler and Tamura 1964. Copyright 1964 by the American Association for the Advancement of Science.)

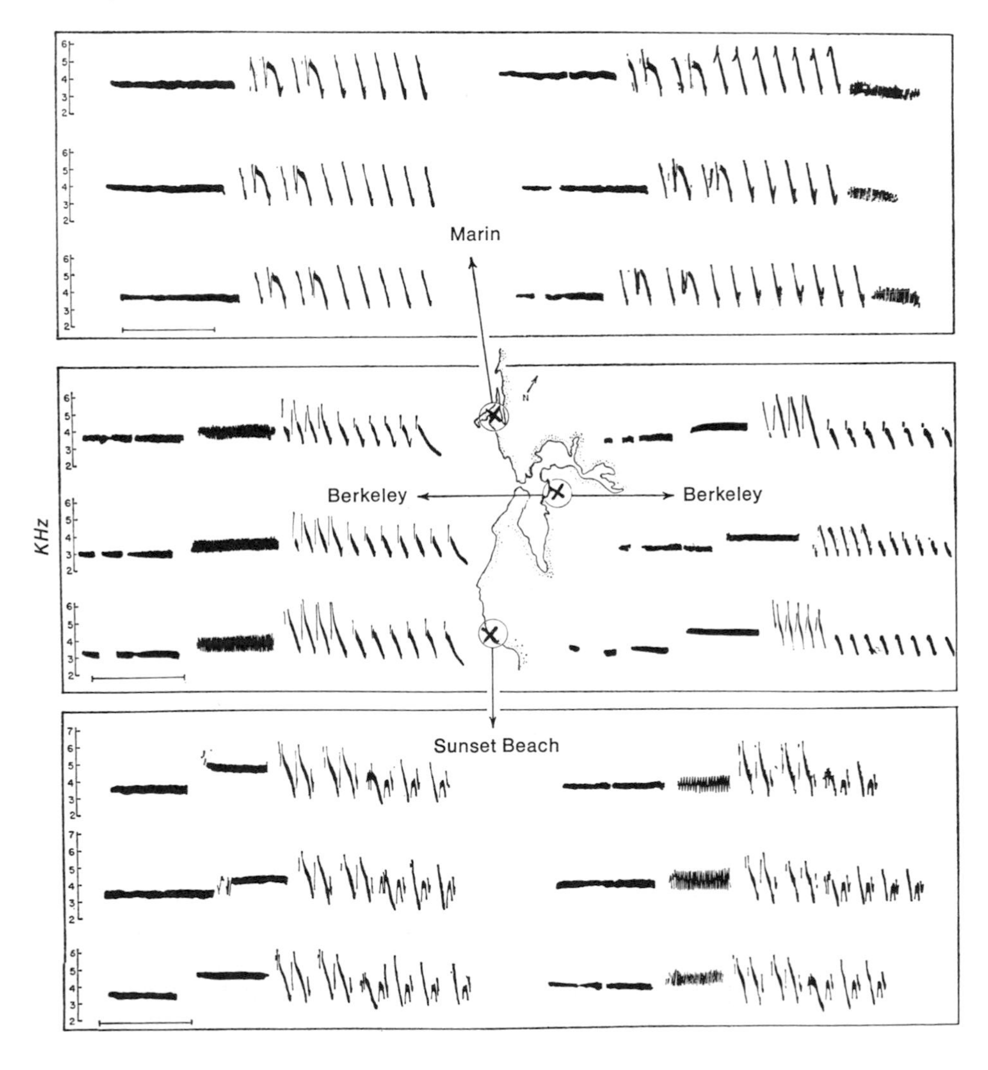

and cultural transmission of the dialect features from generation to generation are important. The white-crowned sparrow provides an excellent example (Marler and Tamura 1964). It has long been known that the various populations in the region of San Francisco Bay could be characterized by their song type. Some songs from three of these populations are illustrated in Figure 28–3. The dialects shown differ in the number of introductory whistles, in their pattern of pitch change, in the presence and location of the buzz, and in the shape of the largely vertical formants, or notes, composing the trill sections of the song. When young were taken from nests in three different dialect areas and raised together in isolation, all developed rather similar songs, none of which resembled any of their three home dialects, as shown in Figure 28–4 in *A*1 through *C*4 (compared to *AN*, *BN*, *CN*). Individual isolates developed somewhat different songs that also failed to resemble any dialect closely (Figure 28–5*A*1, *A*2).

The occurrence of dialects due to cultural transmission introduces more complexity into analyses of geographic variation in mating signals, analyses designed to illuminate the evolution of isolating mechanisms (see Chapter 18). Although virtually nothing is known about the preferences of females for different song types in males, it is conceivable that females may become "imprinted" on the song type of their parent or neighbors during an early critical period. If this preference were strong enough, it might function as an incipient isolating mechanism (Cushing 1941). One might hypothesize that the origin of a new avian species might occur entirely through culturally transmitted traits. In brief, this mode of speciation would require (1) two or more distinct cultural dialects among males, and (2) an "imprinted" dialect preference in females. If two isolated populations with these traits came together, they might fail to interbreed at first (see Chapter 27) and might later develop genetic differences. If, because of differing ecologies in the dialect populations, hybrids were of inferior fitness, selection might then strengthen the isolating mechanisms—as explained in Chapter 18. This hypothesis about speciation differs from the classical one only in that the isolating mechanisms are at first culturally rather than genetically transmitted. Unfortunately, since the role of female song preferences in choice of a mate has not been studied, part of the factual basis for the hypothesis is missing.

Alternatively, the isolating effect of dialects may only impede, but not prevent, gene exchange between populations. The resultant isolation can be expected to be strengthened if the populations differ in their ecologies and their tendencies to leave their familiar habitats. This is consistent with Nottebohm's (1969*b*) finding that dialects in the South American *Zonotrichia capensis* tended to conform with habitat divisions.

28–4 Songs of male white-crowned sparrows from three dialect populations raised together in isolation. These males could learn from each other (explaining the similarity among the songs) but could not learn from other males of their own dialect. These individuals were isolated as a group before the critical period for song learning in this species. Typical songs of the dialect populations from which the individuals came are shown in the boxes labeled *AN, BN,* and *CN.* Songs of individuals are shown from Inspiration Point, 3 km northeast of Berkeley (**A1–3**), from Sunset Beach (**B1–2**), and from Berkeley (**C1–4**). (From Marler and Tamura 1964. Copyright 1964 by the American Association for the Advancement of Science.)

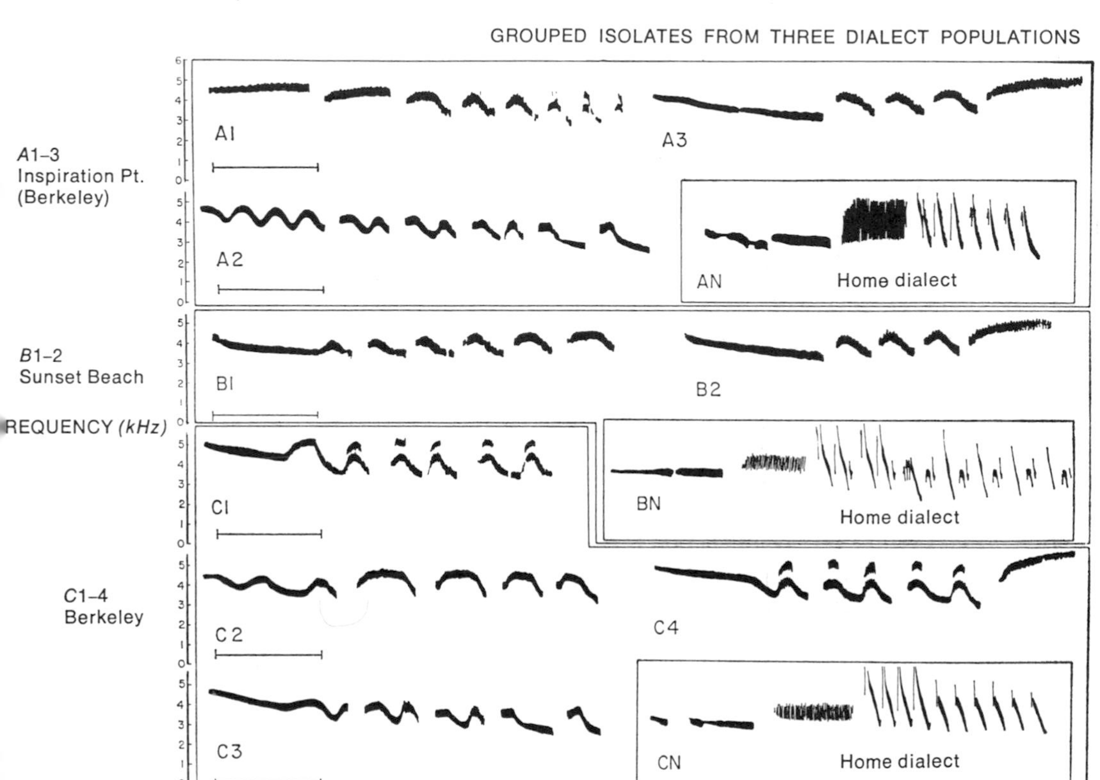

THEORIES OF SONG-LEARNING

The normal pattern of song development in recently studied passerine birds differs from the learning of some other behaviors in adults in that a critical period occurs, and in the apparent "irreversibility" of the final crystallized song patterns. These characteristics are reminis-

28–5 Evidence for constraints on the learning of song by white-crowned sparrows (critical period and preference for song of own species). **A1** and **A2.** Songs of males raised in individual isolation do not resemble those of wild birds, shown in Figure 28–3. **B.** The song of a male isolated at 3 days and tutored from 3 to 8 days (before the critical period, which is about 9 to 50 days of age) resembles neither its home dialect (*Sunset Beach*) nor the dialect of its tutorsong (*Marin*), as can be seen by comparison with Figure 28–3. **C1–4.** Males isolated during the critical period, presumably having been exposed to the song of their own population (*Marin*), develop the dialect of their population with no further training (C1) and in spite of training with another dialect after the critical period (C2–4 were trained on *Sunset Beach* dialect starting at 100, 200, and 300 days of age, respectively.) **D1–2.** Males isolated early and tutored with a *Marin* song during the critical period developed a typical *Marin* song (see Figure 28–3) despite simultaneous exposure to songs of Harris's sparrow for D1 (shown in G) and song sparrow for D2 (shown in F). **E1–3.** Males isolated before the critical period and tutored during the critical period only with the song-sparrow song shown in F failed to learn this song or their own dialect. **F.** The song-sparrow song used for D2, E1–3. **G.** The Harris's sparrow song used for D1. (From Marler and Tamura 1964. Copyright 1964 by the American Association for the Advancement of Science.)

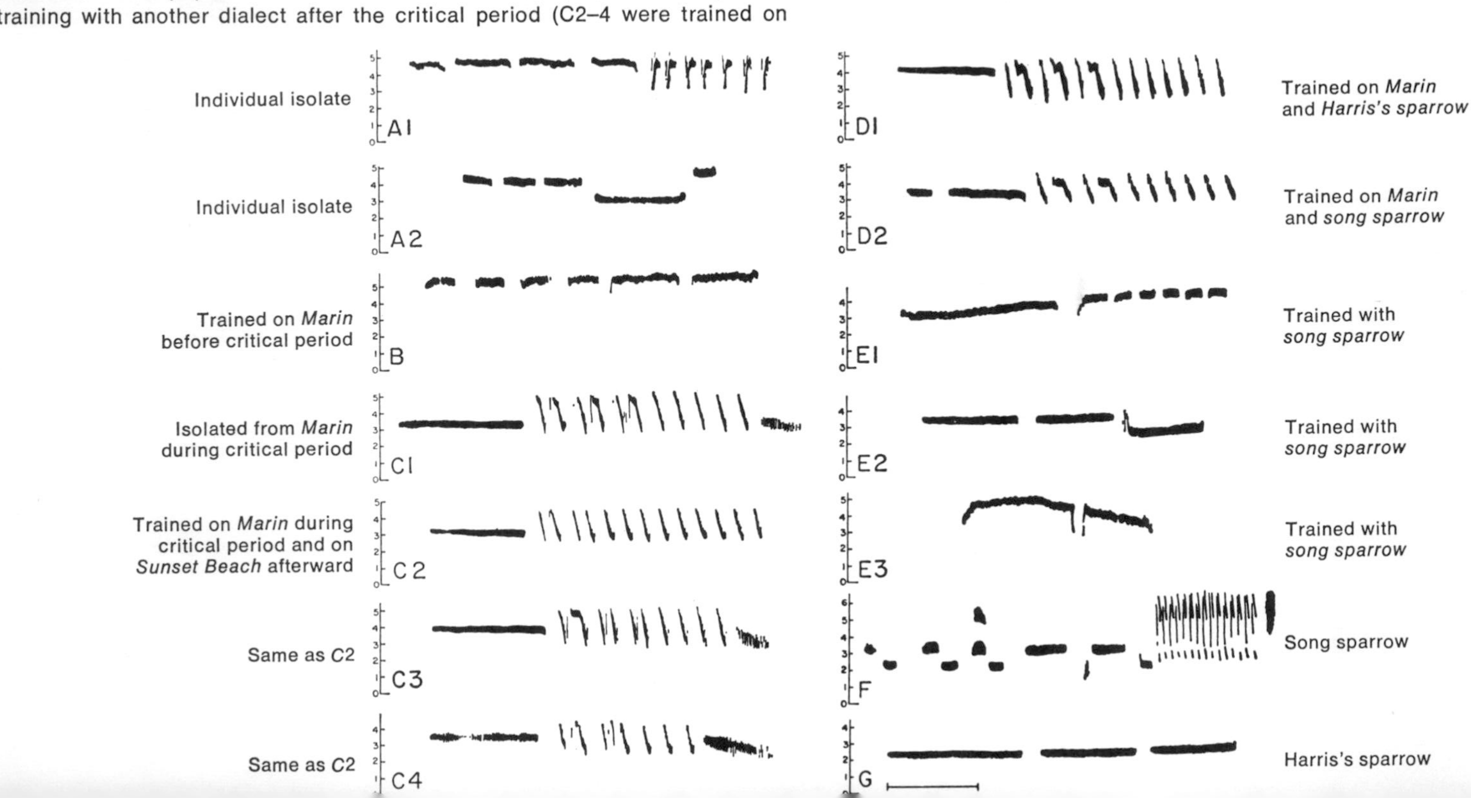

cent of sexual imprinting. Song-learning is also unusual in the constraints on what sounds an individual selects to copy in its song. While not unique to song-learning, these characteristics do impart special interest to it.

A critical period has been described in the song-learning of chaffinches, white-crowned sparrows, zebra finches, and other species (see review of Marler and Mundinger 1971). The period may be brief and very early in juvenile life, as in the zebra finch, or extended over more than one season, as in certain sparrows and finches. After it is over, further song-learning ceases. The timing of the critical period is in part determined by hormonal changes. If the onset of song in the male chaffinch, which is testosterone-dependent, is delayed by castration until two years of age, a bird with no prior singing experience will learn to sing the song of a tutor (Nottebohm 1969*a*). Yet normal chaffinches cease to be able to learn new song elements at ten months of age.

In the chaffinch the end of the critical period coincides with the crystallization of song. But in the white-crowned sparrow, the critical period coincides not with the time when young males first sing, but with an earlier period when they are not singing but can hear older males sing. If exposed to a song tutor at three to eight days of age, before the critical period, no effect is visible (compare *B* in Figure 28–5 with *Marin* in Figure 28–3). If exposed to a tutor only at 100 to 300 days of age, after the critical period spent in the wild, no effect of the tutor can be seen (compare *C*2–4 in Figure 28–5 with *BN* in Figure 28–4 and *Sunset Beach* in Figure 28–3). But if exposed during days 35 to 56, song develops normally (compare *D*1 with *C*1 in Figure 28–5 and with *Marin* in Figure 28–3). Male white-crowns can develop a normal song at about eight months even if they have not heard one since they were about four months old. But if they did not hear a normal song in the earlier (critical) period it is already too late for them at eight months to acquire it. This ability indicates a well-developed auditory memory.

If white-crowns are isolated before starting to sing and after having heard normal songs in the critical period, they retain a memory of the song, and it develops normally (*C*1 in Figure 28–5). But normal development occurs only when they can hear themselves sing (Konishi 1965). If such birds are deafened after exposure to the normal song during the critical period but before having started to sing, they produce a song that differs from those of undeafened isolates in lacking whistles, stability, and dialect properties, as shown in Figure 28–6. This indicates that white-crowns must be able to hear themselves sing to make use of their auditory memory in developing a song to match what they

28–6 Effects of deafening on song development in white-crowned sparrows. These males were taken from a natural population after they had heard their own dialect. They were collected within 100 days of fledging and before attempting to sing themselves. They were deafened surgically within 14 days of capture. **A1–5** and **B1** are songs from 6 different individuals. Males with prefix A are from Inverness and should be compared with natural songs of the Marin dialect; bird B1 is from Berkeley and should be compared with the Berkeley dialect (compare with Figure 28–3). Observe that the songs lack most of the features of natural songs and of the dialects. (From Konishi 1965.)

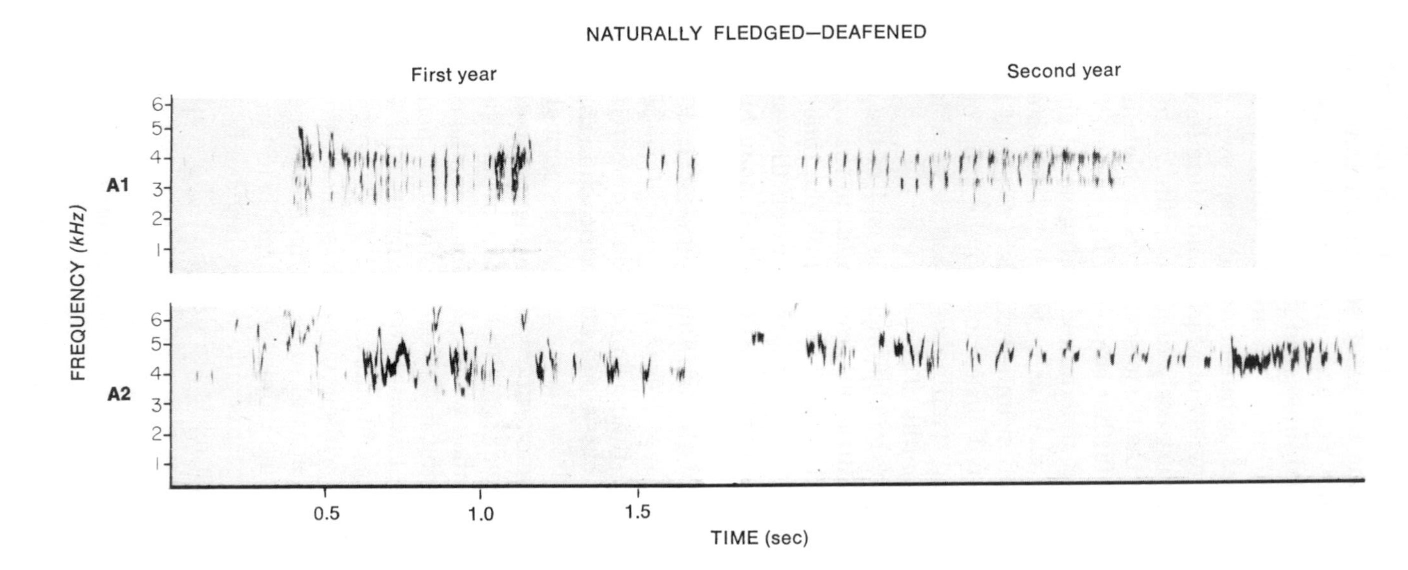

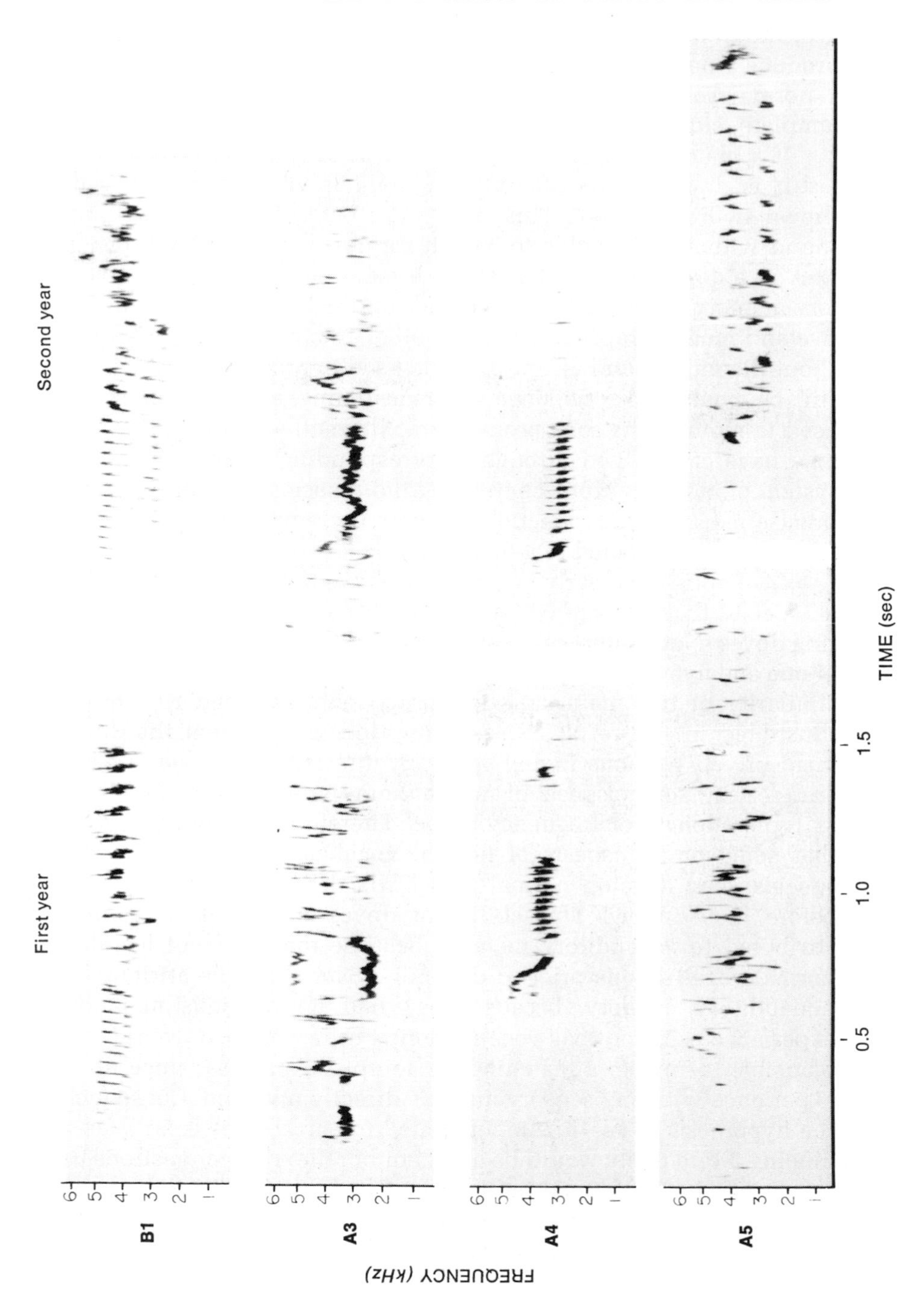
First year
Second year
B1
A3
A4
A5
FREQUENCY (kHz)
TIME (sec)
0.5
1.0
1.5

heard months earlier. Being able to hear what they sing allows them to produce a pure, unquavering whistle, a regular trill, and to crystallize a normal song. This auditory memory is analogous to an auditory template, since it makes possible a matching process.

If a male white-crown is deafened after song crystallization, the result is dramatically different: his song is virtually unchanged, as shown in Figure 28–7. This ability to re-create perfectly a complex sound without being able to hear it cannot depend on auditory feedback in a deafened bird. Therefore, it must depend on some nonauditory memory. There are two principal candidate memories, proprioceptive and motor. Proprioceptive and other internal sensory information about the muscle and air movements used to produce the song should still be available to the singer after deafening, and perhaps could be used to monitor the song production. Alternatively, the motor pattern may have crystallized through a corresponding crystallization in the system of motor neurons controlling the muscles used in singing. The precise nature of this nonauditory memory remains to be established. It too has been likened to a template.

Deafening at an early age does not have the same drastic effect in all species that it has in white-crowns. In chickens (Konishi 1963) and ring doves (Nottebohm and Nottebohm 1971), birds deafened at the age of one and five days, respectively, developed normal vocalizations. The similarity of two main calls in normal and deafened ring doves is illustrated in Figure 28–8. Deafening does not prevent the development of vocalizations in any species yet tested. Even when its effects are greatest, some aspects of normal song persist in developing, such as typical length or frequency range. Therefore, it may be concluded that some or all aspects of normal vocalization, depending on the species, may develop normally in birds deafened very early in life before first singing. This ability of doves and chickens cannot be attributed to an auditory memory because they had not heard their normal vocalizations prior to deafening. Nor can it be attributed to a nonauditory memory, because they had no previous nonauditory experience with normal vocalization. Furthermore, it would not be plausible to invoke a matching (or template) process, since no prior experience with any sensory mode is directly involved. Consequently, the hypothesis of an "innate template" for such cases is far from convincing. More likely would be a system of reflexlike connections based on neural circuits formed independently of learning (in the manner of those controlling coughing, swallowing, and sneezing, but with the addition of spontaneity); however, this is also in the realm of speculation.

In short, we have evidence of contributions to song or call develop-

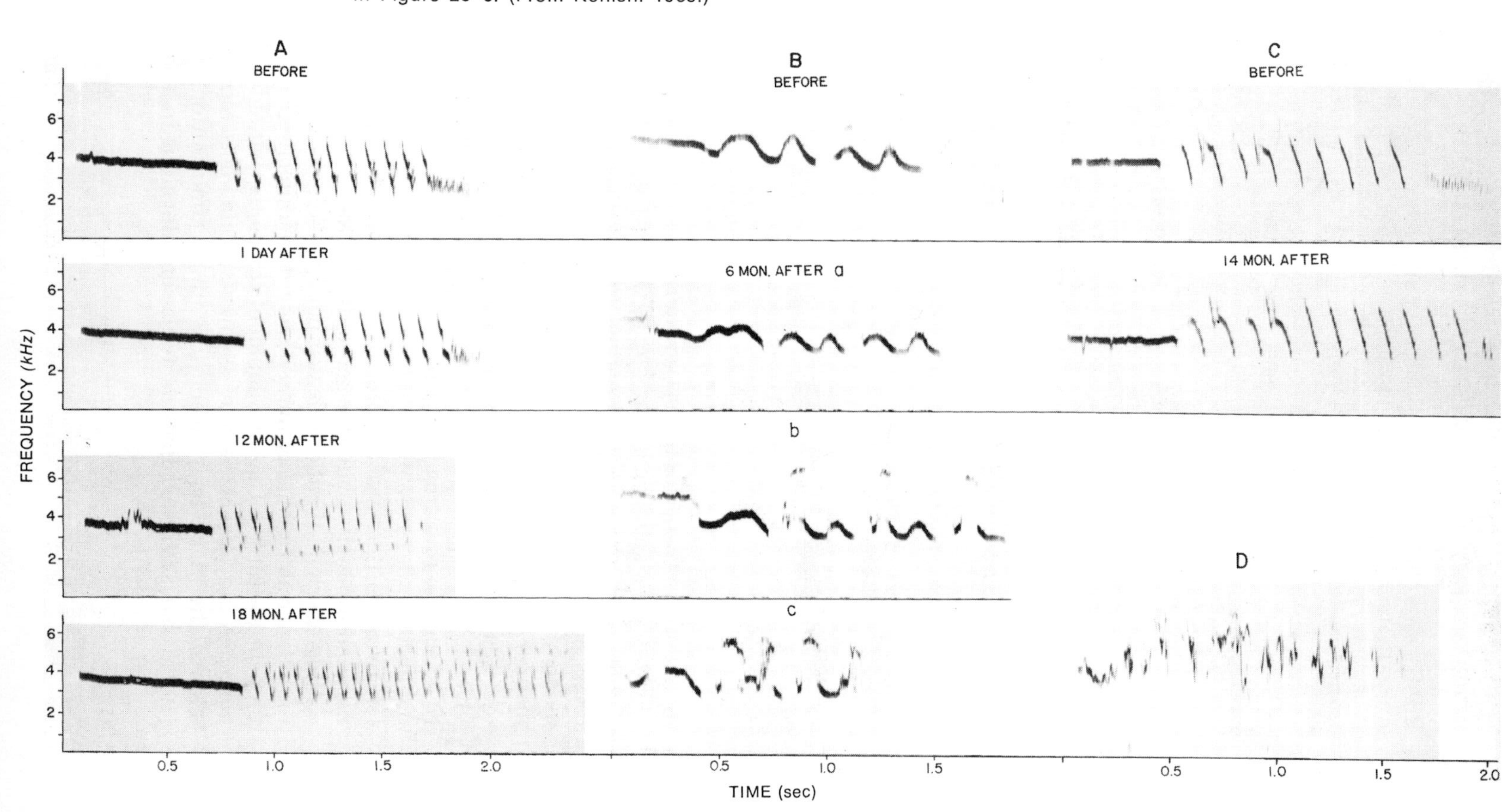

28–7 Effects of deafening on song maintenance in white-crowned sparrows. **A.** Juvenile female induced to sing with testosterone. **B.** Hand-reared bird. **C.** Wild male from Inverness (Marin dialect). **D.** Bird deafened before onset of singing. Observe that the condition shown in A–C does not approach that in D and in Figure 28–6. (From Konishi 1965.)

28–8 Effects of deafening on development of two types of vocalization in ring doves. Deafening was performed at the following ages: *Mag/M,* 6 days; *pBl/Mag,* 5 days; *W/dkG,* 1½ months. All calls were recorded from adults. (From Nottebohm and Nottebohm 1971.)

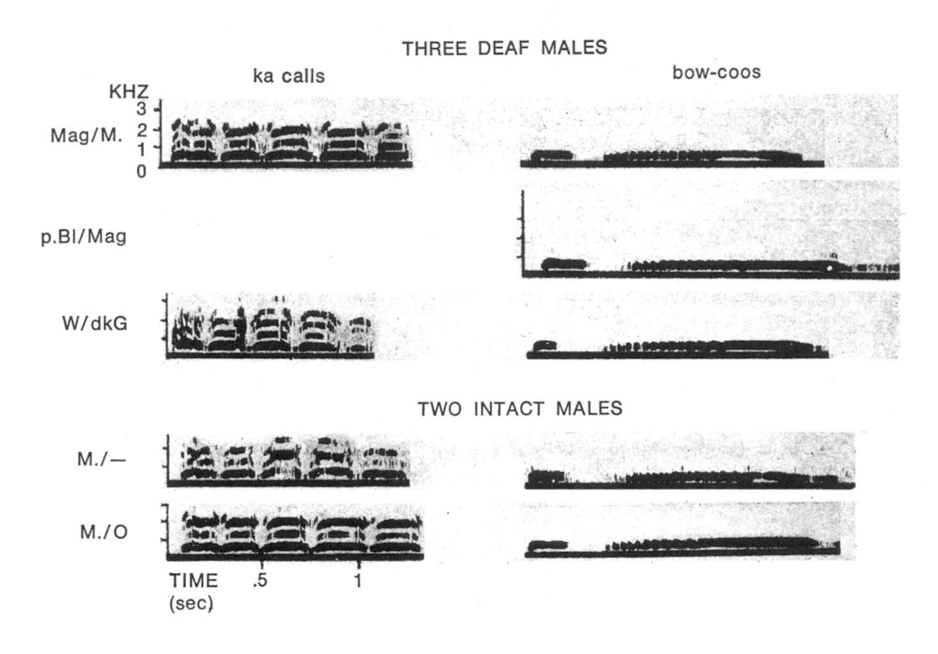

ment in various species from (1) an auditory memory, (2) a nonauditory memory, (3) a nonmemory (or "innate") source of pattern coordination, and (4) a preference for what sounds are to be learned. The existence of these separable contributions has been demonstrated by the experiments described above, but their location in the nervous system and manner of operation remain to be discovered.

The neural aspects of song development are just beginning to be investigated. In the chaffinch, as in many other species, the left and right sound-producing tympaniform membranes of the syrinx can be controlled independently. Nottebohm (1971) showed dominance of the left over the right side by observing that cutting the left hypoglossal nerve to the syrinx caused a more drastic loss of song elements than cutting the right. If either side was cut well before song crystallization, the other was capable of normal song development. The chaffinch is thought to learn during development to use the left side for the more complex and lower-frequency elements of its vocal repertoire and to silence the right one except for minor contributions. Nottebohm (1972*a*) has suggested that the adult song develops in a series of programmed steps from nestling and fledgling begging calls through the chirps and rattles of subsong, and has shown how some of the details of this process can be explained by examining the role of each side of the syrinx separately.

THE EVOLUTION OF VOCAL LEARNING

Having surveyed the incidence and methods of vocal learning in animals, can we deduce anything about the evolution of vocal learning in man and birds? Vocal learning, or the acquisition of elements in the vocal repertoire through learning, is most conspicuous in man and passeriform birds. The evolutionary antecedents of vocal learning in man's predecesors will be difficult to identify using the comparative method because of the present paucity of evidence of vocal learning in all other primates. The failures of attempts to teach chimpanzees to mimic many human sounds probably reflect the normal lack of use of vocal imitation in wild chimpanzees (Marler 1969*b*). Among primates other than man, only in siamangs (*Symphalangus syndactylus*) is there at least indirect evidence of the use of vocal learning. Lamprecht (1970) has observed that siamangs have duetting patterns that are relatively stable and characteristic for each pair, while differing among pairs, much in the manner of duetting birds. Thus, the principal use of vocal learning in nonhuman primates would seem on this slim evidence to be in maintaining the pair bond. Perhaps psychologists and linguists would have more success in their studies of language-learning in primates if they used siamangs instead of chimpanzees.

Among other mammals, vocal imitation and learning has been found among aquatic species and others in which restrictions on other sensory modalities make audition of prime importance. Imitation of sounds has been demonstrated in porpoises (*Tursiops truncatus*); and underwater song of impressive complexity is known in bearded seals (*Erignathus barbatus;* Ray et al. 1969) and humpback whales (*Megaptera novaeangliae;* Payne 1970; Payne and McVay 1971). Dialects in the vocalizations of male elephant seals (*Mirounga occidentalis;* LeBoeuf and Peterson 1969) on their mating grounds have also been described. Although vocal learning has not been experimentally verified in most of these mammals, the complexity and content of the underwater sounds and the occurrence of dialects make it seem likely. In mammals other than man, the principal selective advantages of vocal learning can be very tentatively associated with (1) the advantage given by vocalizations in attracting females to an unmated male (e.g., whales?), and (2) the advantage given by vocalizations in maintaining a stable and reproductive pair or family unit (siamangs; whales? porpoises?).

What can be said of the adaptive value of vocal learning in birds? In most male songbirds, songs are involved in repelling males and in attracting or courting females, often as part of territorial advertisement. Nottebohm (1972*b*) has suggested that male songs have been exaggerated through sexual selection of the female-preference type, much in the way the plumage of birds of paradise and the bowers of bowerbirds are thought to have evolved. If a complex or striking song acts as a supernormal stimulus for a female bird, it will be selected for. If we assume that the songs most effective in attracting females are so complex that they cannot be transmitted from generation to generation as reliably by inheritance as by learning, then individuals able to copy the most effective songs will be selected over individuals who merely inherit a predisposition for development of the song unaided by copying. By favoring males who most effectively attract females, sexual selection would cause the evolution of vocal learning ability together with a preference of males for copying the most effective songs. The principal objection to Nottebohm's hypothesis is that there is a negative correlation between song complexity and mating systems that favor sexual selection. Most of the best singers and versatile vocal learners (mimics, "talking" birds, most passerines) are monogamous. Promiscuous and polygamous species tend to have simple songs, whereas according to Nottebohm's theory they should have the most complex songs and should show more vocal learning ability.

If the songs of passeriform birds are considered as adaptations for improving the efficiency of territorial advertisement, perhaps their principal function would be in repelling males rather than attracting females. It has been shown in studies of red-winged blackbirds (*Agelaius phoeniceus;* Peek 1972*a*, *b;* Smith 1972*a*, *b*) that painting the red wing-patches black or tranquilizing males does not cause males to lose females, but it does make them much less successful in defending their territories against males. Perhaps song functions similarly. If, as seems likely, females of monogamous species choose males more on the basis of the qualities of their territories than the qualities of the males, the role of song in defense against males would seem to be more important than its role in attracting females. Then if the more complex, culturally transmitted songs are the most effective, vocal copying will be selected.

A second adaptive value of vocal learning is probably its role in maintaining pair bonds, as we have discussed with regard to duetting and "talking" birds.

A third adaptive value of vocal learning has been suggested.

Nottebohm (1972*b*) thought that it was selected for its role in maintaining the distinctness of dialect populations. However, implicit in his explanation was the mechanism of population selection; and the conditions for population selection to be effective are unlikely to be operative in most species with known dialects. Dialects may be a result of vocal learning, but they are unlikely to be its evolutionary cause.

CONCLUSIONS

Species that make use of learning in the ontogenetic development of their species-typical or population-typical vocalizations are confined, as far as is known, to a very few mammals and a large number of birds. But even in these groups, isolation experiments suggest that genetic factors are conspicuous in their influence on development. In birds, most calls develop normally in isolates, but the songs of different species vary in their dependence on learning. In many species, songs develop through a variable stage known as subsong by crystallization into a nearly invariant adult pattern. Development is characterized by a hormone-dependent critical period after which further learning typically does not occur. The constraints on what motifs may be learned vary among species; they may be very rigid, or they may permit the learning of a striking variety of sounds of other species, noises in the environment, and even human speech. In some species, song elements are learned only from the father or foster father (bullfinch, zebra finch). In others, calls or song elements are learned only from the mate (American goldfinch, some duet singers, "talking" birds). More commonly, males learn song elements from neighboring rival males. The latter practice is probably responsible for most cases of avian dialects, or population differences in songs or calls. Dialects are of theoretical interest because of their potential role in speciation.

Theories of the mechanisms of song-learning have been derived mainly from experiments employing auditory isolation or deafening. By these means it has been shown that some birds rely on an auditory memory and must be able to hear their own vocalizations to match what they have learned from others. Once crystallized, however, these songs persist for years after deafening, thus implicating a still unidentified nonauditory memory. In some species, deafening naive birds has little or no effect on normal development, and a nonmemory and nonauditory mechanism may be presumed to exist, though it remains to be identified. Of considerable interest is the selectivity by which some species choose to learn the song of their own species in preference to another when both are presented to them only through loudspeakers. The physiological mechanisms of selectivity, song memory, and self-differentiating fixed patterns remain unknown.

Major selective pressures leading to vocal learning ability in mammals other than man are controversial, but sparse evidence suggests the hypotheses that the ability to copy vocal patterns is important in pair maintenance (siamangs), pair formation (cetacea?), and inter-male agonistic encounters (elephant seals). In birds, complex songs, and therefore probably song-learning, are of major importance in territorial advertisement, particularly against males in monogamous species. Vocal copying is probably also adaptive in some birds in maintaining the pair.

Literature Cited

Able, K. P. 1970. A radar study of the altitude of nocturnal passerine migration. Bird-banding 41:282–290.

Abu Gideiri, Y. B. 1969. The development of behaviour in *Tilapia nilotica* L. Behaviour 34:17–28.

Adams, L., and S. D. Davis. 1967. The internal anatomy of home range. J. Mammal. 48:529–536.

Adler, H. E., ed. 1971. Orientation: sensory basis. Ann. N.Y. Acad. Sci. 188:1–408.

Adler, K. 1969*a*. Extraoptic phase shifting of circadian locomotor rhythm in salamanders. Science 164:1290–1292.

Adler, K. 1969*b*. Pineal end organ: role in extraoptic entrainment of circadian locomotor rhythm in frogs, pp. 342–350. *In* Menaker 1971*a*.

Adriaanse, A. 1947. *Ammophila campestris* Latr. and *Ammophila adriaansei* Wilcke. Ein Beitrag zur vergleichenden Verhaltensforschung. Behaviour 1:1–34.

Ahlen, I., and A. Andersson. 1970. Breeding ecology of an eider population on Spitsbergen. Ornis Scand. 1:83–106.

Aho, J., and O. Kalela. 1966. The spring migration of 1961 in the Norwegian lemming (*Lemmus lemmus*) at Kilpisjarvi, Finnish Lapland. Ann. Zool. Fennici 3:53–65.

Alexander, B. K. 1970. Parental behavior of adult male Japanese monkeys. Behaviour 36:270–285.

Alexander, R. D. 1960. Sound communication in Orthoptera and Cicadidae, pp. 38–91. *In* W. E. Lanyon, and W. N. Tavolga, eds., Animal sounds and communication. Amer. Inst. Biol. Sci. Pub. No. 7.

Alexander, R. D. 1962. Evolutionary change in cricket acoustical communication. Evolution 16:443–467.

Alexander, R. D. 1967. Acoustical communication in arthropods. Annu. Rev. Entomol. 12:495–526.

Alexander, R. D. 1968*a*. Life cycle origins, speciation, and related phenomena in crickets. Quart. Rev. Biol. 43:1–41.

Alexander, R. D. 1968*b*. Arthropods, pp. 167–216. *In* Sebeok 1968.

Alexander, R. D. 1974. The evolution of social behavior. Ann. Rev. Ecol. Syst. 5:325–383.

Alexander, R. D., and R. S. Bigelow. 1960. Allochronic speciation in field crickets and a new species *Acheta veletis*. Evolution 14:334–346.

Allee, W. C. 1938. The social life of animals. Beacon Press, Boston.

Allee, W. C. 1952. Dominance and hierarchy in societies of vertebrates. Colloq. Int. Cent. Nat. Recherche Sci. 34:157–181.

Allen, G. R. 1972. The anemonefishes; their classification and biology. T.F.H. Publications, Neptune City, N.J.

Altes, R. A., and E. L. Titlebaum. 1970. Bat signals as optimally doppler tolerant waveforms. J. Acoust. Soc. Amer. 48:1014–1020.

LITERATURE CITED

Altmann, S. A. 1962. A field study of the sociobiology of rhesus monkeys, *Macaca mulatta*. *Ann.* N.Y. Acad. Sci. 102:330–435.

Altmann, S. A. 1965. Sociobiology of rhesus monkeys. II: Stochastics of social communication. J. Theoret. Biol. 8:490–522.

Altmann, S. A. 1967. The structure of primate social communication, pp. 325–362. *In* S. A. Altmann, ed., Social communication among primates. Univ. Chicago Press, Chicago.

Altmann, S. A., and J. Altmann. 1970. Baboon ecology. Univ. Chicago Press, Chicago.

Altum, J. B. T. 1868. Der Vogel und sein Leben. Munster. *See* E. Mayr, 1935, Bernard Altum and the territory theory. Proc. Linn. Soc. N.Y. 45, 46:24–38.

Amadon, D. 1959. Behavior and classification. Festschrift N. Steiner. Vierteljahrschrift naturforschende Gesellschaft Zürich 104:73–78.

Anderson, E. 1949. Introgressive hybridization. Wiley, New York.

Anderson, P. 1964. Lethal alleles in *Mus musculus:* local distribution and the evidence for the isolation of demes. Science 145:177–178.

Anderson, P. K., and J. L. Hill. 1965. *Mus musculus:* experimental induction of territory formation. Science 148:1753–1755.

Anderson, W. W., and L. Ehrman. 1969. Mating choice in crosses between geographic populations of *Drosophila pseudoobscura*. Amer. Midland Natur. 81:47–53.

Andrew, R. J. 1956. Intention movements of flight in certain passerines, and their use in systematics. Behaviour 10:179–204.

Andrew, R. J. 1963*a*. The origin and evolution of the calls and facial expressions of the primates. Behaviour 20:1–109.

Andrew, R. J. 1963*b*. Evolution of facial expression. Science 142:1034–1041.

Andrewartha, H. G., and L. C. Birch. 1954. The distribution and abundance of animals. Univ. Chicago Press, Chicago.

Andrews, R. V. 1968. Daily and seasonal variation in adrenal metabolism of the brown lemming. Physiol. Zool. 41:86–94.

Aneshansley, D. J., T. Eisner, J. M. Widom, and B. Widom. 1969. Biochemistry at 100°C: explosive secretory discharge of bombardier beetles (*Brachinus*). Science 165:61–63.

Archer, J. 1970. Effects of population density on behaviour in rodents, pp. 169–210. *In* J. H. Crook, ed., Social behaviour in birds and mammals. Academic Press, New York.

Ardrey, R. 1966. The territorial imperative. Atheneum, New York.

Armstrong, E. A. 1947. Bird display and behaviour. Lindsay Drummond, London.

Armstrong, E. A. 1955. The wren. Collins, London.

Armstrong, E. A. 1963. A study of bird song. Oxford Univ. Press, Oxford.

Armstrong, E. A. 1964. Distraction display, p. 205. *In* A. L. Thompson, ed., A new dictionary of birds. McGraw-Hill, New York.

Armstrong, J. T. 1965. Breeding home range in the nighthawk and other birds: its evolutionary and ecological significance. Ecology 46:610–629.

Armstrong, P. B., and D. C. Higgins. 1971. Behavioral encephalization in the bullhead embryo and its neuroanatomical correlates. J. Comp. Neurol. 143:371–384.

Arnold, J. M. 1962. Mating behavior and social structure in *Loligo pealii*. Biol. Bull. 123:53–57.

Aronson, L. R., E. Tobach, D. S. Lehrman, and J. S. Rosenblatt, eds. 1970. Development and evolution of behavior. Freeman, San Francisco.

Atz, J. W. 1970. The application of the idea of homology to behavior, pp. 53–74. *In* L. R. Aronson, E. Tobach, D. S. Lehrman, and J. S. Rosenblatt, eds., Development and evolution of behavior. Freeman, San Francisco.

Awbrey, F. T. 1968. Call discrimination in female *Scaphiopus couchii* and *Scaphiopus hurterii*. Copeia 1968:420–432.

Baerends, G. P. 1953. Specialization in organs and movements with a releasing function. Soc. Exp. Biol. Symp. 4:337–360.

Baerends, G. P. 1957. Behavior: the ethological analysis of fish behavior, pp. 229–269. *In* M. E. Brown, ed., The physiology of fishes. Academic Press, New York.

Baerends, G. P. 1958. Comparative methods and the concept of homology in the study of behavior. Arch. Neerl. Zool. 13(Suppl.):401–417.

Baerends, G. P., and J. M. Baerends-van Roon. 1950. An introduction to the study of the ethology of cichlid fishes. Behaviour Suppl. 1:1–242.

Baerends, G. P., R. Brouwer, and H. T. Waterbolk. 1955. Ethological studies on *Lebistes reticulatus* (Peters). I. An analysis of the male courtship pattern. Behaviour 8:249–334.

Baerends, G. P., and N. A. van der Cingel. 1962. On the phylogenetic origin of the snap display in the common heron (*Ardea cinerea* L.) Symp. Zool. Soc. London 8:7–24.

Baker, H. G., and G. L. Stebbins, eds. 1965. The genetics of colonizing species. Academic Press, New York.

Baldwin, P. H. 1953. Annual cycle, environment and evolution in the Hawaiian honeycreepers (Aves, Drepaniidae). Univ. Calif. Pub. Zool. 52:285–398.

Ball, R. W., and D. L. Jameson. 1966. Premating isolating mechanisms in sympatric and allopatric *Hyla regilla* and *Hyla californiae*. Evolution 20:533–552.

Barber, H. S. 1951. North American fireflies of the genus *Photuris*. Smithsonian Misc. Coll. 117(1).

Barbour, R. W., and W. H. Davis. 1969. Bats of America. Univ. Press of Kentucky, Lexington.

Barlow, G. W. 1968. Effect of size of mate on courtship in a cichlid fish, *Etroplus maculatus*. Comm. Behav. Biol. 2:149–160.

Barlow, G. W., and R. F. Green. 1970. Effect of relative size of mate on color patterns in a mouthbreeding cichlid fish, *Tilapia melanotheron*. Comm. Behav. Biol. 4:71–78.

Barlow, H. B., R. Narasimhan, and A. Rosenfeld. 1972. Visual pattern analysis in machines and animals. Science 177:567–575.

Barth, R. 1937. Herkunft, Wirkung und Eigenschaften des weiblichen Sexualduftstoffes einiger Pyraliden. Zool. Jahrb. Allgem. Zool. Physiol. 58:297–329.

Bartholomew, G. A. 1942. The fishing activities of double-crested cormorants on San Francisco Bay. Condor 44:13–21.

Bartholomew, G. A. 1952. Reproductive and social behavior of the northern elephant seal. Univ. Calif. Pub. Zool. 47:369–472.

Bartholomew, G. A. 1970. A model for the evolution of pinniped polygyny. Evolution 24:546–559.

Bastock, M. 1956. A gene mutation which changes a behavior pattern. Evolution 10:421–439.

Bastock, M. 1967. Courtship: an ethological study. Aldine, Chicago.

Bateson, P. P. G. 1966. The characteristics and context of imprinting. Biol. Rev. Cambridge Phil. Soc. 41:177–220.

Bateson, P. P. G. 1971. Imprinting, pp. 369–387. *In* Moltz 1971.

Batham, E. J., and C. F. A. Pantin. 1950. Phases of activity in the sea anemone, *Metridium senile* (L.), and their relation to external stimuli. J. Exp. Biol. 27:377–399.

Batham, E. J., C. F. A. Pantin, and E. A. Robson. 1960. The nerve-net of the sea-anemone *Metridium senile:* the mesenteries and the column. Quart. J. Microscop. Sci. 101:487–510.

Beer, C. G. 1970*a*. Individual recognition of voice in the social behavior of birds. Adv. Behav. 3:27–74.

Beer, C. G. 1970*b*. On the responses of laughing gull chicks (*Larus atricilla*) to the calls of adults. I. Recognition of the voices of the parents. II. Age changes and responses to different types of call. Anim. Behav. 18:652–660, 661–677.

Bellrose, F. C. 1971. The distribution of nocturnal migrants in the air space. Auk 88:397–424.

Bennet-Clark, H. C., and A. W. Ewing. 1968. The wing mechanism involved in the courtship of *Drosophila*. J. Exp. Biol. 49:117–128.

Bennet-Clark, H. C., and A. W. Ewing. 1969. Pulse interval as a critical parameter in the courtship song of *Drosophila melanogaster*. Anim. Behav. 17:755–759.

Bennet-Clark, H. C., and A. W. Ewing. 1970. The love song of the fruit fly. Sci. Amer. 223(1):84–92.

Bennett, E. L., M. C. Diamond, D. Krech, and M. R. Rosenzweig. 1964. Chemical and anatomical plasticity of brain. Science 146:610–619.

Bentley, D. R. 1969. Intracellular activity in cricket neurons during generation of song patterns. Z. vergl. Physiol. 62:267–283.

Bentley, D. R. 1971. Genetic control of an insect neuronal network. Science 174:1139–1141.

Bentley, D. R., and R. R. Hoy. 1970. Postembryonic development of adult motor patterns in crickets: a neural analysis. Science 170:1409–1411.

Bentley, D. R., and R. R. Hoy. 1972. Genetic control of the neuronal network generating cricket (*Teleogryllus gryllus*) song patterns. Anim. Behav. 20:478–492.

Benzer, S. 1967. Behavioral mutants of *Drosophila* isolated by counter-current distribution. Proc. Nat. Acad. Sci. 58:1112–1119.

Bergman, G. 1964. Zum Problem der gemischten Kolonien: Tonband- und Dressurversuche mit Limicolen und Anatiden. Ornis Fennica 4:1–13.

Bergman, G. 1965. Der sexuelle Grossendimorphismus der Anatiden als Anpassung an das Höhlenbruten. Commentationes Biol. 28:1–10.

Berndt, R., and H. Sternberg. 1968. Terms, studies and experiments on the problems of bird dispersion. Ibis 110:256–269.

Berndt, R., and H. Sternberg. 1969. Alters- und Geschlechtsunterschiede in der Dispersion des Trauerschnäppers (*Ficedula hypoleuca*). J. Ornithol. 110:22–26.

Bernstein, I. S. 1968. The lutong of Kuala Selanger. Behaviour 32:1–16.

Beroza, M., and E. F. Knipling. 1972. Gypsy moth control with the sex attractant pheromone. Science 177:19–27.

Bertram, B. 1970. The vocal behaviour of the Indian hill mynah, *Gracula religiosa*. Anim. Behav. Monogr. 3:81–192.

Bigelow, R. S. 1960. Interspecific hybrids and speciation in the genus *Acheta* (Orthoptera: Gryllidae). Can. J. Zool. 38:509–524.

Bigelow, R. S. 1965. Hybrid zones and reproductive isolation. Evolution 19:449–458.

Binkley, S., E. Kluth, and M. Menaker. 1971. Pineal function in sparrows: circadian rhythms and body temperature. Science 174:311–314.

Bjärvall, A. 1967. The critical period and the interval between hatching and exodus in mallard ducklings. Behaviour 28:141–148.

Blair, W. F. 1940. A study of prairie deermouse populations in southern Michigan. Amer. Midland Natur. 24:273–305.

Blair, W. F. 1955*a*. Size difference as a possible isolation mechanism in *Microhyla*. Amer. Natur. 89:297–302.

Blair, W. F. 1955*b*. Mating call and stage of speciation in *Microhyla olivacea–M. carolinensis* complex. Evolution 9:469–480.

Blair, W. F. 1958. Mating call in the speciation of anuran amphibians. Amer. Natur. 92:27–51.

Blair, W. F. 1962. Non-morphological data in anuran classification. Syst. Zool. 11:72–84.

Blair, W. F. 1964. Isolating mechanisms and interspecies interactions in anuran amphibians. Quart. Rev. Biol. 39:334–344.

Blair, W. F. 1968. Amphibians and reptiles, pp. 289–310. *In* T. A. Sebeok, ed., Animal communication. Indiana Univ. Press, Bloomington.

Blair, W. F., ed. 1961. Vertebrate speciation. Univ. Texas Press, Austin.

Blair, W. F., and W. E. Howard. 1944. Experimental evidence of sexual isolation between three forms of mice of the cenospecies *Peromyscus maniculatus*. Contrib. Lab. Vert. Biol. Univ. Michigan 26:1–19.

Blair, W. F., and M. J. Littlejohn. 1960. Stage of speciation of two allopatric populations of chorus frogs (*Pseudacris*). Evolution 14:82–87.

Blest, A. D. 1957*a*. The function of eyespot patterns in the Lepidoptera. Behaviour 11:209–256.

Blest, A. D. 1957*b*. The evolution of protective displays in the Saturnoidea and Sphingidae (Lepidoptera). Behaviour 11:257–309.

Blest, A. D. 1960. The evolution, ontogeny and quantitative control of the settling movements of some New World saturniid moths, with some comments on distance communication by honey-bees. Behaviour 16:188–253.

Blum, M. S., and J. M. Brand. 1972. Social insect pheromones: their chemistry and function. Amer. Zool. 12:553–576.

Boch, R. 1957. Rassenmässige Unterschiede bei den Tänzender Honigbiene (*Apis mellifica L.*). Z. vergl. Physiol. 40:289–320.

Boch, R., D. A. Shearer, and B. C. Stone. 1962. Identification of isoamyl acetate as an active component in the sting pheromone of the honey bee. Nature 195:1018–1020.

Bock, W. J. 1967. The use of adaptive characters in avian classification. Proc. Int. Ornithol. Conf. 14:61–74.

Bock, W. J. 1970. Microevolutionary sequences as a fundamental concept in macroevolutionary models. Evolution 24:704–722.

Boeckh, J., K. E. Kaissling, and D. Schneider. 1965. Insect olfactory receptors. Cold Spring Harbor Symp. Quant. Biol. 30:263–280.

Boehlke, K. W., and B. E. Eleftheriou. 1967. Levels of monoamine oxidase in the brain of C57BL/6J mice after exposure to defeat. Nature 213:739–740.

Bogert, C. M. 1960. The influence of sound on the behavior of amphibians and reptiles, pp. 137–320. *In* W. E. Lanyon and W. N. Tavolga, eds., Animal sounds and communication. Amer. Inst. Biol. Sci. Pub. No. 7.

Boorman, S. A., and P. R. Levitt. 1973. A frequency-dependent natural selection model for the evolution of social cooperation networks. Proc. Nat. Acad. Sci. 70:187–189.

Borror, D. J. 1961. Intraspecific variation in passerine bird songs. Wilson Bull. 73:57–78.

Borror, D. J. 1965. Song variation in Maine song sparrows. Wilson Bull. 77:5–37.

Borror, D. J. 1967. Songs of the yellowthroat. Living Bird. 6:141–161.

Borror, D. J., and W. W. H. Gunn. 1965. Variation in white-throated sparrow songs. Auk 82:26–47.

Bossert, W. H., and E. O. Wilson. 1963. The analysis of olfactory communication among animals. J. Theoret. Biol. 5:443–469.

Bowman, R. I. 1961. Morphological differentiation and adaptation in the Galapagos finches. Univ. Calif. Pub. Zool. 58:1–326.

Boyd, H. 1953. On encounters between wild white-fronted geese in winter flocks. Behaviour 5:85–129.

Brady, J. 1967*a*. Control of the circadian rhythm of activity in the cockroach. I. The role of the corpora cardiaca, brain and stress. J. Exp. Biol. 47:153–163.

Brady, J. 1967*b*. Control of the circadian rhythm of activity in the cockroach. II. The role of the suboesophageal ganglion and ventral nerve cord. J. Exp. Biol. 47:165–178.

Brady, J. 1969. How are insect circadian rhythms controlled? Nature 223:781–784.

Brady, J. 1971. The search for an insect clock, pp. 517–524. *In* Menaker 1971*a*.

Braestrup, F. W. 1966. Social and communal display. Phil. Trans. Roy. Soc. (B) 251:375–386.

Brain, P. F. 1971. The physiology of population limitation in rodents—a review. Comm. Behav. Biol. 6:115–123.

Brander, R. B. 1967. Movements of female ruffed grouse during the mating season. Wilson Bull. 79:28–36.

Breder, C. M., Jr. 1936. The reproductive habits of the North American sunfishes (family Centrarchidae). Zoologica 21:1–48.

Breder, C. M., Jr. 1959. Studies on social groupings in fishes. Bull. Amer. Mus. Natur. Hist. 117:393–482.

Breder, C. M., Jr., and D. E. Rosen. 1966. Modes of reproduction in fishes. Doubleday, Garden City, N.Y.

Bremond, J. C. 1963. Acoustic behaviour of birds, pp. 709–750. *In* Busnel, R. G., ed., Acoustic behaviour of animals. Elsevier, Amsterdam.

Bremond, J. C. 1967. Reconnaissance de schemas reactogenes liés à l'information contenue dans le chant territorial du rouge-gorge (*Erithacus rubecula*). Proc. Int. Ornithol. Congr. 14:217–229.

Bremond, J. C. 1968*a*. Recherches sur la sémantique et les éléments vecteur d'information contenue dans les signaux acoustiques du rouge-gorge (*Erithacus rubecula*). La Terre et la Vie 2:109–220.

Bremond, J. C. 1968*b*. Valeur specifique de la syntaxe dans le signal de defense territoriale du Troglodyte (*Troglodytes troglodytes*). Behaviour 30:66–75.

Bremond, J. C. 1968*c*. Parametres physiques assurant la specificité du chant chez de pouillot siffleur (*Phylloscopus sibilatrix*). Rev. du Comportement Anim. 3:97–98.

Brenner, F. J. 1966. The influence of drought on reproduction in a breeding population of red-winged blackbirds. Amer. Midland Natur. 76:201–210.

Brereton, J. le Gay, and K. Immelmann. 1962. Head-scratching in the Psittaciformes. Ibis 104:167–175.

Brian, A. D. 1949. Dominance in the great tit *Parus major*. Scottish Natur. 61:144–155.

Brian, M. V. 1965. Social insect populations. Academic Press, New York.

Bronson, F. H. 1964. Agonistic behavior in woodchucks. Anim. Behav. 12:470–478.

Bronson, F. H. 1971. Rodent pheromones. Biol. Reprod. 4:344–357.

Brosset, A. 1971. L' "imprinting," chez les Colombidés—étude des modifications comportementales aucours de vieillisement. Z. Tierpsychol. 29:279–300.

Brown, J. L. 1959. Method of head-scratching in the wrentit and other species. Condor 61:53.

Brown, J. L. 1963*a*. Ecogeographic variation and introgression in an avian visual signal: the crest of the Steller's jay, *Cyanocitta stelleri*. Evolution 17:23–39.

Brown, J. L. 1963*b*. Aggressiveness, dominance and social organization in the Steller's jay. Condor 65:460–484.

Brown, J. L. 1963*c*. Social organization and behavior of the Mexican jay. Condor 65:126–153.

Brown, J. L. 1964*a*. The evolution of diversity in avian territorial systems. Wilson Bull. 6:160–169.

Brown, J. L. 1964*b*. The integration of agonistic behavior in the Steller's jay *Cyanocitta stelleri* (Gmelin). Univ. Calif. Pub. Zool. 60:223–328.

Brown, J. L. 1966. Types of group selection. Nature 211:870.
Brown, J. L. 1969*a*. Territorial behavior and population regulation in birds. Wilson Bull. 81:293–329.
Brown, J. L. 1969*b*. The buffer effect and productivity in tit populations. Amer. Natur. 103:347–354.
Brown, J. L. 1969*c*. Neuro-ethological approaches to the study of emotional behavior: stereotypy and variability. Ann. N.Y. Acad. Sci. 159:1084–1095.
Brown, J. L. 1970*a*. Cooperative breeding and altruistic behaviour in the Mexican jay, *Aphelocoma ultramarina*. Anim. Behav. 18:366–378.
Brown, J. L. 1970*b*. The neural control of aggression, pp. 164–186. *In* C. H. Southwick, ed., Animal aggression. Van Nostrand Reinhold, New York.
Brown, J. L. 1972. Communal feeding of nestlings in the Mexican jay (*Aphelocoma ultramarina*): interflock comparisons. Anim. Behav. 20:394–402.
Brown, J. L. 1974. Alternate routes to sociality in jays—with a theory for the evolution of altruism and communal breeding. Amer. Zool. 14:61–78.
Brown, J. L. 1975. Helpers among Arabian babblers *Turdoides squamiceps*. Ibis 117:243–244.
Brown, J. L., and R. W. Hunsperger. 1963. Neuroethology and the motivation of agonistic behaviour. Anim. Behav. 11:439–448.
Brown, J. L., R. W. Hunsperger, and H. E. Rosvold. 1969*a*. Defence, attack, and flight elicited by electrical stimulation of the hypothalamus of the cat. Exp. Brain Res. 8:113–129.
Brown, J. L., R. W. Hunsperger, and H. E. Rosvold. 1969*b*. Interaction of defence and flight reactions produced by simultaneous stimulation at two points in the hypothalamus of the cat. Exp. Brain Res. 8:130–149.
Brown, J. L., and G. H. Orians. 1970. Spacing patterns in mobile animals. Annu. Rev. Ecol. Syst. 1:239–262.
Brown, L. E. 1966. Home range and movement of small mammals. Symp. Zool. Soc. London 18:111–142.
Brown, L. E., and J. R. Brown. 1972. Call types of the *Rana pipiens* complex in Illinois. Science 176:928–929.
Brown, L. N. 1972. Mating behavior and life habits of the sweet-bay silk moth (*Callosamia carolina*). Science 176:73–75.
Brown, R. G. B. 1964. Courtship behaviour in the *Drosophila obscura* group. I. *D. pseudoobscura*. Behaviour 23:61–106.
Brown, R. G. B. 1965. Courtship behaviour in the *Drosophila obscura* group. Behaviour 25:281–323.
Brown, W. L. 1959. Taxonomic problems with closely related species. Annu. Rev. Entomol. 4:77–98.
Brown, W. L. 1968. An hypothesis concerning the function of the metapleural glands in ants. Amer. Natur. 102:188–191.
Brown, W. L., T. Eisner, and R. H. Whittaker. 1970. Allomones and kairomones: transspecific chemical messengers. BioSci. 20:21–22.
Brown, W. L., and E. O. Wilson. 1956. Character displacement. Syst. Zool. 5:49–64.
Bub, H. 1967–1969. Vogelfang und Vogelberingung. Teilen I–IV. A. Ziemsen Verlag, Wittenberg Lutherstadt.
Buchsbaum, R. 1948. Animals without backbones. Univ. Chicago Press, Chicago.
Buck, J., and E. Buck. 1968. Mechanism of rhythmic synchronous flashing of fireflies. Science 159:1319–1327.
Buck, J. B. 1937. Studies on the firefly. II. The signal system and color vision in *Photinus pyralis*. Physiol. Zool. 10:412–419.
Buckley, P. A., and F. G. Buckley. 1970. Color variation in the soft parts and down of royal tern chicks. Auk 87:1–13.

LITERATURE CITED

Buechner, H. K., and R. Schloeth. 1965. Ceremonial mating behavior in Uganda kob (*Adenota kob thomasi* Neuman). Z. Tierpsychol. 22:209–225.

Buettner-Janusch, J. 1963. Hemoglobins and transferrins of baboons. Folia Primat. 1:73–87.

Bullock, T. H. 1959. Neuron doctrine and electrophysiology. Science 129:997–1002.

Bullock, T. H. 1961. The origins of patterned nervous discharge. Behaviour 17:48–57.

Bullock, T. H. 1965. Comparative aspects of superficial conduction systems in echinoids and asteroids. Amer. Zool. 5:545–562.

Bullock, T. H., and G. A. Horridge. 1965. Structure and function in the nervous systems of invertebrates. Vols. I., II. Freeman, San Francisco.

Burckhardt, D. 1944. Möwenbeobachtungen in Basel. Ornithol. Beobacht. 5:49–76.

Burghardt, G. M. 1970. Intraspecific geographical variation in chemical food cue preferences of newborn garter snakes (*Thamnophis sirtalis*). Behaviour 36:246–257.

Burkitt, J. P. 1924–1926. A study of the robin by means of marked birds. Brit. Birds 17:294–303; 18:97–103, 250–257; 19:120–124; 20:91–101.

Burnet, B., K. Conolly, and L. Dennis. 1971. The function and processing of auditory information in the courtship behaviour of *Drosophila melanogaster*. Anim. Behav. 19:409–415.

Burt, W. H. 1943. Territoriality and home range concepts as applied to mammals. J. Mammal. 24:346–352.

Bustard, H. R. 1968. The ecology of the Australian gecko, *Gehyra variegata*, in northern New South Wales. J. Zool., London 154:113–138.

Bustard, H. R. 1969. The population ecology of the gekkonid lizard, *Gehyra variegata* (Dumeril & Bibron) in exploited forest in northern New South Wales. J. Anim. Ecol. 38:35–51.

Bustard, H. R. 1970. The role of behavior in the natural regulation of numbers in the gekkonid lizard *Gehyra variegata*. Ecology 51:724–728.

Butler, C. G. 1970. Chemical communication in insects: behavioral and ecologic aspects. *In* Johnston, et al. 1970, pp. 35–78.

Cade, T. J., and L. I. Greenwald. 1966. Drinking behavior of mousebirds in the Namib Desert, Southern Africa. Auk 83:126–128.

Cade, T. J., E. J. Willoughby, and G. Maclean. 1966. Drinking behavior of sandgrouse in the Namib and Kalahari deserts, Africa. Auk 83:124–126.

Camhi, J. M., and M. Hinkle. 1972. Attentiveness to sensory stimuli: central control in locusts. Science 175:550–553.

Campbell, B., ed. 1972. Sexual selection and the descent of man. Aldine, Chicago.

Campbell, H. W., and W. E. Evans. 1967. Sound production in two species of tortoises. Herpetology 23:204–209.

Capranica, R. R. 1965. The evoked vocal response of the bullfrog: a study of communication by sound. M.I.T. Res. Monogr. 33:1–110.

Capranica, R. R. 1966. Vocal response of the bullfrog to natural and synthetic mating calls. J. Acoust. Soc. Amer. 40:1131–1139.

Carl, E. 1971. Population control in arctic ground squirrels. Ecology 52:395–413.

Carpenter, C. R. 1934. A field study of the behavior and social relations of howling monkeys. Comp. Psychol. Monogr. 10:1–168.

Carpenter, C. R. 1940. A field study in Siam of the behavior and social relations of the gibbon, *Hylobates lar*. Comp. Psychol. Monogr. 16:1–212.

Carrick, R. 1963. Ecological significance of territory size in the Austrialian magpie, *Gymnorhina tibicen*. Proc. Int. Ornithol. Congr. 13:740–53.

Carthy, J. D. 1958. An introduction to the behaviour of invertebrates. Allen & Unwin, London.

Caughley, G. J. 1964. Density and dispersion of two species of kangaroo in relation to habitat. Austral. J. Zool. 12:238–249.

Chance, E. P. 1922. The cuckoo's secret. Sidgwick and Jackson, London.

Chance, M. R. A., and C. J. Jolly. 1970. Social groups of monkeys, apes and man. Dutton, New York.

Chance, M. R. A., and A. P. Mead. 1953. Social behavior and primate evolution. Symp. Soc. Exp. Biol. 7:395–439.

Chappuis, C. 1971. Un example de l'influence du milieu sur les émissions vocales des oiseaux: l'évolution des chants en forêt equatoriale. La Terre et la Vie 118:183–202.

Charles-Dominique, P. 1972. Behavior and ecology of nocturnal prosimians. Advances Ethology 9:7.

Chauvin, R., ed. 1968. Traité de biologie de l'abeille. Vol. 1, Biologie et physiologie generales. Vol. 2, Système nerveux, comportement et regulations sociales. Vol. 3, Les produits de la ruche. Vol. 4, Biologie appliquée. Vol. 5, Histoire, ethnographie et folklore. Masson, Paris.

Childers, W. F. 1967. Hybridization of four species of sunfishes (Centrarchidae). Illinois Natur. Hist. Survey Bull. 29:159–214.

Chitty, D. 1952. Mortality among voles (*Microtus agrestis*) at Lake Vyrnwy, Montgomeryshire in 1936–9. Phil. Trans. Roy. Soc. (B) 236:505–552.

Chitty, D. 1967. The natural selection of self-regulatory behavior in animal populations. Proc. Ecol. Soc. Austral. 2:51–78.

Christian, J. J. 1963. Endocrine adaptive mechanisms and the physiological regulation of population growth, pp. 189–353. *In* W. Mayer and R. van Gelder, eds., Physiological mammalogy, vol. 1. Academic Press, New York.

Christian, J. J. 1970. Social subordination, population density, and mammalian evolution. Science 168:84–90.

Christian, J. J., and D. E. Davis. 1964. Endocrines, behavior, and population. Science 146:1550–1560. Comments in Science 149:376–377.

Christian, J. J., J. A. Lloyd, and D. E. Davis. 1965. The role of endocrines in the self-regulation of mammalian populations. Recent Progr. Hormone Res. 21:501–578.

Clark, E., L. R. Aronson, and M. Gordon. 1954. Mating behavior patterns in two sympatric species of xiphophorin fishes, their inheritance and significance in sexual isolation. Bull. Amer. Mus. Natur. Hist. 103:139–225.

Clark, P. J., and F. C. Evans. 1954. Distance to nearest neighbor as a measure of spatial relationships in populations. Ecology 35:445–453.

Clayton, D. L., and J. V. Paietta. 1972. Selection for circadian eclosion time in *Drosophila melanogaster*. Science 178:994–995.

Cochrane, W. W., G. G. Montgomery, and R. R. Graber. 1967. Migratory flights of *Hylocichla* thrushes in spring: a radio-telemetry study. Living Bird 6:213–225.

Cochrane, W. W., D. W. Warner, J. R. Tester, and V. B. Kuechle. 1965. Automatic radio-tracking system for monitoring animal movements. BioSci. 15:98–100.

Cody, M. L. 1969. Convergent characteristics in sympatric species: a possible relation to interspecific competition and aggression. Condor 71:222–239.

Coghill, G. E. 1929. The early development of behaviour in *Amblystoma* and in man. Arch. Neurol. Psychol. 21:989–1009.

Colbert, E. H. 1958. Morphology and behavior, pp. 27–47. *In* A. Roe and G. G. Simpson, eds., Behavior and evolution. Yale Univ. Press, New Haven.

Cole, L. C. 1954. The population consequences of life history phenomena. Quart. Rev. Biol. 29:103–137.

Collias, N. E. 1943. Statistical analysis of factors which make for success in initial encounters between hens. Amer. Natur. 77:519–538.
Collias, N. E. 1944. Aggressive behavior among vertebrates. Physiol. Zool. 17:83–123.
Collas, N. E. 1964. The evolution of nests and nest-building in birds. Amer. Zool. 4:175–190.
Collias, N. E., and E. C. Collias. 1963. Evolutionary trends in nest building by the weaverbirds (Ploceidae). Proc. Int. Ornithol. Congr. 13:518–530.
Collias, N. E., and E. C. Collias. 1964. Evolution of nest-building by the weaverbirds (*Ploceidae*). Univ. Calif. Pub. Zool. 73:1–239.
Collias, N. E., and E. C. Collias. 1969. Size of breeding colony related to attraction of mates in a tropical passerine bird. Ecology 50:481–488.
Collias, N. E., and E. C. Collias. 1970. The behavior of the West African village weaverbird. Ibis 112:457–480.
Collias, N. E., and E. C. Collias. 1971. Some observations on behavioral energetics in the village weaverbird. I. Comparison of colonies from two subspecies in nature. Auk 88:124–143.
Collias, N. E., J. K. Victoria, E. L. Coutlee, and M. Graham. 1971. Some observations on behavioral energetics in the village weaverbird. II. All-day watches in an aviary. Auk 88:133–143.
Colquhoun, M. K. 1942. Notes on the social behavior of blue tits. Brit. Birds 35:234–240.
Conder, P. 1949. Individual distance. Ibis 91:649–655.
Connolly, A., B. Burnet, and D. Sewell. 1969. Selective mating and eye pigmentation. An analysis of the visual component in the courtship behavior of *Drosophila melanogaster*. Evolution 23:548–559.
Cooch, F. G. 1955. Observations on the autumn migration of blue geese. Wilson Bull. 67:171–174.
Cooch, F. G. 1961. Ecological aspects of the blue–snow goose complex. Auk 78:72–89.
Cooch, F. G. 1963. Recent changes in distribution of color phases of *Chen c. caerulescens*. Proc. Int. Ornithol. Congr. 13:1182–1194.
Cooch, F. G., and J. Beardmore. 1959. Assortative mating and reciprocal difference in the blue–snow goose complex. Nature 183:1833–1834.
Cooke, F., and F. G. Cooch. 1968. The genetics of polymorphism in the goose *Anser caerulescens*. Evolution 22:289–300.
Cory, L., and J. J. Manion. 1955. Ecology and hybridization in the genus *Bufo* in the Michigan-Indiana region. Evolution 9:42–51.
Cott, H. B. 1957. Adaptive coloration in animals. Methuen, London.
Coulson, J. C. 1966. The influence of the pair-bond and age on the breeding biology of the kittiwake gull *Rissa tridactyla*. Ecology 47:269–279.
Coulson, J. C. 1968. Differences in the quality of birds nesting in the center and on the edges of a colony. Nature 217:478–479.
Craig, J. V., A. M. Biswas, and A. M. Guhl. 1969. Agonistic behaviour influenced by strangeness, crowding and heredity in female domestic fowl (*Gallus gallus*). Anim. Behav. 17:498–506.
Crane, J. 1943. Display, breeding and relationships of fiddler crabs (Brachyura, genus *Uca*) in the northeastern United States. Zoologica 28:217–233.
Crane, J. 1952. A comparative study of innate behavior in Trinidad mantids. Zoologica 37:259–293.
Crane, J. 1957. Basic patterns of display in fiddler crabs (Ocypodidae, genus *Uca*). Zoologica 42:68–82.
Crane, J. 1958. Aspects of social behavior in fiddler crabs, with special reference to *Uca maracoani* (Latreille). Zoologica 43:113–130.

Crewe, R. M., and M. S. Blum. 1972. Alarm pheromones of the Attini: their phylogenetic significance. J. Insect. Physiol. 18:31–42.
Crook, J. H. 1962. The adaptive significance of pair formation types in weaverbirds. Symp. Zool. Soc. London 8:57–70.
Crook, J. H. 1963. A comparative analysis of nest structure in the weaver birds. Ibis 105:238–262.
Crook, J. H. 1964. The evolution of social organization and visual communication in the weaverbirds (Ploceidae). Behaviour Suppl. 10:1–178.
Crook, J. H. 1965. The adaptive significance of avian social organizations. Symp. Zool. Soc. London 14:181–218.
Crook, J. H. 1966*a*. Gelada baboon herd structure and movement, a comparative report. Symp. Zool. Soc. London 18:237–258.
Crook, J. H. 1966*b*. Cooperation in primates. Eugenics Rev. 58:63–70.
Crook, J. H. 1970. The socio-ecology of primates, pp. 103–166. *In* J. H. Crook, ed., Social behaviour in birds and mammals. Academic Press, London.
Crook, J. H. 1971. Sources of cooperation in animals and man, pp. 237–260. *In* J. F. Eisenberg and W. S. Dillon, eds., Man and beast: comparative social behavior. Smithsonian Inst. Press, Washington, D.C.
Crook, J. H. 1972. Sexual selection, dimorphism, and social organization in the primates, pp. 231–281. *In* B. Campbell, ed., Sexual selection and the descent of man 1871–1971. Aldine, Chicago.
Crook, J. H., and P. Aldrich-Blake. 1968. Ecological and behavioural contrasts between sympatric ground dwelling primates in Ethiopia. Folia Primat. 8:192–227.
Crook, J. H., and J. S. Gartlan. 1966. Evolution of primate societies. Nature 210:1200–1203.
Cullen, E. 1957. Adaptations in the kittiwake to cliff-nesting. Ibis 99:275–302.
Cullen, J. M. 1960. The aerial display of the Arctic tern and other species. Ardea 48:1–39.
Cullen, J. M., E. Shaw, and H. A. Baldwin. 1965. Methods for measuring the three-dimensional structure of fish schools. Anim. Behav. 13:534–543.
Curio, E., and P. Kramer. 1964. Vom Mangrovefinken (*Cactospiza heliobates* Snodgrass and Heller). Z. Tierpsychol. 21:223–234.
Curry-Lindahl, K. 1962. The irruption of the Norway lemming in Sweden during 1960. J. Mammal. 43:171–184.
Curry-Lindahl, K. 1963. New theory on a fabled exodus. Natur. Hist. 72:46–53.
Curtis, R. F., J. A. Ballantine, E. B. Keverne, R. W. Bonsall, and R. P. Michael. 1971. Identification of primate sexual pheromones and the properties of synthetic attractants. Nature 232:396–398.
Cushing, J. E. 1941. Non-genetic mating preference as a factor in evolution. Condor 43:233–236.
Daanje, A. 1951. On locomotory movements in birds and the intention movements derived from them. Behaviour 3:48–98.
Dales, R. P. 1966. Symbiosis in marine organisms. *In* S. M. Henry, ed., Symbiosis. Academic Press, New York.
Dalquest, W. W. 1953. Mammals of the Mexican state of San Luis Potosi. La. State Univ. Stud., Biol. Sci. Ser. No. 1:1–229.
Dane, B., C. Walcott, and W. H. Drury. 1959. The form and duration of the display actions of the goldeneye (*Bucephala clangula*). Behaviour 14:265–281.
Darling, F. F. 1938. Bird flocks and the breeding cycle: a contribution to the study of avian sociality. University Press, Cambridge, England.
Darling, F. F. 1964. A herd of red deer. Anchor Books, Doubleday, Garden City, N.Y. First published in 1937.

LITERATURE CITED

Darlington, P. J., Jr. 1972. Nonmathematical models for evolution of atruism, and for group selection. Proc. Nat. Acad. Sci. 69:293–297.

Darwin, C. 1871. The descent of man and selection in relation to sex. Modern Library, Random House, New York, pp. 389–1000.

Darwin, C. 1928. The origin of species. Dutton, New York. First published in 1859.

Darwin, C. 1965. The expression of the emotions in man and animals. Univ. Chicago Press, Chicago. First published in 1872.

Davies, S. J. J. F. 1970. Patterns of inheritance in the bowing display and associated behaviour of some hybrid *Streptopelia* doves. Behaviour 36:187–214.

Davis, D. E. 1942. The phylogeny of social nesting habits in the Crotophaginae. Quart. Rev. Biol. 17:115–134.

Davis, D. E., and J. J. Christian. 1957. Relation of adrenal weight to social rank of mice. Proc. Soc. Exp. Biol. Med. 94:728–731.

Davis, D. E., and F. Peek. 1972. Stability of a population of male red-winged blackbirds. Wilson Bull. 84:349–350.

Davis, L. I. 1964. Biological acoustics and the use of the sound spectrograph for detailed analysis of animal sounds, especially of birds. Southwestern Natur. 9:118–145.

Davis, W. C., and V. C. Twitty. 1964. Courtship behavior and reproductive isolation in the species of *Taricha* (Amphibia, Caudata). Copeia 1964:601–610.

Davis, W. J. 1968. Cricket wing movements during stridulation. Anim. Behav. 16:72–73.

Davis, W. J. 1970. Motoneuron morphology and synaptic contacts: determination by intracellular dye injection. Science 168:1358–1360.

Dawkins, R. 1970. Bees are easily distracted. Science 165:751.

Decker, J. D. 1967. Motility of the turtle embryo, *Chelydra serpentina* (Linné). Science 157:952–954.

Decker, J. D., and V. Hamburger. 1967. The influence of different brain regions on periodic motility of the chick embryo. J. Exp. Zool. 165:371–384.

Delacour, J., and E. Mayr. 1945. The family Anatidae. Wilson Bull 57:3–55.

Delco, E. A. 1960. Sound discrimination by males of two cyprinid fishes. Texas J. Sci. 12:48–54.

Delius, J. D. 1965. A population study of skylarks. Ibis 107:466–492.

Delius, J. D. 1967. Displacement activities and arousal. Nature 214:1259–1260.

Delius, J. D. 1970. Irrelevant behaviour, information processing and arousal homeostasis. Psychol. Forsch. 33:165–188.

DeLong, K. T. 1967. Population ecology of feral house mice. Ecology 48:611–634.

Denenberg, V. H. 1972. Readings in the development of behavior. Sinauer, Stamford, Conn.

Dethier, V. G. 1957. Communication by insects: physiology of dancing. Science 125:331–336.

Dethier, V. G. 1963. The physiology of insect senses. Methuen, London.

Dethier, V. G. 1964. Microscopic brains. Science 143:1138–1145.

DeVore, I., ed. 1965. Primate behavior: field studies of monkeys and apes. Holt, Rinehart and Winston, New York.

DeVore, I., and K. R. L. Hall. 1965. Baboon ecology, pp. 20–52. *In* I. DeVore, ed., Primate behavior. Holt, Rinehart and Winston, New York.

Dhondt, A. A., and J. Huble. 1968. Age and territory in the great tit (*Parus m. major* L.). Appl. Ornithol. 3:20–24.

Diamond, J. M. 1972. Further examples of dual singing by southwest Pacific birds. Auk 89:180–183.

Dilger, W. C. 1956. Hostile behavior and reproductive isolating mechanisms in the avian genera *Catharus* and *Hylocichla*. Auk 73:313–353.

Dilger, W. C. 1960. The comparative ethology of the African parrot genus *Agapornis*. Z. Tierpsychol. 17:648–685.

Dilger, W. C. 1962. The behavior of lovebirds. Sci. Amer. 206:88–98.

Dingle, H., and R. L. Caldwell. 1969. The aggressive and territorial behavior of the mantis shrimp *Gonodactylus bredini* Manning (Crustacea: Stomatopoda). Behaviour 33:115–136.

Dingle, H. A. 1969. Statistical and information analysis of aggressive communication in the mantis shrimp *Gonodactylus bredini* Manning. Anim. Behav. 17:561–575.

Dixon, K. L. 1963. Some aspects of social organization in the Carolina chickadee. Proc. Int. Ornithol. Congr. (Ithaca, 1962) 13:240–258.

Dixon, K. L. 1965. Dominance-subordination relationships in mountain chickadees. Condor 67:291–299.

Dobzhansky, T. 1937. Genetics and the origin of species. Columbia Univ. Press, New York.

Dobzhansky, T. 1940. Speciation as a stage in evolutionary divergence. Amer. Natur. 74:312–321.

Dobzhansky, T. 1950. Heredity, environment, and evolution. Science 111:161–166.

Dobzhansky, T. 1968. On some fundamental concepts of Darwinian biology. Evol. Biol. 2:1–34. Appleton-Century-Crofts, New York.

Dobzhansky, T. 1970. Genetics of the evolutionary process. Columbia Univ. Press, New York.

Dobzhansky, T., and O. Pavlovsky. 1971. Experimentally created incipient species of *Drosophila*. Nature 230:289–292.

Dorst, J. 1962. The migrations of birds. Houghton Mifflin, Boston.

Dorst, J., and P. Dandelot. 1970. A field guide to the large mammals of Africa. Houghton Mifflin, Boston.

Dowling, J. E. 1968. Synaptic organization of the frog retina: an electron microscopic analysis comparing the retinas of frogs and primates. Proc. Roy. Soc. (B) 170:205–228.

Dowling, J. E., and B. B. Boycott. 1966. Organization of the primate retina: electron microscopy. Proc. Roy. Soc. (B) 166:80–111.

Drury, W. H., and I. C. T. Nisbet. 1964. Radar studies of orientation of songbird migrants in southeastern New England. Bird-banding 35:69–119.

Duellman, W. E. 1967. Courtship isolating mechanisms in Costa Rican hylid frogs. Herpetologica 23:169–183.

Dumortier, B. 1963. Ethological and physiological study of sound emissions in arthropoda, pp. 583–654. *In* R. G. Busnel, ed., Acoustic behavior of animals. Elsevier, Amsterdam.

Dumpert, K. 1972. Alarmstoffrezeptoren auf der Antenne von *Lasius fuliginosus* (Latr.) (Hymenoptera, Formicidae). Z. vergl. Physiol. 76:403–425.

Dunford, C. 1970. Behavioral aspects of spatial organization in the chipmunk, *Tamias striatus*. Behaviour 36:215–231.

Dunham, D. W. 1963. Head-scratching in the hairy woodpecker, *Dendrocopus villosus*. Auk 80:375.

Dunning, D. C., and K. D. Roeder. 1965. Moth sounds and the insect-catching behavior of bats. Science 147:173–174.

Eastwood, E., and G. C. Rider. 1965. Some radar measurements of the altitude of bird flight. Brit. Birds 58:393–425.

Eaton, R. L. 1970. Group interactions, spacing and territoriality in cheetahs. Z. Tierpsychol. 27:481–491.

Efron, R. 1966. The conditioned reflex: a meaningless concept. Persp. Biol. Med. 9:488–514.

Ehrlich, P. R., and P. H. Raven. 1969. Differentiation of populations. Science 165:1228–1232.

Ehrman, L. 1959. The antennae of *Drosophila* females. Evolution 13:147.

Ehrman, L. 1964. Genetic divergence in M. Vetukhiv's experimental populations of *Drosophila pseudoobscura*. I. Rudiments of sexual isolation. Genet. Res., Cambridge 5:150–157.

Ehrman, L. 1965. Direct observation of sexual isolation between allopatric and sympatric strains of the different *Drosophila paulistorum* races. Evolution 19:459–464.

Ehrman, L. 1966. Mating success and genotype frequency in *Drosophila*. Anim. Behav. 14:332–339.

Ehrman, L. 1969. The sensory basis of mate selection in *Drosophila*. Evolution 23:59–64.

Eibl-Eibesfeldt, I. 1950. Beiträge zur Biologie der Haus- und der Ahrenmaus nebst einigen Beobachtungen an andere Nagern. Z. Tierpsychol. 7:558–587.

Eibl-Eibesfeldt, I. 1959. Der Fisch *Aspidontus taeniatus* als Nachahmer der Putzers *Labroides dimidiatus*. Z. Tierpsychol. 16:19–25.

Eibl-Eibesfeldt. I. 1961. Über der Werkzeuggebrauch des Spechtfinken *Camarhynchus pallidus*. Z. Tierpsychol. 18:343–346.

Eibl-Eibesfeldt, I., and H. Sielmann. 1962. Beobachtungen am Spechtfinken *Cactospiza pallida*. J. Ornithol. 103:92–107.

Eidmann, H. 1926. Die Koloniegrundung der einheimischen Ameisen. Z. vergl. Physiol. 3:776–838.

Eisenberg, J. F. 1963. The behavior of heteromyid rodents. Univ. Calif. Pub. Zool. 69:1–100.

Eisenberg, J. F. 1965. The social organizations of mammals. Handbuch Zool. 8(39):1–92.

Eisenberg, J. F. 1967. A comparative study in rodent ethology with emphasis on the evolution of social behavior. Proc. U.S. Nat. Mus. 122:1–51.

Eisenberg, J. F., and D. G. Kleiman. 1972. Olfactory communication in mammals. Annu. Rev. Ecol. Syst. 3:1–32.

Eisenberg, J. F., N. A. Muckenhirn, and R. Rudran. 1972. The relation between ecology and social structure in primates. Science 176:863–874.

Eisner, T. 1965. Defensive spray of a phasmid insect. Science 148:966–968.

Eisner, T. 1970. Chemical defense against predation in arthropods, pp. 157–217. *In* E. Sondheimer and J. B. Simeone, eds., Chemical ecology. Academic Press, New York.

Eisner, T., and J. Meinwold. 1966. Defensive secretions of arthropods. Science 153:1341–1350.

Eleftheriou, B. E., and K. W. Boehlke. 1967. Brain monoamine oxidase in mice after exposure to aggression and defeat. Science 155:1693–1694.

Eleftheriou, B. E., and R. L. Church. 1968*a*. Brain 5-hydroxy-tryptophan decarboxylase in mice after exposure to aggression and defeat. Physiol. Behav. 3:323–325.

Eleftheriou, B. E., and R. L. Church. 1968*b*. Brain levels of serotonin and norepinephrine in mice after exposure to aggression and defeat. Physiol. Behav. 3:977–980.

Elias, P. 1961. A note on the misuse of "digital" in neurophysiology, pp. 794–795. *In* W. A. Rosenblith, ed., Sensory communication. M.I.T. Press, Cambridge, Mass.

Ellefson, J. O. 1968. Territorial behavior in the common white-handed gibbon, *Hylobates lar* Linn., pp. 180–199. *In* P. C. Jay, ed., Primates. Holt, Rinehart and Winston, New York.

Emlen, J. M. 1970. Age specificity and ecological theory. Ecology 51:588–601.

Emlen, J. M. 1973. Ecology: an evolutionary approach. Addison-Wesley, Reading, Mass.

Emlen, J. T., Jr. 1957. Defended area?—a critique of the territory concept and of conventional thinking. Ibis 99:352.

Emlen, S. T. 1967*a*. Migratory orientation in the indigo bunting, *Passerina cyanea*. I. Evidence for use of celestial cues. Auk 84:309–342.

Emlen, S. T. 1967*b*. Migratory orientation in the indigo bunting, *Passerina cyanea*. II. Mechanisms of celestial orientation. Auk 84:463–489.

Emlen, S. T. 1968. Territoriality in the bullfrog. *Rana catesbeiana*. Copeia 1968:240–243.

Emlen, S. T. 1969*a*. Bird migration: influence of physiological state upon celestial orientation. Science 165:716–718.

Emlen, S. T. 1969*b*. The development of migratory orientation in young indigo buntings. Living Bird 8:113–126.

Emlen, S. T. 1970*a*. Celestial rotation: its importance in the development of migratory orientation. Science 170:1198–1201; 173:460–461.

Emlen, S. T. 1970*b*. The influence of magnetic information on the orientation of the indigo bunting, *Passerina cyanea*. Anim. Behav. 18:215–224.

Emlen, S. T. 1971. The role of song in individual recognition in the indigo bunting. Z. Tierpsychol. 28:241–246.

Emlen, S. T. 1972. An experimental analysis of the parameters of bird song eliciting species recognition. Behaviour 41:130–171.

Emlen, S. T. 1974. Migration: orientation and navigation. *In* D. S. Farner and J. R. King, eds., Avian biology, vol. 3. Academic Press, New York. In press.

Emlen, S. T., and J. T. Emlen. 1966. A technique for recording migratory orientation of captive birds. Auk 83:361–367.

Erlenmeyer-Kimling, L., J. Hirsch, and J. M. Weiss. 1962. Studies in experimental behavior genetics. III. Selection and hybridization analyses of individual differences in the sign of geotaxis. J. Comp. Physiol. Psychol. 55:722–731.

Erulkar, S. D. 1972. Comparative aspects of spatial localization of sound. Physiol. Rev. 52:237–360.

Esch, H. 1961. Über die Schallerzeugung beim Werbetanz der Honigbiene. Z. vergl. Physiol. 45:1–11.

Esch, H. 1964. Beiträge zum Problem der Entfernungsweisung in den Schwänzeltänzen der Honigbienen. Z. vergl. Physiol. 48:534–546.

Esch, H., and J. A. Bastian. 1970. How do newly recruited honey bees approach a food site? Z. vergl. Physiol. 68:175–181.

Eshel, I. 1972. On the neighbor effect and the evolution of altruistic traits. Theoret. Pop. Biol. 3:258–277.

Estes, R. D., and J. Goddard. 1967. Prey selection and hunting behavior of the African wild dog. J. Wildlife Mgmt. 31:52–70.

Evans, H. E. 1953. Comparative ethology of spider wasps. Syst. Zool. 2:155–172.

Evans, H. E. 1962. The evolution of prey-carrying mechanisms in wasps. Evolution 16:468–483.

Evans, H. E. 1963. Wasp farm. Natural History Press, Garden City, N.Y.

Evans, H. E. 1966*a*. The behavior patterns of solitary wasps. Annu. Rev. Entomol. 11:123–154.

Evans, H. E. 1966*b*. The comparative ethology and evolution of the sand wasps. Harvard Univ. Press, Cambridge, Mass.

Evans, H. E. 1966*c*. The accessory burrows of digger wasps. Science 152:465–471.

Evans, H. E., and M. J. W. Eberhard. 1971. The wasps. Univ. Michigan Press, Ann Arbor.

Evans, K. 1966. Observations on a hybrid between sharp-tailed grouse and the greater prairie chicken. Auk 83:128–129.

Evans, R. M. 1970*a*. Parental recognition and the "mew" call in black-billed gulls (*Larus bulleri*). Auk 87:503–513.

Evans, R. M. 1970*b*. Imprinting and mobility in young ring-billed gulls, *Larus delawarensis*. Anim. Behav. Monogr. 3:193–248.

Ewer, R. F. 1968. Ethology of mammals. Logos Press, London.

Ewert, J. P., and F. Hock. 1972. Movement-sensitive neurones in the toad's retina. Exp. Brain Res. 16:41–59.

Ewing, A. W. 1961. Body size and courtship behaviour in *Drosophila melanogaster*. Anim. Behav. 11:93–99.

Ewing, A. W. 1969. The genetic basis of sound production in *Drosophila pseudoobscura* and *D. persimilis*. Anim. Behav. 17:555–560.

Ewing, A. W., and H. C. Bennet-Clark. 1968. The courtship songs of *Drosophila*. Behaviour 31:288–301.

Ewing, A. W., and A. Manning. 1967. The evolution and genetics of insect behaviour. Annu. Rev. Entomol. 12:471–494.

Faber, W. 1967. Beiträge zur Kenntnis sozialparasitischer Ameisen. I. *Lasius* (*Austrolasius* n. sg.) *reginae* n. sp., eine neue temporär sozialparasitische Erdameise aus Österreich (Hym. Formicidae). Pflanzenschutz-Berichte, Wien 36:73–107.

Fabricius, E., and O. Fält. 1969. Sexuell prägling hos gräsandhonor. Zool. Revy 31:83–88.

Falconer, D. S. 1960. Introduction to quantitative genetics. Oliver and Boyd, Edinburgh and London.

Falls, J. B. 1963. Properties of bird song eliciting responses from territorial males. Proc. Int. Ornithol. Congr. 13:259–271.

Falls, J. B. 1969. Functions of territorial song in the white-throated sparrow, pp. 207–232. *In* R. A. Hinde, ed., Bird vocalizations. Cambridge Univ. Press, London.

Farkas, S. R., and H. H. Shorey. 1972. Chemical trail following by flying insects: a mechanism for orientation to a distant odor source. Science 178:67–68.

Farris, H. E. 1967. Classical conditioning of courting behavior in the Japanese quail, *Coturnix coturnix japonica*. J. Exp. Anal. Behav. 10:213–217.

Faugeres, A., C. Petit, and E. Thibout. 1971. The components of sexual selection. Evolution 25:265–275.

Feder, H. M. 1966. Cleaning symbiosis in the marine environment. *In* S. M. Henry, ed., Symbiosis. Academic Press, New York.

Fehmi, L. G., and T. H. Bullock. 1967. Discrimination among temporal patterns of stimulation in a computer model of a coelenterate nerve net. Kybernetik 5:240–249.

Ferguson, G. W. 1971. Variation and evolution of push-up displays of the side-blotched lizard *Uta* (Iguanidae). Syst. Zool. 20:79–101.

Ficken, M. S., and R. W. Ficken. 1962. The comparative ethology of the wood warblers: a review. Living Bird 1:102–122.

Ficken, M. S., and R. W. Ficken. 1968*a*. Head-scratching in wood warblers. Auk 85:136.

Ficken, M. S., and R. W. Ficken. 1968*b*. Reproductive isolating mechanisms in the blue-winged warbler–golden-winged warbler complex. Evolution 22:166–179.

Ficken, M. S., and R. W. Ficken. 1969. Responses of blue-winged warblers and golden-winged warblers to their own and the other species' songs. Wilson Bull. 81:69–74.

Ficken, M. S., and R. W. Ficken. 1970. Comments on introgression and reproductive isolating mechanisms in the blue-winged warbler–golden-winged warbler complex. Evolution 24:254–256.

Ficken, R. W., and M. S. Ficken. 1958. Head-scratching in *Seiurus* (Parulidae) and other passerines. Ibis 100:277–278.

Ficken, R. W., and M. S. Ficken. 1963. The relationship between habitat density and the pitch of songs in Parulidae and other birds. Amer. Zool. 3:500.

Findley, J. S. 1969. Brain size in bats. J. Mammal. 50:340–344.

Fish, M. P., and W. H. Mowbray. 1970. Sounds of western north Atlantic fishes. Johns Hopkins Univ. Press, Baltimore.

Fisher, R. A. 1958. The genetical theory of natural selection. 2nd revised ed. Dover, New York. First published 1930.

Fisler, G. F. 1969. Mammalian organizational systems. Los Angeles Co. Mus. Contrib. Sci. No. 167:1–32.

Fitch, H. S., and H. W. Shirer. 1970. A radiotelemetric study of spatial relationships in the opossum. Amer. Midland Natur. 84:170–186.

Fooden, J. 1969. Color phase in gibbons. Evolution 23:627–644.

Ford, E. B. 1964. Ecological genetics. Methuen, London.

Fox, M. W. 1970. A comparative study of the development of facial expressions in canids; wolf, coyote and foxes. Behaviour 36:49–73.

Fraenkel, G. S., and D. L. Gunn. 1961. The orientation of animals. Dover, New York. First published 1940.

Fraga, R. M. 1972. Cooperative breeding and a case of successive polyandry in the bay-winged cowbird. Auk 89:447–449.

Fredrikson, K. A. 1968. [Observations on parasitic nesting in the tufted duck (*Aythya fuligula*)]. Ornis Fennica 45:127–130.

Fretwell, S. 1968. Habitat distribution and survival in the field sparrow (*Spizella pusilla*). Bird-banding 34:293–306.

Fretwell, S. 1969. Dominance behavior and winter habitat distribution in juncos (*Junco hyemalis*). Bird-banding 40:1–25.

Fretwell, S. D., and H. L. Lucas, Jr. 1969. On territorial behavior and other factors influencing habitat distribution in birds. Acta Biotheoretica 19:16–36.

Friedmann, H. 1929. The cowbirds, a study in the biology of social parasitism. Charles C Thomas, Springfield, Ill.

Friedmann, H. 1955. The honey-guides. Bull. U.S. Nat. Mus. 208:1–292.

Friedmann, H. 1956. Further data on African parasitic cuckoos. Proc. U.S. Nat. Mus. 106(3374):377–408.

Friedmann, H. 1960. The parasitic weaverbirds. Bull. U.S. Nat. Mus. 223:1–196.

Friedmann, H. 1967. Alloxenia in three sympatric African species of *Cuculus*. Proc. U.S. Nat. Mus. 124:1–14.

Friedmann, H. 1968*a*. The evolutionary history of the avian genus *Chrysococcyx*. Bull. U.S. Nat. Mus. 265:1–137.

Friedmann, H. 1968*b*. Additional data on brood parasitism in the honey-guides. Proc. U.S. Nat. Mus. 124(3648):1–8.

Frings, H., M. Frings, B. Cox, and L. Peissner. 1955. Auditory and visual mechanisms in food-finding behavior of the herring gull. Wilson Bull. 67:155–170.

Frisch, K. von. 1967*a*. The dance language and orientation of bees. Harvard Univ. Press, Cambridge, Mass.

Frisch, K. von. 1967*b*. Honeybees: do they use direction and distance information provided by their dancers? Science 158:1072–1076 (comment by Wenner and Johnson 1076–1077).

Frisch, K. von. 1968. The role of dances in recruiting bees to familiar sites. Anim. Behav. 16:531–533.

Frisch, K. von. 1971. Bees: their vision, chemical senses, and language. 2nd ed. Cornell Univ. Press, Ithaca, N.Y.

Frisch, K. von. 1973. The bee language controversy: an experience in science; a review. Anim. Behav. 21:628–630.

Frishkopf, L. S., R. R. Capranica, and M. H. Goldstein, Jr. 1968. Neural coding in the bullfrog's auditory system—a teleological approach. Proc. IEEE 56:969–980.

Frith, C. B. 1970. Sympatry of *Amblyornis subalaris* and *A. macgregoriae* in New Guinea. Emu 70:196–197.

Frochot, B. 1967. Reflexions sur les rapports entre predateurs et proies chez les rapaces. II. L'influence des proies sur les rapaces. La Terre et la Vie 1967:33–62.

Fromme, H. G. 1961. Untersuchungen über das Orientierungsvermögen nächtliche ziehender Kleinvögel (*Erithacus rubecula, Sylvia communis*) Z. Tierpsychol. 18:205–220.

Fry, C. H. 1972. The social organization of bee-eaters (Meropidae) and co-operative breeding in hot-climate birds. Ibis 114:1–14.

Fuller, J. L., and W. R. Thompson. 1960. Behavior genetics. Wiley, New York.

Furuya, Y. 1968. On the fission of troops of Japanese monkeys. I. Five fissions and social changes between 1955 and 1966 in the Gagyusan troop. Primates 9:323–350.

Furuya, Y. 1969. On the fission of troops of Japanese monkeys. II. General view of troop fission of Japanese monkeys. Primates 10:47–69.

Gallup, G. G., Jr. 1970. Chimpanzees: self-recognition. Science 167:86–87.

Gartlan, J. S., and C. K. Brain. 1968. Ecology and social variability in *Cercopithecus aethiops* and *C. mitis*, pp. 253–292. *In* P. C. Jay, ed., Primates: studies in adaptation and variability. Holt, Rinehart and Winston, New York.

Gaston, S. 1971. The influence of the pineal organ on the circadian activity rhythm in birds, pp. 541–546. *In* Menaker 1971*a*.

Gaston, S., and M. Menaker. 1968. Pineal function: the biological clock in the sparrow? Science 160:1125–1127.

Gauthreaux, S. A. 1972. Behavioral responses of migrating birds to daylight and darkness: a radar and direct visual study. Wilson Bull. 84:136–148.

Gauthreaux, S. A., and K. P. Able. 1970. Wind and the direction of nocturnal songbird migration. Nature 228:476–477.

Geist, V. 1966*a*. The evolutionary significance of mountain sheep horns. Evolution 20:558–566.

Geist, V. 1966*b*. The evolution of horn-like organs. Behaviour 27:175–214.

Gerald, J. W. 1971. Sound production during courtship in six species of sunfish (Centrarchidae). Evolution 25:75–87.

Ghent, A. W., and N. E. Gary. 1962. A chemical alarm releaser in honey bee stings (*Apis mellifera* L.). Psyche 69:1–6.

Giesel, J. T. 1970. On the maintenance of a shell pattern and behavior polymorphism in *Acmaea digitalis*. Evolution 24:98–119.

Gifford, E. W. 1919. Field notes on the land birds of the Galapagos Islands and of Cocos Island, Costa Rica. Proc. Calif. Acad. Sci. 2:189–258.

Giles, R. H., Jr., ed. 1969. Wildlife management techniques. 3rd ed. Wildlife Soc., Washington, D.C.

Gill, F. B., and W. E. Lanyon. 1964. Experiments on species discrimination in blue-winged warblers. Auk 81:53–64.

Gilliard, E. T. 1956. Bower ornamentation versus plumage characters in bower-birds. Auk 73:450–451.

Gilliard, E. T. 1958. Living birds of the world. Doubleday, Garden City, N.Y.

Gilliard, E. T. 1959*a*. A comparative analysis of courtship movements in closely allied bowerbirds of the genus *Chlamydera*. Amer. Mus. Novitates 1936:1–8.

Gilliard, E. T. 1959*b*. The courtship behavior of Sanford's bowerbird (*Archboldia sanfordi*). Amer. Mus. Novitates 1935:1–18.

Gilliard, E. T. 1959*c*. Notes on the courtship behavior of the blue-backed manakin (*Chiroxiphia pareola*). Amer. Mus. Novitates 1942:1–19.

Gilliard, E. T. 1963. The evolution of bowerbirds. Sci. Amer. 209:38–46.

Gilliard, E. T. 1969. Birds of paradise and bower birds. Natural History Press, Garden City, N.Y.

Glas, P. 1960. Factors governing density in the chaffinch (*Fringilla coelebs*) in different types of wood. Arch. Neerl. Zool. 13:466–472.

Godfrey, J. 1958. The origin of sexual isolation between bank voles. Proc. Roy. Soc. Edinburgh, 27:47–55.

Goforth, W. R., and T. S. Baskett. 1965. Effects of experimental color marking on pairing of captive mourning doves. J. Wildlife Mgmt. 29:543–553.

Goodwin, D. 1958. The existence and causation of colour-preferences in the pairing of feral and domestic pigeons. Bull. Brit. Ornithol. Club 78:136–139.

Goss-Custard, J. D. 1970. Feeding dispersion in some over-wintering wading birds, pp. 3–35. *In* J. H. Crook, ed., Social behaviour in birds and mammals. Academic Press, New York.

Gottlieb, G. 1968. Prenatal behavior of birds. Quart. Rev. Biol. 43:148–174.

Gottlieb, G. 1970. Conceptions of prenatal behavior, pp. 111–137. *In* L. R. Aronson, E. Tobach, D. S. Lehrman, and J. S. Rosenblatt, eds., Development and evolution of behavior. Freeman, San Francisco.

Gottlieb, G. 1971*a*. Development of species identification in birds. Univ. Chicago Press, Chicago.

Gottlieb, G. 1971*b*. Ontogenesis of sensory function in birds and mammals, pp. 67–128. *In* E. Tobach et al.

Gottlieb, G., ed. 1973. Studies on the development of behavior and the nervous system. Vol. 1, Behavioral embryology. Academic Press, New York.

Gould, E. 1965. Evidence for echolocation in the Tendrecidae of Madagascar. Proc. Amer. Phil. Soc. 109:352–360.

Gould, E. 1970. Echolocation and communication in bats, pp. 144–161. *In* B. H. Slaughter and D. W. Walton, eds., About bats. Southern Methodist Univ. Press, Dallas, Texas.

Gould, E., N. C. Negus, and A. Novick. 1964. Evidence for echolocation in shrews. J. Exp. Zool. 156:19–38.

Gould, J. L. 1974. Honey bee communication. Nature 252:300–301.

Gould, J. L., M. Henerey, and M. C. MacLeod. 1970. Communication of direction by the honey bee. Science 169:544–554.

Grant, V. 1963. The origin of adaptations. Columbia Univ. Press, New York.

Grant, V. 1966. The selective origin of incompatibility barriers in the plant genus *Gilia*. Amer. Natur. 100:99–118.

Greenewalt, C. H. 1968. Bird song: Acoustics and physiology. Smithsonian Inst. Press, Washington, D.C.

Greig-Smith, P. 1964. Quantitative plant ecology. 2nd ed. Butterworth's, London.

Griffin, D. R. 1953. Acoustic orientation in the oil bird. Proc. Nat. Acad. Sci. 39:884–893.

Griffin, D. R. 1958. Listening in the dark. Yale Univ. Press, New Haven. Reprint ed. 1974. Dover, New York.

Griffin, D. R. 1964. Bird migration. Doubleday, New York.

Griffin, D. R. 1969. The physiology and geophysics of bird navigation. Quart. Rev. Biol. 44:255–276.

Griffin, D. R., and R. Galambos. 1941. The sensory basis of obstacle avoidance by flying bats. J. Exp. Zool. 86:481–506.

Griffin, D. R., A. Novick, and M. Kornfield. 1958. The sensitivity of echolocation in the fruit bat, *Rousettus*. Biol. Bull. 115:107–113.

Griffin, D. R., F. A. Webster, and C. R. Michael. 1960. The echolocation of flying insects by bats. Anim. Behav. 8:141–154.

Grinnell, A. D., and V. S. Grinnell. 1965. Neural correlates of vertical localization by echo-locating bats. J. Physiol. 181:830–851.

Grüsser, O. J., and U. Grüsser-Cornehls. 1968. Neurophysiologische Grundlagen visueller angeborener Auslösemechanismen beim Frosch. Z. vergl. Physiol. 59:1–24.

Guhl, A. M. 1953. Social behavior of the domestic fowl. Kansas Agricultural Experiment Station, Manhattan, Kan.

Guhl, A. M. 1956. The social order of chickens. Sci. Amer. 194(2):42–46.

Guhl, A. M., and W. C. Allee. 1944. Some measurable effects of social organization in flocks of hens. Physiol. Zool. 17:320–347.

Guiton, P. 1962. The development of sexual responses in the domestic fowl, in relation to the concept of imprinting. Symp. Zool. Soc. London 8:227–234.

Guthrie, R. D. 1971. A new theory of mammalian rump patch evolution. Behaviour 38:132–145.

Gwinner, E. 1964. Untersuchungen über das Ausdrucks- und Sozialverhalten des Kolkraben (*Corvus corax corax* L.) Z. Tierpsychol. 21:657–748.

Gwinner, E., and J. Kneutgen. 1962. Uber die biologische Bedeutung der "zweckdienlichen" Anwendung erlernter Laute bei Vögeln. Z. Tierpsychol. 19:692–696.

Gwynn, A. M. 1968. The migration of the arctic tern. Austral. Bird Bander, Dec.: 71–75.

Haartman, L. von. 1957. Adaptation in hole-nesting birds. Evolution 11:339–347.

Haartman, L. von. 1969. Nest-site selection and evolution of polygamy in European passerine birds. Ornis Fennica 46:1–12.

Hagiwara, S., and A. Watanabe. 1956. Discharges in motoneurons of cicada. J. Cellular Comp. Physiol. 47:415–428.

Hailman, J. P. 1965. Cliff-nesting adaptations of the Galapagos swallow-tailed gull. Wilson Bull. 77:346–362.

Hailman, J. P. 1967. The ontogeny of an instinct. Behaviour Suppl. 15: 1–159.

Hailman, J. P. 1970. Comments on the coding of releasing stimuli, pp. 138–157. *In* L. R. Aronson, E. Tobach, D. S. Lehrman, and J. S. Rosenblatt, eds., Development and evolution of behavior. Freeman, San Francisco.

Hairston, N. G., D. W. Tinkle, and H. M. Wilbur. 1970. Natural selection and the parameters of population growth. J. Wildlife Mgmt. 34:681–690.

Halberg, F. 1959. Physiologic 24-hour periodicity; general and procedural considerations with reference to the adrenal cycle. Z. Vitamin-, Hormon-, Ferment-forsch. 10:225.

Haldane, J. B. S. 1932. The causes of evolution. Longmans Green, London.

Haldane, J. B. S. 1955. Population genetics. New Biol. 18:34–51. Penguin Books, London.

Haldane, J. B. S. 1959. Natural selection, pp. 101–149. *In* P. R. Bell, ed., Darwin's biological work. Cambridge Univ. Press, London.

Haldane, J. B. S., and H. Spurway. 1954. A statistical analysis of communication in *Apis mellifera* and a comparison with communication in other animals. Insectes Soc. 1:247–283.

Hale, E. B. 1962. Domestication and the evolution of behavior, pp. 21–53. *In* E. Hafez, ed., The behavior of domestic animals. Williams and Wilkins, Baltimore.

Hall, B. P., R. E. Moreau, and I. C. J. Galbraith. 1966. Polymorphism and parallelism in the African bush-shrikes of the genus *Malaconotus* (including *Chlorophoneus*). Ibis 108:161–182.

Hall, K. R. L. 1964. Aggression in monkey and ape societies, pp. 51–64. *In* J. D. Carthy and J. F. Ebling, eds., Academic Press, N.Y. Reprinted in Jay 1968.

Hall, K. R. L. 1965*a*. Social organization of the Old-World monkeys and apes. Symp. Zool. Soc. London No. 14:265–289. Reprinted in Jay 1968.

Hall, K. R. L. 1965*b*. Behaviour and ecology of the wild patas monkey, *Erythrocebus patas*, in Uganda. J. Zool. 148:15–87. Reprinted in Jay 1968.

Hall, K. R. L. 1967. Social interactions of the adult male and adult females of a patas monkey group, pp. 261–280. *In* S. A. Altmann, ed., Social communication among primates. Univ. Chicago Press, Chicago.

Hall, K. R. L., and I. DeVore. 1965. Baboon social behavior, pp. 53–110. *In* I. DeVore, ed., Primate behavior. Holt, Rinehart and Winston, New York.

Hamburger, V. 1963. Some aspects of the embryology of behavior. Quart. Rev. Biol. 38:342–365.

Hamburger, V. 1970. Embryonic motility in vertebrates, pp. 141–151. *In* F. O. Schmitt, ed., The neurosciences second study program. Rockefeller Univ. Press, New York.

Hamburger, V. 1971. Development of embryonic motility, pp. 45–65. *In* E. Tobach, L. R. Aronson, and E. Shaw, eds., The biopsychology of development. Academic Press, New York.

Hamburger, V., and M. Balaban. 1963. Observation and experiments on spontaneous rhythmical behavior in the chick embryo. Developmental Biol. 7:533–545.

Hamburger, V., M. Balaban, R. Oppenheim, and E. Wenger. 1965. Periodic motility of normal and spinal chick embryos between 8 and 17 days of incubation. J. Exp. Zool. 159:1–14.

Hamburger, V., E. Wenger, and R. Oppenheim. 1966. Motility in the chick embryo in the absence of sensory input. J. Exp. Zool. 162:133–160.

Hamilton, W. D. 1963. The evolution of altruistic behaviour. Amer. Natur. 97:354–356.

Hamilton, W. D. 1964. The genetical evolution of social behavior. I, II. J. Theoret. Biol. 7:1–52.

Hamilton, W. D. 1970. Selfish and spiteful behaviour in an evolutionary model. Nature 228:1218–1220.

Hamilton, W. D. 1971*a*. Selection of selfish and altruistic behavior in some extreme models, pp. 59–91. *In* J. F. Eisenberg and W. S. Dillon, eds., Man and beast: comparative social behavior. Smithsonian Inst. Press, Washington, D.C.

Hamilton, W. D. 1971*b*. Geometry for the selfish herd. J. Theoret. Biol. 31:295–311.

Hamilton, W. D. 1972. Altruism and related phenomena, mainly in social insects. Annu. Rev. Ecol. Syst. 3:193–232.

Hamilton, W. J., Jr. 1937. The biology of microtine cycles. J. Agr. Res. 54:779–790.

Hamilton, W. J., III. 1959. Aggressive behavior in migrant pectoral sandpipers. Condor 61:161–179.

Hamilton, W. J., III. 1962. Celestial orientation in juvenal waterfowl. Condor 64:19–33.

Hamilton, W. J., III. 1973. Life's color code. McGraw-Hill, New York.

Hamilton, W. J., III, and G. H. Orians. 1965. Evolution of brood parasitism in altricial birds. Condor 67:361–382.

Hamilton, W. J., III, and K. E. F. Watt. 1970. Refuging. Annu. Rev. Ecol. Syst. 1:263–286.

Hardy, J. W., and R. W. Dickerman. 1965. Relationships between two forms of the red-winged blackbird in Mexico. Living Bird 4:107–129.

Harker, J. E. 1964. The physiology of diurnal rhythms. Cambridge Monogr. Exp. Biol. No. 13:1–114. Cambridge Univ. Press, London.

Harmon, L. D. 1964. Neuromimes: action of a reciprocally inhibitory pair. Science 146:1323–1325.

Harrington, R. W., Jr. 1961. Oviparous hermaphroditic fish with internal self-fertilization. Science 134:1749–1750.

Harris, M. P. 1970. Breeding ecology of the swallow-tailed gull, *Creagrus furcatus*. Auk 87:215–243.

Harris, M. P. 1970. Abnormal migration and hybridization of *Larus argentatus* and *L. fuscus* after interspecies fostering experiments. Ibis 112:488–498.

Harrison, C. J. O. 1969. Helpers at the nest in Australian passerine birds. Emu 69:30–40.

Hartline, H. K. 1949. Inhibition of activity of visual receptors by illuminating nearby retinal areas in the *Limulus* eye. Fed. Proc. 8:69.

Hartzler, J. E. 1970. Winter dominance relationship in black-capped chickadees. Wilson Bull. 82:427–434.

Haugh, J. R., and T. J. Cade. 1966. The spring hawk migration around the southeastern shore of Lake Ontario. Wilson Bull. 78:88–110.

Hauske, G., and E. Neuberger. 1968. Gerichtete Informationsgrössen zur Analyze gekoppelter Verhaltensweisen. Kybernetik 5:171–181.

Hazlett, B. A. 1968. Size relationships and aggressive behavior in the hermit crab *Clibanarius vittatus*. Z. Tierpsychol. 25:608–614.

Hazlett, B. A., and W. H. Bossert. 1965. A statistical analysis of the aggressive communications systems of some hermit crabs. Anim. Behav. 13:357–373.

Hazlett, B. A., and W. H. Bossert. 1966. Additional observations on the communications systems of hermit crabs. Anim. Behav. 14:546–549.

Hediger, H. 1964. Wild animals in captivity. Dover, New York. First published in German in 1942.

Heinroth, O. 1911. Beiträge zur Biologie, namentlich Ethologie und Psychologie der Anatiden. Verh. Int. Ornithol. Kongr. (Berlin 1910) 5:589–702.

Heinroth, O. 1930. Ueber bestimmte Bewegungsweisen der Wirbeltiere. Sitzungsberichte der Gesellschaft naturforschender Freunde zu Berlin. Jahrgang 1929:333–342.

Helmreich, R. L. 1960. Regulation of reproductive rate by intrauterine mortality in the deer mouse. Science 132:417–418.

Helversen, D. V. 1972. Gesang des Männchens und Lautschema des Weibchens bei der Feldheuschrecke *Chorthippus biguttulus* (Orthoptera, Acrididae). J. Comp. Physiol. 81:381–422.

Hensley, M. M., and J. B. Cope. 1951. Further data on removal and repopulation of the breeding birds in a spruce-fir forest community. Auk 68:483–493.

Henson, O. W., Jr. 1965. The activity and function of the middle ear muscles in echolocating bats. J. Physiol. 180:871–887.

Henson, O. W., Jr. 1966. The perception and analysis of biosonar signals by bats, pp. 949–1003. *In* R. G. Busnel, ed., Animal sonar systems, biology and bionics, vol. II. INRA-CNRZ, Jouy-en-Josas, France. Laboratoire de Physiologie Acoustique, Paris.

Hershkovitz, P. 1968. Metachromism or the principle of evolutionary change in mammalian tegumentary colors. Evolution 22:556–575.

Hess, E. H. 1956. Space perception in the chick. Sci. Amer. 195:71–80.

Hess, E. H. 1959. Imprinting. Science 130:133–141.

Hess, E. H. 1964. Imprinting in birds. Science 146:1128–1139.

Hess, E. H. 1973. Imprinting. Van Nostrand Reinhold, New York.

Hinde, R. A. 1956. The biological significance of the territories of birds. Ibis 98:340–369.

Hinde, R. A. 1958. Alternative motor patterns in chaffinch song. Anim. Behav. 6:211–218.

Hinde, R. A. 1966. Animal behaviour. 1st ed. McGraw-Hill, New York.

Hinde, R. A. 1970. Animal behaviour. 2nd ed. McGraw-Hill, New York.

Hinde, R. A., and N. Tinbergen. 1958. The comparative study of species-specific behavior, pp. 251–268. *In* A. Roe and G. G. Simpson, eds., Behavior and evolution. Yale Univ. Press, New Haven.

Hirsch, J., ed. 1967. Behavior-genetic analysis. McGraw-Hill, New York.

Hjorth, I. 1970. Reproductive behavior in Tetraonidae, with special reference to males. Viltrevy 7:271–328.

Hoenigsberg, H. F., A. J. Chejne, and E. Hortobagji-German. 1966. Preliminary report on artificial selection towards sexual isolation in *Drosophila*. Z. Tierpsychol. 23:129–135.

Hoffman, K. 1960. External manipulation of the orientational clock in birds. Cold Spring Harbor Symp. Quant. Biol. 25:379–387.

Hoffman, R. S. 1958. The role of reproduction and mortality in population fluctuations of voles (*Microtus*). Ecol. Monogr. 28:79–109.

Hogan-Warburg, A. J. 1966. Social behavior of the ruff *Philomachus pugnax* (L.) Ardea 54:109–229.

Holtzer, H., and R. P. Kamrin. 1956. Development of local coordination centers. I. Brachial centers in the salamander spinal cord. J. Exp. Zool. 132:391–408.

Homma, K., and Y. Sakakibara. 1971. Encephalic photoreceptors and their significance in photoperiodic control of sexual activity in Japanese quail, pp. 333–341. *In* Menaker 1971*a*.

Hooff, J. A. R. A. M. van. 1962. Facial expressions in higher primates. Symp. Zool. Soc. London 8:97–125.

Hooff, J. A. R. A. M. van. 1967. The facial displays of the Catarrhine monkeys and apes, pp. 7–68. *In* D. Morris, ed., Primate ethology. Aldine, Chicago.

Hooker, T., and B. I. Hooker. 1969. Duetting, pp. 185–205. *In* R. A. Hinde, Bird vocalizations. Cambridge Univ. Press, London.

Hörmann-Heck, S. von. 1957. Untersuchungen über den Erbgang einiger Verhaltensweisen bei Grillen bastarden (*Gryllus campestris* L.- *Gryllus bimaculatus* DeGreer). Z. Tierpsychol. 14:137–183.

Horn, G. 1962. Some neural correlates of perception, pp. 242–285. *In* J. D. Carthy and C. L. Duddington, eds., Viewpoints in biology, vol. 1. Butterworth's, London.

Horn, G. 1965. Physiological and psychological aspects of selective perception. Advances Stud. Behav. 1:155–215.

Horn, H. S. 1968. The adaptive significance of colonial nesting in the Brewer's blackbird (*Euphagus cyanocephalus*). Ecology 49:682–694.

Horridge, G. A. 1954. The nerves and muscles of medusae. I. Conduction in the nervous system of *Aurelia aurita* Lamarck. J. Exp. Biol. 31:594–600.

Horridge, G. A. 1955*a*. The nerves and muscles of medusae. II. *Geryonia proboscidialis* Eschscholtz. J. Exp. Biol. 32:555–568.

Horridge, G. A. 1955*b*. The nerves and muscles of medusae. IV. Inhibition in *Aequorea forskalea*. J. Exp. Biol. 42:642–648.

Horridge, G. A. 1957. The co-ordination of the protective retraction of coral polyps. Phil. Trans. Roy. Soc. London (B) 240:495–529.

Horridge, G. A. 1968. Interneurons. Freeman, San Francisco.

Houlihan, R. T. 1963. The relationship of population density to endocrine size in metabolic changes in the California vole (*Microtus californicus*). Univ. Calif. Pub. Zool. 65:327–362.

Howard, E. 1964. Territory in bird life. Atheneum, New York. First published in 1920.

Howse, P. E. 1964. The significance of the sound produced by the termite *Zootermopsis angusticollis* (Hagen). Anim. Behav. 12:284–300.

Howse, P. E., D. B. Lewis, and J. D. Pye. 1971. Adequate stimulus of the insect typanal organ. Experientia 27:598–600.

Hoyle, G. 1970. Cellular mechanisms underlying behavior-neuroethology. Advances Insect Physiol. 7:349–444.

Hsia, D. Y. 1967. The hereditary metabolic diseases, pp. 176–193. *In* J. Hirsch, ed., Behavior-genetic analysis. McGraw-Hill, New York.

Hubbs, C., and E. A. Delco, Jr. 1960. Mate preference in males of four species of gambusiine fishes. Evolution 14:145–152.

Hubbs, C., and E. A. Delco, Jr. 1962. Courtship preferences of *Gambusia affinis* associated with the sympatry of the parental populations. Copeia 1962:396–400.

Hubby, J. L., and R. C. Lewontin. 1966. A molecular approach to the study of genic heterozygosity in natural populations. I. The number of alleles at different loci in *Drosophila pseudoobscura*. Genetics 54:577–594.

Hubby, J. L., and L. H. Throckmorton. 1968. Protein differences in *Drosophila*. IV. A study of sibling species. Amer. Natur. 102:193–205.

Huber, F. 1962. Central nervous control of sound production in crickets and some speculations on its evolution. Evolution 16:439–442.

Hughes, A., S. V. Bryant, and A. d'A. Bellairs. 1967. Embryonic behaviour in the lizard, *Lacerta vivipara*. J. Zool., London 153:139–152.

Hughes, A. F. W. 1968. Aspects of neural ontogeny. Logos Press, London; Academic Press, New York.

Humphries, D. A., and P. M. Driver. 1967. Erratic display as a device against predators. Science 156:1767–1768.

Hunsaker, D. 1962. Ethological isolating mechanisms in the *Sceloporus torquatus* group of lizards. Evolution 16:62–74.

Hutchinson, G. E., and R. H. MacArthur. 1959. Appendix: on the theoretical significance of aggressive neglect in interspecific competition. Amer. Natur. 93:133–134.

Huxley, J. 1966. A discussion on ritualization of behavior in animals and man. Phil. Trans. Roy. Soc. London (B) Biol. Sci. 251:247–526.

Huxley, J. S. 1934. A natural experiment on the territorial instinct. Brit. Birds 27:270–277.

Iersel, J. J. A. van, and A. C. A. Bol. 1958. Preening of two tern species. A study of displacement activities. Behaviour 13:1–88.

Immelmann, K. 1969*a*. Song development in the zebra finch and other estrildid finches, pp. 61–74. *In* R. A. Hinde, ed., Bird vocalizations. Cambridge Univ. Press, London.

Immelmann, K. 1969*b*. Über den Einfluss frühkindlicher Erfahrungen auf die geschlechtliche Objektfixierung bei Estrildiden. Z. Tierpsychol. 26:677–691.

Immelmann, K. 1970. Zur ökologischen Bedeutung prägungsbedingter Isolationsmechanismen. Verhandlungsbericht der Deutschen Zoologischen Gesellschaft, 64 Tagung, pp. 304–313. Gustav Fischer Verlag.

Immelmann, K. 1972. Sexual and other long-term aspects of imprinting in birds and other species. Advances Stud. Behav. 4:147–174.

Ingolfasson, A. 1969. Behavior of gulls robbing eiders. Bird Study 16:45–52.

Itani, J. 1959. Paternal care in the wild Japanese monkey, *Macaca fuscata fuscata*. Primates 1:61–93.

Iwao, S. 1968. A new regression method for analyzing the aggregation pattern of animal populations. Res. Pop. Ecol. 10:1–20.

Jacklet, J. W. 1969. Circadian rhythm of optic nerve impulses recorded in darkness from isolated eye of *Aplysia*. Science 164:562–563.

Jacklet, J. W., and J. Geronimo. 1971. Circadian rhythm: population of interacting neurons. Science 174:299–302.

Jackson, J. B. C. 1968. Bivalves: spatial and size-frequency distributions of two intertidal species. Science 161:479–480.

Jacobson, M. 1966. Starting points for research in the development of behavior, pp. 341–383. *In* M. Locke, ed., Major problems in developmental biology. Academic Press, New York.

Jacobson, M. 1969. Sex pheromone of the pink bollworm moth: biological masking by its geometrical isomer. Science 163:190–191.

Jacobson, M. 1971. Developmental neurobiology. Holt, Rinehart and Winston, New York.

Jacobson, M., and M. Beroza. 1964. Insect attractants. Sci. Amer. 211:20–27.

James, H., and C. Binks. 1963. Escape and avoidance learning in newly hatched domestic chicks. Science 130:1293–1294.

Jansen, J., and J. K. S. Jansen. 1969. The nervous system of Cetacea, pp. 175–252. *In* H. Y. Anderson, ed., The biology of marine mammals. Academic Press, New York.

Jarman, C. 1972. Atlas of animal migrations. John Day, New York.

Jay, P. C. 1963. Mother-infant relations in langurs, pp. 282–304. *In* H. L. Rheingold, ed., Maternal behavior in mammals. Wiley, New York.

Jay, P. C., ed. 1968. Primates, studies in adaptation and variability. Holt, Rinehart and Winston, New York.

Jenkins, D., A. Watson, and G. R. Miller, 1963. Population studies on red grouse *Lagopus lagopus scoticus* (Lath.) in north-east Scotland. J. Anim. Ecol. 32:317–376.

Jenni, D. A., and G. Collier. 1972. Polyandry in the American jacana (*Jacana spinosa*). Auk 89:743–765.

Jennings, H. S. 1907. Behaviour of the starfish *Asterias forreri* de Loriol. Univ. Calif. Pub. Zool. 4:53–185.

Jennrich, R. I., and F. B. Turner. 1969. Measurement of non-circular home range. J. Theoret. Biol. 22:227–237.

Jenssen, T. A. 1970. Female responses to filmed displays of *Anolis nebulosus* (Sauria, Iguanidae). Anim. Behav. 18:640–647.

Jewell, P. A. 1966. The concept of home range in mammals. *In* P. A. Jewell and C. Loizos, eds., Play, exploration and territory in mammals. Symp. Zool. Soc. London 18:85–109.

Joffe, J. M. 1969. Prenatal determinants of behaviour. Int. Ser. Monogr. Exp. Psychol. 7:1–368. Pergamon, New York.

Johnsgard, P. A. 1960. Pair-formation mechanisms in *Anas* (Anatidae) and related genera. Ibis 102:616–618.

Johnsgard, P. A. 1961. The taxonomy of the Anatidae—a behavioral analysis. Ibis 103*a*: 71–85.

Johnsgard, P. A. 1965. Handbook of waterfowl behavior. Comstock Publishing Associates, Ithaca, N.Y.

Johnsgard, P. A. 1967. Animal behavior. Wm. C. Brown, Dubuque, Iowa.

Johnsgard, P. A. 1973*a*. Grouse and quails of North America. Univ. Nebraska Press, Lincoln.

Johnsgard, P. A. 1973*b*. Natural and unnatural selection in a wild goose. Nat. Hist. 82(10):60–69.

Johnson, C. 1959. Genetic incompatibility in the call races of *Hyla versicolor* LeConte in Texas. Copeia 1959:327–335.

Johnson, C. 1963. Additional evidence of sterility between call-types in the *Hyla versicolor* complex. Copeia 1963:139–143.

Johnson, C. 1966. Species recognition in the *Hyla versicolor* complex. Texas J. Sci. 18:361–364.

Johnson, D. L. 1967. Honey bees: do they use the direction information contained in their dance maneuver? Science 155:844–847.

Johnson, D. L., and A. M. Wenner. 1966. A relationship between conditioning and communication in honey bees. Anim. Behav. 14:261–265.

Johnston, J. W., Jr., D. G. Moulton, and A. Turk, eds. 1970. Advances in chemoreception. Vol. 1. Communication by chemical signals. Appleton-Century-Crofts, New York.

Johnston, R. F. 1961. Population movements of birds. Condor 63:386–389.

Johnston, R. F. 1966. The adaptive basis of geographic variation in color of the purple martin. Condor 68:219–228.

Jolly, A. 1972. The evolution of primate behavior. Macmillan, New York.

Jorgenson, C. D. 1968. Home range as a measure of probable interactions among populations of small mammals. J. Mammal. 49:104–112.

Josephson, R. K. 1965. Three parallel conducting systems in the stalk of a hydroid. J. Exp. Biol. 42:139–152.

Josephson, R. K., and J. Uhrich. 1969. Inhibition of pacemaker systems in the hydroid *Tubularia*. J. Exp. Biol. 50:1–14.

LITERATURE CITED

Kaada, B. 1967. Brain mechanisms related to aggressive behavior. *In* C. D. Clemente and D. B. Lindsley, eds. Aggression and defense. UCLA Forum Med. Sci. 7:95–133.

Kaiser, H. 1968. "Zeitliches Territorialverhalten" bei der Libelle *Aeschna cyanea.* Naturwissenschaften 55:657.

Kaiser, H. 1969. Regulation der Individuendichte am Paarungsplatz bei der Libelle *Aeschna cyanea* durch "zeitliches Territorialverhalten". Zool. Anz. Suppl. 33:79–85.

Kalela, O. 1954. Über den Revierbesitz bei Vögeln und Saugetieren als populationsökölogischer Faktor. Ann. Zool. Soc. "Vanamo" 16:1–48.

Kalela, O. 1957. Regulation of reproduction rate in subarctic populations of the vole *Clethrionomys rufocanus* (Sund.). Ann. Acad. Sci. Fennicae Ser. A., Sect. IV (Biol.), No. 34:1–60.

Kalmus, H., and S. M. Smith. 1966. Some evolutionary consequences of pegmatypic mating systems (imprinting). Amer. Natur. 100:619–635.

Karlson, P., and A. Butenandt. 1959. Pheromones (ectohormones) in insects. Annu. Rev. Entomol. 4:39–58.

Karlson, P., and M. Lüscher. 1959*a*. "Pheromone", ein Nomenklaturvorschlag für eine Wirkstoffklasse. Naturwissenschaften 46:63–64.

Karlson, P., and M. Lüscher, 1959*b*. 'Pheromones': a new term for a class of biologically active substances. Nature 183:55–56.

Kaston, B. J. 1964. The evolution of spider webs. Amer. Zool. 4:175–190.

Kaufman, I. C. 1973. The role of ontogeny in the establishment of species-specific patterns. Proc. Assoc. Res. Nerv. Ment. Dis. 51:381–397.

Kaufman, I. C., and L. A. Rosenblum. 1969. Effects of separation from mother on the emotional behavior of infant monkeys. Ann. N.Y. Acad. Sci. 159:681–695.

Kaufman, J. F. 1966. Behavior of infant rhesus monkeys and their mothers in a free-ranging band. Zoologica 51:17–28.

Kawabe, M. 1970. A preliminary study of the wild siamang gibbon (*Hylobates syndactylus*) at Fraser's Hill, Malaysia. Primates 11:285–291.

Kawai, M. 1958. [On the system of social ranks in a natural troop of Japanese monkeys: basic rank and dependent rank.] Primates 1–2:111–130. Translated in S. A. Altmann, ed., Japanese monkeys. Privately published, Edmonton.

Kawamura, S. 1958. Matriarchal social ranks in the Minoo-B troop: a study of the rank system of Japanese monkeys. Primates 1–2:149–156. Translated in S. A. Altmann, ed., Japanese monkeys. Privately published, Edmonton.

Keenleyside, M. H. A. 1955. Some aspects of the schooling behaviour of fish. Behaviour 8:183–248.

Keenleyside, M. H. A. 1967. Behavior of male sunfishes (genus *Lepomis*) toward females of three species. Evolution 21:688–695.

Keeton, W. T. 1969. Orientation by pigeons: is the sun necessary? Science 165:922–928; 168:153.

Keeton, W. T. 1971. Magnets interfere with pigeon homing. Proc. Nat. Acad. Sci. 68:102–106.

Keeton, W. T. 1974. The orientational and navigational basis of homing in birds. *In* Rubin, ed., Recent advances in the study of behavior. Academic Press, New York.

Kelemen, G. 1963. Comparative anatomy and physiology of the auditory organ in vertebrates, pp. 522–582. *In* R. G. Busnel, ed., Acoustic behaviour of animals. Elsevier, Amsterdam.

Kellogg, P. P., and A. A. Allen. No date. Voices of the night. Houghton Mifflin, Boston. (Phonograph record.)

Kellogg, W. N. 1961. Porpoises and sonar. Univ. Chicago Press, Chicago.

Kemperman, J. H. B. 1967. On systems of mating. Proc. Koninkl. Nederl. Akad. van Wetenschappen-Amsterdam. Series A, 70, No. 3 and Indag. Math. 29:245–304.

Kendeigh, S. C. 1952. Parental care and its evolution in birds. Illinois Biol. Monogr. 22:1–358.

Kennedy, D., W. H. Evoy, and J. T. Hanawalt. 1966. Release of coordinated behavior in crayfish by single central neurons. Science 154:917–919.

Kenyon, K. W., and D. W. Rice. 1958. Homing of Laysan albatrosses. Condor 60:3–6.

Kerr, W. E. 1969. Some aspects of the evolution of social bees (Apidae). *In* T. Dobzhansky, M. K. Hecht, and W. C. Steere, eds., Evolutionary biology, vol. 3, pp. 119–175. Appleton-Century-Crofts, New York.

Kessel, E. L. 1955. Mating activities of balloon flies. Syst. Zool. 4:97–104.

Kessler, S. 1966. Selection for and against ethological isolation between *Drosophila pseudoobscura* and *Drosophila persimilis*. Evolution 20:634–645.

Kettlewell, H. B. D. 1955. Recognition of appropriate backgrounds by the pale and black phases of Lepidoptera. Nature 175:943–944.

Kettlewell, H. B. D. 1956. Further selection experiments on industrial melanism in the Lepidoptera. Heredity 10:287–301.

Key, K. H. L. 1968. The concept of stasipatric speciation. Syst. Zool. 17:14–22.

Kikkawa, J. 1961. Social behaviour of the white-eye *Zosterops lateralis* in winter flocks. Ibis 103*a*:428–442.

King, J. A. 1955. Social behavior, social organization, and population dynamics in a black-tailed prairiedog town in the Black Hills of South Dakota. Contrib. Lab. Vert. Biol. Univ. Michigan No. 67:1–123.

King, J. A., D. Maas, and R. G. Weisman. 1964. Geographic variation in nest size among species of *Peromyscus*. Evolution 18:230–234.

Klein, J. 1970. Histocompatibility-2 (H-2) polymorphism in wild mice. Science 168: 1362–1364.

Klinghammer, E. 1967. Factors influencing choice of mate in altricial birds, pp. 5–42. *In* H. W. Stevenson, ed., Early behavior: comparative and developmental approaches. Wiley, New York.

Klinghammer, E., and E. H. Hess. 1964. Imprinting in an altricial bird: the blond ring dove (*Streptopelia risoria*). Science 146:265–266.

Klomp, H. 1970. The determination of clutch-size in birds. A review. Ardea 58:1–124.

Kluyver, H. M. 1955. Das Verhalten des Drosselrohrsängers, *Acrocephalus arundinaceus* (L), am Brutplatz, mit besonderer Berucksichtigung der Nestbautechnik und der Revierbehauptung. Ardea 43:1–50.

Kluyver, H. M., and L. Tinbergen. 1953. Territory and regulation of density in titmice. Arch. Neerl. Zool. 10:265–286.

Kneutgen, J. 1964. Über die Künstliche Auslösbarkeit des Gesangs der Schamadrossel (*Copsychus malabaricus*). Z. Tierpsychol. 21:124–128.

Knight, G. R., A. Robertson, and C. H. Waddington. 1956. Selection for sexual isolation within a species. Evolution 10:14–22.

Koford, C. B. 1958. Prairie dogs, whitefaces and blue grama. Wildlife Monogr. No. 3:1–78.

Koford, C. B. 1963. Rank of mothers and sons in bands of rhesus monkeys. Science 141:356–357.

Kok, O. B. 1971. Vocal behavior of the great-tailed grackle (*Quiscalus mexicanus prosopidicola*). Condor 73:348–363.

Kok, O. B. 1972. Breeding success and territorial behavior of male boat-tailed grackles. Auk 89:528–540.

Konishi, M. 1963. The role of auditory feedback in the vocal behavior of the domestic fowl. Z. Tierpsychol. 20:349–367.

Konishi, M. 1965. The role of auditory feedback in the control of vocalizations in the white-crowned sparrow. Z. Tierpsychol. 22:770–783.

Konishi, M. 1966. The attributes of instinct. Behaviour 27:316–328.

Konishi, M. 1973*a*. How the owl tracks its prey. Amer. Sci. 61:414–424.

Konishi, M. 1973*b*. Locatable and nonlocatable acoustic signals for barn owls. Amer. Natur. 107:775–785.

Konishi, M., and F. Nottebohm. 1969. Experimental studies in the ontogeny of avian vocalizations, pp. 29–48. *In* R. A. Hinde, ed., Bird vocalizations. Cambridge Univ. Press, London.

Konopka, R. J., and S. Benzer. 1971. Clock mutants of *Drosophila melanogaster*. Proc. Nat. Acad. Sci. 68:2112–2116.

Koopman, K. F. 1950. Natural selection for reproductive isolation between *Drosophila pseudoobscura* and *Drosophila persimilis*. Evolution 4:135–148.

Kortlandt, A. 1940. Wechselwirkung zwischen Instinkten. Arch. Neerl. Zool. 4:442–520.

Kortlandt, A. 1962. Chimpanzees in the wild. Sci. Amer. 206:128–138.

Kortwright, F. H. 1943. The ducks, geese and swans of North America. Am. Wildl. Inst., Wash. D.C.

Koyama, N. 1967. On dominance rank and kinship of a wild Japanese monkey troop in Arashiyama. Primates 8:189–216.

Kramer, G. 1949. Über Richtungstendenzen bei der nächtlichen Zugunruhe gekäfigter Vögel, pp. 269–283. *In* E. Mayr and E. Schüz, eds., Ornithologie als biologische Wissenschaft. Carl Winter, Universitätsverlag, Heidelberg.

Kramer, G. 1951. Eine neue Methode zur Erforschung der Zugorientierung und die bisher damit erzielten Ergebnisse. Proc. Int. Ornithol. Congr. (Uppsala, 1950) 10:269–280.

Kramer, G. 1952. Experiments in bird orientation. Ibis 94:265–285.

Krebs, J. R. 1970. Regulation of numbers in the great tit (Aves: Passeriformes). J. Zool., London 162:317–333.

Krebs, J. R. 1971. Territory and breeding density in the great tit, *Parus major* L. Ecology 52:2–22.

Kreithen, M. L., and W. T. Keeton. 1974. Detection of changes in atmospheric pressure by the homing pigeon. J. Comp. Physiol. 89:73–82.

Krishna, K., and F. M. Weesner, eds. 1969–1970. Biology of termites. Vol. 1:1–600. Vol. 2:1–648. Academic Press, New York.

Kruger, L. 1966. Specialized features of the cetacean brain, pp. 232–254. *In* K. S. Norris, ed., Whales, dolphins, and porpoises. Univ. California Press, Berkeley.

Kruuk, H. 1964. Predators and anti-predator behaviour of the black-headed gull (*Larus ridibundus* L.). Behaviour Suppl. 11:1–130.

Kruuk, H. 1966. Clan-system and feeding habits of spotted hyenas (*Crocuta crucuta* Erxleben). Nature 209:1257–1258.

Kruuk, H. 1972. The spotted hyena. Univ. Chicago Press, Chicago.

Kühme, W. 1964. Die Ernahrungsgemeinschaft der Hanenhunde (*Lycaon pictus lupinus*, Thomas, 1902). Naturwissenschaften 51:495.

Kühme, W. 1965. Communal food distribution and division of labour in African hunting dogs (*Lycaon pictus lupinus*). Nature 205:443–444.

Kullman, E. J. 1972. The convergent development of orb-webs in cribellate and ecribellate spiders. Amer. Zool. 12:395–405.

Kummer, H. 1967. Tripartite relations in hamadryas baboons, pp. 63–71. *In* S. A. Altmann, ed., Social communication among primates. Univ. Chicago Press, Chicago.

Kummer, H. 1968*a*. Two variations in the social organization of baboons, pp. 293–312. *In* P. Jay, ed., Primates; studies in adaptation and variability. Holt, Rinehart and Winston, New York.

Kummer, H. 1968*b*. Social organization of hamadryas baboons. Univ. Chicago Press, Chicago.

Kummer, H. 1971. Primate societies. Group techniques of ecological adaptation. Aldine-Atherton, Chicago.

Kummer, H., and F. Kurt. 1963. Social units of a free-living population of hamadryas baboons. Folia Primat. 1:4–19.

Kuo, Z.-Y. 1932. Ontogeny of embryonic behavior in Aves. I. Chronology and general nature of behavior of chick embryo. J. Exp. Zool. 61:395–430.

Kuo, Z.-Y. 1967. The dynamics of behavior development. Random House, New York.

Kuroda, N. 1966. A note on the problem of hawk-mimicry in cuckoos. Japanese J. Zool. 15:173–181.

Kutter, H. 1969. Die sozialparasitischen Ameisen der Schweiz. Neujahrsblatt der Naturforschenden Gesellschaft in Zürich auf das Jahr 1969, pp. 1–62.

Lack, D. 1947. Darwin's finches. Cambridge Univ. Press, Cambridge, England.

Lack, D. 1953. The life of the robin. Penguin Books, London. First published 1943.

Lack, D. 1954*a*. The evolution of reproductive rates, pp. 143–156. *In* J. Huxley, A. C. Hardy, and E. B. Ford, eds., Evolution as a process. Allen & Unwin, London. Also Collier Books, New York, 1963, 172–187.

Lack, D. 1954*b*. The natural regulation of animal numbers. Oxford Univ. Press, London.

Lack, D. 1966. Population studies of birds. Clarendon Press, Oxford.

Lack, D. 1968. Ecological adaptations for breeding in birds. Methuen, London.

Lade, B. I., and W. H. Thorpe. 1964. Dove songs as innately coded patterns of specific behaviour. Nature 212:366–368.

Landau, H. G. 1951*a*. On dominance relations and the structure of animal societies. I. The effect of inherent characteristics. Bull. Math. Biophysics 13:1–19.

Landau, H. G. 1951*b*. On dominance relations and the structure of animal societies: II. Some effects of possible social factors. Bull. Math. Biophysics 13:245–266.

Landau, H. G. 1965. Development of structure in a society with a dominance relation when new members are added successively. Bull. Math. Biophysics, Special Issue, 27:151–160.

Landau, H. G. 1968. Models of social structure. Bull. Math. Biophysics 30:215–224.

Langley, J. N. 1903. On the sympathetic system of birds and on the muscles which move the feathers. J. Physiol. 30:221–252.

Lamprecht, J. 1970. Duettgesang beim Siamang, *Symphalangus syndactylus* (Hominoidea, Hylobatinae). Z. Tierpsychol. 27:186–204.

Lanyon, W. E. 1957. The comparative biology of the meadowlarks (*Sturnella*) in Wisconsin. Pub. Nuttall Ornithol. Club. 1:1–67.

Lanyon, W. E. 1960. The ontogeny of vocalizations in birds. *In* W. E. Lanyon and W. N. Tavolga, eds., Animal sounds and communication. Amer. Inst. Biol. Sci. Pub. No. 7:321–347.

Lanyon, W. E. 1963. Experiments on species discrimination in *Myiarchus* flycatchers. Amer. Mus. Novitates 2126:1–16.

Layne, J. N. 1969. Nest-building behavior in three species of deer mice, *Peromyscus*. Behaviour 35:288–303.

LeBoeuf, B. J., and R. S. Peterson. 1969*a*. Social status and mating activity in elephant seals. Science 163:91–93.

LeBoeuf, B. J., and R. S. Peterson. 1969*b*. Dialects in elephant seals. Science 166:1654–1656.

Lehrman, D. S. 1953. A critique of Konrad Lorenz's theory of instinctive behavior. Quart. Rev. Biol. 28:337–363.

Lehrman, D. S. 1959. Hormonal responses to external stimuli in birds. Ibis 101:478–496.

Lehrman, D. S. 1961. Hormonal regulation of parental behavior in birds and infrahuman mammals, pp. 1268–1382. *In* W. C. Young, ed., Sex and internal secretions, vol. 2. Williams and Wilkins, Baltimore.

Lehrman, D. S. 1964. Control of behavior cycles in reproduction, pp. 143–166. *In* W. Etkin, ed., Social behavior and organization among vertebrates. Univ. Chicago Press, Chicago.

Lehrman, D. S. 1970. Semantics and conceptual issues in the nature-nurture problem, pp. 17–52. *In* L. R. Aronson, E. Tobach, D. S. Lehrman, and J. S. Rosenblatt, eds., Development and evolution of behavior. Freeman, San Francisco.

Lehrman, D. S., and M. Friedman. 1969. Auditory stimulation of ovarian activity in the ring dove (*Streptopelia risoria*). Anim. Behav. 17:494–497.

Lemon, R. E. 1966. Geographic variation in the song of cardinals. Can. J. Zool. 44:413–428.

Lemon, R. E. 1967. The response of cardinals to songs of different dialects. Anim. Behav. 15:538–545.

Lemon, R. E. 1968*a*. Coordinated singing by black-crested titmice. Can. J. Zool. 46:1163–1167.

Lemon, R. E. 1968*b*. The relation between organization and function of song in cardinals. Behaviour 32:158–178.

Lemon, R. E., and A. Herzog. 1969. The vocal behavior of cardinals and pyrrhuloxias in Texas. Condor 71:1–15.

Lerner, J., S. E. Mellen, I. Waldron, and R. M. Factor. 1971. Neural redundancy and regularity of swimming beats in scyphozoal medusae. J. Exp. Biol. 55:177–184.

Lettvin, J. Y., H. R. Maturana, W. S. McCulloch, and W. H. Pitts. 1959. What the frog's eye tells the frog's brain. Proc. I.R.E. 47:1940–1951.

Lettvin, J. Y., H. R. Maturana, W. S. McCulloch, and W. H. Pitts. 1961. Two remarks on the visual system of the frog. *In* W. Rosenblith, ed., Sensory communication. M.I.T. Press, Cambridge, Mass.

Leuthold, W. 1966. Variations in territorial behavior of Uganda kob *Adenota kob thomasi* (Newmann, 1896). Behaviour 27:215–258.

Levi, W. M. 1965. Encyclopedia of pigeon breeds. T.F.H. Publications, Jersey City, N.J.

Levins, R. 1964. Theory of fitness in a heterogeneous environment III. The response to selection. J. Theoret. Biol. 7:224–240.

Levinton, J. 1972. Spatial distribution of *Nucula proxima* Say (Protobranchia): an experimental approach. Biol. Bull. 143:175–183.

Lewontin, R. C. 1970. The units of selection. Annu. Rev. Ecol. Syst. 1:1–18.

Lewontin, R. C., and J. L. Hubby. 1966. A molecular approach to the study of genic heterozygosity in natural populations. II. Amount of variation and degree of heterozygosity in natural populations of *Drosophila pseudoobscura*. Genetics 54:595–609.

Leyhausen, P. 1956. Verhaltensstudien an Katzen. Z. Tierpsychol. Suppl. 2:1–120.

Leyhausen, P., and R. Wolff. 1959. Das Revier einer Hauskatze. Z. Tierpsychol. 16:666–670.

Li, C. C. 1955. Population genetics. Univ. Chicago Press, Chicago.

Lidicker, W. Z. 1962. Emigration as a possible mechanism permitting the regulation of population density below carrying capacity. Amer. Natur. 96:29–33.

Lill, A., and D. G. M. Wood-Gush. 1965. Potential ethological isolating mechanisms and assortative mating in the domestic fowl. Behaviour 25:16–44.

Lilly, J. C. 1965. Vocal mimicry in *Tursiops*. Science 147:300–301.

Lin, N., and C. D. Michener. 1972. Evolution and selection in social insects. Quart. Rev. Biol. 47:131–159.

Lincoln, F. C. 1950. Migration of birds. U.S. Dept. Interior Fish and Wildlife Service Circular No. 16:1–102.

Lincoln, G. A. 1972. The role of antlers in the behaviour of red deer. J. Exp. Zool. 182: 233–250.

Lindauer, M. 1956. Über die Verständigung bei indischen Bienen. Z. vergl. Physiol. 38:521–557.

Lindauer, M. 1961. Communication among social bees. 1st ed. Harvard Univ. Press, Cambridge, Mass.

Lindauer, M. 1971*a*. Communication among social bees. 3rd ed. Harvard Univ. Press. Cambridge, Mass.

Lindauer, M. 1971*b*. The functional significance of the honeybee waggle dance. Amer. Natur. 105:89–96.

Lindauer, M., and W. E. Kerr. 1958. Die gegenseitige Verständigung bei den Stachellosen Bienen. Z. vergl. Physiol. 41:405–434.

Lindburg, D. G. 1969. Rhesus monkeys: mating season mobility of adult males. Science 166:1176–1178.

Lissaman, P. B. S., and C. A. Shollenberger. 1970. Formation flight of birds. Science 168:1003–1005.

Littlejohn, M. J. 1959. Call differentiation in a complex of seven species of *Crinia* (Anura, Leptodactylidae). Evolution 13:452–468.

Littlejohn, M. J. 1964. Geographic isolation and mating call differentiation in *Crinia signifera*. Evolution 18:262–266.

Littlejohn, M. J. 1965. Premating isolation in the *Hyla ewingi* complex (Anura: Hylidae). Evolution 19:234–243.

Littlejohn, M. J. 1969. The systematic significance of isolating mechanisms. *In* Int. Conf. Syst. Biol, ed., Systematic biology. Nat. Acad. Sci. Pub. 1692:459–482.

Littlejohn, M. J. 1971. A reappraisal of mating call differentiation in *Hyla cadaverina* (= *Hyla californiae*) and *Hyla regilla*. Evolution 25:98–112.

Littlejohn, M. J., M. J. Fouquette, and C. Johnson. 1960. Call discrimination by female frogs of the *Hyla versicolor* complex. Copeia 1960:47–49.

Littlejohn, M. J., and J. J. Loftus-Hills. 1968. An experimental evaluation of premating isolation in the *Hyla ewingi* complex. Evolution 22:659–663.

Littlejohn, M. J., and T. C. Michaud. 1959. Mating call discrimination by females of Strecker's chorus frog (*Pseudacris streckeri*). Texas J. Sci. 11:86–92.

Littlejohn, M. J., and R. S. Oldham. 1968. *Rana pipiens* complex: mating call structure and taxonomy. Science 162:1003–1005.

Lloyd, J. A., J. J. Christian, D. E. Davis, and F. H. Bronson. 1964. Effects of altered social structure on adrenal weights and morphology in populations of woodchucks (*Marmota monax*). Gen. Comp. Endocr. 4:271–276.

Lloyd, J. E. 1966. Studies on the flash communication system in *Photinus* fireflies. Univ. Michigan Mus. Zool. Misc. Pub. 130:1–95.

Lloyd, M., and H. S. Dybas. 1966. The periodical cicada problem. I. Population ecology. II. Evolution. Evolution 20:133–149, 466–505.

Loftus-Hills, J. 1971. Neural correlates of acoustic behavior in the Australian bullfrog *Limnodynastes dorsalis* (Anura: Leptodactylidae). Z. vergl. Physiol. 74:140–152.

Loftus-Hills, J. J., and M. J. Littlejohn. 1971*a*. Pulse repetition rate as the basis for mating call discrimination by two sympatric species of *Hyla*. Copeia 1971:154–156.

Loftus-Hills, J. J., and M. J. Littlejohn. 1971*b*. Mating-call sound intensities of Anuran amphibians. J. Acoust. Soc. Amer. 49:1327–1329.

Loftus-Hills, J. J., and B. M. Johnstone. 1970. Auditory function, communication, and the brain-evoked response in anuran amphibians. J. Acoust. Soc. Amer. 47: 1131–1138.

Löhrl, H. 1959. Zur Frage des Zeitpunktes einer Prägung auf die Heimatregion beim Halsbandschnäpper (*Ficedula albicollis*). J. Ornithol. 100:132–140.

Lorenz, K. 1931. Beiträge zur Ethologie sozialer Corviden. J. Ornithol. 79:67–127.

Lorenz, K. 1932. Betrachtungen über das Erkennen der arteigenen Triebhandlungen der Vögel. J. Ornithol. 80:50–98. Translated in Lorenz 1970.

Lorenz, K. 1935. Der Kumpan in der Umwelt des Vogels. J. Ornithol. 83:137–213, 289–413. Translated: Companions as factors in the bird's environment, pp. 101–258, *in* Lorenz 1970.

Lorenz, K. 1937*a*. Über die Bildung des Instinktbegriffes. Naturwissenschaften 25:289–300, 307–318, 325–331. Translated in Lorenz 1970.

Lorenz, K. 1937*b*. The companion in the bird's world. Auk 54:245–273.

Lorenz, K. 1938. A contribution to the comparative sociology of colonial-nesting birds. Proc. Int. Ornithol. Congr. 8:207–218.

Lorenz, K. 1941. Vergleichende Bewegungsstudien bei Anatiden. J. Ornithol. 89:194–294. Translated: Comparative studies on the behaviour of the Anatinae. Avicultural Magazine 57:157–182; 58:8–17, 61–72, 86–94, 172–184; 59:24–34, 80–91 (1951–1953).

Lorenz, K. 1950. The comparative method in studying innate behavior patterns. Symp. Soc. Exp. Biol. 4:221–268.

Lorenz, K. 1958. The evolution of behavior. Sci. Amer. 199:67–78.

Lorenz, K. 1963. On aggression. Harcourt, Brace & World, New York.

Lorenz, K. 1965. Evolution and modification of behavior. Univ. Chicago Press, Chicago.

Lorenz, K. 1970. Studies in animal and human behaviour. Vol. 1. Translated by R. Martin. Harvard Univ. Press, Cambridge, Mass.

Lorenz, K., and N. Tinbergen. 1939. Taxis und Instinkthandlung in der Eirollbewegung der Graugans. Z. Tierpsychol. 2:1–29.

Losey, G. S. 1972. The ecological importance of cleaning symbiosis. Copeia 1972:820–833.

Louch, C. D., and M. Higginbotham. 1967. The relation between social rank and plasma corticosterone in mice. Gen. Comp. Endocr. 8:441–445.

Lowery, G. H., and R. J. Newman. 1966. A continentwide view of bird migration on four nights in October. Auk 83:547–586.

Lowther, J. K. 1961. Polymorphism in the white-throated sparrow, *Zonotrichia albicollis* (Gmelin). Can. J. Zool. 39:281–292.

MacArthur, R. H. 1962. Some generalized theorems of natural selection. Proc. Nat. Acad. Sci. 48:1893–1897.

MacArthur, R. H. 1968. Selection for life tables in periodic environments. Amer. Natur. 102:381–383.

MacArthur, R. H., and E. O. Wilson. 1967. The theory of island biogeography. Princeton Univ. Press, Princeton, N.J.

MacDonnell, M. F., and J. P. Flynn. 1966. Control of sensory fields by stimulation of hypothalamus. Science 152:1406–1408.

Mackay, R. S. 1970. Bio-medical telemetry. Wiley, New York.

Mackie, G. O. 1960. The structure of the nervous system in *Vellella*. Quart. J. Microscop. Sci. 101:119–131.

Mackie, G. O. 1968. Electrical activity in the hydroid *Cordylophora*. J. Exp. Biol. 49:387–400.

Mackintosh, J. H. 1970. Territory formation by laboratory mice. Anim. Behav. 18:177–183.

MacLean, P. D. 1964. Mirror display in the squirrel monkey, *Saimiri sciureus*. Science 146:950–952.

Maher, W. J. 1970. The pomarine jaeger as a brown lemming predator in northern Alaska. Wilson Bull. 82:130–157.

Mainardi, D. 1964. Effetto evolutivo della selezione sessuale basata su imprinting in *Columba livia*. Rivista Italiana di Ornitologia 34:213–216.

Mainardi, D. 1967. Commento ad un recente studio sugli effetti evolutivi del' accoppioamento preferenziale basato sull' imprinting. Rivista Italiana di Ornitologia 37:51–60.

Mainardi, D., F. M. Scudo, and D. Barbieri. 1965. Assortative mating based on early learning: population genetics. Acta Bio-medica 36:583–605.

Manley, G. H. 1960. The swoop and soar performance of the black-headed gull, *Larus ridibundus* L. Ardea 48:37–51.

Mann, G. 1965. Phylogeny and cortical evolution in Chiroptera. Evolution 17:589–591.

Manning, A. 1959*a*. The sexual behaviour of two sibling *Drosophila* species. Behaviour 15:123–145.

Manning, A. 1959*b*. The sexual isolation between *Drosophila melanogaster* and *Drosophila subobscura*. Anim. Behav. 7:60–65.

Manning, A. 1961. Effects of artificial selection for mating speed in *Drosophila melanogaster*. Anim. Behav. 9:82–92.

Manning, A. 1965. *Drosophila* and the evolution of behaviour, pp. 125–169. *In* J. D. Carthy and C. L. Duddington, eds., Viewpoints in biology. Butterworth's, London.

Manning, A. 1968. An introduction to animal behaviour. Edward Arnold, London.

Manosevitz, M., G. Lindzey, and D. D. Thiessen, eds. 1969. Behavioral genetics; method and research. Appleton-Century-Crofts, New York.

Marchant, S. 1960. The breeding of some S.W. Ecuadorian birds. Ibis 102:349–382.

Markl, H. 1969. Die Verständigung durch Stridulationssignale bei Blattschneiderameisen II. Erzeugung und Eigenschaften der Signale. Z. vergl. Physiol. 60:103–150.

Marks, H. L., P. B. Siegel, and C. Y. Kramer. 1960. Effect of comb and wattle removal on the social organization of mixed flocks of chickens. Anim. Behav. 8:192–196.

Marler, P. 1955*a*. Characteristics of some animal calls. Nature 176:6.

Marler, P. 1955*b*. Studies of fighting in chaffinches. 1. Behaviour in relation to the social hierarchy. Brit. J. Anim. Behav. 3:111–117.

Marler, P. 1955*c*. Studies of fighting in chaffinches. 2. The effect on dominance relations of disguising females as males. Brit. J. Anim. Behav. 3:137–146.

Marler, P. 1956. Studies of fighting in chaffinches. 3. Proximity as a cause of aggression. Brit. J. Anim. Behav. 4:23–30.

Marler, P. 1957. Species distinctiveness in the communication signals of birds. Behaviour 11:13–39.

Marler, P. 1959. Developments in the study of animal communication, pp. 150–206. *In* P. R. Bell, ed., Darwin's biological work. Cambridge Univ. Press, London.

Marler, P. 1963. Inheritance and learning in the development of animal vocalizations, pp. 228–243, 794–797. *In* R.-G. Busnel, ed., Acoustic behaviour of animals. Elsevier, Amsterdam, London, New York.

Marler, P. 1968. Visual systems, pp. 103–126. *In* T. A. Sebeok, ed., Animal communication. Indiana Univ. Press, Bloomington, Ind.

Marler, P. 1969*a*. *Colobus guerza:* territoriality and group composition. Science 163:93–95.

Marler, P. 1969*b*. Vocalizations of wild chimpanzees. Recent Advances Primat. 1:94–100.

Marler, P. 1969*c*. Tonal quality of bird sounds, pp. 5–18. *In* R. A. Hinde, ed., Bird vocalizations. Cambridge Univ. Press, London.

Marler, P. 1970. Birdsong and speech development: could there be parallels? Amer. Sci. 58:669–673.

Marler, P., and W. J. Hamilton III. 1966. Mechanisms of animal behavior. Wiley, New York.

Marler, P., and D. Isaac. 1960. Physical analysis of a simple bird song as exemplified by the chipping sparrow. Condor 62:124–135.

Marler, P., and P. Mundinger. 1971. Vocal learning in birds, pp. 389–450. *In* H. Moltz, ed., The ontogeny of vertebrate behavior. Academic Press, New York.

Marler, P., and M. Tamura. 1962. Song "dialects" in three populations of white-crowned sparrows. Condor 64:368–377.

Marler, P., and M. Tamura. 1964. Culturally transmitted patterns of vocal behavior in sparrows. Science 146:1483–1486.

Marshall, A. J. 1953. Bower-birds: their displays and breeding cycles. Clarendon Press, Oxford.

Marshall, J. T., B. A. Ross, and S. Chantharojvong. 1972. The species of gibbons in Thailand. J. Mammal. 53:479–486.

Martin, A. A. 1970. Parallel evolution in the adaptive ecology of leptodactylid frogs of South America and Australia. Evolution 24:643–644.

Martin, R. D. 1966. Tree shrews: unique reproductive mechanism of systematic importance. Science 152:1402–1404.

Martin, W. F. 1972. Evolution of vocalization in the genus *Bufo*, pp. 279–309. *In* W. F. Blair, ed., Evolution in the genus *Bufo*. Univ. Texas Press, Austin.

Martof, B. S. 1961. Vocalization as an isolating mechanism in frogs. Amer. Midland Natur. 65:118–126.

Martof, B. S., and E. F. Thompson. 1964. A behavioral analysis of the mating call of the chorus frog, *Pseudacris triseriata*. Amer. Midland Natur. 71:198–209.

Mason, L. G. 1964. Stabilizing selection for mating fitness in natural populations of *Tetraopes*. Evolution 18:492–497.

Mason, L. G. 1969. Mating selection in the California oak moth. Evolution 23:55–58.

Mason, W. A. 1965. The social development of monkeys and apes, pp. 514–543. *In* I. DeVore, ed., Primate behavior: field studies of monkeys and apes. Holt, Rinehart and Winston, New York.

Mason, W. A. 1968. Use of space by *Callicebus* groups, pp. 200–217. *In* P. C. Jay, ed., Primates. Holt, Rinehart and Winston, New York.

Masure, R. A., and W. C. Allee. 1934. The social order in flocks of the common chicken and the pigeon. Auk 51:306–327.

Mather, K. 1953. The genetical structure of populations. Symp. Soc. Exp. Biol. 7:66–95.

Mather, K. 1955. Polymorphism as an outcome of disruptive selection. Evolution 9:52–61.

Matthews, G. V. T. 1963. The astronomical bases of "nonsense" orientation. Proc. Int. Ornithol. Congr. 13:415–429.

Matthews, G. V. T. 1968. Bird navigation. 2nd ed. Cambridge Univ. Press, London.

Maturana, H. R., J. Y. Lettvin, W. S. McCulloch, and W. H. Pitts. 1960. Anatomy and physiology of vision in the frog (*Rana pipiens*). J. Gen. Physiol. 43:129–175.

Mayfield, H. 1965. Chance distribution of cowbird eggs. Condor 67:257–263.

Mayr, E. 1942. Systematics and the origin of species. Dover, New York. Reprinted 1964.

Mayr, E. 1950. The role of the antennae in the mating behavior of female *Drosophila*. Evolution 4:149–154.
Mayr, E. 1958. Behavior and systematics, pp. 341–362. *In* A. Roe and G. G. Simpson, eds., Behavior and evolution. Yale Univ. Press, New Haven.
Mayr, E. 1963. Animal species and evolution. Harvard Univ. Press, Cambridge, Mass.
Mayr, E., ed. 1957. The species problem. Amer. Assoc. Advance. Sci. Pub. No. 50.
McAlister, W. H. 1961. The mechanisms of sound production in North American *Bufo*. Copeia 1961:86–95.
McBride, G. 1964. A general theory of social organization and behaviour. Univ. Queensland Papers Faculty of Veterinary Sci. 1:72–110.
McBride, G., J. W. James, and R. N. Shoffner. 1963. Social forces determining spacing and head orientation in a flock of domestic hens. Nature 197:1272–1273.
McCabe, T. T., and B. D. Blanchard. 1950. Three species of *Peromyscus*. Rood Associates, Santa Barbara, Calif.
McClearn, G. E., and J. C. DeFries. 1973. Introduction to behavioral genetics. Freeman, San Francisco.
McDermott, F. A. 1917. Observations on the light-emission of American Lampyridae. Can. Entomol. 49:53–61.
McFarland, D. J. 1969. Mechanisms of behavioural disinhibition. Anim. Behav. 17:238–242.
McFarland, D. J. 1971. Feedback mechanisms in animal behaviour. Academic Press, London.
McFarland, D. J., and E. Baher. 1968. Factors affecting feather posture in the Barbary dove. Anim. Behav. 16:171–177.
McFarland, D. J., and R. Sibly. 1972. 'Unitary drives' revisited. Anim. Behav. 20:548–563.
McNab, B. K. 1963. Bioenergetics and the determination of home range size. Amer. Natur. 97:133–140.
McGeen, D. S. 1972. Cowbird-host relationships. Auk 89:360–380.
McGeen, D. S., and J. J. McGeen. 1968. The cowbirds of Otter Lake. Wilson Bull. 80:84–93.
McKinney, C. 1971. An analysis of zones of intergradation in the side-blotched lizard, *Uta stansburiana* (Sauria: Iguanidae). Copeia 1971:596–613.
McKinney, F. 1965. The comfort movements of Anatidae. Behaviour 25:120–220.
Mech, L. D. 1970. The wolf. Natural History Press, Garden City, N.Y.
Mecham, J. S. 1961. Isolating mechanisms in anuran amphibians, pp. 24–61. *In* W. F. Blair, ed., Vertebrate speciation. Univ. Texas Press, Austin.
Mecham, J. S. 1968. Evidence of reproductive isolation between two populations of the frog, *Rana pipiens*, in Arizona. Southwestern Natur. 13:35–44.
Mecham, J. S. 1971. Vocalizations of the leopard frog, *Rana pipiens* and three related Mexican species. Copeia 1971:504–516.
Meijer, G. M., F. J. Ritter, C. J. Persoons, A. K. Minks, and S. Voerman. 1972. Sex pheromones of summer fruit tortrix moth *Adoxophyes orana*: two synergistic isomers. Science 175:1469–1470.
Menaker, M. 1968. Extraretinal light perception in the sparrow. I. Entrainment of the biological clock. Proc. Nat. Acad. Sci. 59:414–421.
Menaker, M., ed. 1971. Biochronometry. Nat. Acad. Sci., Washington, D.C.
Mendelson, M. 1971. Oscillator neurons in crustacean ganglia. Science 171:1170–1173.
Merkel, F. W., H. G. Fromme, and W. Wiltschko. 1964. Nichtvisuelles Orientierungsvermogen bei nächtlich zugunruhigen Rotkehlchen. Vogelwarte 22:168–173.
Merkel, F. W., and W. Wiltschko. 1965. Magnetismus und Richtungsfinden zugunruhiger Rotkehlchen (*Erithacus rubecula*). Vogelwarte 23:71–77.
Messmer, F., and E. Messmer. 1956. Die Entwicklung der Lautäusserungen und einiger

Verhaltensweisen der Amsel (*Turdus merula merula* L.) unter natürlichen Bedingungen und nach Einzelaufzucht in schalldichten Raümen. Z. Tierpsychol. 13:341–441.

Metfessel, M. 1945. Roller canary song produced without learning from external sources. Science 81:470.

Metzgar, L. H. 1967. An experimental comparison of screech owl predation on resident and transient white-footed mice (*Peromyscus leucopus*). J. Mammal. 48:387–391.

Metzgar, L. H. 1972. The measurement of home range shape. J. Wildlife Mgmt. 36:643–645.

Mewaldt, L. R. 1964. Effects of bird removal on a winter population of sparrows. Bird-banding 35:184–195.

Michael, R. P., and E. B. Keverne. 1968. Pheromones in the communication of sexual status in primates. Nature 218:746–749.

Michael, R. P., E. B. Keverne, and R. W. Bonsall. 1971. Pheromones: isolation of male sex attractants from a female primate. Science 172:964–966.

Michael, R. P., D. Zumpe, E. B. Keverne, and R. W. Bonsall. 1972. Neuroendocrine factors in the control of primate behavior. Recent Progr. Hormone Res. 28:665–706.

Michaud, T. C. 1962. Call discrimination by females of the chorus frogs, *Pseudacris clarki* and *Pseudacris nigrita*. Copeia 1962:213–215.

Michelson, A. 1964. Observations on the sexual behaviour of some longicorn beetles, subfamily Lepturinae (Coleoptera, Cerambycidae). Behaviour 22:152–166.

Michener, C. D. 1964. Evolution of the nests of bees. Amer. Zool. 4:227–239.

Michener, C. D. 1974. The social behavior of the bees. Harvard Univ. Press, Cambridge, Mass.

Michener, C. D., and R. B. Lange. 1958. Distinctive type of primitive social behavior among bees. Science 127:1046–1047.

Miller, H., and A. Dzubin. 1965. Regrouping of family members of the white-fronted goose (*Anser albifrons*) after individual release. Bird-banding 36:184–191.

Miller, H. C. 1963. The behavior of the pumpkinseed sunfish *Lepomis gibbosus* (Linnaeus), with notes on the behavior of the other species of *Lepomis* and the pigmy sunfish, *Elassoma evergladei*. Behaviour 22:88–151.

Miller, L. A., and E. G. MacLeod. 1966. Ultrasonic sensitivity: a tympanal receptor in the green lacewing *Chrysopa carnea*. Science 154:891–893.

Miller, R. E. 1967. Experimental approaches to the physiological and behavioral concomitants of affective communication in rhesus monkeys, pp. 125–134. *In* S. A. Altmann, ed., Social communication among primates. Univ. Chicago Press, Chicago.

Miller, R. E., J. H. Banks, and H. Kuwahara. 1966. The communication of affect in monkeys: co-operative reward conditioning. J. Genet. Psychol. 108:121–134.

Miller, R. E., J. H. Banks, and N. Ogawa. 1962. Communication of affect in "co-operative conditioning" of rhesus monkeys. J. Abnormal Soc. Psychol. 64:343–348.

Miller, R. E., J. H. Banks, and N. Ogawa. 1963. Role of facial expressions in "co-operative avoidance conditioning" in monkeys. J. Abnormal Soc. Psychol. 67:24–30.

Miller, R. E., J. V. Murphy, and I. A. Mirsky. 1959. Relevance of facial expression and posture as cues in communication of affect between monkeys. A.M.A. Arch. Gen. Psychiat. 1:480–488.

Miller, R. S., and W. J. D. Stephen. 1966. Spatial relationships in flocks of sandhill cranes (*Grus canadensis*). Ecology 47:323–327.

Milligan, M. M., and J. Verner. 1971. Inter-populational song dialect discrimination in the white-crowned sparrow. Condor 73:208–213.

Millikan, G. C., and R. I. Bowman. 1967. Observations on Galapagos tool-using finches in captivity. Living Bird 6:23–41.

Minock, M. E. 1971. Social relationships among mountain chickadees (*Parus gambeli*). Condor 73:118–120.

Missakian, E. A. 1972. Genealogical and cross-genealogical dominance relations in a group of free-ranging rhesus monkeys (*Macaca mulatta*) on Cayo Santiago. Primates 13:169–180.

Missakian, E. A. 1973. Genealogical mating activity in free-ranging groups of rhesus monkeys (*Macaca mulatta*) on Cayo Santiago. Behaviour 45:225–241.

Moffat, C. B. 1903. The spring rivalry of birds, some views on the limits to multiplication. Irish Natur. 12:152–166.

Mohr, C. O., and W. A. Stumpf. 1966. Comparison of methods for calculating areas of animal activity. J. Wildlife Mgmt. 30:293–304.

Moltz, H. 1960. Imprinting: empirical basis and theoretical significance. Psychol. Bull. 57:291–314.

Moltz, H. 1963. Imprinting: an epigenetic approach. Psychol. Rev. 70:123–138.

Moltz, H. 1965. Contemporary instinct theory and the fixed action pattern. Psychol. Rev. 72:27–47.

Moltz, H. 1971. The ontogeny of vertebrate behavior. Academic Press, New York.

Moore, R. E. 1965. Olfactory discrimination as an isolating mechanism between *Peromyscus maniculatus* and *Peromyscus polionotus*. Amer. Midland Natur. 73:85–100.

Moreau, R. E. 1972. The Palaearctic-African bird migration systems. Academic Press, London.

Morejohn, G. V. 1968. Breakdown of isolation mechanisms in two species of captive junglefowl (*Gallus gallus* and *Gallus sonneratii*). Evolution 22:576–582.

Morris, D. 1956*a*. The feather postures of birds and the problem of the origin of social signals. Behaviour 9:75–113.

Morris, D. 1956*b*. The function and causation of courtship ceremonies. *In* L'instinct dans le comportement des animaux et de l'homme. Fondation Singer-Polignac. Masson, Paris.

Morris, D. 1957. "Typical intensity" and its relation to the problem of ritualization. Behaviour 11:1–12.

Morse, D. H. 1968. The use of tools by brown-headed nuthatches. Wilson Bull. 80:220–224.

Morse, D. H. 1970. Territorial and courtship songs of birds. Nature 226:659–661.

Moser, J. C. 1964. Inquiline roach responds to trail-marking substance of leaf-cutting ants. Science 143:1048–1049.

Moulton, J. M. 1963. Acoustic behaviour of fishes, pp. 655–693. *In* R.-G. Busnel, ed., Acoustic behaviour in animals. Elsevier, Amsterdam.

Moynihan, M. 1955. Remarks on the original sources of displays. Auk 72:240–246.

Moynihan, M. 1955. Some aspects of reproductive behavior in the black-headed gull (*Larus ridibundus ridibundus* L.) and related species. Behaviour Suppl. 4:1–201.

Moynihan, M. 1959. A revision of the family Laridae (Aves). Amer. Mus. Novitates. 1928:1–42.

Moynihan, M. 1968. Social mimicry: character convergence versus character displacement. Evolution 22:315–331.

Muller, H. J. 1940. Bearing of the "*Drosophila*" work on systematics, pp. 185–268. *In* J. S. Huxley, ed., New systematics. Clarendon Press, Oxford.

Müller-Schwarze, D. 1971. Pheromones in black-tailed deer (*Odocoileus hemionus colombianus*). Anim. Behav. 19:141–152.

Muntz, W. R. A. 1962*a*. Microelectrode recordings from the diencephalon of the frog (*Rana pipiens*), and a blue-sensitive system. J. Neurophysiol. 25:699–711.

Muntz, W. R. A. 1962*b*. Effectiveness of different colors of light in releasing the positive phototactic behavior of frogs, and a possible function of the retinal projection to the diencephalon. J. Neurophysiol. 25:712–720.

Murray, B. G., Jr. 1969. A comparative study of the LeConte's and sharp-tailed sparrows. Auk 86:199–231.

Murray, B. G., Jr. 1971. The ecological consequences of interspecific territorial behavior in birds. Ecology 52:414–423.

Murton, R. K. 1967. The significance of endocrine stress in population control. Ibis 109:622–623.

Murton, R. K. 1968. Some predator-prey relationships in bird damage and population control. *In* R. K. Murton and E. N. Wright, eds., The problems of birds as pests. Symp. Inst. Biol. 17:157–169.

Murton, R. K. 1971. Why do bird species feed in flocks? Ibis 113:534–536.

Murton, R. K., A. J. Isaacson, and N. J. Westwood. 1971. The significance of gregarious feeding behaviour and adrenal stress in a population of wood-pigeons (*Columba palumbus*). J. Zool., London 165:53–84.

Murton, R. K., and E. N. Wright. 1968. The problems of birds as pests. Academic Press, New York.

Myers, K. 1966. The effects of density on sociality and health in mammals. Proc. Ecol. Soc. Austral. 1:40–64.

Myllymaki, A., J. Aho, E. O. Lind, and J. Tast. 1962. Behaviour and daily activity of the Norwegian lemming (*Lemmus lemmus*) during autumn migration. Ann. Zool. Soc. "Vanamo" 24:1–31.

Mykytowycz, R. 1965. Further observations on the territorial function and histology of the subcutaneous (chin) glands in the rabbit, *Oryctolagus cuniculus* (L.). Anim. Behav. 13:400–465.

Mykytowycz, R. 1970. The role of skin glands in mammalian communication, pp. 327–360. *In* Johnston et al. 1970.

Napier, J. R., and P. H. Napier. 1967. A handbook of living primates. Academic Press, New York.

Narayanan, C. H., M. W. Fox, and V. Hamburger. 1971. Prenatal development of spontaneous and evoked activity in the rat (*Rattus norwegicus albinus*). Behaviour 40:100–134.

Narda, R. D. 1968. Experimental evaluation of the stimuli involved in sexual isolation among three members of the *ananassae* species subgroup (*Sophophora*, *Drosophila*). Anim. Behav. 16:117–119.

Negus, N. C., and E. Gould. 1965. Endocrines, behavior and population. Science 149:376.

Neill, W. T. 1964. Isolating mechanisms in snakes. Quart. J. Fla. Acad. Sci. 27:333–347.

Nelson, J. B. 1968*a*. Galapagos; islands of birds. William Morrow, New York.

Nelson, J. B. 1968*b*. Breeding behavior of the swallow-tailed gull in the Galapagos. Behaviour 30:146–174.

Nelson, K. 1964*a*. The temporal patterning of courtship in the glandulocaudine fishes (Ostariophysi, Characidae). Behaviour 24:90–146.

Nelson, K. 1964*b*. The evolution of a pattern of sound production associated with courtship in the characid fish, *Glandulocauda inequalis*. Evolution 18:526–540.

Nero, R. W. 1956. A behavior study of the red-winged blackbird. I. Mating and nesting activities. Wilson Bull. 68:4–37.

Neuweiler, G. 1970. Neurophysiologische Untersuchungen zum Echoortungssystem der grossen Hufeisennase *Rhinolophus ferrumequinum*. Z. vergl. Physiol. 67:273–306.

Nice, M. M. 1941. The role of territory in bird life. Amer. Midland Natur. 26:441–487.

Nice, M. M., and W. E. Schantz. 1959. Head-scratching movements in birds. Auk 76:339–342.

Nichols, D. 1962. Echinoderms. Hutchinson Univ. Libr., London.

Nicol, J. A. C. 1948. The giant axons of annelids. Quart. Rev. Biol. 23:291–324.

Nicolai, J. 1959. Familientradition in der Gesangsentwicklung des Gimpels (*Pyrrhula pyrrhula* L.). J. Ornithol. 100:39–46.

Nicolai, J. 1964. Der Brutparasitismus der Viduinae als ethologisches Problem. Prägungsphänomene als Faktoren der Rassen- und Artbildung. Z. Tierpsychol. 21:129–204.

Nicolai, J. 1967. Rassen- und Artbildung in der Viduinengattung *Hypochera*. J. Ornithol. 108:309–319.

Nicolai, J. 1968. Die Schnabelfarbung als potentieller isolationsfaktor zwischen *Pytilia phoenicoptera* Swainson and *Pytilia lineata* Hueglin (Familie: Estrildidae). J. Ornithol. 109:450–461.

Nisbet, I. C. T. 1963. Measurements with radar of the height of nocturnal migration over Cape Cod, Massachusetts. Bird-banding 34:57–67.

Nisbet, I. C. T., and W. H. Drury. 1968. Short-term effects of weather on bird migration: a field study using multivariate statistics. Anim. Behav. 16:496–530.

Nishiitsutsuji-Uwo, J., S. F. Petropulos, and C. S. Pittendrigh. 1967. Central nervous control of circadian rhythmicity in the cockroach. I. Role of the pars intercerebralis. Biol. Bull. 133:679–696.

Nishiitsutsuji-Uwo, J., and C. S. Pittendrigh. 1968*a*. Central nervous system control of circadian rhythmicity in the cockroach. II. The pathway of light signals that entrain the rhythm. Z. vergl. Physiol. 58:1–13.

Nishiitsutsuji-Uwo, J., and C. S. Pittendrigh. 1968*b*. Central nervous system control of circadian rhythmicity in the cockroach. III. The optic lobes, locus of the driving oscillation? Z. vergl. Physiol. 58:14–46.

Noble, G. K. 1934. Sex recognition in the sunfish, *Eupomotis gibbosus* (Linné). Copeia 1934:151–155.

Noble, G. K. 1936. Courtship and sexual selection of the flicker (*Colaptes auratus luteus*). Auk 53:269–282.

Noble, G. K. 1939. The role of dominance in the social life of birds. Auk 56:263–273.

Noble, G. K., and H. T. Bradley. 1933. The mating behavior of lizards; its bearing on the theory of sexual selection. Ann. N.Y. Acad. Sci. 35:25–100.

Noble, G. K., and B. Curtis. 1939. The social behavior of the jewel fish, *Hemichromis bimaculatus* Gill. Bull. Amer. Mus. Natur. Hist. 76:1–46.

Nogueira-Neto, P. 1970. Behavior problems related to the pillages made by some parasitic stingless bees (Meliponinae, Apidae), pp. 416–434. *In* L. R. Aronson, E. Tobach, D. S. Lehrman, and J. S. Rosenblatt, eds., Development and evolution of behavior. Freeman, San Francisco.

Norris, K. S. 1969. The echolocation of marine mammals. *In* H. T. Andersen, ed., The biology of marine mammals. Academic Press, New York.

Norton, A. L., H. Spekreijse, H. G. Wagner, and M. L. Wolbarscht. 1970. Responses to directional stimuli in retinal preganglionic units. J. Physiol. 206:93–107.

Nottebohm, F. 1969*a*. The "critical period" for song learning. Ibis 111:386–387.

Nottebohm, F. 1969*b*. The song of the chingolo, *Zonotrichia capensis*, in Argentina: description and evaluation of a system of dialects. Condor 71:299–315.

Nottebohm, F. 1970. Ontogeny of bird song. Science 167:950–956.
Nottebohm, F. 1971. Neural lateralization of vocal control in a passerine bird. I. Song. J. Exp. Zool. 177:229–262.
Nottebohm, F. 1972*a*. Neural lateralization of vocal control in a passerine bird. II. Subsong, calls, and a theory of vocal learning. J. Exp. Zool. 179:35–50.
Nottebohm, F. 1972*b*. The origins of vocal learning. Amer. Natur. 106:116–140.
Nottebohm, F., and M. E. Nottebohm. 1971. Vocalizations and breeding behaviour of surgically deafened ring doves (*Streptopelia risoria*). Anim. Behav. 19:311–327.
Novick, A. 1959. Acoustic orientation in the cave swiftlet. Biol. Bull. 117:497–503.
Novick, A. 1963. Orientation in neotropical bats. II. Phyllostomatidae and Desmodontidae. J. Mammal. 44:44–56.
Novick, A. 1965. Echolocation of flying insects by the bat, *Chilonycteris psilotis*. Biol. Bull. 128:297–314.
Novick, A. 1971. Echolocation in bats: some aspects of pulse design. Amer. Sci. 59:198–209.
Novick, A., and J. R. Vaisnys. 1964. Echolocation of flying insects by the bat, *Chilonycteris parnellii*. Biol. Bull. 127:478–488.
O'Donald, P. 1959. Possibility of assortative mating in the arctic skua. Nature 183:1210–1211.
O'Donald, P. 1960. Inbreeding as a result of imprinting. Heredity 15:79–85.
O'Donald, P. 1962. The theory of sexual selection. Heredity 17:541–552.
O'Donald, P. 1963*a*. Sexual selection for dominant and recessive genes. Heredity 18:451–457.
O'Donald, P. 1963*b*. Sexual selection and territorial behaviour. Heredity 18:361–364.
O'Donald, P. 1967. A general model of sexual and natural selection. Heredity 22:499–518.
O'Donald, P. 1972. Sexual selection by variations in fitness at breeding time. Nature 237:349–351.
Odum, E. P. 1942. Annual cycle of the black-capped chickadee—3. Auk 59:499–531.
Odum, E. P., and E. J. Kuenzler. 1955. Measurement of territory and home range size in birds. Auk 72:128–137.
O'Keefe, J., and H. Bouma. 1969. Complex sensory properties of certain amygdala units in the freely moving cat. Exp. Neurol. 23:384–398.
Okon, E. E. 1972. Factors affecting ultrasound production in infant rodents. J. Zool., London 168:139–148.
Orenstein, R. I. 1972. Tool-use by the New Caledonian crow (*Corvus moneduloides*). Auk 89:674–676.
Orgain, H., and M. W. Schein. 1953. A preliminary analysis of the physical environment of the Norway rat. Ecology 34:467–473.
Orians, G. H. 1960. Autumnal breeding in the tricolored blackbird. Auk 77:379–398.
Orians, G. H. 1961. The ecology of blackbird (*Agelaius*) social systems. Ecol. Monogr. 31:285–312.
Orians, G. H. 1966. Food of nestling yellow-headed blackbirds, Caribou Parklands, British Columbia. Condor 68:321–337.
Orians, G. H. 1969. On the evolution of mating systems in birds and mammals. Amer. Natur. 103:589–603.
Orians, G. H. 1971. Ecological aspects of behavior, pp. 513–546. *In* D. S. Farner and J. R. King, eds., Avian biology, vol. I. Academic Press, New York, London.
Orians, G. H., and M. F. Willson. 1964. Interspecific territories of birds. Ecology 17:736–745.
Orr, R. T. 1970. Animals in migration. Macmillan, London.

Otto, D. 1971. Untersuchungen zur zentralnervösen Kontrolle der Lauterzeugung von Grillen. Z. vergl. Physiol. 74:227–271.

Owen, D. F. 1964. Mating preferences in wild birds. Nature 203:986.

Pack, J. C., H. S. Mosby, and P. B. Siegel. 1967. Influence of social hierarchy on gray squirrel behavior. J. Wildlife Mgmt. 31:720–728.

Parsons, P. A. 1967. The genetic analysis of behavior. Methuen, London.

Passano, L. M. 1965. Pacemakers and activity rhythms in medusae: homage to Romanes. Amer. Zool. 5:465–481.

Paterniani, E. 1969. Selection for reproductive isolation between two populations of maize, *Zea mays* L. Evolution 23:534–547.

Patterson, I. J. 1965. Timing and spacing of broods in the black-headed gull *Larus ridibundus*. Ibis 107:433–459.

Patterson, J. T., and W. S. Stone. 1952. Evolution in the genus *Drosophila*. Macmillan, New York.

Patterson, R. L. 1952. The sage grouse in Wyoming. Wyoming Game and Fish Commission, Cheyenne.

Pavlidis, T. 1969. Populations of interacting oscillators and circadian rhythms. J. Theoret. Biol. 22:418–436.

Pavlov, I. P. 1927. Conditioned reflexes. Oxford Univ. Press, Oxford.

Payne, R. 1970. Songs of humpback whales (record and booklet). CRM, Del Mar, Calif.

Payne, R., and S. McVay. 1971. Songs of humpback whales. Science 173:585–597.

Payne, R. B. 1965. Clutch size and numbers of eggs laid by brown-headed cowbirds. Condor 67:44–60.

Payne, R. B. 1967. Interspecific communication signals in parasitic birds. Amer. Natur. 101:363–376.

Payne, R. B. 1968. A preliminary report on the relationships of the indigobirds. Bull. Brit. Ornithol. Club 88:32–36.

Payne, R. B. 1969*a*. Breeding seasons and reproductive physiology of tricolored blackbirds and redwinged blackbirds. Univ. Calif. Pub. Zool. 90:1–137.

Payne, R. B. 1969*b*. Nest parasitism and display of chestnut sparrows in a colony of grey-capped social weavers. Ibis 111:300–307.

Payne, R. B. 1973. Behavior, mimetic songs and song dialects and relationships of the parasitic indigobirds (*Vidua*) of Africa. Ornithol. Monogr. Amer. Ornithol. Union 11:1–333.

Pearce, S. 1960. An experimental study of sexual isolation within the species *Drosophila melanogaster*. Anim. Behav. 8:232–233.

Pearson, E. W., P. R. Skon, and G. W. Corner. 1967. Dispersal of urban roosts with records of starling distress calls. J. Wildlife Mgmt. 31:502–506.

Pearson, K. G., and J. F. Iles. 1970. Discharge patterns of coxal levator and depressor motor neurones of the cockroach, *Periplaneta americana*. J. Exp. Biol. 52:139–165.

Peek, F. W. 1972*a*. An experimental study of the territorial function of vocal and visual displays in the male redwinged blackbird (*Agelaius phoeniceus*). Anim. Behav. 20:112–118.

Peek, F. W. 1972*b*. The effect of tranquilization upon territory maintenance in the male red-winged blackbird (*Agelaius phoeniceus*). Anim. Behav. 20:119–122.

Pelkwijk, J. J. ter, and N. Tinbergen. 1937. Eine reizbiologische Analyse einiger Verhaltensweisen von *Gasterosteus aculeatus* (L.). Z. Tierpsychol. 1:193–200.

Perdeck, A. C. 1958*a*. The isolating value of specific song patterns in two sibling species of grasshoppers (*Chorthippus brunneus* Thunb. and *C. biguttulus* L.). Behaviour 12:1–75.

Perdeck, A. C. 1958*b*. Two types of orientation in migrating starlings, *Sturnus vulgaris* L. and chaffinches, *Fringilla coelebs* L. as revealed by displacement experiments. Ardea 46:1–37.

Perdeck, A. C. 1963. Does navigation without celestial clues exist in robins? Ardea 51:91–104.

Perdeck, A. C. 1964. An experiment on the ending of autumn migration in starlings. Ardea 52:133–139.

Perdeck, A. C. 1967. Orientation of starlings after displacement to Spain. Ardea 55:194–202.

Perrins, C. M. 1965. Population fluctuations and clutch-size in the great tit, *Parus major* L. J. Anim. Ecol. 34:601–647.

Perrins, C. M. 1968. The purpose of the high-intensity alarm call in small passerines. Ibis 110:200–201.

Peters, H. M. 1963. Untersuchungen zum Problem des angeborenen Verhaltens. Naturwissenschaften 22:677–686.

Petit, C. 1951. Le role de l'isolement sexuel dans l'evolution des populations de *Drosophila melanogaster*. Bull. Biol. France et Belgique 85:392–418.

Petit, C., and L. Ehrman. 1969. Sexual selection in *Drosophila*, pp. 177–223. *In* T. Dobzhansky, M. K. Hecht, and W. C. Steere, eds., Evolutionary biology, vol. 3. Appleton-Century-Crofts, New York.

Pfeiffer, W. 1967. Schreckreaktion und Schreckstoffzellen bei Ostariophysi und Gonorhynchiformes. Z. vergl. Physiol. 56:380–386.

Pianka, E. R. 1970. On r and K selection. Amer. Natur. 100:592–597.

Pielou, E. C. 1969. An introduction to mathematical ecology. Wiley-Interscience. New York.

Pigarev, I. N., and G. M. Zenkin. 1970. Dark spot detectors in the frog retina and their role in organization of feeding behavior. Neurosci. Trans. 13:29–33.

Pigarev, I. N., G. M. Zenkin, and S. V. Girman. 1972. Retinal detector activity in frogs behaving freely. Neurosci. Behav. Physiol. 5:325–330.

Pimentel, D., G. J. C. Smith, and J. Soans. 1967. A population model of sympatric speciation. Amer. Natur. 101:493–504.

Pitelka, F. A. 1957. Some aspects of population structure in the short-term cycle of the brown lemming in northern Alaska. Cold Spring Harbor Symp. Quant. Biol. 22:237–251.

Pitelka, F. A. 1959. Numbers, breeding schedule, and territoriality in pectoral sandpipers of northern Alaska. Condor 61:233–264.

Ploog, D., and T. Melnechuk. 1969. Primate communication. Neurosci. Res. Program Bull. 7:419–510.

Ploog, D., and T. Melnechuk. 1971. Are apes capable of language? Neurosci. Res. Program Bull. 9:599–700.

Pollack, G., O. W. Henson, Jr., and A. Novick. 1972. Cochlear microphonic audiograms in the "pure tone" bat *Chilonycteris parnellii parnellii*. Science 176:66–68.

Pomeranz, B., and S. H. Chung. 1970. Dendritic-tree anatomy codes form-vision physiology in tadpole retina. Science 170:983–984.

Porter, K. R. 1968. Evolutionary status of a relict population of *Bufo hemiophrys* Cope. Evolution. 22:583–594.

Porter, K. R. 1969. Evolutionary status of the Rocky Mountain population of wood frogs. Evolution 23:163–170.

Porter, K. R. 1972. Herpetology. Saunders, Philadelphia.

Potter, H. D. 1965. Mesencephalic auditory region of the bullfrog. J. Neurophysiol. 28:1132–1154.

Poulsen, H. 1953. A study of incubation responses and some other behaviour patterns in birds. Vidensk. Medd. fra Dansk. naturh. Foren. 115:1–131.

Poulton, E. B. 1908. Essays on evolution, 1889–1907. Clarendon Press, Oxford.

Quastler, H. 1958. A primer on information theory. *In* H. P. Yockey, ed., Symposium on information theory in biology. Pergamon Press, New York.

Raitt, R. J., and J. W. Hardy. 1970. Relationships between two partly sympatric species of thrushes (*Catharus*) in Mexico. Auk 87:20–57.

Ralls, K. 1971. Mammalian scent marking. Science 171:443–449.

Ralph, C. J., and C. A. Pearson. 1971. Correlation of age, size of territory, plumage, and breeding success in white-crowned sparrows. Condor 73:77–80.

Ramón y Cajal, S. R. 1909. Histologie du système nerveux de l'homme et des vertebres. Vol. I. Maloine, Paris. Republished 1953 by the Instituto Ramón y Cajal, Madrid.

Rao, S. V., and P. DeBach. 1969. Experimental studies on hybridization and sexual isolation between some *Aphytis* species (Hymenoptera: Aphelinidae). IV. The significance of reproductive isolation between interspecific hybrids and parental species. Evolution 23:525–533.

Rashevsky, N. 1959. Mathematical biology of social behavior. Univ. Chicago Press, Chicago.

Raup, D. M., and A. Seilacher. 1969. Fossil foraging behavior: computer simulation. Science 166:994–995.

Rauschert, K. 1963. Sexuelle Affinität zwischen Arten und Unterarten von Rötelmäusen (*Clethrionomys*). Biol. Zentralblatt 82:653–664.

Raveling, D. G. 1970. Dominance relationships and agonistic behavior of Canada geese in winter. Behaviour 37:291–319.

Ray, C., W. A. Watkins, and J. J. Burns. 1969. The underwater song of *Erignathus* (bearded seal). Zoologica 54:79–83.

Recher, H. F., and J. A. Recher. 1969. Some aspects of the ecology of migrant shorebirds. II. Aggression. Wilson Bull. 81:140–154.

Reed, R. A. 1968. Studies of the diederik cuckoo *Chrysococcyx caprius* in the Transvaal. Ibis 110:321–331.

Regnier, F. E. 1971. Semiochemicals—structure and function. Biol. Reprod. 4:309–326.

Regnier, F. E., and E. O. Wilson. 1968. The alarm-defense system of the ant *Acanthomyops claviger*. J. Insect Physiol. 14:955–970.

Regnier, F. E., and E. O. Wilson. 1969. The alarm-defense system of the ant *Lasius alienus*. J. Insect Physiol. 15:893–898.

Regnier, F. E., and E. O. Wilson. 1971. Chemical communication and "propaganda" in slave-maker ants. Science 172:267–269.

Reimer, J. D., and M. L. Petras. 1967. Breeding structure of the house mouse, *Mus musculus*, in a population cage. J. Mammal. 48:88–99.

Remler, M., A. Selverston, and D. Kennedy. 1968. Lateral giant fibers of crayfish: location of somata by dye injection. Science 162:281–283.

Rensch, B. 1925. Verhalten von Singvögeln bei Aenderung des Geleges. Ornithol. Monatsberichte 33:169–173.

Ressler, R. H., R. B. Cialdini, M. L. Ghoca, and S. M. Kleist. 1968. Alarm pheromone in the earthworm *Lumbricus terrestris*. Science 161:597–599.

Reynolds, V. 1967. The apes. Dutton, New York.

Rice, J. O., and W. L. Thompson. 1968. Song development in the indigo bunting. Anim. Behav. 16:462–469.

Richardson, W. J. 1971. Spring migration and weather in eastern Canada: a radar study. Amer. Birds 25:684–690.

Richardson, W. J. 1972. Autumn migration and weather in eastern Canada: a radar study. Amer. Birds 26:10–17.
Ripley, S. D. 1961. Aggressive neglect as a factor in interspecific competition in birds. Auk 78:360–371.
Ritchey, F. 1951. Dominance-subordination, and territorial relations in the common pigeon. Physiol. Zool. 24:167–176.
Robel, R. J. 1966. Booming territory size and mating success of the greater prairie chicken (*Tympanuchus cupido pinnatus*). Anim. Behav. 14:328–331.
Roberts, S. K. deF. 1960. Circadian activity rhythms in cockroaches. I. The free-running rhythm in steady-state. J. Cellular Comp. Physiol. 55:99–110.
Roberts, S. K. deF. 1962. Circadian activity rhythms in cockroaches. II. Entrainment and phase shifting. J. Cellular Comp. Physiol. 59:175–186.
Roberts, S. K. deF. 1965. Photoreception and entrainment of cockroach activity rhythms. Science 148:958–959.
Roberts, S. K. deF. 1966. Circadian activity rhythms in cockroaches. III. The role of endocrine and neural factors. J. Cellular Comp. Physiol. 67:473–486.
Roberts, S. K. deF., S. D. Skopik, and R. J. Driskill. 1971. Circadian rhythms in cockroaches: does brain hormone mediate the locomotor cycle? pp. 505–514. *In* Menaker 1971.
Robertson, A. 1970. A note on disruptive selection experiments in *Drosophila*. Amer. Natur. 104:561–569.
Robertson, D. R. 1972. Social control of sex reversal in a coral-reef fish. Science 177:1007–1009.
Roeder, K. D. 1962. The behaviour of free-flying moths in the presence of artificial ultrasonic pulses. Anim. Behav. 10:300–304.
Roeder, K. D. 1966. Auditory system of noctuid moths. Science 154:1515–1521.
Roeder, K. D. 1971*a*. Insect flight behavior: some neurophysiological indications of its control. Progr. Physiol. Psych. 4:1–36.
Roeder, K. D. 1971*b*. Acoustic alerting mechanisms in insects. Ann. N.Y. Acad. Sci. 188:63–79.
Roeder, K. D., and A. E. Treat. 1961. The detection and evasion of bats by moths. Amer. Sci. 49:135–148.
Roeder, K. D., A. E. Treat, and J. S. Vandeberg. 1970. Distal lobe of the pilifer: an ultrasonic receptor in choerocampine hawkmoths. Science 170:1098–1099.
Roelofs, W. L., and R. T. Cardé. 1971. Hydrocarbon sex pheromone in tiger moths (Arctiidae). Science 171:684–686.
Roelofs, W. L., and A. Comeau. 1969. Sex pheromone specificity: taxonomic and evolutionary aspects in Lepidoptera. Science 165:398–400.
Romanes, G. J. 1885. Jellyfish, starfish, and sea-urchins, being a research on primitive nervous systems. Int. Sci. Ser. 49:1–323.
Rosenblum, L. A., I. C. Kaufman, and A. J. Stynes. 1964. Individual distance in two species of macaque. Anim. Behav. 12:338–342.
Rosin, R., and A. Shulov. 1961. Sound production in scorpions. Science 133:1918–1919.
Ross, D. M., and C. F. A. Pantin. 1940. Factors influencing facilitation in Actinozoa; the action of certain ions. J. Exp. Biol. 17:61–73.
Ross, H. H. 1964. Evolution of caddisworm cases and nets. Amer. Zool. 4:209–220.
Rossi, P. J. 1968. Adaptation and negative aftereffect to lateral optical displacement in newly hatched chicks. Science 160:430–432.
Roth, M., L. M. Roth, and T. E. Eisner. 1966. The allure of the female mosquito. Natur. Hist. 75:27–31.
Rothenbuhler, W. C. 1964. Behavior genetics of nest cleaning in honey bees. IV. Re-

sponse of F_1 and backcross generations to disease-killed brood. Amer. Zool. 4:111–123.

Rothenbuhler, W. C. 1967. Genetic and evolutionary considerations of social behavior of honeybees and some related insects, pp. 61–106. *In* J. Hirsch, ed., Behavior-genetic analysis. McGraw-Hill, New York.

Rothstein, S. I. 1971. Observation and experiment in the analysis of interactions between brood parasites and their hosts. Amer. Natur. 105:71–74.

Rovner, J. S. 1967. Acoustic communication in a lycosid spider (*Lycosa rabida Walckenaer*). Anim. Behav. 15:273–281.

Rowan, M. K. 1966. Territory as a density-regulating mechanism in some South African birds. Ostrich, Suppl. No. 6:397–408.

Rowell, C. H. F. 1961. Displacement grooming in the chaffinch. Anim. Behav. 9:38–63.

Rowell, T. E. 1966. Forest living baboons in Uganda. J. Zool., London 149:344–364.

Rowell, T. E. 1967*a*. A quantitative comparison of the behaviour of a wild and a caged baboon group. Anim. Behav. 5:499–509.

Rowell, T. E. 1967*b*. Variability in the social organization of primates, pp. 219–235. *In* D. Morris, ed., Primate ethology. Aldine, Chicago.

Rowell, T. E. 1972. Social behaviour of monkeys. Penguin Books, Harmondsworth, Middlesex, England.

Rowell, T. E., R. A. Hinde, and Y. Spencer-Booth, 1964. Aunt-infant interactions in captive rhesus monkeys. Anim. Behav. 12:219–226.

Rowley, I. 1965*a*. The life history of the superb blue wren *Malurus cyaneus*. Emu 64:251–297.

Rowley, I. 1965*b*. White-winged choughs. Austral. Natur. Hist. 15:81–85.

Rowley, I. 1968. Communal species of Australian birds. Bonn. zool. Beitr. 19:362–370.

Rudinsky, J. A. 1969. Masking of the aggregation pheromone in *Dendroctonus pseudotsugae* Hopk. Science 166:884–885.

Rumbaugh, D. M., and C. McCormack. 1967. The learning skills of primates, a comparative study of apes and monkeys, pp. 289–306. *In* D. Starck, R. Schneider, and H. J. Kuhn, eds. Neue Ergebnisse der Primatologie. Progress in Primatology. Gustav Fischer Verlag, Stuttgart.

Ryszkowski, L. 1966. The space organization of nutria (*Myocaster coypus*) populations. Symp. Zool. Soc. London 18:259–265.

Sabine, W. S. 1949. Dominance in winter flocks of juncos and tree sparrows. Physiol. Zool. 22:68–85.

Sabine, W. S. 1959. The winter society of the Oregon junco: intolerance, dominance, and the pecking order. Condor 61:110–135.

Sachs, M. B. 1964. Responses to acoustic stimuli from single units in the eighth nerve of the green frog. J. Acoust. Soc. Amer. 36:1956–1958.

Sackett, G. P. 1966. Monkeys reared in isolation with pictures as visual input: evidence for an innate releasing mechanism. Science 154:1468–1473.

Sade, D. S. 1967. Determinants of dominance in a group of free-ranging rhesus monkeys, pp. 99–114. *In* S. A. Altmann, ed., Social communication among primates. Univ. Chicago Press, Chicago.

Salanki, J., and I. Varanka. 1972. Central determination of the rhythmic adductor activity in the fresh-water mussel *Anodonta cygnea* L., Pelycypoda. Comp. Biochem. Physiol. 41A:465–474.

Sales, G. D. 1972*a*. Ultrasound and aggressive behaviour in rats and other small mammals. Anim. Behav. 20:88–100.

Sales, G. D. 1972*b*. Ultrasound and mating behaviour in rodents with some observations on other behavioural situations. J. Zool., London 168:149–164.

Salmon, M. 1965. Waving display and sound production in the courtship behavior of *Uca pugilator*, with comparisons to *U. minax* and *U. pugnax*. Zoologica 50:123–149.

Salmon, M. 1967. Coastal distribution, display and sound production by Florida fiddler crabs (genus *Uca*). Anim. Behav. 15:449–459.

Salmon, M., and S. P. Atsaides. 1968. Visual and acoustical signalling during courtship by fiddler crabs (genus *Uca*). Amer. Zool. 8:623–639.

Salmon, M., and J. F. Stout. 1962. Sexual discrimination and sound production in *Uca pugilator* (Bosc.). Zoologica 47:15–20.

Salomonsen, F. 1967. Migratory movements of the arctic tern (*Sterna paradisaea* Pontoppidan) in the southern ocean. Biol. Medd. Dansk. Vidensk. Selskab 24:1–42.

Sargeant, A. B. 1972. Red fox spatial characteristics in relation to waterfowl predation. J. Wildlife Mgmt. 36:225–236.

Sargent, T. D. 1966. Background selections of geometrid and noctuid moths. Science 154:1674–1675.

Sargent, T. D. 1968. Cryptic moths: effects on background selections of painting the circumocular scales. Science 159:100–101.

Sauer, E. G. F. 1957. Die Sternen orientierung nächtlich ziehender Grasmücken (*Sylvia atricapilla, borin* und *curruca*). Z. Tierpsychol. 14:29–70.

Sauer, E. G. F. 1961. Further studies on the stellar orientation of nocturnally migrating birds. Psychol. Forschung. 26:224–244.

Sauer, E. G. F. 1971. Celestial rotation and stellar orientation in migratory warblers. Science 173:459–460.

Sauer, E. G. F. 1972. Review: "Grundriss der Vogelzugskunde. 2nd Ed." Wilson Bull. 84:353–355.

Sauer, E. G. F., and E. M. Sauer. 1955. Zur Frage der nächtlichen Zugorientierung von Grasmücken. Rev. Suisse Zool. 62:250–259.

Sauer, E. G. F., and E. M. Sauer. 1959. Nächtliche Zugorientierung europäischer Vögel in Südwestafrika. Vogelwarte 20:4–31.

Sauer, E. G. F., and E. M. Sauer. 1960. Star navigation of nocturnal migrating birds; the 1958 planetarium experiments. Cold Spring Harbor Symp. Quant. Biol. 25:463–473.

Schaller, G. B. 1963. The mountain gorilla; ecology and behavior. Univ. Chicago Press, Chicago.

Schaller, G. B. 1965. The behavior of the mountain gorilla, pp. 324–367. *In* I. DeVore, ed., Primate behavior. Holt, Rinehart and Winston, New York.

Schaller, G. B. 1972. The Serengeti lion. Univ. Chicago Press, Chicago.

Schaller, G. B., and J. T. Emlen. 1962. The ontogeny of avoidance behaviour in some precocial birds. Anim. Behav. 10:370–381.

Scharloo, W. 1971. Reproductive isolation by disruptive selection: did it occur? Amer. Natur. 105:83–86.

Schein, M. W. 1963. On the irreversibility of imprinting. Z. Tierpsychol. 20:462–467.

Schein, M. W., and E. B. Hale. 1959. The effect of early social experience on male sexual behaviour of androgen injected turkeys. Anim. Behav. 7:189–200.

Schenkel, R. 1956. Zur Deutung der Balzleistungen einiger Phasianiden und Tetraoniden. Ornithol. Beobacht. 53:182–201.

Schenkel, R. 1958. Zur Deutung der Balzleistungen einiger Phasianiden und Tetraoniden. Ornithol. Beobacht. 55:65–95.

Schiermann, G. 1926. Beitrag zur Schädigung der Wirtsvögel durch *Cuculus canorus*. Beitrage Fortpfl. Vög. 2:28–30.

Schildkraut, J. J., and S. S. Kety. 1967. Biogenic amines and emotion. Science 156:21–30.

Schjelderup-Ebbe, T. 1922. Beiträge zur Sozialpsychologie des Haushuhns. Z. Psychol. 88:225–252.

Schjelderup-Ebbe, T. 1935. Social behavior of birds, pp. 947–973. *In* C. Murchison, ed., Handbook of social psychology. Clark Univ. Press, Worcester, Mass.

Schleidt, W. M. 1964. Über die Spontaneität von Erbkoordinationen. Z. Tierpsychol. 21:235–256.

Schleidt, W. M., M. Schleidt, and M. Magg. 1960. Störung der Mutter-Kind-Beziehung bei Truthühnern durch Gehörverlust. Behaviour 16:254–260.

Schlichter, D. von. 1968. Das Zusammenleben von Riffanemonen und Anemonenfischen. Z. Tierpsychol. 25:933–954.

Schmidt, R. S. 1964. *Apicotermes* nests. Amer. Zool. 4:221–225.

Schneider, D. 1966. Chemical sense communication in insects. Symp. Soc. Exp. Biol. 20:273–297.

Schnitzler, H.-U. 1970. Echoortung bei der Ortungslaute der Hufeisen-Fledermäuse (Chiroptera–Rhinolophidae) in verschiedenen Orientierungssituationen. Z. vergl. Physiol. 68:25–38.

Schoener, T. W. 1967. The ecological significance of sexual dimorphism in size in the lizard *Anolis conspersus*. Science 155:474–476.

Schoener, T. W. 1968. Sizes of feeding territories among birds. Ecology 49:123–141.

Schoener, T. W. 1969. Models of optimal size for solitary predators. Amer. Natur. 103: 277–314.

Schoener, T. W. 1971. Theory of feeding strategies. Annu. Rev. Ecol. Syst. 2:369–404.

Schöne, H. 1968. Agonistic and sexual display in aquatic and semiterrestrial brachyuran crabs. Amer. Zool. 8:641–654.

Schöne, H., and H. Schöne. 1963. Balz und andere Verhaltensweisen der Mangrovekrabbe, *Goniopsis cruentata*, und das Winkverhalten der eulitoralen Brachyuren. Z. Tierpsychol. 20:641–656.

Schönholzer, L. 1959. Beobachtungen über das Trinkverhalten bei Zootieren. Der Zool. Garten. 24:345–434.

Schubert, G. 1971. Experimentelle Untersuchungen über die artkennzeichnenden Parameter im Gesang des Zilpzalps, *Phylloscopus c. collybita* (Veillot). Behaviour 38:289–314.

Schubert, M. 1971. Untersuchungen über die reaktionsauslösenden Signalstruckturen des Fitis-gesanges, *Phylloscopus t. trochilus* (L.) und das Verhalten gegenüber arteigenen Rufen. Behaviour 38:250–288.

Schultze-Westrum, T. 1965. Innerartliche Verständigung durch Düfte beim Gleitbeutler *Petaurus breviceps papuanus* Thomas (Marsupialia, Phalangeridae). Z. vergl. Physiol. 50:151–220.

Schultz, R. J. 1961. Reproductive mechanism of unisexual and bisexual strains of the viviparous fish *Poeciliopsis*. Evolution 15:302–325.

Schutz, F. 1965. Sexuelle Prägung bei Anatiden. Z. Tierpsychol. 22:50–103.

Schutz, F. 1971. Prägung des Sexualverhaltens von Enten und Gänsen durch Sozialeindrücke während der Jugendphase. J. Neuro-visc. Rel., Suppl. 10:339–357.

Schüz, E. 1949. Die Spät-Auflassung ostpreussischer Jungstörche in West-Deutschland durch die Vogelwarte Rossitten 1933. Vogelwarte 15:63–78.

Schüz, E. 1950. Zur Frage der angeborenen Zugwege. Vogelwarte 15:219–226.

Schüz, E. 1963. On the northwestern migration divide of the white stork. Proc. Int. Ornithol. Congr. 13:475–480.

Schüz, E. 1971. Grundriss der Vogelzugskunde. Verlag Paul Parey, Berlin.

Schwartz, R. L., and J. L. Zimmerman. 1971. The time and energy budget of the male dickcissel (*Spiza americana*). Condor 73:65–76.

Schwink, I. 1954. Experimentelle Untersuchungen über Geruchsinn und Strömungswahrnehmung in der Orientierung bei Nachtschmetterlingen. Z. vergl. Physiol. 37:19–56.

Scott, J. P. 1962. Critical periods in behavioral development. Science 138:949–958.

Scott, J. P. 1967. The evolution of social behavior in dogs and wolves. Amer. Zool. 7:353–355.

Scott, J. P. 1968. Evolution and domestication of the dog, pp. 243–275. *In* T. Dobzhansky, M. K. Hecht, and W. C. Steere, eds., Evolutionary biology, vol. 2. Appleton-Century-Crofts, New York.

Scott, J. P., and E. Fredericson. 1951. The causes of fighting in mice and rats. Physiol. Zool. 24:273–309.

Scott, J. P., and J. L. Fuller. 1965. Genetics and the social behavior of the dog. Univ. Chicago Press, Chicago.

Scott, J. W. 1942. Mating behavior of the sage grouse. Auk 59:477–498.

Scott, P. 1951. Key to the wildfowl of the world. Severn Wildfowl Trust, Slimbridge, Gloucestershire, England.

Scott, P. 1970. The wild swans at Slimbridge. Wildfowl Trust, Severn, England.

Scudo, F. 1967. L'accoppiamento assortativo basato sul fenotipo di parenti; alcune conseguenze in popolazioni. Inst. Lombardo (Rend. Sc.) B 101:435–455.

Sebeok, T. A. 1968. Animal communication. Indiana Univ. Press, Bloomington.

Seiger, M. B. 1967. A computer simulation study of the influence of imprinting on population structure. Amer. Natur. 101:47–57.

Seiger, M. B., and R. D. Dixon. 1970. A computer simulation study of the effects of two behavioral traits on the genetic structure of semi-isolated populations. Evolution 24:90–97.

Seilacher, A. 1964. Biogenic sedimentary structures, pp. 296–313. *In* J. Imbrie and N. D. Newell, eds., Approaches to paleoecology. Wiley, New York.

Seilacher, A. 1967. Fossil behavior. Sci. Amer. 217:72–80.

Selander, R. K. 1958. Age determination and molt in the boat-tailed grackle. Condor 60:355–376.

Selander, R. K. 1965. On mating systems and sexual selection. Amer. Natur. 99:129–140.

Selander, R. K. 1970. Behavior and genetic variation in natural populations. Amer. Zool. 10:53–66.

Selander, R. K. 1972. Sexual selection and dimorphism in birds, pp. 180–230. *In* Campbell 1972.

Seliger, H. H., J. B. Buck, W. G. Fastie, and W. D. McElroy, 1964. Flash patterns in Jamaican fireflies. Biol. Bull. 127:159–172.

Sexton, O. J., and H. D. Stalker. 1961. Spacing patterns of female *Drosophila paramelanica*. Anim. Behav. 9:77–81.

Sharma, S. C., R. R. Provine, T. T. Sandel, and V. Hamburger. 1970. Unit activity in the isolated spinal cord of the chick embryo, *in situ*. Proc. Nat. Acad. Sci. 66:40–47.

Sharpe, R. S., and P. A. Johnsgard. 1966. Inheritance of behavioral characteristics in F_2 mallard × pintail (*Anas platyrhynchos* L. × *Anas acuta* L.) hybrids. Behaviour 27:259–272.

Sharpless, S. K., and H. H. Jaspar. 1956. Habituation of the arousal reaction. Brain 79:655–669.

Shaw, E. 1970. Schooling in fishes: critique and review, pp. 452–480. *In* L. R. Aronson, E. Tobach, D. S. Lehrman, J. S. Rosenblatt, eds., Development and evolution of behavior. Freeman, San Francisco.

Shaw, K. C. 1968. An analysis of the phonoresponse of males of the true katydid, *Pterophylla camellifolia* (Fabricius) (Orthoptera, Tettigonidae). Behaviour 31:203–260.

Shelford, V. E. 1943. The abundance of the collared lemming (*Dicrostonyx groenlandicus* (Tr.) var. *richardsoni* Mer.) in the Churchill area, 1929 to 1940. J. Anim. Ecol. 24:472–484.

Sherrington, C. 1906. The integrative action of the nervous system. Yale Univ. Press, New Haven. Revised edition 1947, Cambridge Univ. Press, Cambridge, England. Paperback ed., 1961, Yale Univ. Press.

Sherwood, G. A. 1967. Behavior of family groups of Canada geese. Trans. 32nd N. Amer. Wildlife & Natur. Resources Conf., pp. 340–355.

Shiovitz, K. A., and W. L. Thompson. 1970. Geographic variation in song composition of the indigo bunting, *Passerina cyanea*. Anim. Behav. 18:151–158.

Shorey, H. H., and R. J. Bartell. 1970. Role of a volatile female sex pheromone in stimulating male courtship behaviour in *Drosophila melanogaster*. Anim. Behav. 18:159–164.

Short, L. L., Jr. 1969. "Isolating mechanisms" in the blue-winged warbler–golden-winged warbler complex. Evolution 23:355–356.

Sibley, C. G. 1957. The evolutionary and taxonomic significance of sexual dimorphism and hybridization in birds. Condor 59:166–191.

Sidman, R. L., M. C. Green, and S. H. Appel. 1965. Catalog of the neurological mutants of the mouse. Harvard Univ. Press, Cambridge, Mass.

Siegel, P. B., R. E. Phillips, and E. F. Folsom. 1965. Genetic variation in the crow of adult chickens. Behaviour 24:229–235.

Simmons, J. A. 1971. The sonar receiver of the bat. Ann. N.Y. Acad. Sci. 188:161–174.

Simmons, J. A., E. G. Wever, and J. M. Pylka. 1971. Periodical cicada: sound production and hearing. Science 171:212–213.

Simmons, K. E. L. 1957. The taxonomic significance of the head-scratching methods of birds. Ibis 99:178–181.

Simmons, K. E. L. 1961. Problems of head-scratching in birds. Ibis 103*a*:37–49.

Simonds, P. E. 1965. The bonnet macaque in South India, pp. 175–196. *In* I. DeVore, ed., Primate behavior: field studies of monkeys and apes. Holt, Rinehart and Winston, New York.

Skutch, A. F. 1935. Helpers at the nest. Auk 52:257–273.

Skutch, A. F. 1961. Helpers among birds. Condor 63:198–226.

Skutch, A. F. 1969. A study of the rufous-fronted thornbird and associated birds. II. Birds which breed in the thornbird's nests. Wilson Bull. 81:123–139.

Sladen, F. W. L. 1912. The humble-bee. Macmillan, London.

Smith, C. C. 1968. The adaptive nature of social organization in the genus of tree squirrels *Tamiasciurus*. Ecol. Monogr. 38:31–63.

Smith, D. G. 1972*a*. The red badge of rivalry. Natur. Hist. 81:44–51.

Smith, D. G. 1972*b*. The role of the epaulets in the red-winged blackbird (*Agelaius phoeniceus*) social system. Behaviour 41:251–268.

Smith, J. M. 1964. Group selection and kin selection. Nature 201:1145–1147.

Smith, J. M. 1965. The evolution of alarm calls. Amer. Natur. 94:59–63.

Smith, J. M. 1966. Sympatric speciation. Amer. Natur. 100:637–650.

Smith, L. A. 1928. A comparison of the number of nerve cells in the olfactory bulbs of domesticated albino and wild Norway rats. J. Comp. Neurol. 45:483–499.

Smith, N. G. 1966. Evolution of some arctic gulls (*Larus*): an experimental study of isolating mechanisms. Ornithol. Monogr. 4:1–99.

Smith, N. G. 1968. The advantage of being parasitized. Nature 24:690–694.

Smith, N. G. 1969. Provoked release of mobbing—a hunting technique of *Micrastur* falcons. Ibis 111:241–243.

Snow, B. K., and D. W. Snow. 1968. Behavior of the swallow-tailed gull of the Galapagos. Condor 70:252–264.

Snow, D. W. 1958. A study of blackbirds. Allen & Unwin, London.
Snow, D. W. 1963. The display of the blue-backed manakin, *Chiroxiphia pareola*, in Tobago, W.I. Zoologica 48:167–176.
Snyder, N. F. R. 1967. An alarm reaction of aquatic gastropods to intraspecific extract. Cornell Univ. Agr. Exp. Sta. Memoir 403:1–122.
Snyder, N. F. R., and H. Snyder. 1970. Alarm response of *Diadema antillarum*. Science 168:276–278.
Snyder, R. L. 1968. Reproduction and population pressures. Progr. Physiol. Psych. 2:119–160. Stellar, E., and J. M. Sprague, eds., Academic Press, New York.
Snyder, W. F., and D. L. Jameson. 1965. Multivariate geographic variation of mating call in populations of the Pacific tree frog (*Hyla regilla*). Copeia 1965:129–142.
Sokal, R. R., and T. J. Crovello. 1970. The biological species concept: a critical evaluation. Amer. Natur. 104:127–153.
Southern, H. N. 1954. Mimicry in cuckoos' eggs, pp. 219–232. *In* J. Huxley, ed., Evolution as a process. Allen & Unwin, London.
Southern, H. N. 1966. Distribution of bridled guillemots in east Scotland over eight years. J. Anim. Ecol. 35:1–11.
Southern, H. N., and V. P. W. Lowe. 1968. The pattern of distribution of prey and predation in tawny owl territories. J. Anim. Ecol. 37:75–97.
Southwick, C. H., M. A. Beg, and M. R. Siddiqi. 1965. Rhesus monkeys in North India, pp. 111–159. *In* I. DeVore, ed., Primate behavior: field studies of monkeys and apes. Holt, Rinehart and Winston, New York.
Sower, L. L., L. H. Gaston, and H. H. Shorey. 1971. Sex pheromones of noctuid moths. XXVI. Female release rate, male response threshold, and communication distance for *Trichoplusia ni*. Ann. Entomol. Soc. Amer. 64:1448–1456.
Spalding, D. A. 1873. Instinct with original observations on young animals. MacMillan's Magazine 27:282–293, reprinted in Brit. J. Anim. Behav. 2:2–11.
Spiess, E. B. 1968. Low frequency advantage in mating of *Drosophila pseudoobscura* karyotypes. Amer. Natur. 102:363–379.
Spiess, E. B. 1970. Mating propensity and its genetic basis in *Drosophila*, pp. 315–379. *In* M. K. Hecht and W. C. Steere, eds., Essays in evolution and genetics in honor of Theodosius Dobzhansky. Appleton-Century-Crofts, New York.
Spieth, H. T. 1952. Mating behavior within the genus *Drosophila* (Diptera). Bull. Amer. Mus. Natur. Hist. 99:401–474.
Spieth, H. T. 1968. Evolutionary implications of sexual behavior in *Drosophila*. Evolut. Biol. 2:157–193.
Springer, V. G., and W. F. Smith-Vaniz. 1972. Mimetic relationships involving fishes of the family Blenniidae. Smithsonian Contrib. Zool. No. 112:1–36.
Stärk, A. 1958. Untersuchungen am Lautorgan einiger Grillen und Laubheuschreckenarten zugleich ein Beitrag zum Rechts-Links Problem. Zool. Jahrb. 77:9–50.
Stebbins, G. L., Jr. 1950. Variation and evolution in plants. Columbia Univ. Press, New York.
Steele, R. G., and M. H. A. Keenleyside. 1971. Mate selection in two species of sunfish (*Lepomis gibbosus* and *L. megalotis peltastes*). Can. J. Zool. 49:1541–1548.
Stefanski, R. A. 1967. Utilization of the breeding territory in the black-capped chickadee. Condor 69:259–267.
Stein, R. C. 1956. A comparative study of "advertising song" in the *Hylocichla* thrushes. Auk 73:503–512.
Stein, R. C. 1963. Isolating mechanisms between populations of Traill's flycatchers. Proc. Amer. Phil. Soc. 107:21–50.
Stenger, J. 1958. Food habits and available food of ovenbirds in relation to territory size.

Auk 75:335–346.

Stenger, J., and J. B. Falls. 1959. The utilized territory of the ovenbird. Wilson Bull. 71:125–140.

Sterman, M. B., D. J. McGinity, and A. M. Adinolfi, eds. 1971. Brain development and behavior. Academic Press, New York.

Stevenson, J. G., R. G. Hutchison, J. Hutchison, B. C. R. Bertram, and W. H. Thorpe. 1970. Individual recognition by auditory cues in the common tern (*Sterna hirundo*). Nature 226:562–563.

Stewart, R. E., and J. W. Aldrich. 1951. Removal and repopulation of breeding birds in a spruce-fir forest community. Auk 68:471–482.

Stickel, L. F. 1946. The source of animals moving into a depopulated area. J. Mammal. 27:301–307.

Stickel, L. F. 1968. Home range and travels, pp. 373–411. *In* J. A. King, ed., Biology of *Peromyscus* (Rodentia). Amer. Soc. Mammal. Spec. Pub. No. 2.

Stiles, F. G. 1971. Time, energy, and territoriality of the Anna hummingbird (*Calypte anna*). Science 173:818–821.

Stokes, A. W., and H. W. Williams. 1972. Courtship feeding calls in gallinaceous birds. Auk 89:177–180.

Storer, R. W. 1963. Courtship and mating behavior and the phylogeny of grebes. Proc. Int. Ornithol. Congr. 13:562–569.

Storer, R. W. 1966. Sexual dimorphism and food habits in three North American accipiters. Auk 83:423–436.

Stout, J. F. 1963. The significance of sound production during the reproductive behaviour of *Notropis analostanus* (family Cyprinidae). Anim. Behav. 11:83–92.

Stout, J. F., and F. Huber. 1971. Responses of central auditory neurons of female crickets (*Gryllus campestris* L.) to the calling song of the male. Z. vergl. Physiol. 76:302–313.

Strecker, R. L. 1954. Regulatory mechanisms in house-mouse populations: the effect of limited food supply on an unconfined population. Ecology 35:249–253.

Stresemann, E. 1953. Spielplätze und Balz der Laubenvögel. J. Ornithol. 94:367–368.

Struhsaker, T. T. 1967*a*. Social structure among vervet monkeys (*Cercopithecus aethiops*). Behaviour 29:83–121.

Struhsaker, T. T. 1967*b*. Ecology of vervet monkeys (*Cercopithecus aethiops*) in the Masai-Amboseli game reserve, Kenya. Ecology 48:891–904.

Struhsaker, T. T. 1967*c*. Auditory communication among vervet monkeys (*Cercopithecus aethiops*), pp. 281–324. *In* S. A. Altmann, ed., Social communication among primates. Univ. Chicago Press, Chicago.

Sudd, J. H. 1967. An introduction to the behavior of ants. St. Martin's Press, New York.

Suga, N., and P. Schlegel. 1972. Neural attenuation of responses to emitted sounds in echolocating bats. Science 177:82–84.

Tamarin, R. H., and C. J. Krebs. 1969. *Microtus* population biology. II. Genetic changes at the transferrin locus in fluctuating populations of two vole species. Evolution 23:183–211.

Tan, C. C. 1946. Genetics of sexual isolation between *Drosophila pseudoobscura* and *Drosophila persimilis*. Genetics 31:558–573.

Tavolga, W. N. 1960. Sound production and underwater communication in fishes, pp. 93–136. *In* W. E. Lanyon and W. M. Tavolga, eds., Animal sounds and communications. Amer. Inst. Biol. Sci. Pub. No. 7, with a phonograph record.

Tavolga, W. N. 1964. Symposium on marine bio-acoustics, Lerner Marine Lab. 1963. Pergamon Press, New York.

Tavolga, W. N. 1967. Symposium on marine bio-acoustics, 2nd, Amer. Mus. Natur. Hist., 1966. Pergamon Press, New York.

Tavolga, W. N. 1968. Fishes, pp. 271–288. *In* T. A. Sebeok, ed., Animal communication. Indiana Univ. Press, Bloomington.

Tavolga, W. N., and J. Wodinsky. 1963. Auditory capacities in fishes; pure tone thresholds in nine species of marine teleosts. Bull. Amer. Mus. Natur. Hist. 126:179–239.

Taylor, D. H., and D. E. Ferguson. 1970. Extraoptic celestial orientation in the southern cricket frog *Acris gryllus*. Science 168:390–392.

Teleki, G. 1973. The predatory behavior of wild chimpanzees. Bucknell Univ. Press, Lewisburg, Pa.

Tembrock, G. 1963. Acoustic behaviour of mammals, pp. 751–786. *In* R. G. Busnel, ed., Acoustic behaviour of animals. Elsevier, Amsterdam.

Tenaza, R. 1971. Behavior and nesting success relative to nest location in Adélie penguins (*Pygoscelis adeliae*). Condor 73:81–92.

Terman, C. R. 1966. Population fluctuation of *Peromyscus maniculatus* and other small mammals as revealed by the North American Census of Small Mammals. Amer. Midland Natur. 76:419–426.

Terman, C. R. 1968. Population dynamics. *In* J. A. King, ed., Biology of *Peromyscus* (Rodentia). Amer. Soc. Mammal. Spec. Pub. 2:412–450.

Thielcke, G. 1961. Stammesgeschichte und geographische Variation des Gesanges unserer Baumläufer (*Certhia familiaris* L. und *Certhia brachydactyla* Brehm). Z. Tierpsychol. 18:188–204.

Thielcke, G. 1962. Versuche mit Klangattrappen zur Klärung der Verwandtschaft der Baumläufer *Certhia familiaris* L., *C. brachydactyla* Brehm und *C. americana* Bonaparte. J. Ornithol. 103:266–271.

Thielcke, G. 1965. Gesangsgeographische Variation des Gartenbaumlaüfers (*Certhia brachydactyla*) im Hinblick auf das Artbildungsproblem. Z. Tierpsychol. 22:542–566.

Thielcke, G. 1969. Geographic variation in bird vocalizations, pp. 311–339. *In* R. A. Hinde, ed., Bird vocalizations. Cambridge Univ. Press, London, England.

Thielcke, G. 1970. Die sozialen Funktionen der Vogelstimmen. Vogelwarte 25:204–229.

Thiessen, D. D. 1964. Population density and behavior: a review of theoretical and physiological contributions. Texas Rep. Biol. Med. 22:266–314.

Thiessen, D. D. 1971. Gene organization and behavior. Random House, New York.

Thoday, J. M. 1959. Effects of disruptive selection. I. Genetic flexibility. Heredity 13:187–203.

Thoday, J. M. 1965. Effects of selection for genetic diversity. Genetics Today 3:533–540. Pergamon Press, Oxford.

Thoday, J. M. 1972. Disruptive selection. Proc. Roy. Soc. London (B) 182:109–143.

Thoday, J. M., and J. B. Gibson. 1962. Isolation by disruptive selection. Nature 193:1164–1166.

Thompson, D. H., and J. T. Emlen. 1969. Parent-chick individual recognition in the adelie penguin. Antarct. J. 3:132.

Thompson, T., and T. Sturm. 1965. Classical conditioning of aggressive display in Siamese fighting fish. J. Exp. Anal. Behav. 8:397–403.

Thompson, W. L. 1960. Agonistic behavior in the house finch. II. Factors in aggressiveness and sociality. Condor 62:378–402.

Thompson, W. L. 1970. Song variation in a population of indigo buntings. Auk 87:58–71.

Thorpe, W. H. 1958. The learning of song patterns by birds, with especial reference to the song of the chaffinch *Fringilla coelebs*. Ibis 100:535–570.

Thorpe, W. H. 1961. Bird-song: the biology of vocal communication and expression in birds. Cambridge Monogr. Exp. Biol. 12:1–143. Cambridge Univ. Press, London.

Thorpe, W. H. 1963. Antiphonal singing in birds as evidence for avian auditory reaction time. Nature 197:774–776.

Thorpe, W. H. 1972. Duetting and antiphonal song in birds. Behav. Suppl. 18:1–197.

Thorpe, W. H., and M. E. W. North. 1965. Origin and significance of the power of vocal imitation: with special reference to the antiphonal singing of birds. Nature 208:219–222.

Tinbergen, L. 1960. The natural control of insects in pinewoods. I. Factors influencing the intensity of predation by songbirds. Arch. Neerl. Zool. 13:265–336.

Tinbergen, N. 1940. Die Übersprungbewegung. Z. Tierpsychol. 4:1–40.

Tinbergen, N. 1948. Social releasers and the experimental method required for their study. Wilson Bull. 60:6–52.

Tinbergen, N. 1951. The study of instinct. Oxford Univ. Press, Oxford.

Tinbergen, N. 1952. "Derived" activities; their causation, biological significance, origin, and emancipation during evolution. Quart. Rev. Biol. 27:1–32.

Tinbergen, N. 1953. Social behaviour in animals. Methuen, London.

Tinbergen, N. 1959. Comparative studies of the behavior of gulls (Laridae): a progress report. Behaviour 15:1–70.

Tinbergen, N. 1962. The evolution of animal communication—a critical examination of methods. Symp. Zool. Soc. London 8:1–6.

Tinbergen, N., M. Impekoven, and D. Franck. 1967. An experiment on spacing-out as a defence against predation. Behaviour 28:307–321.

Tinbergen, N., and D. J. Kuenen. 1939. Über die auslösenden und richtungsgebenden Reizsituationen der Sperrbewegung junger Drosseln. Z. Tierpsychol. 3:37–60.

Tinbergen, N., and M. Moynihan. 1952. Head flagging in the black-headed gull; its function and origin. Brit. Birds 45:19–22.

Tinkle, D. W. 1965. Home range, density, dynamics, and structure of a Texas population of the lizard *Uta stansburiana*, pp. 5–29. *In* W. W. Milstead, ed., Lizard ecology. Univ. Missouri Press, Columbia.

Tinkle, D. W. 1967. The life and demography of the side-blotched lizard, *Uta stansburiana*. Misc. Pub. Mus. Zool. Univ. Michigan No. 132:1–182.

Tinkle, D. W. 1969. Evolutionary implications of comparative population studies in the lizard *Uta stansburiana*, pp. 133–154. *In* Int. Conf. Syst. Biol., Univ. Michigan 1967, Systematic biology. National Academy of Sciences, Washington, D.C.

Tischner, H., and A. Schief. 1954. Fluggeräusch und Schallwahrnehmung bei *Aedes aegypti* (Culicidae). Verh. Deut. Zool. Ges. 51:453–460.

Tobach, E., L. R. Aronson, and E. Shaw, eds. 1971. The biopsychology of development. Academic Press, New York.

Tokuda, K., and G. D. Jensen. 1968. The leader's role in controlling aggressive behavior in a monkey group. Primates 9:319–322.

Treat, A. E. 1955. The response to sound in certain Lepidoptera. Ann. Entomol. Soc. Amer. 48:272–284.

Treat, A. E., and J. S. Vandeberg. 1968. Auditory sense in certain sphingid moths. Science 159:331–333.

Tretzel, E. 1965. Artkennzeichnende und reaktionsauslösende Komponenten im Gesang der Heidelerche (*Lullula arborea*). Verh. Deut. Zool. Ges. Jena. 1965:367–380.

Trivers, R. L. 1971. The evolution of reciprocal altruism. Quart. Rev. Biol. 46:35–57.

Trivers, R. L. 1974. Parent offspring conflict. Amer. Zool. 14:249–264.

Tschanz, B. 1965. Beobachtungen und Experimente zur Entstehung der "personlichen" Beziehung zwischen Jungvogel und Eltern bei Trottellummen. Verh. Schweiz. Naturforsch. Ges. 1964:211–216.

Tschanz, B. 1968. Trottellummen. Z. Tierpsychol. Suppl. 4.
Turner, F. B., R. I. Jennrich, and J. D. Weintraub. 1969. Home ranges and body size of lizards. Ecology 50:1076–1081.
Uexküll, J. J. von. 1909. Umwelt und Innenwelt der Tiere. Springer, Berlin.
Underwood, H., and M. Menaker. 1970. Extraretinal light perception: entrainment of the biological clock controlling lizard locomotor activity. Science 170:190–192.
Valenstein, E. S. 1968. Biology of drives. Neurosci. Res. Program 6:1–111.
Valverde, J. A. 1959. Moyens d'expression et hiérarchie sociale chez le vautour fauve *Gyps fulvus* (Hablizl). Alauda 27:1–15.
Vandenbergh, J. G. 1967. The development of social structure in free-ranging rhesus monkeys. Behaviour 29:179–194.
Van Hooff, J. A. R. A. M. 1967. The facial displays of the catarrhine monkeys and apes, pp. 7–68. *In* D. Morris, ed., Primate ethology. Aldine, Chicago.
Van Lawick-Goodall, H., and J. Van Lawick-Goodall. 1971. Innocent killers. Houghton Mifflin, Boston.
Van Tets, G. F. 1965. A comparative study of some social communication patterns in the Pelecaniformes. Ornithol. Monogr. 2:1–88.
Van Tyne, J., and A. J. Berger. 1959. Fundamentals of ornithology. Wiley, New York.
Van Vleck, D. B. 1968. Movements of *Microtus pennsylvanicus* in relation to depopulated areas. J. Mammal. 49:92–103.
Verbeek, N. A. M. 1964. A time and energy budget study of the Brewer blackbird. Condor 66:70–74.
Verbeek, N. A. M. 1972. Daily and annual time budget of the yellow-billed magpie. Auk 89:567–582.
Vereschchagin, S. M., V. P. Lapitskii, and V. P. Tyshehenko. 1971. Functional characteristics of neurons in the central nervous system of *Gryllus domesticus*. Neurosci. Transl. 15:89–98.
Verner, J. 1963. Song rates and polygamy in the long-billed marsh wren. Proc. Ornithol. Congr., Ithaca 13:299–307.
Verner, J. 1964. Evolution of polygyny in the long-billed marsh wren. Evolution 18:252–261.
Verner, J. 1965*a*. Breeding biology of the long-billed marsh wren. Condor 67:6–30.
Verner, J. 1965*b*. Time budget of the male long-billed marsh wren during the breeding season. Condor 67:125–137.
Verner, J., and G. H. Engelsen. 1970. Territories, multiple nest building, and polygyny in the long-billed marsh wren. Auk 87:557–567.
Verner, J., and M. F. Willson. 1966. The influence of habitats on mating systems of North American passerine birds. Ecology 47:143–147.
Verner, J., and M. F. Willson. 1969. Mating systems, sexual dimorphism, and the role of male North American passerine birds in the nesting cycle. Ornithol. Monogr. 9:1–76.
Vince, M. A. 1969. Embryonic communication, respiration and the synchronization of hatching, pp. 233–260. *In* R. A. Hinde, ed., Bird vocalizations. Cambridge Univ. Press, London.
Walcott, C., and R. P. Green. 1974. Orientation of homing pigeons altered by a change in the direction of an applied magnetic field. Science 184:180–182.
Waldron, I. 1964. Courtship sound production in two sympatric sibling *Drosophila* species. Science 144:191–193.
Walker, T. J. 1962. Factors responsible for intraspecific variation in the calling songs of crickets. Evolution 16:407–428.
Walker, T. J. 1964. Cryptic species among sound-producing ensiferan Orthoptera (Gryllidae and Tettigoniidae). Quart. Rev. Biol. 30:345–353.

Walker, T. J., and D. Dew. 1972. Wing movements of calling katydids: fiddling finesse. Science 178:174–176.

Wall, W. van de. 1963. Bewegungsstudien an Anatiden. J. Ornithol. 104:1–15.

Wallace, B. 1950. An experiment on sexual isolation. *Drosophila* Info. Service 24:94–96.

Wallace, B. 1954. Genetic divergence of isolated populations of *Drosophila melanogaster*. Proc. 9th Int. Congr. Genet. 1:761–764.

Wallace, B. 1968*a*. Topics in population genetics. Norton, New York.

Wallace, B. 1968*b*. Polymorphism, population size, and genetic load, pp. 87–108. *In* R. C. Lewontin, ed., Population biology and evolution. Syracuse Univ. Press, Syracuse, N.Y.

Walther, F. 1964. Einige Verhaltensbeobachtungen an Thomsongazellen (*Gazella thomsoni* Gunther, 1884) im Ngorongoro-Krater. Z. Tierpsychol. 21:871–890.

Ward, P. 1965. Feeding ecology of the black-faced dioch (*Quelea quelea*) in Nigeria. Ibis 107:173–214.

Warner, R. W. 1972. The anatomy of the syrinx in passerine birds. J. Zool., London 168:381–393.

Warren, J. M. 1965. Primate learning in comparative perspective, pp. 249–281. *In* A. M. Schrier, H. F. Harlow, and F. Stollnitz, eds., Behavior of nonhuman primates. Academic Press, New York.

Warriner, C. C., W. B. Lemmon, and T. S. Ray. 1963. Early experience as a variable in mate selection. Anim. Behav. 11:221–224.

Wasserman, A. O. 1957. Factors affecting interbreeding in sympatric species of spadefoots (*Scaphiopus*). Evolution 11:320–338.

Watson, A. 1965. A population study of ptarmigan (*Lagopus mutus*) in Scotland. J. Anim. Ecol. 34:135–172.

Watson, A., and D. Jenkins. 1968. Experiments on population control by territorial behaviour in red grouse. J. Anim. Ecol. 37:595–614.

Watson, A., and G. R. Miller. 1971. Territory size and aggression in a fluctuating red grouse population. J. Anim. Ecol. 40:367–383.

Watson, A., and R. Moss. 1970. Dominance, spacing behaviour and aggression in relation to population limitation in vertebrates, pp. 167–218. *In* A. Watson, ed., Animal populations in relation to their food resources. Blackwell, Oxford.

Watson, G. F., and A. A. Martin. 1968. Postmating isolation in the *Hyla ewingi* complex (Anura: Hylidae). Evolution 22:664–666.

Watson, M. 1969. Significance of antiphonal song in the eastern whipbird, *Psophodes olivaceus*. Behaviour 35:157–178.

Webster, F. A., and D. R. Griffin. 1962. The role of the flight membranes in insect capture by bats. Anim. Behav. 10:332–340.

Weeden, J. S. 1965. Territorial behavior of the tree sparrow. Condor 67:193–209.

Weeden, J. S., and J. B. Falls. 1959. Differential responses of male ovenbirds to recorded songs of neighboring and more distant individuals. Auk 76:343–351.

Weinberger, N. M. 1971. Attentive processes, pp. 129–198. *In* J. L. McGaugh, ed., Psychobiology. Academic Press, New York.

Weiss, J. M., B. S. McEwen, M. T. A. Silva, and M. F. Kalkut. 1969. Pituitary-adrenal influences on fear responding. Science 163:197–199.

Welch, A. S., and B. L. Welch. 1968. Reduction of norepinephrine in the lower brain stem by psychological stimulus. Proc. Nat. Acad. Sci. 60:478–481.

Welch, B. L. 1965. Psychophysiological response to the mean level of environmental stimulation: a theory of environmental integration, pp. 39–96. *In* Symposium on medical aspects of stress in the military climate. U.S. Govt. Printing Office Pub.

Weller, M. W. 1959. Parasitic laying in the redhead (*Aythya americana*) and other North American Anatidae. Ecol. Monogr. 29:333–365.

Weller, M. W. 1968. The breeding biology of the parasitic black-headed duck. Living Bird 7:169–207.

Wells, M. J. 1962. Brain and behaviour in cephalopods. Stanford Univ. Press, Stanford, Calif.

Wells, P. H., and A. M. Wenner. 1973. Do honey bees have a language? Nature 241:171–175.

Wenner, A. M. 1962. Sound production during the waggle dance of the honey bee. Anim. Behav. 10:79–95.

Wenner, A. M. 1967. Honey bees: do they use the distance information contained in their dance maneuver? Science 155:847–849.

Wenner, A. M. 1968. Honey bees, pp. 217–243. *In* T. A. Sebeok, ed., Animal communication. Indiana Univ. Press, Bloomington.

Wenner, A. M., P. H. Wells, and D. L. Johnson. 1969. Honey bee recruitment to food sources: olfaction or language. Science 164:84–86.

Werblin, F. S. 1972. Lateral interactions at inner plexiform layer of vertebrate retina: antagonistic responses to change. Science 175:1008–1010.

Wetterberg, L., E. Geller, and A. Yuwiler. 1970. Harderian gland: an extraretinal photoreceptor influencing the pineal gland in neonatal rats? Science 167:884–885.

Wheeler, W. M. 1910. Ants: their structure, development and behavior. Columbia Univ. Press, New York.

White, M. J. D. 1954. Animal cytology and evolution. 2nd ed. Cambridge Univ. Press, Cambridge, England.

White, M. J. D. 1968. Models of speciation. Science 159:1065–1070.

White, S. J. 1971. Selective responsiveness by the gannet to played-back calls. Anim. Behav. 19:125–131.

White, S. J., and R. E. C. White. 1970. Individual voice production in gannets. Behaviour 37:40–54.

White, S. J., R. E. C. White, and W. H. Thorpe. 1970. Acoustic basis for individual recognition in the gannet. Nature 225:1156–1158.

Whitman, C. O. 1899. Animal behavior. Biol. Lect. Marine Biol. Lab., Woods Hole, pp. 285–338.

Whitman, C. O. 1919. The behavior of pigeons. Pub. Carnegie Inst. 257:1–161.

Whittaker, R. H., and P. P. Feeny. 1971. Allelochemics: chemical interactions between species. Science 171:757–770.

Whitten, W. K. 1966. Pheromones and mammalian reproduction. Advance. Reprod. Physiol. 1:155–177.

Whitten, W. K., and F. H. Bronson. 1970. The role of pheromones in mammalian reproduction. Advance. Chemoreception 1:309–326.

Wickler, W. 1961. Über die Stammesgeschichte und den taxonomischen Wert einiger Verhaltensweisen der Vögel. Z. Tierpsychol. 18:320–342.

Wickler, W. 1962. Zur Stammesgeschicte funktionelle Korrelierter Organ- und Verhaltensmerkmale: Ei-Attrappen und Maulbruten bei afrikanischer Cichliden. Z. Tierpsychol. 19:129–164.

Wickler, W. 1963. Zum problem der Signalbildung, am Beispiel der Verhaltensmimicry zwischen *Aspidontus* und *Labroides* (Pisces, Acanthopterygii). Z. Tierpsychol. 20:657–679.

Wickler, W. 1965. Die Evolution von Mustern der Zeichnung und des Verhaltens. Naturwissenschaften 52:335–341.

Wickler, W. 1968. Mimicry in plants and animals. McGraw-Hill, New York.

Wiens, J. A. 1966. On group selection and Wynne-Edwards' hypothesis. Amer. Sci. 54:273–287.

Wiepkema, P. R. 1961. An ethological analysis of the reproductive behaviour of the bitterling (*Rhodeus amarus* Bloch). Arch. Neerl. Zool. 14:103–199.

Wiewandt, T. A. 1969. Vocalization, aggressive behavior, and territoriality in the bullfrog, *Rana catesbeiana*. Copeia 1969:276–285.

Wilbur, H. M. 1971. The ecological relationship of the salamander *Ambystoma laterale* to its all-female gynogenetic associate. Evolution 25:168–179.

Williams, G. C. 1964. Measurement of consociation among fishes and comments on the evolution of schooling. Pub. Mus. Michigan State Univ. 2:351–383.

Williams, G. C. 1966. Adaptation and natural selection. Princeton Univ. Press, Princeton, N.J.

Williams, H. W., A. W. Stokes, and J. C. Wallen. 1968. The food call and display of the bobwhite quail (*Colinus virginianus*). Auk 85:464–467.

Willis, E. O. 1967. The behavior of bicolored antbirds. Univ. Calif. Pub. Zool. 79:1–127.

Willis, E. O. 1968. Studies of lunulated and Salvin's antbirds. Condor 70:128–148.

Willson, M. F. 1966. Breeding ecology of the yellow-headed blackbird. Ecol. Monogr. 36:51–77.

Wilson, D. M. 1964. The origin of the flight-motor command in grasshoppers, pp. 331–345. *In* R. F. Reiss, ed., Neural theory and modeling. Stanford Univ. Press, Stanford, Calif.

Wilson, D. M. 1966. Central nervous mechanisms for the generation of rhythmic behavior in arthropods. Soc. Exp. Biol. Symp. 20:199–228.

Wilson, D. M. 1968. The nervous control of insect flight and related behavior, pp. 219–229. *In* J. E. Treherne and J. W. L. Beament, eds., Recent advances in insect physiology. Academic Press, New York.

Wilson, D. M., and R. J. Wyman. 1965. Motor output patterns during random and rhythmic stimulation of locust thoracic ganglia. Biophys. J. 5:121–143.

Wilson, E. O. 1962. Chemical communication among workers of the fire ant, *Solenopsis saevissima* (Fr. Smith). 1–3. Anim. Behav. 10:134–164.

Wilson, E. O. 1963. Social modifications related to rareness in ant species. Evolution 17:249–253.

Wilson, E. O. 1965*a*. The challenge from related species, pp. 7–24. *In* H. G. Baker and G. L. Stebbins, eds., The genetics of colonizing species. Academic Press, New York.

Wilson, E. O. 1965*b*. Chemical communication in the social insects. Science 149:1064–1071.

Wilson, E. O. 1966. Behaviour of social insects. *In* P. T. Haskell, ed., Insect behaviour. Roy. Entomol. Soc. London Symp. 3:81–96.

Wilson, E. O. 1968. Chemical systems, pp. 75–102. *In* T. A. Sebeok, ed., Animal communication. Indiana Univ. Press, Bloomington.

Wilson, E. O. 1970. Chemical communication within animal species, pp. 135–155. *In* E. Sondheimer and J. B. Simeone, eds., Chemical ecology. Academic Press, New York.

Wilson, E. O. 1971. The insect societies. Harvard Univ. Press, Cambridge, Mass.

Wilson, E. O., and W. H. Bossert. 1963. Chemical communication among animals. Recent Progr. Hormone Res. 19:673–716.

Wilson, E. O., and W. H. Bossert. 1971. A primer of population biology. Sinauer, Stamford, Conn.

Wilson, E. O., and F. E. Regnier. 1971. The evolution of the alarm-defense system in the formicine ants. Amer. Natur. 105:279–289.

Wiltschko, W. 1968. Über den Einfluss statischer Magnetfelder auf die Zugorientierung der Rotkehlchen (*Erithacus rubecula*). Z. Tierpsychol. 25:537–559.

Wiltschko, W., and H. Höck. 1972. Orientation behavior of night-migrating birds (European robins) during late afternoon and early morning hours. Wilson Bull. 84:149–163.

LITERATURE CITED

Wiltschko, W., H. Höck, and F. W. Merkel. 1971. Outdoor experiments with migrating robins (*Erithacus rubecula*) in artificial magnetic fields. Z. Tierpsychol. 29:409–415.

Wiltschko, W., and F. W. Merkel. 1966. Orientierung zugunruhiger Rotkehlchen im statischen Magnetfeld. Verh. Deut. Zool., Jena 1965:362–367.

Wiltschko, W., and R. Wiltschko. 1972. Magnetic compass of European robins. Science 176:62–64.

Windle, W. F. 1940. Physiology of the fetus. Saunders, Philadelphia.

Wine, J. J., and F. B. Krasne. 1972. The organization of escape behavior in the crayfish. J. Exp. Biol. 56:1–18.

Winn, H. E., and B. L. Olla. 1972. Behavior of marine animals. Vol. 2. Vertebrates. Plenum Press, New York.

Wolf, L. L. 1970. The impact of seasonal flowering on the biology of some tropical hummingbirds. Condor 72:1–14.

Wolf, L. L., and F. R. Hainsworth. 1971. Time and energy budgets of territorial hummingbirds. Ecology 52:980–988.

Wolf, L. L., F. R. Hainsworth, and F. G. Stiles. 1972. Energetics of foraging: rate and efficiency of nectar extraction by hummingbirds. Science 176:1351–1352.

Wolfe, J. L., and C. T. Summerlin. 1968. Agonistic behavior of organized and disorganized cotton rat populations. Science 160:98–99.

Wollberg, Z., and J. D. Newman. 1972. Auditory cortex of squirrel monkey: response patterns of single cells to species-specific vocalizations. Science 175:212–213.

Woolfenden, G. E. 1975. Florida scrub jay helpers at the nest. Auk 92:1–15.

Wright, H. O. 1968. Visual displays in brachyuran crabs: field and laboratory studies. Amer. Zool. 8:655–665.

Wright, S. 1945. Tempo and mode in evolution: a critical review. Ecology 26:415–419.

Wynne-Edwards, V. C. 1962. Animal dispersion in relation to social behavior. Hafner, New York.

Wynne-Edwards, V. C. 1963. Intergroup selection in the evolution of social systems. Nature 200:623–626.

Yamagishi, S. 1971. A study of the home range and the territory in meadow bunting (*Emberiza cioides*). Misc. Rep. Yamashima Inst. Ornithol. 6(4):356–388.

Yarnall, J. L. 1969. Aspects of the behaviour of *Octopus cyanea* Gray. Anim. Behav. 17:747–754.

Yoshiba, K. 1968. Local and intertroop variability in ecology and social behavior of common Indian langurs, pp. 217–242. *In* P. C. Jay, ed., Primates: studies in adaptation and variability. Holt, Rinehart and Winston, New York.

Young, J. Z. 1964. A model of the brain. Clarendon Press, Oxford.

Zahavi, A. 1971. The function of pre-roost gatherings and communal roosts. Ibis 113:106–109.

Zahavi, A. 1974. Communal nesting in the Arabian babbler. Ibis 116:84–87.

Zimmerman, J. L. 1966. Polygyny in the dickcissel. Auk 83:534–546.

Zimmerman, J. L. 1971. The territory and its density dependent effect in *Spiza americana*. Auk 88:591–612.

Zucker, R. S. 1972. Crayfish escape behavior and central synapses. I. Neural circuit exciting lateral giant fiber. J. Neurophysiol. 35:599–620.

Zucker, R. S., D. Kennedy, and A. I. Selverston. 1971. Neuronal circuit mediating escape responses in crayfish. Science 173:645–649.

Zuckerman, S. 1932. The social life of monkeys and apes. Routledge and Kegan Paul, London.

Zweifel, R. G. 1968. Effects of temperature, body size, and hybridization on mating calls of toads, *Bufo a. americanus* and *Bufo woodhousii fowleri*. Copeia 1968:269–285.

Appendix: Testing Kinship Theories Ecologically

THE FOLLOWING EXAMPLES illustrate how kinship theory can provide predictions about operational altruism that can be tested using ecological observations of natural populations.

An individual superb blue wren, X, has two alternative strategies open to it at the inception of its first breeding season. It can stay with its parents and help to rear full siblings, the *helping strategy*. Or it can leave its parents and attempt to rear its own offspring, the *selfish strategy*. For selection to favor the helping strategy, this strategy must contribute more genes to the next generation than would the selfish strategy. The genetic contribution in each case is determined by the number of individuals produced to maturity and the probability that each carries a specified gene. The criterion is

$$\begin{matrix}\text{genetic contribution with} \\ \text{helping strategy}\end{matrix} = Gr_{XS} > Lr_{XO} = \begin{matrix}\text{genetic contribution with} \\ \text{selfish strategy.}\end{matrix}$$

G = number of full sibling young raised by X if it helps its parents, above and beyond what the parents could raise unaided

L = number of young raised by X if it attempts to breed on its own in a pair

r_{XS} = probability that a full sibling carries a gene that is identical by descent with a specified gene, A_1, in X

r_{XO} = probability that one of X's own offspring carries gene A_1 identical by descent

Data for offspring surviving to maturity for each strategy are not available, but Rowley's (1965*a*) data for numbers of young at independence can be used as an approximation. Rowley showed that trios (pairs with one helper) produced an average of 2.83 independent young per year, while pairs (without helpers) produced 1.50 young. The number for pairs alone is an estimate of L. The difference between trios and pairs ($2.83 - 1.50 = 1.33$) is an estimate of G.

The criterion for selection to favor the helping strategy can be restated as

$$\frac{G}{L} > \frac{r_{XO}}{r_{XS}}.$$

Since in this case $r_{XO} = r_{XS} = \frac{1}{2}$, $r_{XO}/r_{XS} = 1$. For selection to favor the helping strategy, the criterion is $G/L > 1$.

For females $\frac{G}{L} = \frac{1.33}{1.50} < 1$. Therefore, females are predicted to follow the selfish strategy. In agreement with expectation, female superb blue wrens rarely become helpers.

Since there is a shortage of females, only about 70 percent of the males can breed. For an average male the selfish strategy results in fewer young than for an average female. Specifically, $L = 0.7(1.50) = 1.05$ young. For an average male

$$\frac{G}{L} = \frac{1.33}{1.05} > 1.$$

Therefore, average males are predicted to become helpers. However, a male can be either a helper or a breeder; he cannot be 0.7 of the way in between in this species. If a male can acquire a female, he should keep her, since

$$\frac{G}{L} = \frac{1.33}{1.50} < 1.$$

If he cannot acquire a female, he should become a helper, since

$$\frac{G}{L} = \frac{1.33}{\approx 0} > 1.$$

For male wrens, the optimum strategy seems to be to adopt a contingency plan. Strive to get a female, but leave open the option of helping your parents if you don't get a female. In this case at least, field work has shown that the wrens do what kinship theory predicts they should do.

In scrub jays the situation resembles that of the wrens in that jays can help to rear siblings or leave and attempt to rear their own young. So again the criterion for selection to favor the altruistic strategy is $G/L > 1$. Woolfenden (1975) has shown that in Florida, pairs with one or more helpers produce an average of 1.3 independent young per year, while pairs without helpers produce only 0.5. Therefore,

$$\frac{G}{L} \cong \frac{1.3 - 0.5}{0.5} = \frac{0.8}{0.5} > 1.$$

Kinship theory predicts that both sexes should follow the altruistic strategy. In agreement with expectation, nearly all yearlings of both sexes do stay and help their parents instead of breeding on their own. Most Florida scrub jays cease being helpers and begin to breed in pairs when two years old. The success of such naive pairs is even lower than 0.5, so the actual estimate of G/L should be even larger. The present discussion is necessarily an oversimplification. Further analysis requires data and theory that are not yet published. Although the analysis must be incomplete, it is interesting to note that the greater prominence of operational altruism in the life history of Florida scrub jays is correlated with higher values of G/L than in superb blue wrens, specifically with values greater than 1 for both sexes.

Subject Index

Adaptation, 267–68
  evolutionary, 268
  neuronal, 268
  physiological, 268
  sensory, 268
Adaptive radiation, 10–12
Advertisement songs, 364
Aggression, 110–12, 128–31
  see also Territoriality; Dominance
Agonistic behavior
  defined, 40
  and displays, 323
  in primates, 253–59
Aid-giving behavior, 186–213
  and displays, 323–27
  in primates, 259–62
  in social insects, 214–23
Aid-related behavior, defined, 40
Alarm
  calls, 200, 362–65
  substances, 383–88
Allelochemics, 371
Allen's rule, 18
Allochronic, 405
Allomones, 371–75
Allopatric, 407
Allopolyploidy, 411
Altruism, 196–213
  in social insects, 214–23
Analysis of variance, 25–26
Anemotaxis, 377
Appeasement behavior, 92–94
Assortative mating
  negative, 645–48
  positive, 645–48
Attention, 600–602
Auditory communication, 329–69

Biological rhythms, 588–606
Bowers, 170–71
Brood parasitism, 174, 232–41, 658

Calls, *see* Vocalizations
Central nervous system, 334, 344, 482–86, 503–14, 518–19, 528
Central-pattern-generator hypothesis, 503–14
Character convergence, 437
Character displacement, 436
Chemical communication, 370–99
Cicada principle, 143–44
Circadian rhythms, 592–600
Cline, 18
Clock-shifting, 572–73
Clutch size, 189
Coefficient of genetic relationship, 203–4
Coloniality, evolution of, 134–41
Colonizing species, 192
Communal breeders, 201
Communal group, 198
Communication, 268
  auditory, 329–69
  chemical, 370–99
  visual, 271–328
Competition, 129, 131
Control theory, 602–3
  closed-loop system, 602–4
  feedback loop, 603
  open-loop system, 602–3
Cooperative behavior, 197, 261
Coral colony, 474–76
Crest erection, 302–6
Critical period, 635
Cryptic species, 416

Dances of bees, 449–64
  round, 451
  sickle, 451
  waggle, 451
Deafening, 670–72
Density-intolerant species, 122
Density-tolerant species, 122
Dialects, 432–33, 663–66
Dispersion pattern, defined, 47

Displacement behavior, 296–97
Displays, 170, 276, 282–327
Domestication, 12–13
Dominance, 82–95
  male, 155–61
  matrix, 85–91
Dominion, 59, 61–62
Doppler shift, 553–54
Drinking, 9
Duetting, 359–60, 661–62

Echolocation (EL), 547–55
  and attention mechanisms, 600–601
  by bats, 548–55
  by porpoises, 555
Embryo behavior, 612–23
Emery's rule, 231
Emigration, 113–17, 121–22
Emlen funnel, 576
Energy-budget, 136–37
Ethogram, 2
Ethology, defined, 3
Eusociality, 214–18
Evolution
  convergent, 13–14
  parallel, 13–14
Exclusion effects, 98–112
Exclusive area, 59, 61–62
Extinction, 401

Fitness
  defined, 41
  inclusive, 204
  individual, 196, 204
Fixed action patterns (FAPs), 492
Following response, 628
Food-sharing, 199
Fossil, 18, 41, 315

Ganglia, 499
Genetic drift, 407
Genetics, 21–46, 400–448
  and imprinting, 643–50
  and motor unit firing of hybrid crickets, 513
  and sexual selection, 175–83
  and songs of hybrid crickets, 512
  and songs of hybrid grasshoppers, 418
Giant axon, *see* Giant neuron
Giant fiber, *see* Giant neuron
Giant neuron, 493–503
Gloger's rule, 18
Gluttony principle, 143
Grouping behavior, defined, 40
Groups, 72–96
  coincidental, 73–75
  colonial, 73–74

Groups (*cont.*)
  kin, 72–73
  mating, 72–73
  survival, 73–74
Group selection, 124

Habituation, 600
Haplodiploidy, 220–23
Hardy-Weinberg equilibrium, 42–43
Head-scratching, 8–9
Helpers, 199
Heritability, 25–32
Heterogamy, 433
Homeostatic motivational systems, 604–5
Home range, 50–57, 61–63, 146–49
Homogamy, 433
Homology, 7
Hybrids, 32–34, 409, 411, 418, 430, 440, 512–13, 644
Hybrid swarm, 413, 435

Imprinting, 627–52
  filial, 629
  sexual, 628–34, 643–50
Independent effector, 588–89
Individual distance, 79–82
Individual recognition, 358
Individuation, 623
Innateness, 35–36, 607–11
Innate releasing mechanisms (IRMs), 516
Inquilinism, 225–31
Instinct problem, 466–67
Intention movements, 293–94
Interspecific
  chemical communication, 373–76
  displays, 327
  symbioses, 188, 209–11
Isolating mechanisms, 400–448, 664
  defined, 404
  role of imprinting in, 643–44
Iteroparous reproduction, 191

Kairomones, 371, 376
Kineses, 543–45
  klinokinesis, 545
  orthokinesis, 543
Kin selection, 45–46, 203–7, 261–62
Kinship bond, 76
K-selection, 193–96

Leading lines (migration), 563
Lek, 126, 163–65, 179

Mating systems, 151–85
  and displays, 321–22
  in primates, 243–53

Mauthner cells, 494, 621–23
Migration, 557–87
"Migration-divide," 567–70
Mimicry, 143
  Batesian, 310–11
  of cleaner fish, 311
  of eggs, 232–35
  Müllerian, 310–11
  of nestlings, 235–37
Monogamy, 151–85
Monosynaptic, 483
Morphism, *see* Polymorphism
Motivation, 602–5
Motor neurons, 483, 614–23
Mutualism, 209–10
Myogenic motility, 614

Nassanoff's gland, 451
"Nature-nuture" problem, 33–36, 607–11
Nearest-neighbor distance, 76–78
Nematocyst, 588–89
Nerve nets, 468–82
Neural maturation, 612–23
Neurogenic motility, 614

Opportunistic species, 192
Orientation, 543–56
  echolocation, 547–55
  in invertebrates, 543–47
  migration, 557–87
  one-direction, 564
  transverse, 544
  true goal, 564
Ortstreue, 559–60
Osmotropotaxis, 377

Pacemaker, 591–96
Parasitism, 154
  brood, 232–41, 658
  social, 224–41
Parental care, 166, 189–96, 259–60
Pars intercerebralis (PIC), 593
Peck-dominance, 91–92
Peck order, 83
Peck-right, 91–92
Pedicellariae, 470, 473–74
Phenetics, 403
Pheromones, 399, 450–51
  defined, 370
Phonoresponse, 352
Phonotaxis, 352
Phylogeny, 1–15
Polyandry, 153
Polygamy, 151–85
Polygenes, 25
Polygyny, 153
Polymorphism, 28, 32, 644–50
Polysynaptic, 484
Population selection, 46, 124, 132–34, 147, 209
Predators, 141–46
Prey-carrying, 16–17
Promiscuity, 153
Protogynous hermaphroditism, 154

Receptor cells, 483
  extraretinal, 598
Reciprocal altruism, 208–9
Recognition, 539
  individual, 355, 358–62
  species, 640–52
Reflex, 482–88
  cyclic-reflex hypothesis, 503–14
  defined, 486–87
  monosynaptic, 590
  stretch, 484–85
Reflexogenic motility, 614
Reinforcement, 602
Releaser, 276, 518
Reproductive rate, 189–96
Rhythms, 588–606
  circadian, 592–600
  endogenous, 593
  entrained, 593
  exogenous, 593
  free-running, 593
r-selection, 191–96

Scaphognathites, 591
Schreckstoff, 383
Secondary contact, 409–10
Selection
  artificial, 27
  differential, 28
  directional, 28
  disruptive, 28
  kin, 45, 261–62
  K-selection, 193–96
  natural, 154–55
  on displays, 301–28
  population (deme, interpopulation, interdemic), 46, 124, 132–34, 147
  r-selection, 191–96
  sexual, 151–85
  stabilizing, 27
  units of, 40–41
Selfish behavior, 197
Semelparous reproduction, 191
Semiochemicals, 370
Sensory neurons, 483
Sensory systems, 516–56
Sex attractants, 377–83
Sex change, in *Labroides*, 154

Sexual dimorphism, 170–71, 176
 in primates, 252–53
Sexual preference, 175–76
Sexual selection, 151–85
 in primates, 250–53
Sibling species, 7–8, 416, 442
Slavery in ants, 225–28
Social behavior, defined, 39
Social insects, 214–23, 389–91
Sociality, 214
Social organization (social system), defined, 40
Soliciting or begging behavior, 76
Song
 of birds, 347–49, 354–65, 653–76
 cricket, *see* Cricket
 development, 653–75
 dialects, 432–33, 663–66
 duetting, 359–60, 661–62
 of gibbons, 351
 of grasshoppers, 332, 352–54, 418
 plastic, 656
 primary, 656
 rehearsed, 656
 subsong, 656
 of whales, 351
Spacing behavior, defined, 40, 50
Species, defined, 403
Species recognition, 640–52
Species-specific behavior, defined, 400
Spiteful behavior, 197
Stereotypy, 490
Stimulus-response (S-R) specificity, 539–40
Stridulation, 333
Submissive behavior, 92–94
Subordination effects, 112–22
Survival groups, evolution of, 141–46
Symbiosis, 209–10
Sympatry, defined, 407
Synchronic, 405

"Talking" birds, 661–62
Taxes, 544–45
 klinotaxis, 544–46
 telotaxis, 544, 546
 tropotaxis, 544, 546
Territorial behavior, ontogeny of, 67–69
Territoriality, 58–71
 and diversity of territories, 64–65
 effects of, on reproduction, 106–9
 evolution of, 125–34
 see also Territory
Territory
 all-purpose, 164–65
 artificial, 69–71
 defense of, 65–67
 food, 126
 mating, 126
 quality, 172–73
 size, 172
Thermoregulation, 291–92
Trophallaxis, 218

Visual communication, 271–328
Vocalizations, 34, 169, 337–41, 345–51, 354–65, 653–75
 call, defined, 345
 song, defined, 345

You-first principle, 144–45

Zeitgeber, 593–98
Zooid, 474–76
Zugunruhe, 571–86

Author Index

Able, K. P., 564
Abu Gideiri, V. B., 613–14
Adams, L., 55
Adinolfi, A. M., 612
Adler, H. E., 543
Adler, K., 598
Adriaanse, A., 7
Ahlen, I., 144
Aho, J., 115
Aldrich, J. W., 106
Aldrich-Blake, P., 246
Alexander, B. K., 260
Alexander, R. D., 177, 223, 333, 337, 342–44, 366–67, 405–6, 410, 417, 419, 431, 528, 654
Allee, W. C., 83, 92
Allen, A. A., 339
Allen, G. R., 209
Altes, R. A., 554
Altmann, J., 249–52
Altmann, S. A., 249–52, 255, 258, 269
Altum, J. B. T., 126
Amadon, D., 6
Anderson, E., 405, 410
Anderson, P., 69–71, 117
Anderson, W. W., 434
Andersson, A., 144
Andrew, R. J., 292, 294–95, 309, 320
Andrewartha, H. G., 47
Andrews, R. V., 120
Aneshansley, D. J., 373
Appel, S. H., 35
Archer, J., 49, 115–20
Ardrey, R., 132
Armstrong, E. A., 174, 296, 319, 432, 663
Armstrong, J. T., 56
Armstrong, P. B., 621–23
Arnold, J. M., 296
Aronson, L. R., 33, 625
Atsaides, S. P., 302
Atz, J. W., 7
Awbrey, F. T., 426
Baerends, G. P., 3, 271, 275, 286, 292
Baerends-van Roon, J. M., 271, 275, 292
Baher, E., 261
Baker, H. G., 192
Balaban, M., 616
Baldwin, H. A., 47
Baldwin, P. H., 12
Ball, R. W., 425, 441–42
Ballantine, J. A., 383
Banks, J. H., 275
Barber, H. S., 7
Barbieri, D., 634, 646
Barlow, G. W., 288, 301
Barlow, H. B., 537–38
Bartell, R. J., 426, 446
Barth, R., 380
Bartholomew, G. A., 143, 156–57
Baskett, T. S., 172, 273, 428
Bastian, J. A., 456
Bastock, M., 23, 161, 312–13, 445
Bateson, P. P. G., 627–28, 635, 638
Batham, E. J., 469, 472, 479
Beardmore, J., 429, 445, 646, 649–50
Beer, C. G., 347, 359, 362
Beg, M. A., 258
Bellairs, A. d'A., 614
Bellrose, F. C., 563
Bennet-Clark, H. C., 419, 446
Bennett, E. L., 35
Bentley, D. R., 180, 337, 446, 504, 512–14, 654
Benzer, S., 23, 596, 598
Berger, A. J., 334
Bergman, G., 170, 363
Berndt, R., 49, 558, 560
Bernstein, I. S., 257
Beroza, M., 380
Bertram, B., 362, 660–61
Bigelow, R. S., 180, 405, 410, 446
Binkley, S., 598
Binks, C., 639
Birch, L. C., 47

Biswas, A. M., 78
Bjärvall, A., 634
Blair, W. F., 116, 345, 400, 417, 424–25, 431, 434, 441
Blanchard, B. D., 113, 115
Blest, A. D., 294–95, 311, 313, 462, 464
Blum, M. S., 385, 391
Boch, R., 385
Bock, W. J., 10, 12
Boeckh, J., 380
Boehlke, K. W., 121
Bogert, C. M., 345, 425
Bol, A. C. A., 297
Bonsall, R. W., 377, 383
Boorman, S. A., 209
Borror, D. J., 432, 659
Bossert, W. H., 45, 277, 280–81, 288, 290, 371, 378, 383, 389, 396, 405, 435–37
Bouma, H., 530
Bowman, R. I., 10, 11
Boycott, B. B., 533
Boyd, H., 90
Bradley, H. T., 274
Brady, J., 593
Brain, C. K., 255
Brain, P. F., 119
Brand, J. M., 385, 391
Brander, R. B., 163
Breder, C. M., 47, 143, 412
Bremond, J. C., 347, 356, 366, 421
Brenner, F. J., 104
Brereton, J., 8
Brian, A. D., 87, 102
Brian, M. V., 215
Bronson, F. H., 88, 116, 393–94, 427
Brosset, A., 632, 643
Brouwer, R., 288
Brown, J. L., 8, 18, 45–46, 48, 59, 62, 64, 84–89, 101, 104, 107, 132–33, 200–209, 283, 289, 297, 304–6, 321, 659
Brown, L. E., 52, 55, 425
Brown, L. N., 383
Brown, R. G. B., 445
Brown, W. L., 371, 417, 437
Bryant, S. V., 614
Bub, H., 53
Buchsbaum, R., 589
Buck, E., 590
Buck, J., 590
Buck, J. B., 415
Buckley, F. G., 358
Buckley, P. A., 358
Buechner, H. K., 164
Buettner-Janusch, J., 262
Bullock, T. H., 470, 474–75, 478, 480, 590–91
Burckhardt, D., 79
Burghardt, G. M., 18
Burkitt, J. P., 84
Burnet, B., 34, 419
Burns, J. J., 673
Burt, W. H., 61
Bustard, H. R., 106
Butenandt, A., 370
Butler, C. G., 377, 380–81

Cade, T. J., 9, 563
Caldwell, R. L., 277
Camhi, J. M., 601
Campbell, B., 151
Campbell, H. W., 331
Capranica, R. R., 337, 421, 522–26
Cardé, R. T., 381
Carl, E., 103–4
Carpenter, C. R., 246, 257
Carrick, R., 49
Carthy, J. D., 543
Caughley, G. J., 48
Chance, E. P., 235
Chance, M. R. A., 94, 243
Chantharojvong, S., 341, 351
Chappuis, C., 364
Charles-Dominique, P., 246
Chauvin, R., 214
Chejne, A. J., 445
Childers, W. F., 411
Chitty, D., 119, 147
Christian, J. J., 116, 119–21, 146
Chung, S. H., 534
Church, R. L., 121
Cialdini, R. B., 385
Clark, E., 33
Clark, P. J., 78
Clayton, D. L., 596
Cochrane, W. W., 55, 561
Cody, M. L., 312
Coghill, G. E., 620, 623–24
Colbert, E. H., 18
Cole, L. C., 191–92
Collias, E. C., 18, 136, 163, 307
Collias, N. E., 18, 83–84, 136, 163, 307
Collier, G., 153
Colquhoun, M. K., 87
Comeau, A., 382
Conder, P., 79
Connolly, A., 34
Conolly, K., 419
Cooch, F. G., 411, 429, 559, 645, 649–50
Cooke, F., 411, 646, 649–50
Cope, J. B., 106
Corner, G. W., 348
Cory, L., 405–6
Cott, H. B., 313
Coulson, J. C., 144, 358

Cox, B., 347
Craig, J. V., 78
Crane, J., 302, 313, 321
Crewe, R. M., 385
Crook, J. H., 10, 18, 48, 124, 133, 139, 174, 243, 246, 252, 255, 261–63, 316, 321, 323
Crovello, T. J., 404
Cullen, E., 319
Cullen, J. M., 47, 296
Curio, E., 11
Curry-Lindahl, K., 49
Curtis, B., 273
Curtis, R. F., 383
Cushing, J. E., 664

Daanje, A., 293, 312
Dales, R. P., 209
Dalquest, W. W., 549
Dandelot, P., 428
Dane, B., 282, 320
Darling, F. F., 76, 139, 158–59
Darlington, P. J., 200, 203, 209
Darwin, C., 1, 151, 154–56, 165, 170, 175, 290, 306, 427
Davies, S. J. J. F., 33
Davis, D. E., 104, 116, 119–20, 146, 202
Davis, L. I., 330
Davis, S. D., 55
Davis, W. C., 406
Davis, W. J., 335–36, 502
Dawkins, R., 457
DeBach, P., 410
Decker, J. D., 614, 616
DeFries, J. C., 21
Delacour, J., 6
Delco, E. A., 442
Delius, J. D., 139, 297
DeLong, K. T., 119
Denenberg, V. H., 625
Dennis, L., 419
Dethier, V. G., 334, 462, 464, 604
DeVore, I., 243, 249–50, 252, 255, 258, 262
Dew, D., 335, 368
Dhondt, A. A., 52
Diamond, J. M., 662
Diamond, M. C., 35
Dickerman, R. W., 319
Dilger, W. C., 18, 33, 419, 421–24, 428
Dingle, H. A., 277–80
Dixon, K. L., 87–88
Dixon, R. D., 648
Dobzhansky, Th., 28, 400, 404, 435, 444
Dorst, J., 428, 557, 559
Dowling, J. E., 533
Driskill, R. J., 593
Driver, P. M., 142
Drury, W. H., 282, 320, 564, 577
Duellman, W. E., 425
Dumortier, B., 332–33, 352
Dumpert, K., 385
Dunford, C., 62, 88
Dunham, D. W., 8
Dunning, D. C., 522
Dybas, H. S., 143
Dzubin, A., 91

Eastwood, E., 563
Eaton, R. L., 62
Eberhard, M. J. W., 6, 16, 215
Efron, R., 486
Ehrlich, P. R., 404
Ehrman, L., 29, 178–79, 419, 426, 433–34, 442, 446
Eibl-Eibesfeldt, I., 11, 118, 492
Eidman, H., 229
Eisenberg, J. F., 118, 122, 148, 160, 200, 243, 248, 259, 395
Eisner, T., 373–76, 528
Eleftheriou, B. E., 121
Elias, P., 301
Ellefson, J. O., 246, 341
Emlen, J. M., 195
Emlen, J. T., 59, 362, 572, 574–76, 639
Emlen, S. T., 345, 355–57, 557, 566, 572–82
Engelsen, G. H., 170
Erlenmeyer-Kimling, L., 31
Erulkar, S. D., 547, 555
Esch, H., 454–56
Eshel, I., 133, 203, 209
Estes, R. D., 143
Evans, F. C., 78
Evans, H. E., 6, 16, 18, 33, 215, 232
Evans, K., 33
Evans, R. M., 362
Evans, W. E., 331
Evoy, W. H., 499
Ewer, R. F., 59, 61, 160, 200, 291, 395
Ewert, J. P., 534
Ewing, A. W., 33–34, 180–82, 312, 419, 426, 446

Faber, W., 230
Fabricius, E., 639
Factor, R. M., 591
Falconer, D. S., 25
Falls, J. B., 52, 355–56, 421, 660
Fält, O., 639
Farkas, S. R., 377, 379
Farris, H. E., 318
Fastie, W. G., 415
Faugeres, A., 161
Feder, H. M., 209

Feeny, P. P., 371, 374
Fehmi, L. G., 475
Ferguson, D. E., 598
Ferguson, G. W., 430
Ficken, M. S., 8, 364, 421, 429
Ficken, R. W., 8, 364, 421, 429
Findley, J. S., 35
Fish, M. P., 331
Fisher, R. A., 132, 169–70, 175, 186–87, 191
Fisler, G. F., 200
Fitch, H. S., 52
Flynn, J. P., 602
Folsom, E. F., 181, 655
Fooden, J., 429, 645
Ford, E. B., 32
Fouquette, M. J., 424
Fox, M. W., 76, 323, 619, 624
Fraenkel, G. S., 543–46
Fraga, R. M., 238
Franck, D., 146
Fredericson, E., 40
Fredrikson, K. A., 241
Fretwell, S., 49, 94, 108
Friedman, M., 347
Friedmann, H., 236–37, 240
Frings, H., 347
Frings, M., 347
Frisch, K. von, 214, 449–61
Frishkopf, L. S., 522, 528
Frith, C. B., 317
Frochot, B., 170
Fromme, H. G., 582
Fry, C. H., 206
Fuller, J. L., 12, 13, 21
Furuya, Y., 258

Galbraith, I. C. J., 429
Gallup, G. G., 275
Gartlan, J. S., 255
Gary, N. E., 385
Gaston, S., 380, 598–99
Gauthreaux, S. A., 563–64
Geist, V., 156, 170, 316
Geller, E., 598
Gerald, J. W., 412–13
Geronimo, J., 596
Ghent, A. W., 385
Ghoca, M. L., 385
Gibson, J. B., 445
Giesel, J. T., 32
Gifford, E. W., 10
Giles, R. H., Jr., 53
Gill, F. B., 421, 429
Gilliard, E. T., 162, 171, 302, 316–17
Girman, S. V., 532
Glas, P., 102–3
Goddard, J., 143
Godfrey, J., 434
Goforth, W. R., 172, 273, 428
Goldstein, M. H., Jr., 522, 528
Goodall, J., 258
Goodwin, D., 646, 648
Gordon, M., 33
Goss-Custard, J. D., 65
Gottlieb, G., 347, 612, 620, 625, 636, 640–43
Gould, E., 120, 341, 548
Gould, J. L., 457
Graber, R. R., 561
Grant, V., 400, 435
Green, M. C., 35
Green, R. F., 288, 301
Green, R. P., 580
Greenewalt, C. H., 330, 332, 341
Greenwald, L. I., 9
Greig-Smith, P., 47
Griffin, D. R., 547–50, 557
Grinnell, A. D., 555
Grinnell, V. S., 555
Grüsser, O. J., 535
Grüsser-Cornehls, U., 535
Guhl, A. M., 78, 83
Guiton, P., 634
Gunn, W. W. H., 432, 543–46
Guthrie, R. D., 291
Gwinner, E., 661
Gwynn, A. M., 558

Haartman, L. von, 172, 174, 319
Hagiwara, S., 509
Hailman, J. P., 2, 518
Hainsworth, F. R., 136–37
Hairston, N. G., 193
Halberg, F., 592
Haldane, J. B. S., 45, 203, 277
Hale, E. B., 13, 634
Hall, B. P., 429
Hall, K. R. L., 243, 248–50, 252, 255, 258, 262
Hamburger, V., 613–20, 624
Hamilton, W. D., 41, 45, 141, 144, 146, 196–97, 203, 206–7, 210, 214, 218, 222, 261
Hamilton, W. J., Jr., 65, 119
Hamilton, W. J., III, 126, 143, 241, 313, 378, 543, 572
Hanawalt, J. T., 499
Hardy, J. W., 319, 421
Harker, J. E., 593
Harmon, L. D., 508
Harrington, R. W., Jr., 154
Harris, M. P., 2, 414, 633–34, 643–44
Harrison, C. J. O., 200, 202, 205

Hartline, H. K., 534
Hartzler, J. E., 90
Haugh, J. R., 563
Hauske, G., 277
Hazlett, B. A., 277, 280, 288, 290
Hediger, H., 79
Heinroth, O., 4, 6, 8, 293
Helmreich, R. L., 119
Helversen, D. V., 352–54
Henery, M., 457
Hensley, M. M., 106
Henson, O. W., Jr., 553, 555, 601
Hershkovitz, P., 428
Herzog, A., 660
Hess, E. H., 603, 627–28, 634–35, 638–39
Heywood, V. H., 13
Higginbotham, M., 120
Higgins, D. C., 621–23
Hill, J. L., 69–71
Hinde, R. A., 3, 126, 199, 259, 517, 543, 604, 660
Hinkle, M., 601
Hirsch, J., 21, 25, 31
Hjorth, I., 163
Hock, F., 534
Höck, H., 582
Hoenigsberg, H. F., 445
Hoffman, K., 572
Hoffman, R. R., 119
Hogan-Warburg, A. J., 164, 179–81, 272
Holtzer, H., 623
Homma, K., 598
Hooff, J. A. R. A. M. van, 308
Hooker, B. I., 360, 661
Hooker, T., 360, 661
Hörmann-Heck, S. von, 33
Horn, G., 535, 601
Horn, H. S., 134–36
Horridge, G. A., 334, 470–71, 474–82, 496, 591
Hortobagji-German, E., 445
Houlihan, R. T., 118
Howard, E., 97, 125–26
Howard, W. E., 434
Howse, P. E., 352, 528
Hoy, R. R., 337, 504, 514, 654
Hoyle, G., 504, 509
Hsia, D. Y., 34
Hubbs, C., 442
Hubby, J. L., 35, 416
Huber, F., 509, 511, 530
Huble, J., 52
Hughes, A. F. W., 612, 614
Humphries, D. A., 142
Hunsaker, D., 302, 415–16, 426, 429
Hunsperger, R. W., 289, 297
Hutchinson, G. E., 128
Hutchison, J., 362
Hutchison, R. G., 362
Huxley, J. S., 97, 100, 312

Iersel, J. J. A. van, 297
Iles, J. F., 504
Immelmann, K., 8, 629, 633–34, 640, 659–60
Impekoven, M., 146
Ingolfasson, A., 59
Isaac, D., 330
Itani, J., 260
Iwao, S., 47

Jacklet, J. W., 596
Jackson, J. B. C., 48
Jacobson, M., 382, 612, 624
James, H., 639
James, J. W., 82
Jameson, D. L., 425, 431, 441–42
Jansen, J., 35
Jansen, J. K. S., 35
Jarman, C., 557
Jaspar, H. H., 600
Jay, P. C., 243, 259
Jenkins, D., 49, 108
Jenni, D. A., 153
Jennings, H. S., 470
Jennrich, R. I., 50, 52
Jenssen, T. A., 429
Jewell, P. A., 55
Joffe, J. M., 612
Johnsgard, P. A., 6, 33, 162–63, 297
Johnson, C., 8, 424–25
Johnson, D. L., 457
Johnston, J. W., Jr., 371
Johnston, R. F., 49, 177
Johnstone, B. M., 337, 528–29
Jolly, A., 200, 246, 262
Jolly, C. J., 243
Jorgenson, D. C., 55
Josephson, R. K., 480, 591

Kaada, B., 121
Kaiser, H., 110–12
Kaissling, K. E., 380
Kalela, O., 41, 97, 115, 119, 132
Kalkut, M. F., 121
Kalmus, H., 646
Kamrin, R. P., 623
Karlson, P., 370
Kaston, B. J., 17
Kaufman, I. C., 77
Kaufmann, J. F., 259
Kawabe, M., 246
Kawai, M., 260
Kawamura, S., 260

Keenleyside, M. H. A., 272–73, 412–13
Keeton, W. T., 557, 572, 577–83
Kelemen, G., 332
Kellogg, P. P., 339
Kemperman, J. H. B., 646
Kendeigh, S. C., 196
Kennedy, D., 495, 499
Kenyon, K. W., 566
Kerr, W. E., 449, 451, 457, 461, 463
Kessel, E. L., 15
Kessler, S., 443
Kettlewell, H. B. D., 32
Kety, S. S., 121
Keverne, E. B., 377, 383
Key, K. H. L., 410
Kikkawa, J., 87
King, J. A., 18, 67
Klein, J., 117
Kleiman, D. G., 395
Kleist, S. M., 385
Klinghammer, E., 634
Klomp, H., 189
Kluth, E., 598
Kluyver, H. M., 102, 172
Kneutgen, J., 660–61
Knight, G. R., 44
Knipling, E. F., 380
Koford, C. B., 67, 256, 260
Kok, O. B., 169, 172, 362
Konishi, M., 364, 659, 667–71
Konopka, R. J., 23, 596, 598
Koopman, K. F., 442–44
Kornfield, M., 548
Kortlandt, A., 253, 296
Kortwright, F. H., 562
Koyama, N., 255–56, 260, 262
Kramer, C. Y., 272
Kramer, G., 571–72
Kramer, P., 11
Krasne, F. B., 500–501
Krebs, J. R., 108, 147
Krech, D., 35
Krishna, K., 215
Kruger, L., 35
Kruuk, H., 139, 199, 205
Kuechle, V. B., 55
Kuenen, D. J., 273
Kuenzler, E. J., 50–52
Kühme, W., 199, 325
Kullman, E. J., 17
Kummer, H., 48, 243, 246, 248, 255
Kuo, Z.-Y., 625
Kuroda, N., 241
Kurt, F., 48, 246
Kutter, H., 228
Kuwahara, H., 275

Lack, D., 10, 11, 41, 65, 107–8, 133, 139, 152–53, 164, 172, 175, 189, 200, 202, 205, 237, 241
Lade, B. I., 33, 643
Lamprecht, J., 673
Landau, H. G., 91
Lange, R. B., 217
Langley, J. N., 290
Lanyon, W. E., 355, 421, 429, 655–56, 658, 660
Lapitskii, V. P., 509
Layne, J. N., 18
LeBoeuf, B. J., 157, 673
Lehrman, D. S., 33–36, 272, 347, 610
Lemmon, W. B., 629–30, 639, 659
Lemon, R. E., 355, 432, 660
Lerner, J., 591
Lettvin, J. Y., 530, 532, 534
Leuthold, W., 164
Levi, W. M., 12, 655
Levins, R., 411
Levinton, J., 48
Levitt, P. R., 209
Lewis, D. B., 352
Lewontin, R. C., 35, 41, 46
Leyhausen, P., 62, 272
Li, C. C., 45, 203
Lidicker, W. Z., 50
Lill, A., 634
Lilly, J. C., 654
Lin, N., 223
Lincoln, F. C., 559–61
Lincoln, G. A., 158, 170
Lind, E. O., 115
Lindauer, M., 215–21, 449, 457–61
Lindburg, D. G., 262
Lindzey, G., 21, 31
Lissaman, P. B. S., 143
Littlejohn, M. J., 339, 404, 406, 424–25, 431–32, 436, 438–40, 442
Lloyd, J. A., 116
Lloyd, J. E., 272, 302, 406, 415
Lloyd, M., 143
Loftus-Hills, J., 337, 339, 528–30
Löhrl, H., 566
Lorenz, K., 3, 6, 33, 132, 271, 276, 297, 411, 492–93, 517, 610, 627–29, 634–35
Losey, G. S., 209
Louch, C. D., 120
Lowe, V. P. W., 55
Lowery, G. H., 564
Lowther, J. K., 646
Lucas, H. L., Jr., 108
Lüscher, M., 370

Maas, D., 18
McAlister, W. H., 339

MacArthur, R. H., 41, 128, 148, 192–94
McBride, G., 82
McCabe, T. T., 113, 115
McClearn, G. E., 21
McCormack, C., xvii
McCulloch, W. S., 530, 532, 534
McDermott, F. A., 415
MacDonnell, M. F., 602
McElroy, W. D., 415
McEwen, B. S., 121
McFarland, D. J., 291, 602–4
McGeen, D. S., 238
McGeen, J. J., 238
McGinity, D. J., 612
Mackay, R. S., 55
Mackie, G. O., 480, 591
McKinney, C., 434
McKinney, F., 297
Mackintosh, J. H., 71
Maclean, G., 9
MacLean, P. D., 275
MacLeod, E. G., 520
MacLeod, M. C., 457
McNab, B. K., 56
McVay, S., 347, 351, 673
Magg, M., 347
Maher, W. J., 127–28
Mainardi, D., 634, 646
Manion, J. J., 405–6
Manley, G. H., 296
Mann, G., 35
Manning, A., 29, 34, 181, 312, 419, 426, 434, 446
Manosevitz, M., 21, 31
Marchant, S., 239
Markl, H., 331
Marks, H. L., 272
Marler, P., 12, 79–83, 257, 273, 330, 363–65, 378, 432, 543, 572, 653–67, 673
Marshall, A. J., 6, 171, 316
Marshall, J. T., 341, 351
Martin, A. A., 195, 437, 441
Martin, G., 636
Martin, R. D., 260
Martin, W. F., 340
Martof, B. S., 425
Mason, L. G., 178
Mason, W. A., 257, 259
Masure, R. A., 92
Mather, K., 27–28, 344
Matthews, G. V. T., 557, 572–73
Maturana, H. R., 530–32, 534
Mayfield, H., 238
Mayr, E., 4, 6, 7, 8, 46, 400, 403–4, 407, 410, 416–17, 419
Mead, A. P., 94
Mech, L. D., 143, 395
Mecham, J. S., 425
Meijer, G. M., 382
Meinwold, J., 373, 375
Meisenheimer, J., 15
Mellen, S. E., 59
Melnechuk, T., 654
Menaker, M., 590, 598–99
Mendelson, M., 504, 509, 591–92
Merkel, F. W., 582
Messmer, E., 656
Messmer, F., 656
Metfessel, M., 655
Metzgar, L. H., 50, 117
Mewaldt, L. R., 85
Michael, C. R., 550
Michael, R. P., 377, 383
Michaud, T. C., 425, 436
Michelson, A., 333
Michener, C. D., 17, 217, 223, 231
Miller, G. R., 49, 172
Miller, H., 91
Miller, H. C., 412
Miller, L. A., 520
Miller, R. E., 275
Miller, R. S., 76–78
Milligan, M. M., 660
Millikan, G. C., 11
Minks, A. K., 382
Minock, M. E., 90
Mirsky, I. A., 275
Missakian, E. A., 256, 260, 262
Moffat, C. B., 97
Mohr, C. O., 50
Moltz, H., 492, 628
Montgomery, G. G., 561
Moore, R. E., 427
Moreau, R. E., 429, 559, 563
Morejohn, G. V., 634
Morris, D., 288, 291, 302
Morse, D. H., 12, 347
Mosby, H. S., 95
Moser, J. C., 388
Moss, R., 104, 110
Moulton, J. M., 371
Moynihan, M., 6, 288, 306, 312
Muckenhirn, N. A., 243
Muller, H. J., 427
Müller-Schwarze, D., 391–93
Mundinger, P., 655, 658–59, 667
Muntz, W. R. A., 532
Murphy, J. V., 275
Murray, B. G., 58–59
Murton, R. K., 94, 141, 143–44, 348
Myers, K., 119
Myllymaki, A., 115
Mykytowycz, R., 395

Napier, J. R., 243
Narasimhan, R., 537–38
Narayanan, C. H., 619, 624
Narda, R. D., 426
Negus, N. C., 120, 548
Neill, W. T., 426
Nelson, J. B., 2, 10
Nelson, K., 277, 332
Neuberger, E., 277
Neuweiler, G., 553
Newman, J. D., 530
Newman, R. J., 564
Nice, M. M., 8, 126
Nichols, D., 473
Nicol, J. A. C., 496
Nicolai, J., 237, 428, 659–60
Nisbet, I. C. T., 563–64, 577
Nishiitsutsuji-Uwo, J., 593, 595
Nobel, G. K., 59, 171, 273–74, 412, 428
Nogueira-Neto, P., 231
Norris, K. S., 555
North, M. E. W., 661–62
Norton, A. L., 532
Nottebohm, F., 341, 432, 653–54, 656, 664, 667, 670, 672–75
Nottebohm, M. E., 670, 672
Novick, A., 548, 551–53

O'Donald, P., 126, 165, 176–77, 634, 646
Odum, E. P., 50–52, 87
Ogawa, N., 275
O'Keefe, J., 530
Okon, E. E., 341
Oldham, R. S., 425
Olla, B. L., 345
Oppenheim, R., 616
Orenstein, R. I., 12
Orgain, H., 116
Orians, G. H., 10, 48, 58–59, 64, 104, 134–36, 166–68, 173–74, 241
Orr, R. T., 557
Otto, D., 509, 511
Owen, D. F., 646

Pack, J. C., 95
Paietta, J. V., 596
Pantin, C. F. A., 469, 472, 477, 479
Parsons, P. A., 21, 25–26, 31, 433
Passano, L. M., 591
Paterniani, E., 445
Patterson, I. J., 144
Patterson, J. T., 427
Patterson, R. L., 164
Pavlidis, T., 596
Pavlov, I. P., 486
Pavlovsky, O., 444
Payne, R., 347, 673
Payne, R. B., 135, 236–37, 241, 351
Pearce, S., 444
Pearson, C. A., 52
Pearson, E. W., 348
Pearson, K. G., 504
Peek, F. W., 104, 170, 674
Peissner, L., 347
Perdeck, A. C., 406, 417–18, 564–69, 582
Perrins, C. M., 109, 200, 363
Persoons, C. J., 382
Peters, H. M., 275
Peterson, R. S., 157, 673
Petit, C., 29, 161, 178, 446
Petras, M. L., 117
Petropulos, S. F., 593, 595
Pfeiffer, W., 384–85
Phillips, R. E., 181, 655
Pianka, E. R., 193, 195
Pielou, E. C., 47, 78
Pigarev, I. N., 532, 535–37
Pimentel, D., 411
Pitelka, F. A., 59, 113
Pittendrigh, C. S., 593, 595
Pitts, W. H., 530, 532, 534
Ploog, D., 654
Pollack, G., 553
Pomeranz, B., 534
Porter, K. R., 431, 542
Potter, H. D., 528
Poulsen, H., 9
Poulton, E. B., 427
Provine, R. R., 616
Pye, J. D., 352
Pylka, J. M., 352

Quastler, H., 279

Raitt, R. J., 421
Ralls, K., 395
Ralph, C. J., 52
Ramón y Cajal, S. R., 484–85
Rao, S. V., 410
Rashevsky, N., 91
Raup, D. M., 18
Rauschert, K., 427
Raveling, D. G., 90, 358
Raven, P. H., 404
Ray, C., 673
Ray, T. S., 629–30, 639
Recher, H. F., 65
Recher, J. A., 65
Reed, R. A., 240
Regnier, F. E., 370, 382, 385–87
Reimer, J. D., 117
Remler, M., 495
Rensch, B., 234
Ressler, R. H., 385

Reynolds, V., 247
Rice, D. W., 566
Richardson, W. J., 564
Rider, G. C., 563
Ripley, S. D., 128
Ritchey, F., 92
Ritter, F. J., 382
Robel, R. J., 130, 164, 172
Roberts, S. K. deF., 593–94
Robertson, A., 444–45
Robertson, D. R., 154
Robson, E. A., 469, 479
Roeder, K. D., 519–22
Roelofs, W. L., 381–82
Romanes, G. J., 474
Rosen, D. E., 47
Rosenblum, L. A., 77
Rosenfeld, A., 537–38
Rosenzweig, M. R., 35
Rosin, R., 331
Ross, B. A., 341, 351
Ross, D. M., 477
Ross, H. H., 17
Rossi, P. J., 603
Rosvold, H. E., 297
Roth, L. M., 528
Roth, M., 528
Rothenbuhler, W. C., 23, 33, 216, 219, 231
Rothstein, S. I., 234
Rovner, J. S., 331
Rowan, M. K., 129
Rowell, C. H. F., 297
Rowell, T. E., 199, 249, 255, 258–59, 262–63
Rowley, I., 200, 202, 205, 208
Rudinsky, J. A., 371
Rudran, R., 243
Rumbaugh, D. M., xvii
Ryszkowski, L., 116

Sabine, W. S., 87
Sachs, M. B., 528
Sackett, G. P., 275
Sade, D. S., 256, 260
Sakakibara, Y., 598
Salanki, J., 504
Sales, G. D., 341
Salmon, M., 302, 332
Salomonsen, F., 558
Sandel, T. T., 616
Sargeant, A. B., 54, 62
Sargent, T. D., 32
Sauer, E. G. F., 572, 575, 581
Sauer, E. M., 572
Schaller, G. B., 2, 199–200, 205, 272, 639
Schantz, W. E., 8
Scharloo, W., 445
Schein, M. W., 116, 634
Schenkel, R., 298–99
Schief, A., 528
Schildkraut, J. J., 121
Schlegel, P., 555, 601
Schleidt, M., 347
Schleidt, W. M., 313, 347, 590
Schiermann, G., 234
Schjelderup-Ebbe, 83
Schlichter, D. von, 209
Schloeth, R., 164
Schmidt, R. S., 18
Schneider, D., 380–81, 383–85
Schnitzler, H.-U., 553
Schoener, T. W., 55–57, 59, 137, 143, 170
Schollenberger, C. A., 143
Schöne, H., 302, 316
Schönholzer, L., 9
Schubert, G., 356, 421
Schubert, M., 356, 421
Schultz, R. J., 154
Schultze-Westrum, T., 395
Schutz, F., 630–32, 639
Schüz, E., 566, 570
Schwartz, R. L., 136
Schwink, I., 377
Scott, J. P., 12, 13, 40
Scott, J. W., 164
Scott, P., 311, 645, 647
Scudo, F., 634, 646
Seiger, M. B., 411, 648
Seilacher, A., 18
Selander, R. K., 117, 170–71, 176
Seliger, H. H., 415
Selverston, A., 495
Sewall, D., 434
Sexton, O. J., 78–79
Sharma, S. C., 616
Sharpe, R. S., 33
Sharpless, S. K., 600
Shaw, E., 625
Shaw, K. C., 337
Shearer, D. A., 385
Shelford, V. E., 113
Sherrington, C., 468, 483, 486–88
Sherwood, G. A., 90
Shiovitz, K. A., 432
Shirer, H. W., 52
Shoffner, R. N., 82
Shorey, H. H., 377, 379–80, 426, 446
Short, L. L, 429
Shortt, T. M., 562
Shulov, A., 331
Sibley, C. G., 632
Sibly, R., 604
Siddiqi, M. R., 258
Sidman, R. L., 35

Siegel, P. B., 95, 181, 272, 655
Sielmann, H., 11
Silva, M. T. A., 121
Simmons, J. A., 352, 554
Simmons, K. E. L., 8
Simonds, P. E., 258
Skon, P. R., 348
Skopik, S. D., 593
Skov, H., 570
Skutch, A. F., 199–200, 203, 237
Sladen, F. W. L., 232
Smith, C. C., 127, 136–37
Smith, D. G., 170, 674
Smith, G. J. C., 411
Smith, J. M., 45–46, 200, 203, 363, 411
Smith, L. A., 35
Smith, N. G., 172, 272, 363, 406, 413–15, 428, 442
Smith, S. M., 646
Smith-Vanig, W. F., 210
Snow, B. K., 2
Snow, D. W., 2, 107, 164
Snyder, H., 385
Snyder, N. F. R., 385
Snyder, R. L., 119
Snyder, W. F., 431, 441
Soans, J., 411
Sokal, R. R., 404
Southern, H. N., 55, 234, 429
Southwick, C. H., 258
Sower, L. L., 380
Spalding, D. A., 634
Spekreijse, H., 532
Spencer-Booth, Y., 199, 259
Spiess, B., 29
Spiess, E. B., 179, 181
Spieth, H. T., 6, 445–46
Springer, V. G., 210
Spurway, H., 277
Stalker, H. D., 78–79
Stärk, A., 335
Stebbins, G. L., Jr., 192, 400
Steele, R. G., 413
Stefanski, R. A., 52
Stein, R. C., 421–23
Stenger, J., 52, 126
Stephen, W. J. D., 76–78
Sterman, M. B., 612
Sternberg, H., 49, 558, 560
Stevenson, J. G., 362
Stewart, R. E., 106
Stickel, L. F., 116–17
Stiles, F. G., 136–37
Stokes, A. W., 299, 320, 347
Stone, B. C., 385
Stone, W. S., 427
Storer, R. W., 170, 320
Stout, J. F., 530
Strecker, R. L., 118
Stresemann, E., 6
Struhsaker, T. T., 252, 254–57, 262–63, 349
Stumpf, W. A., 50
Sturm, T., 318
Sudd, J. H., 215, 225
Suga, N., 555, 601
Summerlin, C. T., 118
Sutton, G. M., 315

Tamarin, R. H., 147
Tamura, M., 432, 659, 663–65
Tan, C. C., 446
Tast, J., 115
Tavolga, W. N., 344
Taylor, D. H., 598
Teleki, G., 199
Tembrock, G., 341
Tenaza, R., 139–40, 144
Terman, C. R., 120
Ter Pelkwijk, J. J., 274
Tester, J. R., 55
Thibout, E., 161
Thielcke, G., 347, 421, 425, 432, 442, 663
Thiessen, D. D., 119–20, 211
Thoday, J. M., 28, 411, 445
Thompson, D. H., 362
Thompson, E. F., 425
Thompson, T., 318
Thompson, W. L., 84, 90, 432, 659
Thompson, W. R., 21
Thorpe, W. H., 33, 341, 359, 362, 432, 643, 657, 659, 661–62
Throckmorton, L. H., 416
Tinbergen, L., 102, 602
Tinbergen, N., 3, 6, 139, 146, 271, 273–74, 290, 296–97, 306, 319, 492–93
Tinkle, D. W., 52, 129–30, 193
Tischner, H., 528
Titlebaum, E. L., 554
Tobach, E., 625
Tokuda, K., 256
Topoff, H., 388
Treat, A. E., 519–20
Tretzel, E., 356
Trivers, R. L., 196, 200, 203, 208–9, 363
Tschanz, B., 362
Turk, A., 371
Turner, F. B., 50, 52, 56
Twitty, V. C., 406
Tyshehenko, V. P., 509

Uexküll, J. J. von, 487, 605
Uhrich, J., 591
Underwood, H., 598

Vaisnys, J. R., 553
Valenstein, E. S., 604
Valverde, J. A., 87
Vandeberg, J. S., 520
Vandenbergh, J. G., 258, 262
Van der Cingel, N. A., 286
Van Hooff, J. A. R. A. M., 292
Van Lawick-Goodall, H., 143
Van Lawick-Goodall, J., 143, 258
Van Tets, F. G., 4, 5
Van Tyne, J., 334
Van Vleck, D. B., 116
Varanka, I., 504
Verbeek, N. A. M., 136
Vereschchagin, S. M., 509
Verner, J., 136, 165–66, 169–70, 172, 174, 272, 660
Vince, M. A., 347
Voerman, S., 382
Von Frisch, K., *see* Frisch, K. von

Waddington, C. H., 444
Wagner, H. G., 530
Walcott, C., 292, 320, 580
Waldron, I., 419, 591
Walker, T. J., 335–36, 368, 416–17
Wall, W. van de, 33
Wallace, B., 44, 445
Wallen, J. C., 299, 320
Walther, F., 272
Ward, P., 143
Warner, D. W., 55
Warner, R. W., 341
Warren, J. M., xvii
Warriner, C. C., 629–30, 639
Wasserman, A. O., 426
Watanabe, A., 509
Watkins, W. A., 673
Watson, A., 49, 104, 108, 110, 172
Watson, G. F., 437, 441
Watson, M., 360, 661
Watt, K. E. F., 143
Webster, F. A., 550
Weeden, J. S., 52–53, 355
Weesner, F. M., 215
Weinburger, N. M., 601
Weisman, R. G., 18
Weiss, J. M., 121
Welch, A. S., 121
Welch, B. L., 121
Weller, M. W., 241
Wells, M. J., 497
Wells, P. H., 457
Wenger, E., 616
Wenner, A. M., 449, 454, 456–57
Werblin, F. S., 532
Wetterberg, L., 598
Wever, E. G., 352
Wheeler, W. M., 225, 227
White, M. J. D., 400, 403, 410
White, R. E. C., 361–62
White, S. J., 361–63
Whitman, C. O., 3–4, 643
Whittaker, R. H., 371, 374
Whitten, W. K., 427
Wickler, W., 6, 18, 210, 237, 311, 313
Widom, B., 373
Widom, J. M., 373
Wiens, J. A., 133
Wiepkema, P. R., 283–87
Wiewandt, T. A., 345
Wilbur, H. M., 154, 193
Williams, G. C., 133, 144, 196, 203
Williams, H. W., 299, 320, 347
Willis, E. O., 59, 62, 88
Willoughby, E. J., 9
Willson, M. F., 58, 134, 170, 172, 174
Wilson, D. M., 504–9
Wilson, E. O., 41, 45, 192–94, 214–15, 222–25, 231, 277, 371, 377–78, 380–81, 383, 385–91, 396, 405, 435–37
Wiltschko, R., 582–86
Wiltschko, W., 582–86
Windle, W. F., 623
Wine, J. J., 500–501
Winn, H. E., 345
Wodinsky, J., 344–45
Wolbarscht, M. L., 532
Wolf, L. L., 59, 65, 136–37
Wolfe, J. L., 118
Wolff, R., 62
Wollberg, Z., 530
Wood-Gush, D. G. M., 634
Wright, E. N., 348
Wright, H. O., 302
Wright, S., 41, 203
Wyman, R. J., 508
Wynne-Edwards, V. C., 41, 97, 124, 132, 147, 189

Yamagishi, S., 63
Yarnall, J. L., 62, 88
Yoshiba, K., 247, 262–63
Young, J. Z., 289, 497–98
Yuwiler, A., 598

Zahavi, A., 143, 207
Zenkin, G. M., 532, 535–37
Zimmerman, J. L., 99, 136, 172–73
Zucker, R. S., 499, 501
Zuckerman, S., 76, 243
Zumpe, D., 377
Zweifel, R. G., 339

Species Index

Aardwolf, 291
Acanthomyops claviger, 385–87
Acherontia atropos, 332
Adenota kob, 164
Adoxophyes orana, 382
Aepyceros melampus, 160
Aequorea, 481–82
Aeschna cyanea, 110–11
Agelaius
  phoeniceus, 104, 134–35, 165, 169–70, 173–74, 291, 313, 674
  tricolor, 134–35
Aix galericulata, 297
  sponsa, 298, 631
Alauda, 319
Albatross, 566
Alouatta, 332
Amadina fasciata, 302
Amblyornis, 317
Ambystoma, 154
  maculatum, 623
Ammophila
  adriaansei, 7–8
  campestris, 7–8
Anas
  bahamensis, 297
  flavirostris, 631
  platyrhynchos, 573, 630–32
  querquedula, 297
Anatidae, 4, 6
Anemone, 209, 376, 469, 477, 479
Anhingidae, 6
Ani, 198, 202, 238
Anisomorpha bupestroides, 373
Anser
  albifrons, 90–91
  anser, 492–93
  coerulescens, 32, 411, 429
Ant, 41, 224–31
  fire, 388,
  harvester, 397
  Texas leaf cutting, 389
Antbird, 88
Anthus, 235, 309
Aphelocoma
  coerulescens, 201, 736
  ultramarina, 198, 201–9
Apis
  adreniformis, 460
  dorsata, 458
  florea, 458
  indica, 458
  mellifera, 214–23, 407, 449–64
Aplysia, 596–97
Apostle bird, 203
Apteryx, 332
Apus apus, 189
Archboldia, 317
Ardea cinerea, 286
Armadillidium, 546
Aspidontus taeniatus, 311
Astera chastifex, 388
Asterias forreri, 470
Atta texana, 389
Attophila fungicola, 388
Augochloropsis sparsilis, 223
Auplopus, 17–18
Aurelia aurita, 471, 591
Aythya
  americana, 241
  fuligula, 241

Babbler, Arabian, 207
Baboons, 49
  chacma, 160
  gelada, 48, 246
  hamadryas, 48, 76, 246
  savannah, 198
Bats, 35, 547–55, 557, 600–601
  fruit, 549
  leaf-nosed, 549
  little brown, 547, 550–52, 554
  vampire, 439
Beaver, 332

Bee, 33, 41, 214–23, 449–64
poppy, 215
Beetles, 380
bark, 371
bombardier, 373
cerambycid, 178
lady, 32
Betta splendens, 275, 318
Birds of paradise, 162–63, 291
Bishopbird, 319
Biston betularia, 32
Bitterling, 283–84
Blackbirds
Brewer's, 136
European, 107, 656
red-winged, 104–5, 134, 165, 169–70, 173–74, 291, 313, 674
tricolored, 134
yellow-headed, 165, 172
Blarina, 548
Bluegill, 411–12
Bombus, 231
Bombycilla garrula, 49
Bombyx mori, 380, 382
Bonasa umbellus, 163
Bos taurus, 84
Botfly, 240
Bothriomyrmex decapitans, 229
Bowerbird, 6, 7, 163, 171, 316, 660
Brachinus, 373
Branta canadensis, 76, 90, 358
Bryotopha similis, 382
Bucephala clangula, 282, 334
islandica, 320
Budgerigars, 272
Bufo, 384, 534
americanus, 339
cognatus, 339
hemiophrys, 431
Bullfinch, 660
Bullfrog, northern, 345, 354, 522–28
Bullhead, 621–23
Bumblebee, 231, 462
Bunting
indigo, 355–57, 573–82, 659
meadow, 63
reed, 234
Bush-shrike, 429
Bustard, little, 334
Butterfly, 32

Caddisfly, 17
Cairina moschata, 631
Callicebus moloch, 246, 257
Callorhinus ursinus, 156–57
Callosamia
carolina, 383
angulifera, 383
promethea, 383
Camarhynchus
heliobates, 11
pallidus, 10–11
Camponotus beebei, 388
Canis lupus, 75, 160
Capella, 334
Capreolus capreolus, 158–60
Cardinal, 355, 432, 659–60
Cardinalis cardinalis, 355, 432, 659–60
Carollia, 549
Cassidix mexicanus, 176, 362
Castor, 332
Cat, 62, 601
Catbird, 6, 661
Catharus, 421–24
Catocala ultronia, 32
Centrarchidae, 313
Centrocercus urophasianus, 164
Cerambycidae, 333
Cercopithecus aethiops, 249, 349
Certhia, 432
americanus, 421
brachydactyla, 421
familiaris, 421
Cervus
canadensis, 158
elephas, 76, 158–60
Chaffinch, 79–83, 102–3, 273, 341, 656–59, 667
Chen coerulescens, 32, 411, 429, 559, 562, 645–50
Chickadee, 85
black-capped, 87–88
mountain, 88
Chicken, prairie, 172, 292
Chimpanzee, 199, 259, 275, 673
Chipmunk, 62, 88
Chiroxiphia pareola, 164, 302
Chlamydera
cerviniventris, 316
nuchalis, 316
Chlorophoneus, 429
Choeronycteris, 549
Chordeiles, 334
Chorthippus
biguttulus, 352–54, 417
brunneus, 354, 417
mollis, 354
Chough, Australian white-winged, 198, 203, 205
Cicada, 144–45, 352
Cichlidae, 313

Ciconia ciconia, 566, 570
Clibanarius tricolor, 591–92
Clethrionomys
  glareolus, 434
  rufocanus, 119
Clypeadon laticinctus, 17–18
Cockroach, 388, 592–95
Colaptes auratus, 171, 272–73, 428
Colinus virginianus, 299
Collocalia brevirostris, 548
Colobus, 257
Columba
  livia, 8, 629–30, 646
  palumbus, 94, 141
Columbidae, 4, 9
Condor, California, 2
Connochaetes taurinus, 160
Copsychus malabaricus, 660
Corcorax melanorhamphus, 198
Corvus
  corax, 661
  corone, 139, 146
  monedula, 85
  moneduloides, 12
Coturnix coturnix, 318
Cowbird, 174, 237–38, 658
  giant, 239
Crab
  fiddler, 302, 313, 332, 415
  hermit, 280, 591–92
  horseshoe, 534–35
Crane
  sandhill, 76
  whooping, 2
Crayfish, 495, 499–503, 598
Creeper, tree, 421, 432
Cricetus cricetus, 60
Cricket, 33, 177, 333–36, 342–44, 366–67, 431, 509–14, 654
  Chinese, 12
Crinia
  insignifera, 431–32
  signifera, 432
Crocuta crocuta, 198
Crossbill, red, 49
Crotalus, 345
Crotophaga
  ani, 198, 202
  sulcirostris, 202, 238
Crotophaginae, 203, 238
Crow, 139, 146
  New Caledonian, 12
Cuckoo, 232–35, 238, 658
Cuculidae, 203, 232–35, 662
Cuculus canorus, 232, 658
Cuttlefish, 295, 495
Cyanocitta stelleri, 62, 85–92, 283, 291, 304–6, 321, 659
Cygnus, 332
Cynomys ludovicianus, 67–69, 160

Deer, 272
  black-tailed, 391–93
  Japanese, 120
  red, 76, 158–60
  roe, 158–60
  whitetail, 310
Deermouse, prairie, 116
Dendroctonus pseudotsugae, 371
Dendroica palmarum, 309
  striata, 559–63
Dermasterias imbricata, 376
Diadema antillarum, 385
Dicamptodon ensatus, 345
Dickcissel, 99, 165, 172
Dicrostonyx groenlandicus, 113
Diphylla, 549
Dipodomys nitratoides, 118
Dog, 12–13, 198–99
Dolphin, 555, 673
Domesticated animals, 12–13, 84, 141
  cat, 62, 601
  dog, 12–13
Domestic fowl, 12, 78, 82–84, 181, 272, 670
Dove, 33–34, 632
  Barbary, 34, 291
  mourning, 171, 273, 334, 428
  ring, 34, 670
Dragonfly, 110–11
Drepaniidae, 12
Drosophila, 6, 23, 27, 29–32, 34–37, 178–83, 312–13, 419–20, 426, 434, 442–46, 596–98
  melanogaster, 23, 26, 29–31, 34–35, 180–81, 312, 426
  paramelanica, 78
  paulistorum, 178, 437, 442
  persimilis, 180, 405, 419
  pseudoobscura, 23, 178, 180, 405, 419, 433–34
  simulans, 312, 426
Ducks, 33, 640–43
  Bahama pintail, 297
  Chilean teal, 631–32
  eider, 59
  garganey, 297
  goldeneye, 282, 320, 334
  mallard, 573, 630–32, 634–35, 638
  mandarin, 297
  muscovy, 631
  pekin, 641

Ducks (*cont.*)
redhead, 240
tufted, 241
wood, 298, 631, 636
Dumatella carolinensis, 6, 661

Echidna, spiny, 401
Elk, 158
Emberiza
cioides, 63
schoeniclus, 234
Empidonax spp., 421
Empis, 15
Eptesicus fuscus, 552–54
Equus caballos, 160
Erignathus barbatus, 673
Erithacus rubecula, 65, 169, 308, 580–86
Erolia melanotos, 126
Erythrocebus patas, 248
Estrildidae, 235, 347, 639–40
Etroplus maculatus, 288
Euodice cantans, 633
Euphagus cyanocephalus, 136
Euplectes, 319
Evagetes, 231–232

Ficedula
albicollis, 566
hypoleuca, 153, 558–60
Finches
Bengalese, 633
cutthroat, 302
Galapagos (Darwin's), 10–12
weaver, 235–37
woodpecker, 10
zebra, 12, 633, 660
Firefly, 415–17, 590
Fish, 47, 272
cleaner, 154, 209–10, 311
jewel, 273
Siamese fighting, 12, 275, 318
Flicker, 171, 272–73, 428
Fly, 545
balloon, 15
fruit, *see Drosophila*
Flycatcher, 237, 421
collared, 566
pied, 153, 558–60
Formica
lugubris, 228
rufa, 228
sanguinea, 225, 387
subintegra, 387
subsericea, 385
Fox, 76, 323
arctic, 32
red, 54–55, 62, 168

Fregata, 332
Fregatidae, 6
Frigatebird, 332
Fringilla coelebs, 79–83, 102–3, 273, 341, 356–59
Fringillidae, 10
Frog, 8, 332, 337–39, 424–26, 431, 438–42, 522–29
see also Rana; Hyla
Fulica atra, 631

Galago crassicaudatus, 294
Gallus domesticus, 82–83, 272
Gambusia, 442
Gannet, 360–61
Gasterosteus aculeatus, 195, 274
Gazella granti, 160
Gazelle
Grant, 160
Thompson's, 272
Gehyra variegata, 106–7
Geospizidae, 10
Geryonia, 480–82
Gibbon, 341, 351, 429, 646
lar, 246, 257, 429
Glandulocauda inequalis, 332
Glossophaga, 549
Gonodactylus bredini, 277
Goose
blue-snow complex, 32, 411, 429, 559, 562, 645–50
Canada, 48, 76, 90, 358
graylag, 492–93
white-fronted, 90–91
Gorilla gorilla, 249, 272
Grackle, great-tailed, 176, 362
Grasshopper, 352–54, 417
Grouse, 33, 163
black, 163
red, 49, 172
ruffed, 163, 316
sage, 164
sand, 9
Grus canadensis, 76
Gryllidae, 417
Gryllus veletis, 342, 406, 431
bermudiensis, 431
campestris, 509–12, 530
firmus, 177
pennsylvanicus, 431
Gull, 172, 428, 442, 633
black-backed, lesser, 414, 643
black-billed, 362
black-headed, 139, 286, 288, 307
Galapagos swallow-tailed, 2, 401
glaucous, 413
herring, 414, 643

Gull (*cont.*)
  kittiwake, 319, 358
  Kumlien's, 414–15
  laughing, 362
  ring-billed, 362
  Thayer's, 413–14
Gymnorhina tibicen, 49
Gynmostinops montezuma, 171, 239

Harpobittacus, 381
Heleioporus eyrei, 431
Helisoma spp., 385
Hemichromis bimaculatus, 273
Heron, 32
  common, 286
Heteronetta atricapilla, 241
Hippocampus, 196
Holoquiscalus niger, 171
Homarus, 591
Homo sapiens, 186–87, 242, 244, 557
Honeybee, 23–25, 214–23, 407, 449–64
Honeycreeper, Hawaiian, 12
Honeyguide, 240
Horse, 160
Hummingbird, 162–63, 334
Hyalophora
  cecropia, 405
  promethea, 405
Hyena, 198, 205
Hyla
  cadaverina, 441
  californiae, 425, 441
  chryoscelis, 424–25
  clamitans, 528
  crucifer, 425
  ewingi, 431, 438–41
  regilla, 425, 432, 441–42
  verreauxi, 438–41
  versicolor, 32, 424–25
Hylobates, 341, 645
  lar, 246, 429
  syndactylus, 246
Hymenoptera, 198, 214–32

Ictalurus nebulosus, 621–23
Icteridae, 170
Icterus
  galbula, 171
  spurius, 171
Impala, 160
Indicatoridae, 240
Isoptera, 198

Jacana, 153
Jackdaw, 85
Jaeger
  parasitic, 646
  pomarine, 127
Jay
  Mexican, 198, 201–9
  scrub, 201, 736
  Steller's, 62, 85–92, 282–83, 291, 305–6, 321, 659
Jellyfish, 470, 480–82, 591
Junco hyemalis, 49, 94

Kangaroo, 48
Katydid, 334, 337
Kiwi, 332
Kob, Uganda, 164
Kobus ellipsiprimnus, 160

Labroides dimidiatus, 154, 210, 311
Lacewing, 520
Lagopus
  mutus, 108
  scoticus, 49
Langur, 262–63
Lapwing, 319
Lark, 319
  wood, 356
Larus, 172
  argentatus, 414, 643
  atricilla, 362
  bulleri, 362
  delawarensis, 362
  fuscus, 414, 643
  glaucoides kumlieni, 414–15
  hyperboreus, 413–14, 442
  ridibundus, 139, 286, 288, 307
  thayeri, 413–14, 442
Lasioglossum marginatum, 215
Lasiurus, 553, 557
Lasius spp., 229, 230, 385–86
Lebistes reticulatus, 288
Lemming, 49, 113
  brown, 113, 127
  collared, 113
Lemmus, 113
  lemmus, 49, 113
  trimucronatus, 113
Lepomis, 411–13, 426
  gibbosus, 412
  macrochirus, 411–12
  microlophus, 412
Leucophaea madeirae, 592
Limmulus, 534–35
Limnodynastes tasmaniensis, 339, 529
Lion, African, 198–99, 205

Lizard, 14, 302, 415, 426, 429–30
  Australian, 106–7
  eastern fence, 274
  side-blotched, 129, 430, 434
Lobster, 591–92
Locust, 504
Loligo, 494–95, 497–98
Lonchura striata, 633
Lophophorus impajanus, 299
Lovebird, 33
Loxia curvirostra, 49
Lullula arborea, 356
Lutong, 257
Lycaon pictus, 198
Lyrebird, 660
Lyrurus tetrix, 163

Macaca
  mulatta, 255–56, 258–59, 383
  radiata, 72, 258
Macaque, 72, 275
Macropus cangaru, 48
Magpie, Australian, 49
Malurus cyaneus, 203, 208
Man, 186–87, 242, 244, 557
Manakin, 163, 170
  blue-backed, 164, 302
Marmoset, 428
Marmota monax, 88, 116
Meadowlark
  eastern, 355–56, 660
  western, 355, 660
Medusa, 470
Megachile, 215
Megaleia rufa, 48
Megaptera novaeangliae, 351, 673
Melanerpes formicivorus, 198
Meleagris gallopavo, 590
Meliphagidae, 662
Melospiza
  georgiana, 576
  melodia, 432, 659
Menurus menura, 660
Methoca, 17
Metridium senile, 469, 472, 479
Micrastur, 363
Microhyla
  carolinensis, 441
  olivaceus, 441
Micronycteris, 549
Microtus, 118
  californicus, 118–19
  montanus, 119
  pennsylvanicus, 119
Mimidae, 660–61
Mimus
  longicaudatus, 239
  polyglottos, 661
Minnow, 426
Mirounga occidentalis, 673
Mockingbird, 661
Molossus aztecus, 549
Molothrus, 237
  ater, 238, 658
  badius, 237
  bonariensis, 238
  rufo-axillaris, 238
Monkey, 198
  colobus, 257
  howler, 332
  lutong, 257
  patas, 248
  rhesus, 255–56, 258–59, 383
  squirrel, 275
  titi, 246, 257
  vervet, 248, 257, 349
Mormoops, 549
Moth, 32, 332, 379, 381, 519–22
  bollworm, 377, 382
  California oak, 178
  gypsy, 378, 380
  silkworm, 380, 382
Mouse
  house, 69–71, 117, 119, 393–95
  white-footed, 116–17
Murre, common (guillemot), 362, 429
Mus musculus, 69–71, 117, 119, 393–95
Myiarchus, 421
Mynah, 661
Myotis, 547, 550–54, 600–601

Nesopelia galapagoensis, 632
Nighthawk, 334
Notropis, 426
Nucula proxima, 48
Nyctea scandiaca, 49

Octopus, 88, 499
Odocoileus, 272
  columbianus, 391–93
  virginianus, 310
Oilbird, 548
Onchorhyncus, 195
Onychognathus morio, 129
Opsanus tau, 614
Orangutan, 332
Ornithorhynchus anatinus, 401
Oropendola, 239
Oryctolagus cuniculus, 395
Osmia papaveris, 215
Otis tetrax, 334
Otus asio, 32, 117
Ovenbird, 126, 355–56
Owl
  barn, 364
  screech, 32, 117

Owl (*cont.*)
snowy, 49
tawny, 55
Oxybelus quadrinotatum, 17
Oystercatcher, 348

Pagurus
marshi, 280, 591–92
pollicarus, 591
Panthera leo, 198–99, 205
Pan troglodytes, 199
Papio
anubis, 250, 255
cynocephalus, 198, 250
hamadryas, 48, 76, 246, 250, 255, 262
ursinus, 160, 250, 262
Paramecium, 545
Parus
ater, 102
atricapillus, 87–88
atricristatus, 659
caeruleus, 87
gambeli, 88
major, 87–88, 102, 108–9
Passer domesticus, 8, 596, 599
Passerina cyanea, 355–57, 573–82, 659
Pavo, 178, 299
Peacock, 178, 299, 313
Pectinophora gossypiella, 377, 382
Pelecanidae, 6
Penguin, adélie, 140, 362
Periplaneta americana, 592
Peromyscus, 18, 113–19
californicus, 113–18, 195
leucopus, 116–17
maniculatus, 113–19, 195, 434
polionotus, 434
truei, 113–15
Petaurus breviceps, 395
Phaëthontidae, 4
Phalacrocoracidae, 6
Phalanger, Australian, 395
Phasianidae, 299
Phasianus colchicus, 299
Pheasant
impeyan, 299
peacock, 299
ring-necked, 299
Philohela, 334
Philomachus pugnax, 163, 272
Philornis, 240
Phormia regia, 462
Phryganidia californica, 178
Phylloscopus, 421
collybita, 356
trochilus, 356
Pigeon, 8–9, 12
domestic, 629–30, 646
Pigeon (*cont.*)
homing, 577, 580, 583
wood, 94, 141, 144
Pillbug, 546
Pipilo fuscus, 169
Pipit, 235, 309
Pipridae, 170
Planaria, 546
Platypus, duck-billed, 401
Ploceidae, 235–37, 321, 323
Ploceus cucullatus, 163
Poeciliidae, 33
Poeciliopsis, 154
Pogonomyrmex badius, 397
Polybia atra, 462
Polyergus rufescens, 225–27
Polyplectron bicalcaratum, 299
Pongo pygmaeus, 332
Porcellio, 546
Porpoise, 555, 673
Porthetria dispar, 380
Prairiedog, black-tailed, 67–69, 160
Presbytis
cristatus, 257
entellus, 262–63
Pristella riddlei, 272–73
Procambarus clarkii, 499
Progne subis, 177
Proteles cristatos, 291
Protopolybia, 240
Pseudacris
clarki, 425
nigrita, 425
ornata, 425, 431, 432
streckeri, 425, 431–32
triseriata, 425
Psithyrus, 231
Psophodes olivaceus, 360, 661
Ptarmigan, rock, 108
Pteroclidae, 9
Pteronotus
davyi, 549, 553
parnellii, 549, 553
psilotis, 551, 553
rubiginosa, 549
Pterophylla camellifolia, 337
Ptilonorhynchidae, 6, 171, 660
Ptilonorhynchus violaceus, 171
Pumpkinseed, 412
Pygoscelis adéliae, 362
Pyralidae, 380, 426
Pyrrhuloxia sinuata, 660
Pytilia phoenicoptera, 428
lineata, 428

Quail
bobwhite, 299
Japanese, 318

Quiscalus
mexicanus, 176, 362
quiscula, 171

Rabbit, Australian, 395
Rana
catesbeiana, 345, 354, 522–28
clamitans, 528
pipiens, 425, 530
sylvatica, 431
temporaria, 535
Rat, 35
cotton, 118
Norway, 35, 116, 118
Rattus norvegicus, 35, 116, 118
Raven, 661
Rhinolophus ferrum-equinum, 550, 553–55
Rhodeus amarus, 283–84
Rhopaea, 381
Richmondena cardinalis, see *Cardinalis cardinalis*
Rissa tridactyla, 319
Rivulus marmoratus, 154
Robin
American, 107
European, 65, 169, 308, 356, 580–86
Rousettus, 549
Ruff, 163–64, 179, 181, 272
Rupicola rupicola, 170

Saguinus fuscicollis, 428
Saimiri sciureus, 275
Salamander, 154, 345
red-backed, 32
spotted, 623
Sandpiper
pectoral, 126
ruff, 163–64, 179, 181, 272
Saturniidae, 380–81, 383, 405, 426, 462
Scaphidura oryzivora, 239
Scaphiopus
couchii, 425–26
hurterii, 425–26
Sceloporus, 274, 302, 415–16, 426
Schistocerca gregaria, 504
Sciurus carolinensis, 94–95
Seahorse, 196
Seal
bearded, 673
elephant, 157–58, 292, 673
fur, 156–57
Sea urchin, 376, 385
Seiurus aurocapillus, 126, 355
Sepia, 295, 495
Shrew, 548
tree, 244
Shrimp, mantis, 277
Siamang, 673
Sigmodon hispidus, 118
Silverbill, African, 633
Snail, 385
Snake, rattle, 345
Snipe, 334
Solenopsis
fugax, 230
molesta, 230
saevissima, 388
Sorex, 548
Sparrow
English (house), 8–9, 596, 599
field, 94
song, 432, 659, 666
swamp, 576
tree, 52–53
white-crowned, 85, 432, 576, 656, 659–60, 663–71
white-throated, 355–56, 421, 432, 646
Spermophilus undulatus, 103–4
Sphingidae, 426
Spiza americana, 99, 165
Spizella
arborea, 52–53
pusilla, 94
Squid, 296, 494–95, 497–98, 603
Squirrel
arctic ground, 103–4
gray, 94–95
tree, 127
Starfish, 376, 470, 473
Starling, 41, 189, 565–69, 660
red-winged, 129
Steatornis caripensis, 548
Steganura, 319
Stelopolybia, 240
Stercorarius parasiticus, 646
pomarinus, 127
Sterna hirundo, 362
Stickleback, 195
three-spined, 274, 289
Stomphia coccinea, 376
Stork, white, 566, 570
Streptopelia
risoria, 291, 632
senegalensis, 632
Strongylognathus testaceus, 226, 228
Struthidea cinerea, 203
Sturnella
magna, 355–56, 660
neglecta, 355, 660
Sula bassana, 360–61
Sulidae, 6
Sunfish, 411–13, 426
Swan, 332

Swift, 41
  European, 189
Swiftlet, cave, 548
Sylvia borin, 572
  atricapilla, 572
  curruca, 572
Symphalangus syndactylus, 673

Tachyglossus spp., 401
Tadorna tadorna, 631
Taeniopygia castanotis, 12, 633, 660
Tamiasciurus, 127, 137
Tamias striata, 88
Teleogryllus
  commodus, 514
  oceanicus, 514
Teleutomyrmex schneideri, 225
Telmatodytes palustris, 165, 272
Termite, 462
Tern
  arctic, 558
  common, 362
  royal, 358
Tetraonidae, 170, 316
Tetraopes tetropthalmus, 178
Tettigoniidae, 417
Textor cucullatus, 307, 323
Thalasseus maximus, 358
Theropithecus gelada, 48, 246
Thrasher, brown, 661
Thrush, 419, 421–24
  shama, 660
Tibicen dorsata, 352
Tilapia
  melanotheron, 288
  mossambica, 274
Tiphiidae, 17
Titmouse
  black-crested, 659–60
  blue, 87
  coal, 102
  great, 87–88, 102, 108–9
Toad, 339, 384, 405–6, 425–26, 431, 534, 624–25
Toadfish, 614
Towhee, brown, 169
Toxostoma rufum, 661
Tricoplusia ni, 380, 382
Trigona, 240
  silvestris, 462
Troglodytes troglodytes, 356
Turdoides squamiceps, 207
Turdus
  merula, 107, 656
  migratorius, 107
Turkey, 313, 590
Turnix turnix, 598
Tursiops truncatus, 673
Tyto alba, 364

Uca, 302, 415
  speciosa, 332
Uria aalge, 362, 429
Uta stansburiana, 129, 429–30, 434

Velella, 480
Vermivora
  chrysoptera, 421
  pinus, 421
Vervet, 248, 257, 349
Vespertillionidae, 553
Vicugna vicugna, 160
Vicuña, 160
Viduinae, 235–37, 659
Vole
  bank, 434
  meadow, 119
Vulpes fulva, 54–55, 62, 168, 323

Walkingstick, southern, 373
Warbler, 172
  blackpoll, 559–63
  blue-winged–golden-winged, complex, 421, 429
  European, 575
  ovenbird, 126, 355–56
  palm, 309
  Phylloscopus, 356
Wasp, 7, 231, 462
  Aculeate, 16
  digger, 18
  gall, 16
  solitary, 6
  spider, 17
Waterbuck, 160
Waxwing, Bohemian, 49
Weaver, village, 163, 307, 323
Whale, 555
  humpback, 351, 673
Whipbird, 359–60, 661
Whydah, 319
White-eyes, 85
Wildebeeste, blue, 160
Wolf, 75, 160, 395
Woodchuck, 88, 116
Woodcock, 334
Woodpecker
  acorn, 198
  flicker, 171, 272–73, 428
  ivory-billed, 2
Worm, 496

Wren
European, 356
long-billed marsh, 165, 170, 172, 272
superb blue, 203, 208, 735–36

Xanthocephalus xanthocephalus, 165
Xenopus, 624–25

Zenaidura macroura, 171, 273, 334, 632
Zonotrichia
albicollis, 355, 421, 432, 646
capensis, 664
leucophrys, 85, 432, 576, 656, 659–60, 663–71
Zosterops lateralis, 85